3rd International Conference on Civil and Environmental Engineering for Sustainability (IConCEES 2015)

MATEC Web of Conferences Volume 47 (2016)

Melaka, Malaysia
1 - 2 December 2016

Editors:

Norashidah Abd Rahman
Zainorizuan Mohd Jaini
Riduan Yunus
Siti Nazahiyah Rahmat

ISBN: 978-1-5108-2212-2

Printed from e-media with permission by:

Curran Associates, Inc.
57 Morehouse Lane
Red Hook, NY 12571

Some format issues inherent in the e-media version may also appear in this print version.

Printed by Curran Associates, Inc. (2016)

For additional information, please contact EDP Sciences – Web of Conferences at the address below.

EDP Sciences – Web of Conferences
17, Avenue du Hoggar
Parc d'Activité de Courtabœuf
BP 112
F-91944 Les Ulis Cedex A
France

Phone: +33 (0) 1 69 18 75 75
Fax: +33 (0) 1 69 28 84 91

contact-edps@webofconferences.org

Additional copies of this publication are available from:

Curran Associates, Inc.
57 Morehouse Lane
Red Hook, NY 12571 USA
Phone: 845-758-0400
Fax: 845-758-2633
Email: curran@proceedings.com
Web: www.proceedings.com

TABLE OF CONTENTS

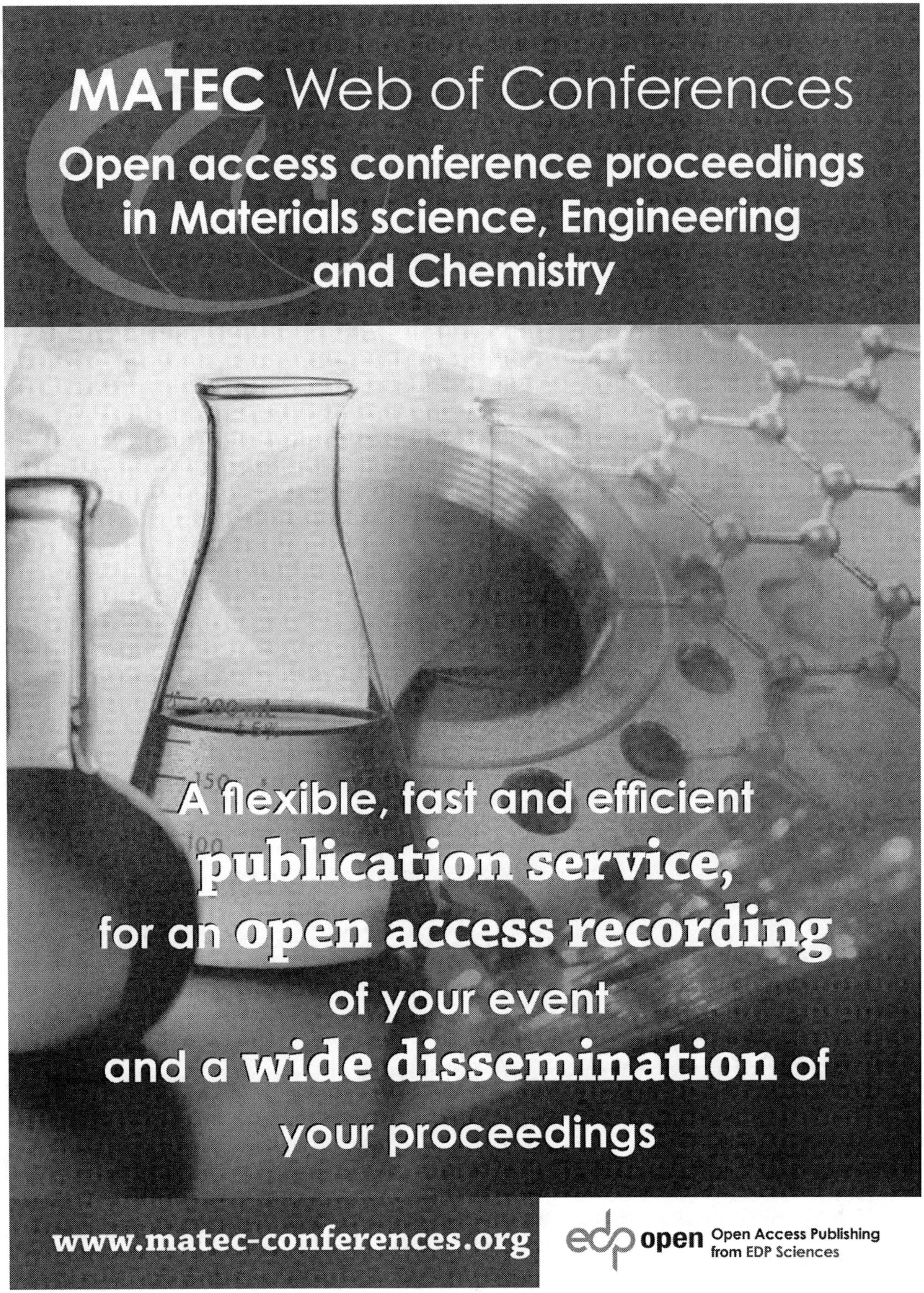
MATEC Web of Conferences
Open access conference proceedings in Materials science, Engineering and Chemistry
A flexible, fast and efficient publication service, for an open access recording of your event and a wide dissemination of your proceedings
www.matec-conferences.org
edp open
Open Access Publishing from EDP Sciences

Aims & scope

MATEC Web of Conferences is an open-access journal devoted to the publishing of conference proceedings dealing with Materials science, Engineering and Chemistry.

The journal aims to make conference proceedings visible, citable and widely usable by the scientific community. It offers outstanding services for the publication and dissemination of contents in order to ensure long-term archiving and maximum exposure of scientific events.

Coverage includes all engineering disciplines: civil, naval, mechanical, chemical, and electrical engineering as well as nanotechnology and metrology. The journal also deals with the physical-chemical characterization of materials, their implementation, their resistance in their environment, etc. Others subdisciplines of chemistry, such as analytical chemistry, petrochemistry, organic chemistry, etc., and even pharmacology, are also welcome.

Why publish with MATEC Web of Conferences?

- **Open access:** conference papers available worldwide.
- **Fast publication:** less than 3 months.
- **Total flexibility:** no page restrictions, free unlimited use of colour in the online version, any kind of scientific document welcomed (traditional articles, posters, abstracts and slideshows).
- **High visibility:** all documents indexed in main databases and easily found online.
- **Fair author rights:** CC-BY Creative Commons license.
- **Range of additional publishing options:** printed proceedings, CD-ROMs, copy-editing, etc.

ISSN (electronic edition): 2261-236X

Contact: Isaline Augusto
matec@webofconferences.org

www.matec-conferences.org

edp open Open Access Publishing from EDP Sciences

Conference Committees

Organizing Committee

Chairman	:	Prof. Dato' Mohd Idrus Mohd Masirin
Vice-Chairman 1	:	Dr. Riduan Yunus
Vice-Chairman 2	:	Assoc. Prof. Norzila Othman
Secretary 1	:	Dr. Shahiron Shahidan
Secretary 2	:	Dr. Noorwirdawati Ali
Treasurer	:	Dr. Hartini Kasmin
Promotions	:	Dr. Basil David Daniel Mrs. Faizah @ Rashidah Dahari
Publication and Indexing	:	Dr. Norashidah Abd Rahman Dr. Zainorizuan Mohd Jaini Dr. Mohd Shalahudin Adnan Assoc. Prof. Noridah Mohamad Dr. Nurazuwa Md Noor Dr. Mustaffa Anjang Ahmad Dr. Norwati Jamaluddin Dr. Saiful Azhar Ahmad Tajudin Dr. Siti Nazahiyah Rahmat Dr. Nickholas Anting Guntor
Sponsorship	:	Dr. Mohd Adib Mohammad Razi Dr. Sasitharan Nagapan Prof. Dato' Ismail Bakar Dr. Ahmad Zurisman Ali
Logistics	:	Dr. Shahrul Niza Mokhatar Dr. Azeanita Suratkon Assoc. Prof. Mohd Yusri Aman Dr. Joewono Prasetijo Dr. Nor Haslinda Abas
Programme and Registration	:	Assoc. Prof. Aeslina Abd Kadir Dr. Rafikullah Deraman Dr. Nor Hazurina Othman Ir. Dr Raha Abd Rahman Dr. Mohd Ariff Ahmad Nazri

Dr. Putera Agung Maha Agung (Politeknik Negeri Jakarta, Indonesia)
Dr. Nasrul Hamidin (Universiti Malaysia Perlis, Malaysia)
Dr. Mohd Remy Rozainy Mohd Arif Zainol (Universiti Sains Malaysia, Malaysia)
Dr. Farah Nora Aznieta Abd Aziz (Universiti Putra Malaysia, Malaysia)
Dr. Noorsuhada Md Noor (Universiti Teknologi MARA, Malaysia)
Dr. Mohd Shalahuddin Adnan (Universiti Tun Hussein Onn Malaysia, Malaysia)
Sr. Dr. Mustaffa Anjang Ahmad (Universiti Tun Hussein Onn Malaysia, Malaysia)
Dr. Nurazuwa Md Noor (Universiti Tun Hussein Onn Malaysia, Malaysia)

MATEC Web of Conferences **47**, 00001 (2016)
DOI: 10.1051/matecconf/20164700001

Preface

We are very grateful that this international conference is convened today to gather more than one hundred speakers presenting papers pertaining to the field of Civil and Environmental Engineering. IConCEES2015 is organized as a platform for scientists, researchers and engineers to share their knowledge and experiences in creating sustainable and environmental friendly world for the present and future. We all know that this world is borrowed from the next generation and in other for us to ensure that it is well looked-after, we have to maintain the three important pillars of sustainability i.e. economics, environment and society. These three pillars are the catalyst to ensure that the world is livable, continue to exist and at the same time is developed to enhance the socio-culture of mankind.

The IConCEES2015 theme, '*Civil and Environmental Engineering for Sustainability: Towards a Better Future*' indicates that the needs to prepare the world for future generation to continue the responsibility to make the world a better place to live. It is the desire of all population and generations that mankind will continue to exist and lead the world. It has been portrayed by many in science fiction movies that the apes will rule the world or the world will be destroyed because of our negligence and too engross in the development of artificial intelligence (AI). It is always possible if we do not care to ensure that what exists can be de-existed and what is impossible is always possible. The best example is the telephone evolution. From wired to wireless and from a typical phone to a smart phone. Constructing buildings was considered difficult and limited to certain people. But now, everybody can construct a house with some skills with the application of industrialized building system (IBS). With research works on reducing the usage of natural materials and increasing the use of recycled materials, pollution and negative environmental impacts can be reduced. Thus, it is our duty to ensure that we conduct the necessary steps towards sustainability of the world in whatever discipline and role that we have.

With Wisdom, We Explore.

Sincerely,

Prof. Dato' Mohd Idrus Mohd Masirin
Chairman

Dr. Riduan Yunus
Assoc. Prof. Norzila Othman
Vice-Chairman

IConCEES 2015
Melaka, Malaysia

The 3rd International Conference on Civil and Environmental Engineering for Sustainability
'*Civil and Environmental Engineering for Sustainability: Towards a Better Future*'
1 & 2 December 2015, Equatorial Melaka, Malaysia

MATEC Web of Conferences **47**, 00002 (2016)
DOI: 10.1051/matecconf/20164700002

Keynote Abstract

Keynote Speaker 1

Sustainability in Built Environment: Green Building Index (GBI) Platinum Certification UCTS Experience

This keynote paper highlights the concept and objectives of green buildings. The fundamental features and characters of green buildings are presented. This paper focuses on the two primary issues; (i) the environmental protection and energy conservation and, (ii) the healthy indoor environment providing residential users with a comfortable and safe living.

The paper also highlights history of green building design with references to a number of green buildings in the world; their primary green features and rating agency.

The Malaysia Green Building Index (GBI), the Malaysia's green rating tools for buildings and complexes is presented in details taking the UCTS platinum rated GBI as an example. The scoring for each six key criteria of the GBI are presented and highlighted.

Prof. Dato' Sabarudin Mohd
Peat Technology Research Center (PETRI)
University College of Technology Sarawak

Keynote Speaker 2

Sustainable Urban Stormwater Management - Key to Future Green-Blue, Resilience and Sustainability of Urban Environments

Rapid urban growth in Malaysia over the last 30 years has resulted in increased stormwater flow into receiving waters, increased flood magnitude and frequency, and degraded water quality. Traditionally, stormwater management focuses on solving flooding through drainage of the urban areas to receiving water bodies. Presently, water supply should have not been an issue in Malaysia, considering the abundance of rainfall in this country. However, the reality is far from this. Malaysia ironically faces water stress in several heavily populated areas. Seasonal rainfall pattern, poor management and ineffectiveness of water supply system have led to occasional and seasonal water supply shortage in some states.

Based on these problems, Malaysian government has introduced a policy changing on Urban Stormwater Management Manual (MSMA). The introduction of Urban Stormwater Management Manual for Malaysia, or MSMA, changed the stormwater management landscape in the country. However, with the increasing demands for green technologies and addressing climate change, stormwater engineers are facing a stiffer challenge to produce effective and sustainable drainage system. This requires the need to inject new technologies or innovations into the design of drainage facilities.

The Sustainable Urban Stormwater System (SUDS) management solutions is consistent with objectives of stormwater management approach, i.e. control at source and stormwater treatment train which focus on both the quantity and quality control of urban runoff. The application of SUDS is a new development attempt to solve three major problems commonly encountered in Malaysia namely flash floods, river pollution and water scarcity during dry periods.

The success of the SUDS implementation proved that with innovation, managing stormwater in resilience and sustainability can be successfully managed without compromising on the overall project cost compared to traditional infrastructure system. A SUDS to the green-blue solutions not only has the potential to reduce the infrastructure costs, but it also reduces maintenance costs. It's also benefit the environment due to its minimise impact on local ecology, encourage the ecosystem values, function as well as preserved and providing benefits to the daily life's.

Prof. Nor Azazi Zakaria
River Engineering and Urban Drainage Research Centre (REDAC)
Universiti Sains Malaysia

Keynote Speaker 3

Green Building Is Not Enough

The definition of "sustainable development" considers three pillars: economic, social and environmental. To these are added the governance or institutional; and cultural pillars. The word 'green' is used to refer to the environmental dimension. In the construction industry, as indicated by the entrenched use of 'green', the approach has narrowly focused on the environmental considerations. This is reflected in the way the industry assesses the performance of buildings which are intended to deal with the negative impacts of buildings on sustainable development and to help to address the issues relating to sustainable development. This is an important subject as buildings are being seen as among major instruments for mitigating against climate change, and are therefore, main components of strategies for attaining national GHG emission-reduction commitments.

Various definitions of "Green Building", the drivers, obstacles and trends are considered. The question is asked: "Is green building the mainstream?" The focus is put on Malaysia and the drivers, key players, attainments and challenges of green building in the country are discussed. Areas which are not yet well covered are highlighted.

The definition of "sustainable development" is applied to formulate an appropriate definition for "sustainable building". Green building is then contrasted with sustainable building. It is shown that the former does not cover some major elements such as the social; economic (project, participants, district, national levels); and institutional and governance. It is suggested that such a broad definition is required in order that the right conceptualization is done, appropriate research is undertaken, suitable policies are formulated and the right inter-related actions are adopted, co-ordinated and integrated. The future of sustainable building in Malaysia is considered. The contributions to be made by these parties are highlighted: government, industry, clients, users and occupiers, and researchers.

Prof. George Ofori
School of Design and Environment
National University of Singapore

edp sciences

MATEC Web of Conferences

Statement of Peer review

In submitting conference proceedings to MATEC Web of Conferences, I certify to the Publisher that I adhere to the **Policy on Publishing Integrity** of the journal in order to safeguard good scientific practice in publishing.

1. All articles have been subjected to peer review administered by the proceedings editors.
2. Reviews have been conducted by expert referees, who have been requested to provide unbiased and constructive comments aimed, whenever possible, at improving the work.
3. Proceedings editors have taken all reasonable steps to ensure the quality of the materials they publish and their decision to accept or reject a paper for publication has been based only on the merits of the work and the relevance to the journal.

Title, date and place of the conference

The 3rd International Conference on Civil and Environmental Engineering for Sustainability (IConCEES2015)

Proceedings editor(s):

Dr. Norashidah Abd Rahman
Dr. Zainorizuan Mohd Jaini
Dr. Riduan Yunus
Dr. Siti Nazahiyah Rahmat

Date and editor's signature

19 December 2015

MATEC Web of Conferences **47**, 01001 (2016)
DOI: 10.1051/matecconf/20164701001

Fresh State and Mechanical Properties of Self Compacting Concrete Incorporating High Volume Fly Ash

N. Mohamad[1,a], M. S. Zulaika[1], A. A. A. Samad[1], W. I. Goh[1], J. Hadipramana[1] and A. Wirdawati[1]

[1]*Jamilus Research Center, University Tun Hussein Onn Malaysia, 86400 Parit Raja, Johor, Malaysia*

Abstract. Self-compacting concrete is considered as a concrete which can be placed and compacted under its own weight without vibration. The elimination of the need for compaction leads to better quality concrete and substantial improvement of working conditions. This paper investigates the fresh state and mechanical properties of self-compacting concrete incorporating high volume fly ash. Fly Ash (FA) was mixed into self-compacting concrete (SCC) as a replacement for cement. Portland cement (PC) was partially replaced with 0%, 20%, 40% and 60% FA. The water to binder ratio was fixed at 0.4 for all mixtures. Tests were carried out on all mixtures to obtain the workability of self-compacting concrete in terms of viscosity by slump flow test (Diameter and T_{500}), V-funnel and J-ring. The mechanical property tests were conducted on SCC cubes and cylinders to determine its compressive strength and modulus of elasticity. The results indicate that replacement of 40% fly ash is the optimum result for the workability and mechanical properties test. The highest compressive strength which is 27.2 MPa was achieved by SCC with 40% FA replacement. Modulus of elasticity increased with the increased percentage of fly ash except for 60% fly ash.

1 Introduction

Conventional concrete has been widely used in construction all over the world. It is the second consumed material after water in the world. The preparation of concrete includes batching, mixing, placing and compacting. It is essential to condense the concrete so that it should completely cover the reinforcement and fill all the space in the formwork to meet the strength and durability requirement. The air entrained in concrete during amalgamation has to be totally expelled out for getting uniform dense mass. If compaction is incomplete, it will lead to reduction in strength and performance of the concrete structure. The compaction becomes complex when percentage of reinforcement is high which does not allow insertion of vibrator at certain places. The vibration will also increase the noise level in and around the construction site [1].

Self-compacting concrete (SCC) is a highly flowable non-segregated concrete which spread into place and does not require vibration. SCC is the modified concrete with the use of chemical (superplasticizers and viscosity modifiers) and mineral (fly ash, slag, silica fume) admixtures in the concrete [2]. It is designed generally with high content of fine material. The three essential properties of self-compacting concrete are its ability to flow under its own mass (filling ability), its ability to pass

[a] Corresponding author : noridah@uthm.edu.my

through congested reinforcement (passing ability), and its ability to resist segregation (segregation resistance).

Advantages of SCC include faster construction, reduction in man power, easier placing, uniform and complete consolidation, better surface finishes, improved durability, increased strength and safer working environment. Recently, many countries report the benefits of SCC by eliminating the need for vibration, shorter construction period, lower overall costs and also improved health and safety conditions [3].

This paper reports the fresh state and mechanical properties of SCC incorporating high volume of fly ash. The fresh state properties of SCC were determined by using mixture designed by Bouzoubaa and Lachemi [4]. The mechanical properties of SCC which include density, compressive strength, and modulus of elasticity were determined by conducting a compressive strength test on SCC cubes and cylincers. All tests were conducted according to British Standard code [5, 6, 7]. Substituting Portland cement with fly ash will make the concrete more workable and suitable for the construction. SCC concrete incorporating high volume of fly ash is able to produce concrete with compressive strength similar or higher than conventional concrete.

2 Experimental Programme

2.1 Mixture design

Generally, batching of materials by volume are often used in the design of concrete mixture due to the importance of the need to fill the voids between the aggregate particles [8]. The hardened properties of SCC are expected to be similar to those of vibrated normal concrete. Laboratory and field tests have demonstrated that the SCC hardened properties are indeed similar to those of vibrated concrete. Table 1 shows the mechanical properties of SCC as obtained from Hajime et al. [3].

2.2 Materials and methods

Self-compacting concrete specimen were formed by applying material such as cement, water, fine aggregate, coarse aggregates, fly ash and superplasticiser. Cube compression tests were conducted at 7, 28 and 56 days of curing ages. Table 2 shows the mixture proportion of the self-compacting concrete incorporating high volume of fly ash.

Table 1. Mechanical properties of SCC [3].

Water-binder ratio (%)	25 to 40
Air content (%)	4.5-6.0
Compressive strength (28 days) (MPa)	40 to 80
Compressive strength (91 days) (MPa)	55 to 100
Splitting tensile strength (28 days) (MPa)	2.4 to 4.8
Elastic modulus (GPa)	30 to 36
Shrinkage strain (x 10^{-6})	600 to 800

Table 2. Mix design for SCC.

Mix	Cement (kg/m^3)	Fly Ash (kg/m^3)	Fly Ash (%)	Fine Aggregates (kg/m^3)	Coarse Aggregates (kg/m^3)	Water (l/m^3)	SP (l/m^3)	W/P ratio
TR1	450	-	-	851	851	180	9	0.4
TR2	180	270	60	851	851	180	9	0.4
TR3	270	180	40	851	851	180	9	0.4
TR4	360	90	20	851	851	180	9	0.4

3 Results and Discussion

3.1 Workability

Table 3 shows results for workability the fresh concrete as obtained from slump flow test (Slump flow diameter and T_{500}), V-funnel, V-funnel at T_{5min} and J-ring for various SCC compositions. The results from the workability tests were obtained as explained in the following sections.

Table 3. Experimental result for workability test.

Mix	Fly ash (%)	Workability Tests				
		Slump flow, SF		V-funnel		J - Ring, (mm)
		Diameter,(mm)	T_{500} (s)	V-funnel (sec)	T_{5min} (sec)	
1	Control SCC	400	-	≥12	-	-
2	60	635	6	10	+3	3
3	40	675	4	8	+2	2
4	20	625	7	9	+3	2.5

3.2 Slump flow test

Results showed that for the SCC control mixture, the slump flow spread diameter was 400 mm while for T_{500} it did not achieve 500mm flow spread. With the presence of 20% fly ash the slump flow spread diameter and T_{500} showed increment up to 625 mm and 7 s, respectively. The 40% fly ash mixture recorded optimum workability compared to 60% fly ash SCC mixture. However, the spreading range obtained was all within the acceptable range which is from 500mm to 700mm [2]. The result are shown in Figure 1 and Figure 2 below.

3.3 V-funnel test

The time measured using the V-funnel was in the range of 8 to 12 second depending mainly on the SCC mixture used. The lowest V-funnel flow time of 8 second was recorded for the SCC with 40% fly ash while the mixture with 60% fly ash had the highest flow time of 10 second. The graphical result is shown in Figure 3.

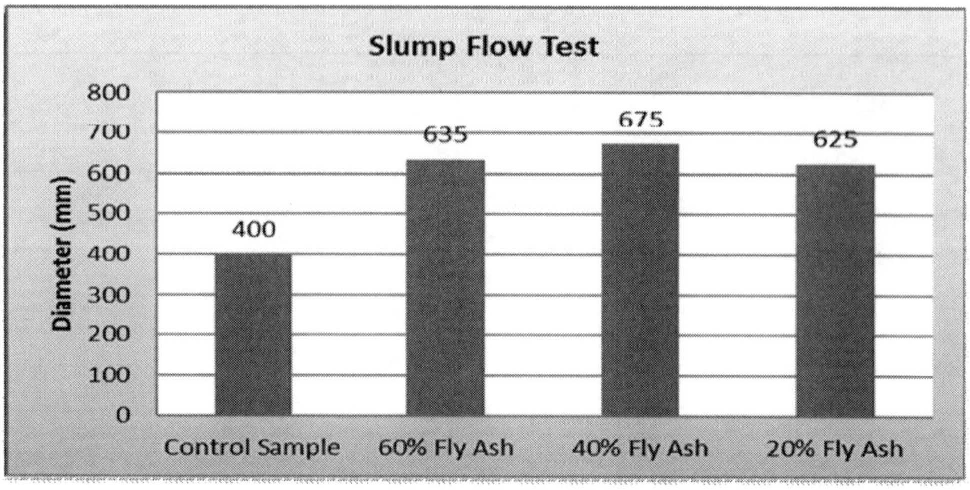

Figure 1. Slump flow test result.

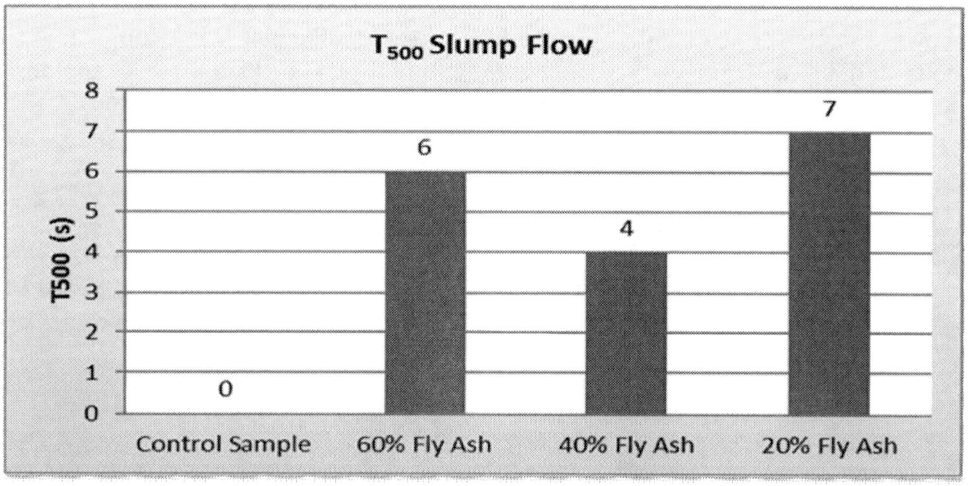

Figure 2. T_{500} slump flow.

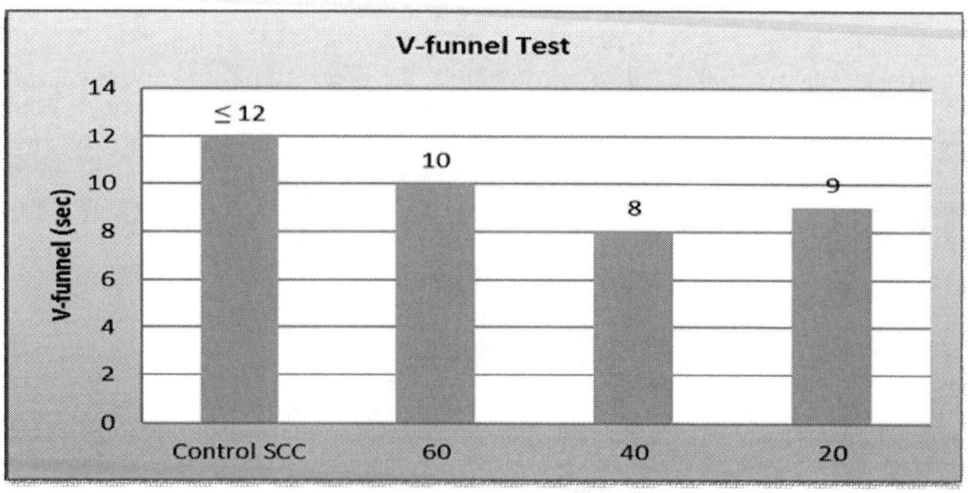

Figure 3. V-funnel result.

3.4 J-Ring test

For J-Ring test, the test is to determine passing ability of the concrete. The blocking step, B_j, was determined by the average of the differences in height at four locations. Figure 4 shows the J-Ring result.

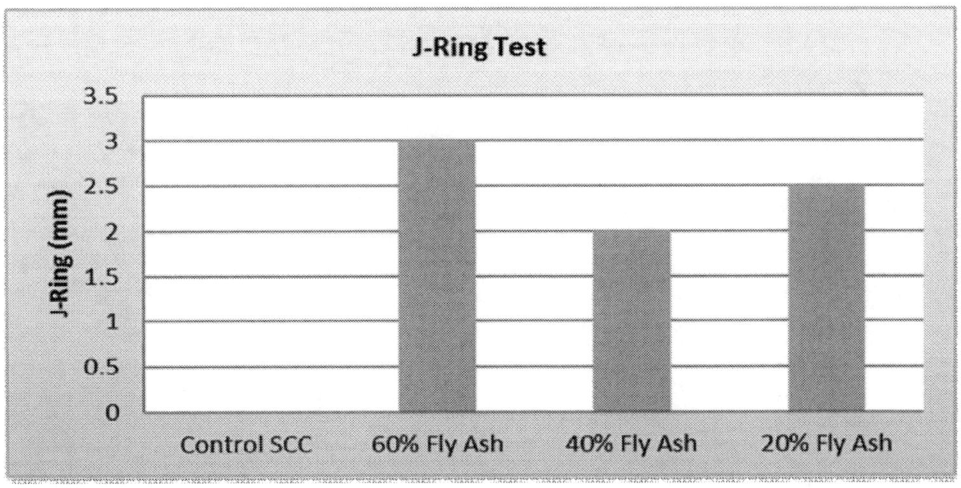

Figure 4. J-Ring results.

4 Mechanical Properties

4.1 Compressive strength

Figure 5 shows the average value of compressive strength for each type of concrete for the respective age of curing. The control SCC recorded compressive strength of 42.7, 46.1 and 50.5 MPa, at 7, 28, and 56 days of age, respectively. SCC contains 40% fly ash recorded the highest value of compressive strength which is 27.2, 39.4 and 43.9 MPa at 7, 28 and 56 days, respectively. The lowest value of compressive strength goes to 60% fly ash which is 17.0, 22.6 and 28.5 MPa at 7, 14 28 and 56 days.

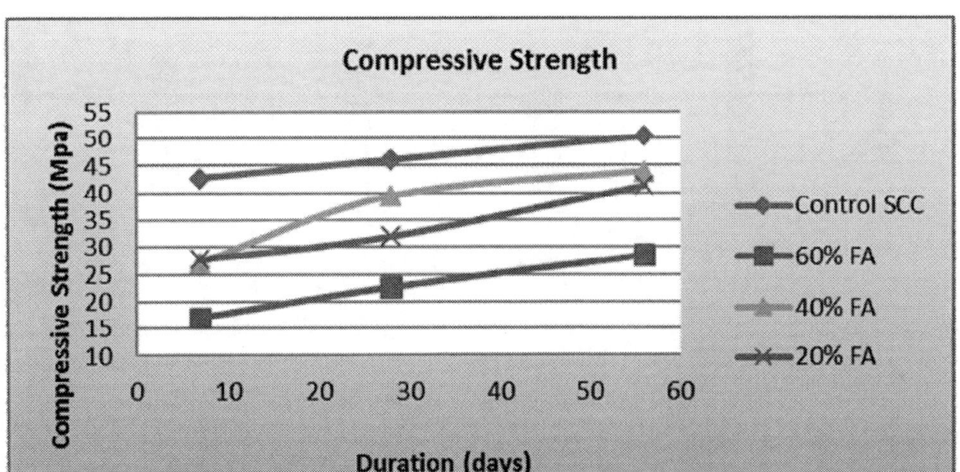

Figure 5. Results for compressive strength.

Figure 6 shows that all of SCC mix design obtained the density within the same range which is 2200kg/m^3-2300kg/m^3. This range of self-compacting concrete density obtained by all of the SCC mix design can be concluded to be in the same range as normal weight concrete which is from 2000 kg/m^3 to 2500 kg/m^3.

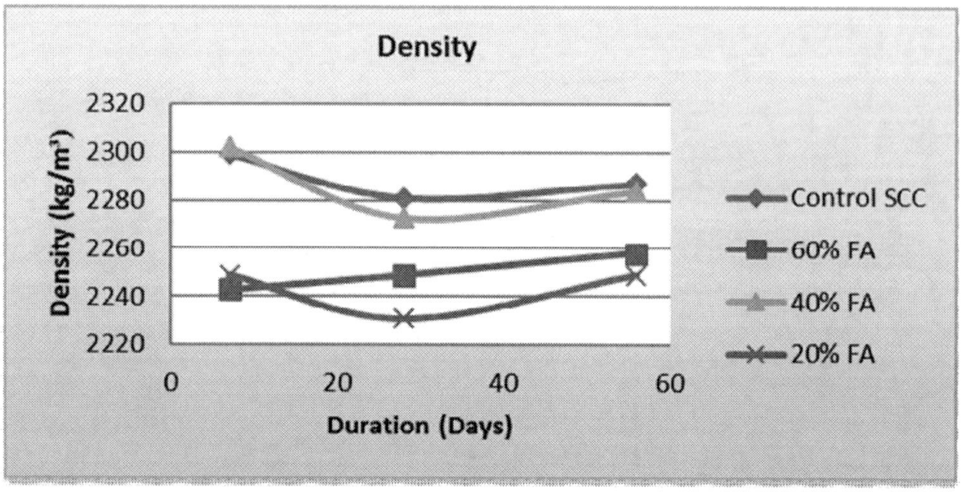

Figure 6. Results for concrete density.

4.2 Modulus of Elasticity

From the results shown in Figure 7, concrete cylinder containing 40% fly ash show the highest value of modulus of elasticity (MOE) which is 31 GPa. In general, results of Modulus of Elasticity for Control SCC and other mixture showed the typical range for the static modulus of elasticity at 28 days based on British Standard, BS 8110 part 2, (1985).

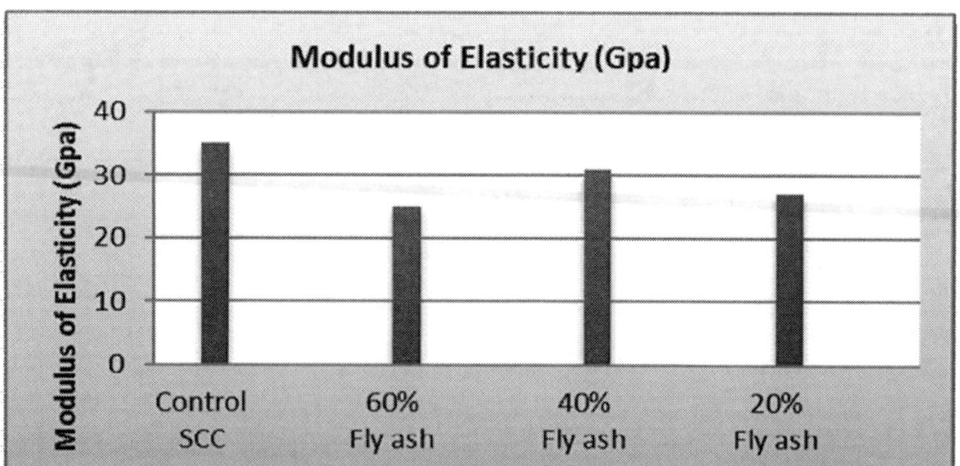

Figure 7. Results of modulus of elasticity.

5 Conclusions

From the analyzed data, the main conclusions for the physical properties are:

i) The workability for fresh concrete increased as the percentage of fly ash increased except for the 60% fly ash. Overall result for Slump flow test (Diameter and T_{500mm}), V-funnel test and J-Ring test showed a stable mixture for concrete was produced from various mixture of SCC with different percentage of fly ash.

ii) A typical concrete mixture has the potential of producing lime. Fly ash chemically reacts with this lime to create more CSH, the same "glue" produced by the hydration of cement and water. 60% of fly ash has higher potential to produce more CSH compared to 40 % of fly ash. This is why the study showed optimum workability was obtained when SCC incorporated with 40% of fly ash.

iii) High percentage of Fly ash can be used to produce SCC with an adequate strength. Using up to 60% fly ash as Portland cement replacement can produce SCC with a strength as high as 30 N/mm^2.

Meanwhile, the main conclusions for the mechanical properties are:

i) The highest compressive strength was recorded by SCC mixture with 40% fly ash with 27.2 MPa, 39.4 MPa and 43.9 MPa at 7, 28 and 56 days, respectively. However, too much percentage of fly ash added will lead to deterioration of strength of the concrete. This is indicated from the results obtained from SCC mixture with 60% fly ash which recorded lowest compressive strength.

ii) The range of concrete density obtained is similar to the range of density in normal concrete which range from 2000 kg/m^3 to 2500 kg/m^3.

iii) Modulus of Elasticity (MOE) is shown to increase with the increase percentage of fly ash except for 60% fly ash. The result of MOE for the all the SCC mixtures are in the typical range for the static modulus of elasticity at 28 days based on British Standard, BS 8110 part 2, (1985).

Acknowledgement

The authors would like to thank Universiti Tun Hussein Onn Malaysia for its support in this research.

References

[1] K.S Sable and M.K. Rathi, Comparison of normal compacted concrete and self compacted concrete in shear and torsion, International Journal of Computer Technology and Electronics Engineering, **2**(4), 74–79, (2012).

[2] N. Bouzoubaa and M. Lachemi, Self-compacting concrete incorporating high volumes of class F fly ash: Preliminary results, Cement and Concrete Research, **31**(3), 413–420, (2001).

[3] H. Okamura and M. Ouchi, Self-compacting concrete, Journal of Advance Concrete Technology, Japan Concrete Institute, **1**(1), 5-15, (2003).

[4] British Standard, Testing Concrete-Part 114: Determination of Density of Hardened Concrete, BS1881-114, (1983).

[5] British Standard, Testing Concrete-Part 116: Method for Determination of Compressive Strength of Concrete Cubes, BS 1881-116, (1983)

[6] British Standard, Testing Concrete-Part 121: Method for Determination of Static Modulus of Elasticity in Compression, BS 1881-121, (1983).

[7] Z. Grdić, I. Despotović and G. Topličić-Ćurčić, Properties of self-compacting concrete with different types of additives, Facta Universitatis - Series: Architecture an Civil Engineering, **6**(2), 173–177, (2008).

MATEC Web of Conferences **47**, 01002 (2016)
DOI: 10.1051/matecconf/20164701002

Flexural Toughness of Ring-Shaped Waste Bottle Fiber Concrete

S. K. Faisal [1,a], J. M. Irwan[1], N. Othman[2] and M. H Wan Ibrahim[1]

[1]*Jamilus Research Centre, Universiti Tun Hussein Onn Malaysia, 86400 Parit Raja, Johor, Malaysia*
[2]*Micropollutant Research Centre, Universiti Tun Hussein Onn Malaysia, 86400 Parit Raja, Johor, Malaysia*

Abstract. Polyethylene terephthalate (PET) bottles are plastic containers that are typically discarded, and thus, cause environmental pollution. To solve this problem, PET bottles are recycled incorporating with concrete. A ring-shaped PET (RPET) fiber are introduced in this study and designed with a special shape to mobilize fiber yielding rather than fiber pullout. Therefore, aim of this paper is to investigate the influence of RPET bottles fibre in terms of toughness strength. The width of RPET fibers is fixed at 5 and 10 mm and the loads were applied to the third points of the specimen. The experiment indicates that RPET-5 and RPET-10 FC presented an increase in the toughness index of I_{20} on averages of 23.1% and 39.9% respectively, compared to normal specimens. It can conclude that incorporating RPET fiber in concrete presents significant improved of concrete properties.

1 Introduction

Polyethylene terephthalate (PET) is one of potential waste which could be used in various applications. Efforts have been made to explore their usage in concrete and one of that is fibre incorporating with concrete [1-3]. Silva et al. [3] mentioned that recycled manufactured monofilament PET fibres increases the toughness indexes of cementitious composites. It is due to the improvement the interfacial bond between fibre and cement. The author used mechanical indentations that create on the fibre surface by pressing the fibre between two hardened steel surfaces. Other authors [4] used embossed type PET fibre reporting that fibre concrete allows a mid-span deflection that is four times higher compared to the plain concrete. This is due to the improvement of interfacial friction energy that gives extra pull-out and friction load resistance in concrete. Ochi et al. [5] mentioned that the used of 30 mm long if intended PET fibres can lead to a flexural strength increase for volume replacements up to 1.5% compared to 20 mm length of fibre. These authors claim that long fibre has a capability interlocking fibre bridge in concrete due to long fibre that could be inserted between aggregates compared to 20 mm length of fibre. Ramadeviet al. [6] studied the possibility of using the waste PET bottle as the partial replacement of aggregate in Portland cement with fibre replacement content 1%, 2%, 4% and 6 %. The finding revealed that the specimens with replacement of fine aggregate with PET bottle fibre increases gradually in flexural strength with increased fibre content. However, the author mentioned that 1% to 2% replacement are reasonable volume fibre content to achieve the optimum strength compared to the other fibre content on average 3.5% to 9.2% significant influence.

[a] Corresponding author : faisalsh@uthm.edu.my

Choi et al. [7] studied performance ring type steel fibre concrete to determine the flexural toughness indices with fibre content from 0.2% to 0.8%. The author mentioned that ring type steel fibre showed effectiveness and superiority to flexural behaviour ring type steel fibre concrete over the straight steel fibre with performs on average 45 %. Pelisseret al.[8] reported that recycle bottle PET with lengths 10, 15, and 20 mm and fibre content of 0.05%, 0.18% and 0.30% presents significant improved properties of concrete. The author founds that 20 mm long PET fibre show significant result on average 10.1% of flexural indices I_{20}.Meanwhile, the addition of recycled PET fibre enhanced absorption and toughness characteristic of concrete on average 4.44% of fibre content from 0.18% to 0.30%.The author also claimed that increased fibre content above than 0.5% indicated that fibre did not affect the flexural toughness compared to the normal concrete. It can be concluded that previous studied [4-6] show capabilities of PET incorporating concrete in term of toughness strength.

Therefore, experiment presents the ring shaped waste bottle (RPET) as fiber concrete. The aim of this experiment is to investigate the influence of ring type of recycled (RPET) bottles fibre in terms of toughness strength according effect of fiber content RPET fiber, and effect of size RPET fiber.

2 Experimental Program

The cement used in all mixtures was a commercially available Portland cement (PC), which corresponds to MS 522: 2007 [9]. The aggregates and fine sand from the same local source were used. The aggregate size was 19 mm and maximum size of fine aggregates conforming to BS EN 12620 [10] is 4.75mm. Fly ash was used in this experiment with 15% additional from Portland cement volume used. The water binder ratio was kept constant at 0.55. RPET FC was basically prepared by adopting the mixing procedure proposed by Liao et al. [11] and Irwan et al. [12]. RPET fibres are made directly from waste bottle as shown in Figure 1.

Waste bottles were cleaned to remove impurities. The waste bottle has been cut manually in a ring shaped. The diameter of the RPET bottles was fixed at 60 ± 5 mm. The width of RPET fiber was fixed at 5 and 10 mm as shown in Figure 2. The concrete mixture was tested to determine the concrete property of toughness strength according to ASTM C1018-97 [13]. All concrete specimens containing seven different content of RPET fiber were tested after 28 days. A steady rate of cross-head displacement at 0.05 mm/min was used. Two rollers were employed as support for the specimens, and the loads were applied on the third points of the specimen. Net deflection was measured with a linear variable differential transducer (LVDT) on both sides of the specimen. The LVDT was connected to data saved in a computer. Figure 3 shows complete setup for the flexural experiment.

Figure 1. The ring bottle of plastic PET bottle.

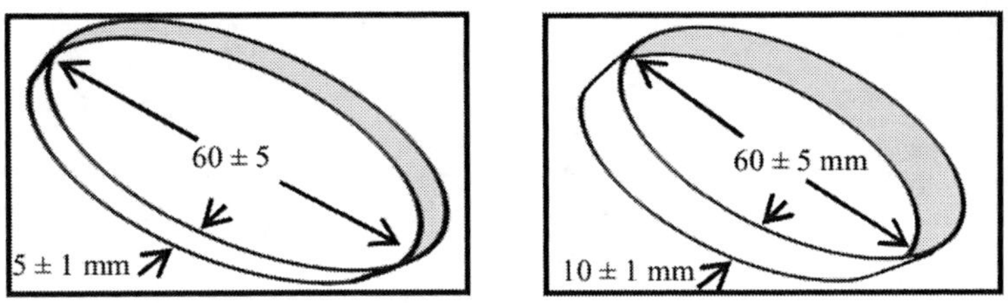

Figure 2. The dimension of ring waste bottles.

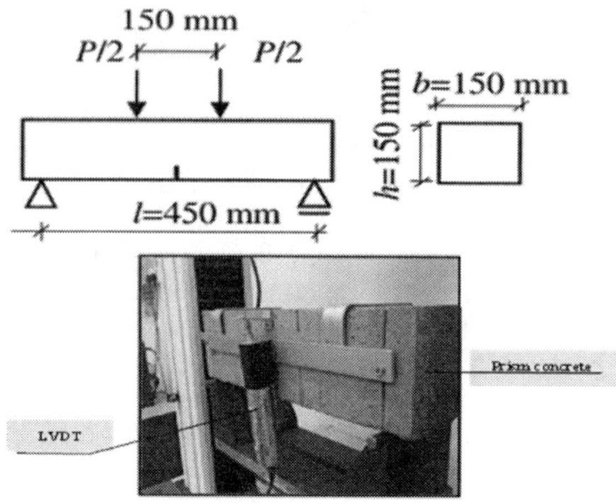

Figure 3. Specimen under third point loading.

3 Result and Discussions

The typical load-deflection response of normal specimen, as well as of RPET-5 and RPET-10 FC, is shown in Figures 4 and 5. A comparison was made between normal and RPET FC specimens, which showed the post-peak response that differentiated normal from RPET FC. Normal specimen behavior in load-deflection is brittle. A crack self-propagates when the strain energy was sufficiently high. Thus, a fracture occurred nearly instantaneously once peak load was reached. This finding is attributed to the tremendous amount of energy being released. RPET fiber in concrete exhibited the fiber bridging effect that helps control the energy release rate. Thus, RPET fiber in concrete can carry the load even after the peak.

The result based on the mentioned figures confirmed that RPET fibers in concrete exhibit a double-peak response in all fiber sizes as shown in Figure 4. Prior to the peak, the load increased proportionally as deflection increased. At the peak (concrete cracking), the load slightly decreased before the fibers began to take over, which caused the load to gradually decrease. In normal specimen, the point of concrete cracking indicates the point of failure or the peak point of the load, and then, the load gradually decreases as deflection increases. This pattern agrees with the finding of Piti [14] who studied the toughness of Polypropylene, PPfiber beams under bending. Piti [14] used 58 mm-long rectangular fibers. The author found that incorporating PP macro fibers into concrete can carry the load after the peak load and double-peak response occurs during load deflection.

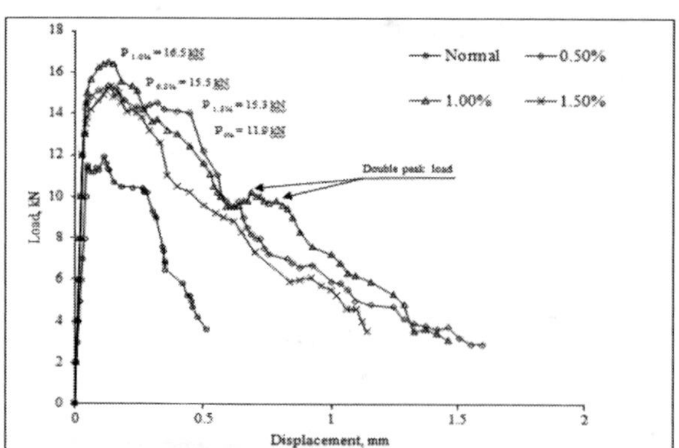

Figure 4. Load-displacement of RPET-5.

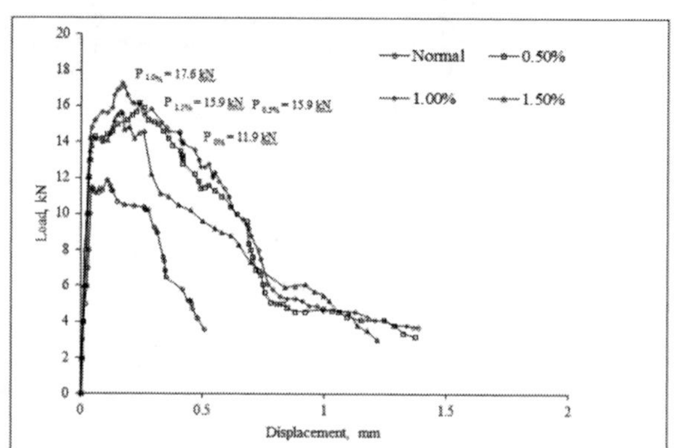

Figure 5. Load-displacement of RPET-10 FC.

3.1 Effect of RPET fiber content on flexural toughness

Figures 6 and 7 show the average value of the toughness indices for RPET-5 and RPET-10 FC, respectively. The toughness indices I_5, I_{10}, and I_{20} for normal specimen are 4.96and 12.87, respectively, as shown in Table 1. Adding 0.5% RPET fiber caused small increase of 5.26 in toughness indices, I_5, I_{10}, and I_{20}, respectively, for RPET-5 FC. Meanwhile, RPET-10 FC obtained 5.29, 10.87, and 14.81. The finding shows a small fiber content similar to that of normal at pre-peak loading or I_5 [7].

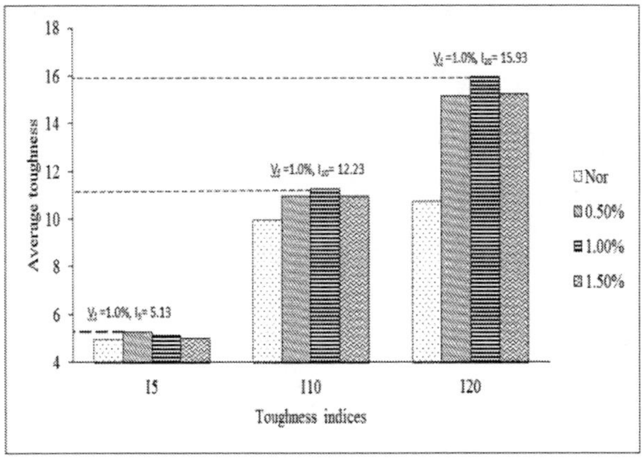

Figure 6. Flexural toughness of RPET-5 FC

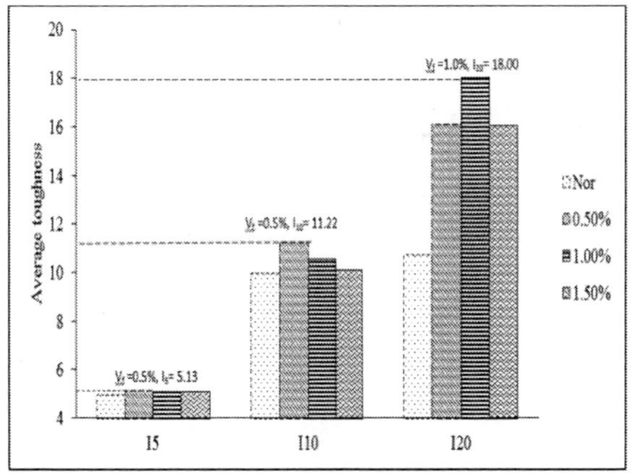

Figure 7. Flexural toughness of RPET-10 FC

The finding also indicated that increasing fiber content in concrete increases the toughness of concrete. The toughness indices I_5, I_{10}, and $I_{20}$for RPET-5, with an average of 1% fiber content, are 5.13, 11.23, and 15.84, respectively compared to normal specimen, whereas RPET-10 FC obtained values of 5.09, 10.19, and 17.26, respectively compared to normal specimen.

The experiment indicates that RPET-5 presented an increase in toughness index,I_{20} of 17.5%, 23.1%, and 17.5% for 0.5%, 1.0% and 1.5% fiber content, respectively, compared with that of normal specimen as shown in Table 1. RPET-10 presented an increase in toughness index, I_{20} of 35.7%, 39.9%, and 24.6% for 0.5%, 1.0% and 1.5% fiber content, respectively, compared with that of normal specimen.

Table 1. Analysis of toughness indices from flexural toughness test.

Fiber	Fiber content, %	Flexural toughness indices					
		I_5	I_{10}	I_{20}	Difference percentage compared to normal, (%)		
					I_5	I_{10}	I_{20}
RPET-5	0	4.96	9.94	12.87	0	0	0
	0.25	5.29	10.81	13.96	6.7	8.8	8.5
	0.50	5.26	10.95	15.12	6.0	10.2	17.5
	0.75	5.09	11.21	15.70	2.6	12.8	22.0
	1.00	5.13	11.23	15.84	3.4	13.0	23.1
	1.25	5.07	11.04	15.76	2.2	11.1	22.5
	1.50	5.02	10.96	15.12	1.2	10.3	17.5
RPET-10	0	4.96	9.94	12.87	0	0	0
	0.25	5.29	10.22	14.81	6.7	2.8	15.1
	0.50	5.13	11.22	17.47	3.4	12.9	35.7
	0.75	5.16	10.89	17.72	4.0	9.6	37.7
	1.00	5.09	10.52	18.00	2.6	5.8	39.9
	1.25	5.12	10.19	17.26	3.2	2.5	34.1
	1.50	5.06	10.09	16.03	2.0	1.5	24.6

Thus, the mechanism of ring-shaped fibers in concrete indicates significant results. Ring-shaped PET fibers provide individual maximum tensile strength by yielding to concrete ruptures [7, 15]. Present study exhibited an increase ranged of 18.6% to 23.8% and 35.7% to 39.9% in the toughness I_{20} index at 1.0% and 1.5% fiber contents, respectively, compared to normal specimen for all size of RPET fiber.

The present study was related to Choi et al. [7] study. Choi et al. [7] presented that ring-shaped steel fibers performed 45% better on average than straight and hooked steel fibers in terms of the toughness index I_{20}. An increase of fiber content leads to improvethe flexural toughness of FC. High fiber content would result in a high amount of fibers crossing a fractured section, and thus, improved failure mechanisms could be mobilized.

3.2 Effect of RPET fiber size on flexural toughness

Based on Figure 8, RPET-10 FC obtained a higher average of 2.9% toughness than RPET-5 FC. It is shows that RPET-10 FC exhibited higher in toughness indices with ranged of 2% to 4.3% compared to RPET-5 FC.

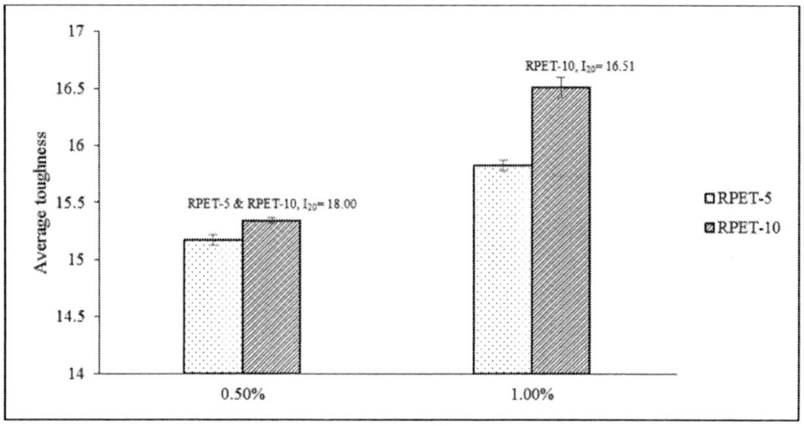

Figure 8. Toughness Indices I_{20} of 0.50 % and 1.00 % fiber content.

Therefore, the present study indicated in line pattern results with Pelisseret al. [8]. The authors found that adding fibers from recycled PET bottles enhances the energy absorption rate and toughness of concrete under flexural load [8]. The 20 mm-long fibers caused an increase of 44% in toughness of the I_{20} index compared with the 10 mm-long fibers at 0.18% fiber content. Meanwhile, the 20 mm-long of fibers resulted in an increase of 30.8% in the I_{20} toughness index compared with the 10 mm-long fibers at 0.3% fiber content. Present study and observed by Pelisseret al. [8] claimed that the size of the fiber area contributes significantly to the toughness indices. Increasing the length of the fibers increases the size of the fiber area that connects to the cement matrix, and thus, contributes to the positive results in the toughness indices, particularly in I_{10} and I_{20}.

4 Conclusions

The experiment indicates that RPET-5 and RPET-10FCpresented an increase in toughness index of I_{20} on averages of 23.1%and 39.9% respectively, compared to normal specimens. This experiment exhibited that the RPET-10 FC presented improvement results compared with RPET-5 because of the size of the surface area that is in contact with concrete. RPET-10 FC presented an increase in toughness index, I_{20} of 7.3% compared to RPET-5 FC.

Acknowledgement

This work was financially supported by University Tun Hussein Onn Malaysia (UTHM) and Minister of Education Malaysia (KPM) through Exploratory Research Grant Scheme (ERGS) Vot. No. E-002.

References

[1] P. Agamuthu and P.N. Faizura, Biodegradability of degradable plastic waste, Waste management, **23**(2), 95-100, (2005).
[2] A.A. Al-Manaseer, T.R. Dalal, Concrete containing plastic aggregates, Concrete International, **19**(8), 47–52, (1997).
[3] D.A. da Silva, A.M. Betioli, P.J.P. Gleize, H.R.Roman, L.A. Gomez and J.L.D. Ribeiro, Degradation of recycled PET fibres in Portland cement-based materials, Cement and Concrete Research, **35**(9), 1741-1746, (2005).
[4] F. Fraternali, V. Ciancia, R. Chechile, G. Rizzano, L. Feo and L. Incarnato, Experimental study of thermo-mechanical properties of recycled pet fiber-reinforced concrete, Composites Structures, **93**(9), 2368-2374, (2011).
[5] T. Ochi, S. Okubo and K. Fukui, Development of recycled PET fibre and its application as concrete-reinforcing fibre, Cement Concrete Composite, **29**(6), 448–455, (2007).
[6] K. Ramadevi and R. Manju, Experimental investigation on the properties of concrete with plastic PET fibre as fine aggregates, International Journal of Emerging Technology and Advanced Engineering, **2**, 2250-2459, (2012).
[7] O.C. Choi and C. Lee, Flexural performance of ring type steel fibre reinforced concrete, Cement and Concrete Research, **33**(6), 841-849, (2003).
[8] P. Fernando, O.R.K. Montedo, P.J.P. Gleize and H. R. Roman, Mechanical properties of recycled PET fibres in concrete, Materials research, **15**(4), 679-686, (2012).
[9] MS EN 196-1, Specification for Portland Cement Composites, Department of Standards Malaysia, (2007).
[10] Eurocode 2, Aggregate for concrete, EN 12620, British Standard Institution, London, (2002).
[11] W. Liao, S. Chao and A. Naaman, Experience self-consolidating high performance fibre reinforced mortar and concrete, American Concrete Institution, **274**(6), 79-94, (2010).
[12] J.M. Irwan, R.M. Asyraf, N. Othman, H.B. Koh, A.K. Aeslina, M.M.K. Annas, and S. K. Faisal, Deflection behaviour of irregular-shaped polyethylene terephthalate fibre reinforced concrete beam, Advanced Materials Research, **911**, 438-442, (2014).
[13] ASTM C1018-97, Standard test method for flexural toughness and first crack load strength of fiber reinforced concrete, The American Society for Testing and Materials Standard, (2006).
[14] S. Piti, Toughness evaluation of steel and polypropylene fiber reinforced concrete beams under bending, Science Technology, **9**(3), 35-41, (2004).
[15] D. Foti, Preliminary analysis of concrete reinforced with waste bottles pet fibers, Construction and Building Materials, **25**, 1906-1915, (2011).

MATEC Web of Conferences **47**, 01003 (2016)
DOI: 10.1051/matecconf/20164701003

A Review on Waste Minimization by Adopting in Self Compacting Concrete

S. Shahidan[1,a], I. Isham[1] and N. Jamaluddin[1]

[1]*Faculty of Civil and Environmental Engineering, Universiti Tun Hussein Onn Malaysia, 86400 Parit Raja, Johor, Malaysia*

Abstract. Self-compacting concrete (SCC) was first developed in late 80's in Japan. SCC is well known for its self-consolidation and able to occupy spaces in the formwork without vibration and become new interesting topic in Construction and Building Materials Research. The aim of this review is to summaries the previous research work related to utilization of waste minimization in SCC from 2009 to 2015 through available literature. It is important to expose new researchers on concept and fundamental theory developed by previous researchers as a reference and guidance in their research. There were a lot of opportunity to be explored in developing SCC especially in utilizing waste material as replacement materials or additives used and mix design method for rheological improvement in SCC. However, these review only focusing on waste materials that have significant to be taken care to reduce environmental impact such as waste product from construction industry and by product industry. As conclusion, this paper will provide significant idea and useful information to those new to SCC and fellow researchers for future studies in utilizing waste materials in SCC mix design.

1 Introduction

Developing Self Compacting Concrete (SCC) was started in late '80s in Japan. It was introduce by Okamura Hajime [1-2] due to shortage of manpower on site and resolve environmental issue such as noise reduction. This concrete technology has received demand from construction industry in Japan to construct bridges, buildings and repair works on site.

In definition, SCC is the technology of concrete that makes use of concrete to flow and fill the gap between reinforcement and fill the void inside the formwork at every corner by it is own weight without vibration process [3-4]. The development of SCC for construction industry in Japan has proved benefit in several factors such as reduce construction duration, required less manpower on site, provide good surface finishes, easier in placing handling, provided freedom idea in structural design, suitable for thinner concrete sections, reduced noise levels and safer working environment [3, 5].

However, the development of SCC still under continues research. This was confirmed by the summary of literature review from published journal since 2009 to 2015 as shown in Table 1. Some of researcher's study on utilizing additive materials, chemical additive, waste materials, light weight aggregate, sand replacement and potential pozzolana materials to improve rheological and hardened properties. However, these summaries only focus on waste materials used in previous research from

[a] Corresponding author : shahiron@uthm.edu.my

Turkey, Greece, India, Thailand, Malaysia, Iran, China, Indonesia, Brazil, Taiwan, Italy, Tunisia and Germany. Hopefully this information can provide an idea for new researcher to explore on SCC by utilizing waste materials that available from construction industry and also from by product industry. The performance of waste materials in SCC is based on fresh state [3, 6-10] and hardened state.

2 Literature from Previous Researches

This review based on information gathered from available literature journal papers since 2009 till 2015. Table 1(a) and (b) have shown the previous research focusing on waste materials in development of sustainable SCC for benefit of economy and environmental issues. It was clearly shown that utilizing waste materials in SCC have significant contribution to environmental solution and indirectly help to contribute in generating economic value to some other industries.

Table 1(a). Summary of previous study base on waste materials used in SCC.

Year	Ref.	Country	Materials	Finding
2009	[11]	Turkey	Mix design using marble powder (MP) (recycle value)	Mix design Monogram. Recycle powder successfully and economically as filler material, f_c (±) 20-60 MPa, using Mix design base on Monogram
	[12]	USA	Mix design using foundry silica dust (SD) as replacement for FA (economic value)	20% SD as economy cost, 30% SD increase demand of HRWA (increase cost of material 14%) , f_c (±) 34-62 MPa
2010	[13]	Greece	Mix design cement paste, SF, Nano SF, LS, P & FA for viscosity effect	Cement paste, 40% LS improved rheological behavior, 20% LS & 20% FA having higher packing density. SF & Pozzolan having lower viscosity
	[14]	India	Mix design cement replacement using SF Vs LS, QD & Cl (comparison water absorption)	Okamura's Method. SF increased dosage SP, improved aggregate matrix bonding & less water absorption (better than QD & LP), LP improved & better workability
2011	[15]	UK	Mix design (Dolomite Powder + FA) Vs (LS + FA). Utilizing DP as alternative materials	Mix by Mass. Utilizing DP with FA can reduced cost SCC & solution to disposal & environmental problem. But, increase DP and decrease FA will reduce compressive strength.
	[16]	Italy	Mix design mineral addition. FA, LP, (Rubble Powder + Fine Recycle) & (RP + Course Recycled)	MP performed better in rheological test. All mix obtained satisfactory in compressive strength (±) 40 MPa.
	[17]	Turkey	Mix design mineral admixtures. FA, GBFS, LP, BP and MP for cement replacement	FA & GBFS increased compressive strength & workability, replacement FA in OPC suitable for HPC
2012	[18]	Algeria	Mix design using various sand (Crushed Sand, River Sand, Dund Sand & mix sand)	Okamura's Method. Increase fine LS, decrease slump flow time & increased V Funnel flow time, 10-15% fine LS improved rheological and compressive strength of mortar but more than that will decreased it strength. DS required higher water demand and larger amount of cement compared with RS & CS
	[19]	Turkey	Mix design MP, LS & FA as filler (Environmental Concern; disposal & recycled materials)	Integration of MP & LS with FA improved workability, MP & LS increase dosage of SP but integration with FA decreased amount of SP.
2013	[20]	Brazil	Mix design using kaolin waste as agg. replacement	Mix using concept of aggregate skeleton (increase compactness will increase density), only focus on fresh test
	[21]	Taiwan	Mix design FA & slag with difference w/c ratio	Densified mixture design algorithm (DMDA), DMDA minimize paste content & overcome aggregate gradation, provide high slump without bleeding & segregation, used less cementitious materials ratio will effect early strength but not for long term strength (after 56 days)
	[22]	Indonesia	Mix design using roof tile powder (RTP)	20% RTP, compressive strength 67.72 MPa, good flow ability
	[23]	UK	Mix design using FA & LP as a filler	30% FA for sustainable SCC with approximately 1.83% SP (by weight), compressive strength 50-60 MPa,
	[24]	India	Mix design using recycled aggregate	10-40% RA, increase RA will decrease compressive, flexural & tensile strength

Table 1(b). Summary of previous study base on waste materials used in SCC.

Year	Ref.	Country	Materials	Finding
2013	[25]	Thailand	Mix design using untreated RHA & PFA	Compressive strength of 40% RHA decreased 1/3 from control sample
	[26]	Thailand	Mix design using cathode ray tube waste	Increase CRT amount will decrease compressive strength
	[27]	Thailand	Mix design using recycled alumina waste (AW) as fine aggregate replacement	AW increased amount of SP, AW increased compressive strength over the period, 75% AW have better compressive strength than control sample
2014	[28]	Greece	Mix design using ladle furnace slag (LFS) and steel fiber reinforcement	LFS as filler improved compressive strength & durability (freezethraw resistance & chloride penetration) with consistence workability, LFS with fibre increase fracture toughness but to many fibre will effect segregation and flowability, increase LFS decrease water cement ratio
	[29]	Turkey	Mix design using Low Lime Fly Ash (LLFA) & GBFS as cement replacement (binder) & Mircronized Calcite (MC) as aggregate replacement	LLFA, GBFS & MC positively improved passing ability & flow ability & reduce viscosity (due to MC), surface roughness LLFA reduce water demand, substitution high volume of GBFS & LLFA up to 60% (compressive strength 42.58 – 52.7 MPa)
	[30]	Iran	Mix design effect of aggregate sizes	Decrease of aggregate size will drop fracture energy
	[31]	Turkey	Mix design utilizing of VMA to reduce fine material	Increase fine material or use VMA to avoid segregation, VA xanthan gum @ 400 kg binder content & VA starch ether (@ 350, 400 & 450 kg binder dosage) reduce bleeding to zero, at some portion segregation will effect compressive strength, W/C ratio significantly effect the compressive strength same as NVC
	[32]	Malaysia	Mix design using Palm Oil Clinker (POC) as LWA	Mix design base on particle packing (PP) can be used to predict hardened properties,
	[33]	UK	Mix design using Recycled Lime Powder (RLP) & Recycled Concrete Aggregate (RCA)	RLP as cement replacement & RCA as aggregate replacement have potential to produce good quality (mechanical & durability performance)
	[34]	Malaysia	Mix design using Metakaolin (MK) with difference course aggregate properties	10 % MK as pozzolan by weight of cement good in workability and strength, same w/b ratio & SP dosage but increase sand content (mortar) will reduce flow ability (lead to bleeding rate of water), SP increase with increasing of MK, higher SP dosage contribute to bleeding, shape & fineness powder effect SP dosage,
	[35]	Malaysia	Mix design using RHA and blended fine aggregate	RHA as supplementary of cement, suitable for normal strength, compressive strength & splitting tensile strength decreased with increasing of RHA, RHA contain silica which not contribute to compressive strength
	[36]	Turkey	Mix design using waste marble (WM) and recycled aggregates (RA)	Crushed limestone agg. as control sample, WM & RA produce slightly lower compressive strength compared with CLS, strength can be improved by reduce w/b ratio
	[37]	Malaysia	Mix design using Bottom Ash	10-15% BA suitable for fine aggregate replacement, more than that it will effect the rheology properties
2015	[38]	India	Mix design using RHA	RHA as supplementary cementitious material, 15% RHA increased compressive strength & tensile strength and also reduced chloride ion penetration
	[39]	China	Mix design using FA & GGBFS	Combination FA & GBFS (20, 30 & 40%) not affect flowability & stability, effective on chloride ion migration, but higher carbonation depth
	[40]	China	Mix design using recycled plastic particles	For light weight SCC, Compressive strength, tensile strength & flexural tensile strength increased with replacement level up to 15%,
	[41]	Germany	Mix design containing various mineral admixtures for high performent concrete (HPC)	Introduced new mix design method containing various mineral admixtures by using Absolute Volume Method (AVM) – adopted from packing theory, RHA improved compressive strength (marginally lower from SF)
	[42]	Tunisia	Mix design using marble and tile waste as filler	Mix design using "Concrete LabPro2" software, compressive strength & tensile strength are sufficient
	[43]	Algeria	Mix design using seashell as fine aggregate for mortar	For fine aggregate by sand substitution for mortar, 100% replacement reduce compressive strength & MOE

3 Mix design methods

SCC has been introduced in Japan since 1986 by Okamura using Empirical Design Method [2, 44]. Through summary review from Table 1, it was shown that development of mix design method has been improvised from time to time by researchers around the world. In 2015, Caijun have summarized a review on mix design method for SCC. Base on his reviewed, mix design method can be classified in five categories which are Empirical Design Method, Compressive Strength Method, Close Aggregate Packing Method, Statistical Factorial Method and Rheology of Paste Model. However, all this method still cannot fully meet requirement for effective cost, sustainability, widely applicable due to variable of alternative raw material used and technical requirement while in fresh and hardened state of SCC [44].

From Table 1, it was confirmed that researchers keep on looking on developing mix design method for SCC. For example, it was found that, 3 new methods have been identified such as Monogram Method [11], Absolute Volume Method [41] and "Concrete LabPro2" software [42]. However, those 3 researchers are using difference types of material in their mix design. They have used MP as filler material [11], RHA, SF, FA, LS powder as mineral admixtures for cementations efficiency [41] and marble and tile waste as filler material[42]. Therefore, the ability of each materials to performance with difference mix design method should be open to evaluate as far as the performance in fresh state are well check with the recommendation by [3, 6-10].

4 Materials

From Table 1, the most common addition materials are FA, SF, GBFS and LS [3, 45]. While currently researchers have looking forward to make used of recycled material such as Quarry Dust (QD), Dolomite Powder (DP), Limestone (LS), Rubble Powder (RP), Recycled Aggregate, Marble Powder (MP), Cement Klin Dust (CKD), Roof Tiles, Rise Husk Ash (RHA), Alumina Waste (AW), Recycled Plastic Particles (RPP), Marble Waste (MW), Tile Waste (TW), Metakaolin (MK), Basalt Powder (BP) and Palm Oil Clinker (POC). All these additions are used as replacement material or substitution materials either for cement, fine aggregate or filler materials.

Some of this material influence SCC performance in positive effect and some of the materials reflect on negative result either for fresh state or hardened state. Majority of researcher having consistent result in their studies such as FA suitable for cement replacement and fine materials will effect amount of SP required, w/b ratio and w/c ratio. However, uncontrolled amount or proportion of replacement material will reduce the compressive strength, tensile strength, MOE and rheology of the SCC. Therefore the performance of fresh state are compulsory to be checked according to the recommendation by [3, 6-10].

Based on Table 1, it shown that utilizing waste material in SCC have getting attention among researchers. Some of potential waste materials either from construction industry or from by product industry and there are commonly available in every country in all over the world. Those waste materials are abundant and causes environmental pollution such as roof tile [22], recycle aggregate [24, 33, 36], waste marble [11, 17, 36, 42], RHA [25, 35, 38, 41] and recycled plastic particle [40]. The performances of utilizing those materials are subjective to the purposed of the study and should be explored and established from time to time for the benefit of all parties. A mechanism to utilize all those potential recycled waste materials should be integrated between all parties either from industry or local authorities. This is important to create a better future to our young generation in order to live in sustainable environment.

As conclusion, no doubt, SCC is a potential concrete technology to be developed by utilizing potential alternative materials or waste materials in all over the world. Utilizing of recycled materials and by product materials should be looking forward as an advantage for new researchers especially in Malaysia. It is not only generate or contribute to new knowledge but it is also helping us to save our environment from destructing due to current generated industrial waste in all over the world.

5 Conclusions

From this summary review, it was found that the development of SCC could be explored by using difference types of waste materials especially using by product waste from industry to reduce environmental issues. Indirectly, this effort will provide better solution in concrete technology and create better benefit in future economic value to the industries and local community. Sustainable values on developing future concrete industry should be expended through quality research among industry and collaborate with higher institutions to create sustainable situation to all parties. Beside that, integration with local authorities laws and regulations will generate better opportunity to industries and researchers to collaborate on waste minimization and utilization. Therefore utilizing of waste materials in developing sustainable SCC should be supported, explored and expended through focus researches. Development of SCC should be verified through performance on fresh state and hardened state to understand the behaviour of each material either in short term or long term effect.

Acknowledgement

This work was supported by Higher Education Ministry Malaysia and Universiti Tun Hussien Onn Malaysia

References

[1] K.O.H. Okamura and K. Maekawa, High Performance Concrete, Gihoubou Pub. (in Japanese), (1993).
[2] H. Okamura, Self-compacting high-performance concrete, Concr. Int., 50–54, (1997).
[3] EFNARC, Specification and Guidelines for Self-Compacting Concrete, United Kingdom, (2002).
[4] H. Okamura and M. Ouchi, Self-Compacting Concrete, Adv. Concr. Technol., **1**(1),. 5–15, (2003).
[5] W. Zhu and P.J.M. Bartos, Permeation properties of self-compacting concrete Wenzhong, Cem. Concr. Res., **21**(7), 921–925, (2003).
[6] BS EN12350-Part 8, Testing Fresh Concrete, Self-Compacting Concrete, Slump Flow Test, (2010).
[7] BS EN12350-Part 9, Testing Fresh Concrete, Self-Compacting Concrete, V Funnel Test, (2010)
[8] BS EN12350- Part10, Testing Fresh Concrete, Self-Compacting Concrete, L Box Test, (2010).
[9] BS EN12350-Part 11, Testing Fresh Concrete, Self-Compacting Concrete, Sieve Segregation Test, (2010).
[10] BS EN12350-Part12, Testing Fresh Concrete, Self-Compacting Concrete, J Ring Test, (2010).
[11] K.E. Alyamaç and R. Ince, A preliminary concrete mix design for SCC with marble powders, Constr. Build. Mater., **23**(3), 1201–1210, (2009).
[12] R.N. Kraus, T.R. Naik, B.W. Ramme and R. Kumar, Use of foundry silica-dust in manufacturing economical self-consolidating concrete, Constr. Build. Mater., **23**(11), 3439–3442, (2009).
[13] N. Diamantonis, I. Marinos, M.S. Katsiotis, A. Sakellariou, A. Papathanasiou, V. Kaloidas and M. Katsioti, Investigations about the influence of fine additives on the viscosity of cement paste for self-compacting concrete, Constr. Build. Mater., **24**(8), 1518–1522, (2010).
[14] C. Selvamony, M.S. Ravikumar, S.U. Kannan and S.B. Gnanappa, Investigations on Self-Compacted Self-Curing Concrete Using Limestone Powder and Clinkers, ARPN J. Eng. Appl. Sci., **5**(3), 1–6, (2010).
[15] S. Barbhuiya, Effects of fly ash and dolomite powder on the properties of self-compacting concrete, Constr. Build. Mater., **25**(8), 3301–3305, (2011).
[16] V. Corinaldesi and G. Moriconi, The role of industrial by-products in self-compacting concrete, Constr. Build. Mater., **25**(8), 3181–3186, (2011).
[17] M. Uysal and M. Sumer, Performance of self-compacting concrete containing different mineral admixtures, Constr. Build. Mater., **25**(11), 4112–4120, (2011).

[18] B. Benabed, E.H. Kadri, L. Azzouz and S. Kenai, Properties of self-compacting mortar made with various types of sand, Cem. Concr. Compos., **34**(10), 1167–1173, (2012).
[19] M. Gesoğlu, E. Güneyisi, M.E. Kocabağ, V. Bayram and K. Mermerdaş, Fresh and hardened characteristics of self compacting concretes made with combined use of marble powder, limestone filler, and fly ash, Constr. Build. Mater., **37**, 160–170, (2012).
[20] G. Azeredo and M. Diniz, Self-compacting concrete obtained by the use of kaolin wastes, Constr. Build. Mater., **38**, 515–523, (2013).
[21] Y.Y. Chen, B.L.A. Tuan and C.L. Hwang, Effect of paste amount on the properties of self-consolidating concrete containing fly ash and slag, Constr. Build. Mater., **47**, 340–346, (2013).
[22] B. Herbudiman and A.M. Saptaji, Self-compacting concrete with recycled traditional roof tile powder, Procedia Eng., **54**, 805–816, (2013).
[23] M.K. Mohammed, A.R. Dawson and N.H. Thom, Production, microstructure and hydration of sustainable self-compacting concrete with different types of filler, Constr. Build. Mater., **49**, 84–92, (2013).
[24] K.C. Panda and P.K. Bal, Properties of self compacting concrete using recycled coarse aggregate, Procedia Eng., **51**, 159–164, (2013).
[25] G. Sua-iam and N. Makul, Use of Unprocessed Rice Husk Ash and Pulverized Fuel Ash in the Production of Self-compacting Concrete, IERI Procedia, **5**, 298–303, (2013).
[26] G. Sua-iam and N. Makul, Use of limestone powder during incorporation of Pb-containing cathode ray tube waste in self-compacting concrete, J. Environ. Manage., **128**, 931–940, (2013).
[27] G. Sua-Iam and N. Makul, Use of recycled alumina as fine aggregate replacement in self-compacting concrete, Constr. Build. Mater., **47**, 701–710, (2013).
[28] E.K. Anastasiou, I. Papayianni and M. Papachristoforou, Behavior of self compacting concrete containing ladle furnace slag and steel fiber reinforcement, Mater. Des., **59**, 454–460, (2014).
[29] H.Y. Aruntas, Workability and mechanical properties of self-compacting concretes containing LLFA , GBFS and MC, **73**, 626–635, (2014).
[30] M.H.A. Beygi, M.T. Kazemi, I.M. Nikbin, J. Vaseghi Amiri, S. Rabbanifar and E. Rahmani, The influence of coarse aggregate size and volume on the fracture behavior and brittleness of self-compacting concrete, Cem. Concr. Res., **66**, 75–90, (2014).
[31] I.E. Isik and M.H. Ozkul, Utilization of polysaccharides as viscosity modifying agent in self-compacting concrete, Constr. Build. Mater., **72**, 239–247, (2014).
[32] J. Kanadasan and H.A. Razak, Mix design for self-compacting palm oil clinker concrete based on particle packing, Mater. Des., **56**, 9–19, (2014).
[33] A. Kanellopoulos, D. Nicolaides and M.F. Petrou, Mechanical and durability properties of concretes containing recycled lime powder and recycled aggregates, Constr. Build. Mater., **53**, 253–259, (2014).
[34] O.R. Khaleel and H. Abdul Razak, Mix design method for self compacting metakaolin concrete with different properties of coarse aggregate, Mater. Des., **53**, 691–700, (2014).
[35] M.E. Rahman, A.S. Muntohar, V. Pakrashi, B.H. Nagaratnam and D. Sujan, Self compacting concrete from uncontrolled burning of rice husk and blended fine aggregate, Mater. Des., **55**, 410–415, (2014).
[36] T. Uygunoğlu, I.B. Topçu and A.G. Çelik, Use of waste marble and recycled aggregates in self-compacting concrete for environmental sustainability, J. Clean. Prod., **84**, (2014).
[37] N.E. Zainal Abidin, M.H. Wan Ibrahim, N. Jamaluddin, K. Kamaruddin and A. F. Hamzah, The Effect of bottom ash on fresh characteristic, compressive strength and water absorption of self-compacting concrete, Appl. Mech. Mater., **660**, 145–151, (2014).
[38] D. Chopra, R. Siddique and Kunal, Strength, permeability and microstructure of self-compacting concrete containing rice husk ash, Biosyst. Eng., **130,** 72–80, (2015).
[39] H. Zhao, W. Sun, X. Wu and B. Gao, The properties of the self-compacting concrete with fly ash and ground granulated blast furnace slag mineral admixtures, J. Clean. Prod.**,** **95**, 66–74, (2015).
[40] S. Yang, X. Yue, X. Liu and Y. Tong, Properties of self-compacting lightweight concrete containing recycled plastic particles, Constr. Build. Mater., **84**, 444–453, (2015).

[41] H. Thanh, M. Müller, K. Siewert, and H. Ludwig, The mix design for self-compacting high performance concrete containing various mineral admixtures, J. Mater., **72**, 51–62, (2015).
[42] M. Tennich, A. Kallel and M. Ben Ouezdou, Incorporation of fillers from marble and tile wastes in the composition of self-compacting concretes, Constr. Build. Mater., **91**, 65–70, (2015).
[43] B. Safi, M. Saidi, A. Daoui, A. Bellal, A. Mechekak and K. Toumi, The use of seashells as a fine aggregate (by sand substitution) in self-compacting mortar (SCM), Constr. Build. Mater., **78**, 430–438, (2015).
[44] C. Shi, Z. Wu, K. Lv and L. Wu, A review on mixture design methods for self-compacting concrete, Constr. Build. Mater., **84**, 387–398, (2015).
[45] ERMCO, The European Guidelines for Self-Compacting Concrete, Eur. Guidel. Self Compact. Concr., (2005).

MATEC Web of Conferences **47**, 01004 (2016)
DOI: 10.1051/matecconf/20164701004

Study on Chloride Ion Penetration Resistance of Rubberized Concrete Under Steady State Condition

Nurazuwa Md Noor[1,a], Daisuke Yamamoto[2], Hidenori Hamada[2] and Yasutaka Sagawa[2]

[1]*Jamilus Research Center, Universiti Tun Hussein Onn Malaysia, 86400 Parit Raja, Johor, Malaysia*
[2]*Department of Civil and Structural Engineering, Kyushu University, Nishi-Ku, Fukuoka 819-0395, Japan*

Abstract. Foamed concrete is a controlled low density ranging from 400kg/m^3 to 1800kg/m^3, and hence suitable for the construction of buildings and infrastructures. The uniqueness of foamed concrete is does not use aggregates in order to retain low density. Foamed concrete contains only cement, sand, water and foam agent. Therefore, the consumption of cement is higher in producing a good quality and strength of foamed concrete. Without the present of aggregates, the compressive strength of foamed concrete can only achieve as high as 15MPa. Therefore, this study aims to introduce the pelletized coconut fibre aggregate to reduce the consumption of cement but able to enhance the compressive strength. In the experimental study, forty-five (45) cube samples of foamed concrete with density 1600kg/m^3 were prepared with different volume fractions of pelletized coconut fibre aggregate. All cube samples were tested using the compression test to obtain compressive strength. The results showed that the compressive strength of foamed concrete containing 5%, 10%, 15% and 20% of pelletized coconut fibre aggregate are 9.6MPa, 11.4MPa, 14.6MPa and 13.4MPa respectively. It is in fact higher than the controlled foamed concrete that only achieves 9MPa. It is found that the pelletized coconut fibre aggregate indicates a good potential to enhance the compressive strength of foamed concrete.

1 Introduction

Concrete structure exposed to aggressive environment such as marine environment and structure exposed to cold weather has tendency to received high penetration of chloride ion into concrete pores which leads to the corrosion of steel reinforcement. Corroded steel reinforcement tends to distress the concrete and high cost of repair and maintenance are required to overcome this problems. In this research, chloride transportation characteristics into concrete was studied on concrete with crumb rubber, CR as a fine aggregate in order to provide good resistance against chloride ion penetration. Study on used tire in concrete namely rubberized concrete started in early 1990's both on strength characteristics and durability performance. With the increasing of population and traffic, the utilization of used tire is important to controlled negative impact to the environment. Thus, this research was conducted to study the effectiveness of crumb rubber as constituent for concrete in providing good resistance against chloride ion penetration.

[a] Corresponding author : nurazuwa@uthm.edu.my

2 Materials Preparation

2.1 Mix design

The mix design was divided into three main group of water to cement ratio, w/c as shown in Table 1. Each group contains CR at 0%, 10%, 15% and 20% and it was added in the concrete as sand replacement by volume. SF was substituted in the mixture with w/c = 35% in order to study the effectiveness of mineral admixture on strength and chloride ion diffusion. Mixing process was done in the controlled temperature room at 20°C. Coarse aggregate was firstly added in the mixer, followed by binder and sand. CR and sand were mixed sufficiently until all CR completely added. All these materials was dry mixed and after 30 seconds, water was added and mixing was continued for an about 90 seconds. Then, mixer was stopped to allow hand mixing to ensure all materials in the mixer were homogeneously mixed. Mixing was finally continued for 60 seconds and stop, which makes total mixing time, became 3 minutes.

Table 1. Mix proportion of rubberized concrete.

Group	CR/(S+CR)	SF/C	w/c	Water	Cement	Silica Fume	Sand	Crumb Rubber	Coarse Aggregates		Chemical Admixture		
				W	C	SF	S	CR	G1	G2	Ether-based polycarboxylate superplasticizers	Air-entrained agent	Air-modifying agent
	(Vol %)	(%)		kg/m^3							%	%	%
Control	0	0	0.50	165	330	0	790	0	636	329	0.5	0.8	
10CR-0SF	10						711	35			0.5	0.8	
15CR-0SF	15						671	53			0.5	0.8	
Control	0	0	0.35	160	457	0	741	0	608	405	0.5	0.8	
10CR-0SF	10						667	34			0.5	0.8	
15CR-0SF	15						629	50			0.7	0.8	
20CR-0SF	20						594	67			0.7	0.7	
10CR-10SF	10	10	0.35	160	457	46	613	34	608	405	0.7		1.5
15CR-10SF	15						575	50			0.7		2.2
20CR-10SF	20						540	67			0.7		2.8

2.2 Steady state condition

Steady state condition is a process where chloride ion is allowed to migrate constantly into the pore solution of concrete under 15V voltage [2]. Figure 1 shows the migration cell in order to measure chloride, Cl^- ion that migrate from cathode towards anode and it achieved steady state condition when increment rate of Cl^- in anode side becomes constant in time as described in Figure 2. The starting sodium hydroxide, NaOH solution in the anode side is 0.5mol/L and 0.3mol/L of sodium chloride, NaCl in the cathode side. Along the testing process, the chloride ion concentration in the anode shall be less than 0.05mol/l, meanwhile as for cathode; the chloride ion concentration shall be more than 0.45mol/l. Solution was completely renewed if any side of anode or cathode solution did not meet this limitation. Test was carried out on Ø100 x 50mm thickness rubberized concrete, cut from cylindrical concrete of size Ø100 x 200mm length.

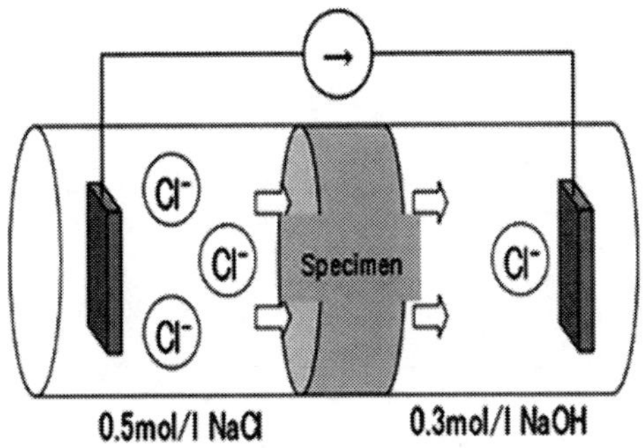

(b) Schematic diagram

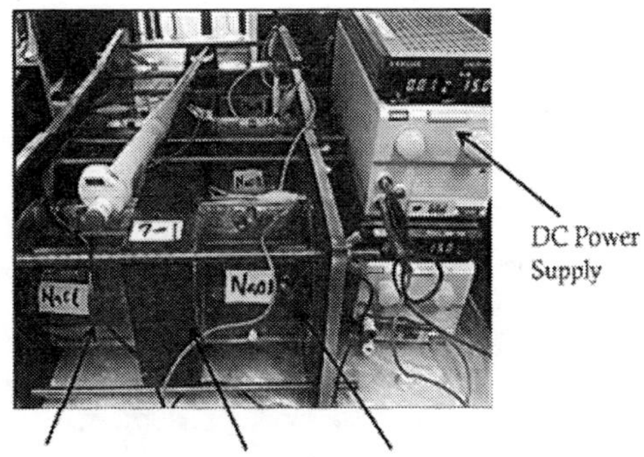

(a) Experimental setup

Figure 1. Migration cell.

Samples from the solution was taken to determine effective diffusion coefficient, D_e by using Nernst-Planck equation,

$$D_e = \frac{J_{Cl} RTL}{|Z_{cl}| FC(\Delta E - \Delta E_C)} .100 \tag{1}$$

where, D_e is effective diffusion coefficient in cm^2/year, R is gas constant = 8.31J/mol K, T is absolute temperature in K units, Z_{Cl} is charge of chloride ion = -1, F is the Faraday constant = 96,500 C/mol, C_{Cl} is measured chloride ion concentration in cathode side in mol/l, ΔE-ΔE_c is electrical potential difference between specimen surfaces in V and L is specimen length in mm.

3 Results and Discussion

Results of D_e are shown in Figure 2 to Figure 5. It was clearly seen that when CR is added in all mixture, resistance against chloride ion migration into the pore solution of concrete was improved. Even though, the effect in each group was slightly reduce due to small amount of CR replacement by volume, but the ability of CR in improving chloride transport characteristics cannot be neglected.

More than 50% of D_e reduction was observed in w/c = 0.50 when 10% CR was added compared to control mixture as shown in Figure 3. When w/c was reduced from 50% to 35%, the resistance against chloride penetration was much improved from 1.538cm^2/year to 0.434cm^2/year due to the increasing of binder in concrete that leads to dense concrete. The D_e in 35% w/c mix kept decreasing slightly with the increasing of CR as shown in Figure 4.

The highest resistance was observed when 10% SF was used as additional binder and the reduction was about 60% to 65% in comparison to control mix as shown in Figure 5. The ultrafine particles of SF allowed voids between cement particles and aggregate to be filled by SF [3]. This led to the porosity reduction and provides denser concrete. In addition, results on compressive strength were presented together with the D_e and according to the literature, relationship between chloride ion migration through pore solution of concrete and compressive strength is linked to the pore structure of the rubberized concrete [4]. From this relationship, it indicates that the pore structure of the rubberized concrete was still under accepted level; where positive improvement in chloride ion transport was observed even though compressive strength was reduced.

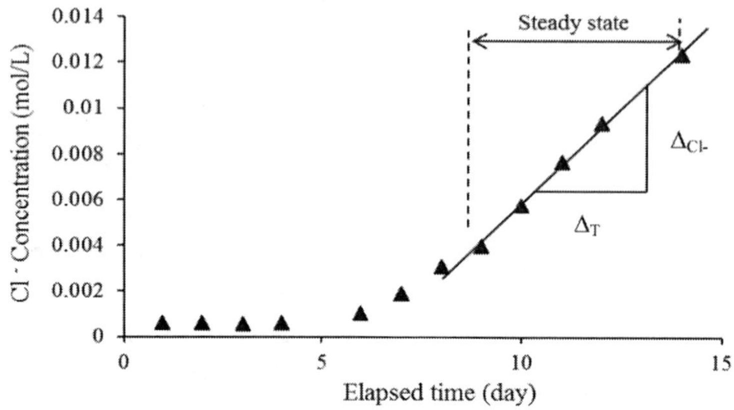

Figure 2. Steady state condition of chloride ion migration test.

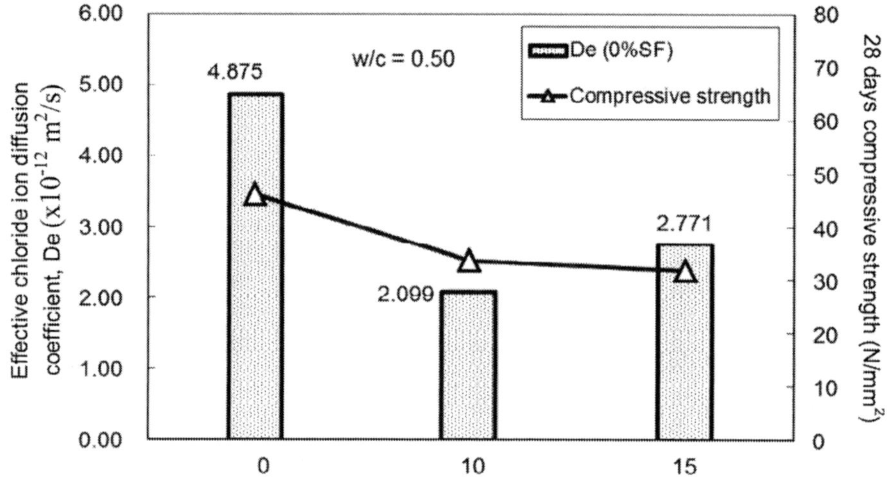

Figure 3. Effective diffusion coefficient and 28 days compressive strength of rubberized concrete with w/c = 0.50 (without SF).

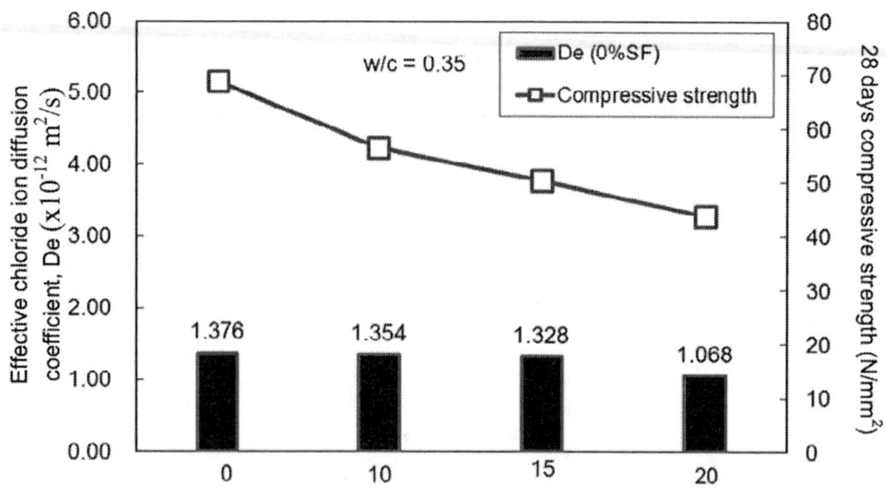

Figure 4. Effective diffusion coefficient and 28 days compressive strength of rubberized concrete with w/c = 0.35 (without SF).

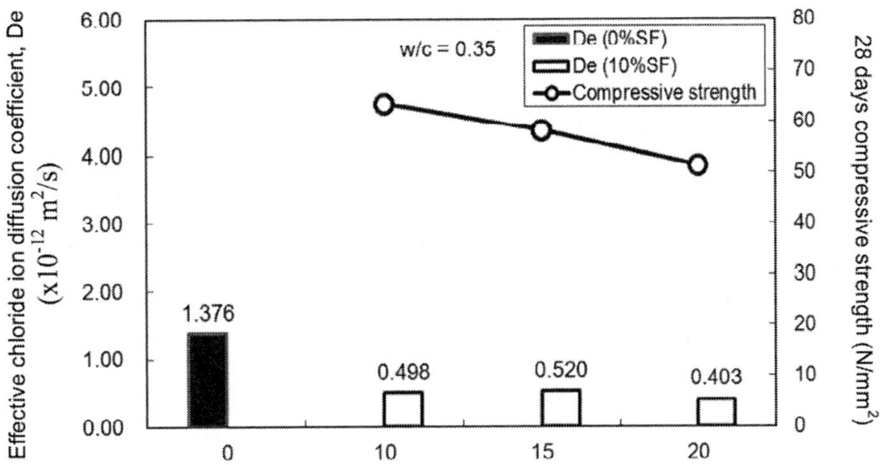

Figure 5. Effective diffusion coefficient and 28 days compressive strength of rubberized concrete with w/c = 0.35 (with 10% SF).

4 Conclusions

Several conclusions can be drawn from this research as follows,

i) Effect of the CR in rubberized concrete was clearly seen in w/c=0.50 and mix with SF in w/c=0.35.
ii) Chloride transport characteristics were improved by increasing amount of CR due to the fact that CR has the ability to repel water.
iii) Resistance of rubberized concrete against chloride ion was much improved when 10% SF was added as a binder.
iv) It was observed that compressive strength was reduce to 43% in 20% CR replacement without SF but it increasing slightly with the present of SF.

Acknowledgement

Special thanks to Hikari World Company Limited for their support in supplying the crumb rubber. Grateful acknowledgement was also dedicated to Malaysian Ministry of Higher Education in providing financial support to the first author.

References

[1] M.N. Nurazuwa, H. Hamada, Y. Sagawa and D. Yamamoto, Strength charactersitics of mortar mixed with tire crumb rubber as fine aggregates, *Proceeding of 67th Conference of Cement Technology*, (2013).
[2] JSCE-G571-2003, Test method for effective diffusion coefficient of chloride ion in concrete by migration, JSCE Guidelines for Concrete No. 2, (2004).
[3] M. Gesoğlu and E. Güneyisi, Strength development and chloride penetration in rubberized concretes with and without Silica Fume, Materials and Structures, **40**, 953-964, (2007).
[4] Cement Concrete and Aggregates, Chloride Resistance of Concrete, Australia, (2009).
[5] G.S. Kumaran, M. Nurdin and M. Lakshmipathy, A review on construction technologies that enables environmental protection: rubberized concrete, AJEAS1, **1**, 41-45, (2008).

MATEC Web of Conferences **47**, 01005 (2016)
DOI: 10.1051/matecconf/20164701005

The Potential of Artificial Polyethylene Coarse Aggregate (APECA) on Compressive Strength of Concrete After Exposed by Temperatures

Shahrul Niza Mokhatar[1,a], Josef Hadipramana[1], Siti Nurhidayah Mohd Isa[2] and Mohamed Muftah Mustafa[2]

[1]*Jamilus Research Center, Universiti Tun Hussein Onn Malaysia, 86400 Parit Raja, Johor, Malaysia*
[2]*Faculty of Civil and Environmental Engineering, Universiti Tun Hussein Onn Malaysia, 86400 Parit Raja,Johor, Malaysia*

Abstract. The strength contribution of Artificial Polyethylene Coarse Aggregate (APECA) as a coarse aggregate replacement in concrete are explored. The compression tests is carried out on twenty-one specimens of concrete cube with 0%, 3%, 6%, and 9% of APECA by utilizing two stages: The concrete specimens containing 6% of APECA as coarse aggregate are heated at different temperatures, the strength of concrete mixes with 0%, 3%, 6% and 9% APECA were compared under normal condition (un-heated) and after exposed by 250°C. Based on literatures, 6% of APECA had an optimal percentage that contribute to the compressive strength of concrete compared to 0%, 3%, and 9% of APECA concrete. Thus, 6% ratio of APECA concrete was applied for continuous observation, which applied various temperatures at 150°C, 200°C, and 250°C. As results, it is observed that the 6% APECA concrete at 150°C had the highest compressive strength than other temperature (200°C and 250°C). Also, it is found that the APECA concrete strength decrease as exposed by higher temperature. However, Scanning Electron Microscope (SEM) results reveals that the APECA melting in the concrete and filled the micro-crack of concrete. It can be concluded that, the APECA has the potential to be a coarse aggregate replacement and more detailed investigations are however demanded to modify the APECA in order to increase its physical properties.

1 Introduction

Nowadays, lightweight concrete still becomes most popular interest among researchers due to its special properties. Lightweight concrete can bring down the weight problem compared to ordinary concrete with 2400 kg/m^3 of density, which contributes to the weight penalty in large volume structures. Besides that, the global economic recession forced materials price increased, consequently the constructions would be high cost, including the cost for coarse aggregate [1]. Hence, to overcome this problem, the ordinary constituent material must be superseded by the low-cost material to cut the expense construction. This occasion provides a chance to waste materials to apply. Re-use waste plastic for replacing ordinary coarse aggregate concrete is the one ways to solve that problem. The waste plastic bag is a non-degradable material and it gives high impact to the environment due to

[a] Corresponding author : shahruln@uthm.edu.my

people are abandoned the plastic waste. Based on this case, it is a severe trouble such that contributes to environmental pollution and wastage of natural resources [2]. In this study, an attempt has been made to re-use the plastic bag as an Artificial Polyethylene Coarse Aggregate (APECA) and replaced the ordinary (natural) coarse aggregate with APECA in the concrete to produce the lightweight concrete.

Fundamentally, the concrete is a high-temperature resistance, which it can conduct heat slowly and able to store a considerable amount of heat from the environment. Thus, a previous investigation [3] stated that the evaluation of fire temperature on the concrete strength was essential for rehabilitation, strengthening, or reconstruction of the structure. Therefore, this study presents the investigation of compressive strength of concrete containing APECA with the various percentage and exposed by different temperatures. APECA is lighter than the natural aggregate, therefore, the APECA concrete will produce lower density than ordinary concrete. The previous investigation [2, 4] observed that the APECA was made from the waste plastic bag and underwent the heating process about 150°C for 10 minutes as shown in Figure 2. The size of APECA based on the standard diameter size of ordinary aggregate by range from 14 to 20 mm in the ball shape. Figure 1 present the compression strength of specimens for each percentage of APECA for 7 and 28 days. It can be seen that the concrete mixes containing 6 % APECA produces highest yet optimum strength about 10 % from normal concrete. Meanwhile, concrete with 9 % APECA will significantly reduce the strength of concrete about 60 %. It is may be due to the bond strength between artificial aggregate and cement paste.

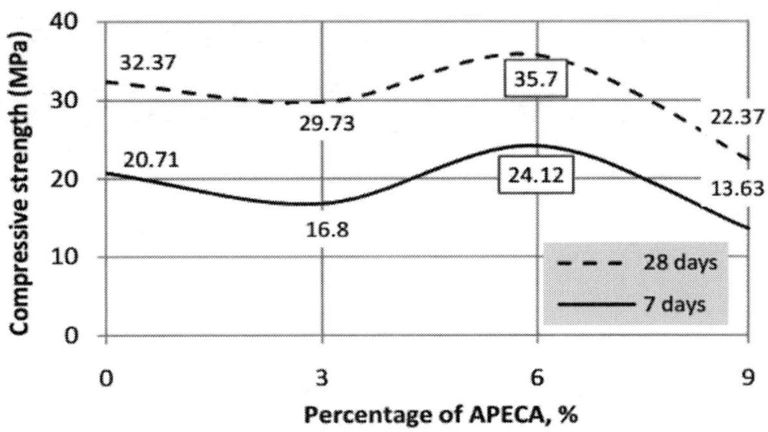

Figure 1. Compressive strength of different percentage of APECA in concrete mix [4].

Basically, the strength of concrete will be decreased when exposed in high temperature, especially on compressive strength. The high-temperature give effects on concrete strength, due to the dehydration of concrete. The dehydration will be occurred by driving out of a free water, interlayer water and chemically combined water [5]. The strength of concrete becomes a loss in the range of the temperature between 400°C and 800°C. At temperature 400°C, the compressive strength of specimens reduced gradually and at 800°C, the concrete strength decreased about 80% [5].

2 Materials and Methods

2.1 Concrete mix design

Table 1 lists the ratio of constituent materials for normal concrete that recommended by Department of Environmental (DOE) to achieve 25 N/mm² of compressive strength on 28 days. In this investigation, the specimen was prepared based on the two stages. The first stage focus on the strength of concrete specimens containing 6% of APECA as coarse aggregate that heated at different temperatures. The second stage involve the strength comparisons of concrete mixes with 0%, 3%, 6% and 9% APECA under normal condition (un-heated) and after exposed by 250°C. The volumes of concrete were determined to ensure the ratio between all the materials were suitable for grade 25 N/mm^2 [6]. This investigation produced totally twenty-one specimens of concrete cubes with 150 mm x 150 mm x 150 of size.

Table 1. Quantities of material based on DOE method.

Quantities	Cement	Water	Fine	Coarse Aggregate (Kg)		
				10 (mm)	20 (mm)	40 (mm)
Per m^3 (to nearest 5 Kg)	340	190	710	-	1157	-
Per trial mix of $3.375x10^{-3}$ m^3	1.145	0.61	2.396	-	3.905	-

(a)

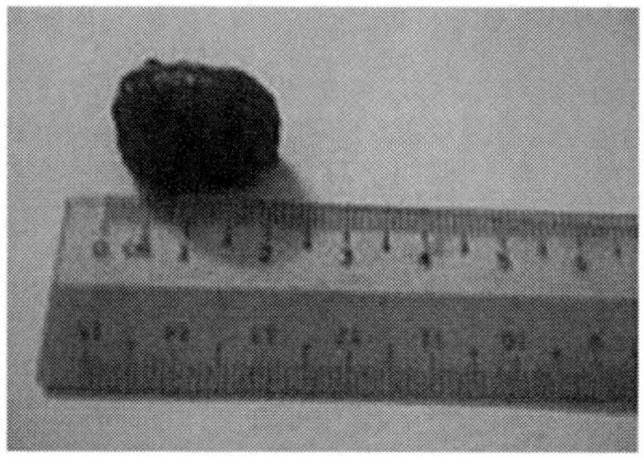

(b)

Figure 2. Condition of APECA (a) before and (b) after heating process.

2.2 Compressive strength test

Basically, the function of concrete is able to support and carry the compression load. This is an important parameter of the investigation was tuned by using the compressive test [6] on the cube concrete specimens. The compressive test was conducted on concrete with optimum percentage (6%) at different temperature (150°C, 200°C and 250°C). Further, the compressive strength of concrete cube containing 0%, 3%, 6% and 9% of APECA were investigated in normal condition (un-heating) and heated at 250°C.

2.3 Heating exposure technique

Generally, the loss rate of concrete strength increased, when gravel concrete was exposed to a temperature of 250°C for more than 1 hour. This agrees with the previous investigation [7], that the reduction in bending strength of concrete was increased when the concrete heated. The bending strength of gravel concrete was extremely affected by heat exposure at 250°C for 2 hours and 3 hours for protected and unprotected specimens, respectively. Therefore, this study applied 250°C for 1 hour to heat the APECA concrete specimens before conduct the compressive test.

At 28 days, all specimens of concrete cubes were removed from the curing tank and were dried in 25–30°C room temperature. The first stage, 6% of APECA concrete were heated at 150°C, 200°C, and 250°C in an oven for about 1 hour. Then, the concrete cube with different percentage of APECA were heated at 250°C for 1 hour as well. Before compression test, all specimens were cooling down in 30 minutes room temperature to ease the handling process on the concrete specimens.

2.4 Microstructure of APECA concrete

To describe the microstructure of APECA concrete, Scanning Electron Microscopy (SEM) has been employed. Changes to the structure caused by the heat need to be investigated and analysed directly with visualization, predominantly, the changes of the microstructure of concrete and APECA due to the heat mechanism. Owing to heat, the matrix concrete would be depreciation and generated the micro-cracks, besides the APECA would be melted. By using SEM, the changes of structure could be analysed, because SEM can achieve better resolution in micrometer of scale. Before sample was analysed under SEM, the sample specimens coated by Platinum (Pt) for 1 minute in sputter coater machine to avoid the bias or excess of reflected light.

3 Results and Discussions

3.1 Compressive strength of 6% APECA concrete exposed by different temperature

As previously mentioned, in the literature review [4], the 6% of APECA used for an optimum percentage in APECA concrete. Therefore, this study implemented 6% of APECA concrete to determine the strength of concrete with different temperatures i.e. 150°C, 200°C and 250°C.

Figure 3 presents a compressive strength of 6 % APECA concrete with different temperature at 150°C, 200°C and 250°C. In general, at 150°C, the 6% of APECA concrete reached a highest compressive strength, which was 31.6 N/mm^2, while, the lowest compressive strength was 28.6 N/mm2 at the 250°C. It is conform to the works presented by Bastami et al. [5] that the higher temperature reduce the strength of concrete. It is interesting to see that the APECA concrete were also reduced its strength when the temperature increased. It is clearly explained by [8] that the high temperature generated the alteration of mechanical properties of concrete due to chemical/physical transformations of aggregate and paste by temperature. Further, the concrete is very risk when it is exposed by fire because concrete is not stable at high temperatures and the concrete can dehydrate [9, 10].

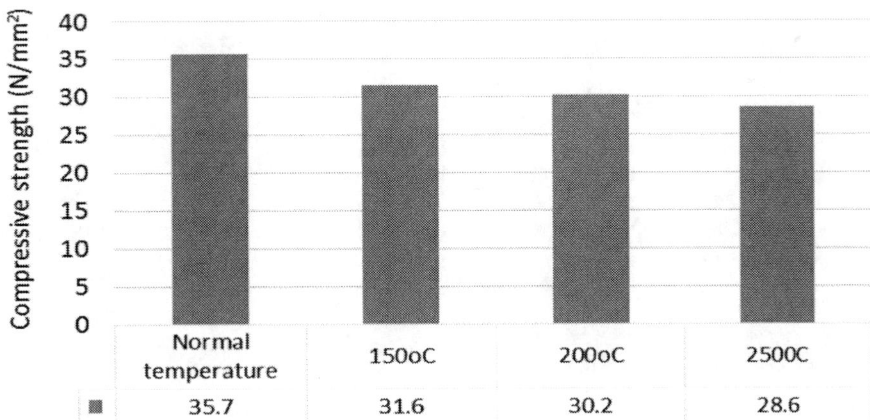

Figure 3. Compressive strength 6% of APECA with different temperature.

3.2 Compression strength comparisons of different percentage of APECA at normal condition (un-heated) and after exposed by 250°C

Figure 4 illustrates the compressive strength of 0%, 3%, 6% and 9% of APECA concrete at two conditions; un-heated condition and after exposed by 250°C. It is important to note that the better compressive strength possessed by APECA concrete with normal condition (un-heated) compared to another, which was exposed by heat at 250°C. This can relates to dehydration process [7, 10].

Focusing and examining the concrete specimens under un-heated condition, the concrete with 6% APECA is the highest compressive strength. Based on comparison of all ratios under heating process, it is clearly shown that the 0% APECA is higher than all ratio. In addition, 6% of APECA concrete is higher than the other two ratio (3% and 9%). It is due to the poor bonding interaction between APECA surface and cement matrix even after exposed by temperature.

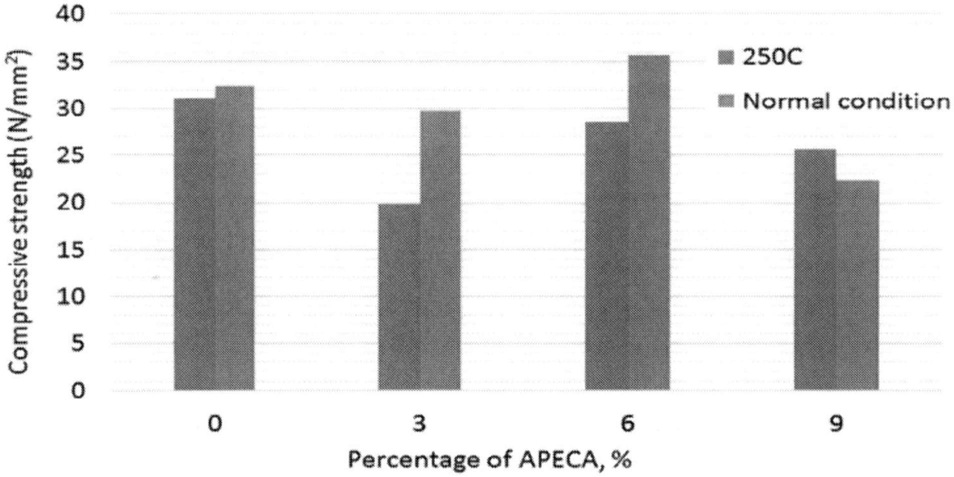

Figure 4. Comparisons of compressive strength of APECA concrete exposed at 250°C and in normal condition (un-heated)

3.3 Microstructure of APECA concrete

Figure 5 reveals the visual result of SEM for APECA concrete before and after being exposed by 250°C for 1 hour. Initially, before heated, the surface of APECA concrete produce micro-crack in cement matrix (refer Figure 5a).

(a) (b)

Figure 5. Microstructure of APECA Concrete before (a) and after (b) exposed to 250°C for 1 hour

While, after exposed to 250°C, it can be seen that the APECA were melted and filled the micro-crack as shown in Figure 5b. The second results contrary to Wasim Khaliq and Hammad Anis Khan [11] that found the development of micro-crack on concrete after exposed by high temperature in the range of 200°C to 600°C. In general, the micro-crack size in concrete reduced after being heated with temperature due to the APECA melted at temperatures above 160°C [2, 4]. It is essential to state that the APECA covering almost the whole micro-crack in concrete. Other than that, the presence of

APECA in concrete under heating delayed the dehydration of concrete. Yuan and Li [12] reported that the coated concrete considerably increased the carbonation resistance under elevated high temperatures. Meanwhile, some materials polymer can be used as a coating of concrete to reduce temperature effect [13].

4 Conclusions

The percentage of APECA plays a vital role in determining the compressive strength of concrete. In general, the result showed that the 6% of APECA concrete are able to produce highest compressive strength at 150°C. Further, APECA covering almost the whole micro-crack in concrete when heated at elevated temperature. However, the results indicate that the APECA concrete reduced its strength when the temperature increased. Nevertheless, APECA has a potential to be a coarse aggregate replacement and more detailed investigations are however demanded to modify the physical properties of APECA in order to increase its capability as a coarse aggregate replacement.

References

[1] A.P. Adewuyi and T. Adegoke, Exploratory study of periwinkle shells as coarse aggregates in concrete works, J. of Engineering and Applied Sciences, **3**(6), 1-5, (2008).
[2] M.M. Bakri A., C.M. Ruzaidi, M.N. Norazian, H. Kamarudin, S.M. Tarmizi and P. Besar, Effects of HDPE plastic waste aggregate on the properties of concrete, J. of Asian Scientific Research, **1**(7), 340-345, (2000).
[3] B. Jonaitis and V. Papinigis, Effect of long-term loading and fire temperatures on mechanical properties of concrete, J. of Civil Engineering and Management, **11**(4), 283-288, (2005).
[4] S.N. Mokhatar, Z.M. Jaini, M.K. Burhanudin, M.L. A. Jeni, and M.N.N Ismail, An experimental study of reinforced concrete beams with artificial aggregate concrete infill under impact loads, *International Civil and Infrastructure Engineering Conference 2014*, 691-701, (2014).
[5] M. Bastami, A. Chaboki-Khiabani, M. Baghbadrani, and M. Kordi, Performance of high strength concretes at elevated temperatures, Scientia Iranica, **18**(5), 1028–1036, (2011)
[6] BS1881, Method for determination of compressive strength, Part 116, (1991)
[7] K. Sakr and E. El-Hakim, Effect of high temperature or fire on heavy weight concrete properties, Cement and Concrete Research, **35**(3), 590–596, (2005).
[8] H. Mohamed, M.N. Ismail and N.S. Ibrahim, Physico-mechanical, microstructure characteristics and fire resistance of cement pastes containing Al_2O_3 nano-particles, Construction and Building Materials, **91**, 232-242, (2015).
[9] M. Heikal, Effect of temperature on the physico-mechanical and mineralogical properties of Homra pozzolanic cement pastes, Cement and Concrete Research, **30**, 1835 – 1839, (2000).
[10] C. Alonso and L. Fernandez, Dehydration and rehydration processes of cement paste exposed to high temperature environments, J. of Materials Science, **39**, 3015 – 3024, (2004).
[11] W. Khaliq and H.A. Khan, High temperature material properties of calcium aluminate cement concrete, Construction and Building Materials, **94**, 475–487, (2015).
[12] G. Yuan and Q. Li, The use of surface coating in enhancing the mechanical properties and durability of concrete exposed to elevated temperature, Construction and Building Materials, **95**, 375–383, (2015).
[13] A.A. Almusallam, F.M. Khan, S.U. Dulaijan, and O.S.B. Al-Amoudi, Effectiveness of surface coatings in improving concrete durability, Cement and Concrete Composites, **25**(4-5), 473–481, (2003).

MATEC Web of Conferences **47**, 01006 (2016)
DOI: 10.1051/matecconf/20164701006

Rubberized Concrete Durability Against Abrasion

Nurazuwa Md Noor[1,a], Daisuke Yamamoto[2], Hidenori Hamada[2] and Yasutaka Sagawa[2]

[1] *Jamilus Research Centre, Universiti Tun Hussein Onn Malaysia, 86400 Parit Raja, Johor, Malaysia*
[2] *Department of Civil Engineering, Faculty of Engineering, Kyushu University, Nishi-Ku, Fukuoka 819-0395, Japan*

Abstract. Durability performance of rubberized concrete against abrasion is presented in this paper. Surface depth loss was measured when abrasion load was constantly applied on concrete surface at each 500 interval rotation. Specimen with water-to cement ratio of 0.50 and 0.35 was prepared and tested at 28 days of curing age. In addition, 10% silica fume, SF was added to provide denser concrete and to understand its effectiveness against wear when added with crumb rubber. Results showed that crumb rubber shows good potential in providing abrasion resistance to concrete mix. However, in the case of rubberized concrete with silica fume, abrasion resistance was found to be slightly decreased with compressive strength more than 50N/mm^2 due to the lack of low elastic modulus of CR particles to accommodate with denser cement matrix.

1 Introduction

Abrasion is described as the behavior of the construction element against destructive influences due to scraping, rubbing, skidding or sliding of the object on the surface which leads to surface erosion. Concrete durability is influenced by several factors such as concrete compressive strength, surfacing finishing, aggregate properties, types of hardeners and curing type [1]. The effectiveness of crumb rubber to improve wear resistance tested by using surface abrasion test is discussed in this paper. Research done by Filipe Valadares et al.[2] on various rubber replacement ratio and size shows significant benefit in enhancing the abrasion resistance. Thus, an experimental study was conducted to see the behavior of crumb rubber with silica fume on abrasion wear resistance under different water to cement ratio. This testing was carried out to explore the possibilities of CR (without any surface treatment) to resists load from abrasion. The objective of the test was to evaluate the resistance of abrasion due to crumb rubber, different water-to-cement ratio and addition of silica fume as binder.

2 Mix Design and Specimen Preparation

In this concrete mix design, CR replacement was 10%, 15% and 20% of sand by volume. Silica fume, SF with 2200kg/m^3 density was added at 10% weight of Ordinary Portland Cement, OPC (density 3160kg/m^3). The mix design was divided into three main group of water to cement ratio, w/c as shown in Table 1. As for workability, ether-based polycarboxylate superplasticizer was added at 0.5%-0.7% of 5% maximum allowable dosage. Air content was controlled by using air-entrained

[a] Corresponding author : nurazuwa@uthm.edu.my

agent (to increase air content) for mixing without SF and air-modifying agent (to reduce air content) for mixing with SF. In this study, air content for all mixes was set from 4% to 5%. Sea sand passing 5mm sieve with density of 2580 kg/m^3 and water absorption of 1.72% which was less than 3.5% as stated in JIS Standard was used as fine aggregate. Combination of 10mm and 20mm crushed stone was used as coarse aggregate with density of 2910kg/m^3. All aggregates were prepared under surface saturated dry condition. Meanwhile, for concrete specimen, diameter, D was 100 mm and height, H between 45 mm to 55 mm. As for example, three small specimens was cut from the origin concrete cylinder (Ø100 x 200 mm) as illustrated in Figure1. Surface for each small specimen should be flat and 90^0 to ensure uniform surface contact with rotating disc.

Table 1. Mix proportion of rubberized concrete.

				Water	Cement	Silica Fume	Sand	Crumb Rubber	Coarse Aggregates		Chemical Admixture		
Group	CR/(S+CR)	SF/C	w/c	W	C	SF	S	CR	G1	G2	Ether-based polycarboxylate superplasticizers	Air-entrained agent	Air-modifying agent
	(Vol %)	(%)					kg/m^3				%	%	%
Control	0	0	0.50	165	330	0	790	0	636	329	0.5	0.8	
10CR-0SF	10						711	35			0.5	0.8	
15CR-0SF	15						671	53			0.5	0.8	
Control	0	0	0.35	160	457	0	741	0	608	405	0.5	0.8	
10CR-0SF	10						667	34			0.5	0.8	
15CR-0SF	15						629	50			0.7	0.8	
20CR-0SF	20						594	67			0.7	0.7	
10CR-10SF	10	10	0.35	160	457	46	613	34	608	405	0.7		1.5
15CR-10SF	15						575	50			0.7		2.2
20CR-10SF	20						540	67			0.7		2.8

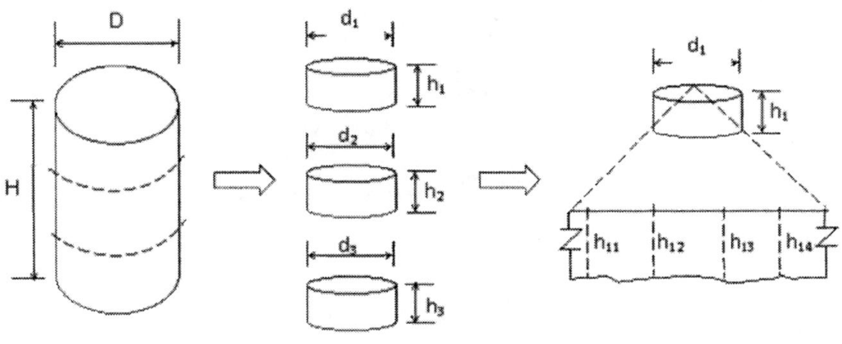

Figure 1. Specimen preparation.

3 Abrasion Test

Abrasion wear resistance of rubberized concrete was performed using Dorry abrasion testing machine as shown in Figure 2. It consists of a disc rotating shaft that connected to a motor to facilitate the removal of abrasive material which flow from above reservoir. Specimen was placed in the grip according to the size and then was tightened to ensure firm condition. Grip was then resting on disc rotating surface. Abrasive material used in this research was silica sand as shown in Figure 3. Silica sand was placed in the reservoir and flow down towards the rotating disc to provide friction on specimen surface. Initial data was recorded before each test and then disc was rotated and stop at each

500 rotation interval. At each interval, the change of weight and depth was recorded; where depth changes were calculated at the average of four points.

Figure 2. Specimen preparation.

Figure 3. Silica sand.

4 Results and Discussion

Figure 4 shows results of surface abrasion wear resistance with the presence of CR at 10%, 15% and 20%. A larger difference was observed between control mix (0CR-0SF) and mix with CR. At 4000 disc rotation the depth loss was 0.85mm compared to mixture with 10% CR addition which was 0.62mm; with 30% reduction difference. Resistance against surface abrasion wear was continuously improving with 15% and 20% CR replacement with depth loss of 0.58 and 0.52 respectively at 4000 disc rotation. Effect of CR added with SF on abrasion resistance is shown in Figure 5. At 4000 disc rotation, the depth loss of rubberized concrete with SF was reducing from 0.85mm for control mix to 0.67mm, 0.61mm and 0.51mm at every 10%, 15% and 20% CR addition as sand replacement.

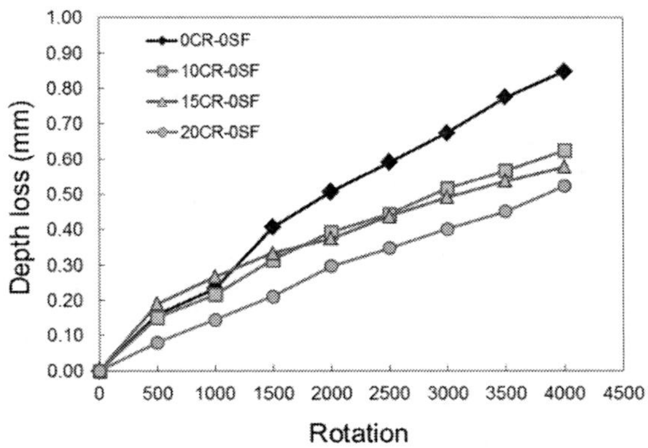

Figure 4. Depth loss of rubberized concrete without silica fume due to abrasion test (w/c = 0.35).

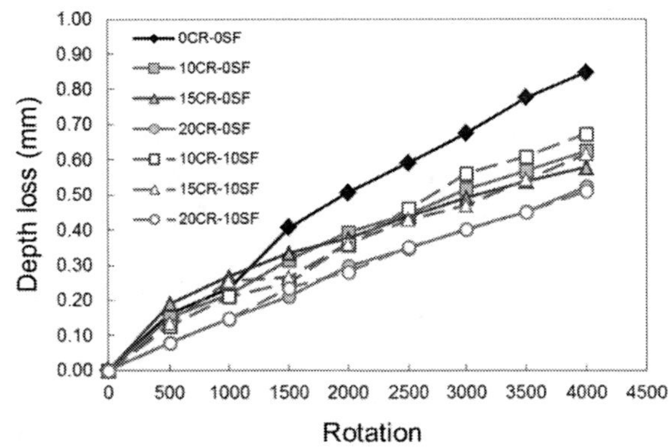

Figure 5. Depth loss of rubberized concrete with silica fume due to abrasion test (w/c = 0.35).

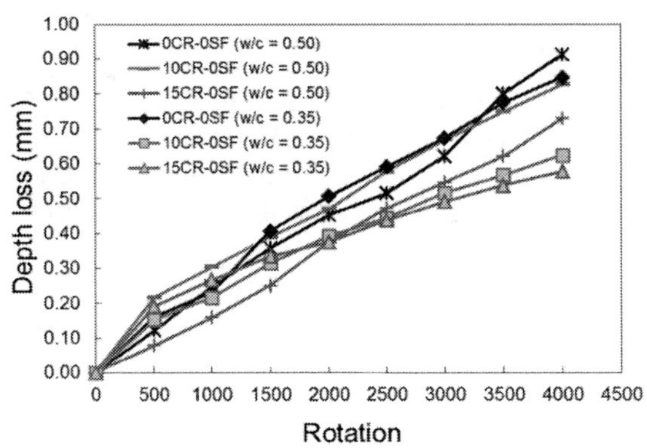

Figure 6. Depth loss of rubberized concrete without silica fume due to abrasion test (w/c=0.50 and 0.35).

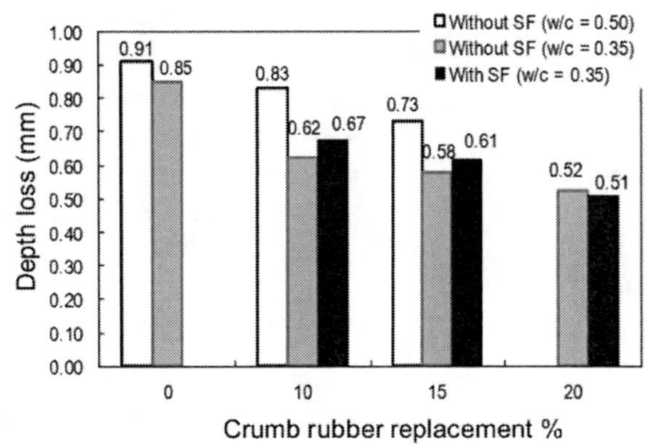

Figure 7. Depth loss of concrete surface at 4000 rotations.

Table 2. Data of abrasion wear resistance at 4000 disc rotations for all mixture.

Abrasion wear resistance at 400 rotation			
Mix		Compressive strength, N/mm2	Depth loss, mm
Without SF (w/c = 0.35)	0CR-0SF	68.8	0.85
	10CR-0SF	56.52	0.62
	15CR-0SF	50.46	0.58
	20CR-0SF	43.81	0.52
With SF (w/c = 0.35)	10CR-10SF	63.43	0.67
	15CR-10SF	58.31	0.61
	20CR-10SF	51.38	0.51
Without SF (w/c = 0.50)	0CR-0SF	46.18	0.91
	10CR-0SF	33.72	0.93
	15CR-0SF	31.89	0.73

However, this reduction was slightly larger than mixture without silica fume at 10% and 15% CR replacement and no difference can be concluded for 20% CR replacement. Effect of water-to-cement ratio of 0.35 and 0.50 on the depth loss was presented in Figure 6. As for mixture with w/c = 0.50 the same reduction pattern with w/c = 0.35 mixture was observed. The depth loss was 0.91mm for conventional concrete and 0.83mm and 0.73mm, respectively, for each 10% and 15% CR replacement. Due to the high w/c, the abrasion resistance was smaller compared to w/c = 0.35.

It was clearly seen that 10% crumb rubber addition as sand replacement provide good resistance than that control mix in all mixes According to Blessen Skariah Thomas et al., this improvement may due to the restriction of the surface grinding due to the CR position which was beyond the smooth surface and acted like a brush [4]. CR with 20% replacement gave much higher resistance against abrasion with 50% depth loss difference for both rubberized concrete with and without SF.

Looking back to the effect of SF, it was found that the inverse behavior was seen; where the abrasion resistance of rubberized concrete was much lower than that mixture without silica fume. Generally, it was stated that, higher compressive strength due to 10% SF addition gave better abrasion resistance as reported by Ghafoori and Diawara in 1999 [5]. But, when CR was added in mixture with higher compressive strength, the bonding between CR particles and cement matrix started to distress when subjected to friction force and may cause certain CR to jut out from the surface. In this research, this fundamental behavior was observed on specimen with compressive strength higher than 50 N/mm^2 (refer Table 2). This situation indicates that the compressive strength was an important factor affecting concrete abrasion resistance. When dealing with water-to-cement ratio effect, it was noted that higher w/c gave higher loss of surface depth due to small compressive strength value compared to mixture with low w/c. However, as discussed above, there was a limitation of abrasion resistance in relation with strength. This limitation was only applicable for CR without any treatment and this fundamental discovery may be further investigated. Abrasion resistance of high strength mixture could be improved by enhancing the bonding of the matrix by applying certain treatment on the rubber surface.

5 Conclusion

Several conclusions can be drawn from this research as follows, (i) The depth loss of concrete surface was decreasing with the increase of crumb rubber volume which indicates that abrasion resistance was improved with the presence of crumb rubber in the mix, (ii) Compressive strength is important factor affecting the abrasion resistance, where abrasion resistance is increase with increasing in compressive strength, and (iii) However in this research, it was found that strength was limited to 50 N/mm^2 due to the lack of crumb rubber elastic modulus to accommodate with high strength of cement paste.

Ackowledgement

The authors would like to express their gratitude to Hikari World Company Limited for their continuous support in supplying the crumb rubber. Grateful acknowledgement was also dedicated to Malaysian Ministry of Higher Education in providing financial support to the first author.

References

[1] A. Çavdar and Ş. Yetgin, investigation of abrasion resistance of cement mortar with different pozzolanic compositions and subjected to sulfate medium, Construction and Building Materials, **24**, 461-470, (2010).

[2] F. Valadares, M. Bravo and J. de Broti, Concrete with used tire rubber aggregates: mechanical performance, ACI Materials Journal, **109**, 283-292, (2012).

[3] G. Sing and R. Siddique, Abrasion resistance and strength properties of concrete containing waste foundry sand (WFS), Construction and Building Materials, **28**, 421-426, (2012).

[4] B.S. Thomas, R.C. Gupta, P. Kalla and L. Cseteneyi, Strength, abrasion and permeation characteristics of cement concrete containing discarded rubber fine aggregates, Construction and Building Materials, **59**, 204-212, (2014).

[5] R. Siddique, Effect of fine aggregate replacement with f fly ash on the abrasion resistance of concrete, Cement and Concrete Research, 33, 1877-1881, (2003).

MATEC Web of Conferences **47**, 01007 (2016)
DOI: 10.1051/matecconf/20164701007

Strength Development and Water Permeability of Engineered Biomass Aggregate Pervious Concrete

S. Shahidan[1,a], H. B. Koh[1], A. M. Sharif Alansi[1] and L. Y. Loon[1]

[1]*Faculty of Civil and Environmental Engineering, Universiti Tun Hussein Onn Malaysia, 86400 Batu Pahat, Johor, Malaysia*

Abstract. Pervious concrete has a high rate of permeability, low strength and high porosity. It is commonly used in the area of storm water management. However, its use has been limited to pavements with low volume traffic. Pervious concrete is different from normal concrete as the mixture contains no fine aggregates. The aggregate is usually of a single size bonded by a cement paste. This study focuses mainly on the effect of Engineered Biomass Aggregate (EBA) on the compressive strength and permeability of previous concrete. Three types of mixtures with 0% natural aggregate (NA), 5% BA and 5% of EBA have been studied in this research. A total of 27 150 mm cubes were casted in the laboratory for compressive strength and permeability test. This research aims to enhance the understanding of engineered biomass aggregate pervious concrete as well as the use of pervious concrete for sustainable construction activities due to the environmental benefits it offers.

1 Introduction

Concrete is considered as the major cause of natural material deployment. However there are many by-products such as waste material which become problematic to the environment. Biomass is conceded as by-product of industrial effluents and is continuously being dumped into landfills. However, research showed that biomass has the potential to be recycled and may be used to replace natural aggregates (NA) in concrete [1-3].

Comparing the properties of normal aggregate with biomass aggregate (BA), it is found that BA shows lower strength. The pervious concrete containing BA allows water to pass through it and is normally used in soft soil due to its lightweight properties [4]. Pervious concrete is also known as "no-fines concrete" and consists of Portland cement, coarse aggregate, water, admixtures, and little or no sand [5-8]. It has high porosity used for concrete flatwork applications that allows water from precipitation and other sources to pass through it, thereby reducing the runoff from a site and recharging ground water levels.

Besides that, Polyethylene or plastic bags also affect the environment due to wide availability and the fact that they are not biodegradable. Plastic bags can be recycled and used to produce Engineered Biomass Aggregate (EBA) [7]. However, engineering technical data on EBA is not well established due to the lack of research on this material. Therefore, this research aims to study the strength

[a] Corresponding author : shahiron@uthm.edu.my

development and water permeability of pervious concrete containing BA and EBA as alternative materials for sustainable development.

2 Biomass Aggregate and Engineered Biomass Aggregate

Biomass Aggregate (BA) is derived primarily from the innovative exploitation of solid wastes from controlled incineration of biomass. The BA used in this research was collected from a factory called Evergreen Fibreboard Sdn. Bhd. located in Parit Raja. The BA is the waste from the furnace which burns wood bark and palm oil fibre as shown in Figure 1. Previous research shows that BA is lightweight and complies with most of the lightweight aggregate properties. The density of BA ranges from 300 kg/m^3 to 800kg/m^3 [8-9]. BA is suitable as construction material for road pavements, interlocking blocks for pathways, lightweight aggregate concrete and etc.

Figure 1. Biomas aggregate.

Engineered Biomass Aggregate (EBA) is a new material derived from combinations of 2 recyclable materials which are biomass aggregate and polyethylene or plastic bag. The BA was wrapped with polyethylene and placed together inside the oven with a temperature of 200°C for half an hour. Figure 2 presents the process of manufacturing EBA.

Figure 2. Engineered biomass aggregate (EBA).

3 Laboratory Work

At the initial stage, all the materials and equipment have to be prepared and checked for availability in order to determine the performance of concrete with BA and EBA. All the testing and laboratory work were carried out according to British Standards.

3.1 Materials and cube preparations

The constituent materials used in this research were composite cement, course aggregate, water, BA, EBA and superplasticizers (SP). A total of 27 pervious concrete cubes with a size of 150 mm x 150 mm x 150 mm each with 0% NA (8-14mm), 5% BA and 5% EBA were prepared with water cement ratio 0.45. Table 1 shows the mix design proportions used in this investigation.

Table 1. Mix design proportions.

Type of Mix	Water (kg/m^3)	Cement (kg/m^3)	Coarse Aggregate(kg/m^3) (8-14)mm	BA (kg/m^3)	EBA (kg/m^3)	SP 1%
100% NA	114	380	1620	-	-	3.8
95%NA +5%BA	114	380	1539	81	-	3.8
95%NA +5%EBA	114	380	1539	-	81	3.8

The materials were stored properly in the laboratory room so that the stability of surface moisture content of the aggregate is maintained. This is because pervious concrete has low and high water content aggregate content. All the materials were mixed together in the mixture machine with the proportion design as shown in Table 1. After completing the mixture process, the pervious concrete was poured into the cube mould and left for a day. On the following day, the pervious concrete was placed into the curing tank for the curing process to take place.

3.2 Testing

Two types of tests were conducted. A compressive strength test, as shown in Figure 3(a), was carried out to determine the short-term strength development (7, 14 and 28 days) of pervious concrete containing BA and EBA. Meanwhile, permeability test was conducted on the 150 mm pervious concrete cubes at 28 days. Permeability is an important parameter of previous concrete since the material is designed to function as a drainage layer in pavement structures. Darcy's law for laminar flow is no longer applicable for previous concrete. In this study, a permeability measurement device and method developed for drainable asphalt mixture similar to pervious concrete in function were used as seen in Figure 3(b).

(a) (b)

Figure 3. Testing method: (a) Compressive strength test, and (b) Permeability test.

4 Results and Discussion

4.1 Compressive strength

Figure 4 shows the result of compressive strength for 7, 14 and 28 days of curing age. The results show that control specimen of 0% NA gained higher value compared to other specimens. NA achieved 15.69 MPa at 28 days but 5% BA and 5% EBA only achieved 13.4 MPa and 10.5 MPa, respectively. Pervious concrete containing 5% EBA achieved the lowest compressive strength. This may be due to the smooth surface of EBA which reduces interlocking between the aggregates. Besides, the mechanical strength of biomass aggregates is generally lower than that of natural aggregates. Overall, the trend of strength development for pervious concrete and pervious concrete containing BA and EBA is similar to normal concrete.

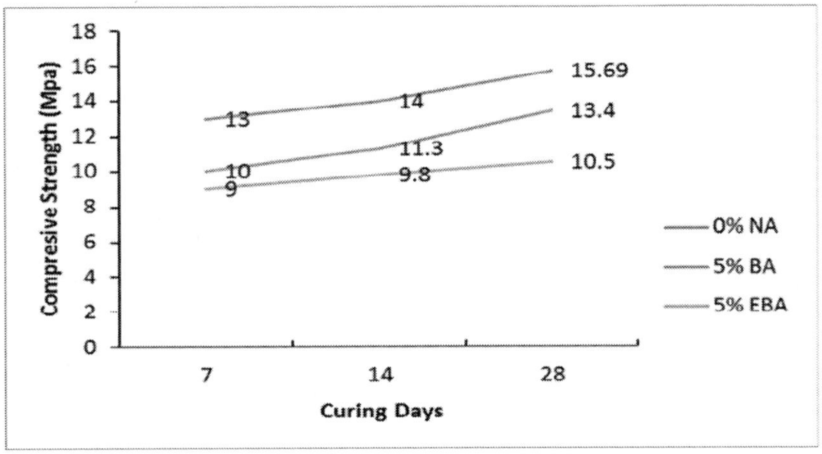

Figure 4. Compressive strength of pervious concrete.

4.2 Water permeability test

The concept of flow rate in this experiment is to determine the duration of water passing through the cubes. The result was obtained by using tubes with specific diminution to calculate the discharge and determine the time of water passing through the specimen. The test was conducted on the cube specimens after 28 days of water curing. Fig 5 shows that the specimen containing 5% of EBA achieved higher water permeability followed by 5% BA and NA. The water permeability of 5% EBA, 5% BA and NA are 14.33 mm/s, 12.15 mm/s and 6.83 mm/s, respectively. Therefore, pervious concrete containing EBA has the highest flow rate compared to normal pervious concrete. The increase in water permeability for pervious concrete containing 5% EBA is due to the wrapping of BA with polyethylene which makes the surface of EBA smooth.

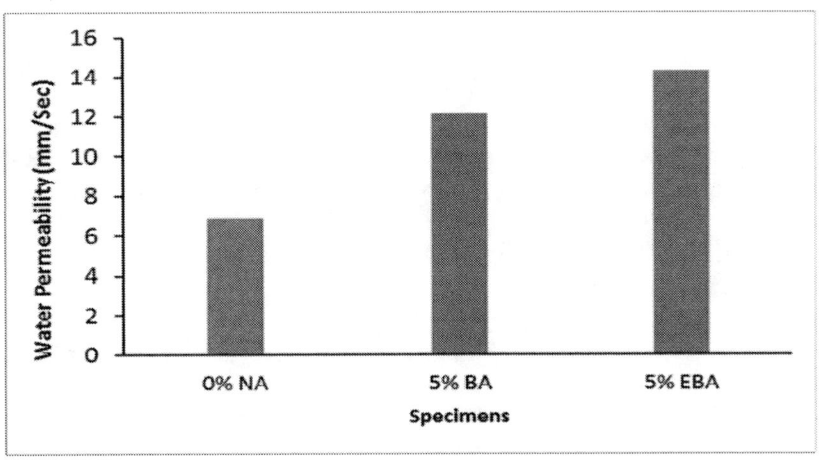

Figure 5. Water permeability test.

4.3 Relationship between water permeability and compressive strength

In general, the compressive strength increases when the water permeability decreases. The relationship between water permeability and 28-day compressive strength for pervious concrete is shown in Figure 6.

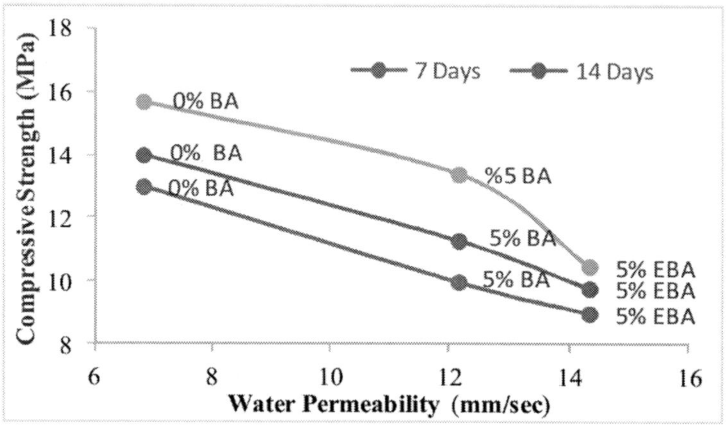

Figure 6. The relationship between water permeability and compressive strength

Figure 6 illustrated that the NA specimen achieved approximately 6.83 mm/sec for water pemeability with a compressive strength of 15.69 MPa. However the strength of concrete decreased slightly by adding 5% of BA and 5% of EBA. From the graph it can be concluded that, the water permeability may be increased by adding 5% of EBA compared to 0% of NA and 5% of BA but the compressive strength is decreased after the addition of BA and EBA.

5 Conclusions

From the above findings, it can be concluded that Biomass Aggregate (BA) and Engineered Biomass Aggregate (EBA) are suitable to be used as the partial replacement for natural aggregate in pervious concrete to increase water permeability. Further research should be carried out to improve the strength of BA and EBA concrete in order to maximize the use of sustainable materials for civil engineering applications.

References

[1] Y.L. Lee, A.K. Ahmad Tarmizi, A.R. Ismail, H.B. Koh, S. Hani, S. Nagapan, R. Fetra Venny and C.T. Jung, Alternative aggregates for sustainable construction, International J. of Zero Waste Generation, **1**(1), 5-10, (2013).

[2] Y.L. Lee, K.O. Khoo, S.S.F. Chong and M.W. Hussin, Strength development and water permeability of high strength TIA concrete, *Proceedings of International Congress on Creating with Concrete*, University of Dundee, 175-181, (1999).

[3] Y.L. Lee, K.S. Goh, K.O. Khoo and I. Bakar, Foamed aggregate pervious concrete – an option for road on peat, *Proceedings of MUCET*, (2009).

[4] Y.L. Lee, T.J. Chai, R. Fetra Venny, H.B. Koh, Y. David and A.K. Ahmad Tarmizi, Biomass aggregate for sustainable geopolymer concrete, International Journal of Civil Engineering and Technology, **5**(3), 340-356, (2014).

[5] F. Montes and L. Haselbach. measuring hydraulic conductivity in pervious concrete, Environmental Engineering Science Journal. November **23**(6), 960-969, (2006).

[6] ASTM C39, Standard Test Method for Compressive Strength of Cylindrical Concrete Specimens, ASTM International, (2009).
[7] BS EN 12390-3, Testing Hardened Concrete, Compressive Strength of Test Specimens, BSI Standards Publication, (2009).
[8] S. Demie, M.F. Nuruddin, M.F. Ahmeed and N. Shafiq, Effects of curing temperature and superplasticizer on workability and compressive strength of self-compacting geopolymer concrete, IEEE, **5**(11), 187-194, (2011).
[9] L.M. Haselbach, S. Valavala and F. Montes, Permeability predictions for sand-clogged Portland cement pervious concrete pavement systems, Journal of Environmental Management, **81**, 42-49, (2006).

MATEC Web of Conferences **47**, 01008 (2016)
DOI: 10.1051/matecconf/20164701008

Identification of Bacteria and the Effect on Compressive Strength of Concrete

L. H. Anneza[1], J. M. Irwan[1,a], N. Othman[2] and A. Faisal Alshalif[1]

[1]*Jamilus Research Center, Universiti Tun Hussein Onn Malaysia, 86400 Parit Raja, Johor, Malaysia*
[2]*Faculty of Civil and Environmental Engineering, Universiti Tun Hussein Onn Malaysia, 86400 Parit Raja, Johor, Malaysia*

Abstract. This paper presents the species of bacteria used in this study as well as the effect of the bacteria on compressive strength of bioconcrete. Bioconcrete is not only more environmentally friendly but it is easy to procure. The objective of this research is to identify the ureolytic bacteria and sulphate reduction bacteria that have been isolated and further use the bacteria in concrete to determine the effect of bacteria on compressive strength. Identification of bacteria is conducted through Polymerase chain reaction (PCR) method and DNA sequencing. The DNA of the bacteria was run through BLAST algorithm to determine the bacterial species.The bacteria were added into the concrete mix as a partial replacement of water. 3% of water is replaced by ureolytic bacteria and 5% of water is replaced by sulphate reduction bacteria. After running BLAST algorithm the bacteria were identified as Enterococcus faecalis (ureolytic bacteria) and Bacillus sp (sulphate reduction bacteria). The result of the compressive strength for control is 36.0 Mpa. Partial replacement of 3% water by ureolytic bacteria has strength of 38.2Mpa while partial replacement of 5% of water by sulphate reduction bacteria has strength of 42.5Mpa. The significant increase of compressive strength with the addition of bacteria shows that bacteria play a significant role in the improvement of compressive strength.

1 Introduction

Bioconcrete is a combination of biology and concrete. In which, the biology part is the bacteria use in this research. Conventional way to increase strength without adding waste in concrete, normally relies on chemical solution or adding more cement. The alternative of using bacteria in concrete is gradually gaining more interest as a natural method to improve concrete properties and durability. Previous research conducted by Ghosh and Mandal [1], Muynck et al. [2], Tittelboom et al. [3] and Arunachalam et al. [4] has shown that adding bacteria into concrete does improve the strength as well as self healing of concrete. As bacteria are able to promote the precipitation of calcium carbonate in the form of calcite. Calcium carbonate precipitation occurs as a by-product of a common microbial metabolic process which would help increase the alkalinity and produce microbial calcite precipitation [5]. Nevertheless, the bacteria used by previous researchers are all from different sources and there are no specific or right bacteria added into the concrete. Therefore, this research aims to identify the bacteria that were previously isolated in [6] and use the bacteria in concrete to determine the effect of

[a] Corresponding author : irwan@uthm.edu.my

the isolated bacteria on compressive strength. This study is based on previous study [7] which determines the optimum percentage of bacteria use.

2 Materials

The aggregate and sand used for this research were all carefully sieved according to BS 882.1992 [8]. Where the maximum size of aggregate used is 20mm and sand is passing the 5mm sieve. The cement used in this research is PCC (Portland composite cement). The ureolytic bacteria and sulphate reduction bacteria that are used in this study are taken from fresh urine and fresh acid mine water in Sg.Pelepah, Kota Tinggi, Johor, Malaysia respectively. The isolation and enrichment process are conducted as in Irwan [6]. After enrichment process, the bacteria are cultivated on agar plates to ensure the growth of bacteria. Figure 1 shows the ureolytic bacteria on the agar plate that is used in this research.

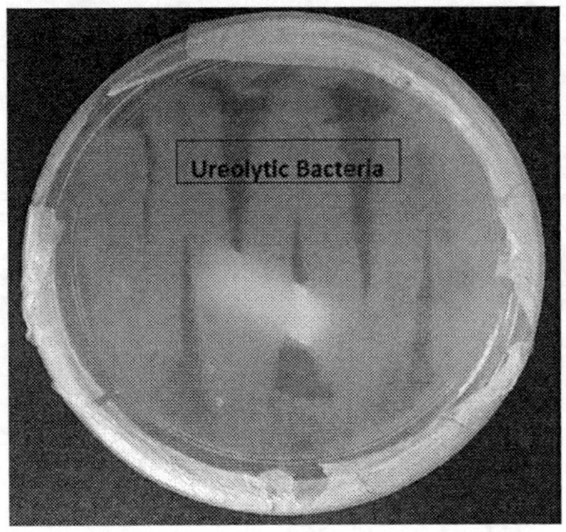

Figure 1. Agar plate with ureolytic bacteria.

3 Methodology

3.1 Determining the type of bacteria

The bacteria sample was first inoculated into Lysogeny Broth (LB) and incubated for 18 hours with temperature of 37° C and 400 rpm. The samples were then centrifuged to collect pellet for genome DNA extraction. The DNA is then quantified spectrophotometrically using a Nanodrop 2000 spectrophotometer (Fisher Scientific). Absorbance measurements were taken at 260nm and 280nm and used to calculate the DNA purity and concentration. 260nm and 280nm are wavelengths which are used to quantify amount of DNA in the samples. DNA samples were considered free from protein contamination if the A_{260} to A_{280} ratio was between 1.8 and 2.0. The result of the quantification is in Table 1. Based on the database using BLAST algorithm the bacteria used in this study which originally named sulphate reduction bacteria is *Bacillus SP.* Whereas, the proper nomenclature of ureolytic bacteria based on result after BLAST algorithm is *Enterococcus faecalis*.

Table 1. The concentration and the quality of DNA using Nanodrop.

No.	Sample ID	Nucleic Acid Conc.	Unit	260/280	260/230
1	S	165.1	ng/µl	1.85	2.15
2	U	159.7	ng/µl	1.84	2.12

3.2 Preparation of inoculum to be casted in bioconcerete

After preparing and sterilizing nutrient broth, 1 colony of bacteria is then added into the sterilized nutrient broth. The solution is then put into an orbital shaker and shaken for 10 days, which is the number of days required to achieve the maximum growth rate of both bacteria. The bacteria solution is then measured and used for casting concrete. Concrete casting is divided into batch 1-4, where each batch uses a different type of bacteria as well as control as a comparison. Figure 2 shows the bacteria solution which has been prepared in order to be used for bioconcrete casting. Solution on the left is ureolytic bacteria (*Enterococcus faecalis*) while the one on the right is sulphate reduction bacteria (*Bacillus SP*).

Figure 2. Ureolytic bacteria (Enterococcus faecalis) and sulphate reduction bacteria (Bacillus SP).

3.3 Preparation and casting of bioconcrete

The grade of concrete use in the design of bioconcrete is a Grade 30, the materials used for casting are calculated based on DOE concrete mix design. Each batch of concrete consists of 3 cubes to determine the average compressive strength. Each batch requires cement (4.26kg), Water (2.1kg), Sand (6.96kg) and coarse aggregate (11.31kg). 3% of partial replacement of water by ureolytic bacteria as in Table 2, batch 2 and 4 consists of 63g of ureolytic bacteria. Whereas, 5% of partial replacement of water by sulphate reduction bacteria as in Table 3, batch 3 and 4 consists of 105g of sulphate reduction bacteria. The dry materials are mixed together in a drum mixer before adding water then bacteria. After adequate consistency and workability, the concrete is put into plastic molds and left to dry for 24hours before removal and dry curing.

3.4 Compressive strength test

Compressive strength test was conducted according to BS EN 12390-3 [9]. The cube specimens of dimensions 150mm × 150mm × 150mm were tested using Universal Testing Machine (UTM). The specimens were tested at the mature age of 28 days after curing. All batches were tested in triplicates and the average results of these triplicates are presented and discussed.

4 Results and Discussion

The average result of the compressive strength of bioconcrete are shown in Table 2 and Figure 3. Where, it shows that concrete in batch 3 has the highest compressive strength which is 42.4 Mpa. This batch contains 5% of *Bacillus SP*. Batch 1 is the batch that does not contain any bacteria as it is the control and therefore use as a comparative among the 4 batches. The maximum strength of batch 1 is 36 MPa. The second highest compressive strength is of Batch 4. This batch contains both 5% of

Bacillus SP and 3% *Enterococcus faecalis*. The maximum strength of batch 4 is 40.6 MPa. As seen in Figure 2, the batch that contains both bacteria does slightly better comparatively to the stand alone bacteria in batch 2 and control. Batch 2 is concrete with 3% of *Enterococcus faecalis*, it has a maximum strength 38.2 MPa. The strength enhancement due to the addition of bacteria is because of the deposition of $CaCO_3$ on the concrete pores. As bacteria has the ability to precipitate calcium carbonate in calcite form.

Table 2. Result of compressive strength.

Batch No.	Description	28th day (MPa)
Batch 1	Control	36.0
Batch 2	Ureolytic bacteria/ *Enterococcus faecalis*	38.2
Batch 3	Sulphate reduction bacteria/ *Bacillus SP*	42.4
Batch 4	Ureolytic + Sulphate reduction bacteria/ *Enterococcus faecalis*+ *Bacillus SP*	40.6

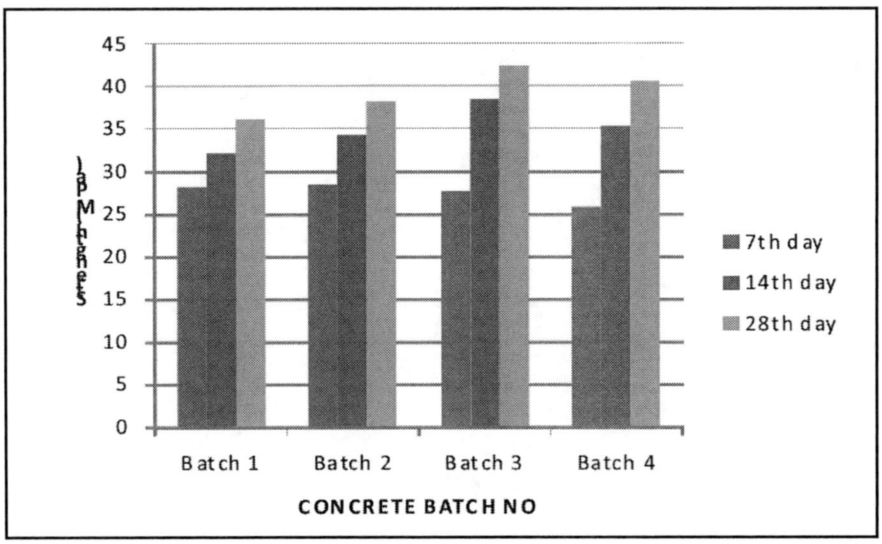

Figure 3. Compressive strength (MPa).

5 Conclusions

In this study, the species of bacteria used has been successfully identified as *Bacillus SP* and *Enterococcus faecalis*. The use of 3% of *Enterococcus faecalis* and 5% of *Bacillus SP* is based on an experimental study of using 1%, 3% and 5% of both bacteria. It was found that 3% of *Enterococcus faecalis* and 5% of *Bacillus SP* has the highest compressive strength. Adding both types of bacteria into a single batch of concrete proves that even though the 2 bacteria are of different sources and therefore have different characteristics. There are still able to precipitate calcium carbonate thus increasing the compressive strength. Further study should be conducted through various other mechanical tests as well as conducting study regarding the healing capability of these bacteria on concrete.

References

[1] P. Ghosh and S. Mandal, Development of bioconcrete material using an enrichment culture of novel thermophilic anaerobic bacteria, Indian J. of Experimental Biology, **44**, 336-339, (2006).

[2] W.D. Muynck, K. Cox, N.D. Belie and W. Verstraete, Bacterial carbonate precipitation as an alternate surface treatment for concrete, Construction and Building Materials, **22**, 875-885, (2008).

[3] T. Kim,Van, N. De Belie, W. De Muynck and W. Verstrate. Use of bacteria to repair cracks in concrete, Cement and Concrete Research, **40**, 157-166, (2010).

[4] K.D. Arunachalan, K.S. Sathyanarayanan, B.S. Darshan and R.B. Raja, Studies on the characterisation of biosealant properties of *Bacillus Sphaericus*, International J. of Engineering Science and Technology, **2**(3), 270-277, (2010).

[5] S.K. Ramachandran, V. Ramakrishnan and S.S. Bang, Remediation of concrete using micro-organisms, American Concrete Institute Materials Journal, **98**(1), 3-9, (2001).

[6] J.M. Irwan, A. Faisal Alshalif, N. Othman and R.M. Asyraf, Isolation of ureolytic and sulphate reduction bacteria: Acclimitize to concrete environment, *International Conference on Civil, Biological and Environmental Engineering*, Istanbul, Turkey, (2014).

[7] J.M. Irwan, A. Faisal Alshalif, N. Othman and L.H. Anneza, Effext of Ureolytic Bacteria on compressive strength and water permeability on bio-concrete, *International Conference on Civil, Architectural, Structural and Construction Engineering*, Seoul, Korea, (2015).

[8] BS 882, Specification for aggregates from natural sources for concrete, British Standard Institution, London, (1992).

[9] BS EN 12390. Part 3: Compressive strength of test specimens, British Standard Institution, London, (2009).

MATEC Web of Conferences **47**, 01009 (2016)
DOI: 10.1051/matecconf/20164701009

Strength of Concrete Containing Rubber Particle as Partial Cement Replacement

Siti Radziah Abdullah[1,a], Wan Raziman Wan Zainal Abidin[1] and Shahiron Shahidan[1]

[1]*Faculty of Civil and Environmental Engineering, University Tun Hussein Onn Malaysia, 86400 Parit Raja, Johor, Malaysia*

Abstract. Uncontrolled issues of disposal waste tire rubber create huge environmental impact and health hazards. An alternative viable solution to minimize these problems is by utilizing the waste rubber tires in construction materials, which in turn can reduce the use of natural resources and eventually lessen the cost of construction. This paper mainly focuses on the use of waste rubber tires particles in concrete with different set of composition ranging from 3 to 12% of cement replacement. Concrete cubes size of 150mm x 150mm x 150mm were prepared for compressive strength test, and concrete cylinders size of 150m x 300mm were prepared for splitting tensile test. The result shows that the compressive and split tensile strength of concrete with rubber as cement replacements is 6-21% lower than the normal concrete.

1 Introduction

Concrete is widely used in the construction worldwide and the depletion of natural resources of concrete mix is an issue to worry about. When the demand of these materials increases the cost of construction also rises. By reducing the usage of natural resources in concrete by incorporating with recycle material, it can help to reduce the cost of construction and hence minimize amount of material dump into landfill. Waste materials such as recycled plastic, rice hush ash, waste glass, wood ash are examples of materials that can be utilized in concrete. Of many recycle material that can be used is vehicle tires where the disposal of this material affects the environment [1]. Rubber particles has acceptable workability when it used to replace total aggregate of up to 50% but the strength is reduced proportionally with substitution of rubber in the mix [2, 3]. Even though compressive strength is reduced, concrete with rubber particles exhibit ductile and plastic failure rather than brittle failure as found for normal concrete [4]. Other than that, concrete with rubber has good resistance and the shrinkage is almost negligible [5]. In order to achieve the acceptable strength of concrete, rubber content should not be more than 20% of total aggregate volume [6]. Through microscopic analysis, heterogenous and hydrophoble rubber particle disturb hydration process of cement that lead to loss in compressive strength [7]. A waste from rubber industry called as carbon black also has been used as additive to increase density of concrete that beneficial in strength. Due to its smaller size, it acts as filler that can reduce the permeability of concrete [8]. There are potential for rubber to be used in construction industry, hence, this study explores the potential application of rubber particle as cement replacement in the concrete.

[a] Corresponding author : radzia@uthm.edu.my

2 Experimental Program

The rubber used in this study was fine rubber from scrap tyres that suitably used as cement replacement in concrete (OPC) due to its fineness as shown in Figure 1. Percentages of rubber particle used to replace the cement by weight were 0%, 3%, 6%, 9% and 12% with 0.5 water-cement ratio. The mix design procedure was for Grade 30. Standard cube specimens of 150mm x 150mm x 150mm were prepared for compressive strength test (BS EN 12390-4) and cylinder specimens with diameter of 150mm and 300mm height were prepared for splitting tensile test (BS EN 12390-6). Prior to concrete casting, slump of each sets of concrete was measured. Each reported compressive and tensile strength is an average of three testing samples that were tested on 7, 28 and 56 days.

Figure 1. Rubber material.

3 Results and Discussion

3.1 Workability

The result of the slump test of all types of concrete with rubber particles is shown in Figure 2. As can be seen, increasing the amount of rubber particles as cement replacement in concrete reduce the workability of the mix. In general, the workability of concrete with rubber is lower compared to normal concrete. As for this study constant water-cement ratio was applied in all mix, therefore modification in this ratio can improve the workability without significantly give adverse effect to its mechanical properties.

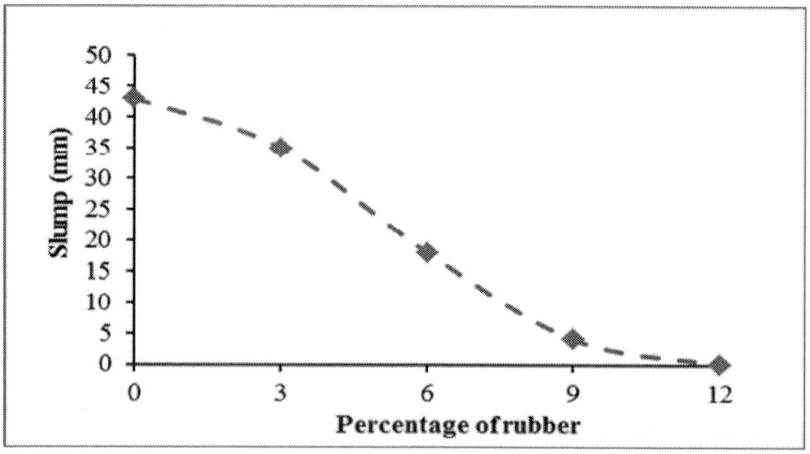

Figure 2. Slump test result.

3.2 Compressive strength test

The compressive strength of concrete cubes containing rubber particle as cement replacements is shown in Figure 3. Concrete with 0% rubber as cement replacement that act as control specimens has compressive strength of 33MPa at 28 days. As seen in the figure, replacing 3 % rubber in cement reduces the strength to 27MPa, but the strength increases when more rubber is replaced in the cement. Among all concrete cubes, 9% rubber shows the best result for 28 days concrete age. The percentage of strength reduction of concrete with rubber is between 6-17 % compared to normal concrete. The compressive strength of these concrete was further assessed for durability. The compressive strength of all the mixtures was increased at 56 days.

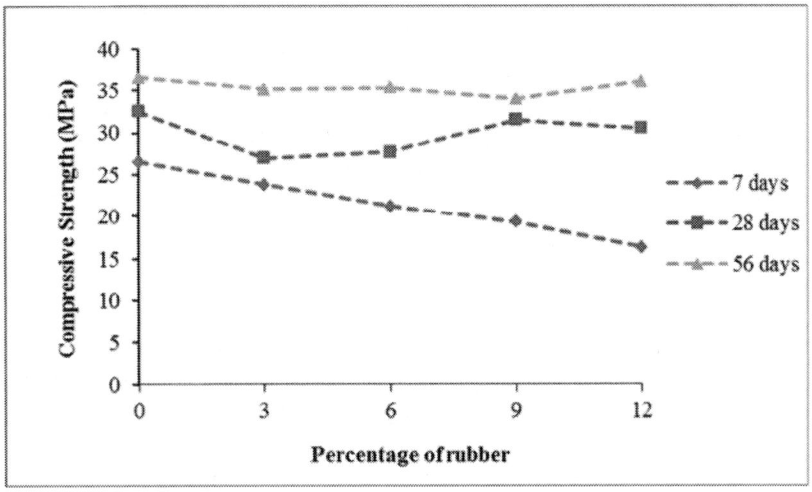

Figure 3. Compressive strength.

3.3 Splitting tensile strength test

Figure 4 shows the result of splitting tensile strength of cylinder made of concrete containing rubber as cement replacement. The control specimen with tensile strength of 2.7 MPa at 28 days is compared with other concrete containing rubber particle.

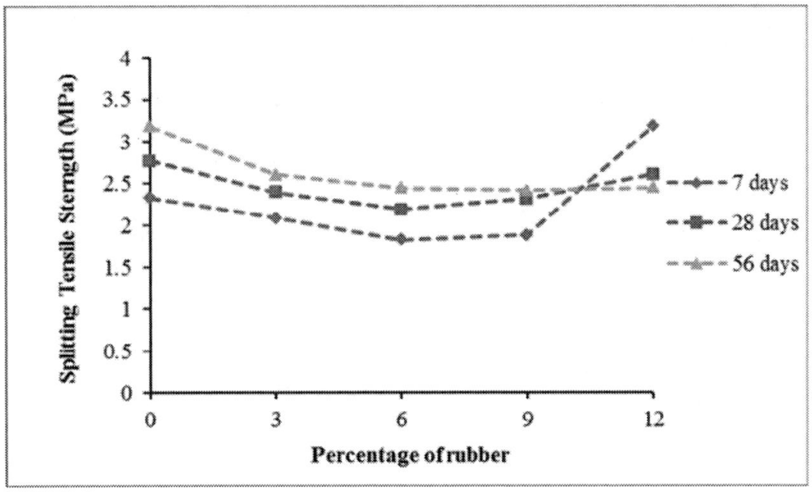

Figure 4. Splitting test strength.

Reduction in tensile strength is observed for 3% and 6% rubber replaced in the cement. After that, tensile strength increases when more percentages of rubber was replaced in the cement. The range of tensile strength differences is between 6% (concrete with 12% rubber) to 21% (concrete with 6% rubber). An increase in strength was recorded for most concrete when it was tested at 56 days. This findings is consistent with the previous research of rubberize concrete where split tensile strength and other properties are lower than normal concrete [9]. It is suggested that pre-treatment of rubber particles can improve the mechanical properties of concrete as well as its durability [2, 10].

4 Summary

This paper provides initial study on the effect of rubber as cement replacement in concrete on the compressive and tensile strength of the concrete. In conclusion, the replacement of cement with suitable percentage of rubber can be used in concrete with acceptable compressive and tensile strength as compared to normal concrete. This is because the range of strength reduction is only 6-21 percentage compare with normal concrete. There is room for improvement for rubber as cement replacement such as pre- treatment of the material to increase its strength without adverse effect on workability, durability and cost of the concrete.

References

[1] V.T. Senthil and R. Thenmozhi, Experimental study on waste tyre rubber replaced concrete - an ecofriendly construction material, Journal of Applied Science Research, **8**(6), 2966-2973, (2012).

[2] K.R. Ali, M. Dehestani and P. Rahmatabadi, Mechanical properties of concrete containing a high value of tire-rubber particles, Waste Management, **28**(12), 2472-2482, (2008).

[3] K.M. Kotresh and M.G. Belachew, Study on waste tyre rubber as concrete aggregate, International Journal of Scientific Engineering and Technology, **3**(4), 433-436, (2014).

[4] A. El-Gammal, A.K. Abdel-Gawad, Y. El-Sherbini and A. Shalaby, Compressive strength of concrete ultilizing waste tire rubber, Journal of Emerging Trends in Engineering and Applied Sciences, **1**(1), 96-99, (2010).

[5] Z. Boudaoud and M. Beddar, Effects of recycled tires rubber aggregate on the characteristics of cement concrete, Open Journal of Civil Engineering, **2**, 193-197, (2012).

[6] K. K. Zaher and F.M. Boyomy, Rubberized portland cement concrete, Journal of Materials in Civil Engineering, **11**(3), 206-213, (1999).

[7] L.H.Chou, C.K. Lu, J.R. Chang and M.T. Lee, Use of waste rubber as concrete additive, Waste Management Resources, **25**(1), 68-76, (2007).

[8] G. Chitra, P.V. Selvi and D. Vijayalakshmi, Carbon black as an additive in conventional concrete, International Journal of Emerging Technology and Advanced Engineering, **4**(2), 194-201, (2014).

[9] R.H. Ghedan and M.D. Hamza, Effect of rubber treatment on compressive strength and thermal conducticity of modified rubberized concrete, Journal of Engineering and Development, **15**(4), 21-29, (2011).

[10] F.P. Torgal, S. Shasavandi and S. Jalali, Tyre rubber wastes based concrete: A review, *First International Conference of WASTES: Solutions, Treatments and Opportunities*, Portugal, (2011).

MATEC Web of Conferences **47**, 01010 (2016)
DOI: 10.1051/matecconf/20164701010

Fresh Properties and Flexural Strength of Self-Compacting Concrete Integrating Coal Bottom Ash

Norwati Jamaluddin[1] , Ahmad Farhan Hamzah[1,5, a], Mohd Haziman Wan Ibrahim[1], Ramadhansyah Putra Jaya[2], Mohd Fadzil Arshad[3], Norul Ernida Zainal Abidin[1] and Nurol Huda Dahalan[4]

[1]*Faculty of Civil and Environmental Engineering, Universiti Tun Hussein Onn Malaysia, 86400 Parit Raja, Johor, Malaysia*
[2]*Faculty of Civil Engineering, Universiti Teknologi Malaysia, 81310 Skudai, Johor, Malaysia*
[3]*Faculty of Civil Engineering, Universiti Teknologi MARA, 40450 Shah Alam, Selangor, Malaysia*
[4]*Faculty of Civil Engineering, Universiti Teknologi MARA, 13500 Permatang Pauh, Pulau Pinang, Malaysia*
[5]*National Hydraulic Research Institute of Malaysia, 43300 Seri Kembangan, Selangor, Malaysia*

Abstract. This paper presents the effect of using coal bottom ash as a partial replacement of fine aggregates in self-compacting concrete (SCC) on its fresh properties and flexural strength. A comparison between SCC with various replacements of fine aggregates with coal bottom ash showed that SCC obtained flexural strength decrease on increase of water cement ratio from 0.35 to 0.45. The natural sand was replaced with coal bottom ash up to 30% volumetrically. The fresh properties were investigated by slump flow, T_{500} spread time, L-box test and sieve segregation resistance in order to evaluate its self-compatibility by compared to control samples embed with natural sand. The results revealed that the flowability and passing ability of SCC mixtures are decreased with higher content of coal bottom ash replacement. The results also showed that the flexural strength is affected by the presence of coal bottom ash in the concrete. In addition, the water cement ratios are influence significantly with higher binder content in concrete.

1 Introduction

Self-Compacting Concrete also referred as SCC is highly workable concrete that can flow through densely reinforced and complex structural elements under its own weight. It can fill all voids with no segregation, excessive bleeding and air migration air-popping, or other separation of materials, and without the need for vibration or other mechanical consolidation [1]. In early 1990's, SCC technology had been developed in Japan. Currently, this concrete has gained extensive use in many countries including Malaysia for different applications and structural configuration [1, 2]. It is agreed that SCC enhanced concrete stability and involves the effect of different mixture variables to certify satisfactory fresh properties and excellent mechanical properties especially compressive strength [2]. The challenge in SCC is to achieve good strength concrete with economical cost, as compared with conventional concrete, which is used high quantity of Portland cement and chemical admixtures [3]. An approach to reduce the cost with mineral admixtures, the utilization of supplementary cementitious materials is applicable in SCC due to durability developments potential in the concrete composite and

[a] Corresponding author : ahmadfarhanhamzah@gmail.com

overall economy. In addition, the inclusion of mineral admixtures is able to sustain workability and long-term properties [4].

In principle, the materials employed in the concrete are appropriate to be utilized in making of SCC including mineral admixture such as silica fume and fly ash, granulated blast furnace slag, etc. Meanwhile, several studies have been found to investigate the performance of SCC incorporating by-product waste material [5–7]. However, the research on bottom ash in the SCC is limited. Though a large number of significant results have been reported on the use of coal bottom ash in concrete, there is not much literature available on the use of coal bottom ash as a fine aggregate replacement material in SCC [8, 9]. Malaysia has established 4 electrical power plants that use a coal fired as power material must have a good dumping system to control the waste which composed of 80% of fly ash and 20% of bottom ash. As the dumping area is limited and become serious problem to environments, coal bottom ash is suggested to be used in concrete making which the result of utilization perhaps positively reduce the particular problem. This paper presents the flexural strength development of SCC with coal bottom ash as a fractional auxiliary of fine aggregates. The authors indicated the results in the form of mechanical strength test with three water cement ratios and different replacement level of coal bottom ash at the age of 28, 90 and 180 days.

2 Experimental Program

2.1 Material

In the current study, Portland Fly Ash cement used as a main binder. The fine aggregates were coal bottom ash and river sand. The coal bottom ash was collected from one of the coal-fired power plant in Malaysia and has a fine classified size distribution within the range of 0.075 mm to 20 mm. The coarse aggregates were graded size sieved through 16 mm and retained at 10 mm, meanwhile for fine aggregates were graded size sieved through 5 mm. Both river sand and coarse aggregate have a specific gravity of 2.61 and 2.67 respectively. The polymer-based superplasticizer has a specific gravity of 1.09 with pH 5.29 which is used to magnify the fresh concrete workability. The physical and chemical properties of cement and coal bottom ash is given in Table 1. The total composition (SiO_2 +Al_2O_3 +Fe_2O_3) in coal bottom ash exceeded 70%, indicating that it can be classified as class F ash as specified in ASTM C 618 [10]. In terms of chemical, coal bottom ash is considered as a pozzolonic material same as Portland and fly ash cement. Figure 1 shows the coal bottom ash particle size spreading and natural sand.

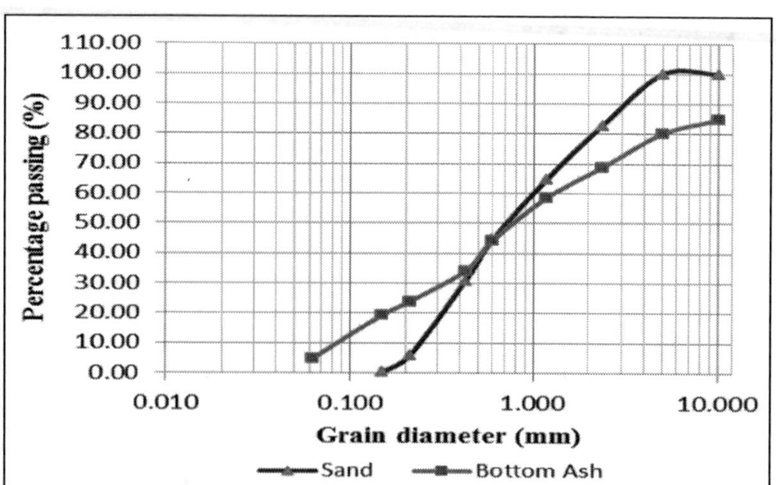

Figure 1. Particle size analysis.

Table 1. Chemical composition of cement and coal bottom ash.

Content	Chemical composition (%)										Loss on ignition	Specific gravity	Moisture Content
	SiO_2	Al_2O_3	Fe_2O_3	CaO	K_2O	TiO_2	MgO	Na_2O	CO_2	MnO			
Coal bottom ash	68.9	18.7	6.5	1.61	1.52	1.33	0.53	0.24	0.1	-	2.68	1.72	40%
Cement	22	8.35	3.92	58.9	1.01	0.72	0.52	0.26	0.1	0.15	1.72	3.0	-

2.2 Mix proportions and preparation of samples

Mix proportion was referred to Jawahar et al. [11]. The mix proportions of the constituent are given in Table 2. The water/cement ratios of 0.35, 0.40 and 0.45 concrete mixtures were considered in six batches volumetric replacement proportion of fine aggregate with coal bottom ash ranged from 0% to 30%. The fine aggregate comprise was preserved at 45% by volume of mortar and coarse aggregate was preserved at 28% by volume of concrete, while air-content being assumed to be 2%. The amount of super plasticizers was adjusted until the concrete achieve self-compatibility.

In order to ensure the concrete mixture in SCC classification, fresh properties of the concrete mixture was tested. The slump flow, slump flow time (T_{500}), L-box blocking ratio and segregation resistance ratio test were executed to determine the fresh properties of the mixtures. All procedure conducting the tests was referred to British Standards [12–14]. The prism size of 100x100x500mm was used and flexural strength was tested at age 28, 90 and 180 days.

Table 2. Mix proportions in kg/m^3.

Mix	w/c ratio	Cement	Coarse Aggregates	Fine Aggregates	Coal Bottom Ash	Water	SP (%)
	0.35	557	715.5	874.50	0	194.95	0.17
BA0	0.40	518	715.5	874.50	0	207.20	0.16
	0.45	485	715.5	874.50	0	218.25	0.16
	0.35	557	715.5	787.05	87.45	194.95	0.23
BA10	0.40	518	715.5	787.05	87.45	207.20	0.20
	0.45	485	715.5	787.05	87.45	218.25	0.19
	0.35	557	715.5	743.33	131.175	194.95	0.28
BA15	0.40	518	715.5	743.33	131.175	207.20	0.26
	0.45	485	715.5	743.33	131.175	218.25	0.21
	0.35	557	715.5	699.60	174.90	194.95	0.36
BA20	0.40	518	715.5	699.60	174.90	207.20	0.32
	0.45	485	715.5	699.60	174.90	218.25	0.30
	0.35	557	715.5	655.88	218.625	194.95	0.17
BA25	0.40	518	715.5	655.88	218.625	207.20	0.16
	0.45	485	715.5	655.88	218.625	218.25	0.16
	0.35	557	715.5	612.15	262.35	194.95	0.23
BA30	0.40	518	715.5	612.15	262.35	207.20	0.20
	0.45	485	715.5	612.15	262.35	218.25	0.19

3 Results

3.1 Fresh properties

The results of fresh characteristic of mixtures such as slump flow, slump spread time (T_{500}), L-box ratio and segregation resistance are presented as shown in Table 3. The slump flow values for all water cement ratios were known within the ranged of 550–750 mm while T_{500} spread times were in ranged between 2.5 and 5 seconds. The flowability of mixtures was reduced significantly with an increase of coal bottom ash. The decreased of slump flow values are due to the porosity of coal bottom ash in concrete, thus absorbed more fluid with higher content of coal bottom ash. The result is affected by irregular shape of coal bottom ash that reducing aggregates inter-particle friction and indicates by combining coal bottom ash reduces the viscosity of SCC mixtures [2, 15, 16]. In the context of passing ability, the L-box blocking ratio decreased when the percentage of coal bottom ash increased from 0% to 30%. The results revealed due to the aggregate crowded in front of openings, which caused from difficulty of particle to flow without obstruction. The trend is parallel as L-box test for sieve segregation test as the ratio decreased with increase coal bottom ash content. The sieve segregation results are indicated by the aggregates settlement which was reduced with higher content of coal bottom ash. As plastic viscosity decreases the settlement rate of coarse aggregate, thus it is improves the segregation resistance of fresh concrete [17].

Table 3. The fresh properties of self-compacting concrete.

	BA0			BA10			BA15		
Water cement ratio	**0.35**	**0.40**	**0.45**	**0.35**	**0.40**	**0.45**	**0.35**	**0.40**	**0.45**
Slump flow *(mm)*	700	720	745	670	710	725	650	688	708
T_{500} *(s)*	2.87	2.59	2.32	3.1	2.84	2.55	3.71	3.31	2.87
L-box ratio	0.97	0.94	0.90	0.92	0.89	0.81	0.92	0.86	0.80
SR (%)	9.61	9.85	10.37	7.93	8.15	9.25	6.61	7.83	8.13
	BA20			**BA25**			**BA30**		
Water cement ratio	**0.35**	**0.40**	**0.45**	**0.35**	**0.40**	**0.45**	**0.35**	**0.40**	**0.45**
Slump flow *(mm)*	615	635	685	588	600	648	545	570	625
T_{500} *(s)*	4.26	3.9	3.05	4.53	4.18	3.77	4.87	4.57	4.02
L-box ratio	0.88	0.80	0.77	0.88	0.80	0.70	0.76	0.69	0.66
SR (%)	5.24	6.06	6.72	5.20	6.28	6.58	5.50	6.20	6.51

3.2 Flexural strength

The results of flexural strength are shown in Figure 2, Figure 3 and Figure 4. It is obvious that the flexural strength of 10% replacement of coal bottom ash is higher compared to control sample for all water cement ratios. The development was observed at the age of 28, 90 and 180 days. As for the sample with 0.35 water cement ratio, the mixture of BA10 shows increment of 1.58%, 2.85% and 9.2% at respective ages.

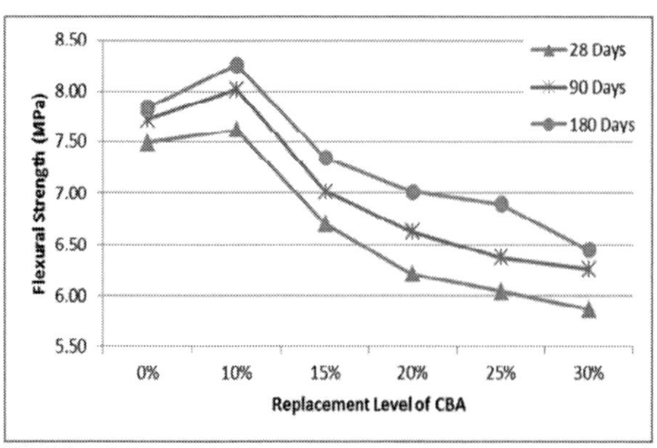

Figure 2. Flexural strength for 0.35 water cement ratios.

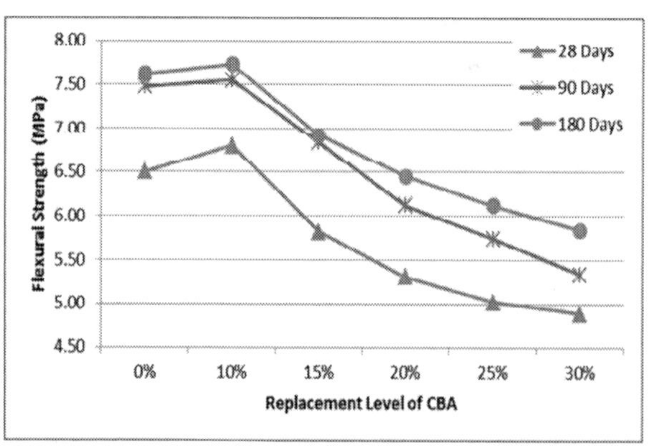

Figure 3. Flexural strength for 0.40 water cement ratios.

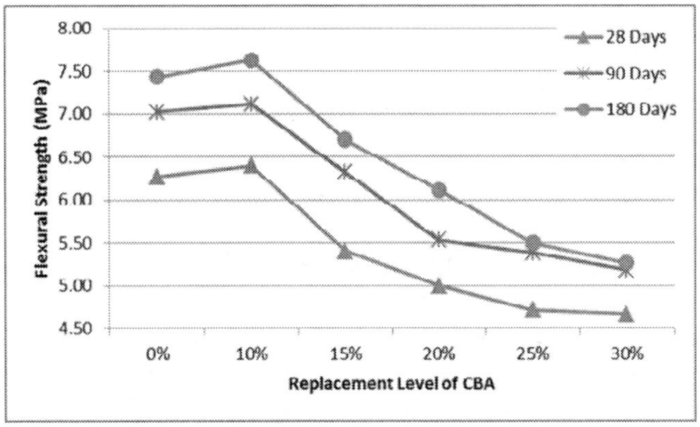

Figure 4. Flexural strength for 0.45 water cement ratios.

The results also determined that the increments of mixture BA10 mixture for 0.40 water cement ratio are 4.41%, 13.91% and 15.91% at 28, 90 and 180 days, correspondingly. Meanwhile for 0.45 water cement ratio, the increments of 2.03%, 11.81% and 17.82% were recorded. Conversely, the flexural strength of samples reduced when 15%, 20%, 25% and 30% replacement of coal bottom ash were used in the concrete for all water cement ratios. It is marked that as the replacement level of coal bottom ash increased, the degree of flexural strength decreased gradually. The decreasing of flexural strength in the particular samples was cause by the poor joining between aggregates. The results of flexural strength in this study were found to be complement to a study conducted by Aggarwal et al. [17] and Zainal Abidin et al. [2].

4 Conclusions

Based on the results obtained, it is confirmed that there are significant effect of coal bottom ash to the fresh properties of the concrete. The fresh properties of SCC mixtures such as slump flow, L-box blocking ratio, and segregation resistance are decreased with the higher content of coal bottom ash replacement. Workability of the concrete mixture was reduced proportionally to the reduction of water cement ratio in the mixture. In addition, the mechanical flexural strength of mixtures decreases as a result of increasing the coal bottom ash replacement and the amount of water cement ratio; thus reduces the viscosity of the mixtures. Meanwhile, the factor of water cement ratios is influenced significantly with higher binder content.

Acknowledgement

The authors thankfully acknowledge the Fundamental Research Grant Scheme (No. Vot 1454) funded by Ministry of Higher Education Malaysia and Faculty Civil and Environmental Engineering, University Tun Hussein Onn Malaysia (UTHM).

References

[1] H. Okamura and M. Ouchi, Self-compacting concrete, Journal of Advanced Concrete Technology, **1**(1), 5-15, (2003).

[2] N.E. Zainal Abidin, M.H. Wan Ibrahim, N. Jamaluddin, K. Kamaruddin and A.F. Hamzah, The effect of bottom ash on fresh characteristic, compressive strength and water absorption of self-compacting concrete, Applied Mechanics and Materials, **660**, 145-151, (2014).

[3] N. Bouzouba and M. Lachemi, Self-compacting concreting high volumes of class F fly ash preliminary results, Cement and Concrete Research, **31,** 413-420, (2001).
[4] A.F. Hamzah, M.H. Wan Ibrahim, N. Jamaluddin, R.P. Jaya and N.E. Zainal Abidin, Cementitious materials usage in self-compacting concrete: A review, In Advanced Materials Research, **1113**, 153-160, (2015).
[5] F.A. Sabet, N.A. Libre and M. Shekarchi, Mechanical and durability properties of self-consolidating high performance concrete incorporating natural zeolite, silica fume and fly ash, Construction Building Materials, **44**, 175–184, (2013).
[6] M. Nepomuceno, L. Oliveira and S.M.R. Lopes, Methodology for mix design of the mortar phase of self-compacting concrete using different mineral additions in binary blends of powders, Construction Building Materials, **26(**1), 317–326, (2012).
[7] M. Najimi, J. Sobhani, B. Ahmadi and M. Shekarchi, An experimental study on durability properties of concrete containing zeolite as a highly reactive natural pozzolan, Construction Building Materials, **35**, 1023–1033, (2012).
[8] R. Kasemchaisiri and S. Tangtermsirikul, Properties of self-compacting concrete in corporating bottom ash as a partial replacement of fine aggregate, Science Asia, **34**, 87–95, (2008).
[9] Mohd Syahrul Hisyam Mohd Sani, Fadhluhartini Muftah and Zulkifli Muda, The properties of special concrete using bottom ash (WBA) as partial sand replacement, International Journal of Sustainable Construction Engineering and Technology, **2**, 65-76, (2010).
[10] ASTM C618-15, Standard Specification for Coal Fly Ash and Raw or Calcined Natural Pozzolan for Use in Concrete, ASTM International, West Conshohocken, (2015).
[11] J.G. Jawahar, C. Sashidhar, I.R. Reddy and J.A. Peter, A simple tool for self compacting concrete mix design, International Journal Advance Engineering Technology, **3**(2), 550–558, (2012).
[12] BS EN12350-Part 8:2010, Testing Fresh Concrete - Self-Compacting Concrete – Slump flow test.
[13] BS EN12350-Part 10:2010, Testing Fresh Concrete - Self-Compacting Concrete – L box test.
[14] BS EN12350-Part 11:2010 Testing Fresh Concrete - Self-Compacting Concrete – Sieve segregation test.
[15] R. Siddique, P. Aggarwal and Y. Aggarwal, Influence of water or powder ratio on strength properties of self-compacting concrete containing coal fly ash and bottom ash, Construction Building Material, **29**, 73–81, (2012).
[16] Abdul Kadir, Aeslina and Mohd Ikhmal Haqeem Hassan, Leachability of self-compacting concrete (SCC) incorporated with fly ash and bottom ash by using synthetic precipitation leaching procedure (SPLP), Applied Mechanics and Materials, **773**, 1375-1379, (2015).
[17] P. Aggarwal, Y. Aggarwal and S.M. Gupt, Effect of bottom ash as replacement of fine aggregates in concrete, Asian Journal of Civil Engineering (Building and Housing), **8(**1), 49-62,(2007).

MATEC Web of Conferences 47, 01011 (2016)
DOI: 10.1051/matecconf/20164701011

Production of Controlled Low Strength Material Utilizing Waste Paper Sludge Ash and Recycled Aggregate Concrete

A. N. Azmi[1,a], M. A. Fauzi[1], M. D. Nor[1], A. R. M. Ridzuan[2] and M. F. Arshad[2]

[1]*Faculty of Civil Engineering, Universiti Teknologi Mara, 13500 Permatang Pauh, Pulau Pinang, Malaysia*
[2]*Faculty of Civil Engineering, Universiti Teknologi Mara, 40450 Shah Alam, Selangor, Malaysia*

Abstract. Recently, the best method to make the concrete industry more sustainable was using the waste materials to replace the natural resources. Currently waste paper sludge is a major economic and environmental problem in this country. In this research, the alternative method is to dwindle the usage of natural resources and the usage of cement in the construction. This method is to replace the usage of cement with the waste paper sludge ash (WPSA) and to use the recycle aggregate collected from the construction is used. The WPSA has ingredient likely cement such as self-cementation but for a low strength. The research was conducted at heavy laboratory UITM Pulau Pinang. Meanwhile, the WPSA is collected at MNI Industries at Mentakab, Pahang. The recycle aggregate is a separated half, which were fine aggregate and the coarse aggregate with the specific size. In this research, the ratio is divided into two (2) which is 1:1 and 1:2 for the aggregate and difference percentage levels of WPSA. The percentage levels of WPSA that use in this research are 10%, 20%, 30%, 40%, 50%, and 60%. A total of 36 cubes were prepared. Aim of this research is to develop a simple design approach for the mixture proportioning of WPSA and recycle concrete aggregate (RCA) within the concrete and to assess the effect of concrete mix with different percentage of WPSA and RCA ratio on the properties. It is found that the best design mix that achieves control low strength material (CLSM) is on 30% of WPSA with the ratio 1:2 on day 28 of compression test.

1 Introduction

CLSM is a self-compacted or cementation material that used to backfill as an alternative to compacted fill. This type of material has several of names like flowable fill, lean-mix backfill, unshrinkable fill, flowable mortar, and controlled- density fill (CDF) but the name of Controlled Low Strength Material (CLSM) has been given by American Concrete Institute (ACI) Committee 229 [14]. CLSM is not a concrete and not also as soil cement, but it has the same properties of both. It cannot be considered as low-strength concrete, but can be considered as self-compacted backfill material that used as compacted fill. The combinations of the mixture to get CLSM are made of Portland cement, water, aggregate, and ash. The normally maximum compressive strength that achieve by CLSM is 8.3 MPa from ACI Committee 229 [14].

The advantages use CLSM can reduced labor, fast construction and reduce the equipment use due to self-leveling properties and no need the compaction. While, for a low strength CLSM that range

[a] Corresponding author : najmifakhrudeen@gmail.com

from 0.3 to 0.7 MPa is allows for future excavation if needed. Other advantages for CLSM that used by product material such as fly ash and foundry sand that can reduce the utilizing of landfills. It was successful and more environmental when using by-product and waste materials to develop a sustainable development. On this paper, WPSA was use to replace the existing by product material to produce CLSM.

Malaysia government has been recently promoted sustainable development to achieve economic growth and a comfortable living for the next future generations. The most active activity in Malaysia that supports the local economic development is construction industries. The construction industries play an important role to develop and to build a comfortable living environment. The production of waste materials and the application of natural sources become negative impacts to this country [3]. The usage of natural sources is needed to be controlled before it's too late. The production of waste material especially from the construction industries need to be governed.

Singapore country nowadays forced to import the natural sources like an aggregate from overseas because their country shortage of the natural sources for their construction industries .Beside that, it would burden to the government and unable to achieve economic growth and comfortable living for the next future generations [3]. The most important material that use in construction industries is aggregates. In this study, the recycled material collected from the construction industries results from the demolited and renovated of buildings were used in the concrete mix. The recycled aggregate (RCA) was used as recycled material.

In this study, the RCA and waste paper sludge ash (WPSA) are used to get low strength called as control low strength material (CLSM). For RCA, it's collected from the batching plant and also can be used from demolition, renovated of building, roads and concrete must through the process that know as crushing.

RCA might be the best alternative solution to create a more sustainable concrete. It is due to the RCA have its own advantages like greater environmental awareness, achieve environmental laws, and keep the construction cost low. In terms of using RCA in structural concrete, Mohd Ridzuan et al. [1] reported that the use of 30% RCA and 70% natural aggregates in high strength concrete produces concrete of similar strength as that containing only natural aggregate. The plant-produced RCA as grainy in texture and later discussed that the RCA has a more rounded, spherical shape which seemed to improve workability [4].

Nowadays, natural construction material like aggregate was deficient. The production of natural aggregates causes of air pollution and environmental problem. WPSA is an innovation material that can be used in construction to replace the waste of cement. By utilizing WPSA, it may contribute Pozzolonic activity dependence on the percentage of WPSA after 7 and 28 days of concreting. Pozzolonic activity is very active where the alkalis and calcium hydroxide results from the mixing of concrete. The incineration of WPSA, when leached with alkali has potential as binder [3]. This study is to determine the strength of CLSM using WPSA in the CLSM mix as replacement of Portland cement. Table 1 show the chemical composition of WPSA and Portland cement based on the previous studies.

Table 1. Chemical Compositions of WPSA and Portland cement [8].

Chemical properties	SiO_2	Al_2O_3	Fe_2O_3	TiO_2	MgO	CaO	Na_2O	K_2O	P_2O_5	MnO	SO_3	LOI
WPSA (%)	15.16	6.06	1.11	0.45	2.00	55.87	0.19	0.34	0.48	0.05	0.78	17.51
OPC (%)	15.05	2.56	4.00	0.12	1.27	72.17	0.08	0.41	0.06	0.06	2.90	1.33

2 Experimental Program

2.1 Preparation of materials

To produce CLSM in this research, materials consist of WPSA, RCAs and soaked water (with RCA) were used. This CLSM unlike the ordinary mix and were produced with low strength same as lean concrete, brick and mortar. WPSA particles have wide range of size and composition and differ to Portland cement [2]. The WPSA was a waste material that collected from the paper industry at Mentakab, Pahang [1, 12, 13]. WPSA is used in this study as a replacement of cement to produce concrete. According to ASTM C618 for classification of fly ash, the WPSA classified as Class-C fly ash because the total combination percentage composition for major constituent components for silicon dioxide (SiO_2), aluminium oxide (Al_2O_3) and iron oxide (Fe_2O_3) more than 50% as shown in Table 1. Furthermore the WPSA contains higher of calcium oxide (CaO) about 56% more than the requirement. Many researchers have proved that the self-cementing of class-C fly ash ability to modify the engineering properties due to its high calcium oxide (CaO) content and pozzolanic characteristics without activators and more economic in stabilization application [9-11]. Hence, the WPSA was used as agent binder in the concrete mix due to pozzolanic composition.

The RCA was collected from recycled aggregates from demolished of and renovated of buildings, road and bridges. The RCA was crushed in the crushing machine and sieved in the sieving machine. The crushing machine was used to separate all materials such as coarse aggregate and fine aggregate. After concrete waste was crushed, the only coarses aggregates and fine aggregates through the next process known as sieving (grading). This process separated (graded) the size of aggregates to 4.75, 2.36, 1.18, 0.6, 0.425, 0.30, 0.15, and 0.075 mm by different of sieving pan. The next process is taking the course aggregates and fine aggregates by any size and quantity that measure it in the design mix to use for this paper study. The RCA size of 5 mm and 20 mm were used as recycled fine aggregate and recycled coarse aggregate respectively.

Prior to use the fine and coarse aggregates, they are soaked in distilled water to obtain the amount of alkali for the production of pozzolanic composition. RCA sizes 20 mm soaked in distilled water for 28days to obtain complete concentration. PH meter was used and identified water containing high alkaline pH 12.2 due to the pozzolanic reaction from the RCA. Alkali RCA released when soaked in water for 28 days the same study by Mohd Ridzuan et al. [1]. The water that containing high alkali was used back when the mixing process do, it can help pozzolonic reaction happen rapidly and have been proven by Abdullah et al. [7] .Two (2) series of concrete mixes was prepared using the WPSA, coarse and fine aggregate. Various ratio of WPSA investigated by batching all mixed with 0%, 10%, 20%, 30%, 40%, 50% and 60% by total weight of aggregates. The mix proportions for all of the mix are based on the weight proportion of 1:1 and 1:2 (Fine RCA: Coarse RCA) on Table 2.

2.2 Slump test

Slump test was performed to measure the workability and consistency of the fresh concrete. Figure 1 shows the slump test that had been conducted. Which based on the BS 12350-2 [17].

2.3 Compressive strength test

The 100mm x 100mm x 100mm sizes of cubes were used on this experiment. A total of 36 cubes were tested at the age of 7 and 28 days with accordance to BS 1881 [18]. A constant load rate of 0.6 ± 0.2 $N/mm^2/s$ was used throughout the test. Two (2) series concrete mixes were prepared using WPSA, coarse aggregate and fine aggregate as shown in Table 2.

Table 2. Mix proportion of CLSM.

Series (Ratio)	WPSA		RCA		Water Content (kg)
	%	Weight (kg/m^3)	Fine (kg/m^3)	Coarse (kg/m^3)	
1:1	10	115	600	550	4
	20	230			
	30	345			
	40	460			
	50	575			
	60	690			
1:2	10	113.3	396	737	4
	20	226.6			
	30	339.9			
	40	453.2			
	50	565.5			
	60	679.8			

Figure 1. Slump Test (ratio 1:2).

3 Results and Discussions

3.1 Slump test

Workability of fresh concrete is influenced by its proportion. Table 3 shows the result of the slump of the fresh concrete. For the first ratio of 1:1, the concrete is totally collapse slump for all percentage of WPSA from 10% to 60% which is in the collapse range. The collapse range is between 150 mm to 250 mm collapse from the height of the cone. For this ratio 1:1, the workability mix is too high and too wet concrete based on the slumps. Figure 2 have shown the slump test ratio 1:1 that totally collapse.

For the second series of mix ratio of 1:2, indicates the fresh concrete is totally true slump as shown in Table 3. At this series, the slump for all percentage of WPSA from 10% to 60% were in the true slump range. The true slump range is between 50 mm and below that collapse from the height of the cone. For this ratio 1:2, the workability mix is too low and not too wet concrete based on the slump type. Figure 1 has shown the true slump of ratio 1:2.

Table 3. Slump test results.

Percentage of WPSA	Series 1 (Ratio 1:1), mm	Series 2 (Ratio 1:2), mm
10	218	45
20	202	44
30	223	31
40	230	35
50	242	42
60	245	43

Figure 2. Slump Test (Ratio1:1).

3.2 Compressive strength of CLSM concrete

Figure 3 depicted the compressive strength of CLSM concrete at 7 and 28 days. It can be seen that the strength for all concrete mix are increased. Table 4 shows the compressive strength and density of the CLSM concrete at 7 and 28 days. It indicates that the density of the CLSM concrete affect to its performance. It is found that the compressive strength of the CLSM concrete is closely related to its weight.

Table 4. Compressive strength and density of CLSM concrete at 7 and 28 days.

Percentage of WPSA (%)	Series 1 (Ratio 1:1)			Series 2 (Ratio 1:2)		
	Compressive strength (kN/m^2)		Density (kg/m^3)	Compressive strength (kN/m^2)		Density (kg/m^3)
	7 days	28 days		7 days	28 days	
10	3.014	3.752	2.13	2.604	3.224	1.90
20	4.343	4.967	1.93	1.770	3.325	1.96
30	3.222	3.855	2.00	6.346	6.425	1.96
40	1.927	2.328	1.80	5.169	5.266	2.00
50	1.362	1.832	1.83	4.972	5.235	1.90
60	1.788	2.146	1.80	4.908	5.081	1.80

For the series 1:1, the lowest strength of CLSM concrete is 1.832 kN/m^2 with the dosage level of 50%. Figure 4 shows the compressive strength of CLSM concrete at different percentage levels of WPSA. The optimum dosage level of WPSA as binder is 20% with the compressive strength of 4.967 kN/m^2. The higher the dosage level reduces the strength of the CLSM concrete. It might be the excess of the binder can reduce the pozzolanic reaction happen between binder and aggregates.

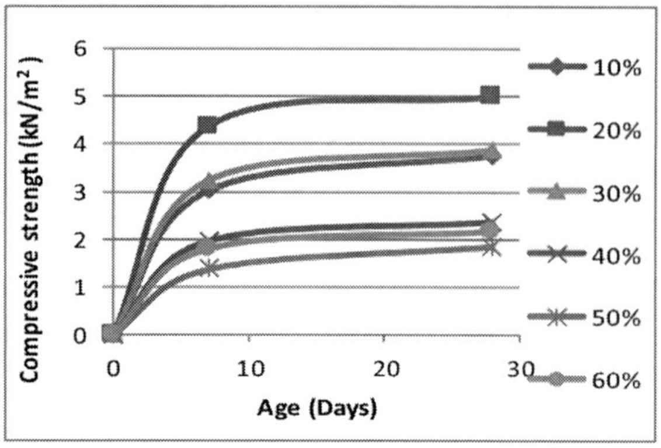

Figure 3. Compressive strength development for series 1 (1:1).

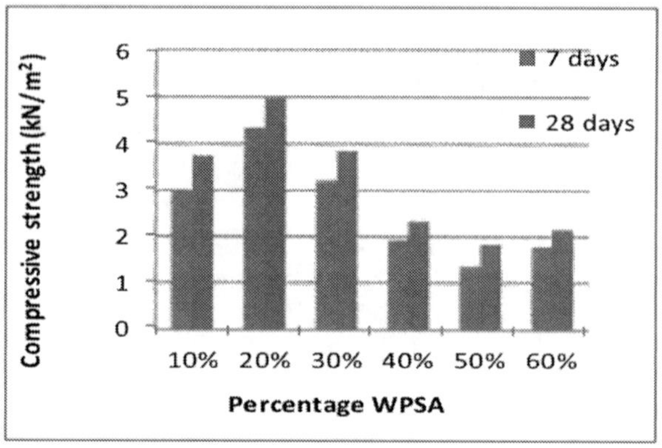

Figure 4. Effect of WPSA percentage for series 1 (1:1).

Figure 5 shows the compressive strength of CLSM concrete at the age of 7 and 28 days for various dosage levels of 10% to 60%. While, the compressive strength of CLSM concrete at different percent for series 2 is presented in Figure 6. The percentage of usage WPSA affects the hardening time process for 7 days until 28 days.

For the ratio 1:2, the lowest strength of CLSM concrete at 28 days is found to be at 10% WPSA of 3.224 kN/m^2. While, the higher strength of this ratio is at 30% WPSA with the strength of 6.425 kN/m^2. The lowest strength of the CLSM concrete for series 1 and series 2 is at dosage level of 10% and 20%, respectively. That means, 30% WPSA mix is the high strength that achieve for the ratio 1:2.

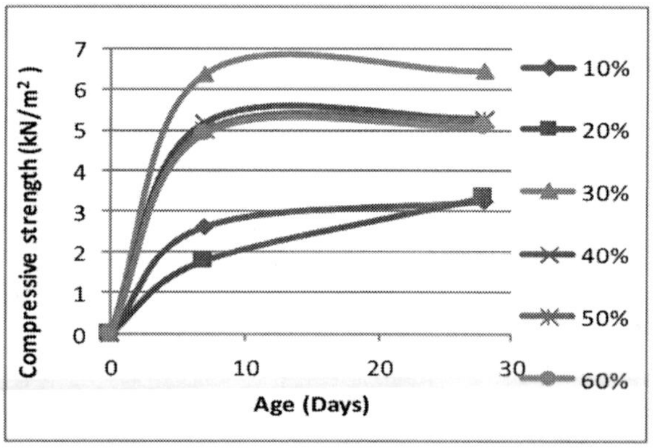

Figure 5. Compressive strength development for series 2 (1:2)

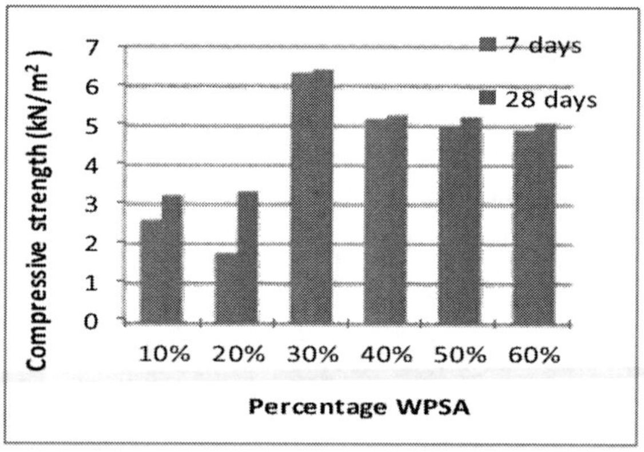

Figure 6. Effect of WPSA percentage for series 2 (1:2)

The higher compressive strength is ratio 1:2 at dosage level of 30% WPSA with the value of 6.425 kN/m^2. While, the lower ratio 1:1 at 50% WPSA with the strength of 1.832 kN/m^2. Overall, the highest compressive strength is obtain at the age of 28 days for the both ratio and the all percentage of WPSA. This is because the concrete has been reached it mature strength on the 28 days after curing process. Besides that, the ratio for the highest strength is obtained. That ratio is 1:2 cubes that contain more coarse aggregate than the fine aggregate, it is more suitable to get the highest strength because this ratio contains more coarse aggregate which it restrains more loading.

4 Conclusions

An investigation has been made of the strength characteristic of controlled low strength material (CLSM) by utilizing of recycled concrete aggregates and waste paper sludge ash. This is done using the compression machine and slump test to get the compressive strength and the slump (workability).

- The highest strength for both series was obtained at 28 days is 6.425 kN/m^2 for 1:2 ratio with 30% of WPSA in the concrete mix.
- The best mix for this research is at the dosage levels of 20% and 30% of WPSA with the ratios of 1:1 and 1:2, respectively.
- It is found that the optimum compressive strength is using WPSA with ratio 1:2. Hence, it becomes the benchmark that would be used throughout the study at dosage level of 30%.
- The workability of concrete mix decreases with the increase in WPSA content, it is suitable for flowable fill application based on ACI standard utilization.
- Utilization of WPSA in concrete would preserve natural resources that can be used for cement manufacture and thus make concrete construction industry sustainable. It also reduces emission of harmful pollutants into environment during the cement manufacturing.
- The compressive strength for ratio 1:1 and 1:2 is achieve the CLSM requirement according to ACI Standard 0.3 to 8.3 MPa for flowable fill application.

References

[1] A.R. Mohd Ridzuan, M.A. Fauzi, E. Ghazali, M.F. Arshad and M.A. Mohd Fauzi, Strength assessment of controlled low strength materials (CLSM) utilizing recycled concrete aggregate and waste paper sludge ash, *IEEE Colloquium on Humanities, Science and Engineering Research*, 208-211, (2011).

[2] E. Mozaffari, M. O'Farrell, J.M. Kinuthia and S. Wild, Improving strength development of waste paper sludge ash by wet-milling, Cement and Concrete Composites, **28**(2), 144-152, (2006).

[3] N.Y. Ho, Y. Pin, K. Lee, W.F. Lim, T. Zayed and M. Asce, Efficient utilization of recycled concrete aggregate in structural concrete, J. of Materials in Civil Engineering, **25**(3), 318-327, (2013).

[4] K. McNeil and T.H.K Kang, Recycled concrete aggregates: A review, Int. J. of Concrete Structures and Materials, **7**(1), 61-69, (2013).

[5] S.F.U. Ahmed, Properties of concrete containing construction and demolition wastes and fly ash, J. of Materials in Civil Engineering, **25**(12), 1864-1870, (2013).

[6] N. Deshpande, S. Kulkarni and H. Pachpande, Strength characteristics of concrete with recycled aggregates and artificial sand, Int. J. of Engineering Research and Applications, **2**(5), 38-42, (2012).

[7] A. Abdullah, M.S Jaafar, Y.H Taufiq-Yap, A. Alhozaimy, A. Al-Negheimish and J. Noorzaei, The effect of various chemical activators on pozzolanic reactivity: A review, Scientific Research and Essays, **7**(7), 719-729, (2012).

[8] S.S. Sharipudin, A.R.M. Ridzuan and H. Mohd Saman, Performance of foamed concrete with waste paper sludge ash (WPSA) and fine recycled concrete aggregate (FRCA) contents, Int. Sustainability and Civil Engineering J., **1**(2), 19-27, (2012).

[9] V. Sata, C. Jaturapitakkul and K. Kiattikomo, Influence of pozzolan from various by-product materials on mechanical properties of development and properties of low-calcium high strength concrete, Construction and Building Material, **21**, 1589-1598, (2007).

[10] M.S.H.M. Sani, F. Muftah and M.A. Rahman, Properties of Waste Paper Sludge Ash (WPSA) as Cement Replacement in Mortar to Support Green Technology Material, *The 3rd International Symposium and Exhibition in Sustainable Energy and Environment*, 94-99, (2011).

[11] A. Misra, D. Biswas and S. Upadhyaya, Psysico-mechanical behaviour of self-cementing class C fly ash-clay mixtures, J. of Fuel, 84, (2005).

[12] A.A. Khairulniza, Effect of alkaline activity to the strength and shrinkage performance of geopolymer concrete. Journal of Sustainability and Civil Engineering, (2012).
[13] A.R.M. Ridzuan, A.A. Khairulniza, M.A. Fadzil and J. Nurliza, Morphology and physical analysis of waste paper sludge ash (WPSA) polymeric mortar, Advanced Materials Research, **701**, 275-279, (2013),
[14] ACI 229 R-99, Controlled low-strength materials, American Concrete Institute, (1999).
[15] BS EN 12390-2, Testing hardened concrete, Making and curing specimens for strength tests, British Standards, (2009).
[16] BS 410-1, Technical requirements and testing, test sieves of metal wire cloth, British Standards, (2000).
[17] BS EN 12350-2, Testing fresh concrete, slump-test, British Standards, (2009).
[18] BS 1881-108, Testing concrete, method for making test cubes from fresh concrete, British Standards, (1983).

MATEC Web of Conferences **47**, 01012 (2016)
DOI: 10.1051/matecconf/20164701012

Lightweight Concrete Using Oil Palm Boiler Clinker (OPBC) – A Review

Herry Hartono[1], Lee Jin Chai[1] and B. C. T. Lee[1,a]

[1]*Department of Engineering, UCSI University, Cheras, 56000 Kuala Lumpur, Malaysia*

Abstract. Lightweight concrete can be effectively produced by replacing normal aggregates (60% to 75% of concrete volume) with a lighter alternative. With depleting natural resources, utilising waste materials, such as oil palm boiler clinker (OPBC), in concrete for structural use is one way to mitigate environmental concerns raised by the construction industry. This paper presents a review of the mechanical properties, structural behaviour and performance of OPBC concrete. Lightweight concrete using OPBC can be designed to achieve different compressive strengths with different mixes. The different OPBC concrete mixes result in different densities and workability. The degree of content and the type of OPBC substitutes used affect the flexural strength and 28-day splitting tensile strength of OPBC concrete. A different effect was observed in the modulus of elasticity as the drying shrinkage and water absorption of OPBC concrete are also impacted. This review study also compares the structural performance of OPBC concrete to that of conventional concrete.

1 Introduction

The need to meet the demand for concrete from the construction sector raises environmental concerns. The resulting carbon dioxide emission is making concrete industry to be not sustainable [1], and construction activities to be labelled as not "green" [2]. Concrete manufacturing also consumes high amount of natural aggregates, accounting for 70% - 80% of total concrete volume [3]. The concrete industry has been known to be the largest contributor to the depletion of limited natural resources i.e. sand, water, rock, and gravel [4].

Because concrete is essentially a blend, the constituents can therefore be selected to enable the maximisation and optimisation of concrete's performance [5]. To curb the issue on environmental concerns, artificial aggregates e.g. fly ash, ground granulated blast furnace slag, and expanded clay, can be used as substitutes to produce an environmentally friendly lightweight concrete [6]. By using lightweight aggregates in concrete, lightweight concrete can be produced.

Concrete is considered as lightweight when its density ranges between 1440 kg/m^3 to 1840 kg/m^3. The density of conventional concrete lies between 2240 kg/m^3 to 2400 kg/m^3. Despite the lower densities, lightweight concrete is still expected to have a compressive strength of at least 17.0 MPa for use on structures [7].

The weight of lightweight concrete is about 25% to 35% less than conventional concrete [8]. With the use of lightweight concrete, the reduction in dead load is approximated to be 20%, which can be

[a] Corresponding author : brendalee@ucsiuniversity.edu.my

translated to about 10% of construction cost saving [9]. The reduction of dead weight consequently allows the reduction in size of load bearing elements as well as the use of steel reinforcement. Lightweight concrete also possesses a low coefficient of thermal expansion and a better fire-resistant as compared to normal weight concrete. It is therefore possible to construct a thinner fire rating slab which leads to a lighter building and lower construction cost [10].

By using low-cost lightweight aggregate, the production cost of lightweight concrete is also reduced. This has been shown possible with the successful substitution of aggregates with agricultural wastes for lightweight concrete production [11]. One of the agricultural wastes is OPBC [12]. OPBC usually has no economic value and are plentifully available [13]. Full aggregate substitution with OPBC has been shown to reduce 30% of manufacturing cost and 22.62% reduction in carbon dioxide emission, with comparable structural efficiency (strength-to-weight ratio) to conventional mix concrete [14].

2 Lightweight Concrete using Agricultural Waste as Aggregates

Lightweight concrete that is made with lightweight aggregate can be categorised into two categories based on its formation: (1) mechanical treatment and (2) thermal treatment. Lightweight concrete that uses agricultural waste as aggregates is grouped in the latter category, where aggregate is produced using combustion of natural or waste material. These aggregates can be made of coconut shell, rice husk, corn cob, tobacco waste, oil palm shell and OPBC as examples [11].

There is always a challenge in attaining high strength in lightweight concrete, i.e. compressive strength of above 55.0 MPa, due to its porosity that tends to restrain the development of mechanical properties of the concrete [15-18]. Despite that, lightweight concrete produced using agricultural wastes as coarse aggregate has been shown to be able to achieve high compressive strength. The 28 days compressive strength of lightweight concrete using crushed oil palm shell in saturated surface dry (SSD) condition was found to be 53.0 MPa [19]. This was achieved by having water-cement ratio of 0.45. When a lower water-cement ratio was used, i.e. 0.35, the 28 days compressive strength was found to be 42.5 MPa [20].

2.1 Lightweight concrete using OPBC as aggregates

OPBC stones are obtained from the combustion of palm oil fibre and palm oil shell. They are irregular in shape and thorny. The OPBC stones are usually grey in colour and are porous. Figure 1 shows the form of OPBC stones in coarse and fine sizes. OPBC particles that pass 4.75 mm sieve are used for fine aggregate replacement whereas the particles that has nominal size of 20 mm (above 4.75 mm) are used for coarse aggregate replacement [12]. Besides coarse and fine aggregates, OPBC can be ground into finer sizes, to be used as binder replacement [14].

Figure 1. Coarse and fine OPBC [14].

2.2 Mix design

In the mix design, normal weight aggregate is usually substituted with OPBC by volume because OPBC is much lighter than normal weight aggregate. Like other lightweight aggregates, OPBC is porous, making it to be sensitive to water content. The use of OPBC in place of coarse and fine aggregates in lightweight concrete has been shown to achieve a 28-day compressive strength of 44.4 MPa in pre-stressed beams. This mix had a water-cement ratio of 0.44 [21]. With a higher water-cement ratio of 0.55, and the use of 10% fly ash, the strength was found to be 42 MPa [12].

At higher water-cement ratios, when OPBC concrete is mixed in accordance to ACI 211.2-98, the expected 28-day compressive strength was in the range of 17.00 to 33.17 MPa [22]. When mixed according to conventional mix design (ACI method), even if OPBC is used only as fine aggregates, it yielded lower compressive strength and workability [23].

In more recent studies, the highest 28-day strength was found in a mix that has even lower water-cement ratio of 0.38, which is 42.56 MPa [24]. For a mix that uses OPBC only as coarse aggregates and water-cement ratio of 0.331, the 28-day compressive strength was found to be about 5% higher at 44.89 MPa [25]. The latter, however, uses plasticizer to improve the workability of the mix.

When designed to be self-compacting concrete mix, OPBC concrete was found to be able to achieve 28-day compressive strength in the range of 50.0 – 60.0 MPa. This mix was designed with water-cement ratios of 0.43 to 0.54, and included a stage in the process where OPBC was immersed for 24 hours for absorption of water [26]. Then, it was shown that it is also possible for self-compacting OPBC concrete to be produced with 28-day compressive strength of between 60.0 to 70.0 MPa [14].

3 Physical and Mechanical Performance

3.1 Slump

For concrete that uses OPBC as coarse and fine aggregates, slump values of 85 to 125 mm could be achieved with 10% inclusion of fly ash as binder material [12]. With the addition of superplasticizer, a high workability mix (slump of 102 mm to 127 mm) was yielded for up to 37.5% of fine OPBC in the mix [25]. However, a reduced slump value of 50 mm to 80 mm was found in concrete incorporating up to 37.5% of OPBC aggregate (SSD condition) [13].

3.2 Density

Weight of concrete was observed to have reduced from 16% - 23% when different mixes of OPBC concrete were tested. The greatest reduction was found in the mix that uses OPBC as coarse and fine aggregates [12].

The use of OPBC in concrete reduces the density of concrete due to it being lighter than any of constituents of a concrete mix. For instance, the weight of OPBC sand is 22% lower than normal sand, that when used, was found to reduce the density of concrete up to 27% [13]. Similar range of densities was also found when OPBC concrete was mixed with oil palm shell as the replacement for the conventional coarse aggregate: 1889 kg/m3 to 1948 kg/m3 [25].

The reduction in density of concrete is directly proportional to the ratio of OPBC being used as replacement. A 100% replacement was found to reduce the density of concrete by 17% [14]. By comparing the effects caused by the use of OPBC in concrete, the use of OPBC as fine aggregate causes more reduction in concrete density. This may be attributed to the greater number of pores that exists in OPBC in the form of fine aggregates.

3.3 Compressive strength

The use of OPBC as fine aggregate was found to increase the compressive strength of concrete even at only 5% of OPBC [27]. However, this was only observed for OPBC concrete mixes under various partial early curing conditions due to the ability of internal curing of OPBC aggregates. When these mixes undergo continuous moist curing, only the mix with 25% of OPBC was able to achieve compressive strength of 37.4 MPa, almost equivalent to that without OPBC, which was 37.8 MPa.

The rest have shown a reduction in compressive strength [13].

When used as coarse aggregate, a high compressive strength was attained, i.e. 44.89 MPa [25]. The difference in effects when OPBC is used as coarse compared to as fine aggregates is due to the greater number of pores in fine OPBC [12].

A different phenomenon was observed when OPBC is used as partial and whole replacement of coarse aggregate, fine aggregate and binder material in self-compacting concrete. The compressive strength was found to be in the range of 60.0 MPa to 70.0 MPa. These values of compressive strength can still be achieved despite the 100% substitution of OPBC as fine and coarse aggregate [14].

3.4 Flexural strength

A similar trend to that of compressive strength is found in the effects that OPBC has in concrete on its flexural strength. It was found that there was an increase of 28% in flexural strength when OPBC is used as 5% of fine aggregates, and an increase of 33% when used as 10% [27]. The opposite effect was found when more than 10% of fine aggregates were replaced with OPBC. The flexural strength of OPBC concrete decreases as the content of OPBC in the concrete increases [13].

The negative effect was also observed when OPBC is used as both coarse and fine aggregates. When used only as coarse aggregates, the reduction in flexural strength was found to be 20.2%. The effect is greater when OPBC is used also as fine aggregates, causing a reduction of 32.3% in flexural strength. This can be attributed to the porous characteristic of OPBC [12].

3.5 Splitting tensile strength

The minimum 28-day splitting tensile strength of structural lightweight concrete is 2.0 MPa [28]. Although the using of OPBC in concrete is expected to reduce the splitting tensile strength of concrete due the weaker grain in OPBC aggregates [29], OPBC concrete is still found to be suitably used as structural lightweight concrete. For concrete that has OPBC as fine aggregates, the splitting tensile strength was only reduced by about 2.7%. This reduction increases as the content of OPBC increases up to 50% [13]. A higher reduction was observed when OPBC was used as both coarse and fine aggregates (up to 32.9%) as compared to when it was used only as coarse aggregates (from 23.5%) [12].

3.6 Modulus of elasticity

The typical modulus of elasticity of structural lightweight concrete is 10.0 GPa to 24.0 GPa [30]. OPBC concrete that uses OPBC as fine aggregates above 37.5% was found to have modulus of elasticity lower than the required values [13, 25]. OPBC concrete that uses OPBC as coarse aggregates, on the other hand, has high modulus of elasticity, 81.6% higher than that of OPBC as 25% of fine aggregate, at 22.95 GPa [25].

3.7 Shrinkage

A higher drying shrinkage in lightweight concrete is expected due to a lower elastic modulus of the aggregate used [31]. This was observed when OPBC is used as coarse and fine aggregates with 10%

of fly ash when compared to conventional concrete mix [32]. The use of OPBC as fine aggregate, on the other hand, was found to have no significant effect on the drying shrinkage of concrete [13].

3.8 Water absorption

The increased porosity in lightweight aggregate causes higher water absorption [33]. This was coherently found in OPBC concrete mix that uses OPBC as fine aggregate and water absorption increases proportionately with the amount of OPBC in concrete [13]. The 24 hours water absorption of fine and coarse OPBC aggregate was 3.60% [13] and 3.56 [34]. As most good performance concrete has water absorption value below 10% by mass [31], the recommended substitution of OPBC is therefore not to exceed 37.5% in oil palm shell concrete [13].

4 Structural Performances

4.1 Flexural behaviour

OPBC concrete beams when cast with the reinforcement ratio less than 0.5% and as pre-stressed were found to behave similarly to normal weight concrete beams. The failure of the beams was due to ductile failure, as expected in normal weight concrete beams [21, 24].

The experimental deflection values for OPBC concrete beams with reinforcement was found to be 10% to 45% less than what was expected according to BS8110. The resulting crack widths for these specimens were also found to be within the acceptable limit in accordance to BS8110 [24].

For OPBC pre-stressed concrete beams, greater deflection was found when compared to that caused in normal weight concrete, as expected. This difference in deflection is translated to only a drop of mere 6.4% in carrying capacity [21].

4.2 Shear behaviour

The shear failure mode in OPBC concrete beam was found to be similar to that of normal weight concrete beams and is in accordance to ASTM: C330 for structural use [24].

5 Summary

The use of OPBC in lightweight concrete has shown promising outcomes. The disadvantages of using OPBC in concrete due to its porosity can be lessened if OPBC is used in certain conditions, size, and percentage of volume. In spite of its sensitivity to water content, its weight brings more advantages to the concrete and construction industry. OPBC concrete, despite being lightweight, can be designed to achieve adequate compressive strength for structural application. The decrease in dead load means cost savings in the construction industry without compromising on structural integrity. The lowered flexural strength in OPBC concrete beams still qualifies itself for use as structural lightweight concrete.

By using OPBC as part of raw materials in concrete also makes the production of OPBC concrete to be more “green’ as it calls for less demand for non-renewable natural resources and on top of that, makes use of waste. The cost of OPBC being very low at RM 0.020 per kg [14], can significantly reduce the cost of concrete.

Lightweight concrete using OPBC can potentially be widely used in the construction industry when optimum ratios of OPBC in different forms can be found to achieve desirable outcomes. More studies could perhaps be conducted to look into the durability of OPBC concrete for these findings to be more conclusive.

References

[1] A.A. Ramezanianpour, M.M. Khani and G.H. Ahmadibeni, The effect of rice husk ash on mechanical properties and durability of sustainable concretes, Int. J. of Civil Engineering, **7**(2), 83-91, (2009).

[2] V.W.Y. Tam, Comparing the implementation of concrete recycling in the Australian and Japanese construction industries, J. of Cleaner Production, **17**, 688-702, (2009).

[3] T.Y. Tu, Y.Y. Chen and C.L. Hwang, Properties of HPC with recycled aggregates, Cement and Concrete Research, **36**, 943-950, (2006).

[4] H. Mefteh, O. Kebaili, H. Oucief, L. Berredjem and N. Nourredine Arabi, Influence of moisture conditioning of recycled aggregates on the properties of fresh and hardened concrete, J. of Cleaner Production, **54**, 282-288, (2013).

[5] R.N. Swamy, Design for sustainable development of concrete construction. *Proc. 4th Int. Structural Engineering and Construction Conference*, Taylor and Francis, London, 1241-1246, (2008).

[6] T.W. Bremner, Environmental aspects of concrete: Problems and solutions. *Proc. of the 1st Russian Conf. on Concrete and Reinforced Concrete Problems,* Moscow, Russia, (2001).

[7] National Ready Mixed Concrete Association, *Structural Lightweight Concrete*, <http://www.nrmca.org/aboutconcrete/cips/36p.pdf> (accessed 31 January 2015).

[8] T.A. Holm and J.P. Ries, *Benefits of Lightweight HPC*, <http://www.escsi.org/uploadedFiles/Technical_Docs/Structural_Lightweight_Concrete/4700.9%20Benefits%20of%20Lightweight%20HPC.pdf> (accessed 3 February 2015).

[9] G.K. Mackie, *Recent Uses of Structural Lightweight Concrete.* <http://www.concreteconstruction.net/Images/Recent%20Uses%20of%20Structural%20Lightweight%20Concrete_tcm45-341131.pdf> (accessed 3 February 2015).

[10] D.P. Martin, A.S. Zimmer, M.J. Bolduc and E.R. Hopps, *Is Lightweight Concrete All Wet*, <http://www.structuremag.org/?p=1163> (accessed 25 July 2015).

[11] P. Shafigh, H.B. Mahmud, M.Z. Jumaat and M. Zargar, Agricultural wastes as aggregate in concrete mixtures – A review, Construction and Building Materials, **53**, 110-117, (2014).

[12] M.H. Ahmad, S. Mohd and M.M. Nurazuwa, Mechanical properties of palm oil clinker concrete, *Proc. of 1st Engineering Conf. on Energy and Environment*, Kuching, Malaysia, (2007).

[13] P. Shafigh, H.B. Mahmud, M.Z. Jumaat, R. Ahmmad and S. Bahri, Structural lightweight aggregate concrete using two types of waste from the palm oil industry as aggregate, J. of Cleaner Production, **80**, 187-196, (2014).

[14] J. Kanadasan and H.A. Razak, Engineering and sustainability performance of self-compacting palm oil mill incinerated waste concrete, J. of Cleaner Production, **89**, 79-86, (2015).

[15] American Concrete Institute, *Guide for Structural Lightweight-Aggregate Concrete*, ACI Committee Report 213R-03, Michigan, United States, (2003).

[16] American Concrete Institute, *Building Code Requirements for Structural Concrete and Commentary*, ACI Committee Report 318-11, Michigan, United States, (2011).

[17] H. Costa, E. Julio and J. Lourenco, New approach for shrinkage prediction of high-strength lightweight aggregate concrete, Construction and Building Materials, **35**, 84-91, (2012).

[18] M. Lopez and L. Kahn, High performance lightweight concrete - A comparison between actual prestress losses and design code estimates, Revista Ingenieria de Construccion, **22**, 59-69, (2007).

[19] P. Shafigh, M.Z. Jumaat, H.B. Mahmud and U.J. Alengaram, A new method of producing high strength oil palm shell lightweight concrete, Materials and Design, **32**, 4839-4843, (2011).

[20] H.B. Mahmud, P. Shafigh and M.Z. Jumaat, Structural lightweight aggregate concrete containing high volume waste materials, Key Engineering Materials, **594-595**, 498-502, (2014).

[21] W. Omar and R.N. Mohamed, The performance of pretensioned prestressed concrete beams made with lightweight concrete, J. Kejuruteraan Awam, **14**(1), (2002).

[22] M. Abdullahi, H.M.A. Al-Mattarneh, A.H.A. Hassan, M.H. Hassan and B.S. Mohammed, Trial mix design methodology for palm oil clinker (POC) concrete, *The International Conference on Construction and Building Technology*, Kuala Lumpur, (2008).

[23] M.A. Mannan and K. Neglo, Mix design for oil-palm-boiler clinker (OPBC) concrete, J. of Science and Technology, **30**, 111-118, (2010).

[24] B. Mohammed, W.L. Foo and M. Abdullahi, Flexural strength of palm oil clinker concrete beams, Materials and Design, **53**(16), 325–331, (2014).

[25] R. Ahmmad, M.Z. Jumaat, S. Bahri and A.B.M.S. Islam, Ductility performance of lightweight concrete element containing massive palm shell clinker, Construction and Building Materials, **53**, 234-241, (2014).

[26] J. Kanadasan and H.A. Razak, Mix design for self-compacting palm oil clinker concrete based on particle packing, Materials and Design, **56**, 9-19, (2014).

[27] E.A.W. Shahreen, R. Omar, S.F.C. Osmi, N.A.M. Khairussaleh and A. Abdullah, Preliminary study on mechanical properties of concrete added with fine palm oil clinker, University Malaysia Pahang, Malaysia.

[28] N.U. Kockal and T. Ozturan, Strength and elastic properties of structural lightweight concretes, Materials and Design, **32**, 2396-2403, (2011).

[29] K.R. Wu, B. Chen, W. Yao, and D. Zhang, Effect of coarse aggregate type on mechanical properties of high-performance concrete, Cement and Concrete Research, **31**, 1421-1425, (2001).

[30] CEB/FIP Manual of Design and Technology, *Lightweight Aggregate Concrete*, The Construction Press, Lancaster, United Kingdom, (1977).

[31] A.M. Neville and J.J. Brooks, *Concrete Technology*, Pearson Education Asia, Kuala Lumpur, Malaysia, (2008).

[32] M.H. Ahmad, L.Y. Loon, N.M. Noor and S.H. Adnan, Shrinkage of Malaysian palm oil clinker concrete, *International Conference on Civil Engineering Practice*, Kuantan, (2008).

[33] T.Y. Lo, H.Z. Cui, W.C. Tang and W.M. Leung, The effect of aggregate absorption on pore area at interfacial zone of lightweight concrete, Construction and Building Materials, **22**, 623-628, (2008).

[34] R. Ahmmad, Z.M. Jumaat , A.U. Johnson, S. Bahri, M.A. Rehman and H.Hashim, Performance evaluation of palm oil clinker as coarse aggregate in high strength lightweight concrete, J. of Cleaner Production, **112**(1), 566-574, (2016).

MATEC Web of Conferences **47**, 01013 (2016)
DOI: 10.1051/matecconf/20164701013

Effect of Pelletized Coconut Fibre on the Compressive Strength of Foamed Concrete

Zainorizuan Mohd Jaini[1,a], Shahrul Niza Mokhatar[1], Ammar Saifuddin Mohd Yusof[2], Syurafarina Zulkiply[2] and Mohd Hadi Abd Rahman[1]

[1]*Jamilus Research Center, Universiti Tun Hussein Onn Malaysia, 86400 Parit Raja, Johor, Malaysia*
[2]*Faculty of Civil and Environmental Engineering, Universiti Tun Hussein Onn Malaysia, 86400 Parit Raja, Johor, Malaysia*

Abstract. Foamed concrete is a controlled low density ranging from 400kg/m^3 to 1800kg/m^3, and hence suitable for the construction of buildings and infrastructures. The uniqueness of foamed concrete is does not use aggregates in order to retain low density. Foamed concrete contains only cement, sand, water and foam agent. Therefore, the consumption of cement is higher in producing a good quality and strength of foamed concrete. Without the present of aggregates, the compressive strength of foamed concrete can only achieve as high as 15MPa. Therefore, this study aims to introduce the pelletized coconut fibre aggregate to reduce the consumption of cement but able to enhance the compressive strength. In the experimental study, forty-five (45) cube samples of foamed concrete with density 1600kg/m^3 were prepared with different volume fractions of pelletized coconut fibre aggregate. All cube samples were tested using the compression test to obtain compressive strength. The results showed that the compressive strength of foamed concrete containing 5%, 10%, 15% and 20% of pelletized coconut fibre aggregate are 9.6MPa, 11.4MPa, 14.6MPa and 13.4MPa respectively. It is in fact higher than the controlled foamed concrete that only achieves 9MPa. It is found that the pelletized coconut fibre aggregate indicates a good potential to enhance the compressive strength of foamed concrete.

1 Introduction

Foamed concrete has received high attention in the construction industry due to its advantageous of lightness and versatility material. The primary characteristic of foamed concrete is cement based mortar that consist at least 20% of air voids. Therefore, foamed concrete has low density, good strength and reliable toughness at minimal consumption of aggregate. Typically, foamed concrete can achieve dry density in the range of 400kg/m^3 to 1600kg/m^3 with compressive strength around 1MPa to 15MPa [1, 2]. The tensile strength of foamed concrete is directly proportion of its compression strength and lower approximately ten times. Meanwhile, the fracture energy of foamed concrete is around 15N/m to 35N/m [3-5]. Although a comparison with normal concrete shows that the strength of foamed concrete is slightly lower. Still, there are demands on the application of foamed concrete as construction material and structural element.

[a] Corresponding author : rizuan@uthm.edu.my

Since the strength of foamed concrete is lower than normal concrete, many attempts have been made to enhance the ability of foamed concrete in carrying meaningful actions. Such attempts include the utilization of supplementary cementitious materials such as rice husk ash, palm oil ash, artificial fibres and natural reinforcements. The fascination toward green and sustainable material in construction industry has prompted the use of natural reinforcements such as coconut and kenaf fibres. Moreover, foamed concrete containing natural reinforcement is an attractive option. Normally, in concrete, steel or polypropylene fibre is used to increase the strength, but it can also increase the weight and cost of the concrete. Therefore, the application of natural fibre is one of the alternatives due to high availability and environmental friendly.

2 Concrete Containing Coconut Fibre

In recent years, the use of coconut fibre as reinforcement on concrete has received high attention because of renewability, low density, high specific properties and low cost. High specific properties include the strength, lightness, stiffness, impact resistance and flexibility [6]. Moreover, coconut fibre in concrete also has become an interesting application towards sustainable construction. Based on an experimental study by Ruben and Braker [7], it was found that concrete containing coconut fibre not just bring several improvement in concrete characteristics but also reduces environment pollution factors. Savastano and Agopyan [8] reported that the combination of short coconut fibres with ordinary Portland cement has presented a significant increase in toughness. Several factors such as properties, volume fraction, distribution and orientation of fibre are identified to be influenced the physical and mechanical properties of concrete. However, the volume fraction appears to be the most dominant factor in influencing the strength of concrete.

The application of concrete containing coconut fibre as structural elements was study by Ali [9]. The study on the behaviour of coconut fibre-concrete columns reveals that non-visible degradation occur prior to the cracking of the column. The use of coconut fibre in cement has increased the flexural strength and consequence improves the crack resistance. Coconut fibre alters the behaviour of concrete when cracks occur across the structure and thus can provide a residual strength of the concrete [10]. Therefore, concrete containing coconut fibre exhibits same performance as that of conventional reinforced concrete produced by steel fibre [11]. In addition, the compressive of strength is also improved up to a certain percentage. Baruah and Talkudar [12] investigated the mechanical properties of concrete containing coconut fibre with different volume fractions ranging from 0.5% to 2%. From the obtained results, it is found that 1.5% of volume fraction of coconut fibre produce optimum strength with increment of 14.66%. A summary of studies about concrete containing coconut fibre is shown in Table 1.

Table 1. Effect of coconut fibre on the strength of concrete.

No.	Reference	Length of Fibre (mm)	Optimum Contain (%)	Increment of Strength (%)
1	Baruah and Talkudar [12]	40	1.5	14.66
2	Abdullah et al. [13]	-	9.0	26.96
3	Domke [14]	50 - 110	3.0	16.50
4	Ruben and Baskar [15]	30	1.0	20.59
5	Adisa et al. [16]	82	1.0	6.70
6	Mydin et al. [17]	34	0.4	21.00

3 Experimental Study

3.1 Material preparation

Main mix proportion consists of cement, sand, foam agent and water. The density of foamed concrete is 1600 ± 5kg/m^3. Therefore, water-cement ratio and cement-sand ratio were set up at 0.55 and 0.5 respectively. Table 2 shows the mix proportion of raw materials. The brown coconut coir was used to produce pelletized coconut fibre aggregate. The coconut coir was compressed into bar with diameter approximately 5mm and thin-coated using epoxy. The bar then cut into pellets with size of 10mm to 20mm. Figure 1 shows the pelletized coconut fibre aggregate.

Table 2. Mix proportion of foamed concrete.

Material	Ratio	Mass (kg)
Cement	1.00	33.00
Sand	2.00	66.00
Water	0.55	18.15
Foam agent	-	26.00

Figure 1. Pelletized coconut fibre aggregate.

Meanwhile, three (3) samples of coconut coir bars were tested under tensile loading to determine the tensile strength and Young's modulus. The results from the test, as shown in Figure 2, gave the average tensile strength and Young's modulus around 3.785MPa and 0.079GPa respectively.

3.2 Specimen preparation

In the experimental study, forty-five (45) cube specimens were prepared for the compression test. Thirty-six (36) of the cube specimens are foamed concrete containing different volume fractions of pelletized coconut fibre aggregate at 5%, 10%, 15% and 20%. Another nine (9) cube specimens were cast for the controlled foamed concrete. Table 3 shows the specification of cube specimens. All cube specimens were placed at the ambient condition for the air curing process at 7, 14 and 28 days before the compression test take place. Figure 3 shows the curing process of the cube specimens and the compression test.

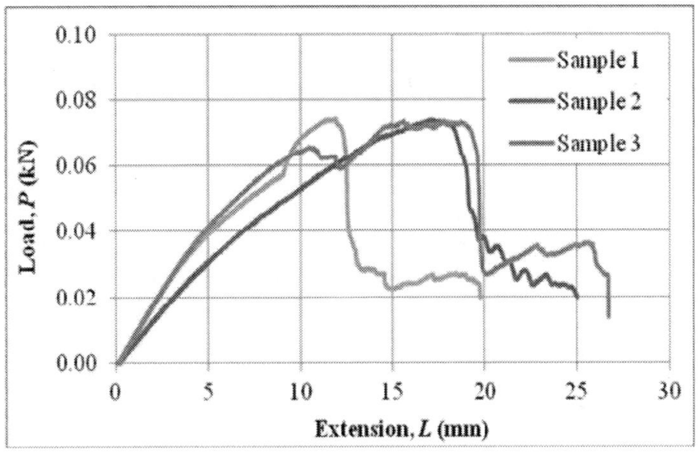

a) Experimental setup b) Tensile strength of coconut fibre bars

Figure 2. Tensile test of coconut fibre bars.

Table 3. Specification of cube specimens for the compression test.

No.	Specimen	Percentage of Pellet (%)	Quantity of Cube Specimen			
			7 days	14 days	28 days	Total
1	FC-A00	0	3	3	3	9
2	FC-B05	5	3	3	3	9
3	FC-C10	10	3	3	3	9
4	FC-D15	15	3	3	3	9
5	FC-E20	20	3	3	3	9
		Total				45

a) Air-curing process b) Compression test

Figure 3. Curing process and compression test of cube specimens at 7, 14 and 28 days.

4 Results and Discussion

The compressive strength of foamed concrete containing pelletized coconut fibre aggregate along with dry density and standard deviation are given in Table 4. The average compressive strength of controlled foamed concrete at 28 days is 9.0MPa. In accordance with Jaini et al. [2] and Aldridge [18], the compressive strength of foamed concrete with density of 1600kg/m^3 is in the range of 7.5MPa to 10.0MPa. Therefore, the controlled foamed concrete showed good agreement and convincing compressive strength.

Table 4. Dry density and compressive strength of cube specimens.

Specimen	Dry Density (kg/m^3)	Curing Age (days)	Compressive Strength (MPa)			Standard Deviation (σ)
			Specimen 1	Specimen 2	Specimen 3	
FC-A00	1611.85	7	6.0	6.2	6.1	0.100
	1605.93	14	8.0	7.8	8.2	0.200
	1602.96	28	9.0	9.2	8.8	0.400
FC-B05	1608.89	7	7.4	7.5	7.3	0.100
	1605.93	14	8.9	8.7	9.0	0.150
	1600.10	28	9.8	9.5	9.6	0.100
FC-C10	1602.96	7	9.0	9.1	9.0	0.050
	1600.96	14	10.5	10.3	10.1	0.200
	1596.23	28	11.5	11.9	10.9	0.500
FC-D15	1600.21	7	9.5	9.7	9.3	0.200
	1597.04	14	11.5	11.3	11.8	0.250
	1586.00	28	14.3	15.0	14.5	0.350
FC-E20	1593.15	7	8.5	8.6	8.8	0.150
	1585.03	14	10.9	10.5	11.3	0.400
	1580.21	28	13.7	13.0	13.5	0.350

Meanwhile, the average compressive strength of foamed concrete containing 5%, 10%, 15% and 20% of pelletized coconut fibre aggregate are 9.6MPa, 11.4MPa, 14.6MPa and 13.4MPa respectively. The pattern of compressive strength due to the volume fraction of pelletized coconut fibre aggregate at 7, 14 and 28 days is shown Figure 4. It can be observed that by increasing the volume fraction of pelletized coconut fibre aggregate on foamed concrete, the compressive strength is escalated significantly. Among the specimens, foamed concrete containing 15% pelletized coconut fibre aggregate showed the highest compressive strength. There is 62.22% of strength increment, as can be referred in Figure 5, compared with that obtained by controlled foamed concrete. Although foamed concrete containing 15% pelletized coconut fibre aggregate has achieved significant increment of compressive strength, however, it is still insufficient for structural application. A comparison with previous studies indicate that the pelletized coconut fibre is sustainable and effective in improving the compressive strength of foamed concrete with high utilization of volume fraction. This condition can contribute to the great saving of cement and consequently promote green concrete in the construction industry. In addition, it is also found that the use of pelletized coconut fibre aggregate tends to reduce the dry density of foamed concrete.

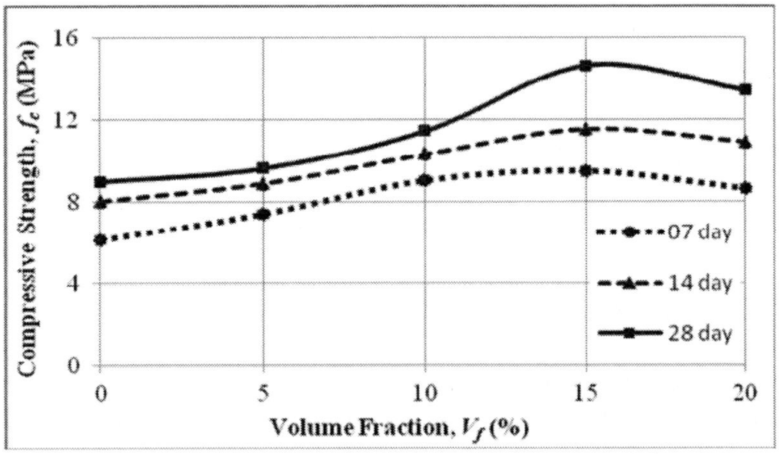

Figure 4. Strength development of foamed concrete based on volume fraction of aggregate.

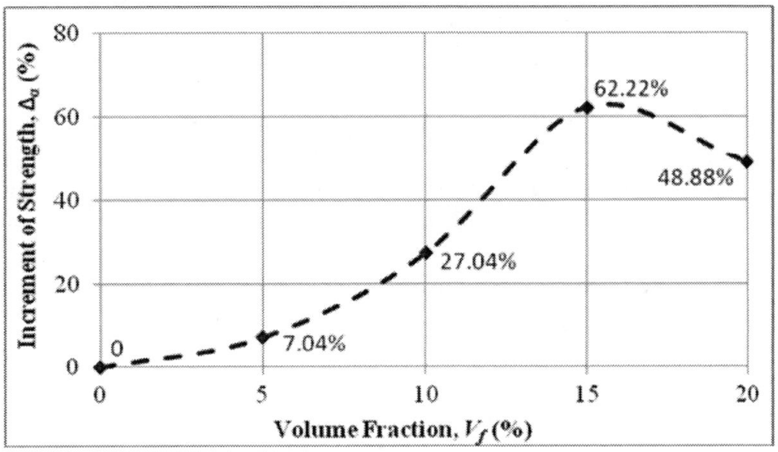

Figure 5. Strength increment of foamed concrete based on volume fraction of aggregate.

5 Conclusions

The study on the effect of pelletized coconut fibre aggregate on the strength of foamed concrete was conducted by using the compression test. The results showed that the compressive strength increases significantly until optimum volume fraction of 15%. It is found that the pelletized fibre aggregate indicates a good potential to enhance the compressive strength of foamed concrete. However, other characteristics of fresh and harden concrete need further investigations.

References

[1] M.R. Jones and A. McCarthy, Preliminary views on the potential of foamed concrete as a structural materials, Magazine of Concrete Research, **7**(1), 21-31,(2005).

[2] Z.M. Jaini, S.N. Mokhatar, Y.T. Feng and M.A. Seman, 2D multi-scale modelling of foamed concrete containing rubber bars, *Proc. of the Int. Civil and Infrastructure Engineering Conference*, Springer, Singapore, 625-638, (2015).

[3] N. Abd Rahman, Z.M. Jaini, N.A. Abd Rahim and S.A. Abd Razak, An experimental study on the fracture energy of foamed concrete using v-notched beams, *Proc. of the Int. Civil and Infrastructure Engineering Conference*, Springer, Singapore, 97-108, (2015).

[4] M. Kozlowski, M. Kadela and A. Kukielka, Fracture energy of foamed concrete based on three-point bending test on notched beams, Procedia Engineering, **108**, 349-354, (2015).
[5] N. Abd Rahman, Z.M. Jaini and N.N.M. Zahir, Fracture energy of foamed concrete by means of the three-point bending test on nocthed beam specimens, ARPN J. of Engineering and Applied Sciences, **10**(15), 6562-6570, (2015).
[6] N. Sgriccia, M.C. Hawley and M. Misra, Characterization of natural fiber surface and natural fibre composites, Composite Part A: Applied Science and Manufacturing, **39**(10), 1632-1637, (2008).
[7] J.S. Ruben and G, Baskar, Experimental study of coir fiber as concrete reinforcement material in cement based composites, J. of Engineering Research and Application, **4**(1), 128-131, (2014).
[8] J.H. Savastano and V. Agopyan, Transition zone studies of vegetable fibre-cement past composites, Cement and Concrete Composites, **21**, 49-57, (1999).
[9] M. Ali, Seismic performance of coconut fibre-reinforced-concrete columns with different reinforcement configurations of coconut-fibre ropes, Construction and Building Material, **70**, 226-230, (2014).
[10] M. Ali, X. Li and N. Chouw, Experimental investigations on bond strength between coconut fibre and concrete, Material and Design, **44**, 596-605, (2013).
[11] K. Bilba, M. Arsene and A. Ouensanga, Study on banana and coconut fibers botanical composition, thermal degradation and textural observations, Bioresource Technology, **98**(1), 58-68, (2007).
[12] P. Baruah and S.A. Talkudar, Comparatuve study of compressive, flexural, tensile and shear strength of concrete with fibres of different originis, Indian Concrete J., **81**(7), 17-24, (2007).
[13] A. Abdullah, S. Jamaluddin and M. Anwar, Assessment of physical and mechanical properties of cement panel influenced by treated and untreated coconut fiber addition, Physics Procedia, **22**, 262-269, (2011).
[14] P.V Domke, improvement in the strength of concrete by using industrial and agricultural waste, IOSR J. of Engineering, **2**(4), 755-759, (2012).
[15] J.S. Ruben and G. Baskar, Experimental study of coir fiber as concrete reinforcement material incement based composites, Int. J. of Engineering Research and Applications, **4**(1), 128-131, (2014).
[16] K.A. Olonade, A.D. Alake and A.G. Morakinyo, Strength development and crack pattern of coconut fibre reinforced strength development and crack pattern of coconut fibre reinforced, Civil and Environmental Research, **4**, 46-53, (2013).
[17] M.A.OMydin, N.A. Rozlan and S. Ganesan, Experimental study on the mechanical properties of coconut fibre reinforced lightweight foamed concrete, J. Material and Environmental Sciences, **6**(2), 407-411, (2015).
[18] D. Aldridge, *Introduction to Foamed Concrete: Use of Foamed Concrete in Construction*, Thomas Telford, London, United Kingdom, (2005).

MATEC Web of Conferences **47**, 01014 (2016)
DOI: 10.1051/matecconf/20164701014

An Experimental Study on Mitigating Alkali Silica Reaction by Using Fly Ash (FA) in Combination with Silica Fume and Expanded Perlite Powder (EPP)

Mohd Isneini[1,a], Yasutaka Sagawa[1], Hidenori Hamada[1] and Daisuke Yamamoto[1]

[1]*Faculty of Engineering, Kyushu University, Nishi-ku, Fukuoka 819-0395, Japan*

Abstract. ASR suppression by FA, SF, EPP, FA in combination with SF and EPP were evaluated by both mortar bar and concrete prism test. Mortar bars were made based on JIS A 1146, meanwhile concrete prism bars were casted in accordance with Rilem AAR-3. Both specimens were stored in 40°C 100% R.H. controlled room. Mortar and concrete mixtures used reactive aggregate in pessimum proportion. The results indicated that FA in combination with SF and EPP showed smaller expansion compared to FA. The best of concrete mixtures in reducing expansion is combination of FA with SF (FA15SF10).

1 Introduction

Fly ash (FA) is a by-product of the combustion of pulverized coal in electric power plants. Silica fume (SF) is a by-product of the smelting process in the silicon and ferrosilicon industry that is used as a pozzolan. SF also referred to as micro silica or condensed silica. SF shows pozzolanic reaction and micro filler effect, and they lead to improve durability of concrete. Moreover, perlite is also a pozzolan due to its glassy structure and high SiO_2 and Al_2O_3 contents. There are studies on the utilization of minerals to inhibit alkali silica reaction (ASR), such as; it is widely accepted that the use of FA at the level of 15% to 25% are considered to be sufficient in mitigating concrete expansion. Tanaka et al. [1] concluded that the significant amount of both alkali and hydroxide ions were consumed by additives in ion analysis of pore fluid [1]. Shehata et al. showed that pastes containing SF increased alkalinity at ages beyond 28 days, while pastes containing ternary blends maintained the low alkalinity of the pore solution throughout 3 years testing period [2].

The objective of this study was intended to determine the performance of mineral combination; FA with SF and FA with expanded perlite powder (EPP). Minerals were combined at 15% and 25% of FA. In this study, mortar bars were made based on JIS A 1146, and mortars were treated under 40°C, and R.H. 100% of controlled room for expansion measurement. Meanwhile, concrete prisms were made according to RILEM AAR-3, and specimens were also cured in a controlled room under 40°C, and R.H. 100% for expansion test. Then, for compressive strength test, the specimens were cured in a controlled room under 20°C, and R.H. 60%.

---

[a] Corresponding author : isneini-m@doc.kyushu-u.ac.jp

2 Material and Mixture Proportions

2.1 Materials

Four types of binder were used in this study, namely; Ordinary Portland Cement (OPC), FA, SF and EPP. Chemical composition of OPC, FA, SF and EPP are shown in Table 1. It can be seen in Table 1 that there is no content of Na_2O and K_2O in SF. However, the content of Na_2O and K_2O in EPP and FA are higher than OPC. OPC, FA, and SF meets the requirements of JIS R 5202, JIS A6201, and JIS A 6207, respectively.

Table 1. Chemical compositions of OPC, FA, SF and EPP.

Constituents (%)	OPC	FA	SF	EPP
SiO_2	20.89	54.96	95.5	65.85
SO_3	2.02	0.49	0.18	-
MgO	0.92	1.30	0.56	0.14
Na_2O	0.35	1.07	-	4.68
K_2O	0.36	1.30	-	4.07
LoI	1.87	2.37	1.22	-

Physical properties of materials are presented in Table 2. Two aggregates were used in this experiment; one was classified as alkali silica reactive (andesite) and the other was classified as non-reactive (limestone).

Table 2. Physical properties of materials.

Material	Properties
Cement (OPC)	Density = 3.16 g/cm^3, Specific Surface Area = 3330 cm^2/g
FA (type II)	Density = 2.26 g/cm^3, Specific Surface Area = 3972 cm^2/g
SF	Density = 2.35 g/cm^3, Specific Surface Area = 18 m^2/g
EPP	Pass sieve No. 200 (0.075 mm), Density = 2.58 g/cm^3
Sand (Limestone)	Density (SSD) = 2.70 g/cm^3, Water absorption = 0.61%, Rc = 8 mmol/l. Sc = 1 mmol/l.
Gravel (1): (Andesite)	Density (SSD) = 2.60 g/cm^3, Water absorption = 2.75%, Rc = 185 mmol/l. Sc = 620 mmol/l.
Gravel (2): (Limestone)	Density (SSD) = 2.70 g/cm^3, Water absorption =0.55%, Rc = 8 mmol/l. Sc = 1 mmol/l.

2.2 Mixture proportions

This study presents an experimental study on mortar and concrete mixtures. Mortar and concrete mixtures were prepared based on the same level of mineral percentage. Mortar bars were casted based on JIS A 1146. Andesite coarse aggregates were crushed to become fine aggregates that meets the requirement of particle size distribution in JIS A 1146. Water to cement ratio for mortar was adjusted to 0.50 and alkali content was set to Na_2O_{eq} of 1.2 weight % cement by adding NaOH solution into mixing water. In addition, concrete prisms were made based on RILEM AAR-3. Concrete mixtures were prepared by using tap water with water binder ratio 0.5. Mixture 1 was made for control, Mixture 2 and 3 used FA at 15%, and 25%. Mixture 4, 5 and 6 used SF at 5%, 10%, and 15%; Mixture 7, 8 and 9 used EPP at 10%, 15%, and 20%. Furthermore, mixture 10 and mixture 11 used combination of FA and SF. Mixture 10 used FA10%, and SF5%, then mixture 11 used FA15% and SF10%. Mixture 12, and mixture 13 were made for combination FA and EPP. Mixture 12 used FA10% and EPP5%, then mixture 13 used FA15%, and EPP10%. Moreover, test results were evaluated to investigate the effectiveness of FA, SF, EPP, and combination of them in mitigating

ASR. Total alkali in concrete were determined around 6 kg/m^3. Mix 1 up to Mix 6, total alkali are 6 kg/m^3. Then, Mix 7 to Mix 13 are: 6.4 kg/m^3, 6.6 kg/m^3, 6.8 kg/m^3, 6.0 kg/m^3, 6.0 kg/m^3, 6.1 kg/m^3, 6.4 kg/m^3, successively. Mixture proportions of concrete are shown in Table 3.

Table 3(a). Mixture proportions of concrete.

Mix	w/b (%)	s/a (%)	(kg/m^3)								(g/m^3)	
			Water	OPC	FA	SF	EPP	Sand	Gravel	NaOH	WR +AE	AE
1	50	45	160	320		-	-	843	992	1.856	1000	12.8
2	50	45	160	272	48	-	-	836	984	2.499	1000	9.60
3	50	45	160	240	80	-	-	831	978	2.928	1000	16.0
4	50	45	160	304	-	16	-	841	990	2.070	1000	12.8
5	50	45	160	288	-	32	-	838	987	2.285	1000	12.8
6	50	45	160	272	-	48	-	836	984	2.499	1760	22.0
7	50	45	160	288	-	-	32	838	987	3.613	1400	17.6
8	50	45	160	272	-	-	48	836	984	4.491	1400	17.6
9	50	45	160	256	-	-	64	833	981	5.370	1400	17.6
10	50	45	160	272	32	16	-	824	970	5.706	880	12.8
11	50	45	160	240	48	32	-	813	957	5.946	880	12.8
12	50	45	160	272	32	-	16	824	970	3.163	880	12.8
13	50	45	160	240	48	-	32	813	957	4.256	880	12.8

(WR+AE)= water reducing and air entraining agent, AE=air entraining agent

Table 3(b). Mixture proportions of concrete.

Mix	w/b (%)	s/a (%)	Temp (°C)	Slump (mm)	Air Content (%)	Compressive Strength (MPa)
1	50	45	19	80	5.0	32.52
2	50	45	23	150	4.0	30.05
3	50	45	22	180	2.8	30.55
4	50	45	22	70	3.8	33.77
5	50	45	21	60	3.5	43.86
6	50	45	30	30	5.0	48.30
7	50	45	24	57	4.0	27.26
8	50	45	25	65	2.9	31.62
9	50	45	22	55	3.9	35.89
10	50	45	21	125	4.5	35.49
11	50	45	21	125	3.1	37.65
12	50	45	20	100	3.3	31.75
13	50	45	20	90	2.9	32.69

3 Expansion Test

3.1 Mortar bar test

The size of the mortar was 40x40x160mm. Samples were wrapped with wet paper and placed vertically in stainless container, then stored in a 40°C, and 100% R.H. of controlled room. Measurement was conducted to observe expansion behavior of mortar at day 1, 2 weeks, 4 weeks, 8 weeks, 12 weeks and 24 weeks.

3.2 Concrete prism test

The size of the prism was 75x75x250 mm. Prisms were wrapped with wet paper and thin plastic sheet. Then, specimens were placed vertically in a polyethylene container with constant level of water at the bottom. Moisture control was important for concrete prisms. A piece of wet unwoven paper (size: 355 x 425 mm) was used to wrap a specimen. Then, prism was covered with thin plastic sheet. Then, polyethylene containers were stored in a 40°C and 100% R.H. of controlled room. Before measuring, the container was taken to 20°C room to cool the specimens for 24 hours. After completing the measurement, the specimens were taken in controlled room. The length of concrete was measured periodically every two weeks. The observation for concrete expansion was conducted in 365 days.

3.3 Compressive strength test

Concrete cylinders were demoulded after 24 hours of casting. Compressive test was conducted after 28 days of curing in accordance with JIS A 1108. The average values of compressive were determined from three specimens for each concrete mixture.

4 Results and Discussion

4.1 Mortar bar

Mortar expansion at the age of 6 months with various aggregate combination ratios is shown in Figure 1. The expansion reached maximum level when the ratio was 30:70 between reactive and non-reactive aggregate. Figure 2 showed that without minerals, expansion increased slowly until eight weeks, then remarkably increased until the end of the observation. It can be seen that expansion was reduced due to usage of FA, SF and EPP. The mortar expansion with FA at level 25% was lower than mortar expansion with FA at level 15%. The mortar expansion with SF at level 5% was lower than mortar expansion without SF. Furthermore, mortar expansion with SF at level 10% was lower than mortar with SF5%. Then, the usage of SF15% showed the lowest expansion. The expansion reduction by SF10%, and SF15% were under 0.1%, that classified as innocuous. Mortar expansion with EPP10% increased remarkably. However, became lower than control at the end of test period. Mortar expansion with EPP15% increased gradually but became lower than EPP10%. Mortar with 20% of EPP showed the lowest expansion. Moreover, expansion reduction by using FA in combination with SF and EPP were under 0.1%, that classified as innocuous.

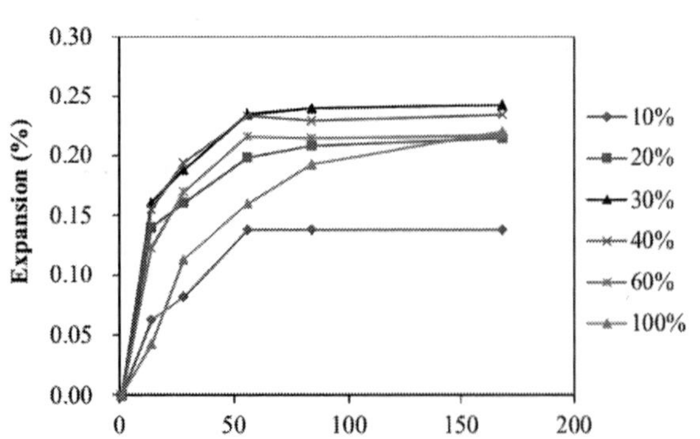

Figure 1. Expansion of mortar at the age of 6 months reactive aggregate ratio.

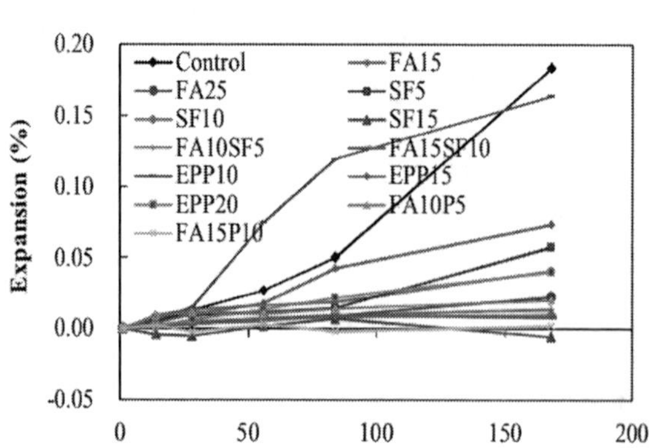

Figure 2. Expansion of mortar with FA, SF, different EPP and mineral combination.

4.2 Concrete prism

Figure 3 showed that without mineral, concrete expansion increased fastly, then enlarge until the end of observation period. It can be seen that there was expansion reduction due to usage of FA. The expansion of concrete with FA15% was lower than control concrete. Then, the expansion of concrete with FA25% was lower than concrete with FA at level 15%. Expansion reduction by using FA 25% was under 0.05%, that classified as non-reactive.

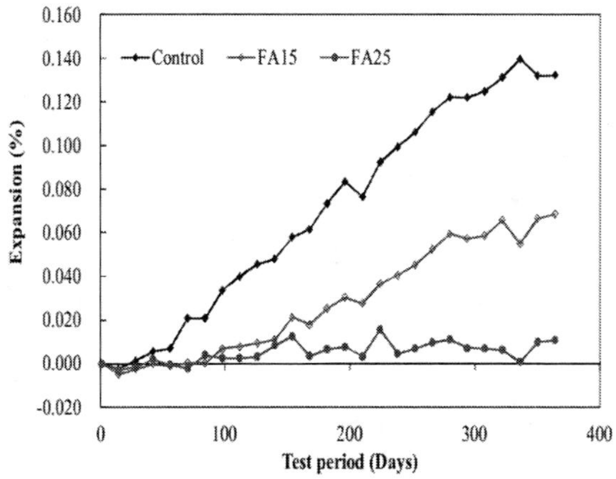

Figure 3. Expansion of concrete with FA.

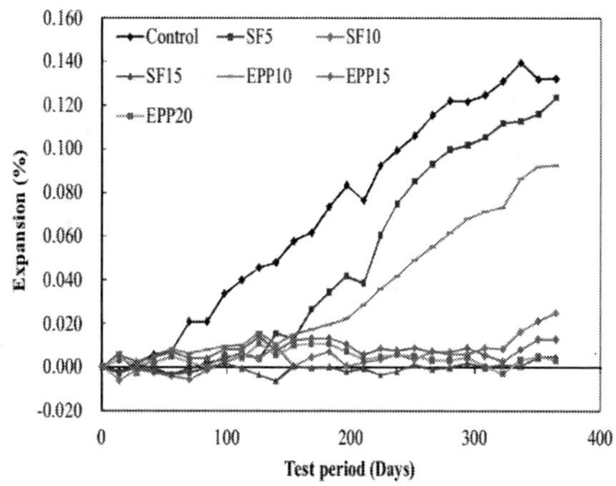

Figure 4. Expansion of concrete with SF and EPP.

Furthermore, concrete expansion with SF, and EPP were shown in Figure 4. Expansion was reduced due to usage of SF and EPP. Control concrete or concrete without mineral, its expansion was large and rapid until the end of the period. The larger the portion of mineral, the lower the expansion will be. As a results, the lowest expansion were achieved at concrete with SF at level 15%, and concrete with EPP at level 20%. Furthermore, expansion reduction by using 10% and 15% of SF were under 0.05%, that classified as non-reactive. Then, expansion of concrete with EPP at level 15% and 20% were under 0.05%, that classified as non-reactive. The expansion comparison of concrete by using FA, SF, EPP, FA in combination with SF and EPP can be seen in Figure 5 and Figure 6. Concrete expansion with combination FA with SF (FA10SF5) and EPP (FA10EPP5) were lower than concrete with FA 15%. Then, concrete expansion of FA in combination with SF (FA15SF10) and EPP (FA15EPP10) were also lower than concrete with FA 25%.

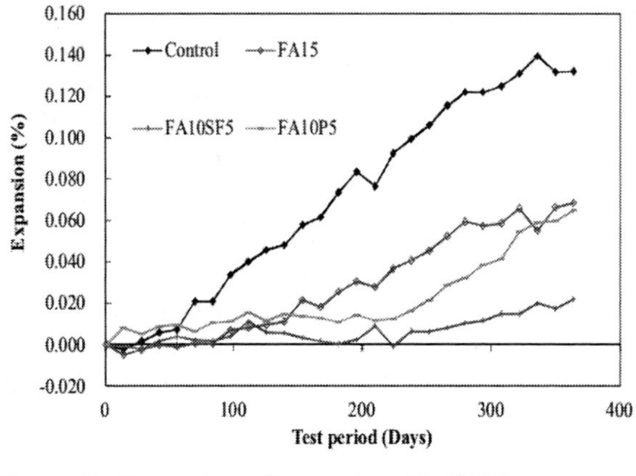

Figure 5. Expansion of concrete with FA15, FA15SF10, FA10P5.

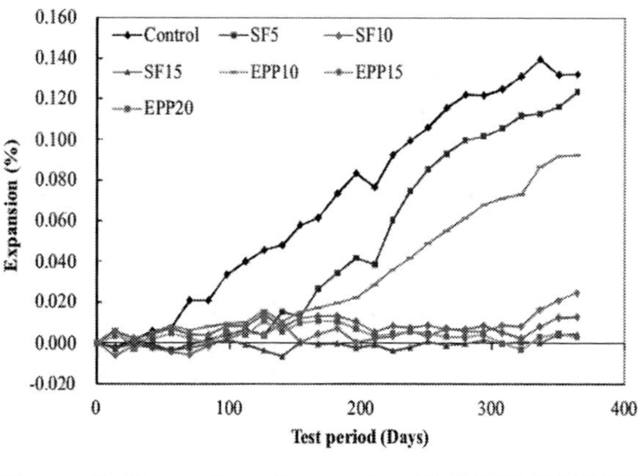

Figure 6. Expansion of concrete with FA25, FA10SF5, FA15P15.

The combination effect of FA with SF and EPP in order to reduce expansion due to ASR were stronger than FA itself. Furthermore, the comparison of expansion suppression by using all minerals used in this study is shown in Figure 7.

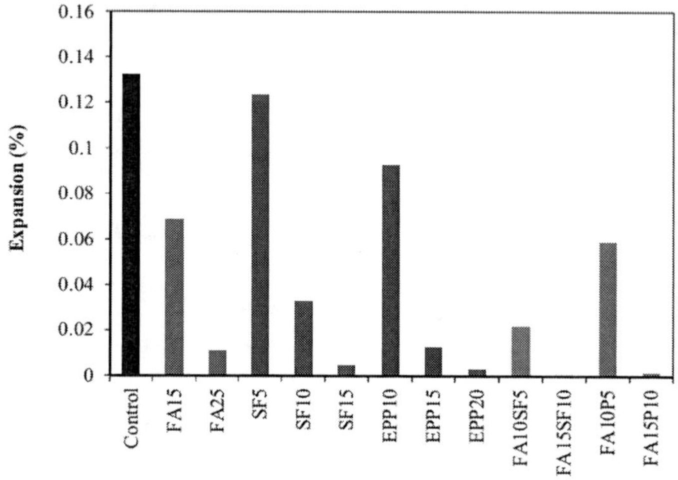

Figure 7. Expansion of concrete with FA, SF, EPP, FA in combination with SF and EPP.

4.3 Compressive strength of concrete

The results of compressive strength test at 28 days are shown in Table 3. It can be observed that the usage of minerals and mineral combination could influence the strength values.

5 Discussion on the Effectiveness of FA, SF and EPP

Finer pozzolans are sufficient in mitigating ASR [3]. SF with a particle size of approximately 0.1 μm is also very effective in preventing ASR. SF is a reactive pozzolanic material because of its fineness and high amorphous silicon dioxide content. SF reacts with calcium hydroxide ($Ca(OH)_2$) to form additional Calcium Silica Hydrate (C-S-H), which is similar to the calcium hydrate formed from the Portland cement. The product of the pozzolanic reaction (secondary C-S-H) can incorporate alkali metal ions into their structure and reduce the alkalinity of the pore solution. At the same time, the secondary C-S-H hydrates fill the adjacent pores and reduce the permeability of concrete, consequently, reducing the free movement of alkali ions[4]. Meanwhile, FA reduces expansion by lowering the alkali concentration and hydroxide ion in the pore solution [5]. Tanaka et al. concluded: the effect of neutralization reaction that the part of alkali ion was consumed in hydrated material while part of FA and SF would react with hydroxide ion. Furthermore, the reduction in alkali concentration of pore fluid was the principal of minerals suppression mechanism [1]. The presence of both FA and SF resulted in low pore solution alkalinity at early and late ages. SF is effective in reducing alkalinity at an early age, meanwhile FA become effective at later ages [2]. The effectiveness of EPP in mitigating ASR can be attributed to pozzolanic activity [6, 7], possessed by the material. In suppressing expansion, the porous nature of EPP might play an important role by providing a distributed air bubble system that accommodates expansive gel in concrete[7]. Furthermore, G.E. Cristidis et al. reported that capable of sodium absorption or posseses binding alkalis property were as a basis for pozzolanic activity of Perlite. FA reduced compressive strength. The compressive strength reduction was related to the properties of FA that declined the heat of hydration. As a result, FA slowed the rate of hardening and reduced the strength [9]. On the other hand, the usage of SF and combination of FA and SF enhanced the strength. The highest strength was achieved on concrete with SF15%. Meanwhile, the combination of FA and EPP were not significant to increase the strength.

Furthermore, the compressive strength of concrete with EPP 20% was larger than control concrete. Demirbŏga et al. attributed that the strength improvement of expanded perlite, related to both the self-curing property and pozzolanic of perlite [10].

6 Summary

The results obtained are summarized as follows:

i) Combination of FA with SF and EPP could reduced expansion lower than FA itself.
ii) Combination of FA with SF (FA15SF10) showed the best concrete mixtures in reducing expansion.
iii) The expansion reduction of concrete with FA10SF5, FA15SF10, and FA15P10 were under 0.05% that classified as non-reactive.
iv) Compressive strength of concrete by using FA in combination with SF showed better than compressive strength of concrete by using combination of FA with EPP.

References

[1] Yasushi Tanaka, Yuji Hayamizu, and Takuya Sudoh, Combination effect of additives on preventives effect of alkali silica reaction, Cement Science and Concrete Technology (in Japanese), **64**, 406-412, (2010).

[2] M.H. Shehata and M.D.A. Thomas, Use of ternary blends containing silica fume and fly ash to suppress expansion due to alkali-silica reaction in concrete, Cement and Concrete Research, **32**, 341-349, (2002).

[3] L.R.L Malvar, efficiency of fly ash in mitigating alkali-silica reaction based on chemical composition, ACI Materials Journal, **September-October**, 319-326, (2006).

[4] G.J.Z. Xu, D.F. Watt and P.P. Hudec, Effectiveness of mineral admixtures in reducing ASR-expansion, Cement and Concrete Research, **25**, 1225-1236, (1995).

[5] H. Takahashi and K. Yamada, Critical characters of fly ash suppressing ASR, *Proceedings of the 14th International Conference on Alkali-Aggregate Reaction*, Austin, (2012)

[6] T.K. Erdem, Ç. Meral, M. Tokyay and T.Y. Erdoğan, Use of perlite as a pozzolanic addition in producing blended cements, Cement and Concrete Composites, **29**, 13-21, (2007).

[7] F. Bektas, L. Turanli and P.J.M. Monteiro, Use of perlite powder to suppress the alkali-silica reaction, Cement and Concrete Research, **35**, 2014-2017, (2005).

[8] G.E. Christidis, I. Paspaliaris and A. Kontopoulus, Zeolitisation of perlite fines, mineralogical characteristics of the end products and mobilization of chemical elements, Appl. Clay Sci., **15**, 305-324, (1999).

[9] R. Demirboğa, İ. Türkmen and M.B. Karakoç, Relationship between ultrasonic velocity and compressive strength for high-volume mineral-admixture concrete, Cement and Concrete Research, **34**, 2329-2336, (2004).

[10] R. Demirbŏga, I. Örüng and R. Gül, Effects of expanded perlite powder and mineral admixtures on the compressive strength of low-density concretes, Cem. Concr. Res., **31**, 1627-1632, (2001).

MATEC Web of Conferences **47**, 01015 (2016)
DOI: 10.1051/matecconf/20164701015

The Effect of Lime Addition on the Setting Time and Strength of Ambient Cured Fly Ash Based Geopolymer Binder

Andi Arham Adam[1,a], Nur Hayati Amiri[1], I. Wayan Suarnita[1] and Nicodemus Rupang[1]

[1]*Department of Civil Engineering, Tadulako University, Palu 94118, Indonesia*

Abstract. One of the limitations of geopolymer as the alternative binders in concrete is the necessity of heat curing. This study aimed to produce fly ash geopolymer binder subjected to ambient curing by adding a small proportion of lime and varying the activator dosage. The Class F fly ash from Mpanau coal-fired power plant was mixed with alkaline solution consists of sodium silicate and sodium hydroxide with Na_2O dosage of 5%, 7%, and 9%. To achieve ambient cured paste, 8%, 9%, and 10% slaked lime was added as the substitute for the fly ash. The setting time test was conducted for each mix and the compressive strength was performed at age of 7, 14 and 28 days. The test result shows that the setting time of the fly ash based geopolymer paste can be controlled by adding a small proportion of slaked lime. The addition of lime increased strength but decreased the setting time.

1 Introduction

Portland cement is the most widely used building material and largely contributes to greenhouse gas emissions. High energy consumption and large amount of raw materials are needed to produce Portland cement, and on the other side significant amount of CO_2 released into the atmosphere. However, as the main ingredient in the manufacture of mortar and concrete, the use of Portland cement in large quantities is unavoidable. Therefore, it is necessary to find an alternative to cement with more environmentally friendly materials.

One alterative is in the form of inorganic polymer binder (geopolymer) that uses materials containing silico-aluminate precursors and inorganic alkali solution as an activator. Geopolymer potentially replace cement as the main ingredient in a concrete mix due to the characteristics of geopolymer concrete, strength and durability which is similar to concrete using Portland cement binder [1]. Low calcium fly ash is a suitable material for geopolymer precursor due to its availability and can be used directly without additional process (burning, grinding etc.)

However previous research shows that one of challenges in the application of low calcium fly ash based geopolymer as binder is its setting process needs relatively high curing temperature 60°C-120°C for 6 - 24 hours [2-7]. This conditions limit the application of geopolymer for precast purposes whilst for normal concrete, the binder should be able to set in room temperature. Addition of calcium to produce ambient cured geopolymer binder has been conducted elsewhere [8-13].

[a] Corresponding author : adam.arham@gmail.com

2 Materials and Methods

Class F fly ash taken from Mpanau power plant and slaked lime were used as binder materials. The chemical composition of the binders can be found in Table 1. Blended sodium silicate solution (15.4% Na_2O ; 32.33% SiO_2) and 15M sodium hydroxide solution were used as chemical activator. The activator dosages (mass ratios of Na_2O in activator solution to binders) were 5%, 7%, and 9% and the modulus of activator (mass ratios of SiO_2 to Na_2O) was kept constant at 1.00. To achieve ambient cured paste, 8%, 9%, and 10% slaked lime was added as the substitute for the fly ash.

Table 1. Properties of the binder materials.

Chemical compositions	Fly ash (%)	Slaked lime (%)
SiO_2	56.04	-
Fe_2O_3	21.27	0.35
AlO_3	12.99	-
CaO	5.18	85.04
K_2O	1.62	1.03
TiO_2	0.90	0.03
MnO	0.26	-
P_2O_5	0.25	-
SrO	-	2.17
Cl	-	1.06
Other oxides	0.57	0.1
LOI	0.94	10.25
Total	100.0	100.0

Initially, water to solid ratio (w/s) of 3.4 were used to prepare the geopolymer paste specimens, however the 7% specimen set too fast and could not be cast into the mold, therefore additional specimens of 7% dosage were prepared using 0.4 w/s. The water content in the w/s calculation is the total water content in sodium silicate solution, sodium hydroxide solution and additional water, whilst the solids are the fly ash, and the solid parts from the sodium silicate and the sodium hydroxide solution. The mix proportions of the geopolymer paste are given in Table 2.

Table 2(a). Mix proportions of geopolymer paste.

Specimens	Na_2O dosage (%)	Binder		Water to solid ratio
		Slaked lime (%)	Fly ash (%)	
G5C8		8	92	0.340
G5C9	5	9	91	0.340
G5C10		10	90	0.340
G7C8		8	92	0.340
G7C9	7	9	91	0.340
G7C10		10	90	0.340

Table 2(b). Mix proportions of geopolymer paste.

Specimens	Na_2O dosage (%)	Binder		Water to solid ratio
		Slaked lime (%)	Fly ash (%)	
G7C8*		8	92	0.400
G7C9*	7	9	91	0.400
G7C10*		10	90	0.400
G9C8		8	92	0.340
G9C9	9	9	91	0.340
G9C10		10	90	0.340

*modified w/s ratio

Fly ash and slaked lime was dry mixed until uniform and mixed with the activator solution for 11 minutes. The paste specimens were then molded into 2.5 cm diameter x 5 cm height plastic cylinders mold. All specimens were cured at an ambient temperature of 30°C and humidity level of 60%. All specimens were demolded and left in the room until the day of testing. The setting time of the geopolymer paste was measured using a Vicat needle according to ASTM C 191-04. Compressive strength tests were conducted in accordance with ASTM C 39 at 7, 14, and 28 days.

3 Results and Discussion

3.1 Effect of lime addition on setting time material

As can be seen in Figure 1, increasing the lime proportion from 8% to 10% significantly decreased the setting time. Different trend was observed on the variation of Na_2O dosages.

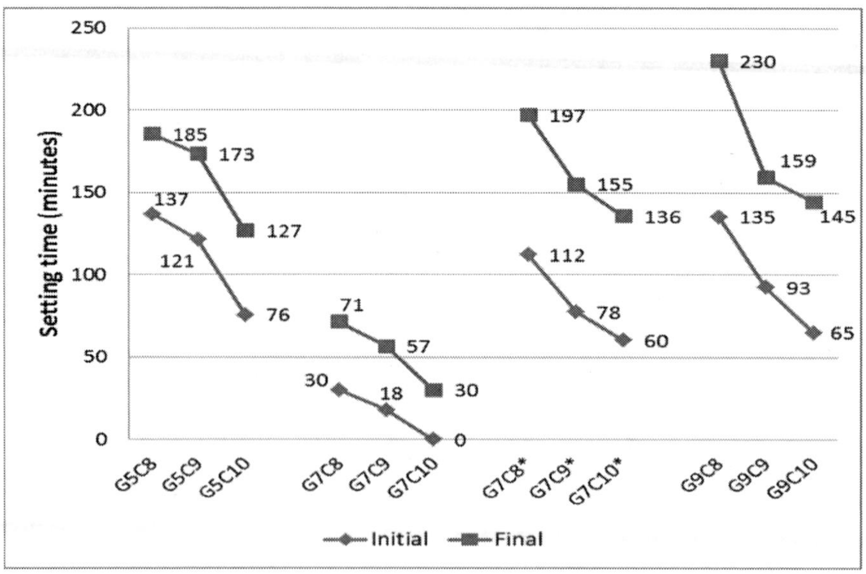

Figure 1. Setting time of geopolymer paste.

At constant w/s ratio, increasing the dosage from 5% to 7% decreased the setting time but further increase from 7% to 9% increased the setting time. The 7% Na_2O specimens with 0.34 w/s ratio had a dry consistency and it was difficult to be molded, as such the w/s ratio was increased to 0.4 which is denoted as G7C8*, G7C9*, and G7C10*. However, even with the increased w/s ratio, the initial setting time was still lower although the final setting time was a little bit higher than the 5% specimens. The cause of this exception in the case of the 7% specimen is not clear. However, it can be said that the setting properties would be influenced in accordance with the interaction between the activator solution and the slaked lime in the binder.

The exact mechanism of hardening of geopolymer with lime addition is not clearly established, however according to Temuujin [12] it depends on many parameters including the aluminosilicate precursors, alkaline activator type and their concentrations and curing temperature. It can be seen from figure 1 that there is a certain value where the increased of Na_2O dosage decreased setting time but beyond the level further increase in Na_2O dosage will increase setting time. The addition of small amount of calcium to the alkali activated binder could accelerate setting, however when the level of addition was high, the effects would be accelerating or retarding depend on the composition of activator, the dissolved species from the solid, and the nature of the calcium [14].

3.2 Effect of lime addition on strength development

Figure 2 shows the compressive strength of fly ash/lime based geopolymer paste with time. At 5% Na_2O, the compressive strength was very low especially for the specimens with 8% and 9% lime. Increasing proportion of lime by 1% increased the strength, although larger increase was obtained by increasing Na_2O dosage. The compressive strength at 7 and 14 days were very low, it was 0 MPa for 5% Na_2O with 8% and 9% lime, 2 – 3 MPa for 7% Na_2O and 4 – 6 MPa for 9% Na_2O dosage. The higher 28 days strength, 8 – 12 MPa was found on the 9% Na_2O specimens. It was also found by other researcher that addition of the hydrated lime up to 10% by weight of binder will improve the mechanical properties of geopolymer paste when cured with water under ambient temperature [12].

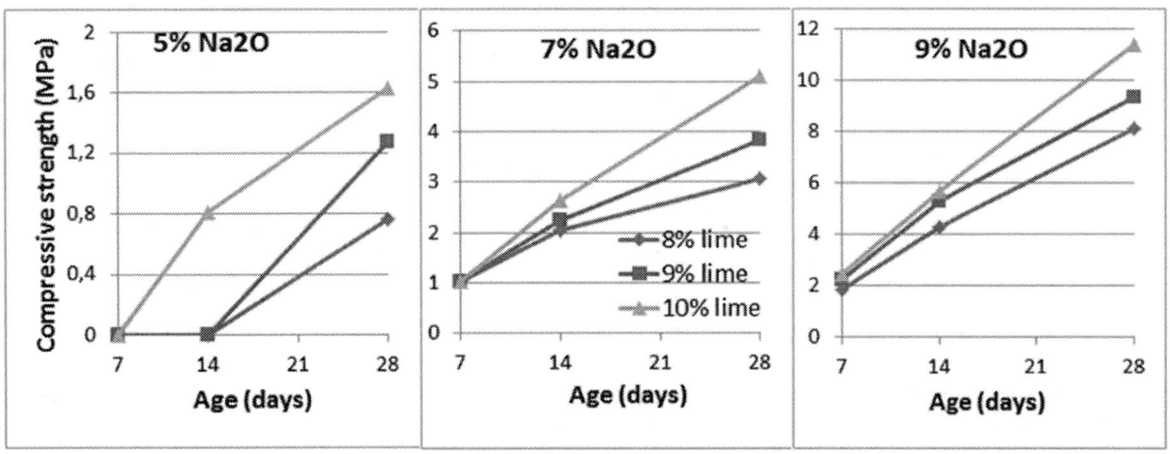

Figure 2. Strength development of geopolymer paste.

4 Conclusions

The test result shows that the setting time of the class F fly ash based geopolymer paste can be controlled by adding a small proportion of slaked lime. The addition of lime increases the strength but decreases the setting time. There is a certain value where the increase of Na_2O dosage decrease setting time. However beyond the level, further increase in Na_2O dosage will increase setting time.

Acknowledgements

The authors wish to thank to Ministry of Research, Technology and Higher Education for funding this research.

References

[1] D. Law, A.A. Adam, T. Molyneaux, I. Patnaikuni and A. Wardhono, Long term durability properties of class F fly ash geopolymer concrete, Materials and Structures, **48**, 721-731, (2015).
[2] A.A. Adam and Horianto, The effect of temperature and duration of curing on the strength of fly ash based geopolymer mortar, Procedia Engineering, **95**, 410-414, (2014).
[3] A.M.M. Al Bakri, H. Kamarudin, M. BinHussain, I.K. Nizar, Y. Zarina and A.R. Rafiza, The effect of curing temperature on physical and chemical properties of geopolymers, Physics Procedia, **22**, 286-291, (2011).
[4] A.R. Brough, A. Katz, G.K. Sun, L.J. Struble, R.J. Kirkpatrick and J.F. Young, Adiabatically cured, alkali-activated cement-based wasteforms containing high levels of fly ash: formation of zeolites and al-substituted C-S-H, Cement and Concrete Research, **31**, 1437-1447, (2001).
[5] A.S. de Vargas, D.C.C. Dal Molin, A.C.F. Vilela, F.J. da Silva, B. Pavão and H. Veit, The effects of Na_2O/SiO_2 molar ratio, curing temperature and age on compressive strength, morphology and microstructure of alkali-activated fly ash-based geopolymers, Cement and Concrete Composites, **33**, 653-660, (2011).
[6] G. Kovalchuk, A. Fernández-Jiménez and A. Palomo, Alkali-activated fly ash: Effect of thermal curing conditions on mechanical and microstructural development - part II, Fuel, **86**, 315-322, (2007).
[7] P.R. Vora and U.V. Dave, Parametric studies on compressive strength of geopolymer concrete, Procedia Engineering, **51**, 210-219, (2013).
[8] K. Dombrowski, A. Buchwald and M. Weil, The influence of calcium content on the structure and thermal performance of fly ash based geopolymers, J. of Materials Science, **42**, 3033-3043, (2007).
[9] F. Puertas, S. Martínez-Ramírez, S. Alonso and T. Vázquez, Alkali-activated fly ash/slag cements: Strength behaviour and hydration products, Cement and Concrete Research, **30**, 1625-1632, (2000).
[10] S. Puligilla and P. Mondal, Role of slag in microstructural development and hardening of fly ash-slag geopolymer, Cement and Concrete Research, **43**, 70-80, (2013).
[11] C.K. Yip, G.C. Lukey, J.L. Provis and J.S.J. van Deventer, Effect of calcium silicate sources on geopolymerisation, Cement and Concrete Research, **38**, 554-564, (2008).
[12] J. Temuujin, A. van Riessen and R. Williams., Influence of calcium compounds on the mechanical properties of fly ash geopolymer pastes, J. of Hazardous Materials, **167**, 82-88, (2009).
[13] H.M. Khater, Effect of calcium on geopolymerization of aluminosilicate wastes, Journal of Materials in Civil Engineering, **24**, 92-101, (2012).
[14] W.K.W. Lee and J.S.J. Van Deventer, The effect of ionic contaminants on the early-age properties of alkali-activated fly ash-based cements, Cement and Concrete Research, **32**, 577-584, 2002.

MATEC Web of Conferences **47**, 01016 (2016)
DOI: 10.1051/matecconf/20164701016

Isolation of Sulphate Reduction Bacteria (SRB) to Improve Compress Strength and Water Penetration of Bio-Concrete

A. Faisal Alshalif[1], J. M.Irwan[1,a], N. Othman[2] and L. H. Anneza[1]

[1]*Jamilus Research Center, University Tun Hussein Onn, Malaysia, 86400 Parit Raja, Johor, Malaysia*
[2]*Faculty of Civil and Environmental Engineering, University Tun Hussein Onn, Malaysia, 86400 Parit Raja, Johor, Malaysia.*

Abstract. The objective of this study is to isolate sulphate reduction bacteria (SRB) from acid mire water collected at Sg Pelepah Kota Tinggi, Johor Malaysia. The isolation process was conducted in high alkaline and anaerobic conditions to sustain the bacteria in concrete environment. Properties tests such as compressive strength and water penetration were conducted. The result showed that optimal growth condition of sulphate reduction bacteria is pH 9-10. It was also observed that the bacteria is a coccus shape after gram staining process. The bacteria was used after 10 days of culturing prior to growth curve measurement. The liquid culture containing sulphate reduction bacteria were used at 1%, 3% and 5% as replacement ratio of water content. Concrete specimens were cured in the air conditions for 7, 14 and 28 days. Maximum increment on compressive strength was 13.0% and decrement in water penetration was 8.5% occurred with 5% of SRB. The enhancement in compressive strength and water penetration performance was due to calcium precipitation within concrete pores. Image of scanning electronic microscopy (SEM) showed bacteria sustained and survived in concrete environment by reducing diameter of pores in concrete specimens.

1 Introduction

Concrete is the first most consumed material in construction projects. The application of concrete is rapidly increasing with the increasing of types of concrete worldwide. Recently, concrete technology presents many types concrete which are economically and friendly with environment. Today innovation is leading to natural solution for environmental impacts of concrete as sustainable alternative. However, the development of a sustainable concrete is urgently needed for environmental reasons. Biotechnology, applied to create new type of concrete namely bio- concrete using microorganism as natural way to improve concrete properties [1].

Therefore, bio-concrete is the used many types of bacteria whereas, the most popular bacteria used in concrete are (Bacillus Pasteurii Pseudomonas Aeruginosa, Bacillus Sphaericus, Ureolytic and Shewanella) to produce carbonate [8-11]. The presence of calcium sources in cement material plays an important role in producing calcium carbonate through enzyme reaction from those bacteria. Metabolic activities of some specific microorganisms in bio-concrete resulted by microbial mineral precipitation which in role improve the overall behavior of concrete and help in healing

[a] Corresponding author : irwan@uthm.edu.my

process of pores and cracks [6]. Bio-concrete application is rapidly increasing worldwide and the development of bacteria mediated bio-concrete is urgently needed for environmental reasons [2, 3].

SRB has the ability to precipitate the dolomite consisting of a carbonate of calcium and magnesium. Precipitation of calcium carbonate is the essentially role of sulphate reduction bacteria which functionally has the ability to create an alkaline environment through various physiological activities [4]. Concrete technology used many types of additional materials to improve concrete properties such as foamed agent, super plasticizers or pozzolanic materials fly ash, biomass, silica fume, and other admixtures materials [5]. In addition, adding bacteria to the conventional concrete or other types of concrete to produce new type of concrete namely bio-concrete. Therefore, bio- concrete can be considered as modern concrete comparing to conventional concrete. The used of bacteria in different types of concrete to improved concrete properties and healing process such as fly ash concrete and concrete incorporating silica fume. Furthermore, the potential of bacteria in bio- concrete to act as self-healing of micro-cracks in concrete has proven to be a promising future [7].

Based on the statements above, it can be seen clearly bacteria have the ability to precipitate calcium which in line with filling pores and enhance concrete properties. But the most problem restricting the growth of bacteria inside the concrete were high alkaline and anaerobic conditions. In this study, (SRB) isolated with high alkaline and anaerobic condition with the aim to survive in the concrete environment. Hence, the evaluation of bio-concrete behavior in compressive strength and water penetration properties should be optimized.

2 Methodology

2.1 Isolation of bacteria

SRB applied to concrete as surface treatment. Based in the previous publications, conducted by other researchers, bacteria with gram positive capable to precipitate carbonate calcium in concrete through the biological process in concrete mix [9, 11-13].

Source of SRB from acid mire water collected from Sg Pelepah Kota Tinggi, Johor Malaysia. Whereas, the isolation process of SRB was following; enrichment, serial dilution, streaking plate, strain purification, gram staining and microscopy. SRB was enriched under special environment to tolerate the high alkalinity and anaerobic conditions for ensuring self sustain in the concrete. In this study magnesium sulphate heptahydrate AR (MgSO4.7H2O) was added to simulate sulphate environment during isolation. The compositions of media for control and SRB were prepared as follow:

a) Composition of sulphate reduction bacteria sample (SRB) consists of 25 ml (nutrient broth +10 ml (Mg $SO_4.7H_2O$) + 1 ml (water from Sg pelepah Kota Tinggi).
b) Composition of control sample for sulphate reduction bacteria consist of 25 ml (nutrient broth) + 10 ml (Mg $SO_4.7H_2O$).

Traditionally, bacteria can't survive in high alkaline environment and anaerobic conditions due to high pH value will effect microbial life which influence the dissociation and solubility of many molecules in bacteria enzyme [14]. Therefore, after samples preparation pH was adjusted in alkaline condition. The pH adjustment was done using NaOH until pH value in the range of 9-10. The other condition that was monitored along this study is anaerobic condition of the enrichment. Nitrogen gas was purged in the enrichment flask before enrichment started. The concentration of oxygen should be at 0 ppm and was measured with dissolve oxygen meter. The flask was shaked at room temperature. The adjustment of pH and anaerobic condition were done every day over the enrichment period of 20 day.

2.2 Concrete mixture proportion and materials properties

Concrete mixtures was designed according to (DOE) Department of Environment for target 28 days compressive strength of 30 MPa. Concrete mix proportion used 0.54 w/c to achieved the target strength as illustrated the details in Table 1. Ordinary portland cement was used and manufactured by Holcim Malaysia which the composition and specification complying with all requirement defined by BS EN 197-1:2000 [17]. The size of fine aggregate was below 5mm while coarse aggregate was in the range (12-20)mm. Whereas, the requirement size of aggregate in concrete used sieve analysis test which conducted according to BS 882-1992 [18].

Table 1. Mix proportion of bio-concrete with different percentages of SRB.

Percentage (%) of bacteria in batch	Cement (kg)	Water (kg)	Fine aggregate (kg)	Coarse aggregate (kg)	Bacteria growth medium (1kg = 1L)
0 %	420	210	685	1115	0
1 %	420	207.9	685	1115	2.1
3 %	420	203.7	685	1115	6.3
5 %	420	199.5	685	1115	10.5

2.3 Preparation of test specimens

Concrete cubes were prepared with different concentration of sulphate reduction bacteria in growth medium which replace 1%, 3% and 5% of water content. The cell concentration was determined from the bacterial growth curve made by observing the maximum turbidity was 710 NTU at day 10. Control concrete specimens were casted with out any additional of SRB. All experiments were performed in triplicates and curried in the dry conditions along the curing period.

2.4 Compressive strength test

Control concrete was designed using (DOE) to reach 30 MPa on compressive strength at 28 day of curing period. For bio-concrete mixes water was partially replace with 1%, 3% and 5% (SRB) liquid culture by weight of water. After casting control sample and different mix of bio- concrete the mortar demoulding in cubs with size of 150mm x 150mm x 150mm to be tested on 7th, 14th and 28th day of curing. The test was carried out using compressive strength Test Machine to conduct according to BS EN 12390- 3:2002 [19].

2.5 Water penetration test

Water penetration test was conducted according to BS EN 12390-8 [20] in ordure to determine the increase in resistance towards water penetration in concrete specimens as shown in Figure 1. The cubes molds of 150mm x 150mm x150mm were prepared both with and without (SRB). Concrete specimens were cured in dry air for 7th, 14th and 28th days. After curing the specimens water penetration test machine siting with constant pressure of 5 KPa for 72 hours to measure the water penetration depth, then specimens is splitting in half and the maximum depth of penetration of water measured in (mm).

Figure1. Water penetration test.

3 Results and Discussion

3.1 Compressive strength

The compressive strength of bio-concrete increased gradually along 7, 14 and 28 days using different ratio of (SRB) liquid culture with 1%, 3% and 5% to replace water. Furthermore, the results emerged the maximum increment of compressive strength were with specimens contained 5% and the lower increment was with 1% of SRB. On the other hand, bio-concrete specimens contained 3% of SRB display good performance in compressive strength whereas; the results was closed to the specimen with 5% of SRB specially at 28 days as shown in Figure 2. The compressive strength result of bio-concrete contains 1% ,3% and 5% of SRB increased comparing to the control samples by 5.9 %, 11.7% and 13.0% respectively at 28 days of curing time. Therefore, the result of compressive strength of concrete increase with the increment of SRB replacement ratio. The observation of the enhancement in compressive strength using (SRB) was due to deposition of calcium which conform the precipitation of CaCO3 on the microorganism cell surfaces and within the pores in side the concrete, which plug the pores within the binder matrix. Similar results were reported by other researchers [3, 15].

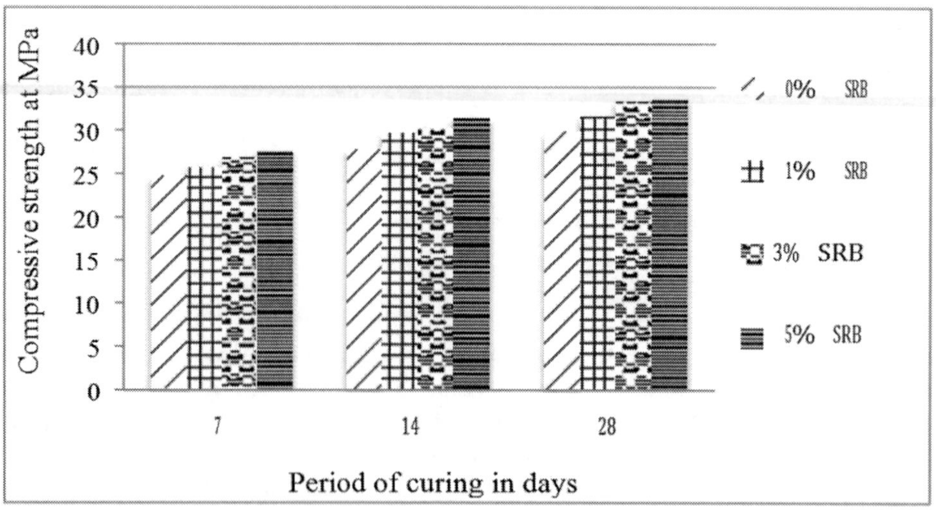

Figure 2. Compressive strength of concrete with different ratio of SRB.

3.2 Water penetration

The results presented reduction of water penetration with all replacements ratio of water 1%, 3% and 5% by SRB in bio-concrete specimens. Also indicated that, the increment of (SRB) had inverse relationship with reduction of water penetration in bio-concrete specimens. It was estimated that, the highest improvement was determined with 5% and the lowest performance was with 1% of (SRB) growth medium in bio-concrete mix. The decrement shows that at 28 day, the water penetration reduce by 5.3% , 7.3% and 8.5% for 1%, 3% and 5% of (SRB) as illustrate in Figure 3.Having bacteria in concrete could reduce the water permeability. Therefore, it increases the durability of concrete as it would prevent water from seeping into concrete to cause carbonation of steel and chlorination which would jeopardize the strength of the structure and cause the whole building to be less durable than intended to [16].

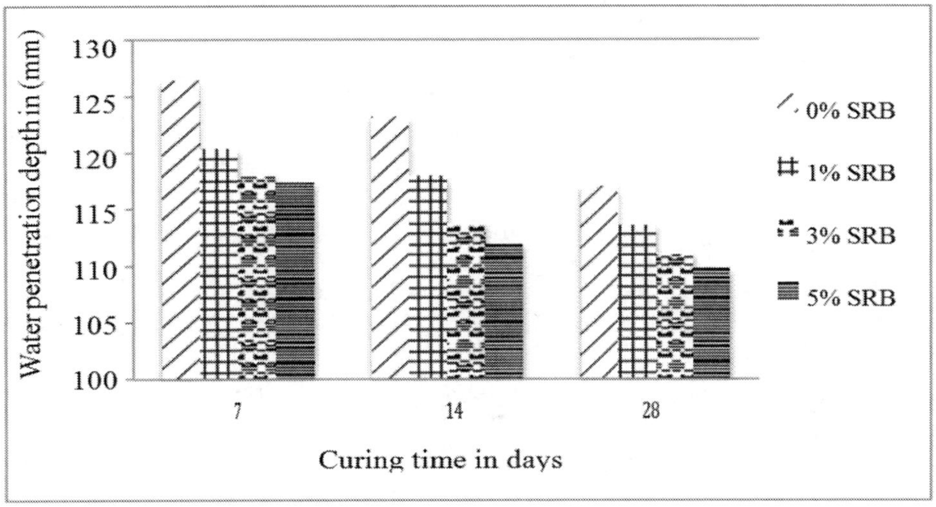

Figure 3. Water penetration depth with different ratio of SRB.

4 Conclusion

Out of all isolated cultures developed and tested of sulphate reduction bacteria it can be concluded that, (SRB) successfully isolated to survive in the alkaline environment in the range of 9-11 pH with anaerobic conditions. Furthermore, using of (SRB) in concrete increase the compressive strength and decrease of water penetration. Whereas, the highest increment in compressive strength of bio concrete was 13% with 5% of (SRB) compared to the control specimens at 28 days with same curing conditions. On the other hand, the maximum decrement of water penetration was 8.5% with replacement ratio 5% of (SRB) compared to the control sample at 28 days.

Acknowledgment

This study has been granted by the Ministry of Education (MOE) under Fundamental Research Grant Scheme (FRGS Vot. 1211) and supported by Universiti Tun Hussien Onn Malaysia (UTHM).

References

[1] G.M. Gonsalves, Bioconcrete- A Sustainable Substitute for Concrete, Master Thesis, Polytechnic University of Catalonia, Barcelona, Spain, (2011).

[2] R. Siddique and N. K. Chahal, Effect of ureolytic bacteria on concrete properties, Construction and Building Materials, **25**, 3791-3801, (2011).

[3] V. Achal, A. Mukerjee and M.S. Reddy, Biogenic treatment improves the durability and remediates the cracks of concrete structures, Construction and Building Materials, **48**, 1-5, (2013).

[4] L.K. Baumgartner, R.P. Reid, C. Dupraz, A.W. Decho, D.H. Buckley and J. R. Spear, Sulfate reducing bacteria in microbial mats: Changing paradigms, new discoveries, Sedimentary Geology, **185**, 131-145, (2006).

[5] R. Rajamma, L. Senff, M.J. Ribeiro, J.A. Labrincha, R.J. Ball and G.C. Allen, Biomass fly ash effect on fresh and hardened state properties of cement based materials, Composites Part B: Engineering, **77**, 1-9, (2015).

[6] F. Nosouhian, D. Mostofinejad and H. Hasheminejad, Influence of biodeposition treatment on concrete durability in a sulphate environment, Biosystems Engineering, **133**, 141-152 (2015).

[7] N. Chahal, R. Siddique and A. Rajor, Influence of bacteria on the compressive strength, water absorption and rapid chloride penetrationof concrete incorporating silica fume, Construction and Building Materials, **37**, 645-651 (2012).

[8] S.K. Ramachandran, V. Ramakrishnan and S.S. Bang, Remediation of concrete using micro-organisms, ACI Materials Journal, **98**(1), 3-9, (2001).

[9] P. Ghosh, S. Mandal, B.D. Chattopadhyay and S. Pal, Use of microorganism to improve the strength of cement mortar, Cement and Concrete Research, **35**, 1980-1983, (2005).

[10] F. Pacheco-Torgal and J.A Labrincha, Biotech cementitious materials: some aspects of an innovative approach for concrete with enhanced durability, Construction and Building Materials, **40**, 1136-1141, (2013).

[11] S.M. Al-Thawadi, Ureolytic bacteria and calcium carbonate formation as a mechanism of strength enhancement of sand, J. Adv. Sci. Eng. Res., **1**(1), 98-114, (2011).

[12] N. Chahal, R. Siddique and A. Rajor, Influence of bacteria on the compressive strength, water absorption and rapid chloride penetrationof fly ash concrete, Construction and Building Materials, **28**, 351-356, (2012).

[13] R. Demirboğa, F. Karagöl, R. Polat and M. A. Kaygusuz, The effects of urea on strength gaining of fresh concrete under the cold weather conditions, Construction and Building Materials, **64**, 114-120, (2014).

[14] R.C.J. Subhashree Parida, Kailash Chandra Samal and Pradeep Kumar Chand, isolation and identification of pathogenic bacteria from brackish waters of Chilika Lagoon, Odisha, India for pharmaceutical use, Malaysian Journal of Microbiology, **8**(3), 197-202, (2012).

[15] H.K. Kim, S.J. Park, J.I. Han and H.K. Lee, Microbially mediated calcium carbonate precipitation on normal and lightweight concrete, Construction and Building Materials, **38**, 1073-1082, (2013).

[16] N. De Belie and W. De Muynck, Crack repair in concrete using biodeposition, *Proceedings of the International Conference on Concrete Repair, Rehabilitation and Retrofitting*, Cape Town, (2008).

[17] British Standards Institution, Cement: Composition, Specifications and Conformity Criteria for Common Cements, London, (2000).

[18] British Standards Institution, Specification for Aggregates from Natural Sources for Concrete, London, (1992).

[19] British Standards Institution, Compressive Strength of Test Cubes, London, (2002).

[20] British Standards Institution, Testing Hardened Concrete, Depth of Penetration of Water under Pressure, London, (2000).

MATEC Web of Conferences **47**, 01017 (2016)
DOI: 10.1051/matecconf/20164701017

A Review on Interlocking Compressed Earth Blocks (ICEB) with Addition of Bacteria

J. M. Irwan[1,a], M. M. Zamer[1] and N. Othman[1]

[1]*Faculty of Civil and Environmental Engineering, Universiti Tun Hussein Onn Malaysia, 86400 Parit Raja, Johor, Malaysia*

Abstract. Interlocking Compressed Earth Block (ICEB) are cement stabilized soil blocks that allow for dry stacked construction. Because of this characteristic, the process of building walls is faster and requires less skilled labour as the blocks are laid dry and lock into place. However there are plenty room for improving the interlocking blocks by increase its durability. Durability of bricks are affected by several factors and one of them are water absorption. High water absorption level will results in low durability. Many studies have been conducted in order to improve the durability of bricks by using environmentally method. One of the method is by inducing bacteria into bricks. Bacteria in brick induced calcite precipitation (calcite crystals) to cover the cracks continuously. Therefore, the issues of durability of bricks can be fully solved. This paper offer a review on interlocking compressed earth blocks and the use of bacteria in reducing water absorption level.

1 Introduction

Interlocking compressed earth block (ICEB) is a cost effective and sustainable construction material which has potential to bring durable and affortable homes to developing counties around the world [1]. ICEB is an earthen bricks compressed with hand-operated or motorized hydraulic machines.

Interlocking Compressed Earth Block (ICEB) are energy efficient because its require anywhere from 1/5 to 1/15 of the energy to make when compared to fired bricks and concrete masonry units [2]. According to Maini [2] there is a large reduction in purchased material since soil is the main ingredient in ICEBs construction. As we know the traditional masonry relies heavily on skilled labor, expensive materials and time consuming. The interlocking nature of ICEBs allows for dry stacked, mortar-less construction, which reduces the need for skilled labor and shortens construction time. These reductions lower the cost of labor by as much as 80% [3]. These advantages make ICEBs a practical and preferred construction form.

However there are problems associated with bricks which are low strength, higher water absorption, low fire resistance and high porosity. Water absorption is a function of clay and cement content and usually related with the strength and durability of earth bricks [4]. Higher rate of water absorption will results in low compressive strength and durability. Numbers and size of pore in brick also affected the rate of water absorption. The bigger size of pore, the faster the brick will absorb

[a] Corresponding author : irwan@uthm.edu.my

water. Therefore this paper will provide an overview on the problem solving method that had been done by previous researcher by introducing the used of bacteria in brick or concrete.

2 Water Absorption and Moisture Content

There is a small but set volume of water present in the form of liquid bridges between soil particles in earth structures when air dried [5]. According to Jaquin [5], "This water is under tension and the magnitude of the pore water pressure (suction) is related to the relative humidity of the surrounding air". It has been identified by [6] that the evaporation of water from unsaturated soils is proportional to the suction. However [7] argued that this evaporation will continue from the earth structure until the relative humidity of the pore air is equal to the humidity of the surrounding air. According to [5], the strength and stiffness of earth structures, and therefore their inherent viability is a function of the mean relative humidity of a region.

Water absorption is a characteristic of clay and cement content and is often related to the strength and durability of earth blocks, therefore it is important to identify the rate of water absorption of earth blocks [4]. According to Oti et al. [8], water absorption rate decreases with increasing age of earth blocks. High rate of water absorption of a specimen may cause swelling of the stabilized clay fraction which will result in losing strength with time [4]. Walker [9] observed that water absorption and porosity increases with clay content and decreasing with cement content. Between cement, lime, cement-lime and cement-resin, combination cement showed the lowest absorption both in capillary absorption and total absorption [10].

Moisture content affects strength development and durability of earth blocks having a significant influence on the long term performance and the effectiveness of the bonding with the mortar at the time of construction [4]. Once the earth block is dry, water is rapidly drawn out of the mortar preventing good adhesion and proper hydration of cement [8]. Due to this Adam et al [11] stated that to achieve maximum strength, compressed earth block need a period of damp curing, where they are kept moist, to prevent rapid moisture evaporation.

3 Compressed Earth Blocks Failure due to Moisture

Moisture related deterioration in compressed earth blocks is due to seasonal or continuous alternate wetting (rainfall) and drying which lead to the block retaining sufficiently high amounts of moisture which leads to destructive effects. The softening and abrasive action of moisture leads to erosion of exposed surfaces [12].

Keefe [13] stated that over the years a building constructed of compressed earth blocks exposed to the elements will be affected by rainwater and due to the ageing of these blocks surface cracks will appear. This will result in water penetration and a build-up of moisture to a critical level [13]. According to Keefe [13] "compressive strength decreases with increasing moisture content and once it reaches its critical moisture content it is likely to fail". Minimal moisture content will results in better strength, water resistance and durability of the blocks.

Actual destructive action of moisture once the block has been penetrated is the dissolution and softening of loose particles and the pore pressure generated will result in a disruptive internal stress. Therefore the capacity of the block to resist the disruptive action of moisture will differentiate the life span of the earth block.

4 Bacteria

Siddique and Chahal [14] has define bacteria as unicellular (single cell) micro-organisms. Bacteria from various natural habitats have frequently been reported to precipitate calcium carbonate both in natural and in laboratory conditions [15]. Different types of bacteria, as well as abiotic factors (salinity and composition of the medium) seem to contribute in a variety of ways to calcium carbonate precipitation in a wide range of environments [16]. Bacillus species are able to precipitate calcium carbonate ($CaCO_3$) on its cell constituents and in its micro-environment by decomposition of urea ($CO(NH_2)_2$) into ammonium (NH_4^+) and carbonate (CO_3^{2-}) [17]. The bacterial degradation of urea locally increases the pH and promotes the microbial deposition of calcium carbonate in a calcium rich environment. In the presence of dissolved calcium ions, the produced carbonate ions precipitate and form calcium carbonate crystals [18]. Microbial mineral precipitation technologies have been successfully demonstrated for consolidation of sand columns, repair of limestone monuments, and to some extent for remediation of cracks in concrete [19].

4.1 Microbiologically induced calcite precipitation (MICP)

Microbiologically induced calcite precipitation (MICP) is the process by which microorganisms deposit carbonates as part of their basic metabolic activities [20]. Precipitation of calcium carbonate crystals occurs by heterogeneous nucleation on the bacterial cell wall. Once super saturation is achieved and these crystals precipitate inside the pore spaces. The bacteria not only initiate calcite precipitation, but also serve as nucleation sites for calcite crystals in association with other factors such as Ca2+ ions, dissolved inorganic carbon, pH, and temperature in the medium [20]. These carbonates/ calcite crystals act as cloggers in the building materials by filling the voids and thereby reducing permeability [21].

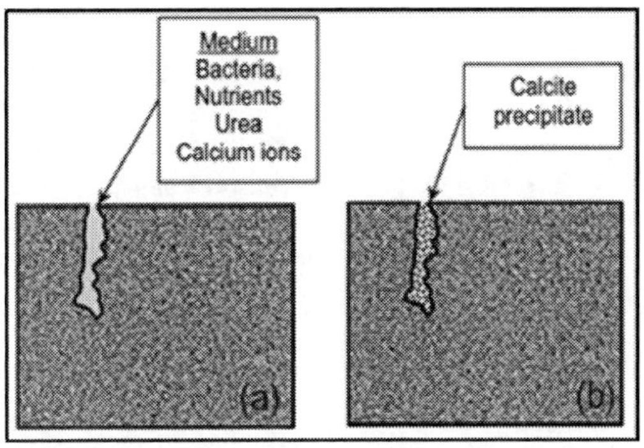

Figure 1. Process of MICP: (a) surface pore filled with medium; (b) calcite precipitates and seals the pore [21].

4.2 Previous research on the use of bacteria

Many research already been done in order to improve the durability and Table 1 provide summary on previous research which more focusing in improving the durability and water absorption level of bricks or concrete.

According to the remark from Table 1, its can be concluded that using bacteria toward brick or concrete can help to improve the water absorption level and increase their durability. Therefore it is possible to solve the ICEB problem by introducing the used of bacteria. Continuous research on the effect of bacteria should be done in future by introducing the used of bacteria toward ICEB.

5 Conclusion

Various methods and bacteria have been used by researchers in order to obtain the best solution on the effect of bacteria toward concrete and bricks as well as to achieve environmentally and sustainable solution. Although there are many researchers do not use the same type of bacteria in their research but the results obtained were very positive and very assuring. Therefore, it can be conclude that by using the bacteria, it can help to improve the durability of concrete or bricks properties. Meaning that, all the bacteria used by the previous researcher manage to undergo MICP process to produce calcite precipitation. Then the calcite will act as protective film coating on the surface of concrete or bricks which prevented ingress of water by filled all the pore with calcite. As the large amount of pores is plugged due to precipitation, the pore connectivity and porosity will be decreased. Hence it will reduce the rate of water absorption in concrete or bricks. Thus the durability problem is fully solved.

Table 1. Various researches works on bacteria.

Researcher	Type of bacteria	Findings	Remark
Shahrood et al, 2015 *"Surface treatment of concrete bricks using calcium carbonate Precipitation"*[22]	-Dimethyl carbonate (DMC)	-water absorption -compressive strength	-the water absorption of concrete bricks significantly reduce and the compressive strength significantly increase.
D. Bernardi et al, 2014 *"Bio-Bricks: Biologically cemented sandstone bricks"*[23]	-Sporosarcina paseurii	-void ratio and dry density -compressive strength	-the void ratio with MICP treatment shows decreasing in results compared to control specimen.
Abhjit et al, 2013 *"Bacterial Calcification for Enhancing Performance of Low Embodied Energy Soil-Cement Bricks"*[24]	- Bacillus megaterium	-water absorption test -wet compressive strength -porosimetry analysis	-the calcite crystal act as biosealant by filling the pores which leads to reduction in water absorption, porosity, permeability and enhance the strength of the bricks.
Navdeep et al, 2012 *"Improvement in strength properties of ash bricks by bacterial calcite"*[21]	-Bacillus megaterium	-microbiological sand plugging -water absorption and initial rate of water absorption -compressive strength	-the bacteria used was found to be very effective in calcite deposition on the surface of bricks which lead to reduction in permeability, decrease in water absorption leading to enhanced its durability.
Willem et al, 2008 *"Bacterial carbonate precipitation improves the durability of cementitious materials"*[25]	-B. sphaericus	-absorption of bacteria -precipitation of carbonate crystal -water absorption	-there are differences between mortar cubes treated with bacteria and a calcium which show less water absorption compared to untreated specimens.

Acknowledgment

This study has been granted by Universiti Tun Hussien Onn Malaysia (UTHM) under GIPS Vot. U264.

References

[1] P.T. Laursen, N.A. Herskedal, D.C. Jansen and B. Qu, Interlocking compressed earth blocks walls: Out-of-plane structural response, *15th World Conference on Earthquake Engineering*, Lisbon, Portugal, 71-81, (2012).

[2] S. Maini, Earthen Architecture in the World, Auroville Earth Institute, (2010). <retrieved from http://www.earth-auroville.com/> (accessed on November, 2015).

[3] K.B. Anand and K. Ramamurthy, Development and evaluation of hollow concrete interlocking block masonry system, The Masonry Society Journal, **23**(1), 11-19, (2005).

[4] F.V. Riza, I.A. Rahman and A.M.A. Zaidi, A brief review of compressed stabilized earth brick (CSEB), *International Conference on Science and Social Research*, Kuala Lumpur, 999-1004, (2010).

[5] P. Jaquin, How mud bricks work using unsaturated soil mechanics principles to explain the material properties of earth buildings, *EWB-UK National Research Conference*, United Kingdom, 49-51, (2010).
[6] G. Wilson, S.L. Barbour and D.G. Fredlund, The prediction of evaporative fluxes from unsaturated soil surfaces, Alonso and Delage (eds), France, 423-429, (1995).
[7] P. Jaquin, Analysis of historic rammed earth construction, PhD Thesis, University of Durham, United Kingdom, (2008).
[8] J.E. Oti, J.M. Kinuthia and J. Bai, Engineering properties of unfired clay masonry bricks, Journal of Engineering Geology, **107**(3-4), 130-139, (2009).
[9] P.J. Walker, Strength and erosion characteristics of earth blocks and earth block masonry, Journal of Materials in Civil Engineering, **16**(5), 497-506, (2004).
[10] A. Guettala, A. Abibsi and H. Houari, Durability study of stabilized earth concrete under both laboratory and climatic conditions exposure, Journal of Construction and Building Materials, **20**(3), 119-127, (2006).
[11] E.A. Adam and A.R.A. Agib, Compressed stabilised earth block manufacture in Sudan, Graphoprint for the United Nations Educational, Scientific and Cultural Organization, France, Paris, (2001).
[12] D.E. Gooding and T.H. Thomas, The potential of cement-stabilised building blocks as an urban building material in developing countries, Overseas Development Administration, United Kingdom, (1995).
[13] L. Keefe, *Earth Building: Methods and Materials, Repair and Conservation*, Taylor and Francis, New York, 145-156, (2005).
[14] R. Siddique and N.K. Chahal, Effect of ureolytic bacteria on concrete properties, Journal of Construction and Building Materials, **25**(10), 3791-3801, (2011).
[15] H.S. Chafetz and C. Buczynski, Bacterially induced lithification of microbial mats, Journal of Palaios, **7**, 277–293, (1992).
[16] H. Knorre and K.E. Krumbein, *Bacterial Calcification, Book of Microbial Sediments*, Springer, Berlin, 25–31, (2000).
[17] F. Hammes, N. Boon, J. de Villiers, W. Verstraete and S.D. Siciliano, Strain-specific ureolytic microbial calcium carbonate precipitation, Applied and Environmental Microbiology, **69**, 4901–4909, (2003).
[18] M.P. Harkes, L.A Van Paassen, J.L. Booster, V.S. Whiffin and M.C.M. Van Loosdrecht, Fixation and distribution of bacterial activity in sand to induce carbonate precipitation for ground reinforcement, Journal of Ecological Engineering, **36**(2), 112–117, (2010).
[19] W. De Muynck, N. De Belie and W. Verstraete, Microbial carbonate precipitation in construction materials: A review, Journal of Ecological Engineering, **36**(2), 118–136, (2010).
[20] S. Stocks-Fischer, J.K. Galinat and S.S. Bang, Microbiological precipitation of $CaCO_3$, Journal of Soil Biology and Biochemistry, **31**(11), 1563-1571, (1999).
[21] N.K. Dhami, M.S. Reddy and A. Mukherjee, Improvement in strength properties of ash bricks by bacterial calcite, Journal of Ecological Engineering, **39**, 31-35, (2012).
[22] S. Amidi and J. Wang, Surface treatment of concrete bricks using calcium carbonate, Journal of Construction and Building Materials, **80**, 273-278, (2015).
[23] D. Bernadi, J.T. Dejong, B.M. Montoya and B.C. Mattinez, Bio-bricks: Biologically cemented sandstone bricks, Journal of Construction and Building Material, **55**, 462-469, (2014).
[24] A. Mukherjee, N.K. Dhami, B.V.V Reddy and M.S. Reddey, Bacterial calcification for enhancing performance of low embodied energy soil-cement bricks, *Third International Conference on Sustainable Construction Materials and Technology*, Kyoto, (2013).
[25] W. De Muynck, D. Debrouwer, N. De Belie and W. Verstraete, Bacterial carbonate precipitation improves the durability of cementitious materials, Journal of Cement and Concrete Research, **38**(7), 1005-1014, (2008).

MATEC Web of Conferences **47**, 01018 (2016)
DOI: 10.1051/matecconf/20164701018

Physical and Mechanical Properties of Compressed Earth Brick (CEB) Containing Sugarcane Bagasse Ash

Noorwirdawati Ali[1,a], Nurul Aini Zainal[1], Mohd Khairy Burhanudin[1], Abdul Aziz Abdul Samad[1], Noridah Mohamad[1], Shahiron Shahidan[1] and Siti Radziah Abdullah[1]

[1]*Faculty of Civil and Environmental Engineering, Universiti Tun Hussein Onn Malaysia, 86400 Parit Raja,Johor, Malaysia*

Abstract. The use of brick in construction is commonly used, especially in the construction of buildings and infrastructure. Various studies have been conducted to produce methods that can increase the strength of brick at the same time can reduce the cost of manufacturing bricks. In order to reduce cost of manufacturing, one of the solution applied was by using waste as part of bricks production materials. In this study, sugarcane bagasse (SuCaB) ash was used as a part of compressed earth brick (CEB) by replacing the quantity of cement for SuCaB ash. The study focused on the physical and mechanical properties of CEB containing SuCaB and the optimum percentage of SuCaB ash as partial cement replacement in CEB. There are 4 types of percentages used; 0%, 20%, 25% and 30% from cement content. All mixed use the same water content of 30% of cement content by weight and the ratio for cement: laterite soil used was 1: 6. A total of 72 specimen with size of 100mm x 50mm x 40mm was produced. The test conducted were Initial Rate Absorption Test (IRA), Density Test, Dimensions Test, Compression Test and Water Absorption Test. From the experimental results, the optimum SuCaB ash percentage as cement replacement in CEB was 20%. It recorded the highest compressive strength of 16.23 MPa at 28 days while for the Initial Rate Absorption test, it lies within the range specified. The density of CEB containing 20% of SuCaB shows slightly lower value where it decreased for about 0.4% from the control specimen. From this study, it can be concluded that waste materials such as sugarcane bagasse can be used as part of construction materials. However, further study needs to be conducted such as on the energy consumption, chemical properties and others to enhance the knowledge on this area before it can be applied into the brick production.

1 Introduction

Due to limited availability of natural resources and rapid urbanization, there is a shortfall of conventional building construction materials. On the other hand, energy consumed for the production of conventional building construction materials pollutes the air, water and land. Accumulation of unmanaged agro-waste, especially from the developing countries has an increased environmental concern [1]. Therefore, development of new technologies to recycle and convert waste materials into reusable material is critically importance for the protection of the environment and sustainable development of the society [2]. The need to conserve traditional building materials that are facing

[a] Corresponding author : wirda@uthm.edu.my

depletion has forced engineers to look for alternative materials. Recycling of such wastes by incorporating them into building materials is a practical solution to the pollution problem [3].

2 Literature

2.1 Compressed earth brick

Compressed Earth Brick (CEB) is combination from three different materials which is Ordinary Portland Cement, soil, and sand that mixed together with addition of water by compression method. It is faster, easier, less skill labor required, good strength, good insulation, less carbon emission, and environmental friendly product. CEB is claimed as a sustainable product that using local material to produce it. However, local soils give a difference performance on their physical and chemical properties such as strength, durability, thermal and water absorption rate. By UNESCO Chair Earthen Architecture (2010), clay laterite soil is suitable for the production of CEB. Since Malaysia has an equatorial climate, the most suitable soils use and easy found is the laterite soil [4].

2.2 Sugarcane bagasse fly ash

In the sugar industry, sugarcane bagasse is more related as the by-product in sugarcane mills. After the sugarcane is pressed to remove sucrose or known as table sugar, the residue is called as sugarcane bagasse which contains highly fibrous residue. After the sucrose removal, the bags is generated and its accumulation presents as a waste problem for the sugar industry. Requirement for economic and environmental friendly materials has extended an interest in natural fibers [5].

The sugarcane bagasse consists of approximately 50% of cellulose, 25% of hemicellulose and 25% of lignin. Each ton of sugarcane generates approximately 26% of bagasse (at a moisture content of 50%) and 0.62% of residual ash. The residue after combustion presents a chemical composition dominates by silicon dioxide (SiO_2). In spite of being a material of hard degradation and that presents few nutrients, the ash is used on the farms as a fertilizer in the sugarcane harvests [6]. Due to limited availability of natural resources and rapid urbanization, there is a shortfall of conventional building construction materials. Therefore, study has been conducted by many researchers on the use of agricultural waste as partial construction materials either cement, sand or aggregate [7-11].

3 Methodology

In this experimental work, a total of 72 brick specimens were produced for the seventh and 28th days. The specimen size was 100 mm x 50 mm x 40 mm. The percentage of SuCaB ash were 0%, 20%, 25% and 30% from cement content. The percentage were chosen based on literature review where beyond 30% replacement, it shows inefficiency to replace cement in normal concrete. All experiments were conducted in accordance with Standards BS 3921: 1985 and MS 76: 1972. Table 1 shows the summary of specimen that produced according to the testing day and types of test conducted.

Table 1. Summary of specimen.

Testing day	7th				28th			
SuCaB(%) / Test	**0**	**20**	**25**	**30**	**0**	**20**	**25**	**30**
Compression	3	3	3	3	3	3	3	3
Water absorption	3	3	3	3	3	3	3	3
IRA **Density**	3	3	3	3	3	3	3	3
Total	9	9	9	9	9	9	9	9
	36				36			

3.1 Material procurement

3.1.1 Laterite soil, cement and water

The dry laterite soil were crushed by using crushing machine and sieve until particle size of soil that past on 1.18mm. In this study, the cement used is Ordinary Portland Cement (OPC) while the water used was available in the laboratory confirming to the requirements of water for mixing and curing.

3.1.2 Sugarcane bagasse (SuCaB) ash

The SuCaB was collected at stall in area Parit Raja, Johor. SuCaB was grinded until fine by using the particle size machine. The bagasse then placed inside electric control furnace and burnt at temperature of 450°C for 2 hours. After that, the ash obtained was grinded again by using planetary micro mill.

3.2 Production of CEB containing SuCaB as partial cement replacement

The brick manufacturing process is a critical element and should be given special attention. The materials used to produce these specimens were measured by volume. Figure 1 show the materials that has been weigh and ready to be mixed. Then the material must first be mixed thoroughly as shown in Figure 2 before pouring the water. The amount of water used was 600ml. After the mix has been thoroughly blend, it was put into a mold as shown in Figure 3. Hydraulic machine as in Figure 4 was used with the pressure of 2000 psi pressed onto the specimen. Figure 5 shows the process of pressing specimen to produces compressed earth brick (CEB).

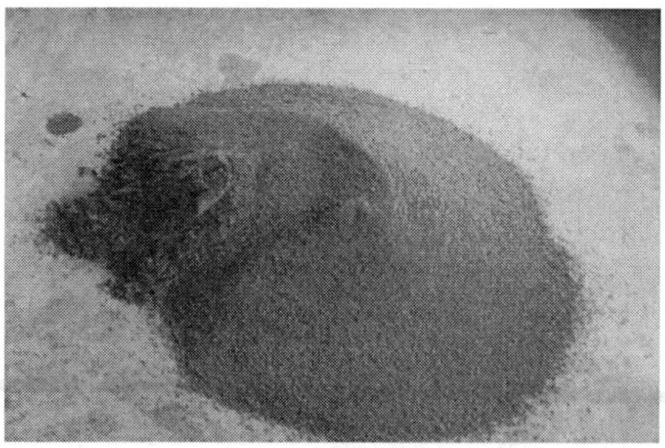

Figure 1. Materials used to produce CEB.

Figure 2. Mix of materials.

Figure 3. Mold used to produce specimen.

Figure 4. Hydraulic Jack with pressure of 2000 psi.

Figure 5. Specimen subjected to pressure.

4 Data Analysis

4.1 Density test

Figure 6 shows the density of bricks for all percentage. For control brick (0% SuCaB ash), the density recorded was 1718.03 kg/m3. For brick containing SuCaB ash of 20%, 25% and 30%, the density slightly decreased to 1710.76 kg/m3 and 1690.73 kg/m3 and 1687.69 kg/m3 respectively. The lowest value was brick containing 30% SuCaB with a decreased of 1.76% compared with the control specimen. From this experimental results, it shows that the density of concrete containing sugarcane bagasse decreased as the percentage increased. This shows that the bagasse makes the bond between aggregate in the concrete weaker than the control specimen.

4.2 Compression test

Figure 7 shows the relationship between the compressive strength and percentage of SuCaB ash on the 7th and the 28th days. Specimen with 20% of SuCaB ash recorded the highest strength compared to other specimens of 16:23 MPa. From the results, it was found that the additional SuCaB ash reduced the strength of the CEB compared to the control brick as the brick become more porous with a higher percentage of SuCaB ash. The additional of exceed 20% of SuCaB ash weaken the bond strength between the brick interface materials. The optimal percentage of SuCaB ash observed was 20% with 16.23 Mpa compressive strength.

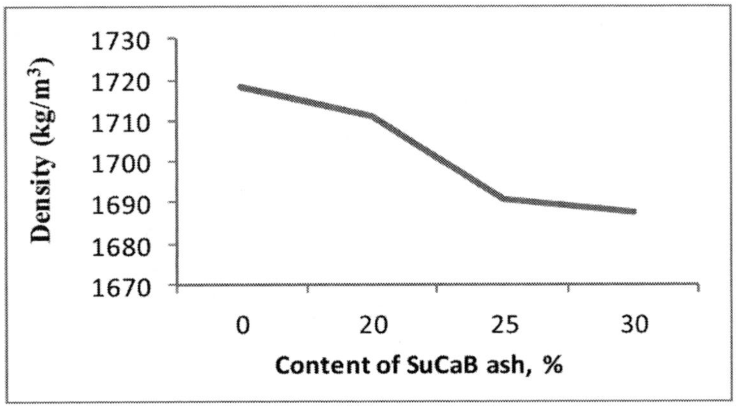

Figure 6. Density of concrete vs percentages of SuCaB ash.

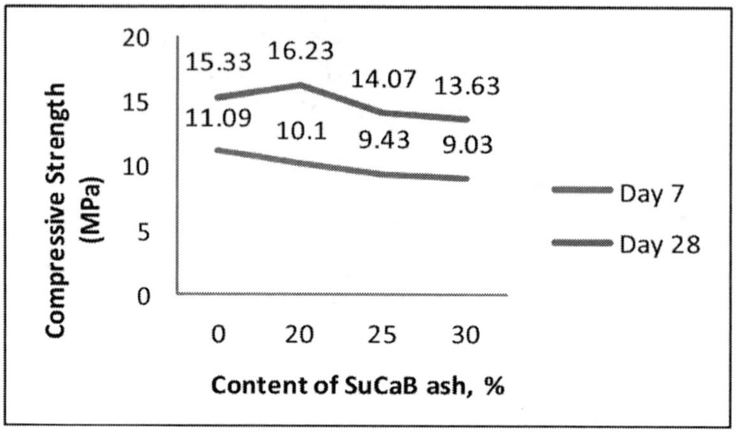

Figure 7. The relationship between the compressive strength and percentage of SuCaB ash on the 7th and the 28th days.

4.3 Water absorption test

Based on Figure 8, the percentage of water content of specimens with 0% SuCaB ash is the lowest with 7.04% on day 7 and 6.04% on day 28 after being soaked for 24 hours. The percentage of water content increased dramatically in the percentage of 20% with 24.17% on day 7 and 23.08% on day 28. Then, it increase slightly on the percentage of 25% and 30% by value of 24.38% and 24.46% at day 7, 23.49% and 23.71% on day 28 respectively.

4.4 Initial rate absorption test

Based on Figure 9, the IRA of 20%, 25% and 30% of SBFA increased compared to control CEB specimen. For 0% and 20% of SuCaB ash, the IRA for 7 and 28 days were within the range specified. While the 25% of SuCaB ash on 28 days, which is within the standard, while the other percentage rest exceed the range of standards.

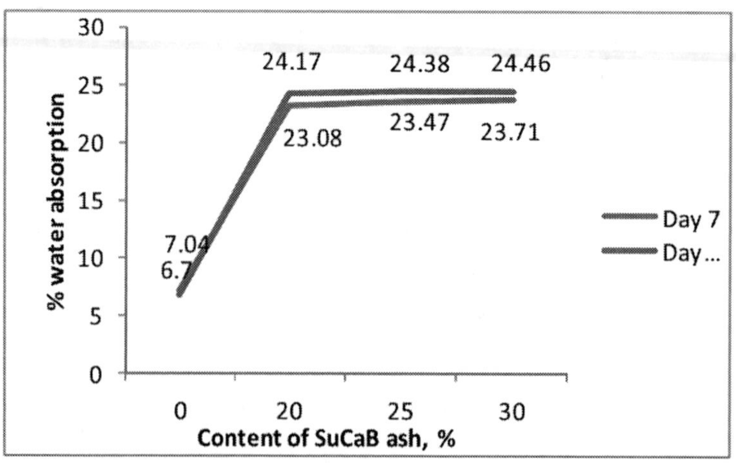

Figure 8. The relationship between the water absorption and percentage of SuCaB ash on the 7th and the 28th days.

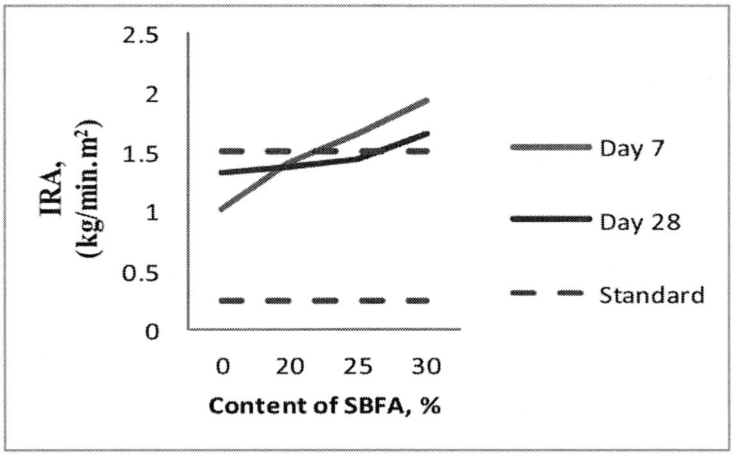

Figure 9. The relationship between the IRA value and SuCaB ash on the 7th and the 28th days.

5 Conclusion

The results show that with the addition of SuCaB ash by 20% of cement content in CEB brick, the compressive strength increased compared with CEB without SuCaB ash. However, higher than that, the compressive strength started to decreased and it shows that the optimum content for SuCaB addition was at 20%. The density of CEB decreased with the increase of SuCaB ash content and thus light weight CEB can be produced with further research.

References

[1] P. Agamuthu, Challenges and opportunities in agro-waste management: An Asian perspective what is agrowaste ?, (2009).

[2] M.V. Madurwar, R.V. Ralegaonkar and S. A. Mandavgane, Application of agro-waste for sustainable construction materials: A review, Constr. Build. Mater., **38**, 872-878, (2013).

[3] S.P. Raut, R.V. Ralegaonkar and S.A. Mandavgane, Development of sustainable construction material using industrial and agricultural solid waste: A review of waste-create bricks Constr. Build. Mater., **25** (10), 4037-4042, (2011).

[4] C. Jayasinghe and R.S. Mallawaarachchi, Flexural strength of compressed stabilized earth masonry materials, Mater. Des., **30** (9), 3859–3868, (2009).

[5] A.A. Kadir and N. Maasom, Recycling sugarcane bagasse waste into fired clay brick, Int. J. Zero Waste Gener., **1**(1), 21–26, (2013).

[6] A. Pandey, C.R. Soccol, P. Nigam and V.T. Soccol, Biotechnological potential of agro-industrial residues. I: Sugarcane bagasse, Bioresour. Technol., **74** (1), 69–80, (2000).

[7] G. Ajay, K. Hattori, H. Ogata and M. Ashra, Processing of sugarcane bagasse ash and reactivity of ash-blended cement mortar, Transactions of the Japanese Society of Irrigation, Drainage and Rural Engineering, **77**(3), 243–251, (2009).

[8] K. Bilba, M.A. Arsene and A. Ouensanga, Sugar cane bagasse fibre reinforced cement composites. Part I. Influence of the botanical components of bagasse on the setting of bagasse/cement composite, Cem. Concr. Compos., **25**(1), 91–96, (2003).

[9] N. Chusilp, C. Jaturapitakkul and K. Kiattikomol, Utilization of bagasse ash as a pozzolanic material in concrete, Constr. Build. Mater., **23**(11), 3352–3358, (2009).

[10] G.C. Cordeiro, R.D. Toledo Filho, L.M. Tavares and E.D.M.R. Fairbairn, Ultrafine grinding of sugar cane bagasse ash for application as pozzolanic admixture in concrete, Cem. Concr. Res., **39**(2), 110–115, (2009).

[11] A. Sales and S.A. Lima, Use of Brazilian sugarcane bagasse ash in concrete as sand replacement, Waste Manag., **30**(6), 1114–1122, (2010).
[12] MS76:1972, Specification for bricks and blocks of fired brick earth, clay or shale, Malaysian Standard, (1972).
[13] BS 3921: 1985. Specification for clay bricks, British Standard, (1985).

MATEC Web of Conferences **47**, 01019 (2016)
DOI: 10.1051/matecconf/20164701019

Tensile Properties of Treated and Untreated Paddy Straw Fiber Using Sodium Hydroxide Strengthened with Polypropylene Resin

Masni A. Majid[1,a], Mimi Attahirah Mohd Hishammudin[1], Noor Azlina Abd Hamid[1], Zalipah Jamellodin[1] and Norhafizah Salleh[1]

[1]*Faculty of Civil and Environmental Engineering, Universiti Tun Hussein Onn Malaysia, 86400 Parit Raja, Johor, Malaysia*

Abstract. This paper focus on the effect of alkaline treatment by using sodium hydroxide on the tensile properties of paddy straw fibers strengthened with polypropylene resin. Two types of paddy straw fibers were used which are MR219 and MR220. The paddy straw fibers were prepared in two conditions as untreated and treated with sodium hydroxide (NaOH) at 5% for 24 hours and then dried at 80°C for another 24 hours. For sampling process, these fibers were weighted according to 5% and 10% mass fraction. A total of 16 samples were prepared for this study based on ASTM D638 and tensile test was conducted by using ASTM D5083. The result showed that paddy straw fiber treated with NaOH had gain higher ultimate tensile strength compared to untreated paddy straw fiber where the highest ultimate tensile strength for the fibers is recorded at 10% of MR219 fiber with value of 2.0230 kN and 3.677 mm displacement were recorded. The highest strain were recorded by the same fibers with an average value of 5.253% and obtained the Young's modulus up to 1110 MPa. However, the Young's modulus which has been obtained by the same fibers was decreased with the percentage difference of 40%.

1 Introduction

The use of synthetic fibers can cause pollution to environment and increment of CO_2 that contribute to effect of global warming [1]. Therefore, researchers nowadays attempted to make changes to the materials that are harmful to humans and environment by finding sustainable and environmental friendly material as an alternative in replacing synthetic fibers.

Natural fibers have the potential to be used as reinforcement in engineering application [2]. Natural fiber is a material that is derived from its own natural resources resulting from plants and animals. There are several types of traditional materials are used as reinforcing material such as wood and steel. Based on previous researcher's studies, natural fibers has advantages of low cost, environmental friendly, renewable, low density, biodegradability and has satisfactory mechanical properties [3]. At present, research is carried out to make the plant natural fibers used as reinforcement fibers replacing iron. In a recent study, composite was produced with the combination of resin to improve the mechanical properties of tensile, impact and flexural properties [4].

[a] Corresponding author : masni@uthm.edu.my

Many researchers have conducted study on the mechanical properties of natural fiber composites using polymers matrix. The polymer matrix or binding agent has ductile more than fibers and it acts as a strengthening of composite. Furthermore, it also serves as to protect the fiber from environmental damage before, during and after the composition process [5]. Among the natural fibers used in previous research are paddy fiber [1, 6], kenaf [2, 4, 7, 8], bamboo [9], abaca [3] and others. Thus this study was proposed to investigate the effect of alkaline treatment on tensile properties of paddy straw fiber.

2 Materials and Methods

2.1 Material preparation

Paddy straw raw fibers used in this study was supplied from paddy industry in Selangor. The supplied fibers were hand-washed, and dried at temperature of 70°C for 24 hours as shown in Figure 1. The sodium hydroxide (NaOH) solution used for fiber treatment and the polymer used for binding agent were polypropylene (PP).

2.2 Fiber treatment

In this study, alkaline treatment has been applied to the paddy straw fibers. Figure 2 shows the fibers were soaked in 5% NaOH solution for 24 hours. The fibers were then washed several times with distilled water to remove the excess of NaOH and dried in an oven at 80 °C for 24 hours.

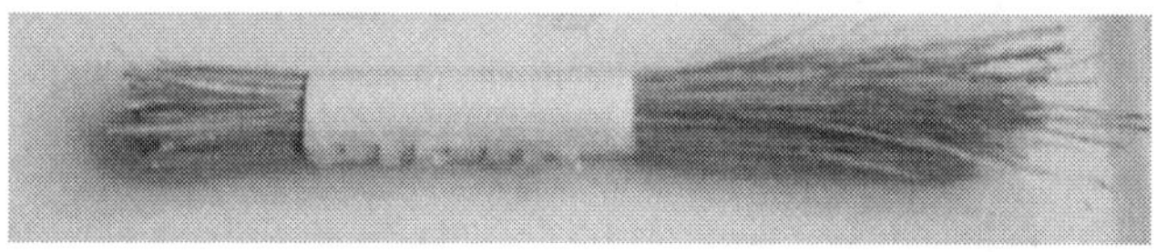

Figure 1. Paddy straw fiber.

Figure 2. Treatment of paddy straw fiber.

2.3 Sample preparation

After alkaline treatment process, the paddy straw fibers and the polypropylene were weighted according to the ratio of (1:9) and (0.5:9.5) for each specimen. In the sampling process, a metal mould (100 x 20 x 6 mm^3) was coated with a layer of release agent (WD-40). The treated fibers were arranged into the mould using sandwich method. Finally, the composite was covered with mould cover and compressed by using Hot Press Machine at 230°C for 20 minutes as can be seen in Figure 3. The same procedure was done for untreated paddy straw fibers as well. The specimen set has been produced for testing purposes as shown in Table 1.

Figure 3. Compression process.

Table 1. Information on specimens produced for testing.

Specimen No	Paddy Fiber Type	Resin
1	Untreated MR219 fibers	Polypropylene
2	Treated MR219 fibers	Polypropylene
3	Untreated MR220 fibers	Polypropylene
4	Treated MR220 fibers	Polypropylene

2.4 Tensile test

In a broad sense, tensile test is measuring the ability to withstand forces; tend to pull it apart and to what extent the material stretches before breaking. Universal Testing Machine was used at cross-head speed of 1m/min. The ultimate tensile strength was obtained from the maximum stress on the stress-strain curve. Besides ultimate tensile strength, the strain at break was also obtained from the stress-strain curve. The modulus elasticity was provided as the slope of the linear portion of stress-strain curve. Three specimens for each case were tested for treated fibers and three specimens for each case tested for untreated fibers as a control. The specimens were pulled until it fractured.

3 Results and Discussion

3.1 Tensile strength

Figure 4 presents the maximum ultimate tensile strength of alkaline treatment and untreated paddy straw fiber composites. For MR220 type of paddy fiber, the impact load is slightly small on both composites because it cannot accommodate higher load. However, the MR219 type of paddy fiber gave higher value of ultimate tensile strength compared with MR220.

Based on the histogram, it is obvious that 10% mass fraction of fibers gave higher value of ultimate tensile strength compared to 5% mass fraction of fibers for MR219 fibers and opposite result for MR220 fibers. For MR219 fibers, an approximation of 40% and 29% increment is noticed for both treated and untreated cases respectively. As for MR220 fibers an approximation of 16% and 26% decrement recorded for treated and untreated case respectively. It is worth nothing that the treated of paddy straw fiber with NaOH solution influences the tensile properties of composites. In particular, the treated fiber of MR219 in composite of 5% mass fraction show ultimate tensile strength of 31% higher than untreated fiber of MR219 in the same composite of mass fraction. It is explained by the increase in uniformity of the yields to enhance strength due to the removal of impurities during treatment. On the other hand, the chemical treatment of paddy fibers leads to increments of 42% in tensile strength of the 10% fibers mass fraction composites.

Moreover, the treated fiber of MR220 also experienced higher ultimate strength of 9% and 21%respectively for both 5% and 10% mass of fractions. However, the MR220 obtained lower ultimate strength value compared to MR219 in 10% mass fractions. Based on research [2], noted that the high impact strength will cause a lower ultimate tensile strength achieved. The result of both type of paddy fibers showed that the 5% NaOH treatment is an evident that the alkali treatment helps in improving the tensile strength and chemical bonding between fibers and resin. It enhances surface adhesion characteristics of fibers by removing impurities and produce rougher fiber surface.

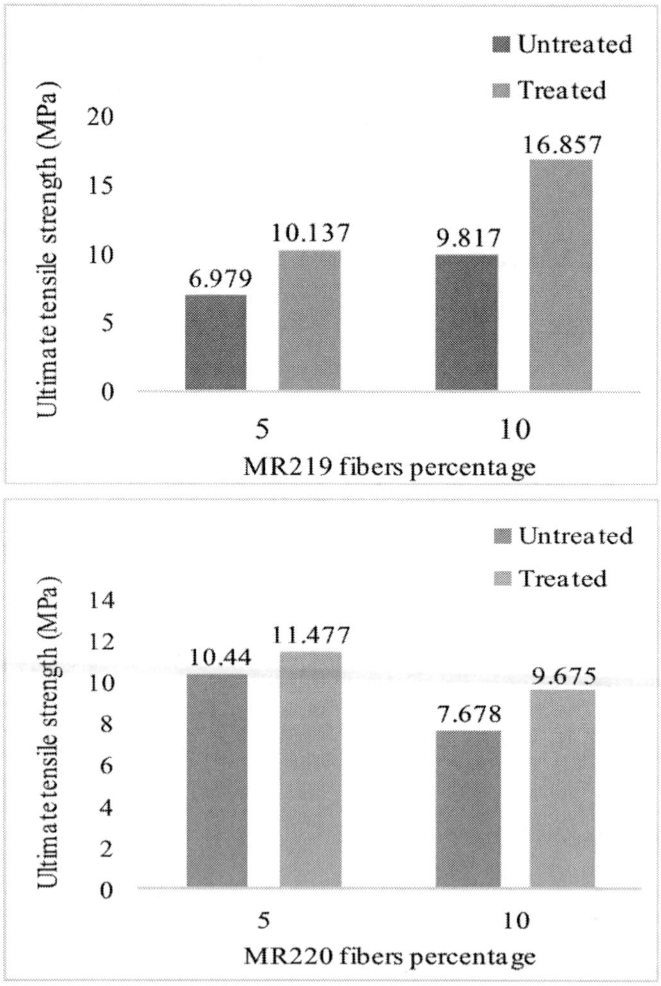

Figure 4. Ultimate tensile strength.

3.2 Modulus elasticity

Based on the graph plotted in Figure 5, the modulus of elasticity was decreases between treated fibers with untreated fibers. This result obtained as opposed to the result recorded by previous researcher [3] using abaca fibers.

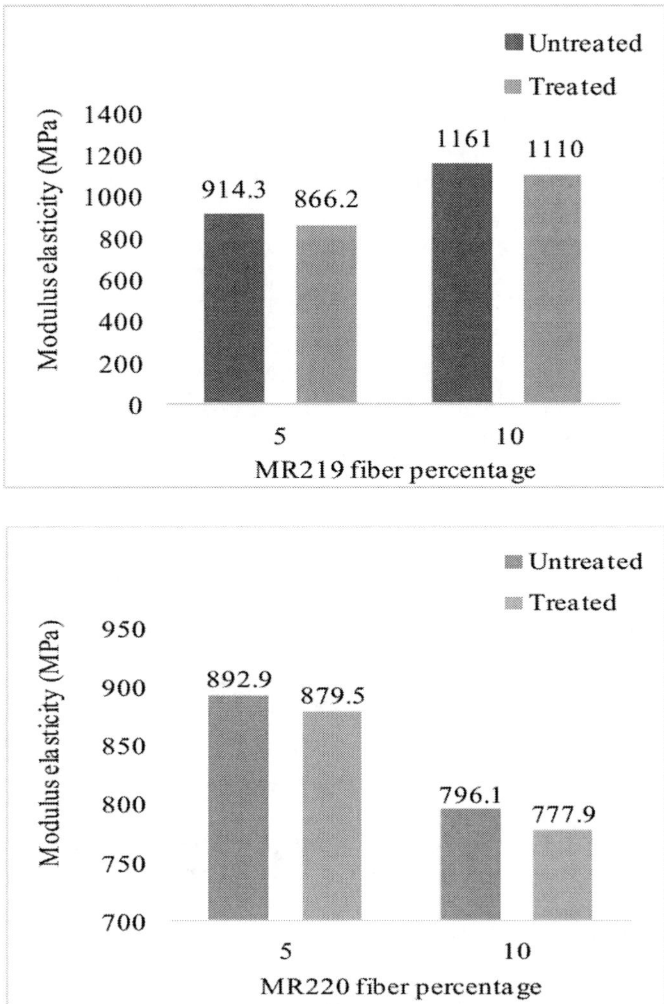

Figure 5. Modulus elasticity.

The values of modulus elasticity recorded are uncertain due to the immersion time, arrangements of fibers are not aligning properly and also the thickness is not the same along the length of specimen. A few researcher reported that this scenario occurred due to longer immersion time that had detrimental effect on the fibers surface and consequently on their mechanical properties [3, 4]. Generally, the existence of alkalization treatment on specimens related to fibers contributed to the value of modulus elasticity.

4 Conclusions

The paddy straw fibers is one of the environmentally friendly natural fiber, easily available, renewable resources, recyclable, widely distributed, good acoustic insulation and economy compared to carbon and glass fibers. In this research, the effect of alkaline treatment (5% NaOH solution) on tensile properties of paddy straw fibers composites was evaluated and it can be summarized that: (i) NaOH produced a better properties compared to the untreated fibers as the NaOH treatment has highly enhanced the interfacial adhesion of the fiber with the resin. MR 219 produce 31% and 42% increment from 5% and 10% mass of treated fiber compared to untreated fiber, while MR220 produce 9% and 21% increment respectively, (ii) The 10% mass of treated paddy fibers produce higher ultimate tensile strength as compared to 5% mass of treated paddy fiber with the difference of 40% increment for MR219 fibers and opposite result for MR220 fibers with 16% decrement, and (iii) The modulus elasticity of treated paddy straw fiber decreased compared to untreated paddy straw fibers. The fibers surface and their mechanical properties effect during the immersion time.

Overall, it can be concluded that the 10% treated MR219 paddy fibers composite gave the highest ultimate tensile strength while the lowest ultimate strength was obtained from 5% untreated MR219 paddy fibers composite. There is a possibility of replacing the synthetic fibers with paddy fibers for tensile applications.

Acknowledgment

The authors wish to thank Universiti Tun Hussein Onn Malaysia (UTHM) for the financial support and not to forget, technical officer at Faculty of Civil and Environmental Engineering, UTHM for their helps in order to complete this project.

References

[1] Y. Abdullah, A. Putra, H. Efendy, W.M. Farid and M.R. Ayob, Investigation on sound absorption coefficient of natural paddy fibers. Int. J. of Renewable Energy Resource, **3**, 8-11, (2013).
[2] M. Mohd Suhairil, O. Muhammad Fauzi, H. Hazran, R. Muhammad Fairuz, M.S. Mohd Syahar, Improving tensile properties of kenaf fibers treated with sodium hydroxide, Procedia Engineering **41**, 1587-1592, (2012).
[3] M. Cai, H. Takagi, A.N. Nakagaito, M. Katoh, T.U. Geoffrey, I.N. Waterhouse and Y. Li, Influence of alkali treatment on internal microstructure and tensile properties of abaca fibers, Industrial Crops and Products, **65**, 27-35, (2015).
[4] V. Fiore, G. Di Bella and A. Valenza, The effect of alkaline treatment on mechanical properties of kenaf fiber and their epoxy composites, Composites: Part B, **68**, 14-21, (2015).
[5] S. Taj, M. Munawar Ali and Shafiullah Khan, Natural fiber-reinforced polymer composites, Proc. Pakistan Academic Science, **44**(2), 129-144, (2007).
[6] R.M. Nasir and N.M. Ghazali, Tribological performance of paddy straw reinforced polypropylene (PSRP) and unidirectional glass-pultruded-kenaf (UGPK) composites. J. Tribologi, 1-17, (2014).
[7] H.M. Akil, M.F. Omar, A.A.M. Mazuki, S. Safiee and Z.A. Mohd Ishak and A. Abu Bakar, Kenaf fiber reinforced composites: A review, Material and Design, **32**, 4107-4121, (2011).
[8] M. Edeerozey ,Chemical Modification of Kenaf Fibers, Material Letters, **61**, 2023-2025, (2007).
[9] K. Murali Mohan Rao, Extraction and tensile properties of natural fibers: Vakka, date and bamboo, Composite Structures **77**, 288-295, (2007).
[10] M. Ali, Natural fibres as construction materials, J. of Civil Engineering and Construction Technology, **3**(3), 80-89, (2012).
[11] S.N. Monteiro, Natural-fiber polymer-matrix composites: Cheaper, tougher, and environmentally friendly, Journal of the Minerals, Metals and Materials Society, **61**(1), 17-22, (2009).
[12] N. Sgriccia, Characterization of natural fiber surfaces and natural fiber composites, Journal of Composites: Part A, **39**, 1632-1637, (2008).

MATEC Web of Conferences **47**, 01020 (2016)
DOI: 10.1051/matecconf/20164701020

The Morphology of Electrospun Titanium Dioxide Nanofibers and Its Influencing Factors

Zi Sheng Tang[1], Nurmin Bolong[1,a], Ismail Saad[1] and Janice Lynn Ayog[1]

[1]*Faculty of Engineering, Universiti Malaysia Sabah, 88400 Kota Kinabalu, Sabah, Malaysia*

Abstract. Titanium dioxide (TiO_2) has high photocatalytic activity and it is extensively applied in solar cell technology and environmental science. Electrospinning is acknowledged as the most versatile technique to fabricate nanofibers such as metal oxide nanofibers. Titanium dioxide nanofibers are generally prepared by electrospinning organic solutions containing alkoxide precursors and a carrier polymer with high voltage supply. The paper discusses on electrospun TiO_2 nanofibers including the spin dopes preparation history, influencing factors on fiber morphology and fiber characterizations. In particular, the parameters such as spin dopes viscosity, supplied voltage, feeding rate and effect of temperature that affect the morphology of the nanofibers are emphasized. Based on several studies, smaller diameter of TiO_2 nanofibers can be produced with lower viscosity solution, higher voltage and lower feeding rate. The heat treatment of 500 °C reduced the fiber size and produces crystallized anatase TiO_2 nanofibers.

1 Introduction

Titanium dioxide (TiO_2) have attracted the interests and attentions of the researchers because it is low cost, environmental friendly and exceptional chemical stability with extraordinary optical and electronic properties [1-5]. TiO_2 nanofibers with wide band gap are promising material in the applications of photocatalysis, solar cell, optical filter and antimicrobial surface coating [5-7]. TiO_2 exists in both amorphous and crystalline forms, in which amorphous form is photocatalytically inactive. There are three crystalline TiO_2 phases, namely anatase, rutile and brookie. Of the phases, anatase and rutile are found to be suitable for photocatalytic applications where as brookie is not being tested for photocatalysis [8]. Anatase TiO_2 is metastable and exhibits high activities for solar cell applications due to its higher band gap (3.2 eV). On the other hand, rutile TiO2 is thermodynamically more stable, and due to the low intrinsic photocatalytic activity with lower band gap (3.0 eV), it can be applied in the cosmetic field and tissue engineering [9-12]. One dimensional nanoarchitectures TiO_2 offers faster electron transportation rate, high porosity and larger surface [13]. One dimensional nanostructures TiO_2 such as whiskers, rods, tubes, wires, belts, spirals and fibers have been produced by sol-gel synthesis, electrochemical fabrication, dip-coating, hydrothermal technique, electroless deposition, crystallize growing method and electrospinning [13-16]. To date, TiO_2 nanofibers are prepared by using crystallize growing method and electrospinning. Crystallize growing method takes

[a] Corresponding author : nurmin@ums.edu.my

extremely long growing time and does not produce uniform fibers [13]. Hence, electrospinning has engrossed the attention of the researchers.

Electrospinning appears as a novel, versatile, simple, cost effective and straightforward technique to produce fibers with diameters down to tens of nanometers, from various materials such as polymers, metal oxides and composites [11, 14, 16]. In a typical electrospinning system, there are three basic components to be used: a high voltage supply, a needle of small diameter with syringe that is connected to a syringe pump and a grounded metal collector [17]. The schematic diagram for a typical electrospinning set up can be referred to [18].

2 Chronology of Spin Dopes Formulation

The titanium dioxide-based spin dopes can be prepared from organic solutions containing alkoxide precursors and a carrier polymer [19]. The alkoxide precursors are titanium tetraisopropoxide (TTIP) or Titanium (IV) n-butoxide (TNBT), the carrier polymers are polyvinylpyrolidone (PVP) or polyvinyl acetate (PVAc). The solvent for the spin are ethanol, dimethyl formamide (DMF), methanol or isopropanol. Acetic acid is a popular choice of catalyst in the spin dopes preparation among many researchers. A summary of the spin dopes which have been productively electrospun into TiO_2 nanofibers by different researchers is presented in Table 1.

Table 1. Spin dopes to fabricate electrospun TiO2/PVP composite nanofibers.

No	Precursor	Carrier Polymer	Solvent	Catalyst	Fiber size (nm)	Ref.
1	Titanium (diisoproproxide) bis (2,4-pentanedionate) 75 wt.% in 2-propanol	PVP (Mw 1,300,000)	Ethanol	Acetic acid	132 ± 24	[7]
2	TNBT	PVP (Mw 1,300,000)	DMF	Acetic acid	50-500	[9]
			DMF/Isopropanol ratio 1/1			
			Isopropanol	Acetic acid	200-2000	
3	TTIP	PVP (Mw 1,300,000)	Ethanol	Acetic acid	200-800	[11]
4	TTIP	PVP	Ethanol	Acetic acid	100-200	[12]
5	TTIP	PVP (Mw 1,300,000)	Ethanol	Acetic acid	200-2000	[16]
6	TNBT	PVP (Mw 360,000)	DMF	Acetic acid	920-1800	[19]
7	TNBT	PVP (Mw 1,300,000)	Ethanol	Acetic acid	200-300	[20]
8	TTIP	PVP (Mw 1,300,000)	Ethanol	Acetic acid	20-200	[21]
9	TNBT	PVP (Mw 1,300,000)	Methanol	Acetyl acetone	150-200	[22]
10	TTIP	PVP (Mw 1,300,000)	Ethanol	Acetic acid	150-200	[23]
11	TTIP	PVAc (Mw 500,000)	DMF	Acetic acid	150-600	[24]
12	TTIP	PVP (Mw 1,300,000)	Ethanol	Acetic acid	70-150	[25]
13	TTIP	PVP (Mw 1,300,000)	Ethanol	Acetic acid	30-200	[26]
14	TTIP	PVP (Mw 300,000)	Ethanol	Acetic acid	184-343	[27]
15	TTIP	PVAc (Mw 800,000)	DMF	Acetic acid	200-600	[28]

Spin dopes consists of the combination of TTIP, PVP (M_w 1,300,000) and acetic acid in ethanol is the most commonly used to fabricate TiO_2 fiber.

In this paper, a systematic review is worked out on the electrospun TiO_2 nanofibers materials selections and parameters that affect the morphology of the nanofibers and the characterization of TiO_2 nanofibers. There are several factors that influence the morphology of the fibers such as applied voltage, flow rate, tip to collector distance and type of collector. However, this paper mainly discusses for the spin dope viscosity, supplied voltage and feeding rate.

3 Main Factors Influencing the Fibers Morphology

3.1 Spin dopes viscosity

The polymer concentration plays a significant role in determining the diameter of the nanofibers. The increase in polymer concentration may consequently elevates the viscosity of the spin dopes significantly [29]. Furthermore, the rise in alkoxide precursor (TTIP) may cause a slight increase in the spin dopes viscosity. In TiO_2 nanofiber fabrication, the increase of polymer concentration cause the expansion of fabricated fiber diameter [21]. Lee et al. [30] found that when the PVP concentration is increased from 0.04 g/ml to 0.10 ml, the fibers size increase significantly from 125 nm to 275 nm while the TTIP concentration raises from 0.12 g/ml to 0.18 g/ml results in increase of the fiber diameter from 200 nm to 280 nm. This has been proved later by [11] and [16]. Thus, the viscosity is greatly affected by the polymer concentration in the spin dopes. When the spin dopes viscosity is higher, under a fixed supplied voltage and feeding rate condition, the fiber formation initiation cannot take place due to the insufficient force to stretch the fibers to the thinner size [11]. At a higher concentration of TTIP, a higher supplied voltage is needed in order to initiate the fiber formation. In order to produce smaller diameter of TiO_2 fibers, it is suggested that the spin dopes should have a lower concentration of PVP in order to obtain smaller size of TiO_2/PVP composite fibers.

3.2 Supplied voltage

Bharwaj and Kundu [18] stated that the fibers can be formed provided the supplied voltage is at the threshold stage which subsequently induces the essential charges on the solution to initiate the electrospinning process. In the fabrication of TiO_2 nanofibers, the fibers' diameter can be reduced by raising the supplied voltage. Li and Xia [21] proved that the fibers size is abruptly decreased from 110 nm to 55 nm as the supplied voltage is raised from 5 kV to 20 kV. In addition, the mean diameter of the fabricated fibers decreases from 110 nm to 80 nm when the supplied voltage of 15 kV is increased to 25 kV [31]. Higher supplied voltage leads to a greater stretching on the spin dopes as the higher electrostatic repulsive force is formed and effectively reduce the fibers size. This causes rapid evaporation of the solvent, which contribute in the fiber size reduction [18]. However, discontinuity in nanofibers occurs when the supplied voltage exceeded 20 kV, causing the occurrence of over pulling effect with the fiber breakage [13, 31-32]. Thus, to fabricate a continuous smaller fibers, the supplied voltage should be within 15-20 kV. The optimized TiO_2 nanofibers can be generated with the supplied voltage of 18.4460 kV based on the morphology analysis [16].

3.3 Feeding rate

Feeding rate, which is tunable by the syringe pump, influences the jet velocity and the transfer rate. Faster feeding rate of spin dopes may result in wider size of fabricated TiO_2 fibers [21, 30]. The study by Jung et al. [13] showed that composite TiO_2/PVP fibers' mean diameter increased from 120 nm to 180 nm with the increment of flow rate from 0.1 mL/hr to 0.3 mL/hr. When the jet velocity is high, the volume ejected is greater and thus it caused the solution is electrospraying sequel of insufficient stretching under the constant supplied voltage [33]. High transfer rate of the jet solution cause the jet to reach the collector rapidly without proper drying time and reflex in larger fibers formation [18]. Thus, the feeding rate should provide adequate time for the solvent to evaporate before reaching collector. Moreover, it is suggested that the optimized morphology of TiO_2 composite nanofibers can be obtained at 0.70 mL/hr [16].

3.4 Heat treatment

Heat treatment is needed after the spin dopes is electrospun on the collector. The electrospun TiO2/PVP composite nanofibers are heated up to 500°C to eliminate PVP and remaining solvent which can reduce the size of fibers [13]. Although calcined at a temperature of 200 °C is enough to eliminate the polymer and solvent, the fibers may stay in amorphous form. Thus, it is suggested that the TiO_2/PVP composite nanofibers is to be sintered up to 500°C to obtain pure TiO_2 nanofibers in crystallize form. However, for sintering more than 600°C, the fibers will transform from anatase to rutile phase [11, 13, 34]. A shrinkage of 35-70% of the diameter when the fibers is annealed at the temperature of 450-500 °C [13-14, 36].

In general, X-ray Diffraction (XRD) is used to determine whether the nanofibers are in amorphous or crystallize form. However, for crystallized TiO_2 nanofibers, XRD is used to identify whether the fibers are either in anatase or rutile phase. Figure 1 shows the heat treatment of TiO_2 nanofibers at different temperature, form and different phase of TiO_2. The nanofibers were converted from amorphous to crystallize form after annealing at 550°C; beyond 400°C, TiO_2 exhibits in anatase phase then at rutile phase when is calcined at 700°C.

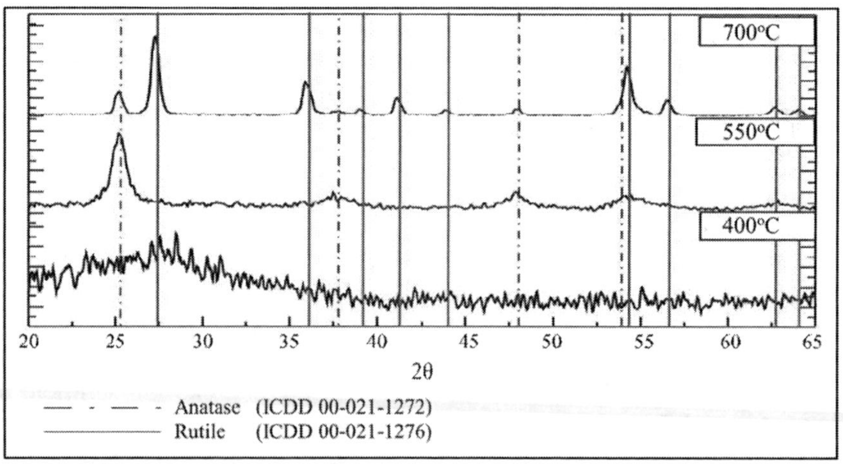

Figure 1. XRD analysis for TiO_2 nanofibers at different temperature of heat treament. TiO_2 is in amorphous form at heat treatment below 400 °C; showing crystallize form at higher temperature with 500°C showing anatase phase and 700 °C showing rutile phase [11].

4 Fibers Characterizations

4.1 Morphology analysis

The morphology of the fabricated composite nanofibers can be revealed by using scanning electron microscope (SEM). The micrograph of SEM provides the diameter of the fibers and the surface condition of the fibers. The diameter of the generated nanofibers as presented in Table 1 was determined and observed by using SEM. The electrospun TiO2/PVP composite nanofiber is straight with smooth surface without heat treatment. The TiO2 fiber is found to have rough surface due to crystallization of TiO_2 when it is calcined. When the fiber is sintered to 700°C and above, the surface structure shows some burl and to have nanocrytallize structure. This is because when the calcination temperature increases, the crystallize growth of the TiO_2 fibers is better [5]. This is further done by [11, 13, 34-35]. The surface structure of the fibers is clearly shown with lower accelerating voltage (e.g. 5-10 kV) of SEM can provide the clearer surface structure image. Meanwhile, in order to measure the diameter of the fiber, the magnification should be range 10k to 200k [11, 16, 35].

5 Conclusions

TiO_2 nanofibers can be produced by using electrospinning process. The diameter of the nanofibers is tunable by changing the parameters especially the applied voltage, feeding rate and PVP concentration. An optimized fiber was suggested to produce at 18 kV, 0.7 mL/hr with 6.5% PVP concentration. Based on the desired applications, the TiO2 nanofibers may transformed from amorphous to crystallize form at 450-500°C. It can be converted from anatase to rutile by calcination of 700 °C and above.

Acknowledgment

The authors would like to acknowledge to Universiti Malaysia Sabah for the financial support under the research grant (SBK0205-TK-2015) and Minister of Education (RACE 0006-TK-2012).

References

[1] H.J. Chen, L. Wang and W.Y. Chiu, Chelation and solvent effect on the preparation of titania colloids, Materials Chemistry and Physics, **101**, 12-19, (2007).

[2] D.S. Xu, J.M. Li, Y.X. Yu and J.J. Li, From titanates to TiO_2 nanostructures: Controllable synthesis, growth mechanism, and applications, Science China Chemistry, **55**(11), 2334-2345, (2011).

[3] H. Gao, J. Tian, H. Zhu, F. Tan and W. Zhang, Effects of Fe doping on the optical and magnetic properties of TiO_2 films deposited on Si substrates by sol-gel route, J. of Sol-Gel Science and Technology, **74**, 521-527, (2015).

[4] A.K. Prasad, F. Nose, N.C. Raut, S. Tripura Sundari, M. Kamruddin, S. Dash and A.K. Tyagi, Phase selective gas sensing properties of nanostructured TiO_2 thin films, Proc. of Int. Conf. on Nanoscience, Engineering and Technology, IEEE, 532-535, (2011).

[5] B. Ding, C.K. Kim, H.Y. Kim, M.K Seo and S.J Park, Titanium nanofibers prepared by using electrospinning method, Fibers and Polymers, **5**(2), 105-109, (2004).

[6] S. Mishra and S.P. Ahrenkiel, Synthesis and characterization of electrospun nanocomposite TiO_2 nanofibers with Ag nanoparticles for photocatalysis applications, J. of Nanomaterials, **2012**(16), 1-6, (2012).

[7] W. Nuansing, S. Ninmuang, W. Jarenboon, S. Maensiri and S. Seraphin, Structural characterization and morphology of electrospun TiO_2 nanofibers, Material Science and Engineering B, **131**, 147-155, (2006).

[8] S. Watson, D. Beydoun, J. Scott and R. Amal, Preparation of nanosized crystalline TiO_2 particles at low temperature for photocatalysis, Journal of Nanoparticle Research, **6**, 193-207, (2004).

[9] R. Chandrasekar, L. Zhang, J.Y. Howe, N.E. Hedin, Y. Zhang and H. Fong, Fabrication and characterization of electrospin titania nanofibers, J. of Materials Science, **44**, 1198-1205 (2009).

[10] B.H.Q. Dang, M. Rahman, D. MacElroy and D.P. Dowling, Conversion of amorphous TiO_2 coating into their crystalline form using a novel microwave plasma treatment, Surface and Coatings Technology, **2015**, 235-240, (2011).

[11] B. Caratao, E. Carneiro, P. Sa, B. Almeida and S. Carvalho, Properties of electrospun TiO_2 nanofibers, Journal of Nanotechnology, **2014**, 1-5, (2014).

[12] M. Mali, S. An, M. Liou, S.S. Al-Deyab and S.S. Yoon, Photoelectrochemial solar water splitting using electrospun TiO2 nanofibers, Applied Surface Science, **328**, 109-114, (2015).

[13] W.H. Jung, N.S. Kwak, T.S. Hwang and K.B. Yi, Preparation of highly porous TiO_2 nanofibers for dye-sensitized solar cells (DSSCs) by electro-spinning, Applied Surface Science, **261**, 343-352, (2012).
[14] T. Mathews, R.P. Antony, P.K. Ajikumar, S. Chakraborty, S. Dash, A.K. Tyagi, Fabrication of TiO2 nanofibers by electrospinning technique, Proc. of Int. Conf. on Nanoscience, Engineering and Technology, IEEE, 540-542, (2011)
[15] M. Boehme and W. Ensinger, Mixed phase anatase/rutile titanium dioxide nanotubes for enhanced photocatalytic degradation of methylene-blue, Nano-Micro Letters, **3**(4), 236-241, (2011).
[16] N. Sarlak, M.A.F. Nejad, S. Shakhesi and K. Shabani, Effects of electrospinning on titanium dioxide nanofibers diameter and morphology: An investigation by Box-Wilson central composite design (CCD), Chemical Engineering J., **210**, 410-416, (2010).
[17] Z.M. Huang, Y.Z. Zhang, M. Kotaki and S. Ramakrishna, A review on polymer nanofibers by electrospinning and their applications in nanocomposites, Composites Science and Technology, **63**, 2223-2253, (2003).
[18] N. Bhardwaj and S.C. Kundu, Electrospinning: A fascinating fiber fabrication technique, Biotechnology Advances, **28**, 325-347, (2010).
[19] N.M. Tikekar, J.J. Lannutti, Effects of humidity on titania-based polyvinylpyrolidone (PVP) electrospun fibers, Ceramics International **38**, 4057-4064, (2012).
[20] C. Wang, X. Zhang, C. Shao, Y. Zhang, J. Ying, P. Sun, X. Liu, H. Liu, T. Xie and D. Wang, Rutile TiO_2 nanowires on anatase TiO_2: A branched heterostructured photocatalysts via interface-assisted fabrication approach, J. of Colloid and Interface Science, **363**, 157-164, (2011).
[21] D. Li and Y. Xia, Fabrication of titania nanofibers by electrospinning, Nano Letters, **4**(3), 555-560, (2003).
[22] D. Sabba, S, Agarwala, S.S. Pramana and S. Mhaisalkar, A maskless synthesis of TiO_2-nanofiber-based hierarchical structures for solid-state dye-sensitized solar cells with improved performance, Nanoscale Research Letters, **9**(14), 1-9, (2014).
[23] R. Nirmala, H.K. Kim, R. Navamathavan, C. Yi, J.J. Won, K. Jeon, A. Yousef, R. Afeesh and M. El-Newehy, Photocatalytic activities of electrospun tin oxide doped titanium dioxide nanofibers, Ceramics Int., **38**, 4533-4540, (2012).
[24] N.A.M Barakat, A.M. Hamza, S.S. Al-Deyab, A. Qurashi and H.Y. Kim, Titanium-based polymeric electrospun nanofiber mats as a novel organic semiconductor, Materials Science and Engineering B, **177**, 34-42, (2012).
[25] J.Y. Park and I.H. Lee, Characterization and morphology of prepared titanium dioxide nanofibers by electrospinning, J. of Nanoscience and Nanotechnology, **10**, 3402-3405, (2010).
[26] H. Jamil, S.S. Batool, Z. Imran, M. Usman, M.A. Rafiq, M. Willander and M.M. Hassan, Electrospun titanium diaoxide nanofibers humidity sensors with high sensitivity, Ceramics Int., **38**, 2437-2441, (2012).
[27] X. Wang, R.A. Gittens, R. Song, R. Tannenbaum, R. Oliver-navarrete, Z. Schwartz, H. Chen and B.D. Boyan, Effects of structural properties of electrospun TiO_2 nano-fiber meshes on their osteogenic potential, Acta Biomater, **8**(2), 878-885, (2012).
[28] M.Y. Song, D.K. Kim, K.J. Ihn, S.M. Jo and D.Y. Kim, Electrospun TiO_2 electrodes for dye-sensitized solar cells, Nanotechnology, **15**, 1861-1865, (2004).
[29] P. Heikkilä and A. Harlin, Parameter study of electrospinning of polyamide-6, European Polymer J., **44**, 3067-3079, (2008).
[30] D.Y Lee, B.Y. Kim, S.J. Lee, M.H. Lee, Y.S. Song and J.Y. lee, Titania nanofibers prepared by electrospinning, J. of the Korean Physical Society, **48** (6), 1686-1690, (2006).
[31] Z.S. Tang, N. Bolong, I. Saad and F.M. Said, Effects of supplied voltage and flow rate on morphology of electrospun titanium oxide nanofibers, *Proceedings of Hong Kong International Conference of Engieering and Applied Sciences*, Hong Kong, (2013).
[32] C. Tekmen, A, Suslu and U. Cocen, Titania nanofibers prepared by electrospinning, Materials Letters, **62**, 4470-4472, (2008).

[33] S. Zargham, S. Bazgir, A. Tavakoli, A.S. Rashidi and R. Damerchely, The effect of flow rate on morphology and deposition area of electrospun nylon 6 nanofibers, J. of Engineered Fibers and Fabrics, **7**(4), 42-49, (2012).
[34] H. Park and W.M. Sigmund, Thermally controlled crystallization of electrospun TiO_2 nanofibers, Advances in Science and Technology, **71**, 80-85, (2010).
[35] J.Y. Park and I.H. Lee, Characterizations and morphology of prepared titanium oxide nanofibers by electrospinning, J. of Nanoscience and Nanotechnology, **10**, 3402-3405, (2010).
[36] L. Francis, A.S. Nair, R. Jose, S. Ramakrishna, V. Thavasi and E. Marsano, Fabrication and characterization of dye-sensitized solar cells from rutile nanofibers and nanorods, Energy, **36**, 627-632, (2011).

MATEC Web of Conferences **47**, 02001 (2016)
DOI: 10.1051/matecconf/20164702001

Impact Resistance of Recycled Aggregate Concrete with Single and Hybrid Fibers

Sallehan Ismail[1,a] and Mahyuddin Ramli[2]

[1]*Faculty of Architecture, Planning and Surveying, Universiti Teknologi Mara, 32610 Seri Iskandar, Perak, Malaysia*
[2]*School of Housing, Building and Planning, Universiti Sains Malaysia, 11800 Pulau Pinang, Malaysia*

Abstract. This paper presents a recycled aggregate concrete (RAC) mix that has been modified by adding treated recycled concrete aggregate (RCA) and various types of fiber-reinforced systems. The effectiveness of these modifications in terms of energy absorption and impact resistance was evaluated and compared with that of the corresponding regular concrete, as well as with unmodified RAC specimens. Results clearly indicate that although modification of the RAC mix with treated RCA significantly enhances the impact resistance of RAC, further diversification with additional fiber, particularly those in hybrid form, can optimize the results.

1 Introduction

Recycling of concrete waste into recycled concrete aggregate (RCA) has been identified as a potential source of alternative aggregates to produce an environmentally benign concrete [1, 2]. However, RCA cannot be simply applied in concrete production because its properties differ from those of natural aggregates. Several researchers [3-5] have found that complete or partial incorporation of RCA as a replacement for natural aggregates in recycled aggregate concrete (RAC) mixes can reduce the mechanical strength of the resulting concrete. Moreover, the presence of old cement mortar in RCA results in weakness in the interface zone between the new cement mortar and the aggregate, thereby crucially affecting the compressive strength of the RAC [6]. In recent years, many researchers have studied the behavior of RAC under static load. However, attempts to study RAC behavior under dynamic loads, such as impact loads, are seldom reported in the literature. In a research conducted by Rao et al. [7] on RAC under drop weight impact load, the impact resistance of RAC was found to decrease proportionally with the increase in RCA replacement level. Ismail and Ramli [8] observed that the impact resistance of RAC beams with treated RCA tends to improve slightly compared with that of RAC beams with untreated RCA. In practice, a concrete structure is designed with high impact resistance criteria to serve in areas subjected to high repeated impact loads, such as during an earthquake, missile impact and explosion, transportation and barrier structures subjected to vehicle crash impact, impact of water wave motions on offshore structure, rock falling and etc. Furthermore, high impact resistance concrete is essential in the production of structure piles, which are designed to sustain being repetitively subjected to impact loading during the pile driving process [9]. Understanding the behavior of concrete subjected under impact loading is important because this behavior differs from that of concrete under quasi-static loading [7].

[a] Corresponding author : salleh_itm@yahoo.com

The present study mainly aims to evaluate the contributions of various fiber-reinforced systems in enhancing the energy absorption and/or impact resistance capacities of RAC beams. The mix proportion of RAC consisted of treated coarse RCA, as developed in our previous research [8]. This mix proportion can further diversify the RAC content through the inclusion of various fiber-reinforced systems. Two different types of synthetic fibers were selected (barchip and polypropylene fibers) and proposed to be added as single and hybrid forms at the appropriate dosage in a series of RAC mixes, respectively. This work is also interested to determine the most suitable fiber composition and the optimum amount of fiber that can significantly enhance the performance of RAC. The results of this study can illuminate the potential benefits of the proposed novel method to develop a comprehensive and systematic approach for RAC production.

2 Experimental Program

2.1 Materials

The cement used in the concrete mixtures in this study was ordinary Portland Cement Type I with a specific gravity of 3.15 g/cm^3. In this study, all of the coarse aggregates used had a maximum size of 20 mm. The natural coarse aggregate used was crushed granite. The coarse RCA used was generated from waste concrete cube collected from the debris collection area. Jaw crushers were used to crush and break down concrete waste into small particle sizes. After crushing, the RCA was graded using a vibrator sieve to obtain aggregates with a maximum size of 20 mm. Two types of coarse RCA were used in this study, treated and untreated. The treated RCA is prepared by modifying the RCA surface structure through the combination of two different surface treatment methods. In this process, the RCA is initially treated by soaking in hydrochloric (HCl) acid at 0.5 molar (M) concentration. They are then impregnated with wollastonite solution. The details on the materials and procedures involved in this treatment process were described in our previous research [8]. Table 1 provides a comparison of the properties of coarse natural aggregate (granite) and coarse treated and untreated RCA in terms of physical and mechanical strength characteristics. The natural fine aggregate or sand used in this study was uncrushed quartzite natural river sand which consisted of particles that passed through a sieve size of 5.00 mm. Two synthetic fibers of different types and geometric properties were used: i) Barchip fiber type 54, which is manufactured in modified olefin synthetic fiber, was supplied by Elasto Plastic Concrete Inc. and ii) polypropylene fibers produced in fibrillated form, supplied by Timuran Engineering Sdn. Bhd. (Table 2). To enhance the workability of the concrete in this study, we used a chloride-free super plasticising admixture based on sulphonated naphthalene polymers.

Table 1. Properties of coarse aggregate [8].

Properties of Aggregate	Sizes of aggregate	Natural Granite	Untreated RCA	Treated RCA
Particle Density - Oven Dry (Mg/m^3)	20 mm	2.60	2.33	2.4
	10 mm	2.58	2.23	2.3
Water absorption (%)	20 mm	0.60	4.44	3.71
	10 mm	0.70	5.58	5.02
Agg. crushing value (%)	14mm	24.32	29.15	28.34
Agg. impact value (%)	14mm	13.98	21.78	19.26
LA abrasion value (%)	14mm	34.76	39.12	36.76

2.2 Mix design

The designated concrete mix proportions were based on a constant effective water/cement ratio of 0.41 for all concrete mixtures in accordance with the British method published by the Department of Environment [10] to achieve a target slump range of 30 mm to 60 mm and a compressive strength of 50 MPa on 28 day. Nine series of mixtures were prepared depending on the type of coarse aggregate and the varying volume fraction of the fiber content, as follows : (i) CO – normal concrete; (ii) RO and TR - refer to RAC composed of untreated and treated coarse RCA, respectively; (iii) BF and PF - RAC composed of treated coarse RCA and incorporated with single barchip and polypropylene fiber, respectively; and (iv) HB1, HB2, HB3 and HB4 - a group of RAC specimens composed of treated RCA and then incorporated with various fractions of fiber volume in hybrid combinations of barchip and polypropylene fiber.

The dosage compositions of the coarse aggregates in this experiment were kept constant by replacing the natural coarse aggregate with untreated or treated RCA at 60% of the weight of the total coarse aggregate content in all RAC mixtures. Meanwhile, the dosage designs of either individual or hybrid fibers added to the respective RAC mix were kept constant at a total volume fraction rate of 1.2% (by volume of cement). This parameter is selected on the basis of our previous research findings [11], in which a similar mix design was used to show that the inclusion of single barchip fiber at a volume fraction of 1.2% (by volume of cement) produces optimum compressive strength of RAC specimens at 28 days. The detailed mix design and proportions in the constituent materials for the overall concrete specimens are presented in Table 3.

2.3 Concrete specimen preparation and curing

All concrete mixes in this study were mixed in accordance to the sequence prescribed in BS1881-125 [12]. A drum mixer was used to prepare all the concrete mixes. For each concrete mix, 100 mm cubes were cast for the compressive strength test, 50 x 100 mm x 500 mm concrete plate were cast for impact resistance test and 100 mm x 100 mm x 500 mm prisms were cast for the flexural strength. All concrete specimens were cast in laboratory conditions, demolded at 24 h after casting, and then fully immersed in water at 25 ± 2 °C until testing age were reached.

2.4 Testing

In this study, the effects of the results of engineering properties of different series modified RAC mix specimens were analyzed and compared with the corresponding to normal aggregate concrete (CO) and also with unmodified RAC specimens (RO) to determine their effectiveness. The parameters such as compressive and flexural strength of hardened concretes were examined at the age of 7, 28, 90, 180 days. The compressive strength test was conducted in compliance with BS EN 12390-3 [13]. Flexural strength test was conducted to measure the ability of a concrete beam to resist bending and the test procedure were performed in accordance to BS EN 12390-5 [14]. An impact resistance test was conducted to gain a better understanding of the load resistance and/or total energy absorption behaviour of concrete when subjected to impact loads. For this test the specimens prepared were in the form of rectangular plates with dimensions of 50 mm × 100 mm × 500 mm. The input energy was provided by a steel ball weighing 450 g, which was released and dropped from the cylindrical steel channel at a height of 800 mm. The steel ball was dropped repeatedly till the ultimate failure of the test specimens. The detail of the instrument involve and test set up of impact resistance test should refer to the references [8]. The impact resistance of the concrete plates was determined by measuring the energy absorbed by the fracture of the specimen, which was impacted by a steel ball through a certain number of blows that cause its ultimate failure. The impact energy absorbed by specimens was calculated from the area under the impact load-deflection curve [15, 16] and was analyzed, moment by moment, after the steel ball was dropped.

Table 2. Specification of barchip and polypropylene fibers.

Item	Specifications of fibers	
	Barchip	Polypropylene
Average Length	27±2mm	15mm
Tensile Strength	640 MPa	45-60 ksi
Average Diameter	0.52 mm	-
Surface Texture	Continuously Embossed	-
No. Fibers per kg	37,000	-
Specific Gravity	0.92 g/cm^3	0.9 g/cm^3
Youngs Modulus	10 GPa	0.5 x 10 ksi
Melting Point	159°C - 179°C	160°C - 170°C
Ignition Point	Greater than 450°C	590°C

Table 3. Mix proportions.

Specimen	Cement kg/m^3	Water kg/m^3	Coarse aggregate kg/m^3		Sand kg/m^3	SP (%)	Vol. fraction of fiber (%)		Total vol. fraction of fiber %
			Gravel	RCA			Barchip	Polypropylene	
CO	512	210	956	-	722	0	-	-	-
RO	512	210	382	574	722	0	-	-	-
TR	512	210	382	574	722	0.2	-	-	-
BF	512	210	382	574	722	0.3	1.2	-	1.2
PF	512	210	382	574	722	0.3	-	1.2	1.2
HB1	512	210	382	574	722	0.3	0.96	0.24	1.2
HB2	512	210	382	574	722	0.3	0.72	0.48	1.2
HB3	512	210	382	574	722	0.3	0.48	0.72	1.2
HB4	512	210	382	574	722	0.3	0.24	0.96	1.2

3 Results and Discussions

3.1 The mechanical properties of modified RAC

The mechanical properties namely bulk density, compressive strength and flexural strength of normal concrete, unmodified and modified recycled aggregate concrete are presented in Table 4. Among all the concrete specimens, the reference concrete (CO) produces the greatest density. Its average bulk densities fall within 2400 kg/m^3, which is considered to meet the mix design target satisfactorily. By contrast, the inclusion of the RCA at 60% replacement of natural coarse aggregate results in the lowering of the density of concrete. The lower density of concrete mixes that contain the RCA is caused by the lower particle density of the RCA than the natural coarse aggregate. The results indicate that effect of the modification on the RAC mix by including the treated coarse RCA is significant in enhancing the engineering properties of RAC. However, further extending the diversity of the production RAC mix by adding fiber, particularly in its hybrid form, can optimize the results. As the results show, the optimum compressive strength is obtained by modified RAC in types HB4, whose produce the highest compressive strength performance compared with those of other specimens, particularly at the long-term curing ages. The results also confirm that modifying RAC using treated RCA and inclusion of single and hybrid fiber found significantly the contribute to the gain in optimum flexural strength. The gain in flexural strength of the specimens with hybrid fiber under the respective curing age is not as obvious as that of the single-fiber specimens. However, using hybrid fiber is found to be beneficial for resisting the flexural load when specimens are exposed to a longer curing period. This result may suggest that the dispersion of different types of fiber strengthens the bond between fibers, and with the cement matrix occurrence, the RAC specimens benefit by reinforcing the effects of hybridisation that can resist and bridge the cracks [17, 18].

Table 4. Mechanical properties of modified RAC.

Specimen	Density at 28 days (kg/m^3)	Compressive strength (MPa) Age at testing (days)				Flexural strength (MPa) Age at testing (days)			
		7	28	90	180	7	28	90	180
CO	2400	45.13	54.54	58.54	62.34	3.86	5.3	5.91	6.27
RO	2346	37.31	46.98	52.24	57.37	3.73	4.65	5.31	5.45
TR	2370	46.61	52.18	57.8	60.98	3.69	4.83	5.75	5.97
BF	2356	48.82	56.13	63.61	66.64	4.92	5.48	6.47	6.71
PF	2352	49.45	52.02	58.89	68.45	4.56	5.38	5.84	6.33
HB1	2354	46.15	53.83	65.6	69.65	5.18	5.35	5.56	6.49
HB2	2355	49.69	53.76	61.28	66.01	4.90	5.25	5.62	6.23
HB3	2354	48.84	52.33	64.42	65.41	4.87	5.04	5.52	6.53
HB4	2348	46.86	53.22	69.28	69.90	4.58	5.05	5.79	6.25

3.2 Impact resistance

The response of all specimens subjected to low-velocity impact loading at ages of 28, 90, and 180 days is plotted in Figures 1. It can seen that the impact energy absorbed by respective specimens at the first crack and at ultimate failure is varied. Investigation of the impact resistance performance of specimens at 28 days shows that the RO specimens present minimal results, followed by the TR specimens. By contrast, the PF specimens demonstrate the highest impact resistance which is approximately 56% higher than that of the CO specimens. These are followed by HB1, H2, BF, HB4, and HB3, which have total impact energy values of approximately 45%, 42%, 34%, 30%, and 29%, respectively, higher than those of the CO specimens. The prolonged curing exposure generally results in an increase in the impact resistance of all specimens. However it could notice that at 90 and 180 days, the impact resistance of the RO specimens on both testing ages is still the lowest, whereas that of the TR specimens is slightly higher than that of the RO specimens. As expected, the impact resistance of the RAC specimens benefited from the reinforcing effect of fiber. This finding is confirmed by the better rate gain in impact energy of the plain concrete at first cracks. Ultimate failure was observed in all RAC specimens that consist of fibers at 90 and 180 days. Additionally, the data plotted in Figure 2 reveal that the maximum gain in total energy is highly present in hybrid fiber specimens, which are HB2 specimens for testing age of 90 days and HB1 specimens for testing age of 180 days.

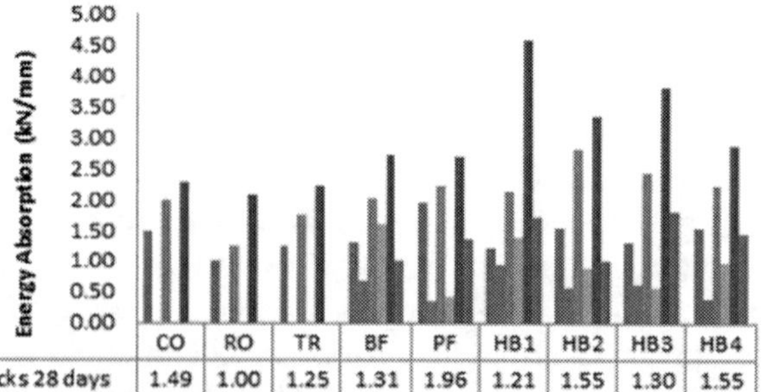

	CO	RO	TR	BF	PF	HB1	HB2	HB3	HB4
Pre cracks 28 days	1.49	1.00	1.25	1.31	1.96	1.21	1.55	1.30	1.55
Post cracks 28 days	0.00	0.00	0.00	0.68	0.36	0.95	0.56	0.62	0.38
Pre cracks 90 days	1.99	1.25	1.75	2.02	2.21	2.13	2.80	2.42	2.21
Post cracks 90 days	0.00	0.00	0.00	1.60	0.41	1.40	0.89	0.56	0.97
Pre cracks 180 days	2.29	2.08	2.22	2.71	2.70	4.58	3.35	3.82	2.87
Post cracks 180 days	0.00	0.00	0.00	1.01	1.37	1.72	1.02	1.81	1.44

Figure 1. Pre-crack and post-crack energy absorption of all specimens exposed at various testing ages.

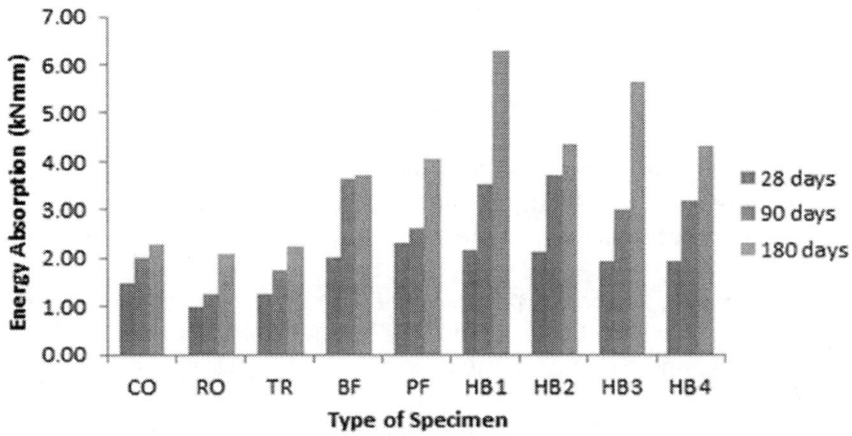

Figure 2. Total energy absorption of all specimens at various age.

Overall, the stiffness of the material is significantly related to the impact resistance of concrete [7, 15] because coarse aggregates are considered dominant constituents of materials during the formation of the final concrete structure. As shown in Table 1, the impact value of the RCA is less than that of natural coarse aggregates. Thus, RCA contributes to reducing the mechanical strength and modulus of elasticity of concrete. In addition, the weak interface between the aggregates and the cement paste may contribute to fracture propagation in the concrete [7]. By contrast, the resistance of concrete to crack propagation under an impact force is enhanced, as indicated in the TR concrete specimens. This phenomenon could be attributed to the improvement of the properties of the RCA and the modification that strengthened the interface between the aggregate–paste bond as a result of surface treatment.

Meanwhile, the most significant property required by a material subjected to impact loading is its energy-absorbing capability. The inclusion of randomly distributed fibers can enhance the toughness and ductility of concrete [19, 20]. Accordingly, based on the data results, the impact resistance of RAC can substantially increase with the addition of randomly distributed fibers. The results suggest that the inclusion of single and/or hybrid fibers into the mixtures results in a conclusive increase of the capability of impact energy absorbed by specimens at first and ultimate cracking. By contrast, in the case of non-fibrous specimens (CO, RO, and TR), the improvement in post-crack resistance was negligible. As observed during testing, the ultimate failure on most of the non-fibrous specimens simultaneously occurred once the first visible crack formed during testing. This observation suggests that the fracture of non-fibrous concretes may suddenly occur because of its brittle behavior. Incorporating different volume ratios of barchip and polypropylene fibers in hybridization formation, such as into the HB1 specimens, is significant in increasing the ductility performance. The capability of fibrous specimens to absorb energy prior to impact loading suggests that the presence of fiber function delays the ultimate failure [9, 19, 20].

4 Summary

The inclusion of treated RCA rather than untreated RCA in RAC specimens may enhance impact resistance. However, the RAC mix diversified with incorporated fibers exhibits higher impact resistance than the non-fibrous specimens. Among the specimens, the total energy absorbed by the modified RAC mix in type HB1 specimens reveals that they exclusively attained the highest value, and these values are estimated to be nearly three times more than those of the CO specimens, particularly in the later testing period after prolonged curing exposure.

References

[1] A. Rao, K.N. Jha and S. Misra, Use of aggregates from recycled construction and demolition waste in concrete, Resources, Conservation and Recycling, **50**(1), 71-81, (2007).

[2] S. Ismail, K.W. Hoe and M. Ramli, Sustainable Aggregates: The potential and challenge for natural resources conservation, Procedia-Social and Behavioral Sciences, **101**, 100-109, (2013).

[3] S. Ismail, K.W. Hoe and M. Ramli, Influence of amount of recycled coarse aggregates and production process on properties of recycled aggregate concrete, Cement and Concrete Research, **37**(5), 735-742, (2007).

[4] S. Ismail and M. Ramli, Engineering properties of treated recycled concrete aggregate (RCA) for structural applications, Construction and Building Materials, **44**, 464-476, (2013).

[5] Ö. Arıöz, A. Özsoy and G. Yılmaz, Concrete with recycled aggregate, Key Engineering Materials, **264-268**, 2145-2148, (2004).

[6] V.W.Y. Tam, X.F. Gao and C.M. Tam, Microstructural analysis of recycled aggregate concrete produced from two-stage mixing approach, Cement and Concrete Research, **35**(6), 1195-1203, (2005).

[7] R.M. Chakradhara, S.K. Bhattacharyya and S.V. Barai, Behaviour of recycled aggregate concrete under drop weight impact load, Construction and Building Materials, **25**(1), 69-80, (2011).

[8] S. Ismail and M. Ramli, Mechanical strength and drying shrinkage properties of concrete containing treated coarse recycled concrete aggregates, Construction and Building Materials, **68**, 726-739, (2014).

[9] W.H. Kwan, M. Ramli and C.B. Cheah, Flexural strength and impact resistance study of fibre reinforced concrete in simulated aggressive environment, Construction and Building Materials, **63**(1), 62-71, (2014).

[10] D.C. Teychenne, R.E. Franklin, H.C. Erntroy, D.W. Hobbs and B.K. Marsh, *Design Of Normal Concrete Mixes*, *Building Research Establishment*, Watford, United Kingdom, (1997).

[11] S. Ismail and M. Ramli, Effects of adding fibre on strength and permeability of recycled aggregate concrete containing treated coarse RCA, Int. J. of Civil, Architectural, Structural and Construction Engineering, **8**(8), 890-896, (2014).

[12] BS 1881-Part 125, *Testing Concrete, Methods for Mixing and Sampling Fresh Concrete in The Laboratory*, British Standards Institution, London, United Kingdom, (1986).

[13] BS 1881-Part 125, *Testing Hardened Concrete, Compressive Strength of The Test Specimens*, British Standards Institution, London, United Kingdom, (2009).

[14] BS 1881-Part 125, *Testing Hardened Concrete, Flexural Strength of The Test Specimens*, British Standards Institution, London, United Kingdom, (2009).

[15] M.M. Al-Tayeb, B.H. Abu Bakar, H. Ismail and H.M. Akil, Effect of partial replacement of sand by recycled fine crumb rubber on the performance of hybrid rubberized-normal concrete under impact load: experiment and simulation, J. of Cleaner Production, **59**, 284-289, (2013).

[16] N. Banthia, P. Gupta and C. Yan, Impact resistance of fiber reinforced wet-mix shotcrete part 1: beam tests, Materials and Structures, **32**(8), 563-570, (1999).

[17] C.X. Qian and P. Stroeven, Development of hybrid polypropylene-steel fibre-reinforced concrete, Cement and Concrete Research, **30**(1), 63-69, (2000).

[18] K. Ramadevi and D.L.V. Babu, Flexural behaviour of hybrid (steel-polypropylene) fibre reinforced concrete beams, European J. of Scientific Research, **70**(1), 81-87, (2012).

[19] M. Nili and V. Afroughsabet, The effects of silica fume and polypropylene fibers on the impact resistance and mechanical properties of concrete, Construction and Building Materials, **24**(6), 927-933, (2010).

[20] S. Erdem, A.R. Dawson and N.H. Thom, Microstructure-linked strength properties and impact response of conventional and recycled concrete reinforced with steel and synthetic macro fibres, Construction and Building Materials, **25**(10), 4025-4036, (2011).

MATEC Web of Conferences **47**, 02002 (2016)
DOI: 10.1051/matecconf/20164702002

Behavior of Reinforced Concrete Box Beam Strengthened with CFRP U-Wrap Strips Under Torsion

Shengqiang Ma[1], N. Muhamad Bunnori [1,a] and K. K. Choong[1]

[1]*School of Civil Engineering, Universiti Sains Malaysia, 14300 Nibong Tebal, Pulau Pinang, Malaysia*

Abstract. The present study focuses on the torsional strengthening behavior of reinforced concrete (RC) box section beams that are widely used in bridges. Four RC box beams were fabricated, and three of them were wrapped by carbon fiber-reinforced polymer (CFRP) U-wrap strips with or without longitudinal strips. The different wrapping configuration, cracking angle, failure pattern, and tensile strain of fibers were investigated and discussed accordingly. The experimental results addressed that U-wrap strips strengthening also can upgrade the ultimate torque of beams moderately. In particular, using U-wrap and longitudinal strips to bond the box beams increased the torsional stiffness slightly. The same equation from different codes for calculating RC specimens can accurately predict the ultimate strength of the control beam, but the calculation of the *fib* model overestimated the torsional strengthening improvement of the wrapped specimens. However, Ghobarah et al. assumed approximately 3000με of the average ultimate fiber strain in calculating the ultimate strength of the wrapped box beams which shows in relatively appropriate agreement with testing results.

1 Introduction

Many structures that have been utilized for years are suffering from accelerated loads, decreased capacity through environmental deterioration and aging, and seismic damage. Strengthening and rehabilitation of the existing structural members would be a cost effectively way with respect to building new structures. Since the past two decades, the materials of fiber-reinforced polymer (FRP) have gradually replaced steel plates to be used in retrofitting reinforced concrete (RC) members. As strengthening materials, the merits of FRP are distinct that they have high tensile strength, lightweight, corrosion immunity, being shaped easily, and high fatigue strength. Although FRP strengthening technique already has been used in upgrading the strength of structures such as buildings and bridges [1], more research work still exists to be studied for further comprehending the behavior of strengthened structures.

Most researchers have paid much attention to flexural and shear strengthening beams bonded with FRP by experimental and numerical studies[2-4], and the investigated achievements have made significant progress. However, few types of research were involved in the torsional behavior of the beams strengthened with FRP. From the present century, Zhang et al. [5] and Ghobarah et al. [6] conducted preliminary and exploratory studies on the torsional strengthening behavior of RC members bonded with FRP. The two experimental investigations showed the ductility and torsional strength of

[a] Corresponding author : cenorazura@usm.my

the strengthened RC rectangular specimens with FRP increased reasonably. The researchers [7-9] did not only perform a torsional test on rectangular beams strengthened with carbon fiber-reinforced polymer (CFRP) and glass fiber-reinforced polymer GFRP but also proposed an analytical method for calculating ultimate torque of strengthened RC specimens. The experimental study proved that the ductility of all strengthened torsional specimens increased, but the level of increment based on the configuration of FRP. The authors suggested that compressive strength of confined concrete should be applied in the proposed model to calculate the ultimate torsional strength more accurately. Hii and Al-Mahaidi [10-12] focused on the experimental and numerical studies of the box and solid cross section RC beams bonded with FRP; the testing results showed an increase in percentage of solid section beam was more than that of the box section beam. Chalioris [13, 14] conducted the experimental study on rectangular and T-beams strengthened with CFRP subjected to torsion; the testing investigation displayed that complete wrapping with continuous sheets was more efficient than the same volume of strips in the torsional strengthening rectangular members, and a limited increase in torsional strength appeared in the T-beams with U-wrap. Thus, the author suggested that rational anchor should be used in T-beams with U-wrap for improving the strengthening effectiveness. In practical projects, it is impossible to access the entire cross section of a beam by fully wrapping due to the presence of the surface of the floor slab or bridge deck [6, 15].

Box section beams are widely used in bridges that are also suffering the problems mentioned above and need to be retrofitted. Few studies have been paid attention to the torsional behavior of RC box beams strengthened with FRP. Therefore, in the current experimental study, CFRP U-wrap strips with or without longitudinal strips were applied in torsional strengthening RC box beams due to inaccessibility of the top face of beam; the objective is to investigate and to evaluate effectiveness of strengthening configuration, failure patterns, twist angle, and fiber strain of the retrofitted box beams .

2 Experimental Programs

Four scaled RC box beams with 350 mm x 400 mm hollow cross-section and 2,000 mm length were cast for a torsional test. Three specimens were externally bonded with CFRP and one specimen was a control beam.

2.1 Materials

The concrete compressive strength obtained by testing cubes cured for 28 days is shown in Table 1. The yield strength of stirrups R6 and R8 was 372 and 472 MPa, respectively. The yield strength of the top longitudinal bar R10 and bottom bar T16 was 466 and 547 MPa. CFRP had a tensile strength of 4,900 MPa, modulus of elasticity of 230 GPa, the thickness of 0.117 mm, and an ultimate elongation of 1.8%.

2.2 Test setup

Special equipment was designed in the torsional experiment. Two fabricated steel end supports were installed on two steel walls, separately. One end of the specimen was fixed in a supportive collar to avoid longitudinal, transverse, and rotational actions. Another supportive collar was designed to twist freely through a rotational shaft whose base was fixed on the steel wall. A rotational cell of 400 kN capacity hydraulic actuator was placed to load on the top of the rotational support of 150 mm perpendicular distance from the loading position to the horizontal center line of the box beam.

2.3 Specimen characteristics

The box beam was 2,000 mm long, 400 mm wide and 350 mm high. The width of the two sides' web and the top thickness were all 50 mm, and the thickness of the bottom was 60 mm. The longitudinal reinforcement bars 4T16 and 3R10 were applied on the bottom and top flange, separately. Apart from the 150 mm range of two ends, 6 mm diameter internal steel stirrups were placed to the whole span at a center to center with the spacing of 100 mm. Two tips were designed to be reinforced with the 8 mm diameter internal transverse steel with a center to center with spacing of 50mm to avoid two supportive ends failure prematurely. The details on the dimensions and reinforcement arrangement of the box beams are presented in Figure 1.

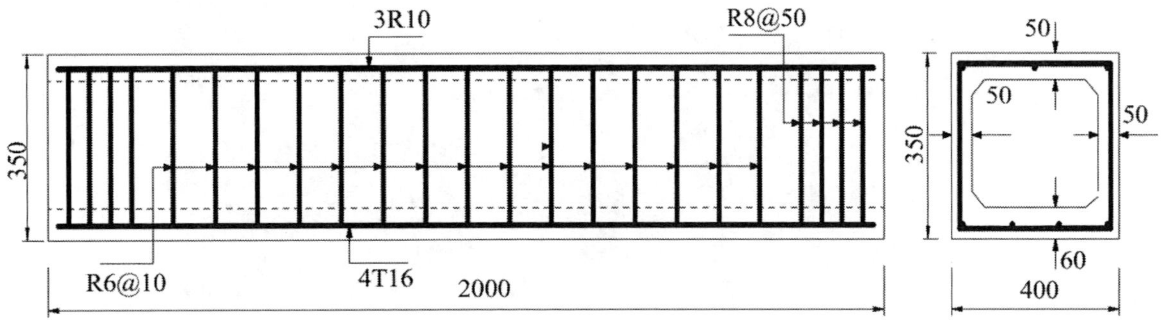

Figure 1. Details on the dimensions and reinforcement arrangement of the box beam: (a) Longitudinal section and (b) Cross section (all dimensions in mm).

2.4 CFRP strengthening configuration

Two webs and the bottom of the beams were strengthened by 100 mm CFRP U-jacketing strips at a center to center of 200 mm spacing along the length of the beams. TBS was only strengthened by one-layer of U-jacketing strips. TBSL1 and TBSL2, however, were wrapped with one and two-layer U-jacketing strips with one-layer longitudinal strips that were used to suppress the ends of U-jacketing to supply nondestructive anchorage. CFRP strengthening configuration is illustrated in Figure 2.

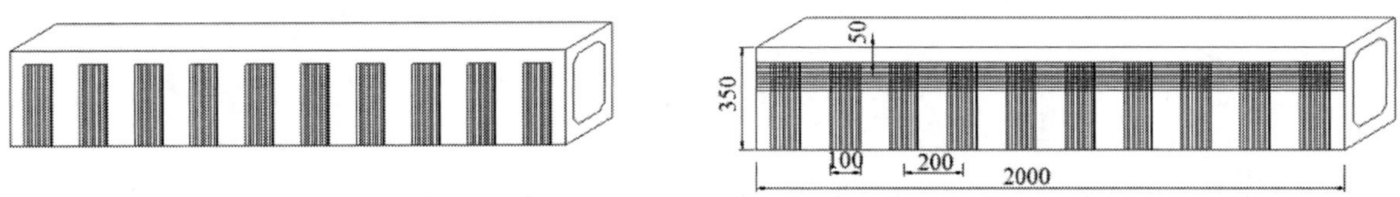

Figure 2. Schematic of the strengthened box beams: (a) wrapped in one-layer strips (TBS), (b) wrapped in one-layer strips with one-layer longitudinal strip (TBSL1) and (c) wrapped in two-layer strips with one-layer longitudinal strip (TBSL2) (all dimensions in mm).

2.5 Instrumentation

The setup of the test instrumentation is presented in Figure 3. A load cell was placed on the top of the rotational end, and the applied load provided by a hydraulic actuator was recorded by a computer-controlled data acquisition system at approximately 5.0 kN loading intervals loaded manually. Two linear variable displacement transducers (LVDTs) were installed at the bottom of the rotational end to calculate the twist angle at every loading level. Out of the five strain gauges, two separate ones were located on the surfaces of CFRP 700 mm and 1200 mm of distance from the rotational end, whereas the others were placed in the same longitudinal location at 900 mm of distance from the rotational end and distributed vertically.

Figure 3. Photograph of the instrument.

3 Experimental Results

The cracking torque, ultimate torque, and twist angle at cracking and ultimate load for all the tested beams are illustrated in Table 1.

Table 1(a). Cracking, ultimate torque, and twist angle of box beams.

Beam	Concrete compressive strength (MPa)	Cracking torque T_{cr} (kN.m)	Increase in T_{cr} (%)	Twist angle at cracking torque (°)	Ultimate torque T_u (kN.m)	Increase in T_u (%)
TBC	38.87	10.29		0.25	22.88	-
TBS	40.37	10.25	-0.4	0.4	26.67	16.6
TBSL1	35.03	12.53	21.7	0.45	26.52	15.9
TBSL2	37.57	12	16.6	0.26	27.56	20.5

Table 1(b). Cracking, ultimate torque, and twist angle of box beams.

Beam	Twist angle at ultimate torque (°)	Increase in torsional stiffness (%)	Calculate ultimate torque T_{uc} (kN.m)	$T_u/T_{u\ cal}$ (*fib*)	$T_u/T_{u\ cal}$ (Ghobarah)
TBC	2.47	-	22.7	1.01	-
TBS	2.85	1	35.99	0.735	0.940
TBSL1	2.71	5.6	35.6	0.745	0.934
TBSL2	2.83	5.1	40.6	0.678	0.810

The curve of the torsional moment versus twist angle of all box beams is drawn in Figure 4. As expected, the ultimate strength of strengthened box beams increases noticeably compared to the control beam TBC. Among all the strengthened beams, TBSL2 wrapped by two-layer U-wrap strips with one-layer longitudinal strip increasing 20.5% has the best strengthening effectiveness at ultimate strength. The maximum increases in cracking torque, 21.7% and 16.6%, are for TBSL1 and TBSL2, which have longitudinal strips bonding over the transverse strips to resist cracks generating. The TBS only with U-wrap strips, however, showed the poorest value at cracking toque. It is an interesting point that the ultimate torque and ultimate twist angle of TBSL1 is slightly lower than that of TBS, which may be explained that the compressive strength of concrete of TBS is higher than the value of TBSL1. It is evidently found that the rise of twist angles is from all strengthened beams, and the torsional stiffness of strengthened box beams showed slight increase, 1% for TBC, 5.6% for TBSL1, and 5.1% for TBSL2.

From the results of the torsional stiffness, CFRP longitudinal strips suppressing transverse strips played a particular action in improving the torsional stiffness of box beams corresponding ultimate torque increasing 15.9% for TBSL1 and 20.9% for TBSL2, respectively. Although TBSL2 with two-layer strips has better strengthening behavior than TBSL1 with the one-layer strip, there is no proportional relationship between the layers and the ultimate strength. It can be seen from Figure 4 that the trend of the torque-twist angle of all beams has not much difference before cracking. In the post-cracking phase, all the curves had a gentle slope to reach the peak torque of beams due to stirrup or stirrup with CFRP strips resisting the torque loaded on the beams. Due to torsional stiffness without remarkable increase, the loading would stop immediately after reaching the peak torque.

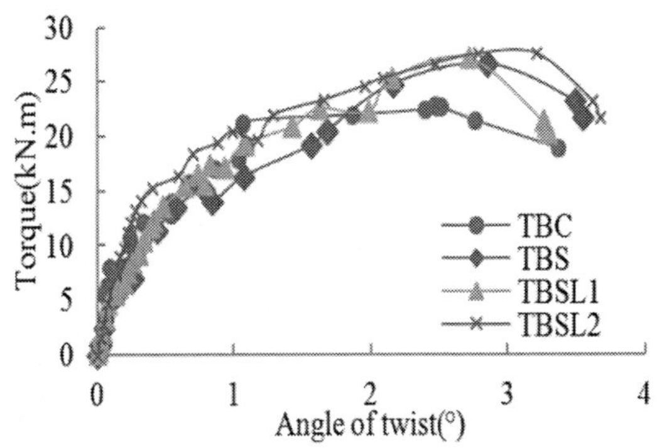

Figure 4. Curve of torque versus twist angle.

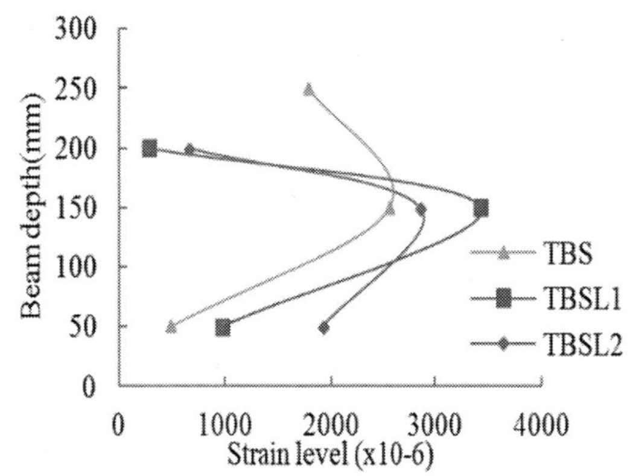

Figure 5. Strain development along depth of beam at ultimate torque.

From Figure 6, the control box beam TBC exhibited typical torsional failure pattern with one widened the inclined spiral crack. Three bonded beams performed different failure patterns. At a high load level, failure of TBS occurred through debonding of the CFRP strips with a wide crack in one strip. To TBSL1 and TBSL2 with CFRP transverse and longitudinal strips, the failure mechanism showed the longitudinal strip rupture that provided physical anchorage to prevent transverse strips debonding. From the investigation in experiment, the primary cracking angles of all beams fluctuated around 45° with respect to the longitudinal axis of beams whereas all the angles varied between 35° and 55°, which is in line with the angle of diagonal crack assumption in *fib* [16]. The distribution of many minor cracks in the unwrapped spacing of strengthened beams was more uniform than that of control beam failed in one main crack. CFRP strips helped distribute the stress along the whole beam of concrete.

Strain gauges were stuck on strips along the depth of each beam to get the strain condition. The even strain of each strengthened box beam versus beam depth is plotted in Figure 5. The general trend of three curves is like a parabola and the maximum strain is less than 3500με at a point near the neutral axis of the beam. The strain of two beams strengthened with CFRP strips and longitudinal strips is higher than TBS only strengthened with transverse strips due to longitudinal strips confining. It also can be discovered from Figure 6 that the strain of TBSL2 is lower than that of TBSL1 despite the same strengthening configuration with different transverse strips plies. It may be considered that rigidity of two layers is greater than that of one layer at approximately the same stress which can be concluded from the ultimate torque.

Figure 6. Failure patterns of the specimens: (a) TBC, (b) TBS, and (c) TBSL1 and TBSL2.

4 Analysis

fib model [16] is an available model in predicting the ultimate torque of FRP contribution, which was calculated as follows:

$$T_f = E_f \varepsilon_{fe} t_f w_f bh \cot(\theta) / s_f \tag{1}$$

$$\varepsilon_{fe} = \min\left[0.17\left(\frac{f_c^{2/3}}{E_f \rho_f}\right)^{0.3} \varepsilon_{fu}, 0.65\left(\frac{f_c^{2/3}}{E_f \rho_f}\right)^{0.56} \times 10^{-3}\right] \tag{2}$$

where E_f is Young's modulus of the FRP; ε_{fe} is the effective strain in the fiber; t_f is the thickness of FRP; S_f is the spacing of FRP strips; w_f is the width of FRP; b and h are the width and height of beam; θ is the diagonal angle of crack; f_c is the concrete compressive strength; $\rho_f = 2t_f w_f / bS_f$ which is FRP reinforcement ratio. Assuming there is no interaction between the internal stirrup and the external FRP strips, the total torque (T_u) of the strengthened RC beam can be calculated by: $T_u = T_{uc} + T_f$. Equation (3) of the ultimate torque is adopted in the ACI318 [17], Australian code [18], and BS8110-2 [19] detailed as:

$$T_{uc} = 2A_0 A_t f_{yt} \cot\theta / s_t \tag{3}$$

where A_0 is the gross area enclosed by shear flow path and is taken as $0.85A_{0h}$, where A_{0h} is the area enclosed inside the hoops; A_t is the area of transverse reinforcement; f_{yv} is the yield strength of transverse reinforcement; S_t is the spacing of stirrups.

The computed ultimate torque using *fib* model [16] was compared to the experimental results as shown in Table.1. In the calculation θ was taken at 45°not only by the angle measured in the experiment but also with the angle provided by *fib* model[16]. From the table, the adopted RC beam model can accurately predict the ultimate torque of the control beam; but the addition of Equation (1) and Equation (3) over-estimated the ultimate torque of the strengthened beams because the calculated strain of fiber was higher than the experimental results. Ghobarah et al. [6] considered that the average ultimate fiber strain can be assumed approximately 3000με; thus for U-jacketing wrapping

$$T_f = 0.003 E_f t_f w_f bh / s_f \tag{4}$$

The ultimate torsional moment of the strengthened beams calculated by Equation (4) shows in relatively good agreement with testing results which is also listed in Table 1.

5 Summary

Three box beams strengthened by CFRP U-wrap strips with or without longitudinal strips were investigated that the ultimate torque increased moderately with respect to the control beam. The increasing percentage of the ultimate torque was 16.6% for TBS, 15.9% for TBSL1, and 20.5% for TBSL2. The beam strengthened with two-layer U-wrap strips and longitudinal strips showed the best improvement. Among of three strengthened beams, TBSL1 and TBSL2 with longitudinal strips showed slightly increase in the torsional stiffness of box beams and also increase in the cracking torque. The two typical failure patterns were observed in the experiment. At a high load level, failure of beam strengthened U-wrap strips was debonding of the strips with a wide crack, but the failure mechanism of beams strengthened with transverse and longitudinal strips showed the longitudinal strip rupture that may provide physical anchorage to avoid the debonding of the transverse strips. The experimental results of diagonal cracks also displayed that the angles mostly fluctuated at around the angle of 45° with the longitudinal axis of beams thus validating the value proposed in the *fib* model[16]. The maximum strain of fiber was less than 3500 $\mu\varepsilon$ and located around the neutral axis of the beam. The ultimate torque of the wrapped beams computed by the *fib* model is higher than the experimental results. However, the computational values using the model proposed by Ghobarah et al. [6] are considerably close to the experimental results.

References

[1] T. Gosbell and R. Meggs, West Gate bridge approach spans FRP strengthening Melbourne, Australia, *IABSE Symposium Report*, International Association for Bridge and Structural Engineering, 55-60, (2002).

[2] T.C Triantafillou and C.P. Antonopoulos, Design of concrete flexural members strengthened in shear with FRP, J. of Composites for Construction, **4**(4), 198-205, (2000).

[3] D. Kachlakev and D.D. McCurry, Behavior of full-scale reinforced concrete beams retrofitted for shear and flexural with FRP laminates, Composites Part B: Engineering, **31**(6-7), 445-452, (2000).

[4] J. Dong, Q. Wang and Z. Guan, Structural behaviour of RC beams with external flexural and flexural–shear strengthening by FRP sheets, Composites Part B: Engineering, **44**(1), 604-612, (2013).

[5] J.W. Zhang, Z.T. Lu and H. Zhu, Experimental study on the behaviour of RC torsional members externally bonded with CFRP, *Proc. of the Int. Conf. on FRP Composites in Civil Engineering*, **1**, 713-722, (2001).

[6] A. Ghobarah, M. Ghorbel and S. Chidiac, Upgrading torsional resistance of reinforced concrete beams using fiber-reinforced polymer, J. of Composites for Construction, **6**(4), 257-263, (2002).

[7] M. Ameli, H.R. Ronagh, M.A. Bradforf and B. Uy, Numerical predictions for the ultimate torque capacity of FRP strengthened reinforced concrete beams, *Proc. of FRP Composites in Civil Engineering*, Balkema, 579-585, (2005).

[8] M. Ameli and H.R. Ronagh, Analytical method for evaluating ultimate torque of FRP strengthened reinforced concrete beams, J. of Composites for Construction, **11**, 384-390, (2007).

[9] M. Ameli, H.R. Ronagh and P.F. Dux, Behavior of FRP strengthened reinforced concrete beams under torsion, J. of Composites for Construction, **11**, 192-200, (2007).

[10] A.K. Hii and R. Al-Mahaidi, Experimental investigation on torsional behavior of solid and box-section RC beams strengthened with CFRP using photogrammetry, J. of Composites for Construction, **10**, 321-329, (2006).

[11] R. Al-Mahaidi and A.K. Hii, Bond behaviour of CFRP reinforcement for torsional strengthening of solid and box-section RC beams, Composites Part B: Engineering, **38**, 720-731, (2007).

[12] A.K. Hii and R. Al-Mahaidi, Torsional capacity of CFRP strengthened reinforced concrete beams, J. of Composites for Construction, **11**, 71-80, (2007).

[13] C.E. Chalioris, Behavioural model of FRP strengthened reinforced concrete beams under torsion, *Asia-Pacific Conf. on FRP in Structures*, Hong Kong, (2007).
[14] C.E Chalioris, Torsional strengthening of rectangular and flanged beams using carbon fibre-reinforced-polymers-experimental study, Construction and Building Materials, **22**(1), 21-29, (2008).
[15] A. Deifalla and A. Ghobarah, Strengthening RC T-beams subjected to combined torsion and shear using FRP fabrics-experimental study, J. of Composites for Construction, **14**, 301-311, (2010).
[16] fib Bulletin 14, *Externally Bonded FRP Reinforcement for RC Structures*, Lausanne, (2001).
[17] ACI 318-08, *Building Code Requirements for Structural Concrete and Commentary*, American Concrete Institute, Michigan, (2010).
[18] Australian Standard, *Concrete Structures*, Standards Association of Australia, Sydney, (2009).
[19] BS 8110, *Structural Use of Concrete Part 2, Code of Practice for Special Circumstances*, British Standards Institution, London, (1985).

MATEC Web of Conferences 47, 02003 (2016)
DOI: 10.1051/matecconf/20164702003

Experimental Strength of Single-Lap Hybrid Joints on Woven Fabric Kenaf Fiber Composites Under Quasi Static Condition

Lee Sim Yee[1,a] and Hilton Ahmad[2]

[1]Faculty of Civil and Environmental Engineering, Universiti Tun Hussein Onn Malaysia, 86400 Parit Raja, Johor, Malaysia
[2]Jamilus Research Center, Universiti Tun Hussein Onn Malaysia, 86400 Parit Raja, Johor, Malaysia

Abstract. For the past decades, usage of natural fiber reinforced composites in low bearing load applications are increasing tremendously due to drawbacks concerning the use of synthetic fibers. Kenaf fibers have a good potential to be used as composite reinforcements as they possesses excellent fiber strength compared to own self-weight. Current work concentrates on mechanical properties of woven fabric kenaf composites with single-lap hybrid joints configurations. Four width to diameter ratio, (W/d) of cross-ply lay-up joints as designed in testing series were tested by using quasi static mechanical testing. Experimental results showed that the failure load increased with the increasing of W/d ratios. Thinner lay-up had better bearing strength compared to thicker lay-up as found in current study.

1 Introduction

Awareness in shortage of non-renewable resources is increasing greatly worldwide and concerns of implementing renewable resources are being developed by many sectors. Compared to synthetic fibers counterparts (e.g. glass fiber reinforced polymer (GFRP) and carbon fiber reinforced polymer (CFRP), natural fiber has superior properties such as higher specific strength [1], biodegradable, low carbon dioxide (CO_2) emissions, environment-friendly, good thermal and acoustic insulation. However, the mechanical properties for natural fibers vary depending on fiber types, region of origin and plantation environment.

Kenaf fiber is an important source of natural fiber as reinforcing substrate in composite materials which demonstrates good specific strength. Compression molded Kenaf-Polypropylene (PP) composites had shown greater tensile strength and flexural strength compared to other natural fibers. Kenaf-PP had lower elongation at break, thus, providing higher value of tensile strength which is about 930 MPa [2].

Hybrid bolted-bonded joints are the combination of mechanically fastened and adhesively bonded joints. Generally, hybrid joint technique has better performance than mechanical fastening and adhesive bonded joint alone. Attaching the mechanical fastening (bolts/rivets) with adhesive bonded joint, may decrease the maximum stress near edges region which primarily decrease the peeling stresses and increase the static strength [3]. There is no standard testing method available as design

[a] Corresponding author : simsim1019@hotmail.com

tools for hybrid joints. Thus, performance of hybrid joints are mainly influenced by the mechanical characteristic of the materials.

In recent years, there are very few available research works on hybrid joints as compared to traditional joining techniques. Fu Maofeng and Mallick [4] had studied on the effect of different washer types on static and fatigue strength for single-lap hybrid joints. Based on their study, the types and shapes of washers had direct influence upon the performance of hybrid joints, whereby the applied stress are transferred through friction due to the bolt tightening. They found that adequate clamping force on the mechanical fastening with appropriate washer type can enhance the mechanical properties of the hybrid joint. Meanwhile, Gordon Kelly [5] concluded that adhesive layer of hybrid joints are able to transfer applied remote loading. Higher load can be transferred with thicker adhesive layer through bolt friction in single-lap CFRP hybrid joint configurations. On the other hand, lesser load was transferred as the overlap length and elastic modulus were increased.

There are three dominant failure modes which occured in bolted joints which are net-tension, shear-out and bearing failures. Karakuzu et al. [6] investigated the effect of joint geometry on woven glass-vinylester composites and found that the end distance to diameter ratio (*e/d*) for cross ply lay-up was 6 to eliminate shear-out failures with W/d ratio ranging from 2 to 5. Single-lap woven fabric CFRP bolted joints had been studied regarding the effects of joint geometry. They found that the bearing strength increases with the *W/d* ratios [7]. However, there are still limited studies on hybrid joint, especially using natural fiber as reinforcement in composite material applications. Hence, this study is designed to investigate the effect of different W/d ratios and laminate lay-up for single-lap hybrid bonded-bolted joint on woven fabric kenaf fiber composite plates. The current testing series were designed and experimental mechanical properties were determined by using quasi-static mechanical machines following ASTM Standard D3039B [8].

2 Experimental Design

2.1 Materials

Kenaf long fibers were used as composite reinforcement. The weaving pattern of kenaf fiber (276 tex) reinforcement was plain weave woven fabric with packing density 4 end per cm (in warp direction) and 4 pick per cm (in weft direction) as shown in Figure 1. Thermoset matrix, epoxy/hardener (SP84/SP76) was used as resin system with density of 1.14 g/cm^3.

Figure 1. Density measurement for woven fabric kenaf fiber.

2.2 Specimen preparation

The laminates were prepared by applying one layer of resin with a ratio of 1 : 2 (hardener : epoxy) on an adjustable aluminum mold. Subsequently, a ply of plain weave (PX) woven fabric was placed on top of it, followed by another layer of resin system. The steps were repeated according to the desired laminate lay-up [PX2 $(0/90)_s$ and PX4 $(0/90)_{2s}$]. Samples were allowed to cure for about 24 hours in a hydraulic compression machine under room temperature. The composite plates obtained were then sectioned according to desired sizes and a hole of 5 mm diameter was drilled on each specimens. Finally, fastening system with finger-tight (FT) clamping torque (~0.5 Nm) and adhesive layer (at overlap region) were applied concurrently. Single-lap kenaf fiber composite plates (FCP) geometries are illustrated in Figure 2.

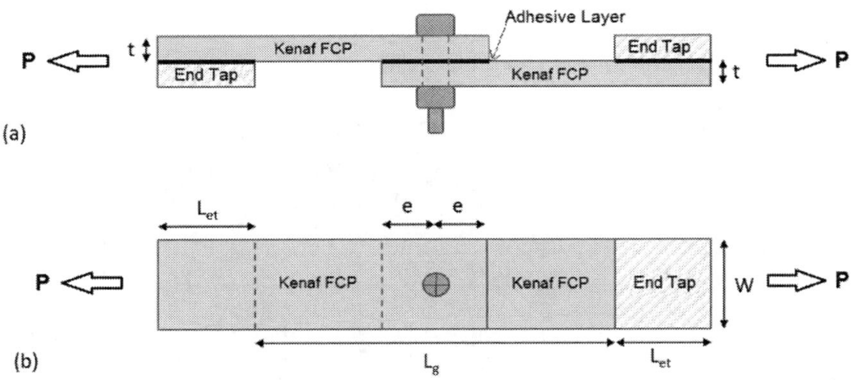

Figure 2. Single-lap hybrid joint test specimen geometry; (a) side view (b) top view.

2.3 Tensile tests

Tensile tests had been conducted based on ASTM Standard D3039B [8] via SHIMAZU AG-I 10kN testing machine at a constant cross-head speed of 0.5 mm/min. All testing series samples and measurements were done in triplicate as listed in Table 1. The end-distance are kept constant (30mm) and the specimens width varied according to W/d ratios.

Table 1. Hybrid joints testing series in current study.

Laminate	Thickness, t [mm]	*W/d* Ratio	*e/d* Ratio	Hole Size, d [mm]	Clamp-up Torques
PX2	2.15 ± 0.01	2	6	5	FT
	2.08 ± 0.01	3	6	5	FT
	2.12 ± 0.09	4	6	5	FT
	2.09 ± 0.08	5	6	5	FT
PX4	3.62 ± 0.01	2	6	5	FT
	3.48 ± 0.10	3	6	5	FT
	3.79 ± 0.05	4	6	5	FT
	3.60 ± 0.16	5	6	5	FT

3 Results and Discussion

This study was carried out to investigate the effect of *W/d* (*W/d* = 2, 3, 4, 5) and different plate thickness (PX2 and PX4) on bearing stress at failure of woven fabric kenaf fiber composite plates under quasi-static loading. The maximum load applied, calculated bearing stress at failure and associated failure modes were determined and stated in Table 2.

Based on visual observation, both PX2 and PX4 lay-ups exhibited two types of failure mode; net-tension and combination bearing/net-tension failures. Net-tension failures occurred in smaller W/d are attribute to the narrow specimen width. Meanwhile, specimens with larger *W/d* ratio portrayed bearing/net-tension failure mode, suggesting similar failure mechanisms as reported by Smith et al. [9] that cross-ply laminates changed from delamination and splitting (full bearing failure) to laminate shear compression failure beneath the pin (due to local tensile failure). It also found that single-lap joint is susceptible to demonstrate bolt pull through and showed washer bending into testing specimen as a result of secondary bending. However, due to adhesively bonded on the shear plane may reduce this effect. The local failure (bearing stress) initially occurred prior to ultimately net-tension failure. Similar findings were found by Ahmad et al. [10] in woven fabric CFRP bolted joints at which intermediate *W/d* net-tension failure was preceded by local bearing damage with unclamped condition. They also suggested full bearing failures seen to be more pronounce in quasi-isotropic lay-ups compared to cross-ply lay-ups. Figure 3 illustrates the photograph of failed PX4 specimens with different W/d ratios.

Table 2. Maximum bearing stress at failures and associated failure modes for all testing series.

Laminate	Width to Diameter ratio, *W/d*	Maximum Load, P_{max} (kN)	Maximum Bearing Stress, σ_{max} (MPa)	Failure Mode
PX2	2	0.78 ± 0.01	73.0 ± 0.4	NT
	3	1.30 ± 0.04	125.1 ± 3.7	NT
	4	1.58 ± 0.11	149.0 ± 6.3	NT
	5	1.90 ± 0.14	181.8 ± 7.1	B-NT
PX4	2	0.77 ± 0.02	42.7 ± 1.3	NT
	3	1.43 ± 0.04	82.2 ± 0.1	B-NT
	4	1.65 ± 0.05	87.3 ± 1.8	B-NT
	5	1.75 ± 0.06	97.3 ± 0.9	B-NT

*NT : Net-Tension failure
*B-NT : Bearing/Net-Tension

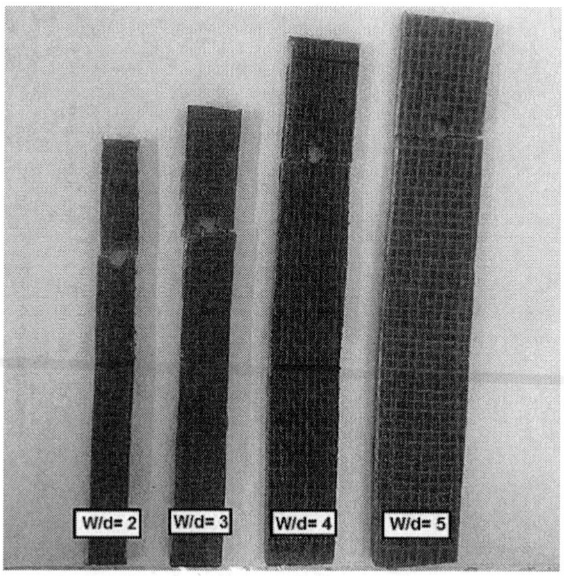

Figure 3. Failed specimens with varied *W/d* ratio on PX4 lay-up.

The thicker specimens exhibited combined local bearing/net-tension at $W/d \geq 3$ compared with PX2 lay-up ($W/d \geq 5$). Correspondingly, bearing strength is dependent upon plate thickness, whereby the bearing strength of PX4 is 97 MPa, 50% lesser than PX2 (182 MPa) at $W/d = 5$. This is presumably due to lack of constraint through the thickness that allow the laminate expansion at unclamped condition. The strength decreased with the increase of bolted joint thickness for net-tension failure. Same findings were reported by Smith et al. [9] on quasi-isotropic bolted joints. In align with previous studies [6], the bearing strength increased with the increase of *W/d* ratio for both PX2 and PX4 specimens. Maximum bearing stresses for all lay-up understudied are shown in Figure 4 and Table 2. Based on the graph plotted, PX4 specimens with *W/d* ratio 3, 4 and 5 slightly increased in maximum bearing stresses which are 82.2 MPa, 87.3 MPa and 97.3 MPa, respectively. Bearing failures showed more progressive failures, therefore constant bearing stress with this failure types perhaps not surprising.

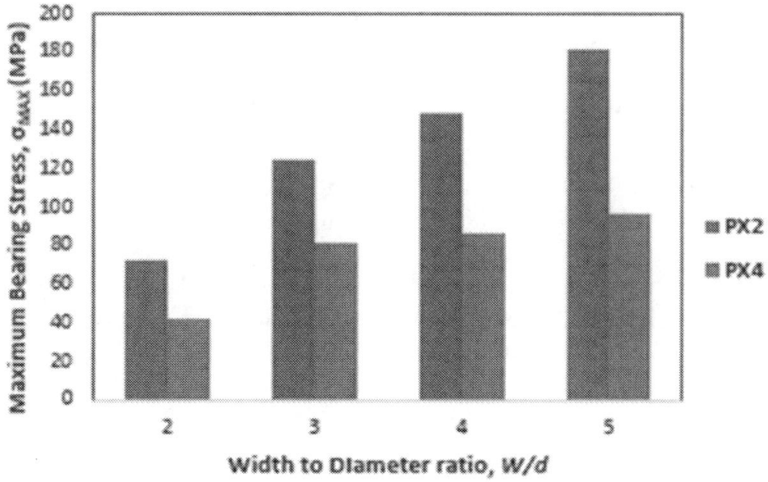

Figure 4. Effect of *W/d* ratio on maximum bearing stress.

In current study, three testing specimens were carried out for each testing series as good reproducibility were found. Figure 5 shows the representative load-displacement curve for PX2 and PX4 with *W/d* = 3. Similar load-displacement profiles were found in other *W/d* ratios. As expected, the specimens showed that the failure loads were increased with W/d ratios which the average maximum load for PX2 and PX4 specimens were 1.30 kN and 1.43 kN, respectively. PX2 specimen exhibited net-tension failure mode and the curve increased gradually up to the ultimate load followed by a small load drop (almost negligible) until breaking load. It is suggested that the adhesive failed first before the joint started to slip. Crack initiated from the hole edge and propagated across the specimen width. On the other hand, the curve for PX4 specimen with bearing/net-tension failure mode was slightly displaced at some points showing more progressive behaviour but ultimately failed in net-tension. However, the secondary bending does not show any significant change in load-displacement profile. The failure load showed increment trends with thicker specimens.

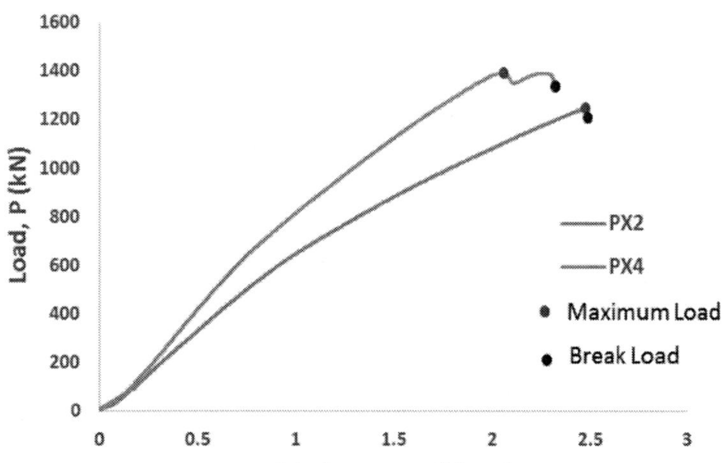

Figure 5. Load-displacement curve of hybrid joints for PX2 and PX4 lay-up with *W/d* = 3.

It is noteworthy that the current work was prepared by using hand weaving and hand lay-up techniques compared to commercially manufacturing production. Therefore, GFRP and CFRP hybrid specimens showed larger failure load than kenaf fiber hybrid specimens, i.e., in the range of 11 kN to 16 kN [11, 12]. However, kenaf fiber still provides good prospect in low bearing stress applications as mentioned previously. Although hand lay-up techniques prone to voids and porosity, current work selects the good batch of testing panels for further mechanical testing. In-depth investigations on natural fibers materials upon hybrid joints behaviour will be carried out in future.

4 Conclusions

The structural behaviour of single-lap hybrid bolted-bonded joints on woven fabric kenaf fiber composite plates were investigated with different lay-up types and *W/d* ratios. Larger *W/d* ratio and thinner specimens were found to provide greater bearing stress at failure. Failure modes in all testing series were net-tension and bearing/net-tension, where critical *W/d* ratios were 5 and 3 for PX2 and PX4 lay-ups respectively. Further investigations on preloaded bolt joints are suggested to study the effect from clamp-up, in which increased joint strength is expected.

Acknowledgements

The authors gratefully acknowledge Universiti Tun Hussein Onn Malaysia (UTHM) and Research Grant Scheme under RAGS R034.

References

[1] H.M. Akil, M.F. Omar, A.A. M. Mazuki, S. Safiee, Z.A. M. Ishak and A. Abu Bakar, Kenaf fiber reinforced composites: A review, Mater. Des., **32**. 4107-4121, (2011).

[2] O. Faruk, A.K. Bledzki, H.P. Fink and M. Sain, Biocomposites reinforced with natural fibers: 2000-2010, Prog. Polym. Sci., **37**, 1552-1596, (2012)

[3] G. Marannano and B. Zuccarello, Numerical experimental analysis of hybrid double lap aluminum-CFRP joints, Compos. Part B Eng., **71**, 28-39, (2015).

[4] M. Fu and P.K. Mallick, Fatigue of hybrid (adhesive / bolted) joints in SRIM composites, Int. J. Adhes. Adhes., **21**, 145-159, (2001).

[5] G. Kelly, Load transfer in hybrid (bonded/bolted) composite single-lap joints, Compos. Struct., **69**, 35–43, (2005).

[6] H. Ahmad, A.D. Crocombe, and P.A. Smith, Strength prediction in CFRP woven laminate bolted double-lap joints under quasi-static loading using XFEM, Compos. Part A Appl. Sci. Manuf., **56**, 192-202, (2014).

[7] R. Karakuzu, T. Gulem and B.M. Icten, Failure analysis of woven laminated glass-vinylester composites with pin-loaded hole, Compos. Struct., **72**, 27-32, (2006).

[8] ASTM D3039/D3039M, *Standard Test Method for Tensile Properties of Polymer Matrix Composite Materials*, ASTM International, Pennsylvania, (2008).

[9] P.A. Smith, K.J. Pascoe, C. Polak and D.O. Stroud, The behaviour of single-lap bolted joints in CFRP laminates, Composite Structures, **6**, 41-55, (1986).

[10] H. Ahmad, A.D. Crocombe and P.A. Smith, Strength prediction in CFRP woven laminate bolted single-lap joints under quasi-static loading using XFEM, Compos. Part A Appl. Sci. Manuf., **66**, 82-93, (2014).

[11] G. Kelly, Quasi-static strength and fatigue life of hybrid (bonded/bolted) composite single lap joints, Compos. Struct., **72**, 119-129, (2006).

[12] J.S. Kim, J.Y. Lim and W.G. Lee, Joining performance evaluation of different types of GEP224 glass/epoxy-to-AZ31B magnesium alloy singlelap joints, Int. J. Precis. Eng. Man., **16**, 1135–1140, (2015).

MATEC Web of Conferences **47**, 02004 (2016)
DOI: 10.1051/matecconf/20164702004

Experimental Study on Impact Load Resistance of RC Beam with Corroded Reinforcement

Isao Kuwahara[1,a], Hiroki Tamai[1] and Yoshimi Sonoda[1]

[1]*Department of Civil Engineering, Kyushu University, 744 Motooka Nishi-ku, Fukuoka 819-0395, Japan*

Abstract. In Japan, many concrete bridges along the coastline have suffered corrosion due to the chloride attack from seawater and airborne salt. Therefore, now, several researches on corroded RC structures have been ongoing around the world. Especially, as a basic research, many researchers have studied on the numerical and experimental method to evaluate static load resistant capacity of RC beams with corroded reinforcements. However, there has been almost no research to evaluate impact load resistant capacity. Of course, it is important that impact load resistant capacity is clarified, if the possibility of acting of impact load including in natural disaster on corroded RC structures is dealt with. With a background like this, this paper describes an experimental investigation of the structural behavior of corroded RC beams. To be precise, in this study, RC beams with corroded reinforcement by using electrolytic corrosion method were made, and both of static loading test and repeated impact test for those RC beams were conducted in order to clarify the static and impact load resistant capacity of them. As a result, we could get some findings. At first, through repeated impact tests (which has 1.0m/s impact velocity and a 100kg steel weight), the number of impacts to the failure of corroded RC beams were less than half of healthy RC beams. It was also found that fracture pattern was completely different due to some axial cracks by corrosion. In addition, even though the corrosion rate is same level, the impact load resistant capacity reduce greatly compared to the static load resistant capacity because of the existence and pattern of cracks due to corrosion.

1 Introduction

The deterioration of reinforced concrete structures due to chloride-induced corrosion of the steel reinforcement is one of the most considerable problems against many infrastructures in civil engineering field all over the world [1]. Especially, the development of the countermeasure against several deteriorations of the structures is an urgent issue in also Japan. Here, the major effect of the corrosion process is the formation of rust and these results in cracking and falling of the cover concrete, which can influence various characteristics such as the mechanical and structural performance and load capacity of the RC structures. To properly maintain some aged RC structures, it is important to quantitatively evaluate the load capacity of them against several loadings such as static and impact loading. So, several researches on corroded RC structures have been ongoing around the world in recent days. Especially, as a basic research, many researchers have studied on the numerical and experimental method to evaluate static load resistant capacity of RC beams with corroded

[a] Corresponding author : kuwahara@doc.kyushu-u.ac.jp

reinforcements [2]. However, there has been almost no research to evaluate impact load resistant capacity. Of course, it is important that impact load resistant capacity is clarified, if the possibility of acting of impact load including in natural disaster on corroded RC structures is dealt with.

In this study, to get fundamental findings regarding impact load resistant capacity of RC beams with corroded reinforcement, the impact loading test was conducted for the damaged RC beams. In addition, the static loading test was conducted to investigate the basic characteristics of the RC beam with corroded reinforcement and to compare the results of the static and impact loading test with corrosion ratio.

2 RC Beams Damaged by Electrolytic Corrosion Method

Figure 1 shows an arrangement of reinforcement and dimensions of the specimen. The length of the RC beam specimens was 1200 mm (span length: 1000 mm), and the cross-sectional dimensions were 120 × 100 mm. The thickness of the cover concrete was 19 mm. The deformed reinforcements (10 mm diameter) were used as the tensile reinforcements in each beam. The round reinforcements (6 mm diameter) were used as the compression reinforcements and as stirrup which were placed with 100 mm intervals in each specimen. Design flexural strength was 18.7kN and design shear strength was 7.6kN. So the specimens were designed the cross-section could lead to the condition that flexural failure was dominant.

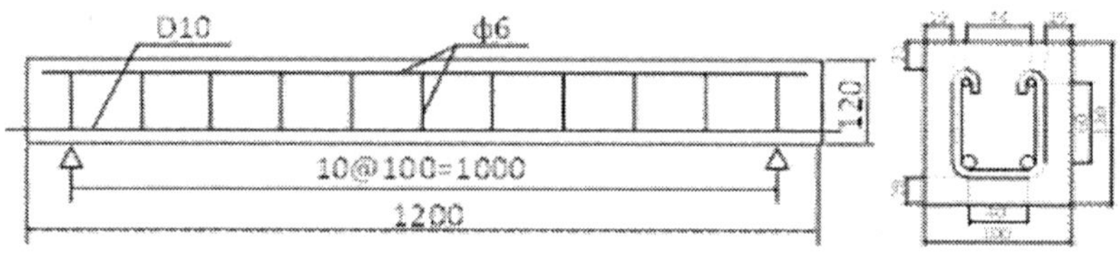

Figure 1. Arrangement of reinforcement and dimensions of the specimen (unit in mm).

In this study, the electrolytic corrosion method was adopted to make the damaged RC beams with corroded reinforcement. This electrolytic corrosion method is one of the methods to accelerate corrosion of reinforcement [3]. In this test, DC power supply was used. In addition, a copper sheet was used as a cathode. The cathode and the specimen were placed in 3% NaCl solution as shown in Figure 2. The corrosion rate of the reinforcements was controlled by energization time.

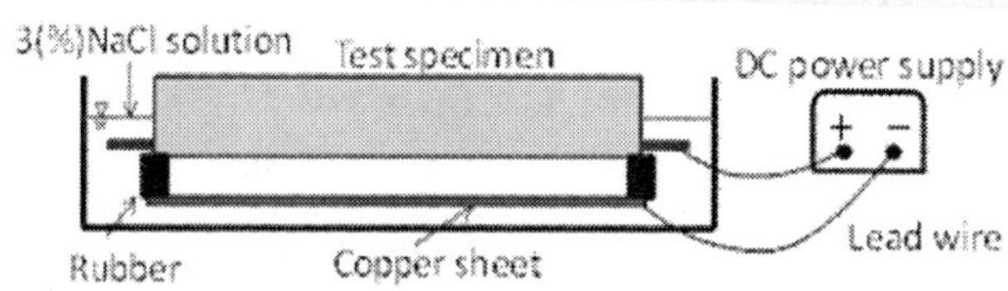

Figure 2. Schematic view of electrolytic corrosion method.

After the loading test, reinforcements were taken out from the specimen and the corrosion rate of the reinforcements was measured. Now, the corrosion rate is calculated according to the following Equation (1).

$$C = \frac{w_0 - w_c}{w_0} \times 100 \tag{1}$$

where C is corrosion rate, w_o is the mass of no corroded reinforcement, and w_c is the mass of corroded reinforcement.

As a result, Table 1 shows measured corrosion rate as any specimens. Further, in specimen's name, D expresses an impact test, S expresses a static loading test and N expresses a no damage specimen, L series expresses a degradation level of specimen. Moreover the crack situations of the specimens after the electrolytic corrosion test were shown in Figure 3. By the expansion of volume due to the generation of corrosion products, cracks in the axial direction of the tensile reinforcements could be confirmed. The corrosion conditions of the tensile reinforcements after derusting are shown in Figure 4. When the corrosion rate increases, a clear bias occurred in the corrosion distribution of tensile reinforcements.

Table 1. Corrosion rate for any specimens.

For any tests	Name of Test Specimen	Corrosion Rate (meV)
For Static Loading Test	S-N	0
	S-L1	0.67
	S-L2	0.94
	S-L3	6.13
	S-L4	11.71
For Repeated Impact Test	D-N	0
	D-L3	7.03
	D-L4	11.04

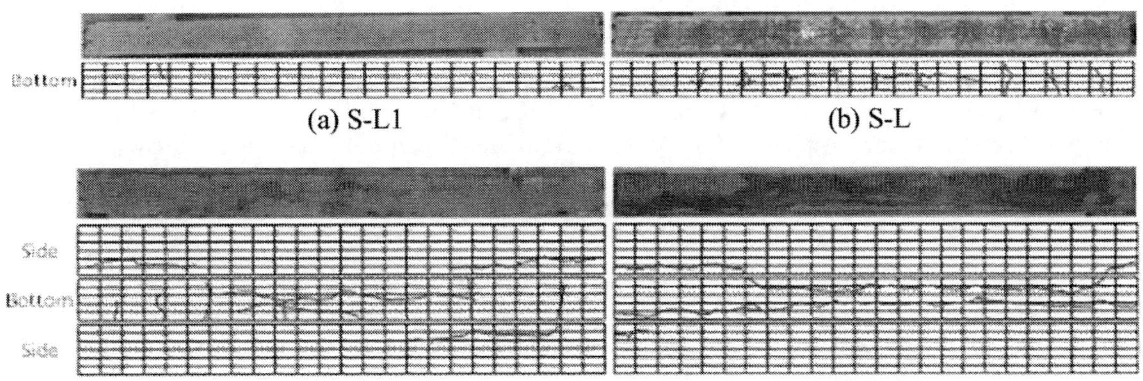

(a) S-L1 (b) S-L

(c) S-L3 (d) S-L4

Figure 3. Crack situations due to corrosion.

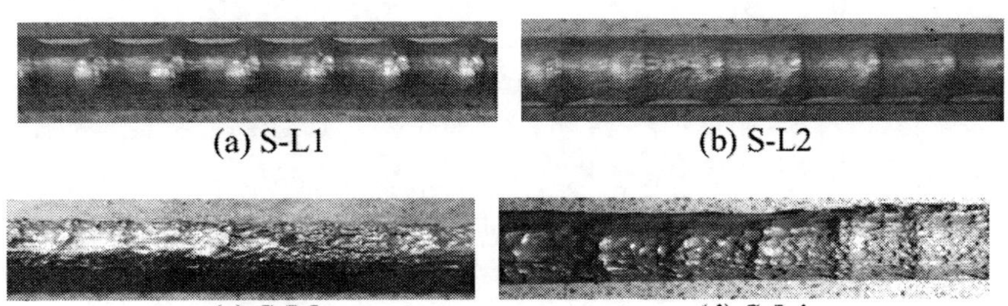

(a) S-L1 (b) S-L2

(c) S-L3 (d) S-L4

Figure 4. Corrosion condition of the tensile reinforcements after derusting.

3 Static Load Capacity

3.1 Experimental procedure

The static loading test was conducted in order to clarify the static flexural load capacity of the RC beams with corroded reinforcement. A universal testing machine was used for acting compressive loading on the beams. As the loading condition, 2 points concentrated load of equal bending moment section 200 (mm) was adopted as shown in Figure 5, and as the loading rate, about 0.5 (mm/min) was controlled because of avoiding the effect of the loading rate. In this test, the acting load and displacement at the mid-span of the RC beams were measured as the measurement item.

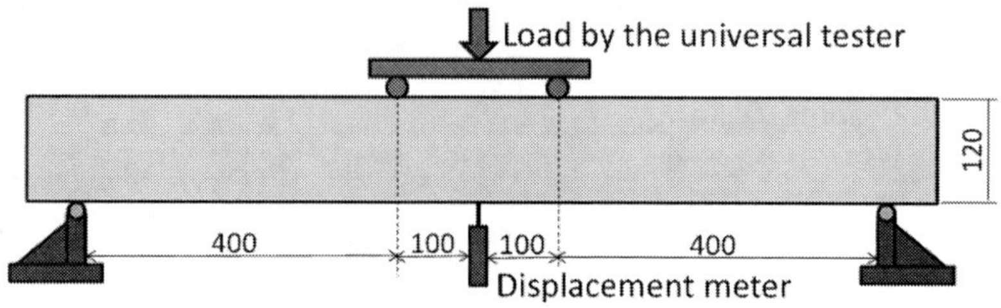

Figure 5. Schematic view of the static loading test (unit in mm).

3.2 Results and discussion

At first, the load-displacement relations are shown in Figure 6 as one of the results of the static loading test. Regarding the process to the fracture in case of S-N (Uncorroded), the flexural cracking occurred at 5kN, after that the reinforcement steel yielded at 23kN and then the beam fractured due to bending at 26kN. Like this, when the displacement at the mid-span of the beams reached 3~4 (mm) in all cases, stiffness fell with a reinforcement yielding, and after the load reached the maximum, the load was indicated the behaviour which was fallen gradually.

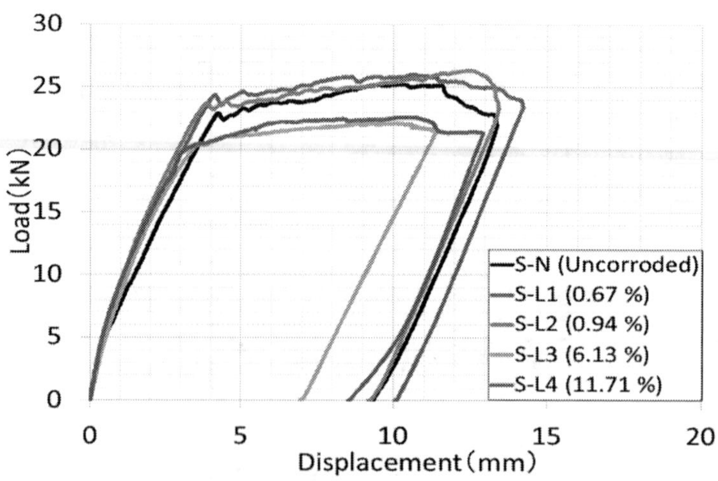

Figure 6. The load-displacement relations.

However, the stiffness degradation due to the occurrence of flexural cracking was not able to be indicated in case of S-L1~L4 (Damaged RC beams). This result could be due to the existence of initial cracks occurred with corrosion. In addition, Table 2 shows the relationship between maximum load and corrosion ratio in all cases. From this table, the maximum load became lower in case of S-L3 (corrosion rate: 6.13%) and S-L4 (corrosion rate: 11.71%) compared to S-N (Uncorroded) as expected, while the maximum load became slightly larger in case of S-L1 (corrosion rate 0.67%) and S-L2 (corrosion rate 0.94%). As these reasons, the influence of the increase of density between rebar and the surrounding concrete by expanded reinforcements due to corrosion becomes more dominant than the influence of the cross-sectional area reduction due to rusting reinforcements in the area with the small corrosion rate. And influence of decrease of a cross-sectional area and a decline of adhesion by corrosion of a reinforcement become dominant in the area with the high corrosion rate.

Table 2. The relationship between maximum load and corrosion ratio in all cases.

Name of Test Specimen	Maximum Load (kN)	Maximum Load Ratio	Corrosion Rate (%)
S-N	25.72	1	0
S-L1	25.98	1.01	0.67
S-L2	26.32	1.02	0.94
S-L3	22.11	0.86	6.13
S-L4	22.6	0.88	11.71

4 Impact Load Resistant Capacity

4.1 Experimental procedure

The impact loading test was conducted in order to clarify the impact load resistant capacity of the RC beams with corroded reinforcement. In this test, a falling weight type impact test indicated in Figure 7 was adopted. No damage specimen (D-N) and reinforcement corrosion specimens (D-L3, corrosion rate: 7.03% and D-L4, corrosion rate: 11.04%) were used. As the loading condition, a 100kg steel weight was dropped freely onto the top surface of the RC beams at mid-span at repeatedly until the fracture of the beams [4]. Here, in this test, fracture of the beams was defined the time when accumulation residual displacement reached 1% of the span length. The impact velocity was set at 1.0m/s in order to make the maximum displacement does not exceed the reinforcement yield displacement that was confirmed in static conditions by tests conducted beforehand. The impact load by a load cell and the residual displacement at the mid-span of the RC beams were measured as the measurement item.

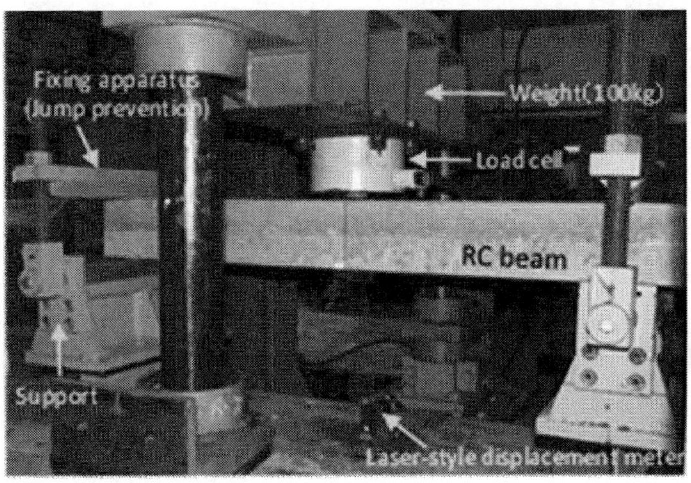

Figure 7. Experimental apparatus (falling weight type impact test machine).

4.2 Results and discussion

First, from the impact force and displacement response of the first impact as shown in Figures 8 and 9, as the level of degradation was increased, impact force response was smaller and displacement response was larger. It could be that resistance against impact force of specimens has declined by corrosion of reinforcements. From the relationship between the accumulation residual displacement and the number of impacts shown in Figure 10, number of impacts to the failure was 75 times in the D-N, 32 times in the D-L3, 16 times in the D-L4 and it was able to confirm that the impact resistance was greatly reduced by corrosion of reinforcements. In addition, the fracture patterns are shown in Figure 11. In D-N, first, flexural cracks were occurred just below of the impact point. Then, with increasing number of impacts, the width of the crack became larger and cracks spread to the entire specimen. On the other hand, in the D-L3 and D-L4, flexural cracks were occurred just below of the impact point and the cracks which had caused by corrosion became larger with increasing number of impacts at the same time.

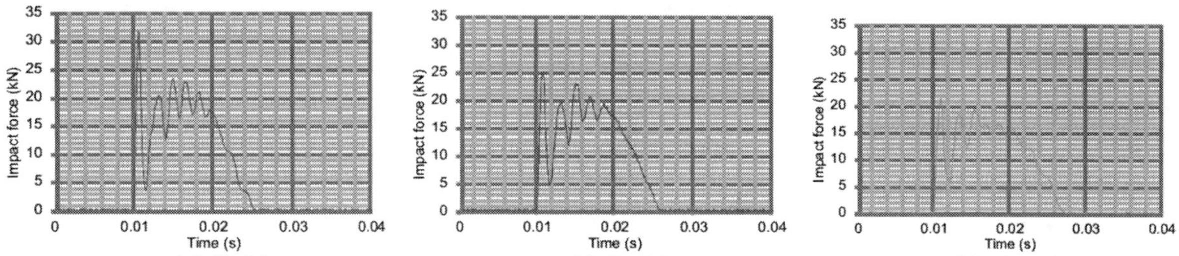

(a) D-N (b) D-L3 (c) D-L4

Figure 8. Impact force response at the first impact.

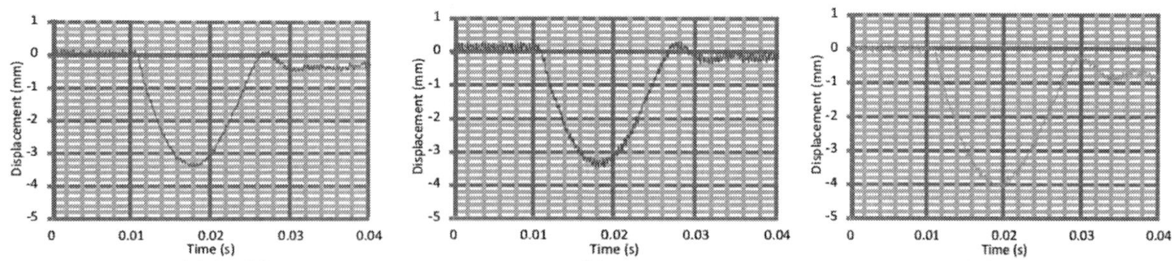

(a) D-N (b) D-L3 (c) D-L4

Figure 9. Displacement response of the beam at mid-span at the first impact.

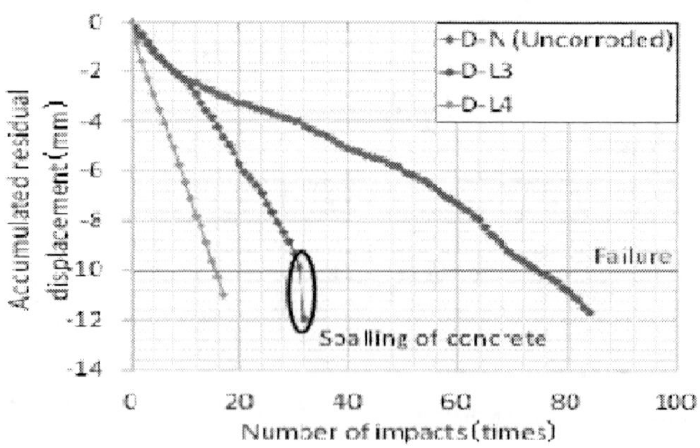

Figure 10. The relationship between the accumulated residual displacement and the number of impacts.

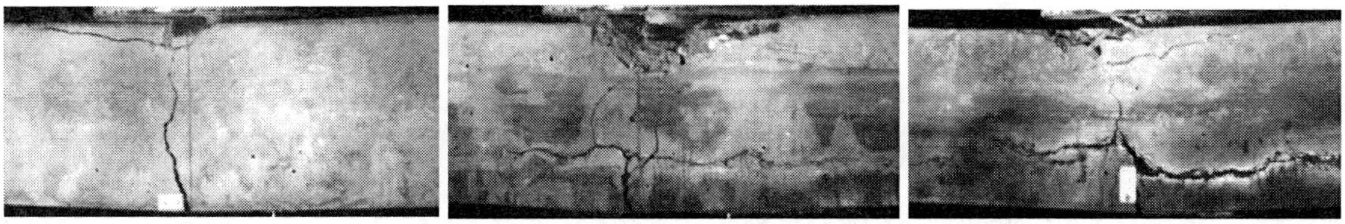

(a) D-N (b) D-L3 (c) D-L4

Figure 11. Fracture pattern.

5 Conclusion

The objective of this study is to get fundamental findings regarding impact load capacity of damaged RC beams with corroded reinforcement through the impact loading test. The followings are confirmed from this study:

i. The impact resistance capacity of the damaged RC beams also decreased with the increasing corrosion ratio through repeated impact test results. It is also found that the existence and pattern of cracks due to corrosion had an obvious effect on the impact resistance capacity based on the number of impacts to fracture.

ii. Even though the corrosion rate is same level, the Impact load resistant capacity reduce greatly compared to the static load resistant capacity because of the existence and pattern of cracks due to corrosion.

References

[1] H. Zhou, J. Lu, B. Dong and F. Xing, Effects of stirrup corrosion on bond–slip performance of reinforcing steel in concrete, Constr. Build. Mater., **93**, 257-256, (2015).

[2] L. Yu, R. Francois, V.H. Dang, V. L'Hostis and R. Gagne, Structural performance of RC beams damaged by natural corrosion under sustained loading in a chloride environment, Eng. Struc., **96**, 30-40, (2015).

[3] O. Michiaki, K. Toshiyuki, Y. Yasuhiko and S. Tsutomu, Influence of corrosion of reinforcing bars on bending performance of reinforced concrete members, J. of Materials, Concrete Structures and Pavements, **62**(39), 542–554, (2006).

[4] I. Mitsuya, S. Yuji and Y. Hiroshi, Structural behavior of reinforced concrete beams subjected to repeated impact loads, *Technical Note of the Port and Airport Research Institute No.1152*, Japan, (2007).

MATEC Web of Conferences **47**, 02005 (2016)
DOI: 10.1051/matecconf/20164702005

The Spalling of Geopolymer High Strength Concrete Wall Panels and Cylinders Under Hydrocarbon Fire

Ahmad Zurisman Mohd Ali[1,a] and Jay Sanjayan[2]

[1]*Department of Structural and Materials, Universiti Tun Hussein Onn Malaysia, 86400 Parit Raja, Johor, Malaysia*
[2]*Centre for Sustainable Infrastructures, Swinburne University of Technology, Australia*

Abstract. Concrete structures were designed to withstand various types of environment conditions from mild to very severe conditions. Fire represents one of the most severe environmental conditions to which concrete structures may be subjected especially in close conduct structure like tunnel. This paper focuses on the spalling of geopolymer high strength concrete exposed to hydrocarbon fire for minimum 2 hours. From the fire test, geopolymer concrete can be classified as a good fire resistance construction materials based on spalling performance of high strength concrete when exposed to hydrocarbon fire. A maximum of 1% (excluding water moisture loss) of spalling recorded for high strength geopolymer concrete wall panel. No explosive spallings were observed for high strength geopolymer concrete.

1 Introduction

The needs for concrete structures to withstand fire condition are far more demanding nowadays with numbers of concrete tunnels, petro-chemical's plant and other mega structures are being built all around the world. For example, a fire in Channel tunnel (35 km railroad tunnel connecting England and France) in 1996 caused severe damage to concrete tunnel rings owing to the spalling of concrete and resulted in six-month closure for repairs costing US$1.5 million per day [1]. On 24 March 1999, a cargo truck carrying margarine and flour in the Mont Blanc tunnel connecting France and Italy caught fire and stopped at Kilometre 6.7 causing 39 human deaths and severely damaging 900 m long tunnel roof due to spalling [2]. A fire occurred in the Great Belt tunnel in Denmark in 1995 also causing severe spalling of concrete tunnel rings [3].Due to high usage of concrete as construction material, it is important to develop high fire resistance concrete that can sustain structural integrity when exposed to fire event. Geopolymer concrete known as very high fire resistance material due to its ceramic-like properties [4]. Lyon et al. [5] reported that geopolymer composites is ideally suited for construction, transportation and infrastructure where fire endurance is part of needed requirement . Low cost geopolymer resins can be produced by activation of fly ash. Geopolymer concrete is produced by combining these resins with coarse and fine aggregates using the conventional concrete technology methods. Since fly ash is an industrial by-product from coal power stations (largely wasted by dumping in landfills), the cost of manufacturing this concrete can be potentially lower than the conventional Portland cement based concrete. Research works reported so far in the literature indicate that geopolymers have superior fire resistance when compared to conventional concretes [6-8]. The

[a] Corresponding author : zurisman@uthm.edu.my

superior fire resistance properties are attributed to the ceramic-like properties of geopolymers, including the way it looks: smooth, glassy and shiny [9] Geopolymer concrete strength after elevated temperature exposure can increase due to further geopolymerisation [6]. However, there are strength decrease of geopolymer concrete after elevated temperature exposure especially geopolymer concrete with coarse aggregate. Thermal incompatibilities due to difference thermal expansions between aggregates and binder contributes to the decline in geopolymer concrete strength [7]. The results in this paper represent the performance of geopolymer high strength concrete panel and cylinder exposed to hydrocarbon fire exposure. The amount of spalling and residual compressive strength after fire exposures are reported in this paper.

2 Experimental Program

2.1 Raw materials

Aluminosilicate source for making geopolymer concrete in this study is fly ash. The fly ash was sourced from Gladstone power station in Queensland. Chemical composition of Gladstone fly ash is shown in Table 1. Sand for the geopolymer high strength concrete was sourced from Langwarrin quarry, Melbourne. The sand absorbed 0.3% of water. Coarse aggregate with the maximum diameter of 14 mm sourced from Oakland Junction, Hornsfel and supplied by Readymix^{TM}. Coarse aggregate water absorption was 0.8%. Sodium silicate and sodium hydroxide were used as alkaline activator for the geopolymer high strength concrete. D grade sodium silicate was supplied by PQ Australia. Sodium hydroxide was supplied by Orica Chemical with 99% purity. Sodium hydroxide of 8.0M solution was made by mixing 26.2% of NaOH solids and 73.8% of water [10].

Table 1. Fly ash chemical compounds.

Chemical Compounds	Weight (%)
Al_2O_3	25.56
BaO	0.09
CaO	4.3
Fe_2O_3	12.48
K_2O	0.7
MgO	1.45
MnO	0.15
Na_2O	0.77
P_2O_5	0.885
SO_3	0.24
SiO_2	51.11
TiO_2	1.32
LOI*	0.57
Total	99.62

2.2 Geopolymer concrete mix

Targeted compressive strength for geopolymer high strength concrete is 50 MPa. Geopolymer concrete was made with alkaline activator to fly ash ratio of 0.5. The $Na_2SiO_3/NaOH$ ratio for alkaline solution was 2.5. Average compressive strength of geopolymer concrete was 64.1 MPa and it can be classified as geopolymer high strength concrete according to EN 206:2013 [11].

2.3 Geopolymer concrete casting

Geopolymer concrete was casted at Commonwealth Scientific and Industrial Research Organization (CSIRO) laboratory in Highett, Australia. The geopolymer concrete mixing procedure was according to Hardjito et al. [12]. Sodium silicate and sodium hydroxide were mixed together for 10 minutes prior to concrete mixing to form an alkaline activator. Geopolymer concrete was mixed in high shear concrete mixer.. Dry components of concrete (coarse aggregates + sand) were mixed for 3 minutes. One percent of water was added into the dry mixture to make the coarse aggregates and sand in saturated surface dry condition. Fly ash was added and mixed with aggregates for further 3 minutes. Alkaline activator was then added into the mixer for further 5 minutes. The fresh geopolymer concrete had a stiff consistency and was glossy in appearance. The geopolymer concrete was then cast and compacted in 1.075 m x 1.075 m x 0.2 m (thickness), 150 mm (diameter) x 300 mm and 100 mm (diameter) x 200 mm moulds. The cast geopolymer concrete panels and cylinders were cured using steam curing at 60°C temperature for 24 hours before demoulded. Compressive strength test for geopolymer concrete was carried out at the age of 1 day, 7 days and 28 days.

2.4 Fire testing regime

Hydrocarbon fire temperature versus time in accordance to EN 1991-1-2 was used for exposing the specimens. Standard fire curve is commonly used for testing for fire exposures in buildings whereas the Hydrocarbon fire test is more suitable for infrastructure applications where the concrete is likely to be exposed to fire originated from hydrocarbons (eg. tunnels). Equation (1) is the hydrocarbon fire temperature [13]. The fire tests were conducted for minimum of 120 minutes.

$$T = 1080\left(1 - 0.325e^{-0.167t} - 0.675e^{-2.5t}\right) + 20 \tag{1}$$

where T = Temperature (°C) and t = time (minute).

3 Result and Discussions

3.1 Spalling

Figure 1 shows the geopolymer concrete condition after exposed to hydrocarbon fire for 2 hours for panels and cylinders respectively. No explosive spalling noises were heard during the whole fire test period. After the fire test, there were some minor surface spalling was observed on 2 out of 4 geopolymer concrete panels with maximum of 4.65% concrete spalling. Geopolymer concrete cylinder did not have any spalling at all and all cylinders remained in their original shape. However, there was a loss in weight for all specimens due to loss of moisture content. Based on weight loss on unspalled panels, 3.57% - 3.75% of moisture evaporated from the concrete panel during the fire test. Therefore, it can be estimated that less than 1% of geopolymer concrete panel spalled when exposed to hydrocarbon fire. For geopolymer concrete cylinders, since there is no any spalling observed, average of 2.86% of weight loss was purely the loss of moisture content inside the concrete.

Figure 1. Fire tested Geopolymer concrete specimens.

3.2 Residual strength

Geopolymer concrete achieved its strength at very early age. Under curing regime of 60°C steam curing for 24 hours, the compressive strength of geopolymer concrete was 54.1 MPa. All geopolymer concrete panels and cylinders remained intact after 2 hours of hydrocarbon fire exposure. Therefore, residual compressive strength of geopolymer concrete could be measured. Residual compressive strength of a structure after fire event is an important factor in assessing the safety and integrity of whole structure. The decision for appropriate repair works on structure after fire is based on a reliable assessment of residual strength [14]. Table 2 summarises the average residual compressive strength of geopolymer concrete panel, 150 mm diameter cylinder and 100 mm diameter cylinder. From Table 2, it can be seen that geopolymer concrete panel has average residual compressive strength which was the highest (35.1 MPa). Cylinders with 150 mm diameter and 100 mm diameter have residual compressive strength of 13.3 and 6.7 MPa respectively. These results show that geopolymer concrete is structurally viable even after fire exposure. The specimen size effect was also exhibited with larger specimen's size showing lesser strength loss due to fire exposure [15].

Table 2. Residual compressive strength summary.

Conrete Type		**Residual Compreesive Strength (MPa)**	
Panel (exposed surface)	Panel 1	38.9	38.5
	Panel 2	35.6	
	Panel 3	34.3	
	Panel 4	45.2	
Cylinder (150 mm diameter)		12.1	13.3
		14.3	
		13.5	
Cylinder (100 mm diameter)		7.9	6.7
		8.2	
		4.1	

4 Conclusion

i) There were no explosive spalling observed during the whole length of hydrocarbon fire test on geopolymer specimens. Panel 2 (P2) recorded the highest geopolymer concrete spalling with 4.65% spalling percentage. However, it is estimated that 3.57% - 3.75% of the weight loss was due to the loss of moisture content during fire test. The estimation was based on loss of weight for unspalled panels (P1 and P4). Thus, the spalling percentage of geopolymer concrete was actually less than 1%. There was no spalling observed on geopolymer concrete cylinders even though there were 2.84% and 2.86% weight loss for 150 mm diameter and 100 mm diameter respectively. The weight loss was due to loss of moisture during fire test.

ii) There was residual compressive strength on geopolymer concrete after hydrocarbon fire test. The fire tested geopolymer concrete panel, 150 mm diameter cylinder and 100 mm diameter cylinder have 38.5, 13.3 and 6.7 MPa compressive strength respectively. It proved that geopolymer concrete structurally can withstand load even after being exposed to 2 hours of hydrocarbon fire exposure. It can be concluded that geopolymer concrete is good fire resistant construction material.

References

[1] F.J. Ulm, P. Acker, and M. Lévy, The "Chunnel" fire. II: Analysis of concrete damage, J. of Engineering Mechanics, **125**(3), 283-289, (1999).

[2] J.S. Roh, S.S. Yang, H.S. Ryou, M.O. Yoon and Y.T. Jeong, An experimental study on the effect of ventilation velocity on burning rate in tunnel fires—heptane pool fire case, Building and Environment, **43**(7),1225-1231, (2008).

[3] K.D. Hertz, Limits of spalling of fire-exposed concrete, Fire Safety J., **38**(2), 103-116, (2003).

[4] J. Davidovits and M. Davidovics. Geopolymer: Ultra-high temperature tooling material for the manufacture of advanced composites, SAMPE, **2**, 1939-1949, (1991).

[5] R.E. Lyon, P.N. Balaguru, A. Foden, U. Sorathia, J. Davidovits and M. Davidovics, Fire-resistant aluminosilicate composites, Fire and Materials, **21**(2), 67-73 (1997).

[6] Z. Pan, J.G. Sanjayan and B.V. Rangan, An investigation of the mechanisms for strength gain or loss of geopolymer mortar after exposure to elevated temperature, J. of Materials Science, **44**(7), 1873-1880 (2009).

[7] D.L.Y. Kong and J.G. Sanjayan, Damage behavior of geopolymer composites exposed to elevated temperature, Cement and Concrete Composites, **30**(10), 986-991(2008).

[8] D.L. Kong, J.G. Sanjayan, and K. Sagoe-Crentsil, Factors affecting the performance of metakaolin geopolymers exposed to elevated temperatures, J. of Materials Science, **43**(3), 824-831(2008).

[9] A. Palomo, M.W. Grutzeck, and M.T. Blanco, Alkali-activated fly ashes: A cement for the future, Cement and Concrete Research, **29**(8), 1323-1329 (1999).

[10] B.V. Rangan, Fly ash-based geopolymer concrete, Indian Concrete J., **80**(2), 35-41, (2006).

[11] IS EN 206:2013, *Concrete-Specification, Performance, Production and Conformity*, European Committee for Standardization, Brussel, (2013).

[12] D. Hardjito, S.E. Wallah, D.M.J. Sumajouw and B.V. Rangan, On the development of fly ash-based geopolymer concret,. ACI Materials J., **101**(6), 467-472, (2004).

[13] BS EN 1991-1-2, *Part 1 -2: General Actions-Actions on Structures Exposed to Fire*, European Committee for Standardization, Brussels, (2005).

[14] E. Annerel and L. Taerwe. Approaches for the assessment of the residual strength of concrete exposed to fire, Concrete Repair, Rehabilitation and Retrofitting II, *Proc. of the 2nd Int. Conf. on Concrete Repair, Rehabilitation and Retrofitting*, 245-246, (2009).

[15] M. Li, C. Qian and W. Sun, Mechanical properties of high-strength concrete after fire, Cement and Concrete Research, **34**(6),1001-1005, (2004).

MATEC Web of Conferences **47**, 02006 (2016)
DOI: 10.1051/matecconf/20164702006

Evaluation of Ultimate Strength of Reinforced Concrete Beams Strengthened with FRP Sheets under Torsion

Shengqiang Ma[1], N. Muhamad Bunnori[1,a] and Kok Keong Choong[1]

[1]*School of Civil Engineering, Universiti Sains Malaysia, Engineering Campus, 14300 Nibong Tebal, Penang, Malaysia*

Abstract. The ultimate torque of reinforced concrete (RC) members strengthened with fiber reinforced polymer (FRP) sheets does not only depend on the torque of RC members, but also on the FRP contribution to the torque. For structural design, predicting the accurate torsional capacity of the strengthened beams is considerably important. Three existing models for calculating the ultimate torsional moment of RC beams and two existing models for computing the FRP contribution to the ultimate torque are described and combined. Based on an experimental database collected from existing literature, six combinations were discussed and evaluated from the calculative values compared with the experimental results. The comparison shows that the combination of ACI 318 and *fib* Bulletin 14 models (Group 2), as well as Chinese and Ghobarah models (Group 6), can reasonably and accurately predict the ultimate torque of beams strengthened with FRP sheet. Furthermore, the ultimate torque of six box-section beams strengthened with fully wrapping or U-wrap calculated by the Group 6 shows closely to the experimental results.

1 Introduction

Since the beginning of the present century, researchers have paid much attention to the torsional behavior of reinforced concrete (RC) beams strengthened with fiber-reinforced polymer (FRP) sheets by experimental investigation [1-3]; however, very few models exist to predict the ultimate strength of strengthened RC beams. The *fib* Bulletin 14[4] proposed equations to calculate the ultimate torsional moment of strengthened beams involving two typical failure patterns, namely, FRP fracture and debonding. Ghobarah [2] proposed a model of the torsional moment contributed by externally bonded FRP and assumed that the mean ultimate strain of fiber was approximately 0.003, as indicated by experimental records. These two models assumed that no interaction exists between the RC and FRP sheets, which contributed to the torsional capacity of the beam. Deifalla and Ghobarah [5] used the compression field theory to propose a series of equations to predict the complete behavior of strengthened RC beams with FRP. On the basis of the softened membrane model for torsion, Zojaji and Kabir [6] developed a computational procedure to predict the full torsional behavior of strengthened beams subjected to torsion. Although complete procedure analysis can accurately predict the torsional behavior, the models need a trial-and-error algorithm for iteration to calculate each point of the torsional moment versus the angle of twist curve. For obtaining the torsional capacity of beams strengthened with FRP in design, the full computational procedure is made to be considerably complicated but difficultly accessible. The use of available models to calculate the precise torque of

[a] Corresponding author : cenorazura@usm.my

strengthened beams does not only depend on the contribution of RC beam but also on the contribution of FRP. To determine which model for computing the torsional strength of RC beam combined with existing torsional contribution of wrapped FRP is more accurate in calculating the ultimate torque of strengthened beam with FRP, this study presents three models to compute the ultimate torque of RC beams and two models to calculate the FRP contribution for torque. On the basis of the experimental data collected from the literature, the combination of models and comparison of ratios are discussed and evaluated to identify the appropriate models for reasonably and accurately predicting the ultimate torque of strengthened beams.

2 Models of Torsional Strength of Reinforced Concrete Beams

2.1 Rahab model [7]

The model was proposed by hollow tube analogy, which was a simple method to predict the ultimate torque of RC beam subjected to torsion. This model showed that the torsional strength was correlated with the volume of the stirrup and longitudinal reinforcement of beams. The equation was deduced from the relationship between the ultimate torque and the ultimate shearing stress in the walls of the equivalent tube as follows:

$$T_{uc} = 0.67(A_c^2 / p_c)v_u \tag{1}$$

where T_{uc} is the ultimate torque of the RC beam section; A_c and p_c are the outer area and the perimeter of the beam cross section, respectively; and v_u is a nominal shear stress related to transverse and longitudinal steel and concrete compressive strength, which can be calculated from the curve based on the results of modified compression field theory[7].

2.2 ACI model [8]

The equation designed to calculate the ultimate torque of RC beams recommended by ACI 318-2011[8] is:

$$T_{uc} = \frac{2A_{cor}A_t f_{yv}}{s_t}\cot(\theta) \tag{2}$$

where A_{cor} is the gross area enclosed by shear flow path; A_t is the area of stirrup; f_{yv} is the yield strength of stirrup; S_t is the spacing of stirrups, and θ is the angle of the diagonal crack with respect to the horizontal axis of the beam. θ shall represent 45° for RC members and 37.5° for pre-stressed members. The equation assumed that the torsional resistance is afforded by mainly closed transverse reinforcement. The outside concrete of these stirrups is comparatively invalid.

2.3 Chinese code model [9]

The torsional strength of RC beams is obtained by the contribution of concrete tensile strength and the contribution of transverse and longitudinal reinforcement in the following models:

$$T_{uc} = 0.35 f_t W_t + 1.2\sqrt{\zeta}\, f_{yv}\frac{A_t A_{cor}}{s_t} \tag{3}$$

$$\zeta = \frac{f_y A_{sl} s_t}{f_{yv} A_t u_{cor}} \tag{4}$$

where f_t is the tensile strength of concrete; W_t is the plastic resistant moment of the cross section of torsional members; ζ is the strength ratio of longitudinal reinforcement to transverse reinforcement; f_y and A_{sl} are the yield strength and area of longitudinal steel, respectively; and u_{cor} is the perimeter of the center line of the shear flow. In the subsequent equations, W_t should be considered when a different cross section of members uses different equations to calculate the rectangular and box beams as follows:

$$W_t = \frac{b^2}{6}(3h - b) \quad \text{for rectangular beam} \tag{5}$$

$$W_t = \frac{b^2}{6}(3h - b) - \frac{(b - 2t)^2}{6}\left[3(h - 2t) - (b - 2t)\right] \quad \text{for box beam} \tag{6}$$

where b, h, and t corresponding to the width, height, and web thickness of the box beam.

3 Models of FRP Contribution for Torsional Strength

3.1 *fib* model [4]

The externally bonded FRP to beam provides the contribution to the torsional capacity of beams, of which the contributions of full wrapping and U-jacketing wrapping are calculated as follows:

$$T_f = \frac{2E_f \varepsilon_f t_f w_f bh \cot(\theta)}{s_f} \quad \text{or} \quad T_f = \frac{E_f \varepsilon_f t_f w_f bh \cot(\theta)}{s_f} \tag{7}$$

For full wrapping with CFRP in fracture controls, effective strain is computed as follows:

$$\varepsilon_{fe} = 0.17\left(\frac{f_c^{2/3}}{E_f \rho_f}\right)^{0.3} \varepsilon_{fu} \tag{8}$$

For U-jacketing CFRP wrapping, the effective strain depended on fracture and peeling off, which is computed as follows:

$$\varepsilon_{fe} = \min\left[0.17\left(\frac{f_c^{2/3}}{E_f \rho_f}\right)^{0.3} \varepsilon_{fu}, 0.65\left(\frac{f_c^{2/3}}{E_f \rho_f}\right)^{0.56} \times 10^{-3}\right] \tag{9}$$

where T_f is the torsional strength of FRP contribution, E_f is Young's modulus of the FRP, ε_{fe} is the effective strain in the fiber, t_f is the thickness of FRP, s_f is the spacing of FRP strips, w_f is the width of FRP, f_c is the concrete compressive strength, and $p_f = 2t_f w_f / bs_f$, which represents the FRP reinforcement ratio.

3.2 Ghobarah model [2]

This model assumed that the average ultimate strain of fiber was approximately 0.003; thus, the equation was proposed by simplifying the *fib* model as torsional capacity of full and U-jacketing FRP contribution, which is computed as follows:

$$T_f = 0.006 w_f t_f E_f bh / s_f \quad \text{or} \quad T_f = 0.003 w_f t_f E_f bh / s_f \tag{10}$$

The equation shows that that the angle of the diagonal crack with respect to the horizontal axis of the beam θ is at 45° for RC members, and the equation is only applicable for the vertical fiber wrap of transverse fiber with respect to the longitudinal axis of the beam. For U-jacketing FRP torsional contribution, the torque is equal to half of full wrapping. The total torsional moment of the beam strengthened with FRP can be computed by combining the torsional contribution of the RC beam and the contribution of the externally bonded FRP as follows:

$$T_u = T_{uc} + T_f \tag{11}$$

4 Database Collection

A full database consisting of 28 experimental beams (Table 1) was utilized to compare the theoretical torsional moment of FRP strengthened beams with the experimental results. The database was collected from the existing literature [1, 2, 10-12], which included different cross-sections, such as rectangular and box beams, strengthening configuration of full wrapping with strips or continuous sheets and U-shaped beams, and beams without internal stirrups. The values of the database were obtained from 6 beams with box cross section and 22 beams with rectangular cross-section. In the database, 5 beams strengthened with U-jacketing strips or continuous sheets, 6 beams strengthened with complete wrapping, and 17 beams wrapped with full strips were clearly identified. 6 rectangular beams without transverse reinforcement were also included in the database. The properties of the strengthened RC beams such as cross section dimension, concrete compressive strength, stirrup and longitudinal steel properties, FRP strengthening configuration, and FRP properties are listed in Table 1.

5 Comparisons and Discussions

Assuming that no correlation exists in the contribution of the torsional moment between RC beam and FRP wrapping [2], the total ultimate torque should be computed by two-part addition. Therefore, the discussion above described the equations to calculate the torque of non-strengthened beam and those for calculating the torsional contribution of FRP wrapping. The experimental ultimate torque for calculating ratios, average values, and standard deviations through six combinations are listed in Table 2. As shown in the table, Group 5 gives the poorest calculation among the experimental results, exhibiting the highest average and standard deviation values. Therefore, the combination of Group 5 is inappropriate in calculating the ultimate torque of strengthened beams. The ACI and *fib* models (Group 2), as well as the Chinese and Ghobarah models(Group 6), present good results in average ratios, which are 1.097 and 1.099, respectively; however, their standard deviations are higher than those of other groups, except Group 5. Almost all numbers calculated from Group 1 and Group 3 are noticeably greater than the experimental numbers, excluding the results of several beams, showing that the combinations have remarkably overestimated the calculation of the ultimate torque. The experimental ultimate torsional values of strengthened beams from the literature[2] are all remarkably lower than the computational values in all combinations. Although Group 3 presents the lowest standard deviation 0.156, it also displays the same case that most computed values with Group 1 and Group 4 are overestimated. Groups 1 and 4 of the Rahal model, combined with two equations of FRP torsional contribution calculating the strengthened box beams, show twice the experimental values probably because the area of the box beams used in the equation is not subtracted a hollow part.

According to all the group comparisons, Groups 2 and 6 showed relatively reasonable and close results compared with the experimental values, which are particularly dotted in Figure 1. Two groups display the same case with the ratio distributed between two lines, but the ratio is very close to the line of T_{exp} / T_{cal} = 1. The calculation values are fully conservative at the high level of the torsional moment, but at the low level, some values are conservative and some are overestimated. Removed the highly overestimated calculative values of the literature [2], Group 2 and Group 6 can more accurately estimate the ultimate torque of strengthened beams. However, the combination of Chinese model and Ghobarah model (Group 6) can predict the ultimate strength of RC box-section strengthened with fully wrapping or U-wrap precisely which can be clearly found from the results of six strengthened box-section beams in Table 2.

Table 1. Details of beams collected from the literature and tested by the authors.

Ref.	Beams	Section (mm)	fc (MPa)	Ø* (mm)	S_s (mm)	f_{yv} & f_y (MPa)	t_f & layer (mm)	w_f (mm)	s_f (mm)	E_f (GPa)
Zhang et al. [1]	L2	150x250	16.75	6.5,10	120	256,446	0.111	60	120	235
	L3	150x250	16.75	6.5,10	120	256,446	0.111	60	120	235
	L5	150x250	18.72	6.5,10	120	256,446	0.111	60	120	235
	L6	150x250	18.72	6.5,10	120	256,446	0.111	60	150	235
	L7	150x250	18.72	6.5,10	120	256,446	0.111	60	150	235
	L10	150x250	16.75	6.5,10	120	256,446	0.111	60	120	235
Ameli et al. [11]	CFE	150x350	39	6,16	80	251,502	0.165	1[d]	1[d]	244
	CFE2	150x350	39	6,16	80	251,502	0.165	1	1	244
	CFS	150x350	39	6,16	80	251,502	0.165	100	200	244
	CJS[c]	150x350	39	6,16	80	251,502	0.165	100	200	244
	CJE[c]	150x350	39	6,16	80	251,502	0.165	1	1	244
Ghobarah et al. [2]	C1	150x350	37	6.32,15	70	457,409	0.165	1	1	235
	C2	150x350	37	6.32,15	70	457,409	0.165	100	200	235
	C4	150x350	37	6.32,15	70	457,409	0.165	250	325	235
	C5	150x350	37	6.32,15	70	457,409	0.165	100	250	235
Chalioris [12]	Ra-F1	100x200	27.5	0,8		0,560	0.111	1	1	230
	Ra-F2	100x200	27.5	0,8		0,560	0.112	1	1	230
	Ra-FS1502	100x200	27.5	0,8		0,560	0.112	150	300	230
	Rb-F1	150x300	28.5	0,8		0,560	0.111	1	1	230
	Rb-F2001	150x300	28.5	0,8		0,560	0.111	200	400	230
	Rb-F3001	150x300	28.5	0,8		0,560	0.111	300	600	230
Hii et al. [10]	FS050D2	350x500	56.4	6,25	125	426,395	0.176	50	175	240
	FH075D1	350x500x50[b]	48.9	6,25	125	426,395	0.176	50	262.5	240
	FH050D2	350x500x50[b]	52.8	6,25	125	426,395	0.176	50	175	240
	FH050D1	350x500x50[b]	56.4	6,25	125	426,395	0.176	50	175	240
Authors[a]	TBS[c]	400x350x50[b]	40.37	6,16	100	372,502	0.117	100	200	230
	TBSL1[c]	400x350x50[b]	35.03	6,16	100	372,502	0.117	100	200	230
	TBSL2[c]	400x350x50[b]	37.57	6,16	100	372,502	0.117	100	200	230

[a] beams were tested by authors; [b] for box beams; [c] U-jacketing wrapping; [d] for FRP complete wrapping.
*refer to the stirrup and longitudinal steel diameters.

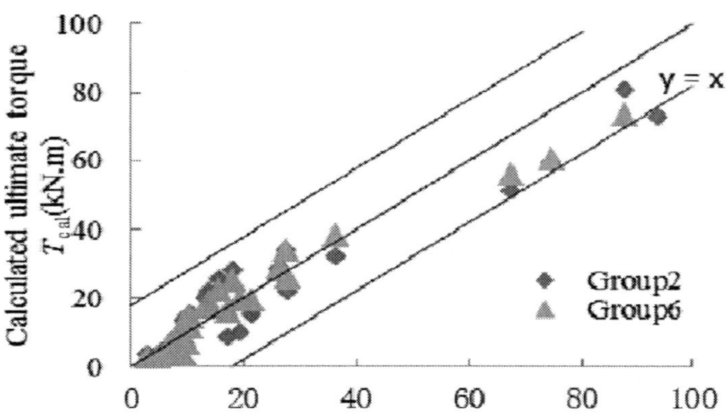

Figure 1. Group 2, 6 with experimental results comparison.

Table 2. Comparison of experimental results with calculated results with different model combinations.

Beams	Section (mm)	Ultimate Torque, T_{exp} (kN.m)	Group 1	Group 2	Group 3	Group 4	Group 5	Group 6
				T_{exp} / T_{cal}			T_{exp} / T_{cal}	
L2	150x250	8.21	0.775	0.974	0.752	1.018	1.391	0.979
L3	150x250	10.23	0.605	0.693	0.593	0.734	0.869	0.718
L5	150x250	10.53	0.615	0.701	0.598	0.759	0.895	0.733
L6	150x250	9.09	0.923	1.174	0.879	1.225	1.711	1.149
L7	150x250	9.55	0.621	0.719	0.602	0.788	0.954	0.758
L10	150x250	10.51	0.75	0.888	0.734	0.956	1.19	0.929
CFE	150x350	28	0.996	1.287	0.945	1.157	1.572	1.09
CFE2	150x350	36.5	0.948	1.136	0.913	0.99	1.197	0.951
CFS	150x350	21.7	0.998	1.412	0.934	1.215	1.89	1.121
CJS	150x350	17.4	1.161	2.022	1.056	1.184	2.093	1.075
CJE	150x350	19.5	1.203	1.982	1.101	1.092	1.698	1.007
C1	150x350	18.1	0.568	0.651	0.632	0.643	0.751	0.726
C2	150x350	14.1	0.547	0.65	0.626	0.639	0.784	0.749
C4	150x350	15.83	0.542	0.63	0.61	0.624	0.744	0.716
C5	150x350	13.4	0.54	0.66	0.632	0.63	0.799	0.759
Ra-F(1)	100x200	4.87	1.024	1.29	1.031	1.212	1.604	1.604
Ra-F(2)	100x200	6.65	0.935	1.084	0.939	0.943	1.095	1.095
Ra-FS150(2)	100x200	3.02	0.635	0.8	0.639	0.751	0.995	0.995
Rb-F(1)	150x300	10.05	0.767	1.04	0.778	0.979	1.471	1.471
Rb-F200(1)	150x300	9.315	0.993	1.566	1.012	1.359	2.727	2.727
Rb-F300(1)	150x300	7.52	0.801	1.264	0.817	1.097	2.202	2.202
FS050D2	350x500	93.8	1.069	1.291	0.699	1.58	2.12	0.886
FH075D1	350x500x50	67.5	0.475	1.566	0.942	0.534	2.468	1.208
FH050D2	350x500x50	74.8	0.467	1.451	0.932	0.534	2.369	1.24
FH050D1	350x500x50	87.7	0.462	1.207	0.864	0.543	1.982	1.199
TBS	400x350x50	26.67	0.423	0.886	0.74	0.484	1.201	0.948
TBSL1	400x350x50	26.52	0.428	0.902	0.744	0.485	1.194	0.932
TBSL2	400x350x50	27.56	0.41	0.8	0.678	0.454	0.989	0.809
Average Ratio			0.739	1.097	0.801	0.879	1.463	1.099
Standard Deviation			0.241	0.381	0.156	0.3	0.571	0.443

* Note:

Group 1 - (Rahal)Eq.1 + (fib 14)Eq.7
Group 2 - (ACI)Eq.2 + (fib 14)Eq.7
Group 3 - (China)Eq.3 + (fib 14)Eq.7
Group 4 - (Rahal)Eq.1 + (Ghobarah)Eq.10
Group 5 - (ACI)Eq.2 + (Ghobarah)Eq.10
Group 6 - (China)Eq.3 + (Ghobarah)Eq.10

6 Summary

Three models for calculating the torsional strength of RC beams and two models for computing FRP torsional contribution to strengthened members have been described. Assuming that no relationship exists between these two parts, the total ultimate torque of strengthened members can be obtained by the torsional strength of the RC beams and the torque of the FRP contribution. Six groups were combined using different models from two parts. Group 5 presents the highest mean and standard deviation values, which makes the combination inappropriate to calculate the ultimate torque. Although Group 1 and Group 3 reveal the lowest standard deviation, the most calculated results are more overestimated than the experimental values the same case to Group 4. The combination of ACI 318 and *fib* Bulletin 14 models (Group 2), as well as the combination of Chinese and Ghobarah models (Group 6), predict relatively reasonable and accurate results compared with the experimental values. For prediction of the ultimate torque of RC box-section beams strengthened with FRP completely wrapping or U-wrap, the results computed by Group 6 are in good agreement with the experimental values.

References

[1] J.W. Zhang, Z.T. Lu and H. Zhu, Experimental study on the behaviour of RC torsional members externally bonded with CFRP, *Proc. of the Int. Conf. on FRP composites in Civil Engineering*, **1**, 713-722, (2001)

[2] A. Ghobarah, M. Ghorbel and S. Chidiac, Upgrading torsional resistance of reinforced concrete beams using fiber-reinforced polymer, J. of Composites for Construction, **6**(4), 257-263, (2002).

[3] S. Panchacharam and A. Belarbi, Torsional behavior of reinforced concrete beams strengthened with FRP composites, *Proc. of 1st FIB Congress*, Osaka, (2002).

[4] fib Bulletin 14, Externally bonded FRP reinforcement for RC structures, Lausanne, (2001).

[5] A. Deifalla and A. Ghobarah, Full torsional behavior of RC beams wrapped with FRP: Analytical model, J. of Composites for Construction, **14,** 289-300, (2010).

[6] A.R. Zojaji and M.Z. Kabir, Analytical approach for predicting full torsional behavior of reinforced concrete beams strengthened with FRP materials, Scientia Iranica, **19**(1), 51-63, (2012).

[7] K.N Rahal, Torsional strength of reinforced concrete beams, Canadian J. of Civil Engineering, **27**, 445-453, (2000).

[8] ACI 318-2011, *Building Code Requirements for Structural Concrete*, American Concrete Institute, Michigan, (2011).

[9] China Structural Industry Press, Code for design of concrete structures GB 50010-2010, China Structural Science Academe Beijing, 64-73, (2010)

[10] A.K. Hii and R. Al-Mahaidi, Experimental investigation on torsional behavior of solid and box-section RC beams strengthened with CFRP using photogrammetry, J. of Composites for Construction, **10**, 321-329, (2006).

[11] M. Ameli,, H.R. Ronagh and P. F. Dux, Behavior of FRP strengthened reinforced concrete beams under torsion, J. of Composites for Construction, **11**, 192-200, (2007).

[12] C.E. Chaliori, Torsional strengthening of rectangular and flanged beams using carbon fibre-reinforced-polymers-experimental study, Construction and Building Materials, **22**, 21-29, (2008).

MATEC Web of Conferences **47**, 02007 (2016)
DOI: 10.1051/matecconf/20164702007

Potential Effects of Corrosion Damage on the Performance of Reinforced Concrete Member

Hamidun Mohd Noh[1,2,a] and Yoshimi Sonoda[1]

[1]*Department of Civil Engineering, Kyushu University, 744 Motooka, Nishi-ku, Fukuoka 819-0395, Japan*
[2]*Faculty of Technology Management and Business, Universiti Tun Hussein Onn Malaysia, 86400 Parit Raja, Johor, Malaysia*

Abstract. Corrosion of steel in reinforced concrete is one of the most serious issues for the owners. The main consequences of this phenomenon include loss of cross section of steel area, induced the expansive pressure which caused cracking of concrete, spalling and delaminating of the concrete cover. Thus, it reduces the bond strength between steel reinforcing bar and concrete, and deteriorates the strength of whole structure. In this study, a non-linear finite element analysis was conducted to investigate the potential effects of corrosion damage. A series of corroded reinforced concrete cylinder is used in the parametric study to assess the influence of different level of corrosion to the structural performance. As a result, a clear picture of corrosive expansion was investigated and the relationship between the expansive pressure and the degree of corrosion was identified to be the important factors of the structural deterioration.

1 Introduction

Corrosion reinforcement is one of the main causes of structural deterioration and the most predominant degradation mechanisms in the reinforced concrete structures. It is usually associated with carbonation phenomenon or chlorides penetration, which generally induces uniform and localized, attacks respectively. Corrosion of steel in concrete causes internal damage to reinforced concrete elements, owing to the loss of steel area and the formation of associated expansive corrosion products.

The properties of oxide layers as the corrosion products have a great influence on corrosion crack in the concrete. The most influential factor is the expansion ratio of the oxide which depends on the specific type of oxide formed [1, 2]. Depending on the level of oxidation, the volume increase due to rebar corrosion is commonly around 2.0 [3, 4] and would be up to 6.5 times than the original iron volume which get consumed by the corrosion process because of the formation of various corrosion products [5]. Nevertheless Molina [4] claimed that the effective expansion ratio maybe less than that corresponding to a given type if the oxide diffuse in the porous structure of the concrete.

When corrosion of reinforcement develops significantly, the corrosive products expand continuously and generate internal pressure to the concrete surface around the steel bar. The continuous process of reinforcement corrosion does not only affects in structural serviceability by

[a] Corresponding author : hamidun@doc.kyushu-u.ac.jp

cracking, or even spalling the concrete cover, but also give serious impact on the structural safety by decreasing the load-bearing capacity. Besides, the physical effects of corrosion include loss of steel area, loss of bond strength between steel reinforcing bars and concrete, and reduce the concrete strength due to cracking.

In the present study, the potential effects of corrosion damage on the performance of reinforced concrete cylinder were investigated. The main effects of corrosion are recognized and considered in the analysis. A reinforced concrete cylinder with concrete diameter of 100mm and 200mm length was modelled consisting of 19mm diameter of reinforcement bar. The detail dimension is shown in Figure 1.

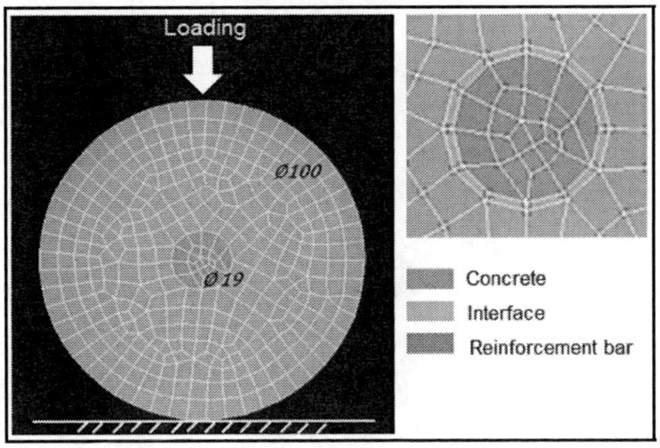

Figure 1. Beam cross-section and dimension.

2 Numerical Simulation

2.1 Finite element modelling of beam specimen

A three-dimensional finite element model is adopted in this study using the non-linear finite element code of MARC 2010. By using the 8 nodes of hexahedral elements, concrete and the main reinforcements was classified as solid. By considering the changes in the adhesion strength between the reinforcement bar and concrete, the interface element was applied by using 8 nodes hexahedral elements.

2.2 Corrosion damage model

Figure 2 shows the material properties of concrete and the bond-slip relationship. The rebar was assumed as complete elasto-plastic body and constitutive law based on the von Mises yield condition is adopted. As for the concrete, elastic-plastic constitutive law based on the Drucker-Prager yield condition is applied. The degradation ratio due to the corrosion is defined according to the degree of damage in three principal axis direction as $\varnothing_x$, $\varnothing_y$ and $\varnothing_z$ by assuming the tension softening. To adequately assess the progress of cracks in the concrete, it is necessary to make the tension softening stress released by subsequent tensile strength. Then, only the extracted positive principal strain component, the degree of damage in principal strain direction D^L_x, D^L_y and D^L_z are converted to the whole coordinate axes of D^G_x, D^G_y and D^G_z. The degradation ratio of each coordinate axis direction was calculated by the following Equation (1).

$$\varnothing_x = \sqrt{1 - D_x^G}, \left(D_x^L = \frac{\varepsilon_x^L}{\varepsilon_u} \right); \varnothing_y = \sqrt{1 - D_y^G}, \left(D_y^L = \frac{\varepsilon_y^L}{\varepsilon_u} \right); \varnothing_z = \sqrt{1 - D_z^G}, \left(D_z^L = \frac{\varepsilon_z^L}{\varepsilon_u} \right) \quad (1)$$

Equation (1) shows the relationship between $Ø_x$, $Ø_y$, $Ø_z$ and $D^G{}_x$, $D^G{}_y$, $D^G{}_z$ with the definition of $D^L{}_x$, $D^L{}_y$, $D^L{}_z$. Here, ε_u is defined as tensile strain limit. In addition, the relationship between the stress rate and the degradation ratio associated with the tension softening is defined in Equation (2).

$$\alpha_{xx} = {\phi^2}_x, \quad \alpha_{yy} = {\phi^2}_y, \quad \alpha_{zz} = {\phi^2}_z \tag{2}$$

Furthermore, in this analysis the degradation ratio was multiplied by elastic stiffness matrix and anisotropic constitutive law with damage influence is obtained as shown in Equation (3).

$$\{\sigma_{ij}\} = \{[D^e]-[D^p]\}\{\varepsilon_{kl}\} \tag{3}$$

where D^e is derived as following equation :

$$[D^e] = \begin{bmatrix} \varphi_x^2(\lambda+2\mu) & \varphi_x\varphi_y\lambda & \varphi_x\varphi_z\lambda & 0 & 0 & 0 \\ \varphi_y\varphi_x\lambda & \varphi_y^2(\lambda+2\mu) & \varphi_y\varphi_z\lambda & 0 & 0 & 0 \\ \varphi_z\varphi_x\lambda & \varphi_z\varphi_y\lambda & \varphi_z^2(\lambda+2\mu) & 0 & 0 & 0 \\ 0 & 0 & 0 & \varphi_x\varphi_y 2\mu & 0 & 0 \\ 0 & 0 & 0 & 0 & \varphi_y\varphi_z 2\mu & 0 \\ 0 & 0 & 0 & 0 & 0 & \varphi_z\varphi_x 2\mu \end{bmatrix} \tag{4}$$

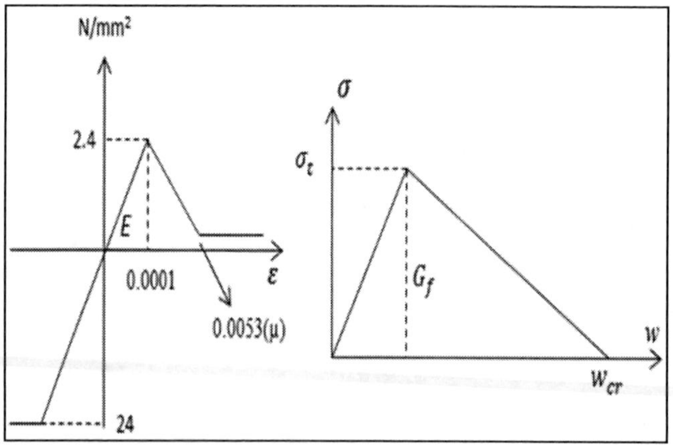

Figure 2. Material properties and bond-slip.

3 Mechanism of Deformation Process Due To Steel Corrosion

3.1 Corrosion and expansion pressure

The corrosion in concrete causes volume expansion of steel bar. Generally, the expansive pressure due to the volumetric change of the steel bar in concrete induces the tensile strains in the surrounding concrete and they are proportional to the degree of corrosion. As the expansion process begins continuously, the tensile strains developed and induces cracks in the surrounding concrete and also at the surface of concrete cover. In order to analyze the cracking of concrete cover due to steel corrosion, it is necessarily to know the relation between the amount of corrosion of steel bar and the internal pressure arises from the corrosion. Therefore, a realistic relation between the amount of corrosion and the internal expansion pressure can be investigated in this study.

3.2 Deformation of corrosion layer

Kim [6] proposed a deformation model of steel bar corrosion. It was detailed out in Figure 3 which consist r_0 as the initial radius of steel bar and x_p is the loss radius of steel bar due to corrosion. Meanwhile, the radius increment of steel bar due to corrosive expansion under restraint and unrestraint condition (mm) represent by U_r and Δr_b, respectively while the degree of corrosion of the steel bar which based on the ratio of weight loss due to corrosion to the initial weight of steel bar is known as w_{corr}.

The corrosion product layer is compressed by the strain ε_{rust} due to expansive pressure, *P*. Lundgren [7] derived the relationship between normal stress versus strain of rust evaluated from a combination of experimental results and analysis together with the chosen parameter K_{rust} and *p*. According to their research, analysis results have a good agreement with the experimental results using the value of K_{rust} = 7.0 GPa and p = 7.0. Thus the expansive pressure P is defined in Equation (5) as follows:

$$P = K_{rust}\left[\frac{u_r - r_b(\sqrt{1+(v_r-1)w_{corr}}-1)}{r_b[(1-\sqrt{1-w_{corr}})+(\sqrt{1+(v_r-1)w_{corr}}-1)]}\right] \quad (5)$$

3.3 Mechanical behavior of the rust

The mechanical behavior of the corrosion products needs to be known in modelling the corrosion layer. Molina [4] assume that the rust is elastic but the mechanical properties of the rust should be replaced by initial property of steel. Meanwhile, a scratching test was used by Petre-Lazar and Gerard [8] to explore the mechanical properties of the corrosion products. They conclude that rust is cohesion-less assemblage of incompressible crystals. Al-Sulaimani [9] and Cabrera and Ghoudoussi [10] have carried out pullout tests on corroded reinforcement bars concentrically placed in concrete blocks. In their study, the reinforcement corroded until the specimen was cracked and corrosion penetration was measured by the weight loss method. In addition, they performed axisymmetric finite element analyses of the test specimens and only the concrete was modelled with expansive normal stress. In this study, we also considered only the concrete domain with a constitutive model based on elastic-plastic theory.

Since the Young's modulus of steel is much larger than concrete, Lundgren [7] assumed that the deformation of the reinforcement bar is negligible. Furthermore, the stiffness of the bond layer is large enough so that the deformation of the bond layer is also negligible. Thereby, Lundgren [7] concluded that the deformation at the hole approximately equals to the deformation in the corrosion layer, U_r. Therefore, the value of U_r can be calculated as the value of strain in the rust.

4 Results and Analysis

In general, cracks and rebar corrosion of the reinforced concrete structures are caused by various environment factors and reduction of adhesion strength is observed. In this study, the adhesion strength of rebar was also considered with reference to the research by Lee [11]. The relationship of the corrosion rate and the adhesion strength of the corroded reinforcing bars were stated in Equations (6) and (7).

$$\tau_{\max} = 0.34\sigma_c - 1.93 \quad (6)$$

$$\tau_{\max} = 5.21e^{-0.0561\Delta w} \quad (7)$$

Here, σ_c is the compressive strength of the concrete and Δw represents the corrosion rate of reinforcing steel (%). Corrosion of the rebar was analyzed with respect to the 2 levels of degradation state which are 0%, 10%, 20%, 40% and 60%. The bond strength for each corrosion rate are 6.23MPa, 2.97MPa, 1.70MPa, 0.55 MPa and 0.18MPa, respectively.

4.1 Volume expansion of rust in corroded concrete

When a rebar starts to corrode, a gradual decrease of its diameter is produced. In this analysis, the gradual decrease of diameter, *xp* is calculated at 0.4875mm. Together with the generation of rust, the increases in volume induce the formation of cracks when the concrete's tensile strength is surpassed. The cracks are generated first at the bar-concrete interface and then propagate radially.

As can be seen in Figure 3, it is presented the rust expansion of 19mm diameter steel bar at 20% of corrosion degree. Under the unrestraint condition, Δr_b at α value of 0.523 and 0.622 correspond to the free expansion of 1.6303mm and 1.2544mm, respectively. Meanwhile, under restraint concrete condition, the value of U_r at α value of 0.523 and 0.622 are 1.6209mm and 1.2463mm, respectively.

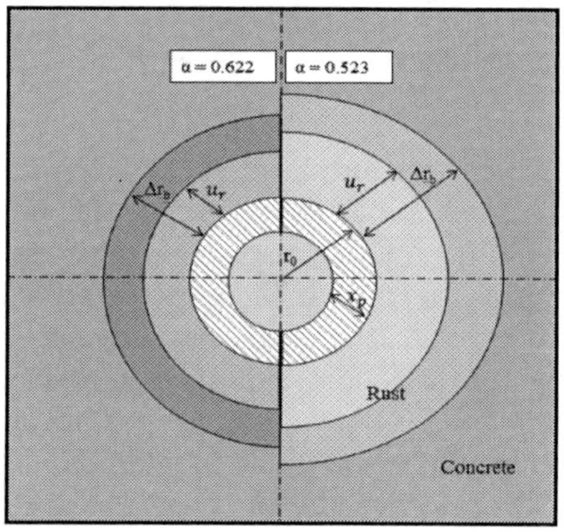

Figure 3. Expansion of rust in concrete.

4.2 Expansive pressure in corroded concrete

Once corrosion is induced in a structure, the expansive pressure will be generated which caused cracking to the concrete. As can be seen in Figure 4, the expansive pressure is increasing when the amount of corrosion increase.

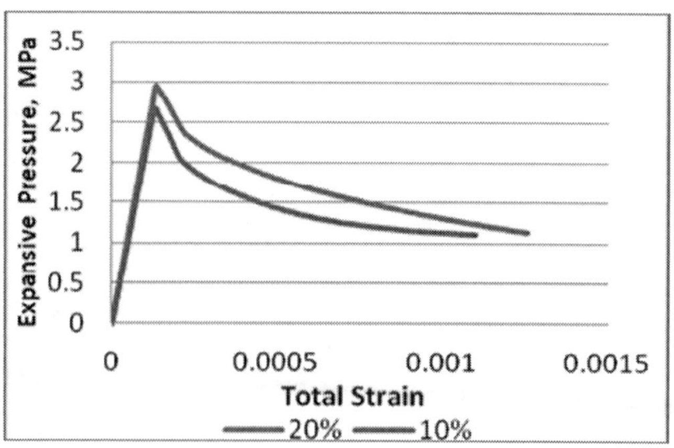

Figure 4. Expansive pressure relationship.

Structure with 20% corrosion generates higher expansive pressure with 2.94MPa. Meanwhile, the pressure value drop until 2.62MPa on the 10% corrosion curve. The distribution of strain at the cross section of the beam is presented in Figure 5. Since the corroded areas consist of the reinforcement only, the strain distribution is concentrated at the bar-concrete area as the stress level is increasing when the corrosion degree increase.

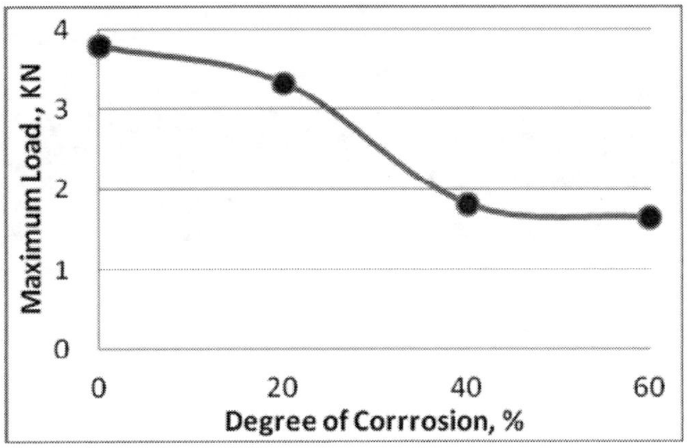

Figure 5. Structural strength reduction.

4.3 Structural strength reduction due to rebar corrosion

Further analysis was conducted to investigate the influence of corrosion to the structural strength. The corrosion is considered along the reinforcement bar in the concrete cylinder. Figure 6 presents the relationship between load bearing capacity and the degree of corrosion of the concrete cylinder. It was shown that 0% corrosion have the higher bearing force with 3.79kN. Meanwhile, the value of the bearing force for corrosion degree of 20% dropped slightly to 3.33kN. Then, the gradient of the graph change drastically to the 40% corrosion degree with bearing capacity of 1.81kN before decrease to 1.64kN at 60% degree of corrosion. The results illustrate that the corroded structure generates expansive pressure and caused cracking to the concrete surface. After reach the maximum bending capacity, the structure get off as the corrosion cause severe condition since the stiffness decrease due to corrosion and caused noticeable strength reduction as shown in Figure 7.

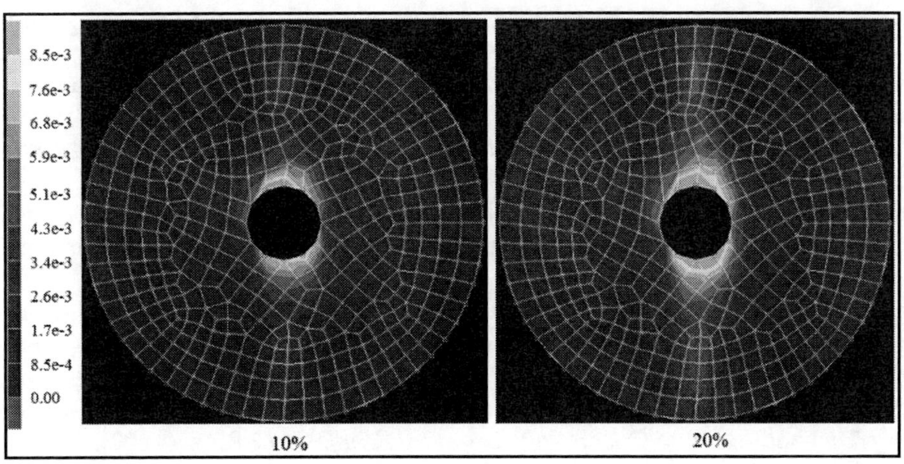

Figure 6. Strain distribution.

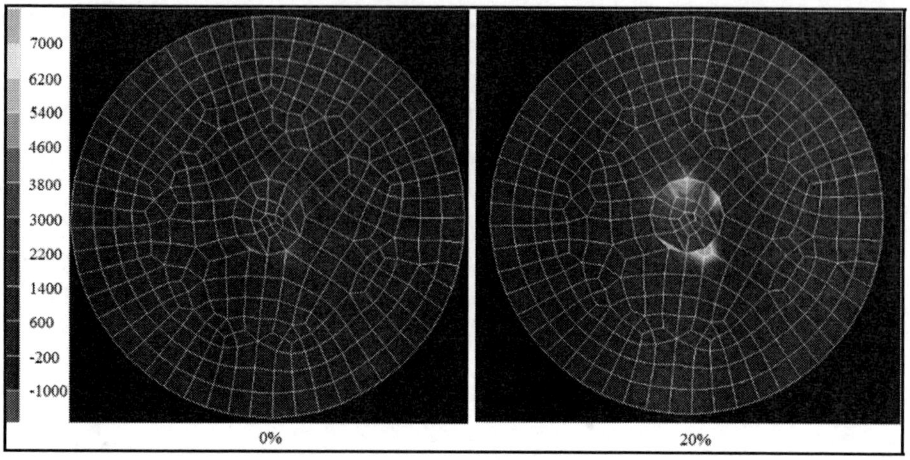

Figure 7. Stress distribution.

5 Conclusions

In this study, we could predict the residual strength of reinforced concrete cylinder under the corrosion condition using the finite element analysis. However, the quantitative relation between the degree of corrosion and its influence on the mechanical properties are still needed to investigate. Thus, we are going to do the experimental validation in the next stage.

References

[1] K. Tutti, Corrosion of steel in concrete. Report No. 4 of Swedish Cement and Concrete Research Institute, Stockholm, Sweden, (1982).

[2] P.K. Mehta and P.J.M. Monteiro, *Concrete Microstructure, Properties and Materials*, Indian Concrete Institute, Chennai, India (1997).

[3] C. Andrade, C. Alonso and F.J. Molina, Cover cracking as a function of rebar corrosion: Part I-experimental test, Mater. Struct., **26**, 453-464, (1993).

[4] F.J. Molina, C. Alonso and C. Andrade, Cover cracking as a function of rebar corrosion: part II-numerical model, Mater. Struct., **26**, 532-548, (1993).

[5] K. Bhargava, A.K. Ghosh, Y. Mori and S. Ramanujam, Model for cover cracking due to rebar corrosion in RC structures, Eng. Struct., **28**, 1093-1109, (2006).

[6] K.H. Kim, S.Y. Jang, B.S. Jang and B.H. Oh, Modeling mechanical behavior of reinforced concrete due to corrosion of steel bar, ACI Material J., **107**(14), 106-113, (2010).

[7] K. Lundgren, Bond between corroded reinforcement and concrete, Technical Report, Chalmers University of Technology, Göteborg, Sweden, (2001).

[8] I. Pètre Lazar and B. Gerard, Mechanical behaviour of corrosion products formed at the steel concrete interface: Testing and modelling, Conditions Monitoring of Materials and Structures, *Proc. of 14th Engineering Mechanics Conf.*, United States, (2000).

[9] Al-Sulaimani, G.J. Kaleemullah, M. Basunbul and I.A. Rasheeduzzafar, Influence of corrosion and cracking on bond behaviour and strength of reinforced concrete members, ACI Struct. J., **87** (2), 220-231,(1990).

[10] J.G. Cabrera and P. Ghoddoussi, The effect of reinforcement corrosion on the strength of the steel-concrete bond, Bond in Concrete, *Proc. of Int. Conf. CEB*, Riga, (1992).

[11] H.S. Lee, T. Noguchi and F. Tomosawa, Evaluation of the bond properties between concrete and reinforcement as a function of the degree of reinforcement corrosion, Cement and Concrete Research, **32**(8), 1313-1318, (2002).

MATEC Web of Conferences **47**, 02008 (2016)
DOI: 10.1051/matecconf/20164702008

Prestress Loss and Bending Capacity of Pre-cracked 40 Year-Old PC Beams Exposed to Marine Environment

Amry Dasar[1,a], Rita Irmawaty[2], Hidenori Hamada[3], Yasutaka Sagawa[3] and Daisuke Yamamoto[3]

[1]*Graduate School of Engineering, Kyushu University, 744 Motooka, 819-0395 Nishi-ku, Fukuoka, Japan*
[2]*Civil Engineering Department, Hasanuddin University, Perintis Kemerdekaan KM. 10 Makassar, 90245 South Sulawesi, Indonesia*
[3]*Faculty of Engineering, Kyushu University, 744 Motooka, 819-0395 Nishi-ku, Fukuoka, Japan*

Abstract. Six prestressed concrete beams (PC beam) were used for evaluation, consist of four post-tension beams (PC-O) and two pre-tension beams (PC-R). In order to investigate the effect of crack on prestress loss and bending capacity after long-term exposed, prestressed concrete beams were pre-crack and then exposed to marine environment. Experimental work was carried out to evaluate PC beams performance after long-term exposed. In addition, visual observations and load bearing capacity test was carried out. Furthermore, evaluation of prestress loss conducted using three-point loading bending test and the remaining tendon forces in the beam were determined using Crack Re-opening Method. The experimental results revealed that prestress loss was increased due to corrosion of strand/wire which affected by the pre-crack on the prestressed beams. Approximately a prestress loss around 26% and 30% was recorded for post-tension and pre-tension beams respectively.

1 Introduction

Reinforced concrete and prestressed concrete structures exposed to aggressive environments, cracks play an important role in accelerating the corrosion process [1]. In contrary, rather than crack widths, the surrounding concrete (i.e., concrete resistance, oxygen permeability, and chloride ion content) more contributed to the corrosion rate of steel bars in concrete, it is expected that with any size of cracks [2]. On the other hand, high strength prestressing steels show a far more sensitive reaction to corrosion attack than reinforcing bars. As the result, the loss of an equivalent sized reinforcing bar have not given a more significant effect on the structural capacity compared if the loss of one prestressing strand or wire will be. Several researchers found that corrosion of prestressing steel reduced in flexural capacity. If one 12.7 mm diameter prestressing strand is assumed to fail in the 100% U PS (prestressed based on nominal strengths requirements) section the nominal flexural capacity is reduced by 13.5% [3]. Furthermore, bridge was assessed to be up 10% loss in live load capacity in the individual beam which has lost the two strands [4].

[a] Corresponding author : amrydasar@doc.kyushu-u.ac.jp

2 Experimental Program

2.1 Beam details

Six prestressed concrete beams (PC beam) were used for evaluation, consist of four post-tension beams (PC-O) and three pre-tension beams (PC-R). All the beams were cast in 1975, and have 2400 mm in length, with identical cross-section of 150x300 mm. Variation of bar arrangements and cover depth are adjusted to the type beam designated. Cover depths were 30mm and 35mm for the PC-O beams and PC-R beams respectively. Prestressing tendons were round wires of 2-2.9 mm in diameter with yield strengths 1148N/mm^2 for PC-R beams and round bars of 17mm in diameter with yield strengths 1795N/mm^2 for PC-O beam respectively. Deformed bars of 10 mm in diameter were embedded as stirrups with spacing of 100mm. High early strength Portland cement was used as a cement for making concrete. The properties of the aggregate with fine-river sand (FM = 2.84; SG = 2.25) and crushed stone (FM = 6.63; SG = 2.25). Mix proportion of concrete was designed with water/cement ratio (w/c) and sand/total aggregate ratio (s/a) of 40.7% and 37.0%, respectively. The composition of material in unit weight (kg/m^3) for water, cement, sand and gravel was 167, 410, 640 and 1175 respectively [5]. The concrete was designed with a slump target of 5 ± 1 cm and air content about 3 ± 1%. After placing of concrete until the start of exposed, PC beams were steam cured (max. temperature was 60°C) for about 10 hours followed by curing in air after the beams were moisture cured for one day and demoulded. Prestress was introduced by the pre-tensioned method for PC-R beam and the post-tensioned method for PC-O beam. Cement paste was grouted between the bars and the sheath after post-tensioning. Also, the ends of each beam were covered with a pad of cement mortar to prevent corrosion of the ends of strands or ends anchorages.

Table 1. Summary of PC beam specimens.

Beam	Type	Cover Depth (mm)	Cracking Condition	Age of Specimen (years)	Effective Pretsress (N/mm^2)
PC-O-1	Post-tensioning	30	No Pre-cracked	35	13.7
PC-O-2			No Pre-cracked	35	
PC-O-3			Pre-cracked	40	
PC-O-4			Pre-cracked	40	
PC-R-1	Pre-tensioning	35	No Pre-cracked	35	12.7-13.2
PC-R-2			Pre-cracked	35	

2.2 Exposed condition

All beams have been exposed to actual marine environments at the Sakata Port for 20 years, which is located in north-west of Japan (38°56'N, 139°47'E) and faces the Sea of Japan. Two months after placing of concrete, the beams were placed in the tidal zone just in front of a caisson-type quay wall, where the beams have been subjected to wet and dry conditions alternately due to tidal action. The beams can be categorized into two conditions in term of pre-cracking, i.e. 'A' and 'B'. Condition A refers to pre-cracked, for which beams were pre-cracked by bending moment until 65% of the ultimate bending moment [6]. After cracking, the bending moment was released, and no continuous load was applied during exposed. Condition B refers to no pre-cracks. Summary of the beam specimens were presented in Table 1. After 20 years of exposed, the beams were transferred to the PARI (Port and Airport Research Institute) laboratory, stored at a constant temperature and sheltered from the rain over 15 years. Then beams were taken out to Kyushu University site for laboratory tests [7].

2.3 Loading procedure

Three-point loading with a distance of support 2100 mm for bending test was carried out. Bending test conducted in three separates stage was applied to each beam. In the first stage of testing, strain gauges, crack gauges and displacement transducers (LVDT, linear variable differential transformer) installed after load was applied to create and locate a series of flexural cracks. The second stage of testing conducted based on strain and displacement measurements of crack openings for the crack identified and instrumented in the first stage of testing. Then, the second stage was used in order to determine the decompression load and first cracking load in each beam. Stage three each test involved the loading until the ultimate failure reached to each beam.

2.4 Prestress loss evaluation

Nowadays, prestress loss can be determined analytically and experimentally. Analytically techniques such as lump-sum methods, refined or detailed methods and time-step methods. Also, there are many ways to determine the loss of prestress in experimentally. Although experimental techniques, still require a back-calculation of the prestress loss from the test data using theory of mechanics concepts. The experimentally method that will use in this research is load testing to determine crack initiation and/or crack re-opening loads to obtain the available compressive stress in the bottom flange of a member. This method is destructive test and provide information only on existing prestressing tendon stress at testing times. Some investigators define the decompression load as the load at which the first crack appears at the extreme fibers of the critical section, such us the bottom midspan fibers of a simply supported beam [8]. Then, effective prestress force (P_e) was calculated by using the formula as shown as below:

$$f = -\frac{P_e}{A_C}\left(1+\frac{ey_t}{r^2}\right)+\frac{My_t}{I_g} \tag{1}$$

where f is stress at the center of gravity of prestressing force, P_e is effective prestress after losses, A_c is cross sectional area, e is eccentricity, y_t is neutral axis, r^2 is radius of gyration, M is moment and I_g is moment of inertia.

3 Results and Discussion

3.1 Visual observation

The cracking traces of appearance beams are shown in Figure1. Flexural crack was not found in PC-O-3 although it was applied for pre-cracked before exposed. It implys that, narrower cracks heal in the natural marine environment after long-term exposed [9]. In addition, if at the flexural crack to be expected to corrosion firstly appears, the phenomenon a self-healing [10] will occurred due to slow down of chlorides migration generated from the crack filling by the corrosion products [11], and corrosion activity will be decreased [12]. This tends to prove that for longitudinal corrosion cracks, there is no self-healing as in the case of flexural cracks [13]. Moreover, longitudinal cracks pattern was found along the beams with maximum crack widths for PC-O-3 showed larger than PC-O-2 with 2.2 mm and 0.5 mm, respectively. It indicates that the passive film breaks down already since the chloride threshold reached and corrosion likelihood started inspite narrower cracks healed. The cracks position showed that the crack occurred due to internal pressure resulted from corrosion of tendon. It shows that pre-cracked promote more corrosion activity and affected to increase size of crack width in spite of the fact that the tendon covered by grouted mortar and sheet. Based on JSCE standard for both type beam conditions can be categorized as second half of acceleration stage (e.g., numerous cracks occur due to corrosion, leaching of rust is observed, partial spalling is found and the corrosion amount of steel increased) for grades of structure appearance and deterioration state [14].

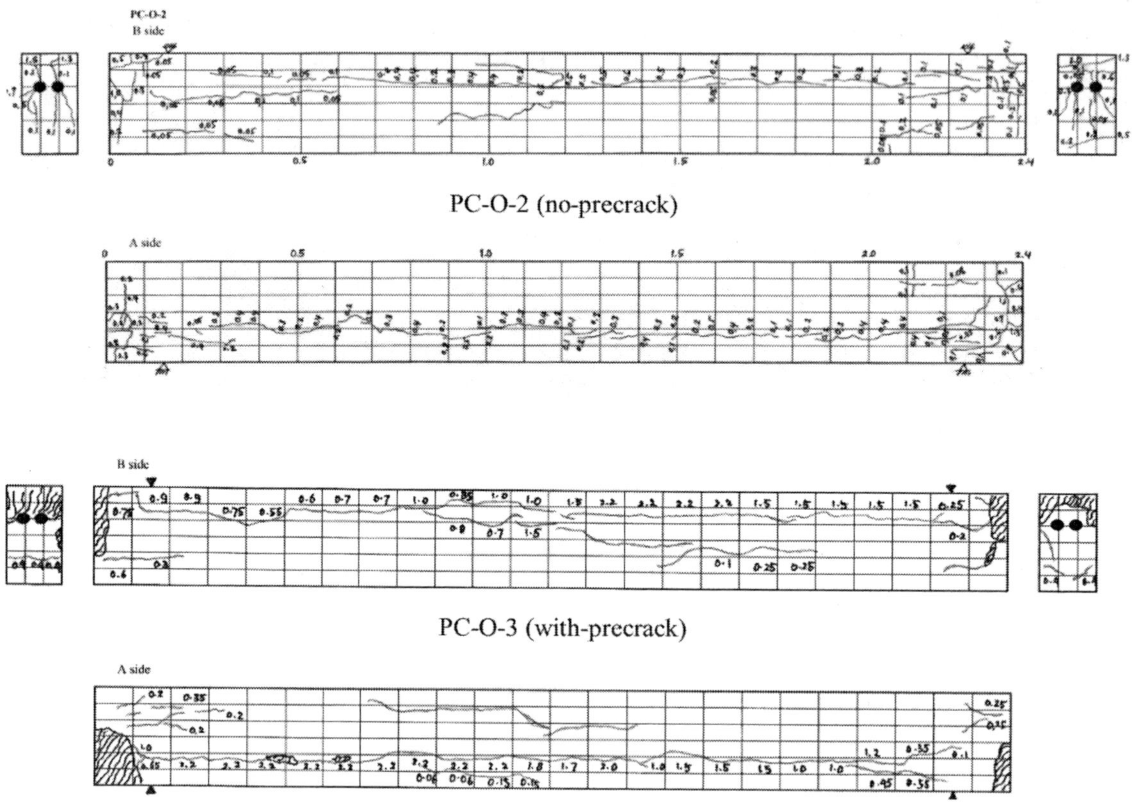

Figure 1. Crack pattern of beams over 40 years of exposed.

Extensive damage visually showed by PC-O beams compared to PC-R beams. In post-tensioned structures the anchorage zone is subdivided into the local zone and general zone, where local zone refers to the volume of concrete surrounding and immediately ahead of a mechanical anchorage device that transfers the prestress force to the member. While, pre-tensioned structures the local zone term is not applicable since these members do not contain mechanical anchorage device. The stress resulted from mechanical anchorage device perform the number of cracks in the end part of PC-O beams.

3.2 Decompression load and prestress loss

The decompression load was determined by intersection the two slopes of the load-strain curves. The relationship between applied load and strain is presented in Figure 2. Using the decompression load and theory of mechanics concepts then the effective prestress can be calculated (Figure 3). In addition, by comparing with the initial prestressing force (P_i), prestress loss was calculated (Table 2). It seems that pre-cracked affected to the decompression load as the results reducing in effective prestressed. Post-tensioning and pre-tensioning beams with pre-cracked showed smaller decompression load obtained compared than without pre-cracked. This means that pre-cracked could considered to reduce decompression load about 16% for post-tensioning and 19% for pre-tensioning, respectively. The most probable explanations for above conditions are loss of bond between the strand/wire and concrete or through cracking and spalling of the concrete may give significantly affect as the result from corrosion product. Also, by introducing pre-cracked before exposed, means allow more aggressive agents penetrate through inside concrete and decreased life time of structure become shorter to enter deterioration stage.

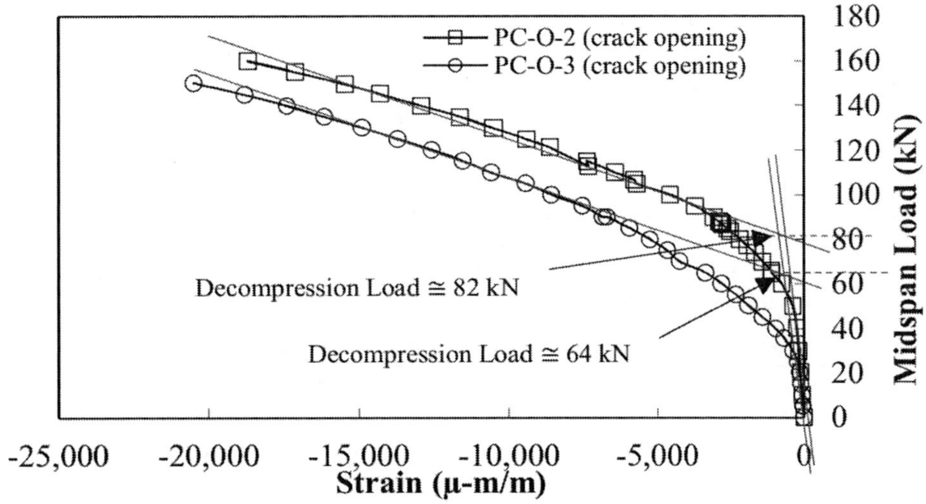

Figure 2. Determination of decompression load from the load-strain data.

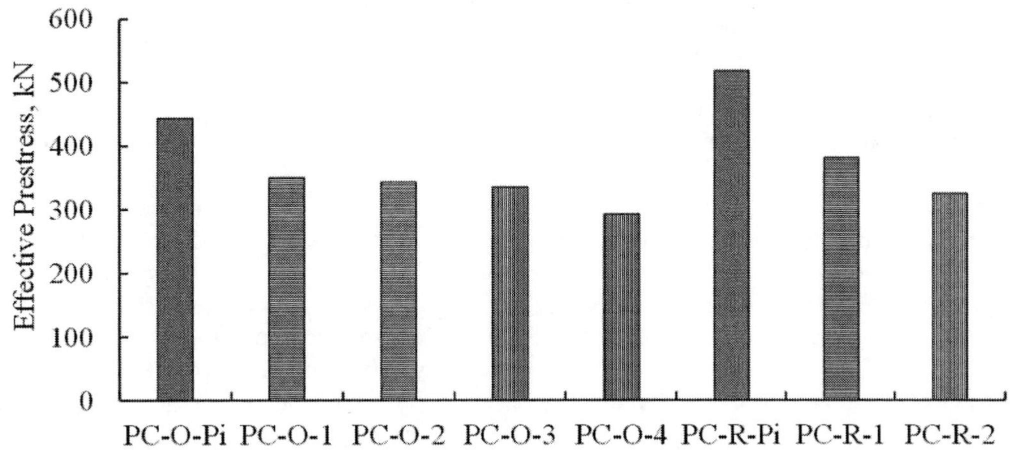

Figure 3. Effective prestress of PC beams.

The effect of pre-cracked for both PC-O and PC-R is presented in Table 2. The prestress loss increased due to pre-cracked 24% for PC-O and 31% for PC-R respectively. Post-tensioning type, pre-cracked also contributed of prestress loss due to produce deteriorated a part of anchorage zone. During loading for pre-cracked (i.e., 0.65 ultimate moment) anchorage zone have an increasing local stress and produced numerous of crack. In addition, for post tensioning type, the anchors moved inwards during the pre-cracked, which cause a shortening of the tendon thus increasing the prestress loss. One reason of prestressed loss for pre-tension is due to the bond between wire and concrete interrupt due to pre-cracked by bending, which is produce bond loss between wire and concrete. Further, prestress loss for pre-tension type also contributed from elastic shortening due to prestressing introduced to wire before casting which cause large of prestress loss. The prestress loss perhaps may be produced by accumulative of deterioration of the beams (i.e., spalling, longitudinal cracks, anchorage zone cracks, debonding) as the result of the corrosion products.

Table 2. Summary of several points during loading procedure.

Name	Applied Load, kN			Neutral Axis, mm	Initial Prestress, P_i (kN)	Effective Prestress, P_e (kN)	Prestress Loss, (%)
	Dec. Load	First Crack	Ultimate Load				
PC-O-1	80	92	215	22.8	443.5	350.8	20.9
PC-O-2	82	90	236	24.6		342.8	22.7
PC-O-3	64	50	227	62.4		334.8	24.5
PC-O-4	72	74	231	36.1		292.0	34.2
PC-R-1	89	90	192	22.2	517.3	381.3	26.3
PC-R-2	72	50	227	33.2		324.5	37.3

Summary of several points during loading procedure is presented in Table 2. The first crack observed both PC-O and PC-R beams with and without pre-cracked at load 50 kN and 90 kN respectively. It seems that pre-cracked reduced load-carrying capacity by promote early cracks during loading. The probable explanations why specimen with pre-cracked cracks occurred earlier, because higher of prestress loss make prestressed concrete starts to behave like a reinforced concrete beam when the value of the flexural moment is well beyond the cracking moment. This explanation was linked to the neutral axis data during loading, by reduced the tensile region means crack will occurred sooner. The effect of pre-cracked reduced the effective of tensile region about 70 mm and 10 mm for PC-O and PC-R beams respectively.

3.3 Ultimate flexural moment capacity

During load, flexural cracks appeared in the tension zone (i.e., both existed and new crack) and failure occurred by crushing of the concrete in the top face of compression region. Thus, the failure mode of beams was failure by bending. The ultimate flexural moment is expressed as the ratio to the initial (before exposed) flexural moment was shown in Figure 4. It seems pre-cracked affected to the flexural moment both PC-O and PC-R beams. The change of flexural moment affected due to pre-cracked about 5.58% and 4.44% for PC-O and PC-R respectively. The moment capacity decrease as loss of prestress in spite reached the large ultimate load enough. This considered performance degradation in sequence due to pre-cracked in the view point of load bearing capacity.

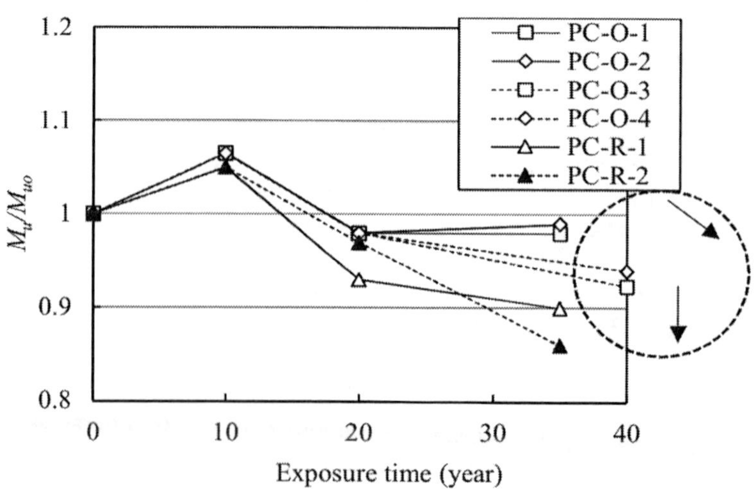

Figure 4. Change in flexural moment capacity.

4 Summary

Several comprehensive test results showed that pre-cracked promote more deterioration of the beams (i.g., spalling, longitudinal cracks, anchorage zone cracks) as the result of the corrosion products. In addition, load test results of post-tensioned and pre-tensioned concrete beams with corroded tendons/wires are presented also. The prestress loss increased due to pre-cracked for PC-O and PC-R about 26% and 30% respectively. Further, change of flexural moment capacity reduced due to pre-cracked about 5.58% and 4.44% for PC-O and PC-R consecutively. This considered degradation of structural performance contributed by pre-cracked more reasonable by increased prestress loss and reduced bending capacity

Acknowledgement

The authors thanks to PARI (Port and Airport Research Institute) for offering PC beam specimens tested in this experiment. The first author also wishes to thank Indonesia Endowment Fund for Education (LPDP) for financial support during study.

References

[1] H.S. Lopez-Calvo, T.W. Bremner, M.D.A. Thomas and P. Montes-Garcia, Effect of corrosion on tension bars in cracked HPC containing CNI, *Concrete Under Severe Condition*, Taylor and Francis,, 397-404, (2010).
[2] T.U. Mohammed, N. Otsuki, M. Hisada and S. Tsunenori, Effect of crack width and bar types on corrosion of steel in concrete, Material Civil Engineering, **13**(3), 194-201, (2001).
[3] J.S. West, C.J. Larosche, B.D. Koester and M.E. Kreger, State of the art report about durability of post-tensioned bridge substructures, Research Report 1405 (1), The University of Texas, Austin. (1999).
[4] S.M. Bruce, P.S. McCarten, S.A. Freitag and L.M. Hasson, Deterioration of prestressed concrete bridge beams, Report of Land Transport New Zealand Research, 337, 72, (2008).
[5] H. Hamada, N. Otsuki, and M. Haramo, Durabilities of concrete beams under marine environments exposed in port of Sakata and Kagoshima (after 10 years` exposed), Technical Note of The Port and Airport Research Institute, **614**, 3-43, (1988).
[6] H. Yokota, T. Fukute and H. Hamada, Structural assessment of deteriorated RC and PC beams exposed to marine environment for more than 20 years, Wakachiku Kensetsu Doboku Gijutsu Nenpo, **8**, 78-84, (1999).
[7] Rita Irmawaty, A study on long-term performance of concrete over 50MPa in compressive strength, PhD Thesis, Kyushu University, Japan, (2013).
[8] E.G. Nawy, *Prestressed Concrete: A Fundamental Approach*, Prentice Hall, New Jersey, (2009).
[9] T.U. Mohammed, N. Otsuki and H. Hamada, Corrosion of steel bars in cracked concrete under marine environment, Journal of Materials in Civil Engineering, **15**(5), 460-469, (2003).
[10] S. Jacobsen, J. Marchand and L. Boisvert, Effect of cracking and healing on chloride transport in OPC concrete, Cement and Concrete Research, **26**(6), 869-881, (1996).
[11] C.Q. Li, Corrosion initiation of reinforcing steel in concrete under natural salt spray and service loading – results and analysis, ACI Materials Journals, **97**(6), 690-697, (2000).
[12] P. Schiessl and M. Raupach, Laboratory studies and calculations on the influence of crack width on chloride-induced corrosion in concrete, ACI Materials Journals, **94**(1), (1997).
[13] T. Vidal, A. Castel and R. François, Corrosion process and structural performance of a 17 year old reinforced concrete beam stored in chloride environment, Cement and Concrete Research, **37**, 1551–1561, (2007).
[14] JSCE, Standard specification for concrete structures, Japan Society of Civil Engineers, (2007).

MATEC Web of Conferences **47**, 02009 (2016)
DOI: 10.1051/matecconf/20164702009

Optimum Shapes and Sizes on Torsion Behaviour of I-Beam with Web Opening

F. De'nan[1] and N. S. Hashim[1,a]

[1]School of Civil Engineering, Universiti Sains Malaysia, 14300 Nibong Tebal, Pulau Pinang, Malaysia

Abstract. Numerical study on torsion behaviour of I-beam with web opening was carried out and presented in this paper. Optimum shapes and sizes which enhance the torsion behaviour of the I-beam with web opening as well as lead to economic design in terms of both manufacture and usage.Comparison is made between I-beam without web opening and I-beam with various shapes and sizes of web opening. Different types of finite elements model, sizes and types of web opening were used to investigate the effect of various shapes and sizes of web opening on the torsion behaviour. The results are expressed in terms of torsional rotation. It is concluded that I-beam without web opening has lower torsional resistance compared to that of I-beam with various shapes of web opening. Meanwhile, the model 2 is the optimum model compare to model 1.The size of opening has slightly effect on the torsional angles values. It was noted that the optimum web opening size is 0.5D due to the low values of the torsional angle values compared with 0.6D and 0.7D.

1 Introduction

Nowadays, the provision of beams with web openings more reliable engineering practice, and eliminates the probability to cut the holes subsequently in inappropriate locations. This form of construction maintains a smaller construction depth with placement of services within the girder depth, at the most appropriate locations. In the 1960s, 1970s, and 1980s, studies on different web opening configurations were completed in the United States and Canada, including square, rectangular, circular, concentric, and eccentric openings in both non-composite and composite steel beams [1-4].

However, in modern composite structures, the behaviour of statically indeterminate castellated composite beams is more complex than that of simply supported beams [5]. This is because instability effects of the castellated composite beam may be subjected to the negative moment regions where the bottom compression flange of the beam is unrestrained. In other countries, this type of open-web beams has found widespread use, primarily in buildings because of great savings in materials and construction costs when beam spans exceed forty feet [6]. Furthermore, for large projects requiring more than one hundred beams, by using castellated beams some savings may be achieved. However, the provision of these web openings has a significant effect on the stress distribution and deformation characteristics [7]. Studies on the behaviour of steel beams with web opening using finite element analysis have been conducted in detail over the last three decades [8].

[a] Corresponding author : wani_a02@yahoo.com

Situations arise where torsional effects are significant, typically where the demands of practical construction result in eccentrically applied loads. For instance, precast units are often supported on one side of a flange or on a shelf angle; in the temporary condition, with one side loaded, most of the load is applied eccentrically. Besides that, if for architectural reasons, a beam cannot be placed concentrically under the wall it supports [9].

Therefore, the designer has to evaluate the magnitudes of the torsional effects and consider the resistances of the members under the combined bending and torsion. In some circumstances, the designer may choose to used 'closed' structural hollow sections, which have a much better performance in torsion; effects and resistances for these have to be evaluated. The effect of warping in all thin-walled cross section loaded by torsion must be assumed in stress and deformation analyses of structures [10]. During the analysis, the secondary torsion moment deformation effect was included into the stiffness matrix [11].

2 Torsion Analysis

Torsion is seldom relied upon as a significant load path in the design of steel buildings. This is primarily because open shapes are torsionally very flexible and twists readily when torque is applied. In uniform torsion, the applied torque is resisted by a single set of shear stresses distributed around the cross section. The ratio of the applied torque to the twist rotation per unit length is defined as the torsional rigidity, GJ of the member, where G is the shear modulus and J is the torsional constant [12, 13]. In many cases, only uniform (or St. Venant's) torsion is applied to the ends of a member and the rate of change of angle of twist is constant along the member and the ends are free to warp. In this case the applied torque is resisted entirely by shear stresses and no warping stresses result. The total angle of twist, ϕ is given by :

$$\phi = \frac{TL}{GJ} \tag{1}$$

where : T is torque applied, L is span length, J is torsional constant, and G is shear modulus. The maximum shear stress in the element of thickness, t is given by:

$$\tau_t = Gt\phi' \tag{2}$$

However, in non-uniform torsion, the second component of the resistance to torsional loading may act when the rate of change of the angle of twist rotation varies along the member. Therefore, an additional set of shear stresses may act in conjunction with those due to uniform torsion to resist the torque acting. The stiffness of the member associated with these additional stresses is proportional to the warping rigidity, EI_w, where E is the modulus of elasticity and I_w is the warping section constant [13].

When warping deformation is constrained, the member undergoes non-uniform torsion. Non-uniform torsion such as of an I-section fixed at one end is subjected to torsion at the other end. Here the member is restrained from warping freely as one end is fixed. The warping restraint causes bending deformation of the flanges in their plane in addition to twisting. The bending deformation is accompanied by a shear force in each flange. In general cases of loading and boundary condition, the warping is non-uniform along the axis of a-beam. This leads to a beam mechanical behavior that may be different from that predicted by Saint Venant beam theory or other theories which are restricted to uniform warping [14].

For thin-walled open section member, the torsion is resisted by a combination of the resistance to uniform torsion developed by shear stresses. This is almost nearly across the thickness of the section wall, and the resistance to warping torsion developed by equal and opposite flange bending and shear actions [15]. The effects of material and geometric nonlinearities on the strength s of I-section members in torsion have been clearly considered. It showed that linear first yield analyses significantly underestimate the real strengths because the favourable effects of strain-hardening, inelastic behaviour including torque redistribution, and the longitudinal Wagner stresses that develop at large rotations have been ignored [16]. The inelastic behaviour was modelled using trilinear stress-strain curves, a combines stress yield criterion, a flow rule and a hardening rule. It was found that the present model can predict more realistic results at higher rotations than other models which ignore the tranverse uniform torsion shear stresses [17].

The method of plastic collapse analysis provides the most logical method for strength design when the strength of the cross-section member is reduced by local buckling. Such member cross-sections are usually described as 'plastic' or 'compact' and width-thickness limits for such members in bending or shear are specified in design codes [18]. Corresponding method were also developed for first hinge design (without plastic redistribution), first yield design, and local buckling design [13]. The use of plastic design should be limited to plastic members which have sufficient ductility to reach the plastic collapse mechanism.

In most engineering type structures, displacement will occur due to an applied torque. The out-of–plane distortions do not induce any normal stresses providing these displacements are not restrained along the axis of the section. If the warping is restrained, warping normal stress will be induced [19]. In elastic non-uniform torsion, both the rate of change of the angle of twist $d\theta/dz$ and the longitudinal warping deflections, w vary along the length of the member. The varying warping deflections induce longitudinal strains and stresses.

In this paper, torsional loading, P is applied at the end of cantilever beam. Linear analysis was carried out to obtain the torsional rotation, ϕ for both models. From the displacement value obtained from finite element analysis, the torsion rotation of each model can be calculated. Theoretically, the equation for torsion rotation is given in Equation (3). For open cross-sections, the general formula is :

$$J = \sum\left(\frac{dt^3}{3}\right) \tag{3}$$

where: b is plate lengths between points of intersection of the axes, and t is plate thickness.

From Finite Element Analysis, the angle of rotation:

$$\phi_{fea} = \tan^{-1}\frac{dy}{dz} \tag{4}$$

where: δ is displacement from the calculated node from centre rotation, and r is depth of beam ($r = D/2$ where D is the height of web).

In design process, various strength properties of steel beam need to be taken into consideration. One of them is torsional behavior of steel beam. Therefore, it is essential to develop the knowledge about torsional behavior of the section including the torsional properties such as torsional constant (J) and warping constant (H). A general idea of torsional properties i.e. torsional constant (J) and warping constant (H) can be obtained through Appendix B BS5950: Part 1: 2000. However, the effect of torsional behavior due to openings is not stated in BS 5950. In this paper, the theoretical and finite element analysis of I-beam steel section with and without openings subject to torsion behavior was presented. The effects of various shapes and sizes of web openings and also the effect of section properties were included. Warping normal stress, displacements on longitudinal axis and angle of twist obtained through finite element analysis were used as comparison parameters between I-beam with and without web opening.

3 Numerical Analysis

3.1 Process of analysis

The first part of this analysis is to find the optimum shape and size of web opening for 1.1m beam length. Then, the optimum size and shapes of web opening will be applied to different beam lengths and structural properties to study the structural behavior of a I-beam with certain web opening. Two types of model, five shapes and three sizes of web opening have been used. Figure 1 shows the dimensions of a typical section of I-beam without web opening steel section.

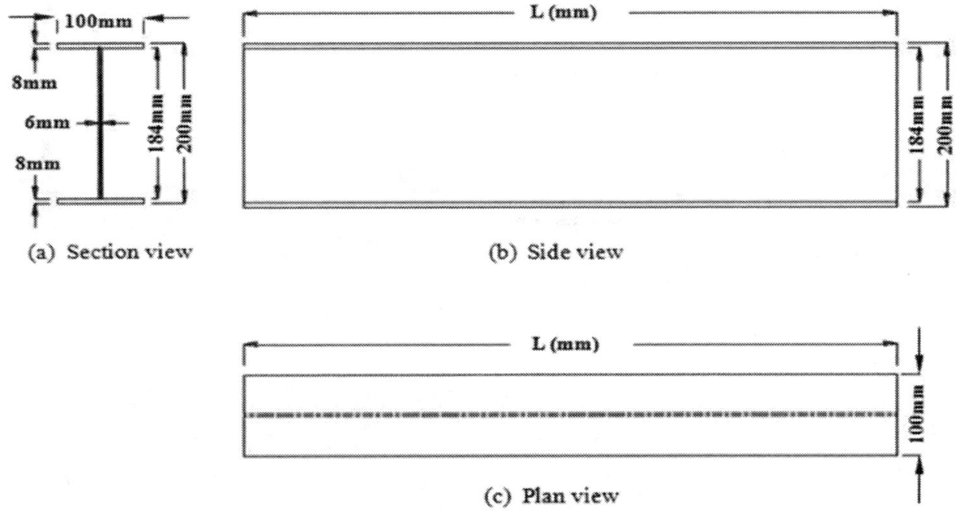

Figure 1. Shape and dimensions of a typical I-beam without web opening.

3.2 Types of model

There are two types of model are used namely Model 1 and Model 2. For the Model 1, the distance between two openings is equal to 150 mm center to center and 200 mm center to center for Model 2. Meanwhile, the edge length is 50 mm for Model 1 and 100 mm for Model 2. Figure 2 shows the details summarize of the model.

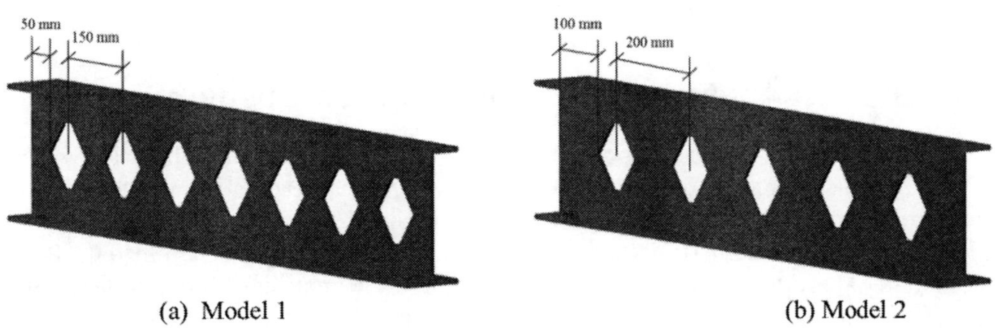

Figure 2. Types of model.

3.3 Sizes of opening

To investigate the torsion behavior of the I-beam with web opening, there are three sizes of opening involved; the first size of opening is equal to 0.5D which the diameter of the opening is 100 mm, the second size of opening is equal to 0.6D which the diameter of the opening is 120 mm and the third size of opening is equal to 0.7D which the diameter of the opening is 140 mm where D is the height of the web.

3.4 Types of opening

Five types of opening have been adopted, which are c-hexagon, octagon, hexagon, circle, square and I-beam without web opening as shown in Figure 3(a) to (f) respectively. The numerical results of I-beam with web opening are compared with corresponding I-beam without web opening, to know better understanding of torsion behaviour of this type of beams.

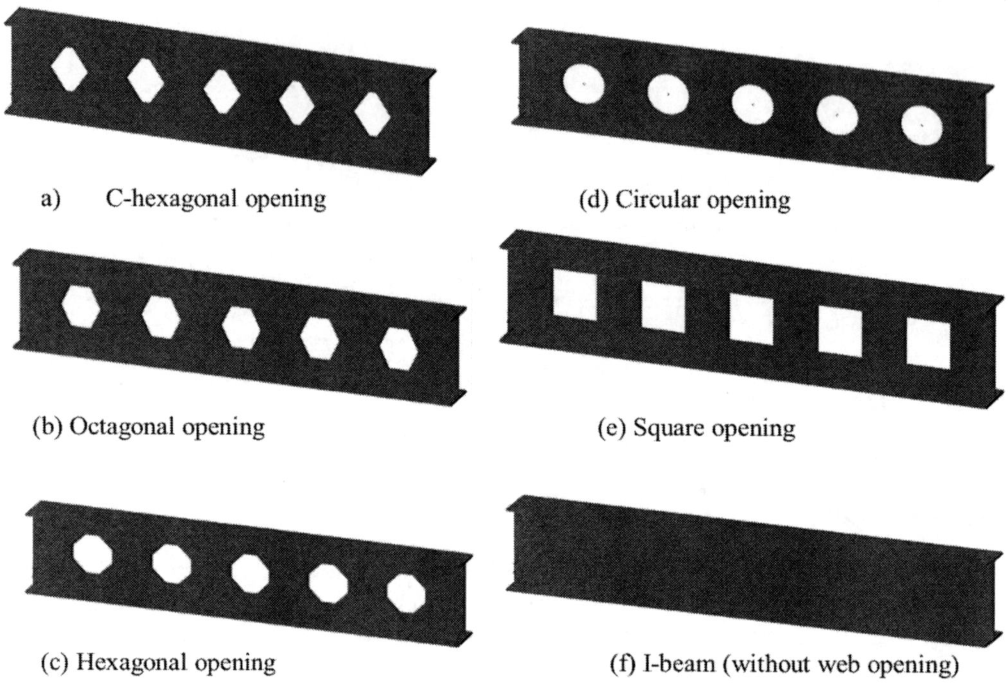

a) C-hexagonal opening

(d) Circular opening

(b) Octagonal opening

(e) Square opening

(c) Hexagonal opening

(f) I-beam (without web opening)

Figure 3. Types of opening.

4 Finite Element Analysis of I-Beam with Web Opening

LUSAS 14.0 was used to discuss the pure torsion behavior of steel beam with various shapes and sizes of opening. Quadrilateral thick shell elements of quadratic interpolation order are chosen for the analysis purpose. The whole model are assigned as ungraded mild steel for its material property with Young's Modulus, E, of 209×103 N/mm^2, shear modulus, G of 79×10^3 N/mm^2 and Poisson ratio of 0.3. These material properties remain constant throughout the analysis. All the nodes at the support are constrained the translation and rotation in x, y and z directions at one side of the beam model. The loading condition is modeled by applying one coupled point load at the edge of upper and bottom flange in opposite direction at one side of the beam model.

4.1 Convergence study

A convergence study was carried out to determine suitable mesh for I-beam without web opening model. The convergence of the mesh is established by independently increasing the mesh density in each part of the model beam section. The model is given increased mesh density in all parts of the section simultaneously, and with higher-order elements (QTS8). It is observed that the increment in displacement becomes smaller from Model 1 to Model 8. The results clearly indicate that a convergence solution has been obtained when the number of elements is 2228. Elements size 20 is used in subsequent analyses.

convergence study was carried out to determine suitable mesh for I-beam without web opening model. The convergence of the mesh is established by independently increasing the mesh density in each part of the model beam section. The model is given increased mesh density in all parts of the section simultaneously, and with higher-order elements (QTS8). It is observed that the increment in displacement becomes smaller from Model 1 to Model 8. The results clearly indicate that a convergence solution has been obtained when the number of elements is 2228. Elements size 20 is used in subsequent analyses.

4.2 Result of numerical analysis

The numerical investigations on the behaviour of I-beam with and without web opening profile section were carried out in this study. The angle of twist result of I-beam without web opening under 10kN loads is shown in Table 1 and Table 2 to Table 4 for I-beam with different types of web opening.

Based on the results from Table 3, Table 4 and Table 5 it shows that the angle of twist for model 2 is lower than model 1. This is because the differences between edge and centre to centre distance of the web opening. However, the I-beam with the opening size of 0.7D has higher torsional rotation value compared to the I-beam. The size of opening has slightly effect on the torsional angles values. It was noted that the optimum web opening size is 0.5D due to the low values of the torsional angle values compared with 0.6D and 0.7D.

Table 1. Torsional angle values of I-beam without web opening.

Types of opening	Model 1=Model 2
	Angle of twist ϕ (rads)
I-beam	0.05954

Table 2. Torsional angle values of I-beam with different types of 0.5D web opening sizes.

Types of opening	Model 1	Model 2
	Angle of twist ϕ (rads)	Angle of twist ϕ (rads)
C-hexagon	0.06098	0.06061
Hexagon	0.06122	0.06081
Octagon	0.06129	0.06085
Circular	0.06125	0.06084
Square	0.06172	0.06123

Table 3. Torsional angle values of I-beam with different types of 0.6D web opening sizes.

Types of opening	Model 1	Model 2
	Angle of twist ϕ (rads)	Angle of twist ϕ (rads)
C-hexagon	0.06165	0.06111
Hexagon	0.06197	0.06136
Octagon	0.06208	0.06145
Circular	0.06201	0.06141
Square	0.06263	0.06194

Table 4. Torsional angle values of I-beam with different types of 0.7D web opening sizes

Types of opening	Model 1	Model 2
	Angle of twist ϕ (rads)	Angle of twist ϕ (rads)
C-hexagon	0.06258	0.06171
Hexagon	0.06304	0.06208
Octagon	0.06326	0.06219
Circular	0.06311	0.06214
Square	0.06439	0.06289

From Figure 4 it was noted that C-hexagon and hexagon shapes have a torsional angles values very close to each other. The closeness of angle of twist values could be attributed to the similarity of the form. Besides that, the differences in angle of twist values increase when the opening becomes more different in form such as circle and square. However, the values of torsional angle still close to each other as tabulated in Table 1 to Table 3. While, the model 2 is the optimum model compare to model 1. From the results, it is also clear that the different sizes of opening have close values of torsional angles to each other. In conclusion, the differences between torsional angles values of different shapes and sizes are close to each other for 1.1m length

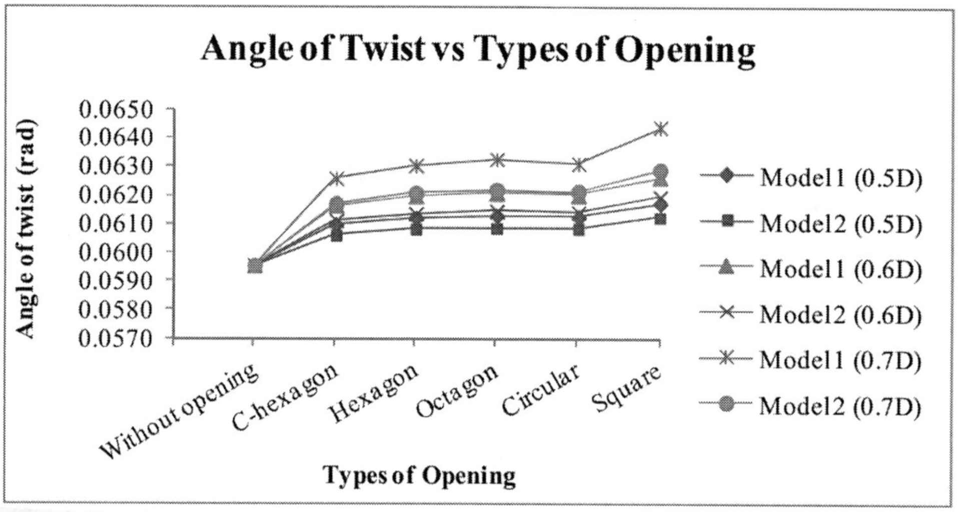

Figure 4. Angle of twist, ϕ versus types of opening for I-beam with different types of web opening and without web opening.

5 Conclusions

In this paper, I-beam with web opening is developed in order to compare the torsion resistance with the I-web without web opening. Three variables such as types of model, sizes and types of web opening were tested to determine their influence to the torsional rotation of the model. LUSAS software was fully used throughout this study because it provided good prediction and it is more economical for designing a structural member. From the results, it was noted that C-hexagon and hexagon shapes have a torsional angles values very close to each other. The closeness of displacement values could be attributed to the similarity of the form. Besides that, the differences in displacement values increase when the opening becomes more different in form such as circle and square. However, the values of torsional angle still close to each other. While, the model 2 is the optimum model compare to model 1.

Acknowledgement

The authors gratefully acknowledge the financial support of Universiti Sains Malaysia (USM), during the course of this research. This research was also made possible by the Fundamental Research Grant Scheme (FRGS) (Account Number: 203/PAWAM/6071239) provided by the Ministry of Higher Education (MOHE).

References

[1] D. Darwin and R.C. Donahey, LFRD for composite beams with unreinforced web openings, J. of Structural Engineering, **114**, 535-552, (1988).
[2] D. Darwin and W.K. Lucas, LFRD for steel and composite beams with web openings, J. of Structural Engineering, **116,** 1579-1593, (1990).
[3] D. Darwin, Design of steel and composite beams with web openings, Steel Design Guide Series American Institute of Steel Construction 2, (1990).
[4] G.D.S Verissimo, R.H. Fakury and J.C.L. Ribeiro, Design aids for unreinforced web openings in steel and composite beams with W-Shapes, Eng. J., **20**, 1-14, (2011).
[5] M.A. Gizejowski and W.A. Salah, Numerical modeling of composite castellated beams, *International Conference on Composite Construction in Steel and Concrete 2008-Composite Construction in Steel and Concrete VI*, (2011).
[6] H.P.E. Estrada, J.J. Jimenez and F. Aguiniga, Cost analysis in the design of open-web castellated beams, *Architectural Engineering Conference (AEI) Omaha*, Nebraska, United States, (2006).
[7] N.K. Chhapkhane and S.R. Kamble, Analysis of stress distribution in castellated beam using finite element method and experimental techniques, International Journal of Mechanical Engineering Applications Research, **3**, 190-197, (2012).
[8] T.C.H Liu and K.F. Chung, Steel beams with large web openings of various shapes and size: Finite element investigation, Journal of Constructional Steel Research, **59**, 1159-1176, (2003).
[9] Steel Construction Institute, *Design of Steel Beams in Torsion*, Berkshire, (2011).
[10] J. Murin and V. Kutis, An effective finite element for torsion of constant cross-sections including warping with scondary torsion moment deformation effect, Engineering Structures, **30**, 2716-2723, (2008).
[11] J. Murin., V. Kutis, V. Kralovic and T. Sedlar, 3D Beam finite element including non uniform torsion, Procedia Engineering, **48**, 436-444, (2012).
[12] C.G. Salmon and J.E. Johnson, *Steel Structures, Design and Behaviour, Emphasizing Load and Resistance Load Factor Design*, Harper Collins College Publishers, (1996).
[13] N.S. Trahair, M.A. Bradford and D.A. Nethercot, *The Behaviour and Design of Steel Structures to BS 5950*, Spon Press, (2001).
[14] R.E. Fatmi, Non-uniform warping including the effects of torsion and shear forces, Part I: A general beam theory, International J. of Solids and Structures, **44**, 5912-5929, (2007).
[15] N.S. Trahair and Y.L. Pi, Torsion, bending and buckling of steel beams, Engineering Structures, **19**(5), 372-377, (1997).
[16] Y.L. Pi and N.S. Trahnair, Inelastic torsion of steel i-sections beams, Journal of Structural Engineering, **121**(4), 609-620, (1995).
[17] G. Chen and N.S. Trahair, Inelastic nonuniform torsion of steel I-Beams, J. of Constructional Steel Research, **23**, 189-207, (1992).
[18] N.S. Trahair and M.A. Bradford, *The Behaviour and Design of Steel Structures*, Chapman and Hall, London, (1991).
[19] C.P. Heins, *Bending and Torsional Design in Structural Members*, Health and Company, Canada, (1975).

MATEC Web of Conferences **47**, 02010 (2016)
DOI: 10.1051/matecconf/20164702010

Average Frequency – RA Value for Reinforced Concrete Beam Strengthened with Carbon Fibre Sheet

M. Z. Mohamad[1,a], N. Md Noor[1] and W. N. A. Wan Ahmad[1]

[1]*Faculty of Civil Engineering, Universiti Teknologi Mara, 13500 Permatang Pauh, Pulau Pinang, Malaysia*

Abstract. Acoustic Emission (AE) is one of the tools that can be used to detect the crack and to classify the type of the crack of reinforced concrete (RC) structure. Dislocation or movement of the material inside the RC may release the transient elastic wave. In this situation, AE plays important role whereby it can be used to capture the transient elastic wave and convert it into AE parameters such as amplitude, count, rise time and duration. Certain parameter can be used directly to evaluate the crack behavior. But in certain cases, the AE parameter needs to add and calculate by using related formula in order to observe the behavior of the crack. Using analysis of average frequency and RA value, the crack can be classified into tensile or shear cracks. In this study, seven phases of increasing static load were used to observe the crack behavior. The beams were tested in two conditions. For the first condition, the beams were tested in original stated without strengthened with carbon fibre sheet (CFS) at the bottom of the beam or called as tension part of the beam. For the second condition, the beams were strengthened with CFS at the tension part of the beam. It was found that, beam wrapped with CFS enhanced the strength of the beams in term of maximum ultimate load. Based on the relationship between average frequency (AF) and RA value, the cracks of the beams can be classified.

1 Introduction

In building construction industries, reinforced concrete (RC) is usually applied for construction of foundation, beam, column and slab. Failure may take place if there is no maintenance action taken. It might be the worst moment of loss of human life [1]. Those moments can be avoided if prevention measurement takes place. In order to avoid worst cases happened, structural health monitoring (SHM) must be considered. Non-destructive test (NDT) may be used to evaluate the damage because it did not damage the structure. One of reliable technique for NDT is Acoustic Emission (AE). This technique can be used to evaluate the damage such as crack. Dislocation or fracture inside harden concrete will produce transient elastic wave [2]. This transient elastic wave can be captured by transducer and transform the transient elastic wave into electrical wave form [3]. AE parameters can be produce in AE data acquisition system such as count, rise time, energy, duration time and amplitude [4-6].

There are two types of analyses that can be made based on AE parameters such as parametric analyses and signal waveform analyses [4]. Md Nor et al. [2] stated that there are five types of

[a] Corresponding author : zakariamohamad91@gmail.com

analyses can be made such as intensity analyses, b-value, RA value, calm ratio and relaxation ratio. All these kind of analyses can be used to classify the type of cracks and crack zones.

One of the methods is using average frequency (AF) versus RA value analyses (AF-RA value). The cracks can be classified into tensile crack so called as Mode Type I and shear crack so called Mode Type II [7]. Aggelis et al. [8] stated that, tensile crack or Mode Type I can be determined when the waveforms leading in faster ways together with short rise time (RT) and high rise angle. RT is the first waveform that crossing the threshold until it reaches the peak of amplitude (A). Rise angle is A per RT. It is contras if compare with shear crack whereby waveform travel slower with longer rise time and less rise of angle. Figure 1 shows the detail for tensile and shear crack according to AE waveform.

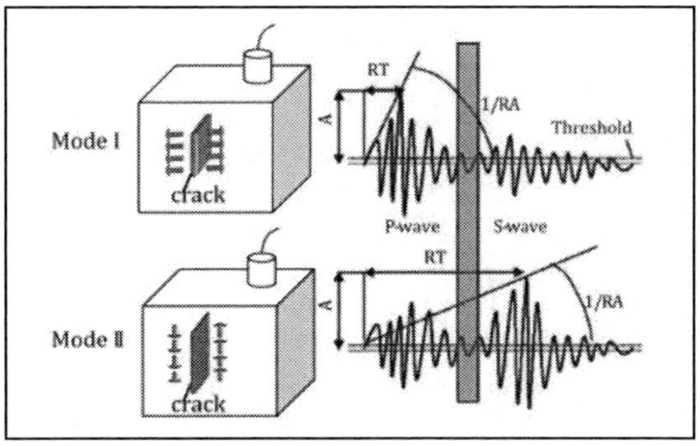

Figure 1. Detail AE waveform [8].

For the present study, two types of beam were tested and evaluated the cracks using AF-RA value analyses. One of the beams was tested in original condition with out strengthened using CFS at tension part and the rest was strengthened using 1 layer of CFS at the tension part. The AE technique and visual observation were implemented to classify the type of crack of RC beam specimen.

2 Experimental Programme

2.1 Sample preparation

Four RC bean specimens were manufactured with diamension size of 150 mm X 150 mm X 750 mm and classified into two types of condition. For each beam specimen, two mild steels with diameter of 8 mm were used at compression part and two high tensile steels were used at tension part with 6 mm diameter of shear link tie 100 mm center to center. Figure 2 shows detail reinforcement of the RC beam. The same type of beams were used as represented by Md Nor et al. [9].

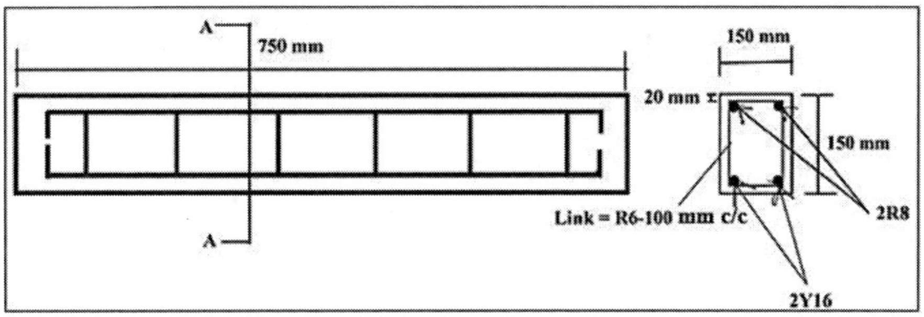

Figure 2. Detail reinforcement of RC beam.

Each condition consists of two beam specimens. The first condition was the beam without strengthened with CFS and designated as B1 and B2. The second condition was the beam strengthened with one layer of CFS at tension part and designated as B1 1 CFS and B2 1CFS.

2.2 Experimental work

Three-point bending test was used to determine the ultimate load P_{ult} for both conditions of beams. Three-point bending was carried out using 1000 kN UTM machine that located at heavy structure laboratory Universiti Teknologi MARA Pulau Pinang. Increasing static loads were developed based on each ultimate load. Table 1 and Table 2 show P_{ult} and load phases for each beam condition, respectively.

Table 1. Ultimate load for each beam condition.

Beam	Average ultimate load, P_{ult} (kN)
B1	114.5
B1 1 CFS	138

Table 2. Load phase for each condition of beam.

Load phase	Load ratio	B2 (kN)	B2 2CFS (kN)
Phase 1	0.1 P_{ult}	11.5	13.8
Phase 2	0.2 P_{ult}	23.0	27.6
Phase 3	0.3 P_{ult}	34.5	41.4
Phase 4	0.4 P_{ult}	46.0	55.2
Phase 5	0.5 P_{ult}	57.5	69.0
Phase 6	0.6 P_{ult}	69.0	82.8
Phase 7	0.7 P_{ult}	80.5	96.6

2.3 Af – Ra value

Average frequency against RA value was used to determine the mitigation of crack from flexural crack to shear crack. This analyses allows characterisation of tensile crack and shear crack [7, 10]. AF is counts (CNTS) divided by the duration (D), measured in kHz as shown in Equation (1). RA values is the rise time (R) divided the amplitude (A), measured as shown in Equation (2).

$$\text{Average frequency} = \text{Counts} / \text{Duration} \tag{1}$$

$$\text{RA values} = \text{Rise time} / \text{Amplitude} \tag{2}$$

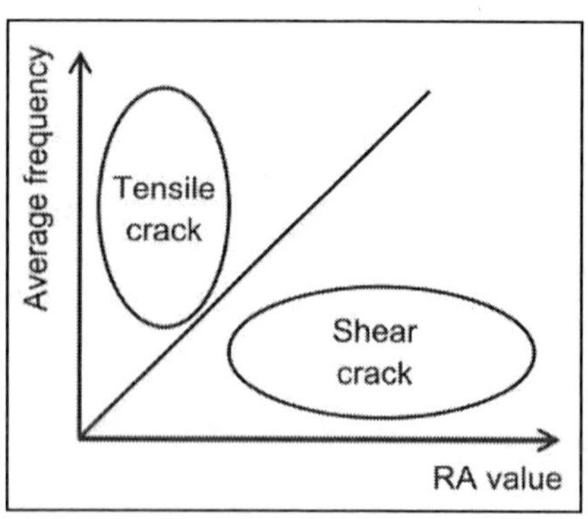

Figure 3. Location of tensile and shear crack based on AF-RA value [2].

2.4 Visual observation

Visual observation was carried out to confirm the initiation of crack and at the same to observe the type of crack [2]. Tourch light was used to confirm the initiation of crack especially hairline crack. The cracks were marked by using black marker pen for each load phase. Crack that appears was marked along the crack line and marked according to load phases.

3 Results and Discussion

3.1 Comparison between B2 and B2 1CFS

At load Phase 1 (0.1 Pult) based on Figure 4 and Figure 6, AF-RA value for B2 shows more plots at tensile zone if compared to B2 1CFS respectively. No cracks were visually observed when load Phase 1 (0.1 Pult) (see Figure 5 and Figure 7) applied on the both beams. The plots in graph AF-RA value indicate the presence of visible crack in the both beam. At this load Phase, the maximum value of 56 kHz than the beam B2 1CFS of 45 kHz. From these values indicate that the CFS strengthened the beam performance when load applied. At the same time, the CFS reduces the development of crack when the same load phase applied on both beams.

Based on Figure 5, the first crack appeared on the beam surface when load Phase 2 (0.2 Pult) applied for beam B2. However, for beam B2 1CFS (see Figure 7), the first crack was fully appeared on the beam surface when load Phase 3 (0.3 Pult) applied on the beam. It indicates that, the CFS attached on the soffit of the beam strengthened the beam and reduced the development of crack. At this load phase, the plots in the AF-RA value drastically increases for B2 compared to beam B2 1CFS.

Concentration of plotted point at Figure 4 higher if compared with plotted point at Figure 6 especially at tensile zone from Phase 1 until Phase 7. For shear zone, there were less plotted point for both Figure 4 and Figure 6 even though shear crack was observed by using visually observation. From visually observation, shear crack keep happened. At the same time, concentration of plotted point at tensile zone keep increased but at shear zone it was slightly less increased. This situation happened due to tensile crack that first initiate before it continues to transform into shear crack [14]. It can be conclude that the shear developed through extension of flexural crack for RC beam specimen without strengthened with CFS.

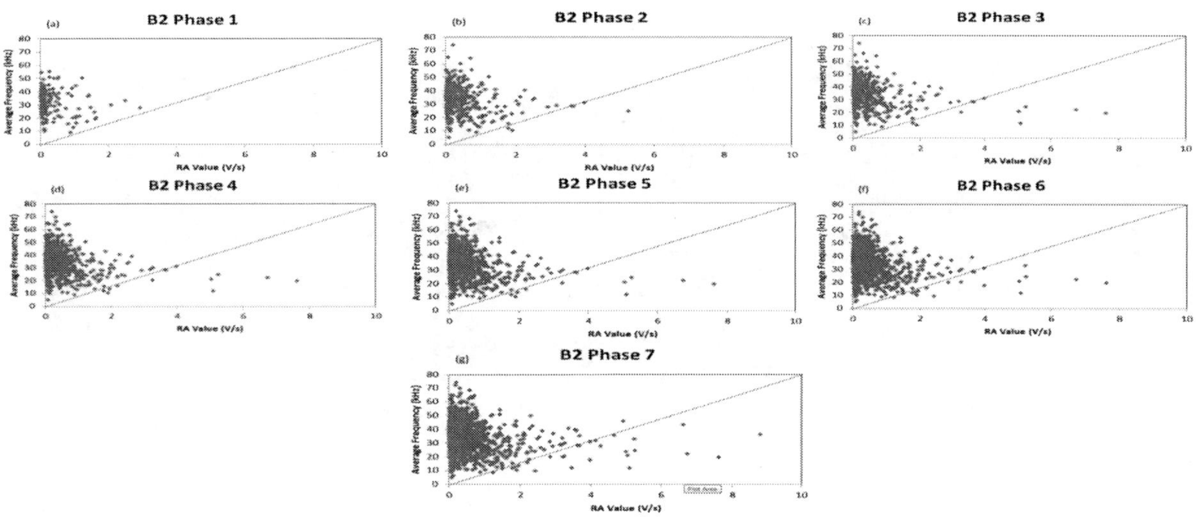

Figure 4. Relationship between average frequency and RA value for Beam B2 (a) Phase 1 (0.1 P_{ult}), (b) Phase 2 (0.2 P_{ult}), (c) Phase 3 (0.3 P_{ult}), (d) Phase 4 (0.4 P_{ult}), (e) Phase 5 (0.5 P_{ult}), (f) Phase 6 (0.6 P_{ult}) and (g) Phase 7 (0.7 P_{ult})

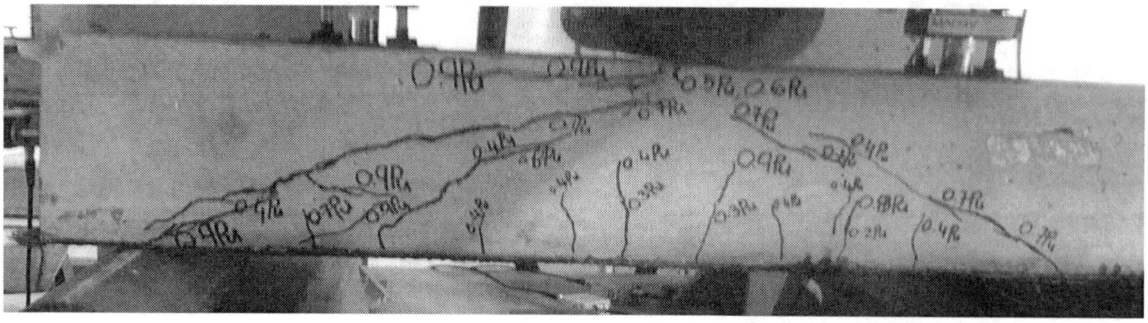

Figure 5. Crack pattern of RC beam specimen without strengthened with CFS on the soffit

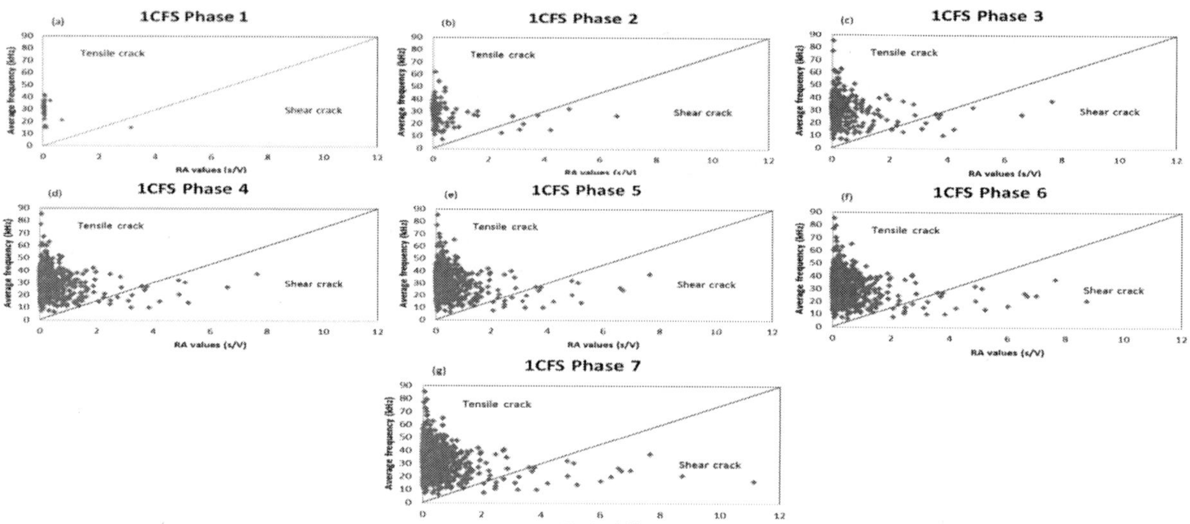

Figure 6. Relationship between average frequency and RA value for Beam B2 1CFS (a) Phase 1 (0.1 Pult), (b) Phase 2 (0.2 Pult), (c) Phase 3 (0.3 Pult), (d) Phase 4 (0.4 Pult), (e) Phase 5 (0.5 Pult), (f) Phase 6 (0.6 Pult) and (g) Phase 7 (0.7 Pult).

Figure 7. Crack pattern of RC beam specimen strengthened with CFS on the soffit.

4 Conclusion

The relationship between AF and RA value from AE signal collected from channel 6 was investigated. The crack classifications of RC beam specimen with and without strengthened using CFS on the soffit were identified. Two types of cracks namely tensile crack and shear crack were recognized. The result from the relationship between AF and RA value are well matched with the crack pattern appeared on the RC beam surface. At the early stage, AF-RA value graph shows the concentration of plots at tensile zone for both types of RC beams and no crack was visually observed. This situation happened at Phase 1 for B2 and Phase 1 until Phase 2 for B2 1CFS. The plots might be from the invisible crack that cannot be seen by naked eyes. First initiation of crack happened at Phase 2 for B2 and Phase 3 for B2 1CFS. The crack start to transform into shear crack at Phase 2 for B2 and Phase 4 for B2 1CFS. Shear crack occurred due to increasing of static load. This indicates that shear crack is extension process from tensile crack. This shows that, AF-RA value can be used to evaluate type of crack according to increment of loading phase. Overall, the attachment of CFS on the soffit reduces the development of crack in the RC beam.

References

1. M.R. Esfahani, M.R. Kianoush and A.R. Tajari, Flexural behaviour of reinforced concrete beams strengthened by CFRP sheets, Engineering Structures, **29**, 2428-2444, (2007)
2. N. Md Nor, A. Ibrahim, N. Muhamad Bunnori and H. Mohd Saman, Acoustic emission signal for fatigue crack classification on reinforced concrete beam, Construction and Building Materials, **49**, 583-590, (2013)
3. D.G. Aggelis, Classification of cracking mode in concrete by acoustic emission parameters, Mechanics Research Communications, **38**(3), 153-157, (2011)
4. A. Behnia, H.K. Chai and T. Shiotani, Advanced structural health monitoring of concrete structures with the aid of acoustic emission, Construction and Building Materials, **65**, 282-302, (2014)
5. J. Lee and J. Lee, Non-destructive evaluation on damage of carbon fiber sheet reinforced concrete, Composite Structures, **58**, 139-147, (2002)
6. G. Ma, H. Li, M. Asce and Z. Duan, Repair effects and acoustic emission technique-based fracture evaluation for predamaged concrete columns confined with fiber-reinforced polymers, J. of Composites for Construction, **16**(6), 626-639, (2012)
7. S. Shahidan, R. Pullin, K.M. Holford, N.M. Bunnori and N.M. Noor, Quantitative evaluation of the relationship between tensile crack and shear movement in concrete beams, Advanced Materials Research, **626**, 355-359, (2012)
8. D.G. Aggelis, A.C. Mpalaskas, D. Ntalakas and T.E. Matikas, Effect of wave distortion on acoustic emission characterization of cementitious materials, Construction and Building Materials, **35**, 183-190, (2012)
9. N. Md Nor, A. Ibrahim, N. Muhamad Bunnori, H.M. Saman, S.N. Mat Saliah and S. Shahidan, Diagnostic of fatigue damage severity on reinforced concrete beam using acoustic emission technique, Engineering Failure Analysis, **41**, 1-9, (2014)
10. M. Ohtsu, T. Isoda and Y. Tomoda, Acoustic emission techniques standardized for concrete structures, J. Acoustic Emission, **25**, 21-32, (2007)
11. T. J. Hadzor, R.W. Barnes, P.H. Ziehl, J. Xu and A.K. Schindler, *Development of acoustic emission evaluation method for repaired*, Highway Research Center, (2011)
12. M.A.A. Aldahdooh, N.M. Bunnori and M.A. Megat Johari, Damage evaluation of reinforced concrete beams with varying thickness using the acoustic emission technique, Construction and Building Materials, **44**, 812-821, (2013)
13. B.B. Adhikary and H. Mutsuyoshi, Behavior of concrete beams strengthened in shear with carbon-fiber sheets, J. of Composites for Construction, **8**, 258-264, (2004)
14. N. Md Nor, A. Ibrahim, N.M. Bunnori, H.M. Saman, S. Noor, M. Saliah and S. Shahidan, Fatigue crack inspection and acoustic emission characteristics of precast RC beam under repetition loading, **774**, 1022-1026, (2015)

MATEC Web of Conferences **47**, 02011 (2016)
DOI: 10.1051/matecconf/20164702011

Identification of Dynamic Characteristics of Bridge Crossing Sungai Simpang Kiri Using Free Vibration Analysis

Siti Mas Ayu Mat Yusoff[1,a], Sherliza Zaini Sooria[2], Ahmad Fahmy Kamarudin[1] and Koh Heng Boon[1]

[1]*Jamilus Research Center, Universiti Tun Hussein Onn Malaysia, 86400 Parit Raja, Johor, Malaysia*
[2]*Public Works Department, 50480 Kuala Lumpur, Malaysia*

Abstract. This paper presents a free vibration analysis (FVA) of a reinforced concrete bridge crossing Sungai Simpang Kiri, located at Batu Pahat. The bridge consists of three simple spans with a total length of 70meter (20 m + 30 m + 20 m). The concrete deck of the bridge is supported by concrete tensioned pre-stressed T-beam. The girders are sitting on two abutments at both ends and two piers as internal supports. The base of the piers is considered as fixed base and the abutments were free to move. The structural dynamic characteristics of the bridge in terms of fundamental frequency and mode shapes were obtained analytically using SAP2000 three-dimensional finite element modeling software. It is very important to evaluate the dynamic characteristics of reinforced concrete bridges that can lead to the detection of stiffness reduction or damage of the structure. From the analysis, the fundamental frequency of the bridge was 1.94 Hz with fundamental mode shape is critical in transverse bending mode.

1 Introduction

Bridges are important elements in a transportation system which span over aquatic obstacles, valleys and gorges. Bridges also act as a key elements in the lifeline of a country and play a fundamental role economically, politically and militarily [1]. Indeed they have always occupied a special place in the attentions of structural designers because their structural form tends to be a simple expression of their functional requirement [2]. These bridge designs which comply with current codes may satisfy safety and strength requirements; however, they may not have the capacity to withstand undesirable dynamic responds in the form of large displacements and consequently increased stress, which may affect the long-term performance of the bridge.

Therefore, there is a need to incorporate dynamic characteristics of modal parameter as an important parameter for the design. The measured fundamental frequency is useful information as they relate closely to the stiffness of the bridges [3].

This paper discussed the dynamic characteristics in terms of fundamental frequency and fundamental mode shape of the non-seismically designed bridges bridge crossing Sungai Simpang Kiri located at Batu Pahat by free vibration analysis (FVA) to make a start toward the seismic evaluation or retrofit of the bridge. These dynamic characteristics used as a basis for the finite element model updating, structural control, damage detection, condition assessment and long-term health monitoring of the structure [3].

---

[a] Corresponding author : ayuyusoff90@gmail.com

2 Description of the Bridge Crossing Sungai Simpang Kiri

The bridge crossing Sungai Simpang Kiri is a two carriageway, three-span structure with a total length of 70 meter (20 m + 30 m + 20 m). The deck was supported on two piers. The pier end is considered as fixed base whereby soil- structure interaction is ignored. The abutments of the bridge were free ends on both sides. The real view of the bridge is shown in Figure 1.

Figure 1. The bridge crossing Sungai Simpang Kiri.

3 Bridge Modelling

The bridge crossing Sungai Simpang Kiri was modeled using SAP2000 software in 3- dimensional finite element model (FEM) as shown in Figure 2 and the geometrical details of the bridge are as tabulated in Table 1.

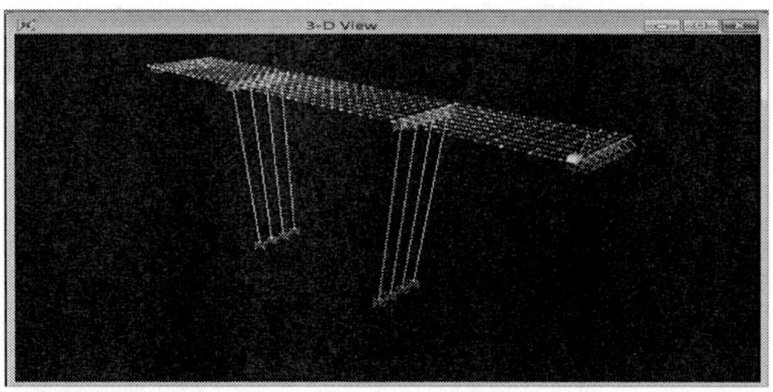

Figure 2. 3D model of the bridge.

Table 1. Geometrical details of the bridge.

Type of bridge	Concrete tensioned pre-stressed T-beam
Material	Reinforced concrete
No of span(s)	3
Support condition	Fixed at pier, roller at abutment
Length	70m (20m+30m+20m)
Deck width	12m
Slab thickness	200mm
No of intermediate piers	4

4 Free Vibration Analysis (FVA) Methodology

The bridge model was analyzed using free vibration analysis (FVA) or the so called Eigen value analysis without excitation of external load. The importance of the analysis is to identify the dynamic characteristics which include fundamental frequency and fundamental mode shapes. This method of analysis has been choosing as mentioned in [4] dynamic test can be classified according to the three types which are Forced Vibration Test, Ambient Vibration Test and Free Vibration Test.

5 Results and Discussion

5.1 Fundamental frequency

During the analysis, the fundamental frequency or first frequency was considered without diaphragm effect and excitation of external forces. The frequencies for the six modes shapes selected were tabulated in Table 2.

Table 2. Frequencies of the bridge crossing Sungai Simpang Kiri.

Mode	Frequency (Hz)
1	1.94
2	2.24
3	3.39
4	6.36
5	6.52
6	6.60

The measured fundamental frequency is useful information as they relate closely to the stiffness of the bridges [5]. The fundamental frequency of bridges is in general rather low and depends of course strongly on the span length and there type of bridge construction [5]. For Highway Bridge, the fundamental frequency may be simply estimates as:

$$f = \frac{100}{L(m)} \tag{1}$$

where L= span length in meter.

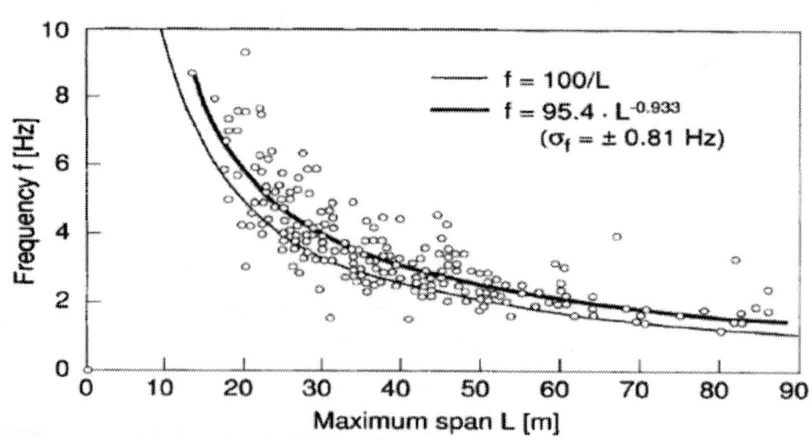

Figure 3. Frequency, f as a function of the maximum span L, derived from 224 bridges [6].

From Figure 3, many of bridge lie within range 2 to 4 Hz. The fundamental frequency of bridge depends on the span length and type of bridge construction [6]. Based on Figure 3, the longer the maximum span length produced lower frequency value. The maximum span length for RC bridge crossing Sungai Simpang Kiri is 30 meter. Estimated frequency value is 3.3 Hz. The fundamental frequency obtained from free vibration analysis is slightly different when compared to the empirical equations. It is due to the fundamental frequency of bridge depending on the span length and the type of bridge construction [7]. Furthermore, empirical relationships exist in codes to estimate elastic fundamental frequency but they miss experimental data to validate them accounting for national feature of building design [8].

5.2 Mode shapes

Fundamental mode shape is the mode having the lowest fundamental frequency [9]. Figure 4 illustrates the corresponding mode shapes of or bending mode of the bridge crossing Sungai Simpang Kiri. The first six modes were selected to limit the mode of vibration since only the dominant mode of fundamental frequency will be considered that relate closely to the stiffness of the bridge. Moreover, mode 6 shows changes in bending mode from vertical to transverse which reverse back to the dominant mode. From the illustration by software, the first, second and sixth modes are identified as transverse bending mode whiles the others in vertical bending mode.

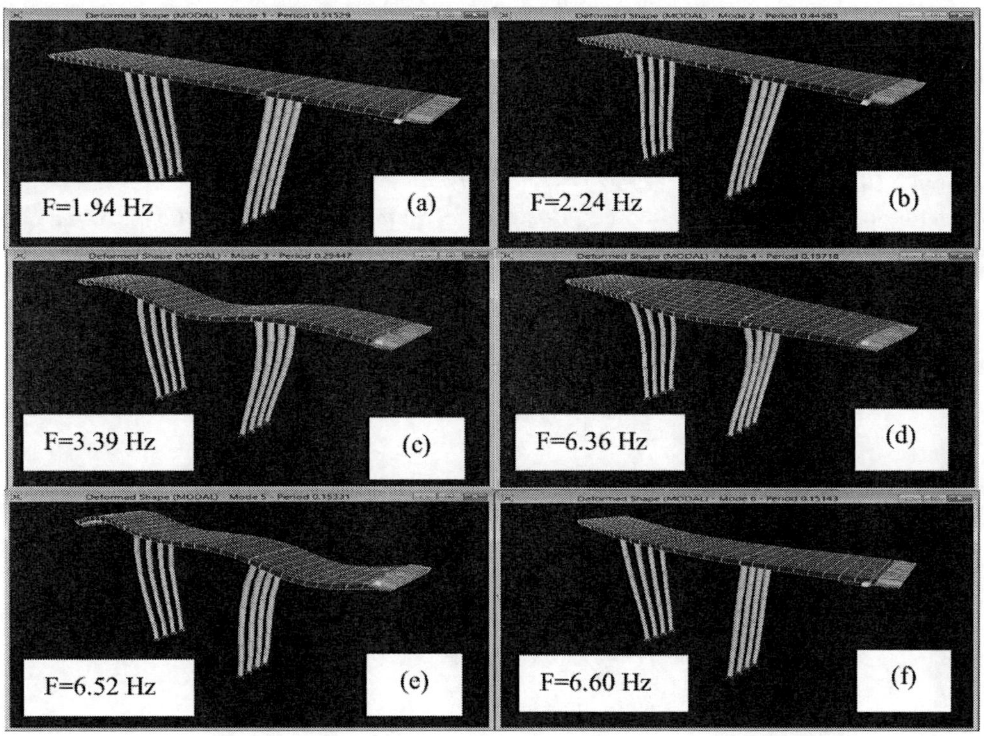

Figure 4. Mode shapes direction of the bridge crossing Sungai Simpang Kiri by FVA (a) Mode 1(Transverse) (b) Mode 2(Transverse) (c) Mode 3(Vertical) (d) Mode 4(Vertical) (e) Mode 5(Vertical) (f) Mode 6 (Transverse).

6 Conclusion

This paper aimed to determines the dynamic characteristics using FVA by SAP200 software. Based on the results obtained from the analysis, the following conclusions can be made: (i) The fundamental frequency is 1.94 Hz bending in transverse fundamental mode, (ii) This method of analysis was the fastest and inexpensive method to evaluate and localize damage using the change in dynamic parameters, and (iii) If the external load or dynamic load excited on the same direction of the dominant mode, the frequency of the bridge will be amplified and causing damage.

References

[1] A. Nicknam, A. Mosleh and H. Hamidi Jamnani, Seismic performance evaluation of urban bridge using static nonlinear procedure, case study: Hafez bridge, Procedia Engineering, **14**, 2350-2357, (2011).

[2] T.S. Najmi and M. Desai, Feasibility of capacity spectrum method for seismic analysis of concrete bridges for Indian terrain as a step towards performance based design philosophy, (2004).

[3] W.X. Ren, X.L. Peng and Y.Q. Lin, Experimental and analytical studies on dynamic characteristics of a large span cable-stayed bridge, Engineering Structures, **27**(4), 535-548, (2005).

[4] B.A. Cunha, E. Caetano and R. Delgado, Dynamic tests on a large cable-stayed bridge an efficient approach, (2001).

[5] C.W. Lin and Y.B. Yang, Use of a passing vehicle to scan the fundamental bridge frequencies: An experimental verification, Engineering Structures, **27**(13), 865-878, (2005).

[6] W. Amman, Vibration induced by traffic and construction activity – bridge, *Vibration Problems in Structures: Practical Guidelines*, Birkhausel, Boston, 125-127 (1995).

[7] S.O. Citak, M. Ohori and T. Nakamura, Microtremor measurements and earthquake response analysis on Urado Bridge, *The 15th World Conferences of Earthquake Engineering LISBOA*, Kochi, Japan, (2012).

[8] C. Micheal, P. Gueguen, Lestuzzi and P.Y. Bard, Comparison between seismic vulnerability models and experimental dynamic properties of existing buildings in France, Bulletin of Earthquake Engineering, **8**(6), 1295–1307, (2010).

[9] S.R. Damodarasamy and S. Kavitha, *Basic of Structural Dynamics and Seismic Design*, PHI Learning, New Delhi, (2009).

[10] V. Gioncu and F.M. Mazzolani, *Earthquake Engineering for Structural Design*, Spon Press, USA, (2011).

[11] G. Lorant, Whole Building Design Guide-Seismic Design Principles, (2012). <retrieved from http://mceer.buffalo.edu>

MATEC Web of Conferences **47**, 02012 (2016)
DOI: 10.1051/matecconf/20164702012

Ambient Vibration Test on Reinforced Concrete Bridges

Nurul Shazwin Idris[1,a], Koh Heng Boon[1] and Ahmad Fahmy Kamarudin[1] and Sherliza Zaini Sooria[2]

[1]*Jamilus Research Center, Universiti Tun Hussein Onn Malaysia, 86400 Parit Raja, Johor, Malaysia*
[2]*Public Works Department, 50480 Kuala Lumpur, Malaysia*

Abstract. An investigation was carried out to determine dynamic characteristic of reinforced concrete (RC) bridges by using ambient vibration test (AVT). The ambient vibration sources on bridges may come from traffic, wind, wave motion and seismic events. AVT describes the dynamic characteristics of the bridge and ground by measuring the natural frequencies using highly sensitive seismometer sensor. This test is beneficial due to light weight equipment and smaller number of operator required, cheap and easy to be handled. It is able to give a true picture of the bridge dynamic behavior without any artificial force excitation when vibration data is recorded. A three-span reinforced concrete bridge located in Sri Medan, Batu Pahat, Johor was measured by using microtremor equipment consist of three units of 1 Hz eigenfrequency passive sensors used in this test was performed in normal operating condition without excitation required from any active sources or short period noise perturbations. Ten measurements were conducted on the bridge deck and ten measurements on the ground surface in order to identify the natural frequencies of the bridge. Several peak frequencies were identified from three components of Fourier Amplitude Spectra (FAS) in transverse (North-South), longitudinal (East-West) and vertical (Up-Down) direction as well as squared average Horizontal to Vertical Spectral Ratio (HVSR) of ground response, computed by using Geopsy software. From the result, it was expected the bridge have five vibration modes frequencies in the range of 1.0 Hz and 7.0 Hz with the first two modes in the transverse and longitudinal direction having a frequency 1.0 Hz, the third mode is 2.2 Hz in transverse direction, fourth and fifth mode is 5.8 Hz and 7.0 Hz. For ground natural frequencies are in range 1.0 Hz to 1.3 Hz for North-South direction and 1.0 Hz to 1.6 Hz for East-West direction. Finally the results are compared with several empirical formulas for simple verification.

1 Introduction

Ambient vibration also known as microtremor is the excitation experienced by a structure under its normal operating condition [5]. Examples of ambient vibration exerted on the bridge are traffic vibration, wind, wave motion and seismic excitation. This vibration can be determined by measuring the vibration behavior using highly sensitive acceleration sensor [16]. This test would give the true picture of the behavior of the structure without artificial excitation. In the past research, ambient vibration test widely applied to different types of bridges such as a double-deck bridge [9], pre-stressed concrete box girder bridge [3, 6], continuous girder bridge [15] , reinforced concrete bridge [6], cable stayed bridge [15] and stone masonry bridge [12] in order to determine the dynamic

[a] Corresponding author : nurulshazwin@gmail.com

characteristics of the bridge. Many researchers have used this technique because it is economic and practical to conduct.

It is important to measure the dynamic characteristics for both newly constructed and existing bridge to understand better their dynamic behavior under normal traffic loads and also during extreme loads such as those causes by seismic events or high wind [5]. Dynamic characteristics in terms of natural frequencies, mode shape and damping ratios are important for accurately predicting the dynamic response of the structure at design and reassessment stage [2]. This study only focused on field measurement of ambient vibration from free vibration sources. This information is important for prediction in seismic design and also for reassessment purpose.

2 Methodology

The ambient vibration test (AVT) was conducted on a three-span, two-carriageway of total length of 70 meter with a main span of 30 meter and two side span of 20 meter. The bridge width is 12 meter. The bridge is semi integral bridge and the deck made of pre-stressed concrete T-beam supported by four intermediate and two abutments. The bridge constructed by Jabatan Kerja Raya (JKR) located in Sri Medan, Batu, Johor as shown in Figure 1(a). The bridge was opened to public used in 2013, linked Kampung Parit Laman, Sri Medan to Kangkar Merlimau, Parit Sulong.

Microtremor equipment consists of three units of 1 Hz Lennartz Triaxial seismometer and CityShark II data logger as shown in Figure 1(b). Ten measurements were conducted on bridge deck and also ground, with 15 minutes of recording length for each measurement. The seismometers were portable data and placed at different points on the bridge deck. All real response data recorded using the data logger with 100 Hz sampling of frequency.

Figure 1. (a) Location of Sungai Simpang Kiri Bridge, (b) Setting up data logger and seismometer for deck measurement

The sensor positions on the bridge are illustrated in Figure 2. Black dot is the permanent position of a reference sensor (Sensor 1) while the red and purple dots are the moveable sensors points (Sensor 2 and Sensor 3), which microtremor measurements were conducted on both bridge edges. Three sensors are used simultaneously for every measurement. Five measurements were conducted on each side. Sensors are also placed on expansion join as proposed by Wenzel and Pichler, 2005. By referring to Figure 3, expansion joint of the bridge is located at R5, R9, L5 and L9.

3 Ambient Vibration Results

The response of the bridge due to ambient vibration is analyzed using (GEOPSY) software version 2.6.3. Natural frequencies evaluated in NS, EW and UD directions of the bridge. Bridge responses were analyzed by applying FAS and HVSR computational. Figure 3 shows example of recorded ambient vibration signal from one of the measurement on the bridge deck taken from GEOPSY software. FAS curve were computed based on AVT input motions taken on the bridge deck whereas HVSR from the ground surfaces. Each spectrum curve is obtained by the following steps:

i) Selection of 25-second stable signal windows, using an anti-trigger (STA=1s, LTA=30s, with low and high threshold of 0.2 and 2.5, respectively) to reject transients,
ii) A 5% cosine taper is applied on both of the ends of the selected window,
iii) Calculation of the Fourier amplitude velocity spectra for each selected window,
iv) Smoothing of each individual window spectrum by Konno and Omachi (1998) smoothing constant of 40. The mean Fourier amplitude spectra curve is obtained by averaging the individual window spectral curves in respective axes of NS, EW and UD. The natural frequencies from the mean NS, EW and UD of FAS curved were picked automatic by GEOPSY software at significant maximum peak amplitude from respective FAS components.

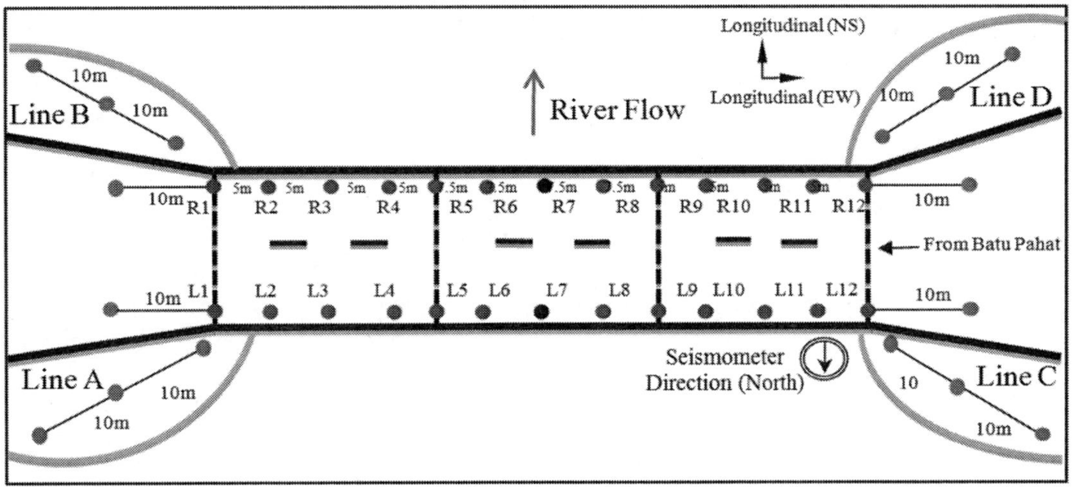

Figure 2. Position of sensor from plan view of the bridge.

Figure 4 shows spectra amplitude recorded at reference point R7, the curve shows six repeated measurements on the point. Six vibration modes of the bridge were identified in the range of 1.0 to 7.0 Hz. The first mode vibration of the bridge is transverse mode with a frequency of 1.0 Hz, first longitudinal mode was found to be occurring with a frequency of 1.0 Hz, while the first vertical mode appeared at a frequency of 5.8 Hz. The main frequencies recorded on the bridge is summarize in Table 1.

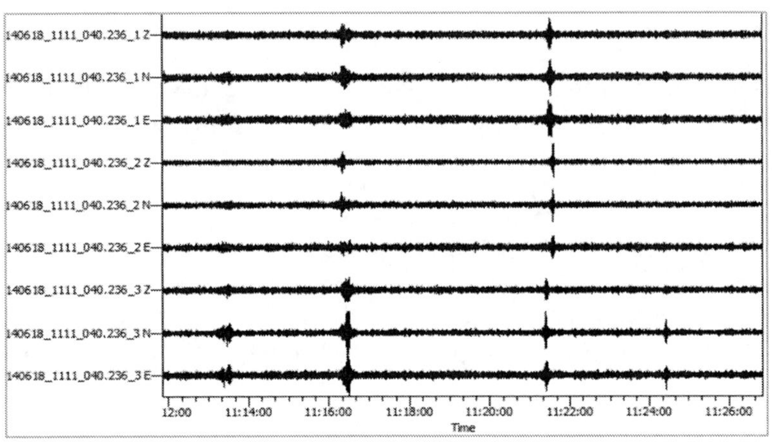

Figure 3. Ambient vibration time histories in NS, EW and UD from three sensors.

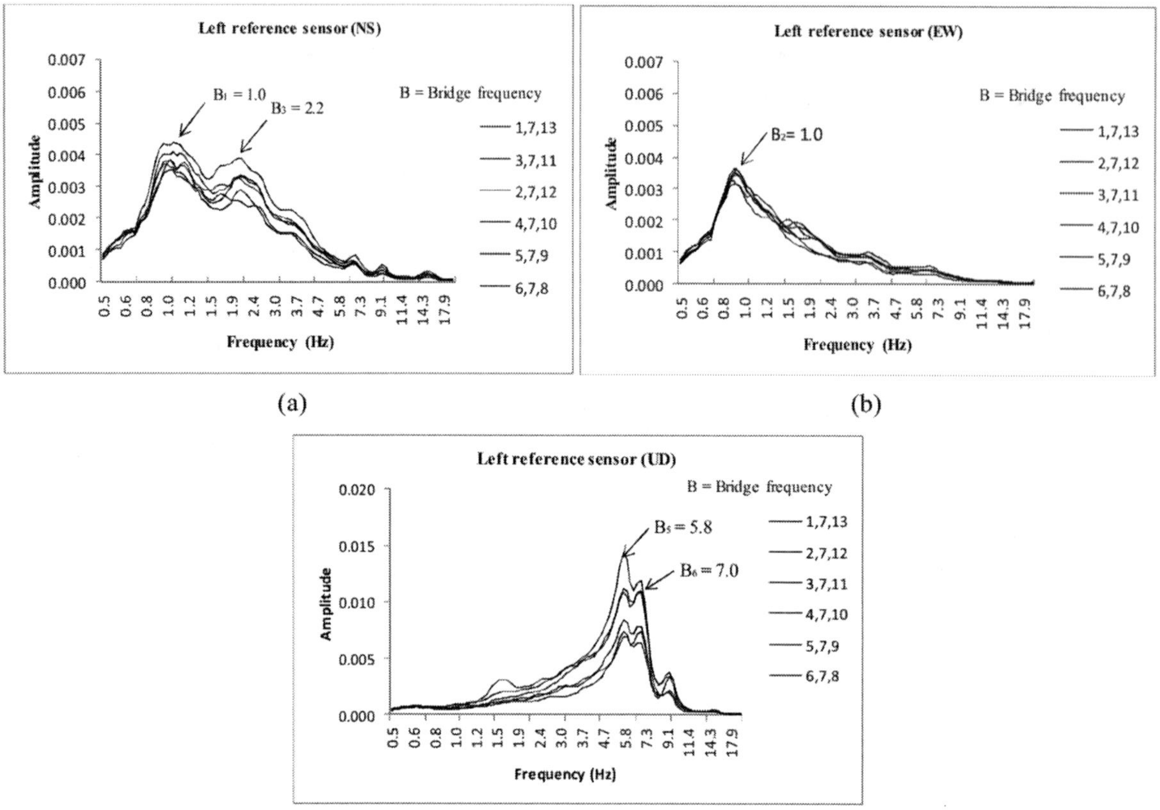

(c)

Figure 4. Fourier spectra on the bridge for three directions a) NS direction b) EW direction c) UD direction.

Table 1. Identified natural frequencies of the bridge

Mode	Frequency (Hz)	Direction
1	1.0	Transverse
2	1.0	Longitudinal
3	2.2	Transverse
4	2.6	Transverse
5	5.8	Vertical
6	7.0	Vertical

Fundamental mode of vibration is the mode having the lowest natural frequency [4]. The fundamental frequency of bridges is in general rather low and depends of course strongly on the span length and there type of bridge construction [1]. For a highway bridge, the fundamental frequency can be simply estimated as:

$$f=100/L(m) \quad (1)$$

where L= span length in meter. From Figure 5, many of bridge lie within range 2 to 4 Hz. The fundamental frequency of bridge depends on the span length and type of bridge construction [1]. Based on Fig. 6, the longer the maximum span length produced lower frequency value. The maximum span length for RC bridge crossing Sungai Simpang Kiri is 30 meter. Estimated frequency value is 3.3 to 4.0 Hz. The natural frequency obtained from AVT is slightly different when compared to the empirical equations. The measured natural frequency is useful information as they relate closely to the stiffness of the bridges [13]. The measured natural frequencies and the corresponding mode shape of a structure are primarily a function of the mass and stiffness characteristics and their distribution in the structure, the boundary conditions of the structure and the connectivity or continuity conditions between the various subcomponents of the structure [8].

Furthermore, empirical relationships exist in codes to estimate elastic fundamental frequency but they miss experimental data to validate them accounting for national feature of building design [14]. Even if the natural frequency extracted from AVT may be larger than the elastic frequency (at yield) generally used in earthquake engineering, ambient vibration recoding may provide large set of data for statistical analysis of periods versus building characteristic relationship [14].

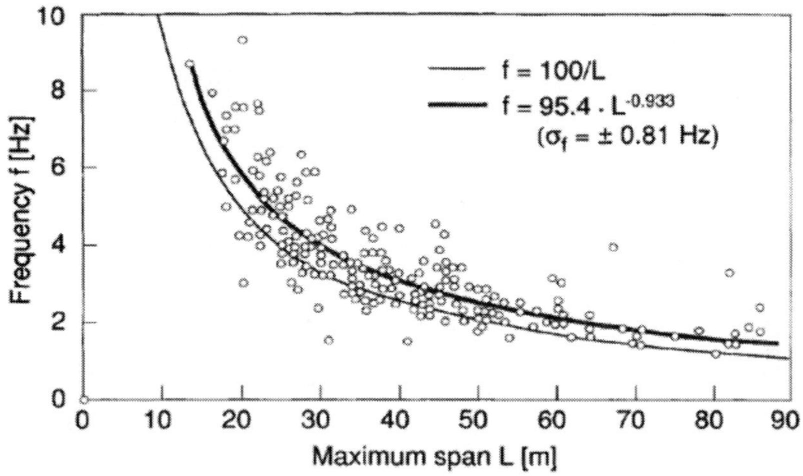

Figure 5. Fundamental frequency, f as a function of the maximum span, L derived from 224 bridges [1].

4 Conclusion

Based on the results obtained from the testing, the bridge has six vibration modes in range 1.0 to 7.0 Hz. AVT is able to be used for prediction of natural frequency of the existing bridges.

References

[1] W. Amman, Vibration induced by traffic and construction activitiy - bridge, *Vibration Problems In Structures: Practical Guidelines*, Birkhausel, Boston, 125-127, (1995).

[2] G. Chen and S. Beskhyroun, Dynamic properties of an eleven-span motorway bridge at different levels of excitation, *Proc. of 2014 New Zealand Society for Earthquake Engineering Int. Conference*, (2014).

[3] S.O. Citak, M. Ohori and T. Nakamura, Microtremor measurements and earthquake response analysis on Urado Bridge, *The 15th World Conferences of Earthquake Engineering LISBOA*, Kochi, Japan, (2012).

[4] S.R. Damodarasamy and S. Kavitha, *Basic of Structural Dynamics and Seismic Design*, PHI Learning, New Delhi, (2009).

[5] C.R. Farrar and G.H. James III, System identification from ambient vibration measurements on a bridge, J. of Sound and Vibration, **205**(1), 1–18, (1997).

[6] C.R. Farrar, T.A. Duffey, P.J. Cornwell and S.W. Doebling, Excitation methods for bridge structures, 1063–1068, (1984).

[7] L. Fyba, *Dynamics of Railway Bridges*, Thomas Telford, London, (1996).

[8] K.A. Grimmelsman, Experimental characterization of towers in cable-supported bridges by ambient vibration testing, PhD Thesis, Drexel University, (2006).

[9] I.E. Harik, D.L. Allen, R.L. Street, M. Guo, J. Harison and M.J. Gawry, Free and ambient vibration of Brent-Spence bridge, J. of Structural Engineering, **123**, 1262–1268, (1997).
[10] A. Kibboua and M.N. Farsi, Modal analysis and ambient vibration measurements on Mila-Algeria cable stayed bridge, Structural Engineering and Mechanics, **29**(2), 171–186, (2008).
[11] K. Konno and T. Ohmachi, Ground-motion characteristics estimation from spectral ratio between horizontal and vertical component microtremor, Bulletin of the Seismological Society of America, **88**(1), 228-241, (1998).
[12] L.S. Krstevska, M. Kustura and L.A. Tashkov, Experimental in-situ testing of reconstructed old bridge in Mostar, *The 14th World Conference On Earthquake Engineering*, Beijing, China, 1-8, (2008).
[13] C.W. Lin and Y.B. Yang, Use of a passing vehicle to scan the fundamental bridge frequencies: An experimental verification, Engineering Structures, **27**(13), 1865–1878, (2005).
[14] C. Micheal, P. Gueguen, Lestuzzi and P.Y. Bard, Comparison between seismic vulnerability models and experimental dynamic properties of existing buildings in France, Bulletin of Earthquake Engineering, **8**(6), 1295–1307, (2010).
[15] W. Ren, X. Peng and Y. Lin, Experimental and analytical studies on dynamic characteristics of a large span cable-stayed bridge, Engineering Structures, **27**(4), 535–548, (2005).
[16] H. Wenzel and D. Pichler, *Ambient Vibration Monitoring*, John Wiley and Sons, England, (2005).

MATEC Web of Conferences **47**, 02013 (2016)
DOI: 10.1051/matecconf/20164702013

Amplitude Distribution of Emission Wave for Cracking Process

Shahiron Shahidan[1,a], Sharifah Salwa Mohd Zuki[1], Mustaqqim Abdul Rahim[2], Noorwirdawati Ali[1] and Noridah Mohamad[1]

[1]*Department of Structural and Material Engineering, Faculty of Civil and Environmental Engineering, Universiti Tun Hussien Onn Malaysia,86400, Parit Raja, Johor, Malaysia*
[2]*Civil Engineering Department, School of Environmental Engineering, Universiti Malaysia Perlis, 02600 Arau, Perlis, Malaysia*

Abstracts: Acoustic emission technique is a method of assessment for structural health monitoring system. This technique is an effective tool for the evaluation of any system without destroying the material conditions. It enables early crack detections and has very high sensitivity to crack growth. The crack patterns in concrete beam have been identified according to the type of cracking process and the crack classifications using the AE data parameters are mainly based on the AE amplitude, rise time, and average frequency. These data parameters have been analysed using statistical methods of b-value analysis. This research paper will mainly focus on the utilization of statistical b-value analysis in evaluating the emission amplitude distribution of concrete beams. The beam specimens (150 X 250 X 1900) mm were prepared in the laboratory system and tested with the four point bending test using cyclic loading together with acoustic emission monitoring system. The results showed that this statistical analysis is promising in determining the cracking process in concrete beams.

1 Introduction

Structural concrete bridges are highly at risk in terms of structural conditions due to over loading compared to other structures. This condition occurs due to the increase of daily traffic loading on bridges. The most common defects found on superstructure concrete bridges are micro-cracking and macro-cracking [1-3]. These types of damage are becoming serious and have caused great concern among the public [4]. The AE technique evaluation method has been found to be more effective and accurate for developing damage grading systems. This system is very useful and significant for assessing concrete structures such as bridges for daily use and deterioration in buildings to classify the differences of class condition in concrete structures [3-5].

In the AE evaluation method, many researchers have developed a technique to identify the damage mechanism using AE signal parameters such as signal strength, hits, amplitude and wave energy [6-7]. These AE parameters have been used extensively in the evaluation system. One approach being used to evaluate the damage mechanism in concrete material is b-value analysis [8]. *b*-value analysis is a distribution of the frequency of AE hits with amplitude as well as with cumulative frequency. This method of analysis has been widely utilized in seismology to characterize the earthquake population

[a] Corresponding author : shahiron@uthm.edu.my

[9-10] and composite material. However, the principle of AE signal data acquisition system has several limitations. Due to this matter, the objective of this paper is to explore the scaling of the AE amplitude distribution waves generated during the cracking process based on the loading system.

2 B Value Analysis

This statistical method is widely used in seismology to characterize earthquake populations. This relationship was established and developed by Gutenberg and Richer in 1949.In terms of AE techniques, the Gutenberg-Ritcher formula had been modified as [10-13]:

$$\log_{10} N = a - b' A_{dB} \tag{1}$$

whereas A_{dB} = peak amplitude of the AE events in dB.

$$A_{dB} = 10 \log_{10} A_{\max}^2 = 20 \log_{10} A_{\max} \tag{2}$$

Then, the b-value is calculated as the slope of log linear plot of the AE frequency magnitude distribution as seen in Figure 1.

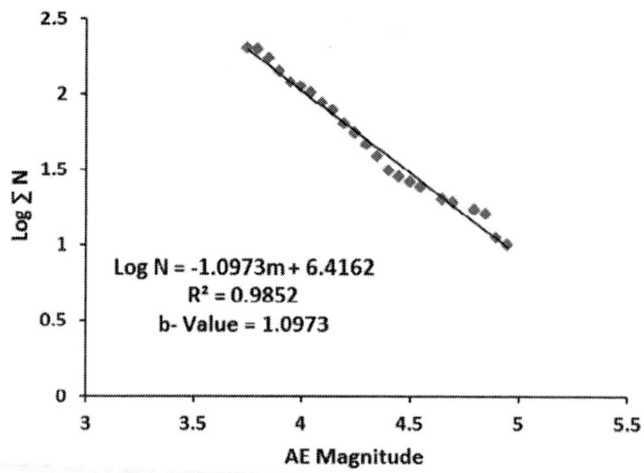

Figure 1. Sample calculation of B-value analysis.

Finally, the values are compared to define and summarize the type of cracking and cracking process (refer to Table 1).

Table 1. Quantitative results analysis [11].

Value	Description
Value > 1.7	Micro-cracks are dominant or Macro-cracks are opening
1.2 < value < 1.7	Uniformly distributed cracking
Value <1.2	Macro-crack forming

3 Laboratory Work

The beam specimens with the following dimensions (150 X 250 X 1900) mm were prepared according to BS8110 for this research as seen in Figure 2. The specimens were tested using a four-point bending setup under cyclic load test (CLT). All the beams specimens were reinforced by 2T12 and 4T12 longitudinal rebar for compression and tension respectively. The designed concrete strength is 40MPa with a curing process of 28 days according to the DOE method. The beam specimens were placed on a steel support with a neoprene pad to reduce acoustic noise during the test. The beams tested were loaded under a cyclic load test (CLT) system with a four-point bending tests and were monitored throughout the test using an AE monitoring system as shown in Figure 3.

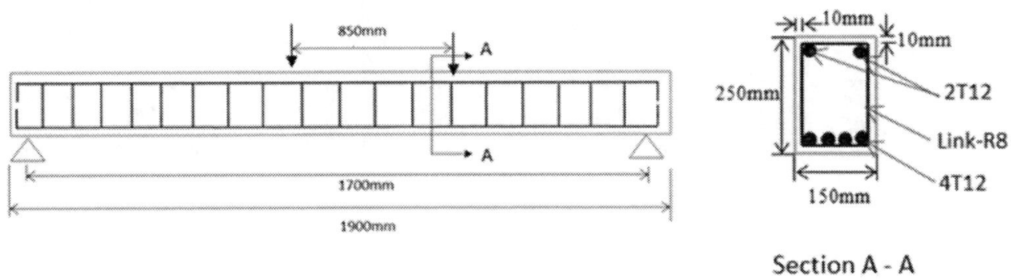

Figure 2. Beam specimens.

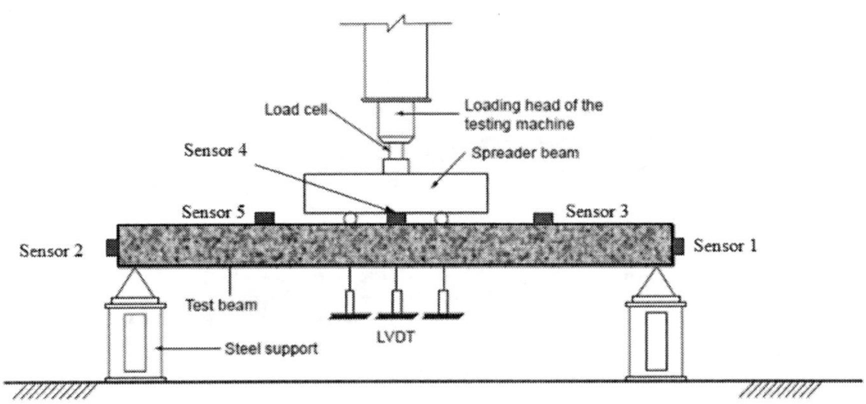

Figure 3. Experimental setup with four point bending test and AE system.

The beam specimens were loaded at a rate of 0.5mm/min using a hydraulic jack system. The Cyclic Load Test system was applied with a starting value 0.5kN and the first loading cycle was increased up to 20% of the ultimate load and held for three minutes before releasing 10% of the calculated ultimate load. This was held for a further three minutes before the next cycle was applied. These procedures were repeated in the second cycle with a similar load level to complete the loading set (LS). This loading and hold procedure was continued with increasing load until ultimate failure.

4 Result Analysis

Figure 4 presents amplitude versus time of each load set for the CLT method. The highest amplitude began from service load level (LS4), where the amplitude of hits ranged from 45dB (threshold level) to 96dB. In addition, the range value of amplitude from 95dB to 100dB kept increasing when the load set was raised up until failure. Most the high amplitude values indicate a higher value throughout the fracture process. This is clearly seen in LS5 to LS8. The AE amplitude raw data is calculated for *b*-value analysis to determine the type of cracks and fracture process. A detailed equation is shown in Equation (2).

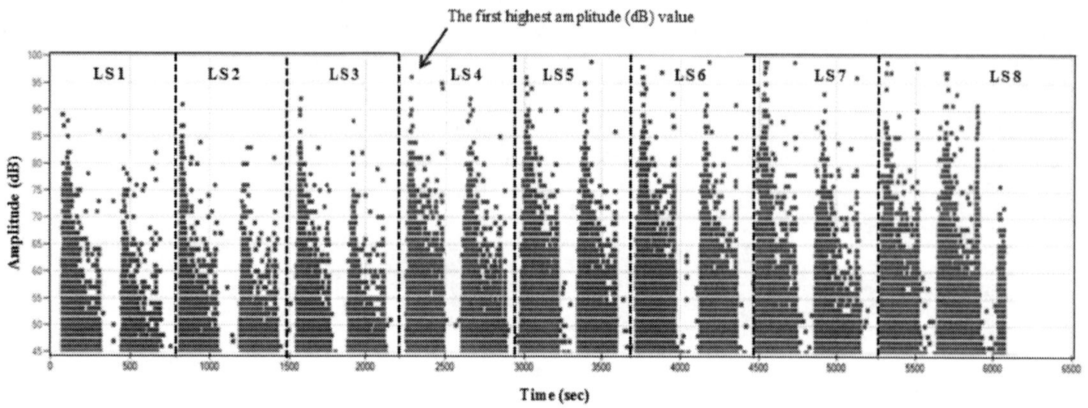

Figure 4. AE amplitude.

The trends of the b-value analysis were then plotted after using statistical analysis (seen in Figure 5). The pattern of the development of fracture process could be clearly observed in terms of micro-cracks and macro-cracks in concrete structure for each load set. From Figure 5, three phases of the fracture process were identified namely the initial load phase, service load phase and failure phase.

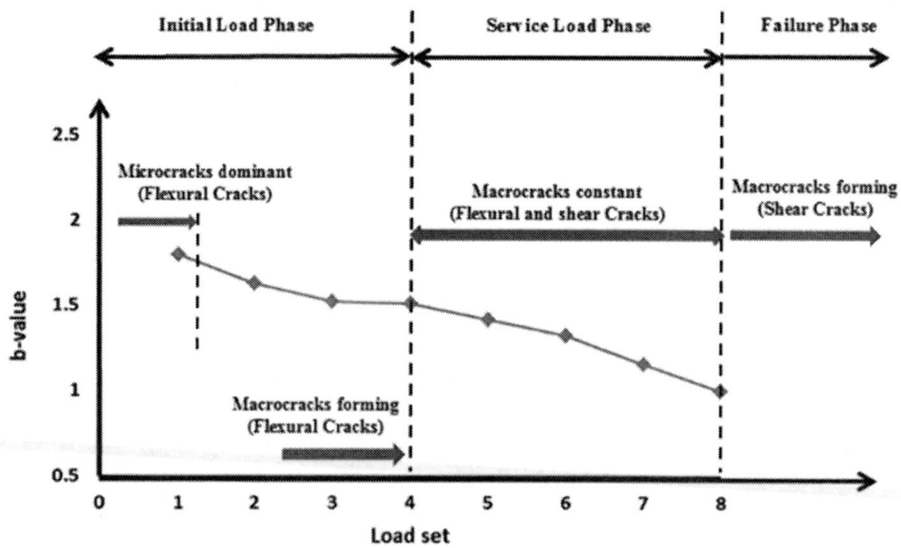

Figure 5. B-value analysis.

From Figure 5, micro-cracks were considered as first cracking and were dominant during the early stage of deformation at LS1. Therefore by increasing the loading to LS3, the macro-cracks began to form and grow as well due to the deformation process and movement from initial load phase and service load phase. At this time, the cracks were classified as flexural cracks and the cracks developed mostly at the bottom of the mid span between two supports.

At the second phase of service load, the macro-cracks were constant and uniformly distributed along the beam which can be recognized during this phase. Based on the visual observation, no new cracks appeared on the beam structure. The flexural cracks continued to grow with some diagonal shear. Subsequently, the macro-cracks were formed and opened up as the beam was failing. At this failure phase, the flexural cracks did not grow to the top, but the diagonal shears at the beam support spread upwards toward the loading point. This analysis was supported by real observation as presented in Figure 6.

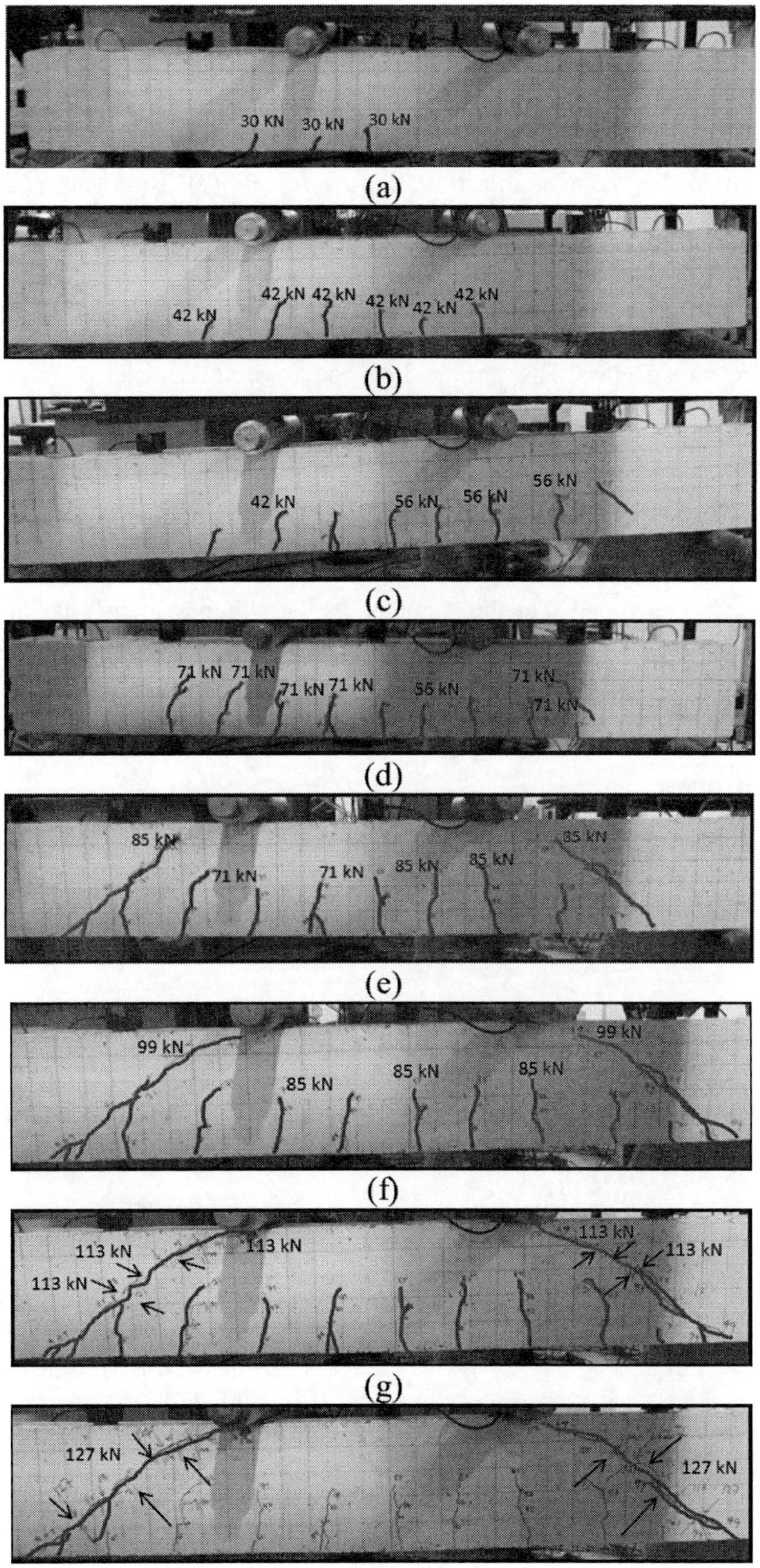

Figure 6. Real observation.

Figure 6 (a) – (h) presented the typical development of cracking and fracture process from LS1 to LS8. During the initial stage, micro-cracks appeared on the middle part of the beam from a load set of 32kN to 43kN as shown in Figure 6 (a) – (b). This type of crack is known as flexural cracking. At this time, the tension steel began to bear the load and contributed to the overall stiffness.

Nonetheless, the loading increased from approximately 60kN to 71kN which is above service load level. The flexural cracks continue to grow and simultaneously some diagonal cracks were formed between support and loading point as shown in Figure 6 (c) and (d). Throughout this point, the stirrup carried shear load and contributed to the global stiffness of the beam. This type of crack is classified as mixed mode cracking (flexural and shear). When the loading increased to the next LS5, the flexural cracks grew quickly between two points loading to form major cracks and spread to the compression zone. In addition, the shear cracks started to form near both supports when the loading was increased to approximately 85kN as shown in Figure 6 (e). From loading of approximately 100kN-120kN, both diagonal shears are spread upwards to the loading point, while the flexure cracks show no notable change as presented in Figure 6(f) and (g).

This period is known as diagonal crack development. Eventually, the flexural and shear cracks become widely open and are connected together when the loading reaches the final point, Figure 6 (h).

5 Conclusion

From the analysis, it was found that the trend of AE amplitudes in b-value method managed to develop the process of micro-cracking and macro-cracking for RC beams subjected to the CLT method. In addition, the correlation between the fracture processes in concrete and the localization of damage was established

Acknowledgement

This work was supported by Universiti Tun Hussein Onn Malaysia's Short Term Grant (U114) and Universiti Sains Malaysia

References

[1] N. Md Nor, A. Ibrahim, N. Muhamad Bunnori, H. M. Saman, S. N. Mat Saliah and S. Shahidan, Diagnostic of fatigue damage severity on reinforced concrete beam using acoustic emission technique, Journal Engineering Failure Analysis, **41**, 1-9, (2013).

[2] S. Shahidan, R. Pulin, N. Muhamad Bunnori and K. M. Holford, Damage classification in reinforced concrete beam by acoustic emission signal analysis, Construction and Building Material, **45**, 78–86, (2013).

[3] S. Shahidan, N.M. Nor and N.M. Bunnori, Overview of moment tensor analysis of acoustic emission signal in evaluation concrete structure, *IEEE 7th International Colloquium Signal Process*, 15–18, (2011).

[4] D.G. Aggelis, A.C. Mpalaskas and T.E. Matikas, *Acoustic Emission and Related Non-Destructive Evaluation Techniques in the Fracture Mechanics of Concrete*, Elsevier, (2015).

[5] K.M Holford, R. Pullin, S.L. Evans, M. J. Eaton, J. Hensman and K. Worden, Acoustic emission for monitoring aircraft structures, Proceedings of the Institution of Mechanical Engineers, Part G: Journal of Aerospace Engineering, **223**(5), 525-532, (2009).

[6] C. Grosse, M. Ohtsu and T. Shiotani, *Acoustic Emission Testing*, Springer, Berlin, (2008).

[7] G.K Kocur and T. Vogel, Classification of the damage condition of preloaded reinforced concrete slabs using parameter-based acoustic emission analysis, Construction and Building Materials, **24**(12), 2332-2338, (2010).

[8] W.C. Choi and H.D. Yun, Acoustic emission activity of CFRP-strengthened reinforced concrete beams after freeze–thaw cycling, Cold Regions Science and Technology, **110**, 47–58, (2015).

[9] S. Gresta and G. Patane, Variation of b value before the etnean eruption of March 1981, Pure and Applied Geophysics, **121**(2), 287-295, (1983).

[10] I.S. Colombo, I.G. Main and M.C. Forde, Assessing damage of reinforced concrete beam using "*b-Value*" analysis of acoustic emission, Journal of Material in Civil Engineering, **15**(3), 899-1561, (2003).

[11] R.V. Sagar, B.K.R. Prasad and S.S. Kumar, An experimental study on cracking evolution in concrete and cement mortar by the b-value analysis of acoustic emission technique, Cement and Concrete Research, **42**(8), 1094-1104, (2012).

[12] S.G. Shah and J.M. Chandra Kishen, Use of acoustic emissions in flexural fatigue crack growth studies on concrete, Engineering Fracture Mechanics, **87**, (2012) .

[13] S. Shahidan, N.M. Bunnori, N. Md Nor and S.R. Basri, Damage severity evaluation on reinforced concrete beam by means of acoustic emission signal and intensity analysis, *IEEE Symposium on Industrial Electronics and Applications*, 337–341, (2011).

MATEC Web of Conferences **47**, 02014 (2016)
DOI: 10.1051/matecconf/20164702014

Non-Destructive Concrete Strength Evaluation Using PZT Based Surface Wave Propagation Technique – A Comparative Study

Kok Zee Kwong[1,a], Yee Yan Lim[2] and Willey Yun Hsien Liew[3]

[1]*Civil Engineering Program, Faculty of Engineering, University Malaysia Sabah, 88400 Kota Kinabalu, Sabah, Malaysia*
[2]*Civil Engineering, School of Environment, Science and Engineering, Southern Cross University, Lismore NSW 2480, Australia*
[3]*Mechanical Engineering Program, Faculty of Engineering, University Malaysia Sabah, 88400 Kota Kinabalu, Sabah, Malaysia*

Abstract. Surface wave propagation (SWP) technique employing Lead Zirconate Titanate (PZT) transducer has recently been found to be a useful tool in concrete hydration monitoring. In this paper, the performance of PZT based SWP technique was compared with the conventional techniques such as ultrasonic pulse velocity (UPV) test, rebound hammer (RH) test and concrete compression test. Results showed that the SWP technique, in addition to its inherent advantages, performed equally well as the conventional counterparts in concrete strength prediction.

1 Introduction

Structural safety is the ultimate concern of engineering design. Failure due to insufficiently cured concrete during construction as well as concrete weakening from aging and deterioration occurred every now and then. As a result, an effective concrete strength monitoring technique is highly sought after to ensure safety and cost effectiveness.

A great variety of conventional techniques are available in the market for the monitoring of concrete strength development and continuous health assessment of concrete. These techniques are broadly classified into destructive evaluation technique and non-destructive evaluation (NDE) technique. Destructive evaluation techniques usually involve core test and loading test. These techniques induce damage to the existing structure as they require extraction of cylindrical specimens from the structure, which is often undesirable. On the other hand, NDE techniques include resonant frequency technique, hydration heat-based monitoring technique, ultrasonic pulse velocity (UPV) test and rebound hammer (RH) test.

Recent development of smart material, namely piezoelectric (Lead Zirconate Titanate, PZT) transducer is capable in overcoming some shortcomings of the conventional structural health monitoring (SHM) techniques. Initially, the PZT transducer is used primarily for damage monitoring involving thin metallic or composite structures. It has later found its application in concrete hydration

[a] Corresponding author : charleskwongkz@gmail.com

and strength monitoring. The advantages of employing PZT transducer based SHM techniques include autonomous, continuous and online remote monitoring, low power consumption and cost effective.

Surface wave based concrete hydration monitoring using conventional techniques has been well established. For instance, Luo and Bungey introduced a method to determine the dynamic modulus of elasticity for concrete using compression wave [1]. Moreover, Song et al. conducted a study on SWP in concrete structures using a PZT actuator/sensor system [2]. To date, research on WP techniques focused mainly on SHM of metallic and composite structures [3-5].

PZT based SWP technique has shown its capability in predicting the dynamic Young's modulus [6] and compressive strength of concrete [7]. This study aimed at comparing the performances of different concrete strength monitoring techniques, namely the PZT based SWP technique, UPV test, RH test and compression test. Study was conducted on a series of lab-sized mortar specimens with different strength.

2 Physical Principle of Wave Propagation

In SWP technique, two or more PZT transducers are permanently bonded on the surface of concrete specimen using high strength epoxy at predetermined distances. One of the patches is assigned as actuator to excite a five-peak tone burst signal. Another patch acts as a sensor to pick up the wave from the actuator. In this study, actuation of PZT transducer on the surface of thick concrete structure is expected to predominantly excite surface wave (Rayleigh or R-wave). R-wave propagates mainly along the surface of the concrete and decreases rapidly in magnitude with depth below the surface [8]. Weaker pressure wave (P-wave) may also be present [9].

It is noteworthy that the P-waves are characterized with high velocity and low amplitude, while the R-waves have lower velocity and higher amplitude [10]. Therefore, the first wave packet is induced by the propagation of P-wave while the second wave packet is induced by R-wave as shown in Figure 1. Waves following these two wave packets are complex reflections of bulk waves and noises.

Velocity of P-wave and R-wave can be calculated by measuring the time of flight (TOF) for any distance between actuator and sensor. TOF is the time calculated from wave packet of the actuator to the arrival of the wave packet of the sensor (Figure 1). As the concrete hardens, it becomes stiffer and thus affecting the velocity of the wave. Consequently, the TOF of the wave which can be measured from the electrical signatures, is altered. The data acquired is analyzed using various signal processing techniques such as Hilbert transform and cross-correlation analysis to achieve higher accuracy.

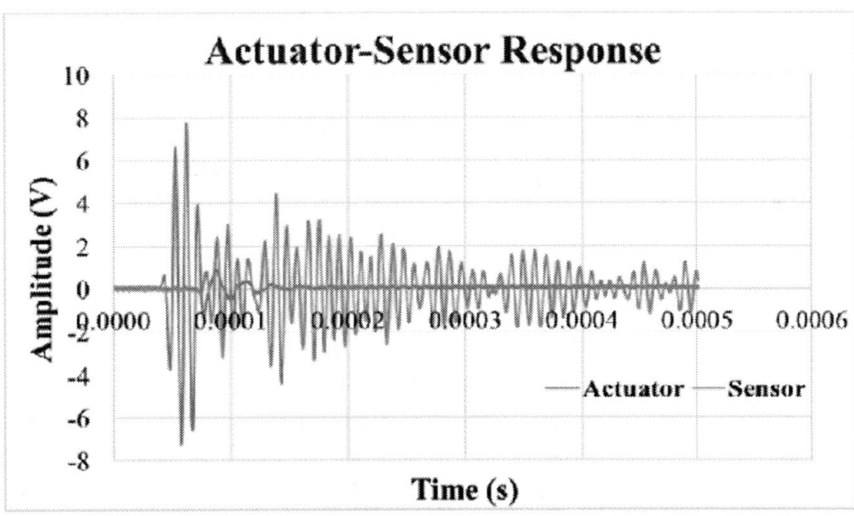

Figure 1. Wave packet and TOF for Prism H at 120 kHz and 120 mm actuator-sensor distance.

The fundamental theories of WP techniques were described by Jones [11]. The relationship between the modulus of elasticity of the solid medium and the wave velocity in homogeneous and isotropic solids of infinite dimensions can be expressed as:

$$V_P = \sqrt{\frac{E_d(1-v)}{\rho(1+v)(1-2v)}} \tag{1}$$

$$V_R = \frac{0.87+1.12v}{1+v}\sqrt{\frac{E_d}{\rho}\frac{1}{2(1+v)}} \tag{2}$$

where, E_d is the dynamic modulus of elasticity, v the Poisson's ratio, ρ the density, V_P the velocity of pressure wave, V_R the velocity of surface wave.

3 Experimental Setup

Two batches of cement mortar were mixed using cement to sand ratio of 1:2. First batch of cement mortar (Prism H) was mixed with water to cement ratio (W/C ratio) of 0.3. The second batch of cement mortar (Prism L) was mixed with W/C ratio of 0.4. Two cement mortar prisms with the dimension of 100 mm x 100 mm x 500 mm were prepared for each batch of casting. Cement mortar cubes with the dimension of 100 mm x 100 mm x 100 mm were also prepared from the same batch of mix for the evaluation of compressive strength after 1 day, 3 days, 7 days, 14 days and 28 days of curing. Three pieces of PZT patches with the dimension of 20mm x 20mm x 0.2mm (PI ceramics) as shown in Figure 2 were surface bonded using two parts quick set epoxy onto the smooth surface of Prism H and Prism L respectively. The distances between Patch 1 and Patch 2, and Patch 2 and Patch 3 were 80mm and 120mm, respectively as depicted in Figure 3.

A Waveform Generator (Agilent 33500B Series) was used to generate a 10V peak-to-peak Hanning windowed five-peak tone burst at a frequency of 120 kHz. Due to the high rate of energy dissipation in concrete, the signal input on the actuation must be strong enough to generate appreciable output from the sensors. The signal was amplified through a Power Amplifier (Trek PZD350A) to 150 V peak-to-peak which was then applied to the actuating PZT patch. The signal from the sensor was measured using a Digital Oscilloscope (YOKOGAWA DL 1620) and recorded on a personal computer. The setup is schematically presented in Figure 4.

For both specimens, Patch 1 was first actuated, and this was followed by Patch 2 and Patch 3 acting as sensors. Patch 2 was later actuated with Patch 3 acting as sensor. Readings were taken immediately after the setting of epoxy and monitored continuously for the first 7 days and thereafter at 14 days, 21 days and 28 days. UPV, rebound hammer and compression tests were performed at 1 day, 7 days, 14 days, 21 days and 28. Both specimens (Prism H and L) were subjected to the same compaction and curing condition. The cubes and samples were cured outdoor under a roof, simulating site curing condition.

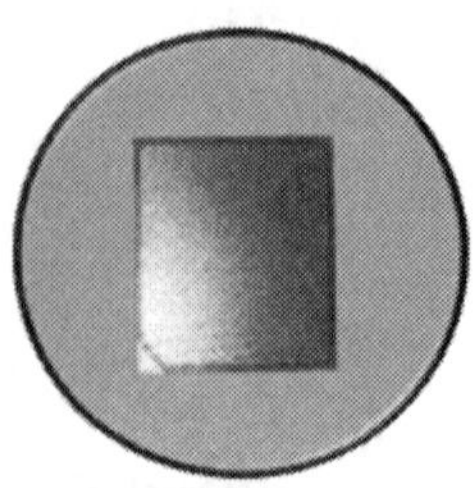

Figure 2. PZT patch.

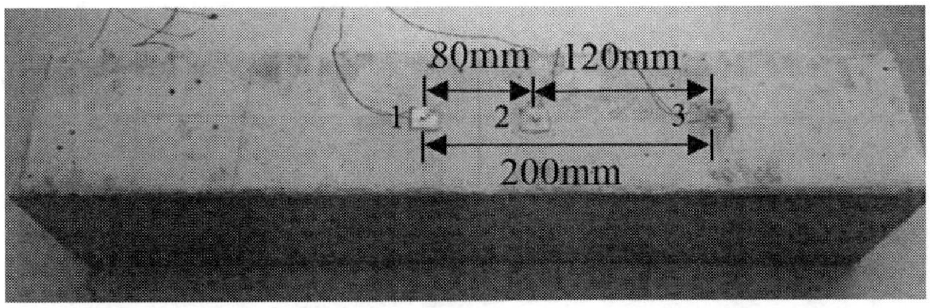

Figure 3. Configurations of PZT patches.

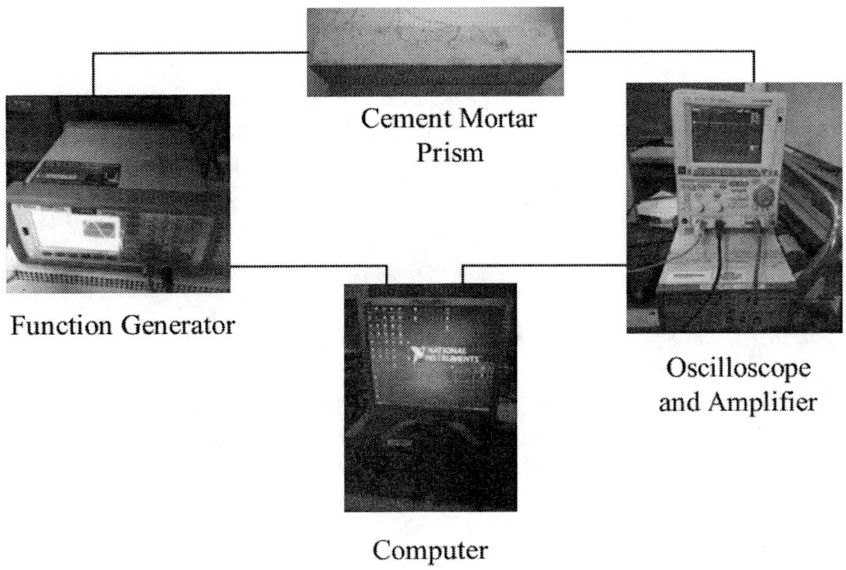

Figure 4. Experimental setup.

4 Results and Discussions

The sensor signal obtained from early age of curing (less than 9 hours) was very weak for both Prism H and Prism L indicating that the concrete was in a "soft" and "semi-solid" state. High damping and low stiffness prevented the P-wave and R-wave from reaching the sensor effectively. Moreover, the sensor signals acquired were seriously contaminated by noise. As a result, the sensor signal's TOF for P-wave and R-wave could only be measured after 19 hours of curing.

Figure 5 illustrates the relationship between the wave velocity and the curing time for P-wave and R-wave, respectively. The results showed that all velocities increased with curing time. According to the elastic wave theory, the velocity of wave is proportional to the stiffness of the medium. In this case, as the concrete hardened, and its stiffness and strength increased, these was reflected in the increase in the velocity. On the other hand, the P-wave velocities and R-wave velocities acquired from Prism H were consistently higher than those from Prism L. This indicated that the stiffness and the strength of Prism H was higher than Prism L.

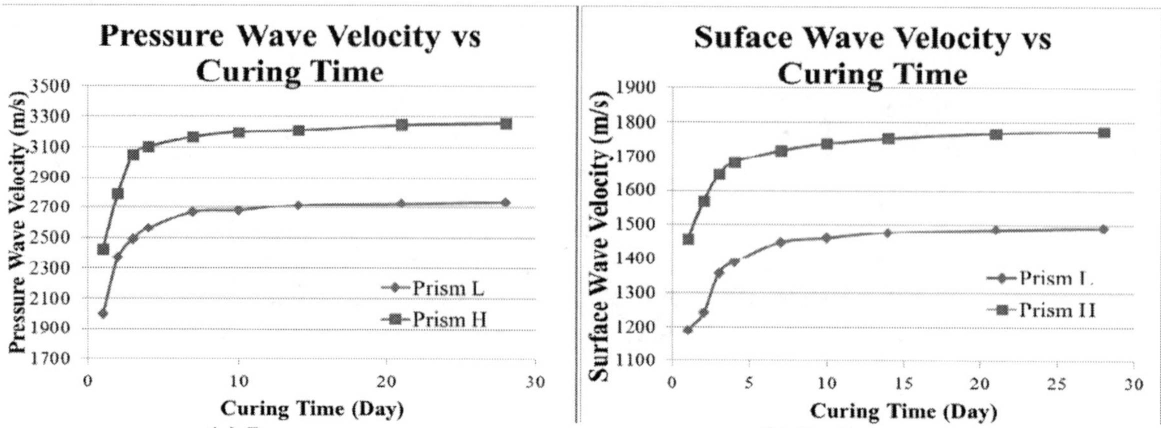

(a) Pressure wave (b) Surface wave

Figure 5. Wave velocity versus curing time.

Knowing the velocities of P-wave and R-wave, the dynamic modulus of elasticity and Poisson's ratio can be evaluated using Equation 1 and Equation 2. The ACI equation was widely adopted in the engineering community to empirically relate the dynamic modulus of elasticity and the compressive strength [12]:

$$E = k\sqrt{f_{cu}} \tag{3}$$

where E is the Young's modulus of concrete, f_{cu} the compressive strength of concrete, k the curve fit proportionality constant.

Straight line curve fitting was plotted using the least square method from relationship between dynamic Young's modulus of concrete and square root of concrete strength. The gradients for the straight line are the P-wave velocity and R-wave velocity. In this study, k value was found to be 3.24 for both Prism H and L.

Combining Equation (2) and Equation (3), the concrete strength can be correlated to the R-wave velocity through a semi-empirical relationship as presented schematically in Figure 6:

$$f_{cu} = \left[\frac{2\rho(1+\nu)\left(\frac{V_S(1+\nu)}{0.87+1.12\nu}\right)^2}{k}\right]^2 \tag{4}$$

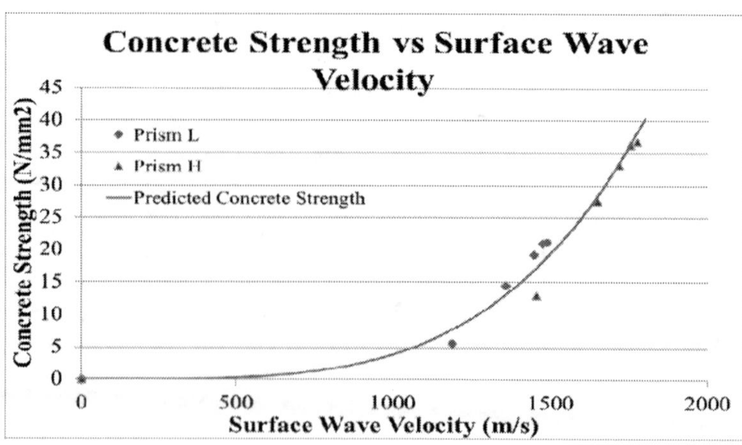

Figure 6. Strength prediction chart (SWP technique).

5 Comparison with Conventional Techniques

Figure 7 shows the comparison of the concrete strength obtained from the SWP technique, and the compression, rebound hammer and UPV tests. In general, the rebound hammer test yielded to a more conservative outcome for both specimens. Most of the predicted concrete strength were lower than the actual concrete strength obtained from the compression test. The standard deviation with the concrete strength obtained from compression test is 4.38.

The predicted strength obtained from the SWP techniques and UPV was similar, indicating that both techniques work equally well. They also showed superiority over the Rebound Hammer test in terms of accuracy of prediction. The standard deviation for UPV is 1.83 while the standard deviation for SWP technique is 2.66.

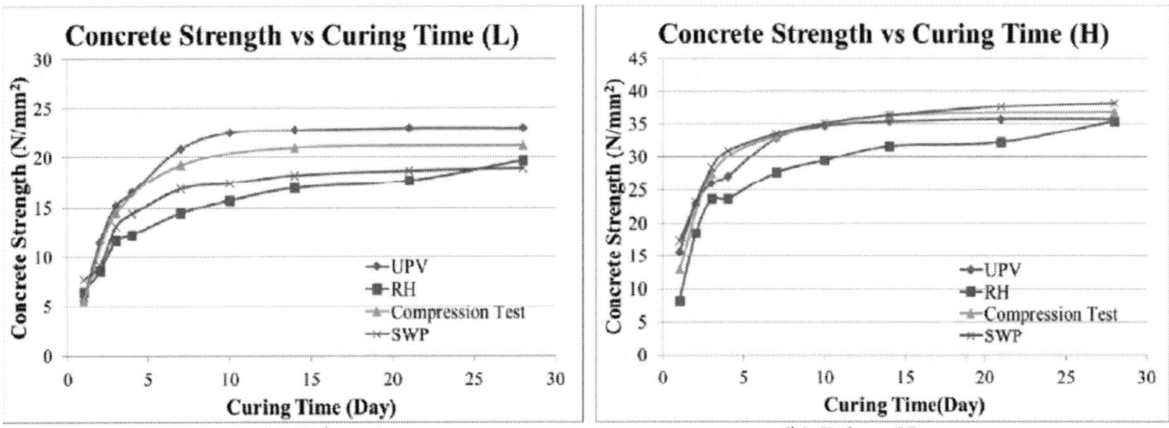

(a) Prism L (b) Prism H

Figure 7. Concrete strength versus curing time.

6 Conclusions

This paper presents a comparative study into the concrete strength prediction capability of SWP technique and similar conventional techniques, namely the UPV, the rebound hammer and the compression tests. It was shown that the SWP technique has the capability of predicting the strength of different types of concrete. It also performs equally well as the UPV test, yielding reasonably accurate predictions as compared to the compression test.

SWP technique enables users to measure both the R-wave and P-wave simultaneously. Hence, both the dynamic Young's modulus and Poisson's ratio of concrete can be evaluated.

As the results are promising, SWP technique could potentially replace the conventional NDE counterparts in providing an autonomous, real-time and remote concrete strength prediction technique. Future studies should focus on the investigation of the consistency, repeatability and reliability of this technique as well as the FE modelling on the PZT based SWP technique. On the other hand, a handy list of correction/modification factors accounting for varying conditions (e.g. temperature, humidity, curing condition, surface roughness, presence of aggregate, size of aggregate, types of cement etc.) should be established before real life application becomes possible.

References

[1] Q. Luo and J.H. Bungey, Using compression wave ultrasonic transducers to measure the velocity of surface waves and hence determine dynamic modulus of elasticity for concrete, Journal Construction and Building Materials, **10**,237-242, (1996).

[2] F. Song, G.L. Huang, J.H. Kim and S. Haran, On the study of surface wave propagation in concrete structures using a piezoelectric actuator/sensor system, Journal Smart Material and Structure, **17**(5), 1-8, (2008).

[3] V. Giurgiutiu, J. Bao and W. Zhao, Active sensor wave propagation health monitoring of beam and plate structures, *Proceeding SPIE's 8th Annual International Symposium on Smart Structures and Materials*, 234-245, (2001).

[4] M. Gresil and V. Giurgiutiu, Guided wave propagation in carbon composite laminate using piezoelectric wafer active sensors, *Proceeding SPIE Smart Structures and Materials+ Nondestructive Evaluation and Health Monitoring*, 869525-869525, (2013).

[5] J.B. Ihn and F.K. Chang, Detection and monitoring of hidden fatigue crack growth using a built-in piezoelectric sensor/actuator network: I. Diagnostics, Journal Smart Material Structure, **13,** 609, (2004).

[6] Y.Y. Lim, N. Majain, K.Z. Kwong and W.Y.H. Liew, On the study of concrete hydration process using piezoelectric based surface wave propagation technique, ACMSM23, 1121-1128, (2014).

[7] Y.Y. Lim, K.Z. Kwong, W.Y.H. Liew and C.K. Soh, A feasibility study into concrete strengthprediction using PZT based surface wave propagation technique, Journal NDT&E International, (2014).

[8] Y. Lu, J. Li, L. Ye and D. Wang, Guided waves for damage detection in rebar-reinforced concrete beams, Journal Construction and Building Materials, **47**, 370-378, (2013).

[9] L. Yu, In-situ structural health monitoring with piezoelectric wafer active sensor guided-wave phased arrays, PhD Thesis, University of South Carolina, (2006).

[10] M. Sun, W.J. Staszewski, R.N. Swamy and Z. Li, Application of low-profile piezoceramic transducers for health monitoring of concrete structures, Journal NDT&E International, **41,** 589-595, (2008).

[11] R. Jones, *Non-Destructive Testing of Concrete*, Cambridge University Press, London, (1962).

[12] ACI Committee: 318-1995 Building code requirements for structural concrete and commentary, American concrete Institute, Michigan, (1995).

MATEC Web of Conferences **47**, 02015 (2016)
DOI: 10.1051/matecconf/20164702015

Effect of Large Negative Phase of Blast Loading on Structural Response of RC Elements

Zubair Iman Syed[1,a], Priyan Mendis[2] and Shaikh Atikur Rahman[3]

[1]*Department of Civil Engineering, Abu Dhabi University, UAE*
[2]*Department of Infrastructure Engineering, The University of Melbourne, Australia*
[3]*Department of Civil Engineering, Universiti Teknologi PETRONAS, Malaysia*

Abstract. Structural response of reinforced concrete (RC) elements for analysis and design are often obtained using the positive phase of the blast pressure curve disregarding the negative phase assuming insignificant contribution from the negative phase of the loading. Although, some insight on the effect of negative phase of blast pressure based on elastic single-degree-of-freedom (SDOF) analysis was presented before, the influence of negative phase on different types of resistance functions of SDOF models and on realistic finite element analysis has not been explored. In this study, the effects of inclusion of pulse negative phase on structural response of RC elements from SDOF analysis and from more detailed finite element analysis have been investigated. Investigation of SDOF part has been conducted using MATLAB code that utilizes non-linear resistance functions of SDOF model. Detailed numerical investigation using finite element code DIANA was conducted on the significance of the negative phase on structural response. In the FE model, different support stiffness was used to explore the effect of support stiffness on the structural response due to blast negative phase. Results from SDOF and FE analyses present specific situations where the effect of large negative phase was found to be significant on the structural response of RC elements.

1 Introduction

Due to the high degree of uncertainty involved in blast load prediction, complex time consuming detailed structural analysis is often not justified. Single-degree-of-freedom (SDOF) analysis has the potential of giving a good prediction of final response of the critical point of a structure. All commonly used design manuals for blast loading like PDC-TR 08-07 [1] and UFC 3-340-02 [2], and widely used design tools such as Single-Degree-of-Freedom Blast Effects Design Spreadsheets (SBEDS) [3] and Facility and Component Explosive Damage Assessment Program (FACEDAP) [4], are based on the SDOF response calculation of structural components. Since initiation of using SDOF approach for structural response calculation for impulsive loads, significant advances have been achieved in the application of SDOF models, hence the use of SDOF models has also increased significantly in area of protective design.

Depending on the location of blast and type of explosive, the negative phase of a blast pressure curve can be significant and can contribute to the damage of RC elements. Literature related to the

[a] Corresponding author : zubair.syed@adu.ac.ae

development of the damage assessment tool FACEDAP has acknowledged the fact that the negative phase can be "in-phase" with the component response. In CEDAW [1], it has been indicated that the negative phase loading only affects component response when scaled pressure is high relative to scaled impulse, which corresponds to cases where the positive phase blast load durations are relatively short compared to the component response time.

In this study, the effect of inclusion of pulse negative phase on structural response of RC elements obtained using SDOF models has been investigated. Non-linear transient analysis was also conducted to explore the significance of the negative phase on structural response using finite element code DIANA [5]. Different support stiffness was used for the FE model to explore the effect of support stiffness on the structural response due to blast load with large negative phase. Results from both SDOF model and FE analysis are presented to establish the effect of large negative on structural response of RC elements.

2 Blast Pressure Idealisation and Negative Phase

Selection of appropriate blast loading is critical for realistic and reliable structural response of reinforced concrete elements under blast loading. Chemical investigations and experimental data have shown that the evolution of blast load pressure (P) with time (t) can be simulated more accurately by a quasi-exponential distribution [6] as a modified Friedlander equation as given below:

$$p(t) = p_{so}(1 - \frac{t}{t_d}) exp\left\{ -\frac{\alpha t}{t_d} \right\} \quad (1)$$

where α is the dimensionless waveform parameter also known as decay coefficient. The parameter alpha (α) determines the extent of negative phase in the exponential distribution of Equation (1). If α is smaller than one, there is a larger negative phase in the pressure curve, while if α is larger than one, the negative phase becomes less significant. The blast pulse profile for different values of waveform parameter alpha (α) in Friedlander equation is shown in Figure 1.

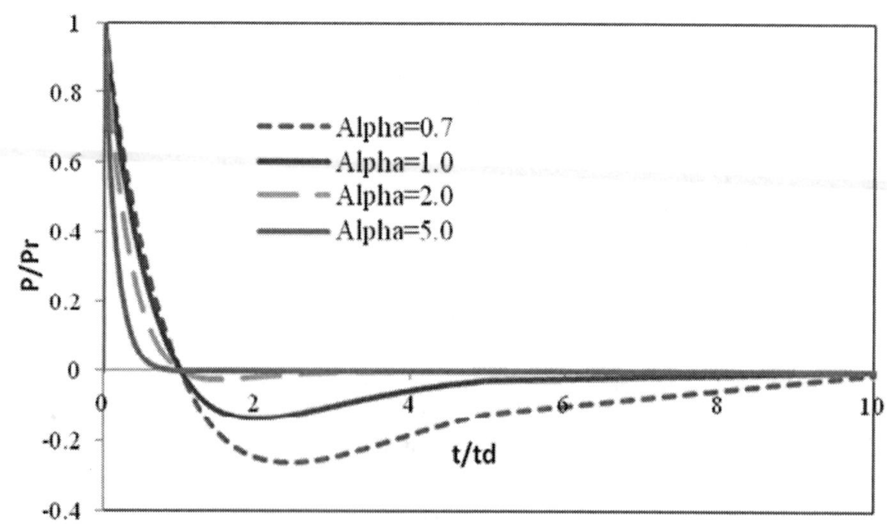

Figure 1. Blast load profile obtained using Friedlander equation with different values of waveform parameter (α).

In this figure, the ratio between pressure at a given time (P) and the peak pressure (P_r) has been plotted against the ratio of corresponding time (t) to the positive duration value (t_d). The blast pulse profile presented by the Friedlander equation with an α value of unity has been found to match well with some experimental results [7]. However, depending on various factors such as change weight, scaled distance, type of blast and other factors, the blast pressure can have a negative phase much higher than the phase represented by the α value of one. In those situations, to idealise the blast pulse profile, α values smaller than unity are required. Values of α higher than unity correspond to smaller negative pressure, and values above five produce virtually zero negative phase. The duration of the negative phase is usually double that of the positive phase [6].

3 Single Degree of Freedom Models

The equivalent SDOF method, which is widely used in protective design practices, has been used in this study to convert the real reinforced concrete elements into equivalent SDOF models. The transformation factors that are applied to the distributed values to calculate the equivalent lumped values are a function of the distribution of mass and loading over the element and the shape function of the deflected shape [8].

Numerical time integration procedures were implemented in MATLAB [9] code to obtain solutions for SDOF models with advanced non-linear resistance functions. Non-linear resistance functions were developed using the moment-curvature relationships which were constructed using material constitutive models of concrete and reinforcement bars. Numerical approximation method proposed by Newmark with modified Newton-Raphson iteration was adopted in the MATLAB codes for dynamic response. The response of the SDOF model with non-linear resistance function was calculated using incremental time-stepping approach. Details on these steps can be found in Chopra [10]. For each load value, the corresponding maximum deflection at the mid-span was obtained by integration of the curvature along the half-length of the element. These codes were developed and verified against numerical FE results and other known experimental results. The details about the selected material models and the code verification can be found in Syed et al. [11, 12].

4 Response of SDOF Models under Blast with Negative Phase

Response spectra for SDOF models can readily be used for predicting the maximum response of the system. Widely referred response spectra for blast design given in UFC 3-304-2 [2] are for elastic-plastic resistance functions and for idealised blast load with positive phase only. Gantes and Pnevmatikos [13] employed an elastic-perfectly plastic resistance function to compare response spectra obtained using exponential load. In the present research, response spectra were obtained for elastic-plastic hardening using exponential blast load with negative phase to mainly investigate the effect of negative phase on a realistic idealised structure. Figure 2 shows comparison of the response obtained from elastic-plastic ($K_s = 0$), elastic-plastic hardening with secondary stiffness 10% of initial stiffness ($K_s = 0.1K$) and elastic-plastic hardening with secondary stiffness 20% of initial stiffness ($K_s = 0.2K$) for exponential loading. Detail on how to obtain system stiffness value for equivalent SDOF system is presented in Syed et al. [11]. The responses for various ratios of maximum response to elastic response of the system (X_{max}/X_{el}) are presented in Figure 2 in terms of the ratios of maximum resistance to peak load (R_m /P). Variation between maximum responses of a SDOF model depends on the ratio of peak resistance to peak load (R_m/P) and the ratio of the blast duration to SDOF system period (t_d/T). Figures 3 and 4 show the comparison of responses for from triangular and exponential load loads.

It can be observed from the response spectra of elastic-plastic hardening system that when R_m/P is larger (between 1 and 2), the response for exponential load is smaller compared to that obtained from triangular load. This can be explained as the exponential load decreases faster than the triangular one, and this has more influence in elastic-plastic situations than in purely elastic cases. The duration of blast loading (t_d) usually varies within a small range of values; large values of t_d/T correspond to small values of the fundamental period (T), therefore, to stiff systems. Commonly used assumptions of triangular blast load can sometimes be slightly under-conservative, particularly for flexible structural systems, but can also be significantly over-conservative for stiffer structures.

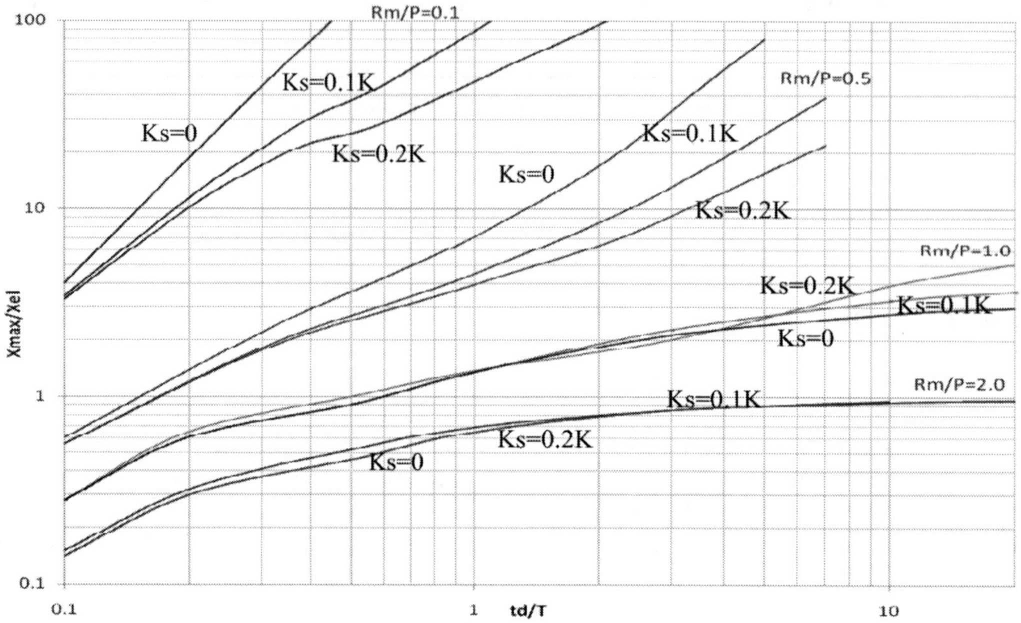

Figure 2. Response spectra for elastic-perfectly plastic and plastic hardening SDOF system under exponential loading with negative phase.

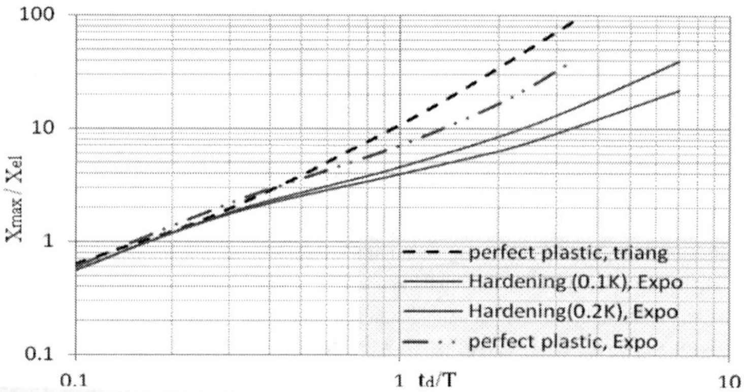

Figure 3. Comparison between response under triangular and exponential load with negative phase for $R_m/P = 0.5$.

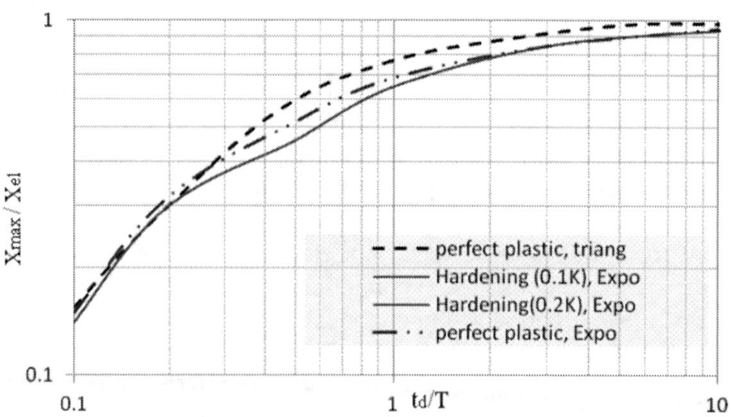

Figure 4. Comparison between response under triangular and exponential load with negative phase for $R_m/P = 2.0$.

If rebound of the system occurs in a period of negative phase of the blast load, then this may lead to maximum displacement of the system due to the negative phase. Thus, the negative plastic deformations due to negative plastic forces may exceed in magnitude the positive plastic deformations caused by the larger forces of the triangular pulse.

If t_d/T is longer than 0.5, then the maximum deformation occurs during the pulse and the shape of loading is of great significance. If t_d/T is less than 0.5, then the maximum deformation occurs during the free vibration and is mainly controlled by the time integral of the phase, independent of the shape of loading. Similar conclusions can also be found in Chopra [10] and Gantes and Pnevmatikos [13].

5 Structural Response of RC Beam to Negative Phase

Non-linear FE analyses were performed using FE code DIANA to explore the effect of negative phase on the overall structural response of a reinforced concrete beam for different support conditions. A normal strength reinforced concrete beam (B40-D4) tested by Magnusson and Hallgren [14] under blast load was modelled. Dimensions of the beam were 300 (W) x 160 (D) x 1500 (L) and had a static compressive strength of concrete of 43 MPa. The beam was reinforced with 5H16 mm bars with yield strength of 555 MPa.

The beam was modelled using beam-elements, CL18B (a class-III beam-element from the DIANA library), based on the Mindlin-Reissner beam theory utilizing total strain-based modelling approach. In total strain-based modelling of concrete elements, the stress-strain relationship of the concrete by Thorenfeldt et al. [15] has been used for modelling the compressive behaviour. This model has been found to be in good correlation with the experimental data and has been used by a large number of researchers in the area of impact and blast loading on concrete elements [16]. Non-linear tension softening model proposed by Hordijk and in build in DIANA material library has been used in this study. A rotating axis approach has been adopted that does require an explicit value of shear retention during modelling. The constitutive model of steel reinforcement is based on the von Mises plasticity model with strain hardening. The plastic modulus of the material (E_{sh}) is assumed to be 1.5% of its elastic modulus (E_s) as similar values is found to be used successfully in the past studies [17]. Mid-span deflection-time histories obtained from finite element simulation was compared with the experimental results for verification purposes. In the FE simulation, the blast pressure on the beam was applied as a time-dependent line load.

To study the effect of negative phase, a blast load with a peak pressure of 0.25 MN/m and positive phase duration of 8 msec was used. The negative phase of the selected load followed the modified Friedlander equation as defined previously. The response of the modelled beam was calculated with four different support spring stiffness values and also for a simply supported condition. The maximum inward and outward deflections of the mid-span for different support conditions under blast load with and without negative phase are presented in Table 1. It was found that the inclusion of negative phase reduced the inbound displacements by a small amount for some support conditions and the rebound displacements were found to increase in all the support conditions. For all types of support stiffness values, similar changes were observed. Although the less stiff support springs showed a higher displacement under blast load with and without negative phase compared to the stiffer supports, it did not demonstrate any significant higher rate of change in inbound and rebound displacements for the inclusion of the negative phase.

The responses of the beam with four types of support conditions under different peak pressures were also calculated. In all cases, the positive duration was selected as 8 ms and the waveform parameter or decay coefficient (α) of the modified Friedlander equation was selected as 0.7. The objective of this investigation was to explore the effect of negative phase of blast pressure on the structural response under different peak pressures. Three different peak load values were selected: 0.15 Mn/m, 0.25 MN/m and 0.40 MN/m. Maximum mid-span deflections calculated for load with negative phase for different peak pressures for different support conditions are tabulated in Table 2.

Table 1. Maximum mid-span deflections for the beam with different support conditions under blast load with and without negative phase.

Support condition notation	Support spring–beam stiffness ratio (K_{sp}/K_{el})	Maximum mid-span deflection (mm)			
		Blast load without negative phase		Blast load with negative phase	
		Inbound	Rebound	Inbound	Rebound
SP1	0.7	8.12	–3.22	7.42	–4.33
SP2	1	8.28	–2.32	8.12	–2.38
SP3	10	7.71	–2.21	7.81	–3.20
SP4	100	7.57	–2.11	7.57	–3.04
S-S	Simply supp.	7.55	–2.14	7.54	–3.18

Table 2. Summary of the responses calculated for load with negative phase for different peak pressures for different support conditions.

Support condition notation	Peak load 0.15 N/m		Peak load 0.25 N/m		Peak load 0.40 MN/m	
	Maximum mid-span deflection (mm)		Maximum mid-span deflection (mm)		Max mid-span deflection (mm)	
	Inbound	Rebound	Inbound	Rebound	Inbound	Rebound
S-S	3.70	–3.97	7.54	–4.01	19.20	–4.72
SP2	3.61	–3.96	8.18	–3.78	17.80	–4.75
SP3	3.73	–3.98	7.88	–3.89	17.85	–4.74
C-C	1.09	–0.91	1.97	–1.69	4.09	–1.71

The results in Table 2 indicates that for low peak load values, the rebound of the structure is significant compared to the inbound displacement. The results obtained for the peak load value of 0.15 MN/m, clearly indicate that the rebound displacements of the beam are higher than the inbound displacements for the simply supported case and also for the cases where springs were used to provide support flexibility. The results obtained for a peak load of 0.15 MN/m, reinforce the previous findings for single degree of freedom (SDOF) models in previous section, where it was found that if rebound of the system occurs in a period of negative phase of the blast load, then this may lead to maximum displacement of the system due to the negative phase. Thus, the negative plastic deformations due to negative plastic forces may exceed in magnitude the positive plastic deformations caused by the larger forces of the triangular pulse.

The displacement-time history of the beam with simply supported conditions under a peak load of 0.15 MN/m is given in Figure 5 and the support reaction is given in Figure 6. The figures show that the maximum support reaction is occurring with the maximum displacement at the rebound phase. The results in Table 2 shows that the blast loads with high peak pressure value increase the inbound displacements significantly and reduce the rebound displacements. Hence, the effect of the negative phase diminishes with the increase in peak pressure value.

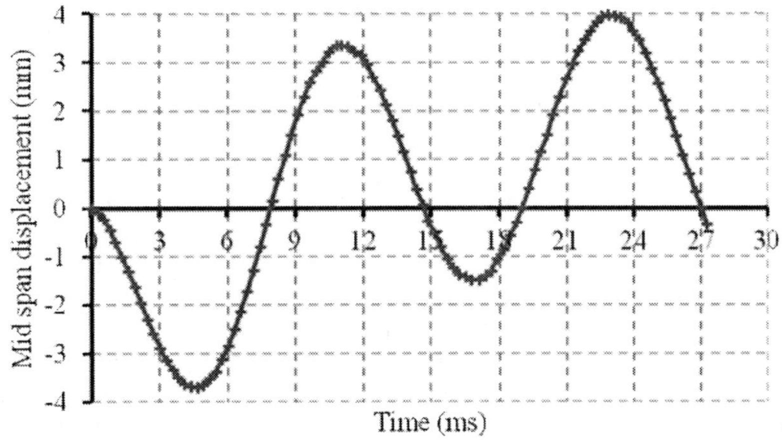

Figure 5. Mid-span deflection-time history of the simply supported beam under a blast load with a peak of 0.15 MN/m and duration of 8 ms.

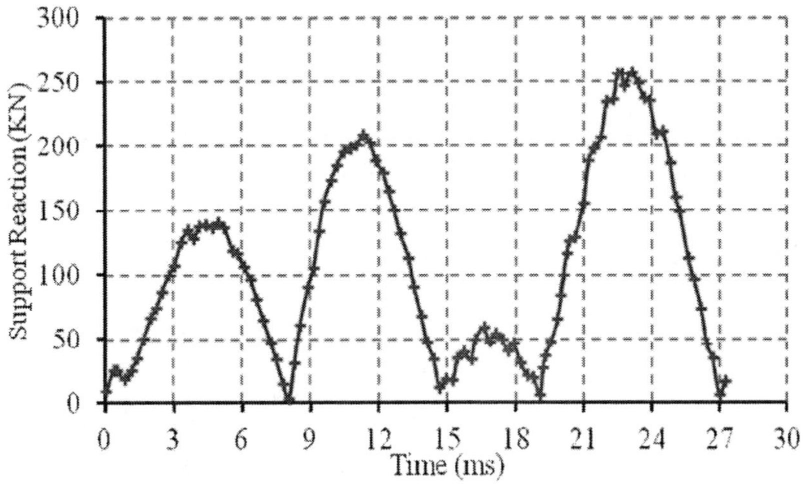

Figure 6. Support reaction-time history of the simply supported beam under a blast load with a peak of 0.15 MN/m and duration of 8 ms .

The mid-span maximum displacements calculated for blast loads with the same peak pressure and positive phase duration, but with different values of waveform parameter of the modified Friedlander equation are given in Table 3. The results indicate that the negative phase has a significant effect when the blast waveform parameter is less than unity and becomes insignificant in contribution for the cases were that negative phase can be presented by the Friedlander equation with an α-value larger than one.

Table 3. Summary of the responses calculated for loads with different amounts of the negative phase part for two support conditions.

Support condition notation	Maximum mid-span deflection (mm)					
	Waveform parameter α = 0.7		Waveform parameter α = 1.0		Waveform parameter α = 2.0	
	Inbound	Rebound	Inbound	Rebound	Inbound	Rebound
S-S	3.70	-3.97	3.70	-3.51	3.70	-3.19
C-C	1.09	-0.91	1.09	-0.89	1.09	-0.66

6 Conclusion

Effects of the negative phase of blast pressure curve on the structural response of RC elements were investigated. The effect of support flexibility on structural response under blast load with negative phase is also explored. Both SDOF analysis approach and detailed FE analysis was conducted. The study showed that when the negative phase of the blast loading was included in the analysis, no significant change was observed in the positive (inbound) displacement, but an increase in the negative (rebound) displacement was observed. It was also found that under low peak load, if the negative phase of blast load was considered, the rebound of the system could occur in a period of negative phase of the blast load, and then this would cause the maximum displacement to occur at rebound. However, the effect of negative phase of blast loading was found to be insignificant for beams with fixed support conditions. The flexibility of supports was found to increase the mid-span deflections when the support stiffness was lower than or close to the stiffness of the concrete element. The support stiffness was found to be insignificant to influence the overall structural response when the stiffness of support was few times higher than the stiffness of the concrete element.

References

[1] PDC-TR 08-07, Methodology Manual for Component Explosive Damage Assessment Workbook (CEDAW), Protective Design Center Technical Report, (2008).
[2] UFC 3-340-02, Structures to Resist the Effects of Accidental Explosions (Unified Facilities Criteria), US Department of Defense, (2008).
[3] PDC-TR 06-08, Single Degree of Freedom Structural Response Limits for Anti-terrorism Design: U.S. Army Corps of Engineers, (2008).
[4] C.J. Oswald and D. Skerhuf, FACEDAP User's Manual. Omaha District: U.S. Army Corps of Engineers, (1993).
[5] DIANA, User's Manual, Release 9.3. Delft, The Netherlands: TNO DIANA BV, (2008).
[6] G.F. Kinney and K.J. Graham, Explosive Shocks in Air, Springer, New York, (1985).
[7] A.C. Jacinto, R.D. Ambrosini and R.F. Danesi, Experimental and computational analysis of plates under air blast loading, Int. J. of Impact Engineering, **25**(10), 927-947, (2001).
[8] C.M. Morison, Dynamic response of walls and slabs by single-degree-of-freedom analysis-A critical review and revision, Int. J. of Impact Engineering, **32**, 1214, (2006).
[9] Mathworks, MATLAB Version 7.10 Natick, MathWorks, MA, (2010).
[10] A.K. Chopra, *Dynamic of Structures: Theory and Applications to Earthquake Engineering*, Pearson Prentice Hall, (2007).
[11] Z.I Syed, P. Mendis, N.T. Lam and T. Ngo, Concrete damage assessment for blast load using pressure-impulse diagrams, *Proc. of Annual Technical Conference of the Australian Earthquake Engineering Society*, Canberra, New Zealand, (2006).
[12] Z.I. Syed, P. Mendis, T. Ngo and N. Lam, *Deflection Damage Criteria for Concrete Elements under Blast Loading Incorporating Sustainable Practice in Mechanics and Structures of Materials*, CRC Press, 609-614, (2010).
[13] C.J. Gantes and N.G. Pnevmatikos, Elastic-plastic response spectra for exponential blast loading, Int. J. of Impact Engineering, **30**(3), 323-343, (2004).
[14] J. Magnusson and M. Hallgren, Reinforced high strength concrete beams subjected to air blast loading, Structures and Materials, **15**, 53-62, (2004).
[15] E. Thorenfeldt, A. Tomaszewicz and J.J. Jensen, Mechanical properties of high-strength concrete and applications in design, *The Utilization of High-Strength Concrete*, Stavanger, (1987).
[16] Elvira, Assessment of Progressive Collapse in Reinforced Concrete Frame Structures, PhD Thesis, The University of Melbourne, Melbourne, (2008).
[17] T. Ngo, Behaviour of High Strength Concrete Subject to Impulsive Loading, PhD Thesis, The University of Melbourne, Melbourne, (2005).

MATEC Web of Conferences **47**, 02016 (2016)
DOI: 10.1051/matecconf/20164702016

An Orthotropic Membrane Model for the Large Deformation Analysis and Snapping Phenomena of the Dome Inflated

Shohei Yamada[1,a], Katsushi Ijima[1], Hiroyuki Obiya[1] and Muhammad Nizam Zakaria[2]

[1]*Department of Civil Engineering and Architecture, Graduate School of Science and Engineering, Saga University, Honjo, Saga 840-8502, Japan*
[2]*Department of Structural and Material Engineering, Faculty of Civil and Environmental Engineering, University Tun Hussein Onn Malaysia, 86400 Parit Raja, Johor, Malaysia*

Abstract. The paper proposes a method to enable analyzing any large deformation of membrane structures. If the analysis uses even very small rigidity against compression in the structures, the computation becomes multi-bifurcated problem and unstable. The original compression-free model used in the method keeps the structures in tension field and this makes computing the deformation always very stable. The paper uses the advantage of the method and shows a snapping phenomenon during constructing a membrane dome. Though the snapping phenomenon is not so public, the phenomenon occasionally happens in the construction site and it causes the destruction.

1 Introduction

Since membrane structures are light, easily expandable and compactly folded, the structures are widely used in various ways. The mechanical characteristic of the membrane structure is that tension in the membrane keeps the structural shape and that the structure in small tension is so flexible that is easily deformed. Therefore, when the structure is constructed or expanded, the characteristic brings unexpected phenomena, such as snapping of a membrane dome shown in the paper. In order to estimate exactly those phenomena, we need a method that can stably compute any state of the membrane structure.

The authors are proposing a method that uses the compression-free unit composed by cables [1], [2]. The unit in tension has almost the same stiffness as that of the triangular element with uniform strain in it. If compressive force in the element occurs even very small, computing the equilibrium necessarily becomes a multi-bifurcated problem. Since the compressive force in the membrane is essentially very small, using the compression-free unit is appropriate and besides the use makes the analysis completely free from the multi-bifurcated problem.

However, the compression-free assumption occasionally brings unpractical solutions as a result from setting a certain boundary condition of the membrane structure. The unpractical solution is, for example that the area of a membrane element becomes to end zero. The modification to the fault is settled with taking into considering bending moment in a side of a triangular element. Using the end shear forces balanced with the end moment in the element keeps only the translational degrees of freedom at each node without increasing a degree of nodal freedom. The idea has been already

[a] Corresponding author : 15577018@edu.cc.saga-u.ac.jp

published by many studies on analyzing shell structures [3-5]. Membrane material generally has so small rigidity against the bending that the end moment in the element side is very small. However, the end shear force derived from the small end moment can resist shrinking the element into ending zero area.

The method can stably compute equilibrium solutions in any state of membrane structures. An experiment of large deformation of a suspended membrane has verified the validity of the computation by using the method [6]. As an example of analyzing the large deformation, the paper shows snapping phenomena of a membrane dome occurring in the inflation. There was an actual accident that the phenomenon extended to a tense state before the occurrence of the snapping during inflating a large-scale membrane dome. If the dome came to just before the snapping, the dome surly ended broken.

2 Concept of the Theory

The method in the paper uses different procedure from that of the popular FEM. The strain energy of an element is defined by the positions of the nodes connected to the apexes of the element. In the method, the energy principal is applied to each element, and it gives the element end forces. The end forces in this paper are classified into the forces in the plane of the compression-free element and the end shear forces in the normal direction to the element plane. The method finds the nodal positions that satisfy the equilibrium equation of those element end forces and the nodal forces working at each node.

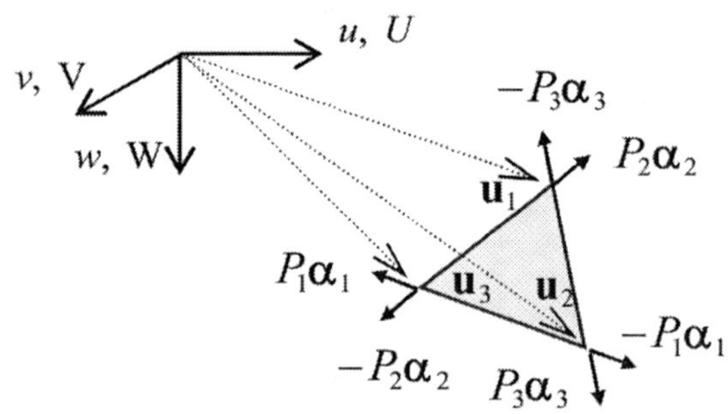

(a) End forces and the position vectors in the element

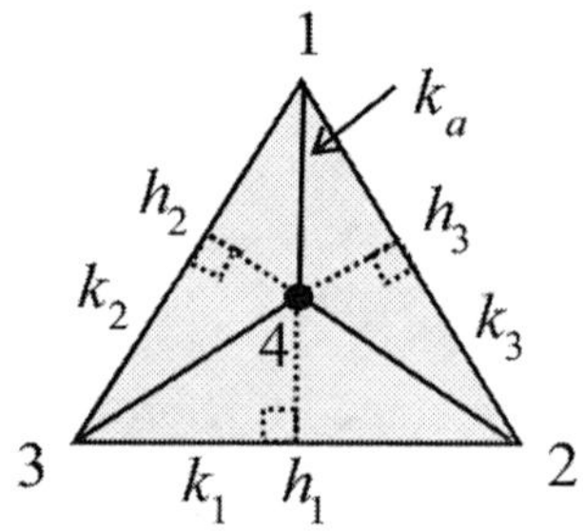

(b) Compression-free unit composed by 6 axial members

Figure 1. Element end forces in the plane of the element

3 Compression-Free Model in the Plane of the Membrane Element

The deformation of the triangular element can be defined as the changes, Δl_1, Δl_2, Δl_3 from the lengths of the three sides in the non-stress element, l_{10}, l_{20}, l_{30}. The deformations of the side lengths are given from the nodal position vectors, $\mathbf{u}_1$, $\mathbf{u}_2$, $\mathbf{u}_3$ shown in Figure 1(a) with geometric exactitude, as follows.

$$\begin{aligned}\Delta l_1 &= l_1 - l_{10} = \sqrt{(\mathbf{u}_3-\mathbf{u}_2)\bullet(\mathbf{u}_3-\mathbf{u}_2)} - l_{10}\\ \Delta l_2 &= l_2 - l_{20} = \sqrt{(\mathbf{u}_1-\mathbf{u}_3)\bullet(\mathbf{u}_1-\mathbf{u}_3)} - l_{20}\\ \Delta l_3 &= l_3 - l_{30} = \sqrt{(\mathbf{u}_2-\mathbf{u}_1)\bullet(\mathbf{u}_2-\mathbf{u}_1)} - l_{30}\end{aligned} \quad (1)$$

In the element of an orthotropic membrane, when the strain in the element is assumed to be uniform, the relation of the end force vector, $\boldsymbol{P}_e = \{P_1 \ \ P_2 \ \ P_3\}^{\mathrm{T}}$ and the elongations of the element sides, $\boldsymbol{\Delta l} = \{\Delta l_1 \ \ \Delta l_2 \ \ \Delta l_3\}^{\mathrm{T}}$ can be expressed by the following equation.

$$\mathbf{P}_e = \mathbf{a}^{\mathrm{T}}\boldsymbol{\omega}^{\mathrm{T}}\mathbf{E}\boldsymbol{\omega}\mathbf{a}\Delta\mathbf{l} = \mathbf{k}_e\Delta\mathbf{l} \tag{2}$$

Where **E** is the matrix expressing the relation of the stress and the strain in the two directions of the fibers making a right angle. **ω** is the translation matrix based on the direction of the fiber in the triangular element and **a** is the translation matrix from the rectangular coordinates to the coordinates based on the direction of each side.

The method replaces the membrane element in uniform strain with the compression-free unit composed by 6 of axial members, as shown by Figure 1(b). The 6 of stiffness coefficients in $\mathbf{k}_e$ of Equation (2) provide the elongation stiffness per unit length, k_1, k_2, k_3 of the three members that form the element shape, the stiffness k_a of the three members connecting each apex and the sub-node, that position is also given by the coefficients in $\mathbf{k}_e$. In Figure 1(b), h_1, h_2 and h_3 are the distance between the sub-node and each side.

Hyperbolic function is applied to expressing the axial forces of the 6 members in the unit. The use keeps the compression-free characteristic from extremely slack state that the three apexes converge at a point to tense state with smoothly change of tensions.

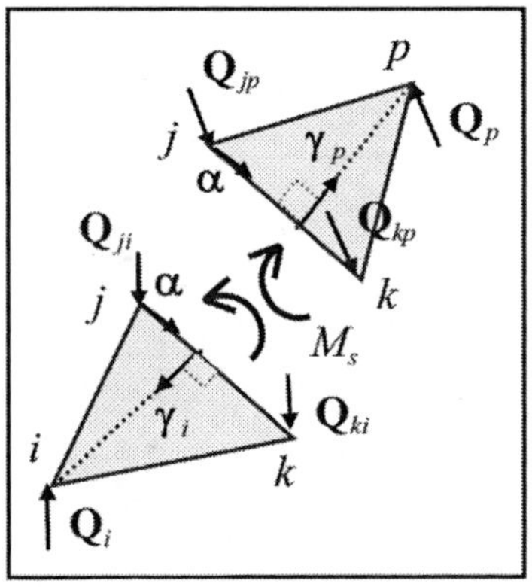

(a) End shear forces balanced with the end moment

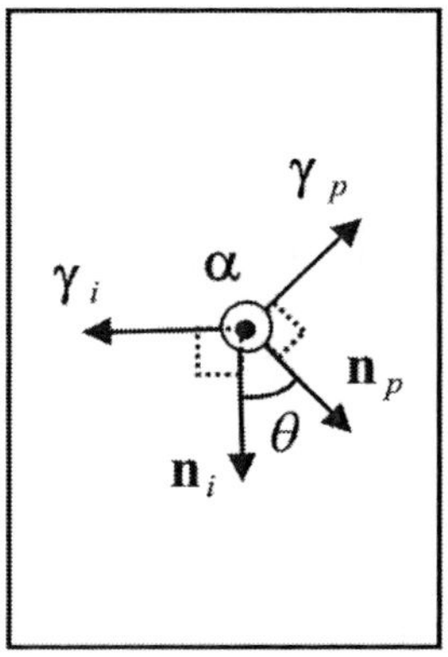

(b) Direction vectors

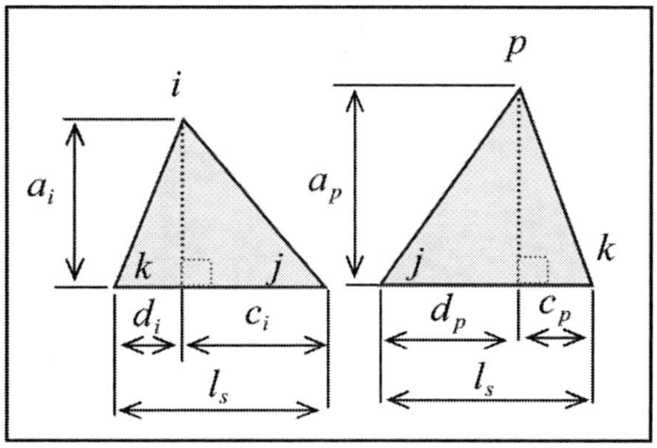

(c) Lengths in the two elements

Figure 2. Element end forces out of the plane of the element

4 End Shear Forces Based on the End Moment at the Element Side

Though bending rigidity of general membrane material is very small, considering the bending moment generated at a side of an element can prevent the element from shrinking occasionally into zero area because of the compression-free assumption. Moreover the consideration does not bring a multi-bifurcated problem.

The bending moment, M_s generating at the side can be converted into the two of the end moments working on each side. The sum of the rotational angles of the two ends by the end moments is defined as $\Delta\theta$. The rotational angle can be exactly given from the four positions of the apexes in the two elements. That is $\Delta\theta = \theta - \theta_0$, where θ_0 is the angle between the two planes of the elements in non-stress and θ is that of the two elements in a certain state of the structure. Using the direction vectors normal to the two planes, as shown in Figure 2(b), gives θ with geometric exactitude, as follows.

$$\cos\theta = \mathbf{n}_i \bullet \mathbf{n}_p, \qquad \sin\theta = \boldsymbol{\alpha} \bullet \mathbf{n}_i \times \mathbf{n}_p \tag{3}$$

When the two elements in Figure 2 are regarded as the two beams with the spans, a_{i0} and a_{p0}, that are the lengths of the perpendiculars in the two elements in non-stress, the widths of the two beams linearly change from zero to the side length, l_s along the perpendiculars. Therefore, the curvatures induced by the end moment, M_s are uniform along the perpendiculars on the assumption that the deformations of the two beams are very small, and then the relation of the end moment M_s and the changed angle $\Delta\theta$ becomes,

$$M_s = \frac{2EI_s}{a_{i0} + a_{p0}} \Delta\theta \tag{4}$$

In Equation (4), EI_s is the bending rigidity of the side jk, E is the Young's modulus in the direction normal to the side, and I_s is the moment of inertia of the cross section in the side.

5 Equilibrium Equations and the Tangent Stiffness Equations Including Geometric Stiffness

The nodal forces balanced with the end forces in the plane of the compression-free element shown in Figure 1(a) are defined as $\mathbf{U}_1^{in}$, $\mathbf{U}_2^{in}$, $\mathbf{U}_3^{in}$. In the universal coordinates, the end forces have the directions of the sides expressed by $\boldsymbol{\alpha}_1$, $\boldsymbol{\alpha}_2$, $\boldsymbol{\alpha}_3$ and the equilibrium equations of those forces at the three nodes of the apexes become Equation (5).

$$\begin{Bmatrix} \mathbf{U}_1^{in} \\ \mathbf{U}_2^{in} \\ \mathbf{U}_3^{in} \end{Bmatrix} = \begin{bmatrix} \mathbf{0} & \boldsymbol{\alpha}_2 & -\boldsymbol{\alpha}_3 \\ -\boldsymbol{\alpha}_1 & \mathbf{0} & \boldsymbol{\alpha}_3 \\ \boldsymbol{\alpha}_1 & -\boldsymbol{\alpha}_2 & \mathbf{0} \end{bmatrix} \begin{Bmatrix} P_1 \\ P_2 \\ P_3 \end{Bmatrix} = \mathbf{C}_e^{in} \mathbf{P}_e \tag{5}$$

On the other hand, when the nodal forces balanced with the end shear forces shown in Figure 2(a) are defined as $\mathbf{U}_i^{out}$, $\mathbf{U}_j^{out}$, $\mathbf{U}_k^{out}$, $\mathbf{U}_p^{out}$, the equilibrium equation of those forces becomes the following.

$$\begin{Bmatrix} \mathbf{U}_i^{out} \\ \mathbf{U}_j^{out} \\ \mathbf{U}_k^{out} \\ \mathbf{U}_p^{out} \end{Bmatrix} = \begin{Bmatrix} \mathbf{Q}_i \\ \mathbf{Q}_{ji} + \mathbf{Q}_{jp} \\ \mathbf{Q}_{ki} + \mathbf{Q}_{kp} \\ \mathbf{Q}_p \end{Bmatrix} = \begin{Bmatrix} -\dfrac{\mathbf{n}_i}{a_i} \\ \dfrac{d_i \mathbf{n}_i}{a_i l_s} + \dfrac{c_p \mathbf{n}_p}{a_p l_s} \\ \dfrac{c_i \mathbf{n}_i}{a_i l_s} + \dfrac{d_p \mathbf{n}_p}{a_p l_s} \\ -\dfrac{\mathbf{n}_p}{a_p} \end{Bmatrix} M_s = \mathbf{C}_s^{out} M_s \tag{6}$$

When the nodal forces working at the all of the nodes in the structure are defined as the force vector of $\mathbf{U}$ where $\mathbf{U}$ is the total sum of $\mathbf{U}_1^{in}$, $\mathbf{U}_2^{in}$, $\mathbf{U}_3^{in}$ in the all of the elements and $\mathbf{U}_i^{out}$, $\mathbf{U}_j^{out}$, $\mathbf{U}_k^{out}$, $\mathbf{U}_p^{out}$ in the all of the sides in the structure. Therefore, when the nodal forces balance with the end forces in the compression-free elements and the end shear forces at the all of the nodes, the unbalanced force vector of $\Delta\mathbf{U}$ in Equation (7) becomes zero vector.

$$\mathbf{U} - \sum_{e=1}^{m_e} \mathbf{C}_e^{in} \mathbf{P}_e - \sum_{s=1}^{m_s} \mathbf{C}_s^{out} M_s = \Delta\mathbf{U} \tag{7}$$

where m_e is the total number of the elements in the structure and m_s is the total number of the sides.

When the nodal forces are optionally given to the structure, the unbalanced forces in Equation (7) do not generally become zero. The method is to find the nodal positions fulfilling that the all of the unbalanced forces are less than the allowable value. Iterative methods such as the Newton-Raphson's method or the Dynamic Relaxation method can be applied to the computation.

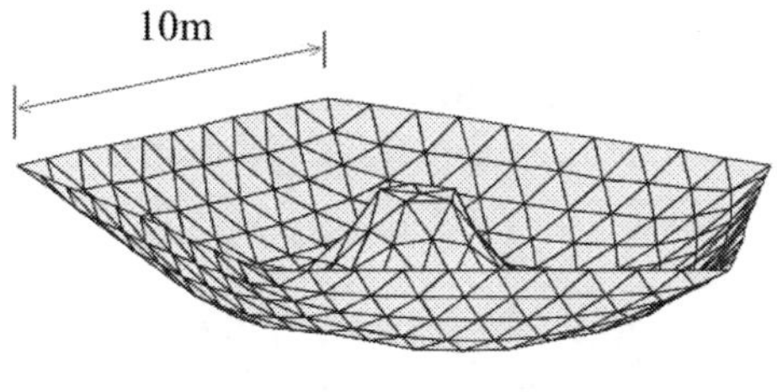

(a) Initial shape of the membrane suspended from the circumference and the six nodes at the crown fixed.

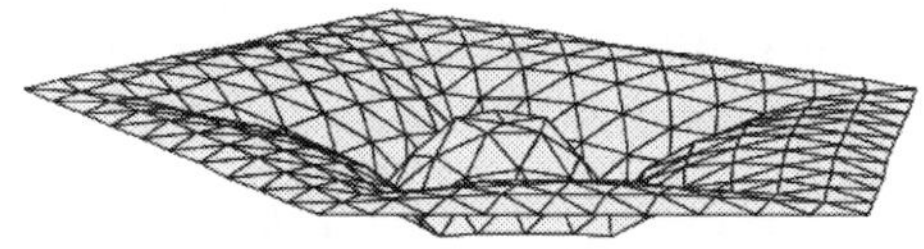

(b) Shape under 13Pa in inner pressure; the six nodes at the crown are still fixed.

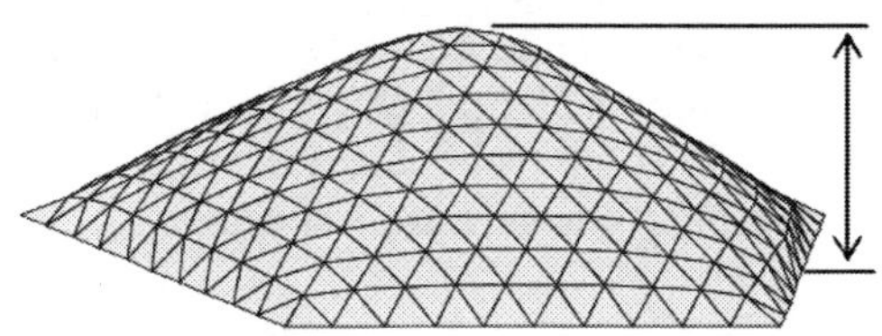

(c) Shape under 15Pa in inner pressure.

Figure 3. Inflating a membrane dome and the snapping phenomenon.

If the tangent stiffness equation shown by Equation (8) derived from differentiating the equilibrium equations of Equation (5) and Equation (6) is used in the iterative methods, the iteration quickly converges to the equilibrium solution.

$$\begin{aligned}\delta\mathbf{U} &= \sum_{e=1}^{m_e}\left(\delta\mathbf{C}_e^{\mathrm{in}}\mathbf{P}_e + \mathbf{C}_e^{\mathrm{in}}\delta\mathbf{P}_e\right) + \sum_{s=1}^{m_s}\left(\delta\mathbf{C}_s^{\mathrm{out}} M_s + \mathbf{C}_s^{\mathrm{out}}\delta M_s\right) \\ &= \left(\mathbf{K}_{\mathrm{G}}^{\mathrm{in}} + \mathbf{K}_{\mathrm{O}}^{\mathrm{in}}\right)\delta\mathbf{u} + \left(\mathbf{K}_{\mathrm{G}}^{\mathrm{out}} + \mathbf{K}_{\mathrm{O}}^{\mathrm{out}}\right)\delta\mathbf{u},\end{aligned} \tag{8}$$

6 Computational Example of Inflating a Membrane Structure

Figure 3 is a computational result of inflating a membrane dome with the pentagonal boundary fixed. The membrane used in the computation is an isotropic material, the elongation stiffness: E_t = 882kN/m, the Poisson ratio: $v = 0.4$ and the weight per unit area: $w = 9.8\text{N/m}^2$, where t is the thickness of the membrane. Those material coefficients roughly correspond to a membrane made from polyester textile coated with vinyl chloride. Incidentally, the snapping phenomena of membrane domes occur during the inflation without regard to isotropic material or orthotropic one. The dome is a three-dimensional structure formed by seaming the five sheets of the membrane of the isosceles triangle with the base of 10m and with the angle of 58° at the vertex in shape. In the computation, 500 elements of the triangle similar to the shape of the sheet of the membrane compose the dome. The meshing keeps enough accuracy in the computation. The circumference of the dome is fixed through the inflation, and the six nodes at the crown are fixed at the position of the same level as the circumference until the inner pressure increases to 13Pa. The six nodes are released over the inner pressure of 13Pa.

The initial shape of the membrane structure before the inflation is the suspended form, as shown in Figure 3(a), calculated on the condition of fixing both of the circumference and the crown. The inflation proceeds by increasing the inner pressure of the dome. When the inner pressure reaches to 13Pa, the membrane forms into Figure 3(b). At that moment, the curvatures at the seams become very large and the tension in the membrane become also so large. When the phenomenon occurs in a real structure of membrane, the structure almost collapses by cracks at the seams. Even if the part of the crown is not fixed, the phenomenon occurs under smaller pressure than 13Pa.

While the inner pressure increases from 13Pa, the part of the crown is released from the fixing. When the inner pressure reaches 15Pa, the shape becomes Figure 3(c). The process of the deformation from Figure 3(b) to Figure 3(c) is just a snapping phenomenon, and we should avoid the phenomenon in a real membrane structure.

7 Summary

The paper proposed a method that can stably analyze any large deformation of membrane structures. The method is to find the nodal positions fulfilling the equilibrium equations at the nodes expressed by the external nodal forces and the end forces derived from the deformation of the elements. The end forces are the two kinds of that in the plane of the compression-free element and the end shear forces normal to the element. By using the method, the paper explained snapping phenomena occurring in the inflation of a membrane dome. Since the phenomena bring destruction of the structure, the prediction of the occurrence and the countermeasure are important.

References

[1] K. Ijima, H. Obiya and N. Kawasaki, Large displacement analysis of compression-free membrane structures replaced by elastic cable units, *International Conference on Textile Composites and Inflatable Structures, Structural Membranes*, (2009).

[2] K. Ijima, H. Obiya and K. Kido, An orthotropic membrane model replaced with line-members and the large deformation analysis, *International Conference on Textile Composites and Inflatable Structures, Structural Membranes*, (2011)

[3] R. Phaal and C.R. Calladine, A simple class of finite elements for plate shell problems, II: An element for thin shells with only translational degrees of freedom, International Journal Numerical Method Engineering, **35**(5), 979-996, (1992).

[4] E. Oñate and M. Cervera, Derivation of thin plate bending elements with one degree of freedom per node, Engineering Computations, **10**, 553-561, (1993).

[5] F.G. Flores and E. Onate, A rotation-free shell triangle for the analysis of kinked and branching shells, International Journal Numerical Method Engineering, **69**(7), 1521-1551, (2007).

[6] M. Tomioka, K. Ijima and H. Obiya, Large deformation analysis of flexible membrane structure and the experiment, *International Conference on Textile Composites and Inflatable Structures, Structural Membranes*, (2015).

MATEC Web of Conferences **47**, 02017 (2016)
DOI: 10.1051/matecconf/20164702017

A Numerical Case Study on Contact Analysis with Large Displacement

A. Yamasaki[1,a], H. Obiya[1], K. Ijima[1] and Z. M. Nizam[2]

[1]*Department of Civil Engineering and Architecture, Saga University, Honjo, Saga 840-8502, Japan*
[2]*Faculty of Civil and Environmental Engineering, Universiti Tun Hussein Onn Malaysia, 86400 Parit Raja, Johor, Malaysia*

Abstract. In case of geometrically non-linear analysis of plane frame structures, Bernoulli-Euler beam elements considering the shear deformation to be zero are generally used. But, in node-element contact case, when a contact point is approaching to one of both element ends, diverging of unbalanced forces makes the analysis impossible because of neglecting of the shear deformation. However, in such a case, the element force equation with Timoshenko beam theory which considers the shear deformation is quite effective. In this study, the contact element consistent of both ends and a contact node derived by Timoshenko beam theory is applied to several case of numerical examples. As a result, its convergent performance can be verified in the almost part of element and got accuracy of the solutions.

1 Introduction

In case of geometrically non-linear analysis of frame structures, the general formulation of the finite element method (FEM) is to derive geometrical stiffness from the relation between strain inside of element's body and displacements of node. On the other hand, the tangent stiffness method (TSM), proposed by Goto, Hane and Tanaka [1], defines elements' behavior in the element force equation. Therefore, the geometrical stiffness, derived by the differential of the equilibrium equation between element edge forces and nodal forces, can express the non-linearity caused by rigid body displacement of each element.

In contact case of node-element or element-element with large displacement, FEM may require so difficult and complex procedures to create a new node when the occurrence of contact. For example, Konyukhov et al. [3] had to use solid elements to simulate a contact model of simple 2-dimensional beam. On the other hand, TSM uses the element force equation, defined in the element coordinate system with a statically determinate and stable support condition. Therefore, it is so easy to add one or more contact points to an element, and to obtain solutions of perfect equilibrium as well as in case of using ordinary element with two nodes of both ends. Nizam et al. [4] applied a contact beam element with three nodes using Timoshenko beam theory to 2-deimensional frame structures. This "Timoshenko contact element" makes it easy to obtain solutions even if when the contact node approaches to an element end, because of effect of shear deformation. In this study, several numerical examples are shown and the rationality of the Timoshenko contact elements is verified.

[a] Corresponding author : 15577017@edu.cc.saga-u.ac.jp

2 Tangent Stiffness Method

Tangent Stiffness Method (TSM) is a method to evaluate a geometric nonlinearity caused by the rigid body displacement of element, when the element deformation is defined in the element force equation. This makes the unbalanced forces converge by the iteration process that is equal to Newton Raphson method. In addition, strict compatibility and an equilibrium equation are applicable in the iteration process to converge the unbalanced force. Therefore, TSM has an extremely high convergence performance.

2.1 General formulation

Here, an element has two ends and the force element vector of both edges are represented by **S**. Considering a plane coordinate system, if the external force vector is represented by **U**, and the equilibrium matrix by **J**, the equilibrium condition can expressed by the following equation:

$$\mathbf{U} = \mathbf{JS} \tag{1}$$

By differentiation of Equation (1), the tangent stiffness equation can expressed as:

$$\boldsymbol{\delta}\mathbf{U} = \mathbf{J}\boldsymbol{\delta}\mathbf{S} + \boldsymbol{\delta}\mathbf{J}\mathbf{S} = (\boldsymbol{K_O} + \boldsymbol{K_G})\boldsymbol{\delta u} \tag{2}$$

Here, the differentiation of Equation (1) simultaneously extracts $\boldsymbol{\delta}\mathbf{S}$ and $\boldsymbol{\delta}\mathbf{J}$ which enables the expression of a linear function of the displacement vector $\boldsymbol{\delta}\mathbf{u}$ in the local coordinate system. Meanwhile, $\boldsymbol{K_o}$ represents the element matrix, which also simulates the element behavior corresponding to the element stiffness. Meanwhile, $\boldsymbol{K_G}$ is the tangent geometrical stiffness.

2.2 Equilibrium condition in case of contact problem

The tangent geometrical stiffness of contact problem can be obtained by differentiating the equilibrium equation. Figure 1 shows the element edge forces for a single contact element. On the other hand, Figure 2 shows the nodal forces. The rotation of point **c** is neglected, which the degree of freedom of the node is two.

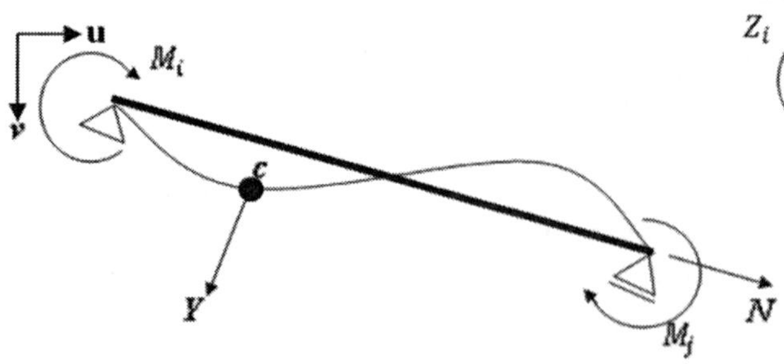

Figure 1. Element edge forces for contact element.

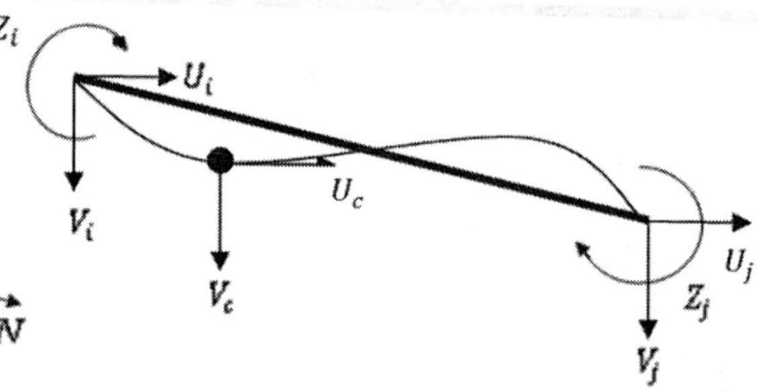

Figure 2. Nodal forces for contact element.

The vector of the element edge forces separate each other and are defined by the following equation:

$$\mathbf{S} = \begin{bmatrix} N & M_i & M_j & Y \end{bmatrix}^T \tag{3}$$

Furthermore, the vector **U** can be expressed as following:

$$\mathbf{U} = \begin{bmatrix} U_i & V_i & Z_i & U_j & V_j & Z_j & U_c & V_c \end{bmatrix}^T \tag{4}$$

In other words, by differentiating Equations (3) and (4), the tangent geometrical stiffness can be obtained from the equilibrium between **S** and **U**.

2.3 Timoshenko contact element

Timoshenko beam elements are effective in case that the shear deformation can not be neglected. Figure 3 shows the equilibrium condition of a beam with elastic and stable support condition under the action of the axial force N, edge moments M_i and M_j, and contact force Y_c. Figure 3 shows the location of the geometric and kinematic variables, and it is assumed that the contact force Y_c is within the range of the beam. This local coordinate is a simple but accurate idealization for frictionless node-element contact problem. The element force equation of the Timoshenko contact element is given as Equations (5) to (8).

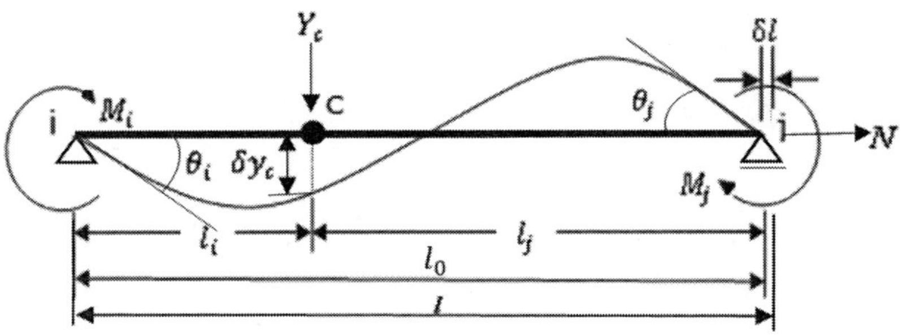

Figure 3. Timoshenko contact element in simply supported beam coordinate.

$$\begin{bmatrix} N \\ M_i \\ M_j \\ Y_c \end{bmatrix} = \begin{bmatrix} \frac{EA}{l} & 0 & 0 & 0 \\ 0 & \left(4\Omega + 3l_i l_j^3 - 108\Psi^2\right)k_c & \left\{9\left(l_0^2 - l_j\Psi\right) - \Omega\right\}k_c & -3l_0\left(l_j l_0 + 6\Psi\right)k_c \\ 0 & \left\{9\left(l_0^2 - l_j\Psi\right) - \Omega\right\}k_c & \left(4\Omega + 3l_i^3 l_j - 108\Psi^2\right)k_c & 3l_0\left(l_i l_0 + 6\Psi\right)k_c \\ 0 & -3l_0\left(l_j l_0 + 6\Psi\right)k_c & 3l_0\left(l_i l_0 + 6\Psi\right)k_c & \frac{3l_0\left(l_0^3 + 12l_0\Psi\right)}{l_i l_j}k_c \end{bmatrix} \begin{bmatrix} \delta l \\ \theta_i \\ \theta_j \\ \delta y_c \end{bmatrix} \tag{5}$$

$$k_c = \frac{EI}{l_0\Omega} \tag{6}$$

$$\Psi = \frac{EI}{GA_e} \tag{7}$$

$$\Omega = \left(l_i l_j\right)^2 + 3\Psi\left(l_0^2 + l_i l_j\right) + 36\Psi^2 \tag{8}$$

3 Numerical Examples

3.1 Contact of a couple of cantilevers

As shown in Figure 4, there are two cantilevers face each other, which are in contact beforehand. Both cantilevers have 8 equal divisions. Node 1 and node 18 are fixed perfectly, and compulsory displacement is given to V direction at node 1. Then, every contact point begins to slide. The material parameters in this example are $E = 2.0\times 10^{7}\text{N/m}^2$, $A = 3.0\times 10^{-4}\text{m}^2$, $I = 2.2\times 10^{-8}\text{m}^4$, $\nu = 0.4$ and $G = 7.142\times 10^{6}\text{N/m}^2$ respectively. Increment per 1 step of compulsory displacement is 0.208m.

Figure 5 shows the shape of solution at 10th incremental step, and Figure 6 shows the stable convergent process by TSM. It takes only 9 times of iteration to reduce the unbalanced forces 8 places on its convergent process, in spite of the case that many contact incidents occur simultaneously. Figure 7 shows the solution at 18th incremental step, in which every node approaches to a node on another cantilever extremely. The ratio of $[lj/l]$, which is expresses the position of the contact node on an element (see Figure 3), is 0.9739. Even in such a complex condition, the solution can be obtained with stable convergent process of only 12 times iterations as shown in Figure 8.

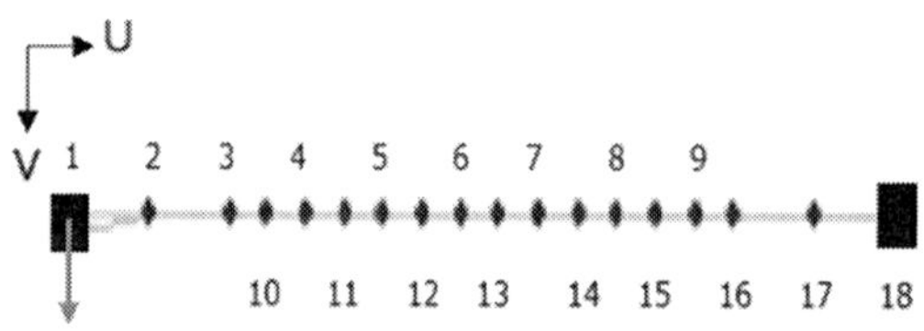

Figure 4. Initial shape.

Figure 5. Shape of solution at 10th step.

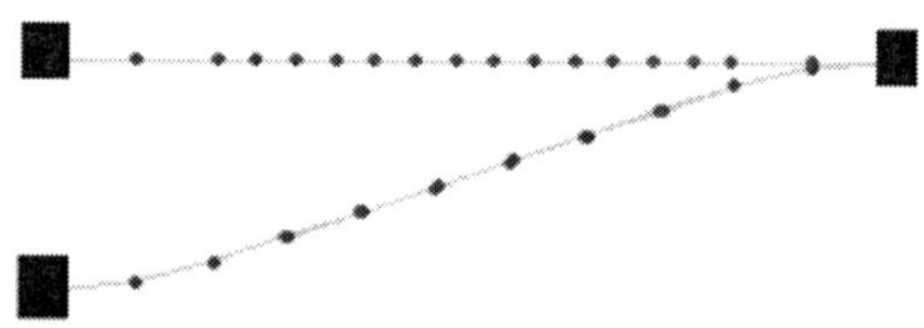

Figure 7. Shape of solution at 18th step.

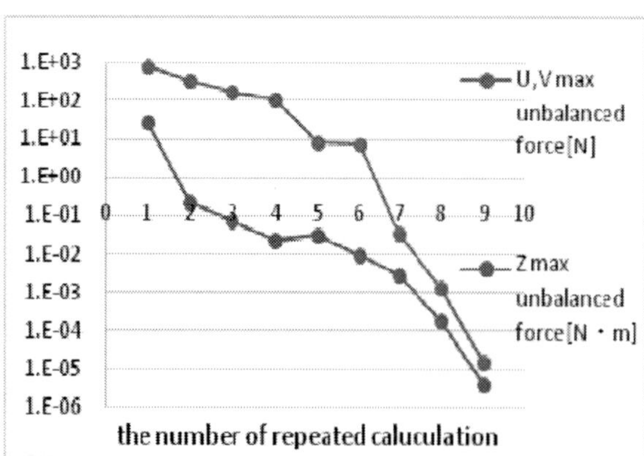

Figure 6. Process of convergence.

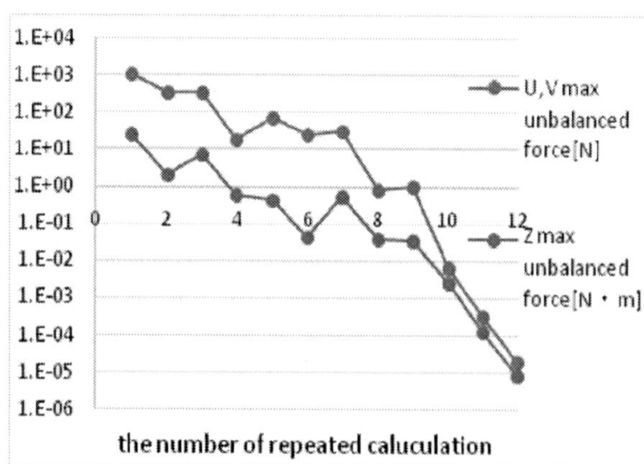

Figure 8. Process of convergence.

3.2 Three cantilevers and one contact point

As shown in Figure 9, this example is consist of three independent cantilevers with 11 nodes and a contact node. Node1, node22, node23 and the contact node are fixed perfectly. The compulsory displacement is given to V direction at the contact node. The material parameters in this example are $E = 2.0 \times 10^7 \text{N/m}^2$, $A = 3.0 \times 10^{-4} \text{m}^2$, $I = 2.2 \times 10^{-8} \text{m}^4$, $\nu = 0.4$ and $G = 7.142 \times 10^6 \text{N/m}^2$. Figure 10 shows the shape of solution when the contact node reached to the cantilever A. Following to it, Figure 11 is the shape of solution when the node 11 on cantilever A contact to the cantilever B, and Figure 12 is when the node 12 on cantilever B contact to cantilever C. When the node12 approaches to the node 30 (Figure 13), the unbalanced force diverged as shown in Figure 14. In such a case, if the incremental calculation is re-started from the previous equilibrium solution, the converged solution can be obtained as shown in Figure 15 by giving larger compulsory displacement and switching the configuration of the contact element to the next door element. This is the simplest procedure to be possible to continue the analysis.

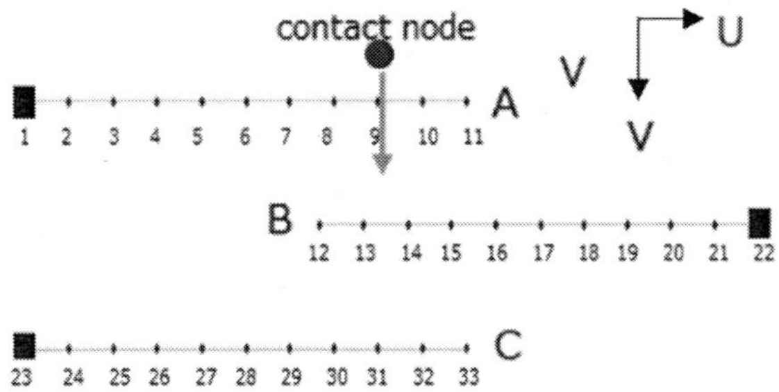

Figure 9. Initial shape.

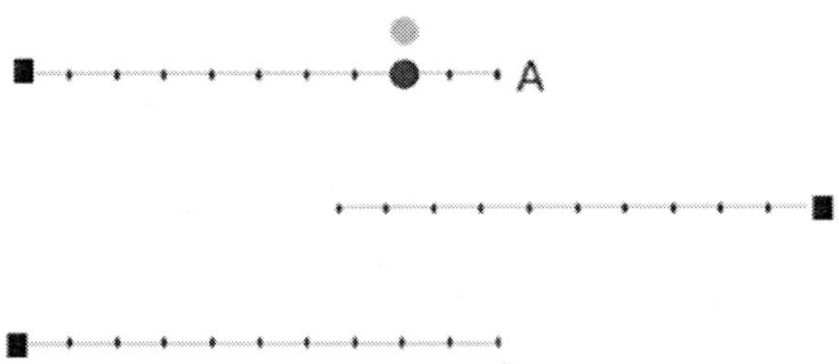

Figure 10. First contact to cantilever A.

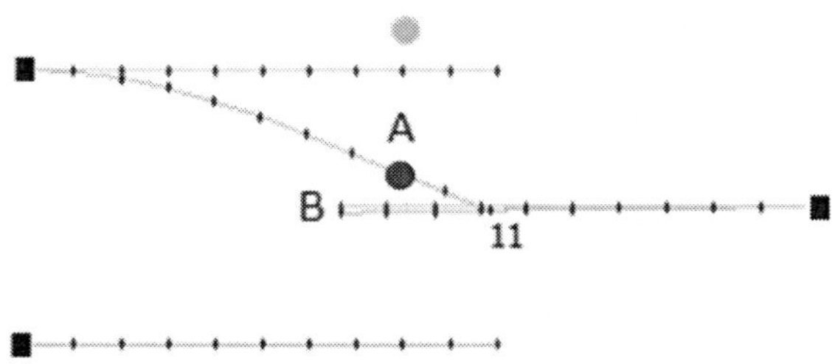

Figure 11. Second contact of A to B.

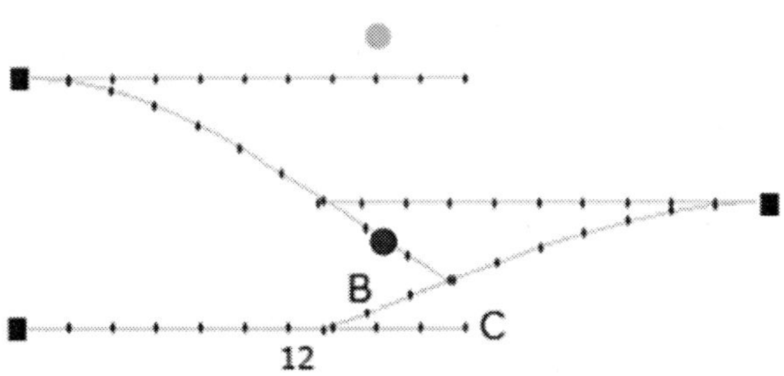

Figure 12. Third contact of B to C.

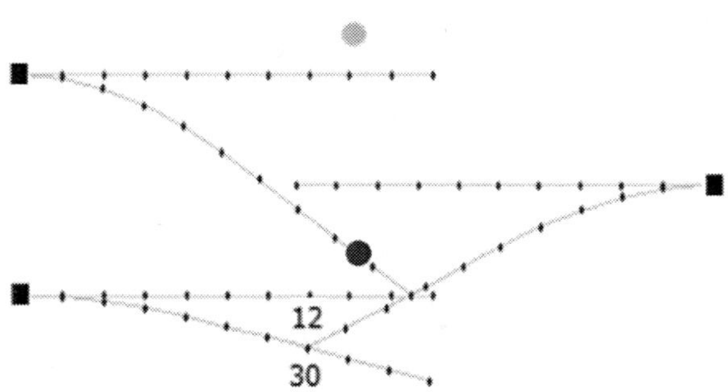

Figure 13. Node12 approaching to Node30.

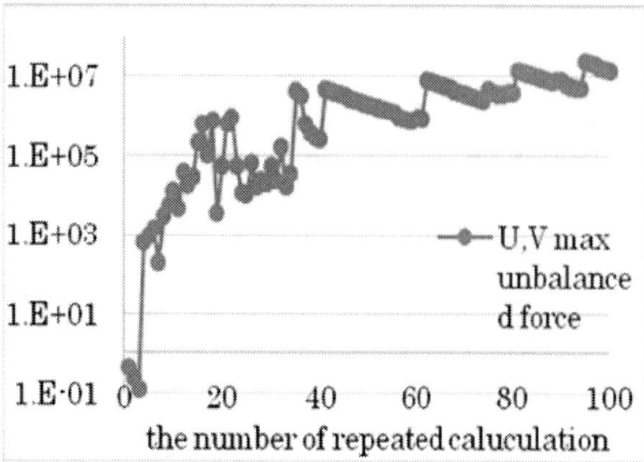

Figure 14. Process of divergence.

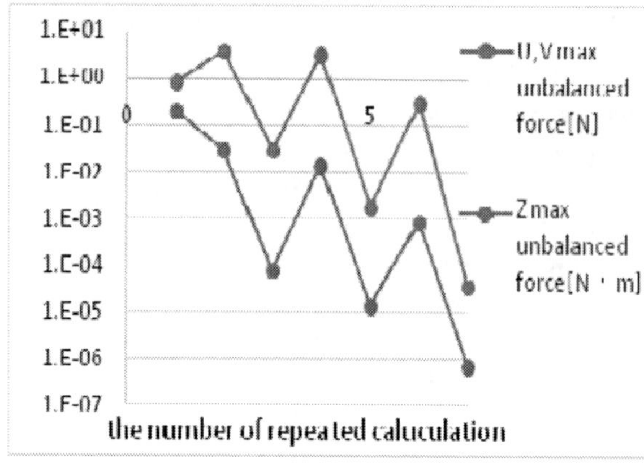

Figure 15. Process of convergence.

The contact node approaches to the end of the cantilever A as shown in Figure16, and then jumps from cantilever A to cantilever B as shown in Figure 17. At the same time, the contact between the contact node and the cantilever A was canceled. On the same process as above, the contact node approaches to the end of the cantilever B as shown in Figure18, and then jumps from cantilever B to cantilever C as shown in Figure 19. Also in this case, the contact between the contact node and the cantilever B was canceled at the same time. Figure 20 is the solution that the contact node approaches to the end of beam C, and all the contact is canceled in Figure 21.

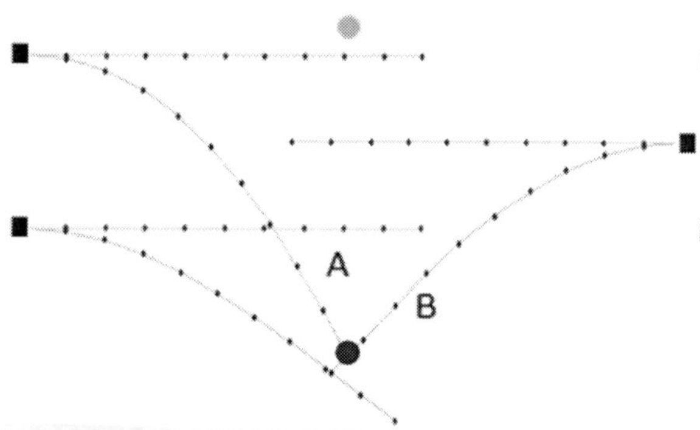

Figure 16. Contact node approaching to end of cantilever A.

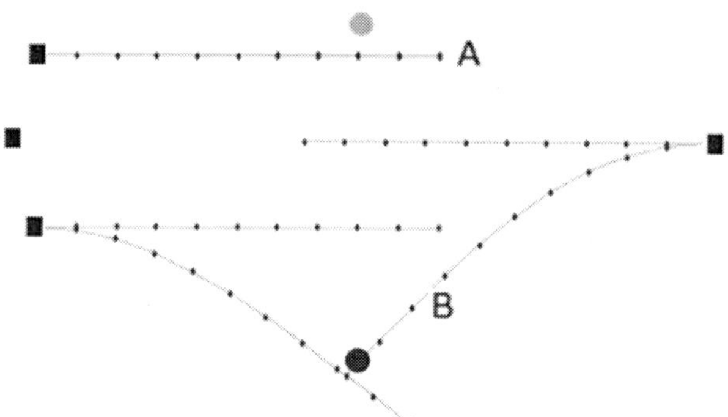

Figure 17. Contact node jumping from A to B.

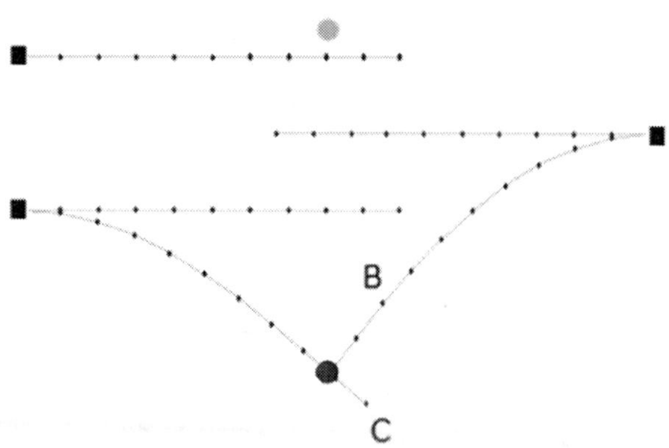

Figure 18. Contact node approaching to end of cantilever B.

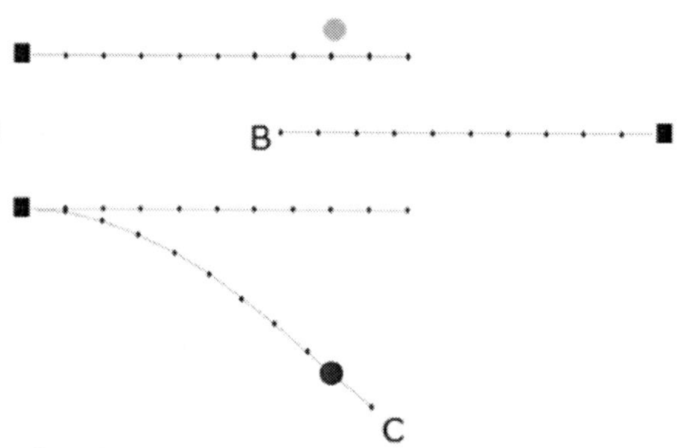

Figure 19. Contact node jumping from B to C.

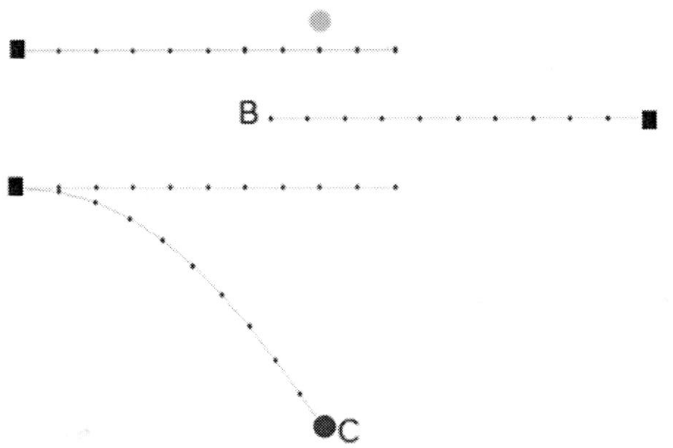

Figure 20. Contact node approaching to end of cantilever B.

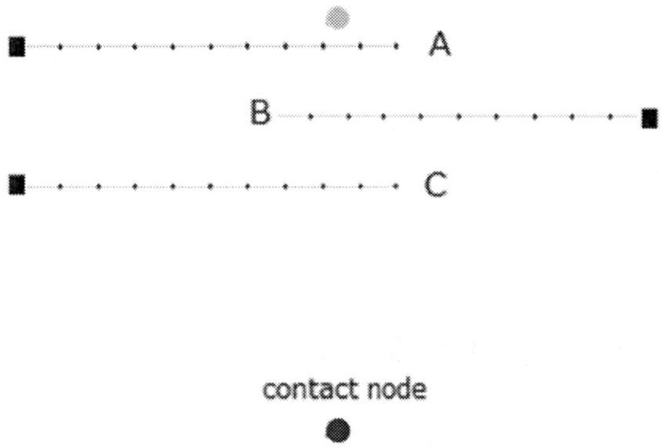

Figure 21. Final solution (non-stressed state).

4 Conclusions

In this study, two numerical examples are shown and the rationality of the Timoshenko contact elements is verified. In the first example, Even if when the contact node approaches to the end of element and the shear deformation can not be neglected, the converged solution can be obtained by using Timoshenko contact element considered the shear deformation. In the second example, when the contact node approaches to an element end, the unbalanced force diverged. In such a case, the converged solution can be obtained by giving the contact node larger compulsory displacement at the previous equilibrium solution and making it move to the next door element. In another words, distortion problem becomes possible to prevent by using the numerical technique above. As a result, the convergent performance of Timoshenko contact element can be verified in the almost part of element. For the future prospects, it is considered that the converged solution can be obtained on contact analysis with large displacement in 3-dimensional frame structures.

References

[1] S. Goto, G. Hane and T. Tanaka, The large deformation analysis of plane frame structures by the tangent stiffness method, J. of Japan Society of Civil Engineer, **238**, 31-42, (1975).

[2] S. Goto, The pluses and minuses of finite element method in a geometrically non-linear analysis method of structures, <http://www.forum8.co.jp/forum8/ronbun1.htm 1998>.

[3] A. Konyukhov, and K. Schweizerhof, Geometrically exact covariant approach for contact between curves, Computer Methods in Applied Mechanics and Engineering, **199**(36-40), 2510-2531, (2010).

[4] Z.M. Nizam, Effect of shear deformation on convergability of simple contact analysis with large deformation, *Proc. Asia Pacific Congress on Computational Mechanics and 4th Int. Symposium on Computational Mechanics*, Singapore, (2013).

MATEC Web of Conferences **47**, 02018 (2016)
DOI: 10.1051/matecconf/20164702018

The Modification of Boundary Treatment in the Incompressible SPH for Pressure Calculation Accuracy on the Solid Boundary

Nur`Ain Idris[1,a], Mitsuteru Asai[1] and Yoshimi Sonoda[1]

[1]*Department of Civil Engineering, Graduate School of Engineering, Kyushu University, Motoka 744, Nishi-ku, Fukuoka 819-0395, Japan*

Abstract. The Incompressible Smoothed Particle Hydrodynamic (ISPH) is one of the particle methods and commonly used to solve some complicated physical problems including free surface flow problems. The study regarding the boundary treatment has become an active research area in the mesh-free or particle method recently for measuring the accurate and robust pressure near the boundary. The penetrations of fluid particles may be happened if the adequate pressure boundary condition on the solid boundary cannot be satisfied. In this paper, a simple boundary treatment, which can be satisfied the non-homogenous Neumann boundary condition on the solid boundary and Dirichlet condition on the water surface, is proposed. The key point of our proposed treatment is that these boundary conditions are automatically satisfied by solving a modified pressure Poisson equation. Lastly, the effectiveness and accuracy of boundary treatment proposed are then authenticated with couples of numerical analysis and compared with the experimental tests.

1 Introduction

Smoothed particle hydrodynamics (SPH) is a Lagrangian meshless particle method that has been popular and actively used by researchers especially in computational fluid dynamics nowadays. This SPH method was first developed by Lucy [1] and Gingold [2] which originated from simulation of the astrophysical problems about three decades ago. SPH method is a unique particle method which does not required any grid and has some advantages over grid based method such as simple in implementation and also eases of handling even for complex fluid and larger deformations. Then, the SPH method was extended to incompressible viscous flow. Nowadays, a stabilized Incompressible Smoothed Particle Hydrodynamic (ISPH) approach is one of the particle methods and commonly used to solve some of complicated physical problems including free surface flow problems [3]. In addition, boundary treatment of the particle method is one of the important issues for accurate solutions. Many researchers tried to improving the boundary treatment from the ghost particle then fixed ghost particle by Marrone [4], and virtual marker by Asai [5] and until know this research topic still become a vigorous research area. Therefore, a modified boundary treatment is proposed, which resolves the problem of pressure accuracy near the boundary including the prevention the penetration.

[a] Corresponding author : nurain@doc.kyushu-u.ac.jp

2 A Stabilized Incompressible Smoothed Particle Hydrodynamics (ISPH)

In this paper, the basic of stabilized ISPH which introduced by Asai [3] was adopted. The following sections will be described the governing equations used and the concept of Smoothed Particle Hydrodynamics (SPH) with the modification for incompressible flow.

2.1 Governing equation

The continuity equation and the Navier- Stokes equation in the Lagrange description are given as

$$\frac{D\rho}{Dt} + \rho\nabla \cdot \mathbf{u} = \mathbf{0} \tag{1}$$

$$\frac{D\mathbf{u}}{Dt} = -\frac{1}{\rho}\nabla p + \nu\nabla^2\mathbf{u} + \frac{1}{\rho}\nabla \cdot \boldsymbol{\tau} + \mathbf{F} = \mathbf{0} \tag{2}$$

where ρ and ν are density and kinematic viscosity of fluid, $\mathbf{u}$ and p are the velocity vector and pressure of fluid respectively. $\mathbf{F}$ is an external force, and t indicates time. The turbulence stress $\boldsymbol{\tau}$ is necessary to represent the effects of the turbulence with coarse spatial grids. In the most general incompressible flow approach, the density is assumed by a constant value with its initial value. Then the above mentioned governing equations lead to

$$\nabla \cdot \mathbf{u} = \mathbf{0} \tag{3}$$

$$\frac{D\mathbf{u}}{Dt} = -\frac{1}{\rho^0}\nabla p + \nu\nabla^2\mathbf{u} + \frac{1}{\rho^0}\nabla \cdot \boldsymbol{\tau} + \mathbf{F} = \mathbf{0} \tag{4}$$

2.2 Stabilization of pressure in PPE

Generally, in SPH approach, there some difficulty occurred due to the accuracy of the density representation in the SPH formulation and the treatment of pressure Poisson equation (PPE). It is difficult for the numerical densityto keep its initial value. Therefore, here, the PPE is reconsidered to overcome the error of artificial pressure fluctuation in the ISPH. In density invariance scheme, the particle position updates after each predictor step and the particle density is updated on the intermediate particle positions. While in the correction step, the pressure term is used to update the particle velocity obtained from the intermediate step. Then, the final form of PPE in ISPH can be approximately redefined by the following equation whereby they are introducing the relaxation coefficient, α $(0 \le \alpha \le 1)$ and usually the small value is used around 1% or less.

$$< \nabla^2 p_i^{n+1} > = \frac{\rho^0}{\Delta t} < \nabla \cdot \mathbf{u}_i^* > + \alpha \frac{\rho^0 - < \rho_i^n >}{\Delta t^2} \tag{5}$$

3 Boundary Treatment

3.1 Slip and Non-Slip condition with virtual marker

The virtual marker proposed to be located in a position which is symmetrical to the wall particle across it solid boundary whereas the wall particles is placed on a grid like structure. Normally, the velocity and pressure on the marker are interpolated based on the concept of weighted average of neighboring particles, which is the fundamental equation of SPH as in the Equation (6). However, the portion of the particles within the compact support might be in the wall domain or unoccupied domain such as air phase. For that case, the interpolation approximation will used the modified weight function.

$$\phi(\boldsymbol{x}_i,t) \approx \langle \phi_i \rangle = \sum_j \frac{m_j}{\rho_j} \widetilde{W}(r_{ij},h)\phi_j(\boldsymbol{x}_j,t) \tag{6}$$

The virtual marker is not directly associated with the SPH approximation but only a computational point to give the wall particle accurate physical properties. Then, the velocity on the virtual marker which is corresponding to the wall particle across its solid boundary is interpolated by the SPH approximation. Practically, it is necessary to discuss an optimized condition between slip and no-slip condition to refer the resolution of the simulation model. Therefore, the following equation is proposed which can be consider as mixed condition whereby β represents the ratio between slip and no-slip condition. Where **M** is a second order tensor to implement the mirroring processing, and it is symbolized by using inward normal vector of the wall and the kronecker delta and **R** is a mirror symmetric tensor

$$\mathbf{v}'_w = \beta \mathbf{M}\mathbf{v}_v + (1-\beta)\mathbf{R}\mathbf{v}_v \quad ,(0 \le \beta \le 1) \tag{7}$$

3.2 Pseudo Neumann boundary condition: Dirichlet conditions

Using the similar approach as introduced by Asai [5], the pressure on each virtual marker and the external force will evaluated by using the physical quantity in the particles neighborhood with the normalized weight function as in Equation (6). Thus, the pressure on the walls particles in the border processing technique does not directly evaluated by solving the modified pressure Poisson, Equation (5). By introduced solving the PPE will gave entire circumference of Dirichlet condition whereby approximation of the value of the wall particles pressure by referring the value of the virtual marker as to satisfy the Neumann boundary condition.

3.3 Neumann boundary treatment (completely satisfy the pressure Neumann conditions)

In this section, the Modified Virtual Marker with Regular grid (MVMRG) as new boundary treatment which satisfied the non-homogenous Neumann boundary condition will be described. The concept of this treatment is to give a wall particle accurate physical properties, velocity and pressure. As concern, the virtual marker used in pseudo Neumann condition is convenient to perform the analysis including the rate of real condition as slip or non-slip condition. However, by solving PPE with the giving the entire circumference Dirichlet condition causes a lower pressure value than the actual hydrostatic pressure distribution in the vicinity of wall particles. Therefore, the MVMRG is proposed in order to overcome this problem. The procedures will be described briefly. First, the wall particles placing similarly as in Pseudo Neumann condition method which placed equally spaced structural grid-like inside solid. Then, the location of the virtual markers are located on the boundary surface which now no longer in the fluid domain. The process of calculated the value of the inner speed on the virtual marker is similar with the previous section which using the SPH approximation as in Equation (6). Therefore, the modification of the coefficient matrix and the source term need to be conducted due to the flow velocity is not strictly mirror-symmetrical. Then, pressure Dirichlet conditions at the stage of solving the Poisson equation was applied only to the place given originally and the pressure approximation of wall particles is evaluated by using the following equation:

$$P'_{\mathrm{w}} = P_{\mathrm{w0}} + d\rho\, \boldsymbol{f}_{\mathrm{w0}} \cdot \boldsymbol{n} \quad \text{whereby } P_{\mathrm{w0}} \cong \overline{P}_{\mathrm{v}} = \frac{\sum_{j \in water}^{n_W} P_j}{n_w} ;\quad \boldsymbol{f}_{\mathrm{w0}} \cong \overline{\boldsymbol{f}}_{\mathrm{v}} = \frac{\sum_{j \in water}^{n_W} \boldsymbol{f}_j}{n_w} \tag{8}$$

By using the above techniques, the pressure solved by the PPE can be limited to the water particles. Hence, the special modification of the coefficient matrix and source terms when the particles were discretized. The relationship of the concept mentioned above between the virtual marker, water particles and the wall particles as shown in Figure 1.

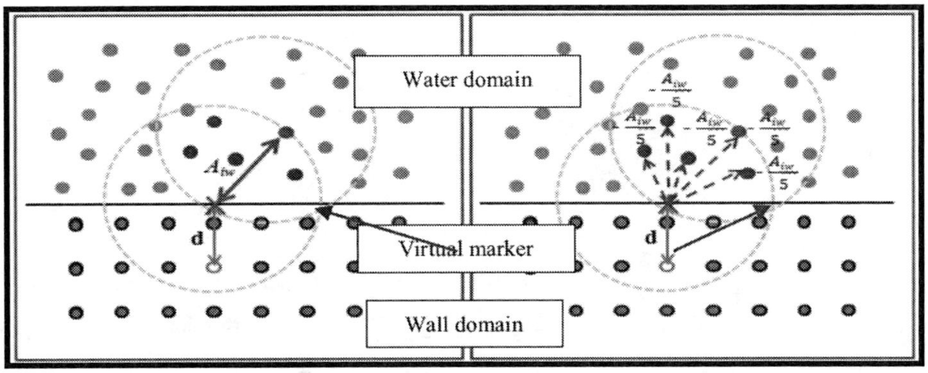

Figure 1. The relationship of the modified virtual marker on the boundary with the wall particles and water particles.

After solving the PPE abovementioned, the pressure needs to go through mapping process whereas the pressure and the external force will be analyzed by SPH formulation.

$$P'_{w} = < P_{v} > + d\rho < \boldsymbol{f}_{v} > \cdot \boldsymbol{n} \tag{9}$$

Hence, there are some verification and validation test was made using this new modification which is used MVMRG to see the performance and also in comparison with experimental test that will be discussed in the following sections.

4 Verification and Validation

4.1 Verification with hydrostatic test

The first test is the simplest hydrostatic problem which the rectangular tank with compatible surface. The comparison of the pressure distribution was made between different boundary treatments which are using pseudo Neumann condition and proposed MVMRG boundary treatment. This test was conducted to show the robustness of our modified boundary treatment. The tank was filled up with the water height of 0.2m and the particles size is 0.01m. Where, the point of pressure measurement is located in the middle of bottom tank. Figure 2(a) shows the different results on the contour pressure distribution especially the effects near the bottom tank which obtained using different treatment. It is clearly show that there are high pressure distribution using the proposed modified treatment, using MVMRG compared with the Pseudo Neumann boundary condition that give lower pressure distribution. Then, the numerical result of the hydrostatic pressure were used to compared with the theoretical of value of hydrostatic pressure as shown in Figure 2(b). Obviously, the result obtained by using modified boundary treatment of Neumann condition, MVMRG gives almost similar result with the theoretical. However, the result found from Pseudo Neumann treatment gives the lower value especially near the bottom surface due to the treatment used at the boundary surface not perfectly treat as non-homogenous of Neumann condition. Next, the same setup was used but with the additional of several geometrical object submerged into the water with size of the particles is reduced to 0.005m. Generally, Figure 3(a) shows the pressure distribution inside the water domain and also at the concave-convex shape boundary surface without any penetration occurred. Then, the numerical results of the pressure distribution were compared with the theoretical hydrostatic pressure as shown in

Figure 3(b). Based on the result obtained, the result using MVMRG gives higher value compared with the Pseudo Neumann condition.

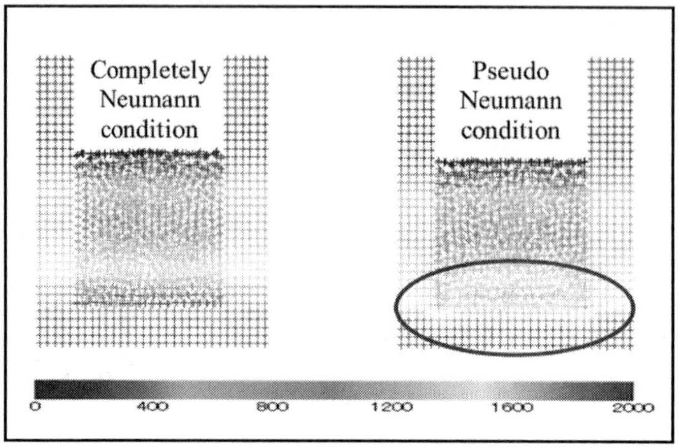

(a)

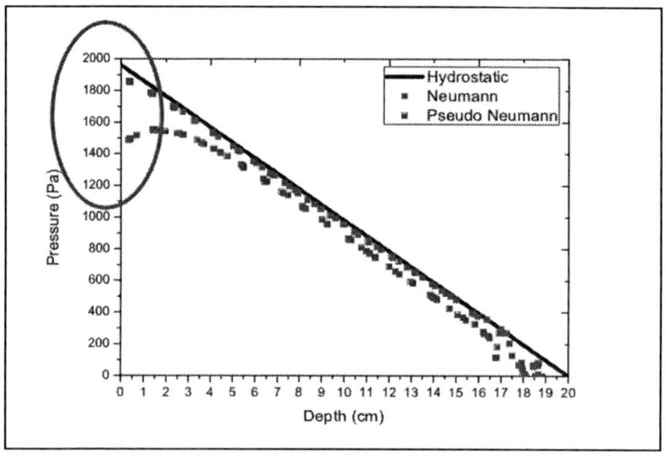

(b)

Figure 2. (a) The pressure distribution by using different boundary condition (b) the comparison of pressure distribution with the depth by using hydrostatic pressure, perfectly Neumann boundary treatment and Pseudo Neumann boundary condition for compatible surface boundary.

4.2 Validation with 3D dam break flow with the opening gate

The simulation of three dimensional dam break flow is used to validate the versatility of our proposed method and compared with the experimental results. The experimental test on the investigation of dynamic pressure loads during dam break with an opening gate that was carried out at Technical University of Madrid (UPM) as reported by Lobovsky [6] is used as the reference for simulate purposed and design the analysis model as shown in the schematic diagram (Figure 4). The clear comparison of free surface profile evolution snapshot between the numerical simulation and the experimental at certain times as shown in Figure 5 and the effect of the dam gate removal looks similar at every snapshot on the free surface shape is observed. Then, the observation of the location of the wave front also shows the same tendency with the experiment. The pressure contour is also including in this figure shows the pressure gradient from the static condition before an opening gate and after gate's removal and the impact after hit the wall which shows higher pressure.

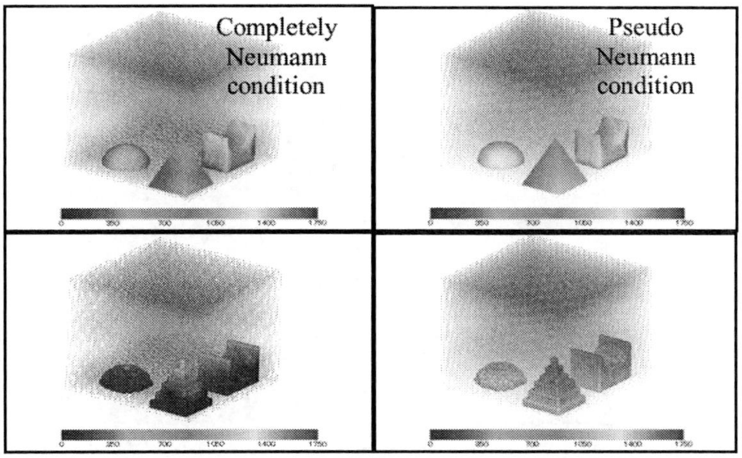

(a)

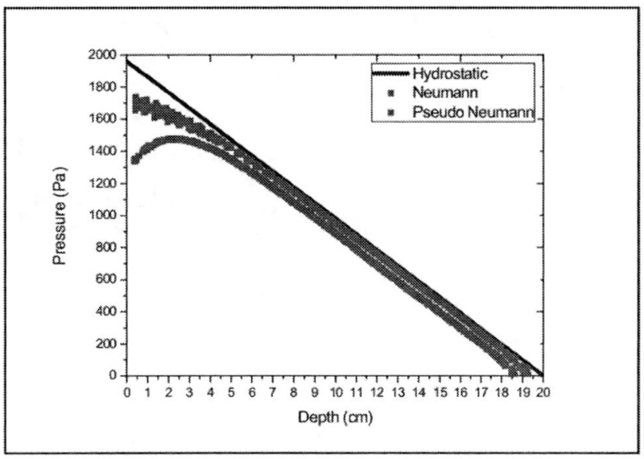

(b)

Figure 3. (a) The pressure distribution on the water particles and wall boundary by using different boundary condition (b) the comparison of pressure distribution with the depth by using hydrostatic pressure, MVMRG with perfectly Neumann condition and Pseudo Neumann condition for compatible surface with complex geometrical objects.

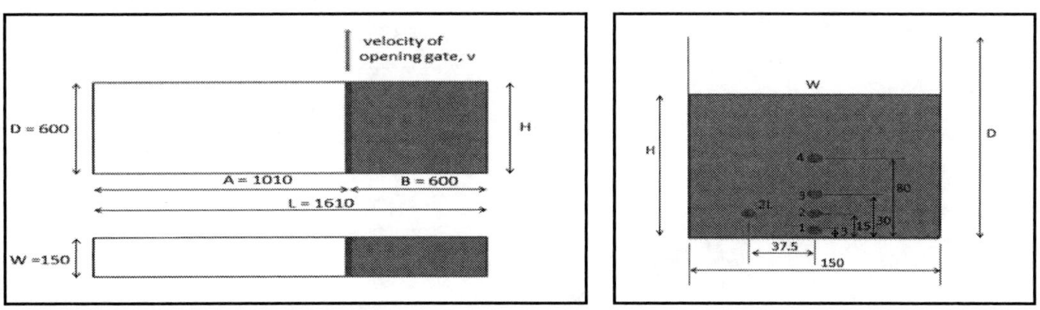

Figure 4. The dam break simulation with an opening gate which corresponding to the experimental [6].

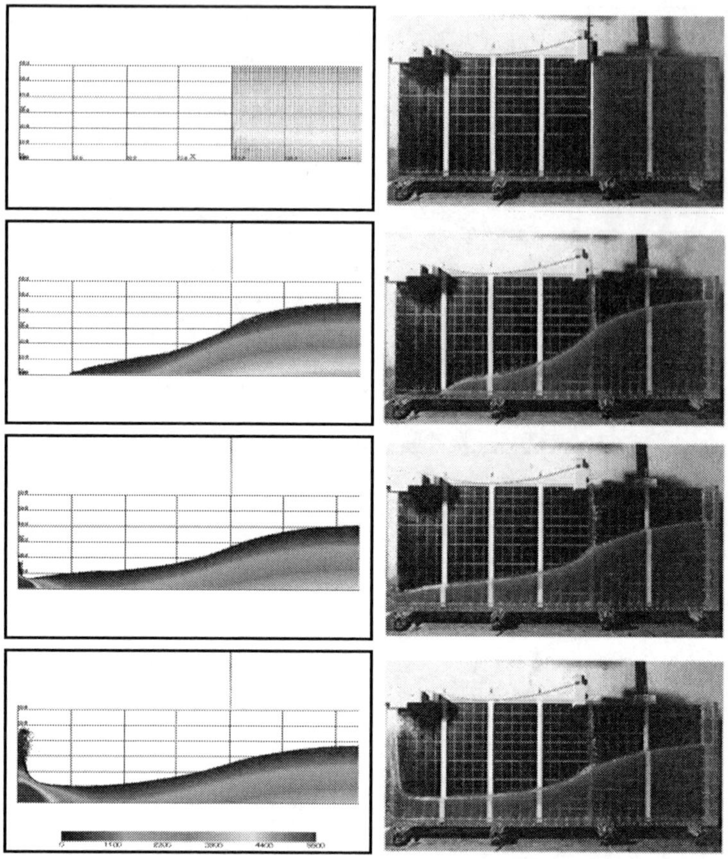

Figure 5. The comparison between simulation and experimental of the free surface evolution at times 0.0, 316.7, 413.4, 463.3, and ±3.3ms.

Furthermore, from the snapshot, it is clearly shows that the lowered sensors are exposed to the high pressure as the red color of contour is observed after the water smashed on the wall surface. Therefore the results obtained at the lowered sensor from the simulations are compared with the experimental data generally. It indicates the significant results, where the simulation results fall within the range of 100 tests of experimental results from sensor 1 with the simulation's result as shown in Figure 6. From the comparison, the arrival time from the numerical simulation is almost same with the arrival time in the experiment and the tendency of the pressure time history is similar.

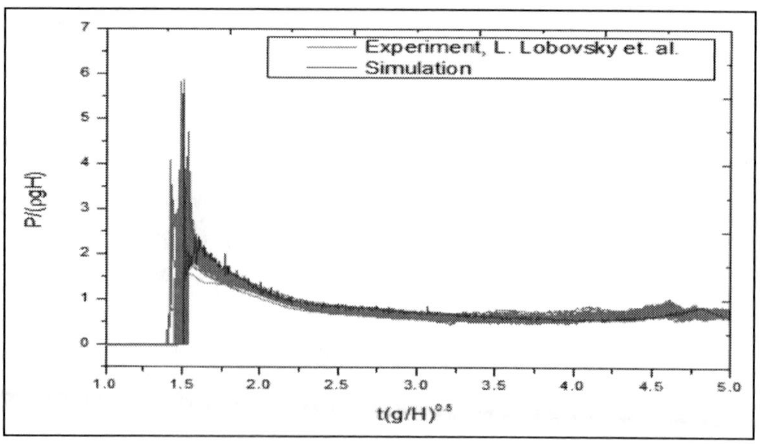

Figure 6. The comparison of the non-dimensional pressure at sensor 1 between the simulation results with 100 tests of experiment data.

5 Conclusions

The numerical simulation that been conducted using the proposed boundary treatment, MVMRG with perfectly Neumann condition shows the good tendency and more accurate pressure value near the boundary surface based on the hydrostatic tests and also the comparison with the experimental test for hydrodynamic pressure evaluation. The profile of the free surface also shows the similarity with experiment results. In addition, the comparison between both boundary condition show clear discrepancy in the pressure especially at the edge of bottom surface and this strongly proves that the proposed modified treatment gives the better treatment.

Acknowledgement

The numerical simulation that been conducted using the proposed boundary treatment, MVMRG with perfectly

References

[1] M.RL.B. Lucy, A numerical approach to the testing of the fusion process, Astron J., **88**, 1013-1024, (1977).
[2] R.A. Gingold and J.J. Monaghan, Smoothed particle hydrodynamics: Theory and application to non-spherical stars, Mon. Not. R. Astron Soc., **181**, 375-389, (1977).
[3] M. Asai, A.M. Aly, Y. Sonoda and Y. Sakai, A stabilized incompressible SPH method by relaxing the density invariance condition, Int. J. of Applied Mathematics, **2012**, 1-24, (2012).
[4] S. Marrone, M. Antuono, A. Colagrossi, G. Colicchio, D.L. Touze and G. Graziani, δ-SPH model for simulating violent impact flows, Comput. Methods Appli. Mech. Eng., **200**, 1526-1542, (2012).
[5] M. Asai, K. Fujimoto, S. Tanabe and M. Beppu, Slip and no-slip boundary treatment for particle simulation model with incompatible step-shaped boundaries by using a virtual marker, J. of JSCES, Paper No.20130011, (Japanese), (2013).
[6] L. Lobovsky, E.B. Vera, F. Castellana, J.M. Soler and A.S. Iglesias, Experimental investigation of dynamic pressure loads during dam break, J. of Fluids and Structures, **48**, 407-434, (2014).

MATEC Web of Conferences **47**, 02019 (2016)
DOI: 10.1051/matecconf/20164702019

Multi-Scale Bridge Wash Out Simulation During Tsunami by Using a Particle Method

Yoshiya Miyagawa[1,a] and Mitsuteru Asai[1]

[1]*Department of Civil Engineering, Graduate School of Engineering, Kyushu University, Motoka 744, Nishi-ku, Fukuoka 819-0395, Japan*

Abstract. In 2011, the huge tsunami caused by the great east Japan earthquake devastated many infrastructures in pacific coast of north eastern Japan. Particularly, collapse of bridges caused a traffic disorder and these collapse behaviors led to delay of recovery after the disaster. In this study, the bridge wash away accident is selected as a target issue, and it is represented by a numerical simulation. For this purpose, Smoothed Particle Hydrodynamics (SPH) Method, which is one of the pure mesh free methods, is utilized for the rigid body motion simulation. In this study, rigid body motion is introduced for the fluid-rigid interaction behavior during bridge wash away simulation. In the numerical analysis, the upper bridge structure is washed away by receiving an impact fluid force. The wash away simulation of two types of the bridge girder showed good agreement with the real accident on the great east Japan earthquake tsunami.

1 Introduction

On March 11, 2011, the huge tsunami caused by the great east Japan earthquake devastated many infrastructures in pacific coast of north eastern Japan. Particularly, the damage of outflow of bridge girders caused a traffic disorder and these collapse behaviors led to delay of recovery after the disaster. After 2011 tsunami, disaster prevention and mitigation techniques are actively developing in coastal infrastructures and establishing prediction method for tsunami disaster is one of the severe issues toward the next millennium tsunami.

In this study, the bridge wash out accident is selected as a target issue, and we try to represent these accidents by using a numerical analysis. For this purpose, one of the mesh free methods; Smoothed Particle Hydrodynamics (SPH) Method [1, 2] is utilized for Tsunami flow. The main advantage of SPH and the other particle simulation is the absence of a computational grid or mesh since it is spatially discretized into Lagrangian moving particles. This allows the possibility of easily modeling flows with a complex geometry or flows where large deformations or the appearance of a free surface occurs. Recently, this method is widely used in field of fluid and solid dynamics. A stabilized ISPH [3], which is one of the modified versions of the SPH and can evaluate much smoother pressure distribution, has been developed by our research group. Then a fluid-solid interaction algorithm including rigid body motion is developed in this study.

[a] Corresponding author : y_miyagawa@doc.kyushu-u.ac.jp

2 SPH Formulation

In this paper, SPH method was adopted. A basic concept in SPH method is that any function φ attached to particle "i" at a position r_i is written as a summation of contributions from neighbor particles

$$\phi(r_i) \approx \langle \phi_i \rangle = \sum_j \frac{m_j}{\rho_j} \phi_j W(r_{ij}, h) \tag{1}$$

Note that, the triangle bracket $<\phi_i>$ means SPH approximation of a function φ. The divergence of a vector function can be assumed by using the above defined SPH approximation as follows:

$$\nabla \cdot \bar{\phi}(r_i) \approx \langle \nabla \cdot \bar{\phi}_i \rangle = \frac{1}{\rho_i} \sum_j m_j (\bar{\phi}_j - \bar{\phi}) \cdot \nabla W(r_{ij}, h) \tag{2}$$

and the expression for the gradient can be represented by:

$$\nabla \phi(r_i) \approx \langle \nabla \phi_i \rangle = \rho_i \sum_j m_j \left(\frac{\phi_j}{\rho_j^2} + \frac{\phi_i}{\rho_i^2} \right) \nabla W(r_{ij}, h) \tag{3}$$

In this study, a stabilized ISPH method developed in the incompressible fluid analysis was adopted. In this method, the pressure is calculated implicitly and the velocity is calculated explicitly.

3 Fluid-rigid Interaction Formulation

3.1 Governing equations

The governing equations are the continuum equation and the Navier-Stokes equation. These equations for the flow are represented as:

$$\frac{D\rho}{Dt} + \rho \nabla \cdot \boldsymbol{u} = 0 \tag{4}$$

$$\frac{D\boldsymbol{u}}{Dt} = -\frac{1}{\rho} \nabla p + \nu \nabla^2 \boldsymbol{u} + \frac{1}{\rho} \nabla \cdot \boldsymbol{\tau} + \boldsymbol{F} = \boldsymbol{0} \tag{5}$$

where, ρ and ν are density and kinematic viscosity of fluid, $\boldsymbol{u}$ and p are the velocity and pressure vectors of fluid respectively. $\boldsymbol{F}$ is external force, and t indicates time. The turbulence stress τ is necessary to represent the effects of turbulence with coarse spatial grids. In the most general incompressible flow approach, the density is assumed by a constant value with its initial value.

3.2 Modification in the source term of pressure Poisson equation

The main concept in an incompressible SPH method is to solve a discretized pressure Poisson equation at every time step to get the pressure value. In a sense of physical observation, physical density should keep its initial value for incompressible flow. However, during numerical simulation, the 'particle' density may change slightly from the initial value because the particle density is strongly dependent on particle locations in the SPH method. If the particle distribution can keep almost uniformity, the difference between 'physical' and 'particle' density may be vanishingly small. In other words, accurate SPH results in incompressible flow need to keep the uniform particle distribution. For this purpose, the different source term in pressure Poisson equation can be derived using the 'particle' density. The SPH interpolations are introduced into the original mass conservation law before the perfect compressibility condition is applied.

$$< \nabla \cdot \boldsymbol{u}_i^{n+1} > = -\frac{1}{\rho^0}\frac{< \rho_i^{n+1} > - < \rho_i^* >}{\Delta t} \tag{6}$$

Then, the pressure Poisson equation reformulated as:

$$< \nabla^2 p_i^{n+1} > = \frac{\rho^0}{\Delta t} < \nabla \cdot \boldsymbol{u}_i^* > + \alpha\frac{\rho^0 - < \rho_i^* >}{\Delta t^2} \tag{7}$$

where, α is relaxation coefficient, $\boldsymbol{u}_i^*$ is temporal velocity and triangle bracket < > means SPH approximation. Note that this relaxation coefficient is strongly dependent on the time increment and the particle resolution. Then, the reasonable value can be estimated by the simple hydrostatic pressure test using the same settings on its time increment and the resolution.

3.3 Treatment of moving rigid body

In this study, the general momentum conservation law of the rigid body is solved numerically with the external forces including hydrodynamic as a fluid-rigid interaction formulation. Figure 1 shows the rigid motion algorithm. The rigid body is discretized to particles at the beginning. Next, it is postulated that the external force is calculated from the gravity g, hydrodynamic force at the rigid surface $\boldsymbol{F}_f$ and contact force between rigid body and fixed boundary $\boldsymbol{F}_e$, and these value are calculated by following equations.

$$\boldsymbol{F}_f = \sum_i^{\text{on the surface}} \boldsymbol{P}_i \Delta S_i \boldsymbol{n}_i \tag{8}$$

$$\boldsymbol{F}_e = k\delta^{3/2} \tag{9}$$

$$\delta = r - 1 \tag{10}$$

$$k = \frac{4\sqrt{r}}{3}\frac{E_i E_j}{\left(1-v_i^2\right)E_j + \left(1-v_j^2\right)E_i} \tag{11}$$

where, $\boldsymbol{P}_i$ is the pressure of surface rigid particle, and ΔS_i is the area of surface rigid particle. And the contact force is modeled by the elastic contact theory by Hertz. In the above equation, l is the distance between rigid particle and the fixed boundary. In above equation, v_i, E_i, E_j is Poisson ratio, elastic modulus of the particle coming into contact, elastic modulus of the contacted particle. And then the hydrodynamic moment $\boldsymbol{M}_f$ and contact moment $\boldsymbol{M}_e$ are calculated by following equation.

$$\boldsymbol{M}_f = \sum_i^{\text{on the surface}} (\boldsymbol{r}_i - \boldsymbol{r}_c) \times \boldsymbol{P}_i \Delta S_i \boldsymbol{n}_i \tag{12}$$

$$\boldsymbol{M}_e = \sum_i^{\text{on the surface}} (\boldsymbol{r}_i - \boldsymbol{r}_c) \times \boldsymbol{F}_e \tag{13}$$

And then, translational velocity $\boldsymbol{T}$ and angular velocity vector ω are evaluated by following equations.

$$M\frac{d\boldsymbol{T}}{dt} = M\mathbf{g} + \boldsymbol{F}_f + \boldsymbol{F}_e \tag{14}$$

$$\boldsymbol{I}\frac{d\boldsymbol{\omega}}{dt} = \boldsymbol{M}_f + \boldsymbol{M}_e \tag{15}$$

Finally, the velocity of each particle in the solid body can be expressed as :

$$\boldsymbol{v}^{n+1} = \boldsymbol{T} + \boldsymbol{\omega} \times \boldsymbol{r}_k \tag{16}$$

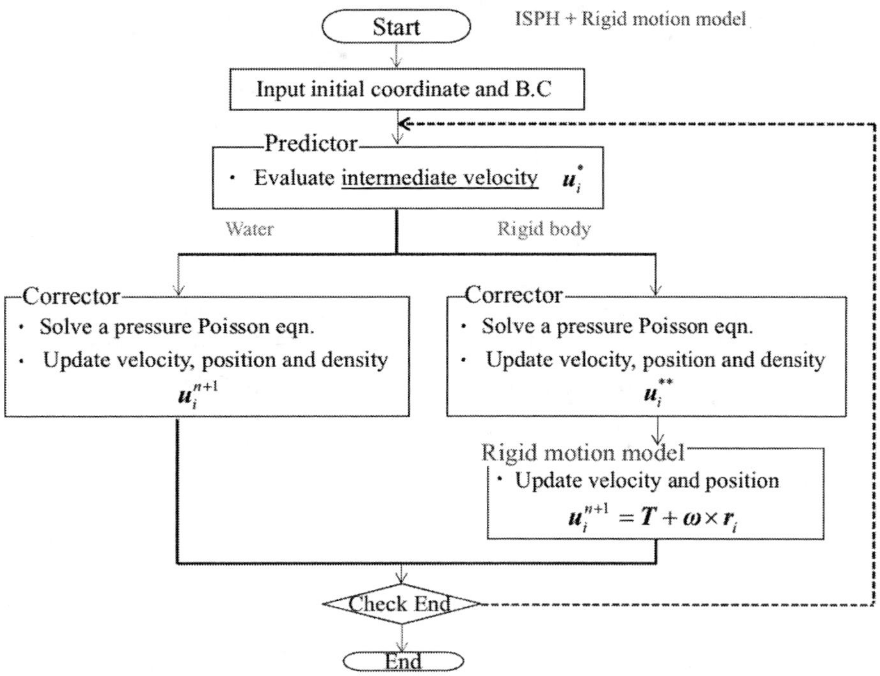

Figure 1. Rigid motion algorithm.

4 Bridge Wash Out Simulation

Figure 2 shows Utatsu Bridge after the Tsunami impact, the bridge is in Minamisanriku-cho, Miyagi prefecture. From this picture, high height girders turned over and low height girders are pushed out without rotation. In this study, we have selected two types of bridge girder model; high girder and low girder model as a target model and compare the simulation results and disaster report. The detail of analysis model and girder model are shown in Figure 3 to Figure 5. The density of particles of the girder model is 24.5 kg/m^3 referring to the density of reinforced concrete. And the wave is modeled for a gentle stream. Initial water level is set to be 8m and the initial velocity of the wave is set 10m/s. In addition, 10m/s is continuously given at the position of 20m from the left corner of the water storage. The depth in the models is 11m. The particle distance d_0= 0.2m, time increment Δt = 0.001s and the total number of particles is about one million.

The results of bridge wash out behavior are shown in Figure 6. From these results, the high girder model case shows a higher rotation than the low girder model case. The disaster report say that the high height girders turned over and low girders just wash out without rotation, and our simulation results show a similar tendency to this disaster report. Our fluid-rigid coupling algorithm can expressed the change of wash out behavior depending on the shape of the bridge girder. However, turn-over of the bridge girder cannot be represented because of lack of several numerical settings. One of the causes is considered that the inflow condition is not correct as a real tsunami accident. Then, we have conducted a tsunami innundation simulation as a next topics.

Figure 2. The damage at Utatsu Bridge.

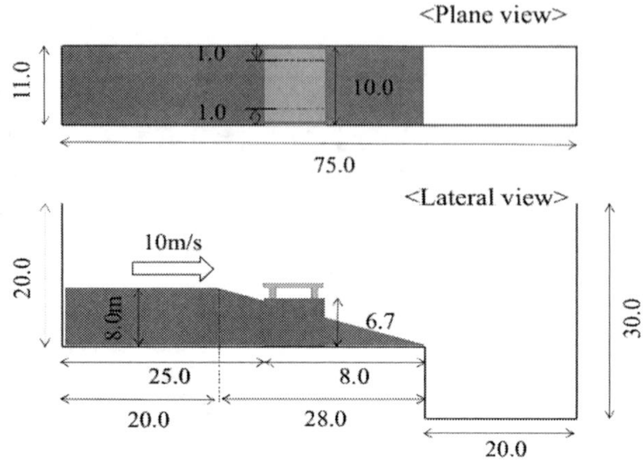

Figure 3. Analysis model.

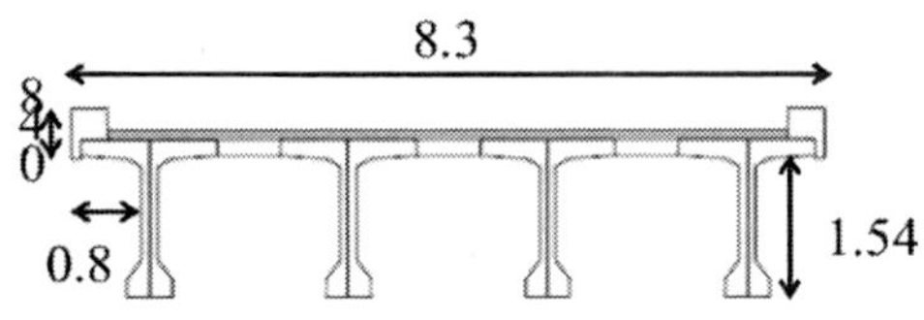

a) High girder

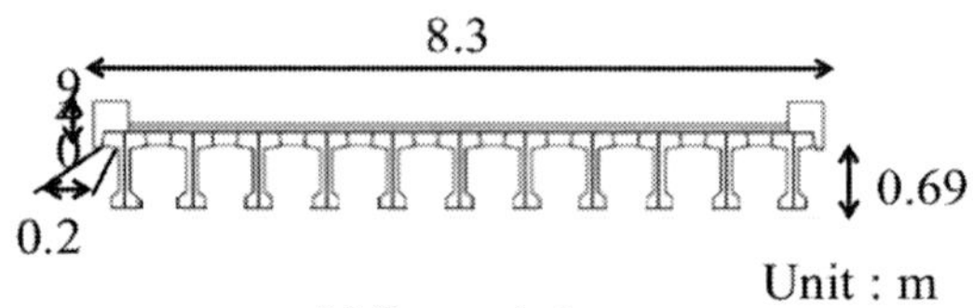

b) Low girder

Figure 4. Cross section of Utatsu Bridge.

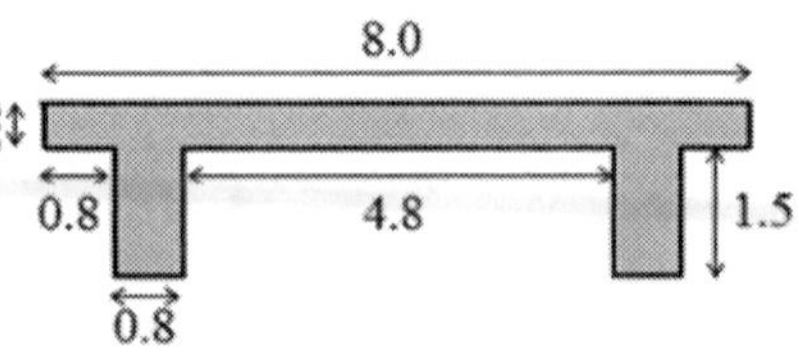

a) High girder model

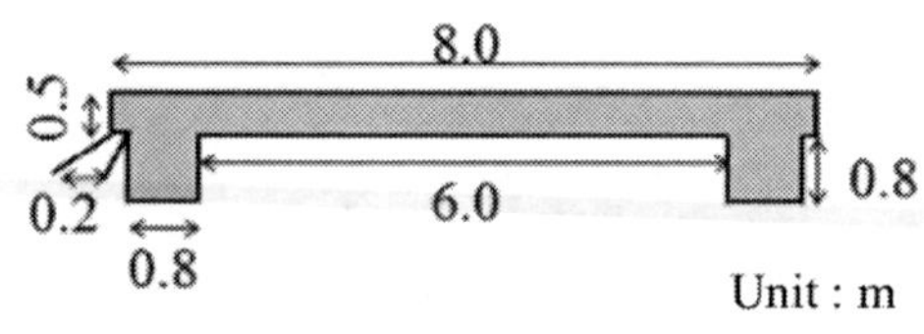

b) Low girder model

Figure 5. Girder model.

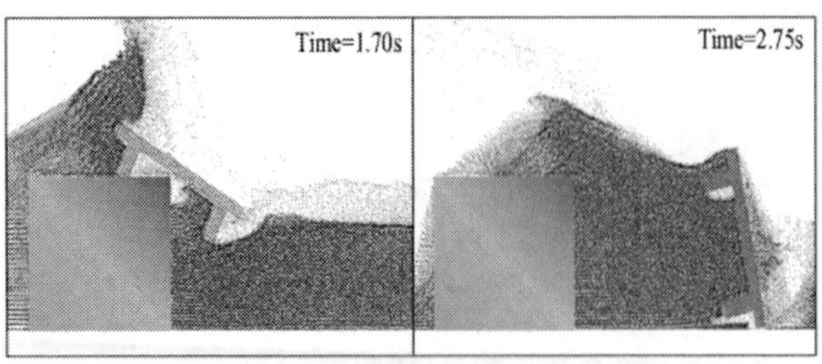

a) High girder model

b) Low girder model

Figure 6. Snapshot of moving girder model.

5 Tsunami Run-up Simulation

According to the conclusion at previsou section, the tsunami run-up simulation is conducted to reconsider the inflow condition. The Utatsu area where Utatsu Bridge is in is selected as a simulation area.

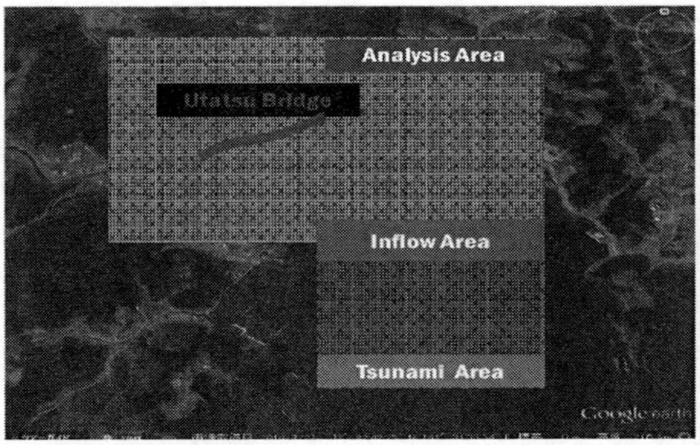

Figure 7. Analysis area and Tsunami are at Utatsu.

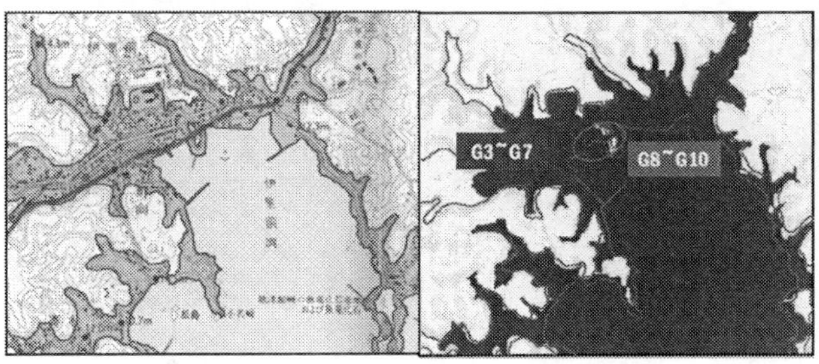

Figure 8. Comparison of tsunami inundation between disaster report and simulation result.

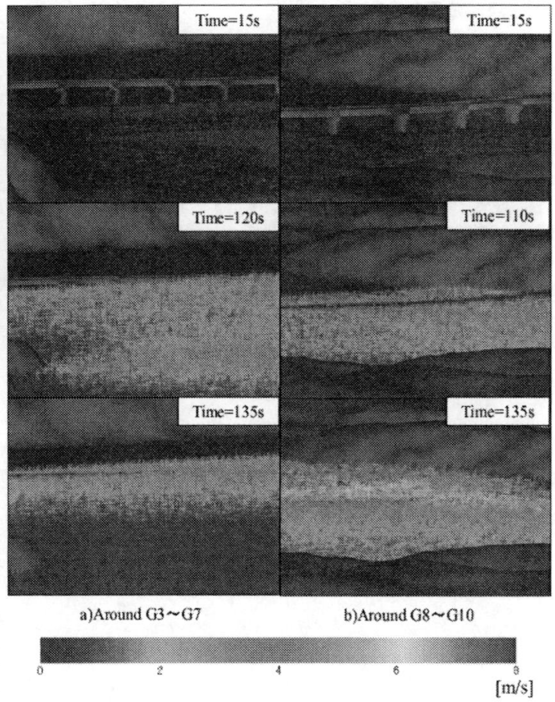

Figure 9. The velocity distribution around the Utatsu Bridge

Figure 7 shows the analysis area and Utatsu Bridge. In addition, the Tsunami area is modeled and initial water level is set to be 10m. And then, 10m/s is continuously given for water particles which pass through the inflow area. The particle distance d_0= 3.0m, time increment Δt = 0.001s and the total number of particles is about seven million.

Figure 8 shows the comparison of tsunami inundation area between disaster report and the simulation result. Although the simulation result shows a similar tendency in the comparison of inundation area, the details of inundation area shows several mismatch between them. The simulated result is overestimated into the northern direction, the other hand the inundation area into the western direction is smaller than disaster report [4]. It's because the inflow velocity into northern direction is faster than actual velocity in this simulation. Figure 9 shows velocity distribution around the Utatsu Bridge by our simulation. In comparison in these two areas, the Tsunami velocity around the high height girder (G8-G10) is larger than the velocity around the low height girder (G3-G7). The above simple inflow conditon may be replaced by a consistent inflow condition based on a large area tsunami propagation simulation based on the shallow water equation and/or observation data.

6 Conclusion

In this study, fluid-rigid interaction formulation is conducted by introducing rigid motion algorithm into ISPH. From bridge wash out simulation, we could express the change of wash out behavior depending on the shape of the bridge girder and it shows good agreements with the disaster report on

March 11, 2011. Then Tsunami run-up simulation is conducted for the improvement of inflow condition of bridge wash out simulation. In the future work, the development of boundary treatment which can manipulate measured values in each analysis with high accuracy is unavoidable, before the numerical simulation apples to the bridge design against to huge Tsunami.

References

[1] L.B. Lucy, Numerical approach to the testing of the fusion process, Astronomical J., **88**, 1013-1024, (1977).

[2] R.A. Gingold and J.J. Monaghan, Smoothed particle hydrodynamic: theory and application to non-spherical stars, Royal Astronomical Society, **181**, 375-389, (1977).

[3] M. Asai, A.M. Aly, Y. Sonoda and Y. Sakai, A stabilized incompressible SPH method by relaxing the density invariance condition, J. of Applied Mathematics, **2012**, 1-24, (2012).

[4] K. Maruyama, Y. Tanaka, K. Kosa, A. Hosoda, T. Arikawa, N. Mizutani and T. Nakamura, Evaluation of tsunami force acting on bridge structures, Japan Society of Civil Engineers, Japan, (2013).

MATEC Web of Conferences **47**, 02020 (2016)
DOI: 10.1051/matecconf/20164702020

Finite Element Analysis of Lateral Torsional Buckling Behaviour of Tapered Steel Section with Perforation

Nurfarhah Naaim[1], Fatimah De'nan[1,a], Choong Kok Keong[1] and Faiz Azar[1]

[1]*School of Civil Engineering, Universiti Sains Malaysia, 14300 Nibong Tebal, Pulau Pinang, Malaysia*

Abstract. This paper presents the finite element analysis of lateral torsional buckling of tapered steel section with perforation. The numerical analysis consists of four main models with a ratio of tapering 0.3 for 3.5m and 5.5m span and ratio of tapering 0.5 for 3.5m and 5.5m span. This study has 94 models in order to investigate the effect of thickness and the openings on the buckling moment. LUSAS software was used to carry out the finite element analysis to study the effect of web opening on the buckling behavior which had five variables such as thickness of web and flange, opening size, ratio of tapering, and opening arrangement. The opening size from 0.2D to 0.4D is recommended because buckling moment capacity difference only around 2% was observed for the opening size exceeds 0.4D. Ratio of tapering 0.3 and ratio of tapering 0.5 was observed increase with the web opening area. The deformation patterns are almost similar for the different sizes of web openings and ratio of tapering. Based on the finite element results, tapered steel section without perforation has higher buckling moment capacity. The comparison of the buckling moment between non-perforated sections with the recommended section is around 2% to 4%.

1 Introduction

Tapered steel section is a section consisting of two flanges connected by tapered web as shown as in Figure 1. Lateral buckling of tapered members were first investigated by Boley and Zimnoch [1] in 1952. This paper indicated the investigations of the application of a numerical method that had developed for the problem of the lateral buckling. Trahair [2] carried out research on the interaction buckling of tapered beams. The results show the elastic lateral buckling of braced and continuous beams is influenced by the separate effects of moment distribution, taper, and restraints between adjacent segments. Elastic lateral-torsional buckling behavior of discretely restrained tapered beams show inaccurate to model a web-tapered I-beam as an assembly of prismatic segments through theoretical and numerical results [3].

Lagaros et al. [4] stated that the arrangement of multiple openings in the beams has also been adopted for aesthetical considerations. The effects of circular or square web openings on the ultimate strength of horizontally curved composite plate girders had conducted by Basher et al. [5]. They found that the reduction of ultimate shear strength is small and can be neglected when the opening size is small. Abidin and Izzudin [6] carried out meshless local buckling analysis of steel beams with irregular web openings. The combination of the element-free Galerkin method with the rotational

[a] Corresponding author : cefatimah@usm.my

spring analogy yields accurate and efficient buckling predictions. Liu and Chung [7] used finite element method to analyze the characteristics of steel beams with large openings of various shapes and sizes. The result shows that perforated section under a global moment Mo,Sd and a global shear force Vo,Sd, three local actions are induced in the tee-sections above and below the web opening.

Lateral torsional buckling strength of unsymmetrical plate girders with corrugated webs subjected to uniform moment demonstrated a good agreement between the finite element results and the proposed formula [8]. The un-symmetrical corrugated web girder has an elastic lateral torsional buckling capacity up to 11% more than the corresponding plate girder with the conventional flat web. Lateral torsional buckling analysis of steel frames with corrugated webs has been studies by Sabat [9]. Finite element analysis and Eurocode 3 show very close value of the result and similar lateral-torsional buckling behavior is obtained for both of frame with corrugated webs and the frame with plane webs. Moon et. al [10] had investigated theoretical and finite element analysis of the lateral torsional buckling of I-girder with corrugated webs under uniform bending. They had found the elastic lateral-torsional buckling strength is increased up to 10% for I-girders with corrugated webs compared to I-girder with flat webs.

In this regard, this paper studies the lateral torsional buckling behavior of tapered steel section with perforation using finite element analysis. The objectives of this study are to study the behavior of tapered steel section with perforation subjected to lateral torsional buckling, to determine the lateral torsional buckling moment for tapered steel section with perforation and to investigate the effect of section properties on the buckling moment resistance of tapered steel section with perforation.

2 Finite Element Analysis

In this research analysis, concentrated load, P is applied at the end of flange at the top part of the flange in vertical y-axis direction. The web is restrained at one end so it acts as cantilever or fixed-free. The following material properties and loading are kept constant throughout the analysis:

a) Ungraded Mild Steel
   - Young's Modulus, $E = 209$ GPa
   - Poisson Ratio, $v = 0.3$

b) Loading, $P = 10$kN

A convergence study had been done to get the optimum meshing that will reduce the computational effort to render the surface. The details of the model used to get the appropriate meshing size in convergence study is shown in Table 1. The convergence study was done by decreasing the element size to obtain increasing number of element of the model simultaneously will get the deflection value. Both number of element and deflection value was used to plot a graph, the graph converged and element size of 25 was used.

Table 1. Section properties of tapered I-beam.

Section	Tapered I-Beam
D_{max}	250 mm
B	100 mm
t_w	13 mm
t_f	9 mm
L	1000 mm
Opening Dimension	0.5D
Tapered Ratio (D_{min}/D_{max})	0.5

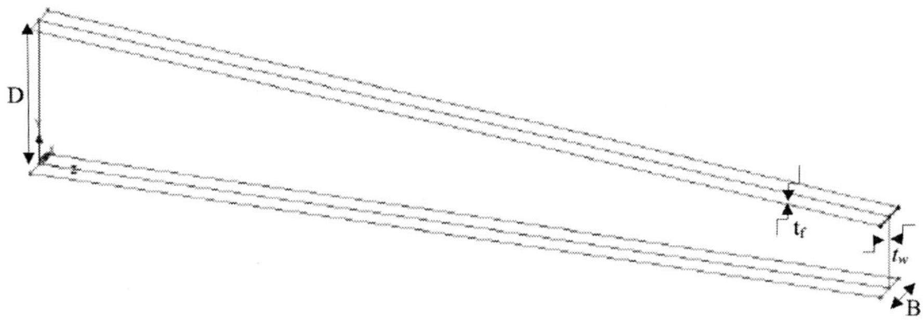

Figure 1. Three dimensional view of tapered steel beam section.

The formula used to calculate the buckling moment is:

$$P_b = \text{eigenvalue x concentrated load} \quad (1)$$

$$M_b = P_b \text{ x } L/4 \quad (2)$$

3 Results and Discussion

Finite element analysis was used to assess the engineering properties and performance of tapered steel section with perforation. The buckling moment capacity with different opening size and area, thickness of web and flange, arrangement and number of openings and ratio of tapering was found from eigenvalue buckling analysis. When the ratio of tapering and opening size increases, the buckling moment capacity decreases as can be seen in Figure 2 to Figure 5.

Buckling moment capacity reduces when ratio of tapering changed from 0.3 to 0.5. The results shows around 30% of decreasing moment of buckling for 3.5m span and decreases around 40% for 5.5m span. Ratio of tapering 0.3 was recommended because it can withstand buckling failure more than ratio of tapering 0.5. The opening areas are inversely proportional to the buckling moment. The rate of decrease of buckling moment is around 3% to 6%. Opening size from 0.2D to 0.4D is recommended because when opening size exceeds 0.4D, buckling moment capacity difference of only 2% is observed.

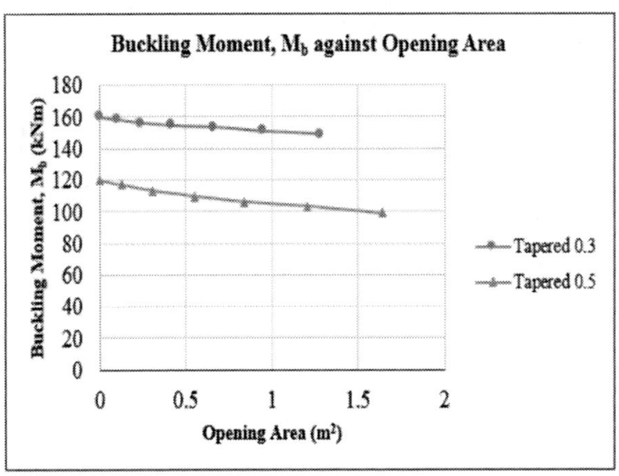

Figure 2. Graph of buckling moment, M_b against opening area for 3.5m span (effect of tapering ratio).

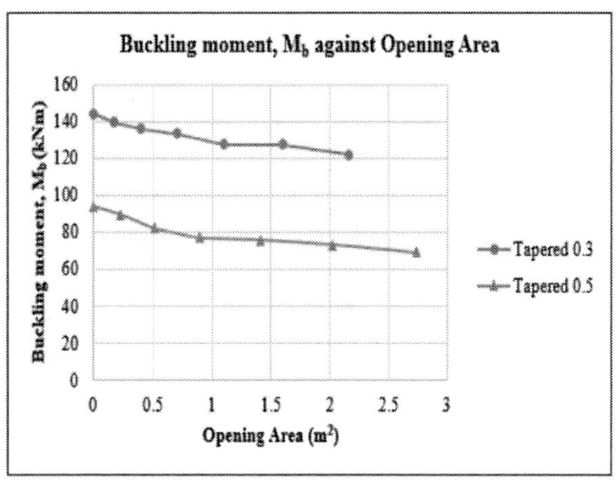

Figure 3. Graph of buckling moment, M_b against opening area for 5.5m span (effect of tapering ratio).

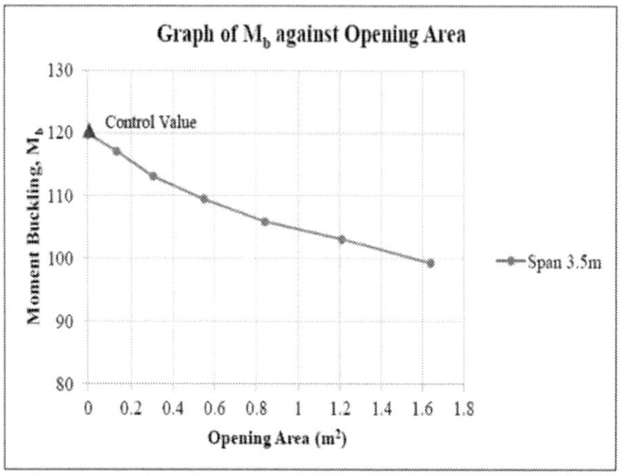

Figure 4. Graph of buckling moment, M_b against opening area for span 3.5 m.

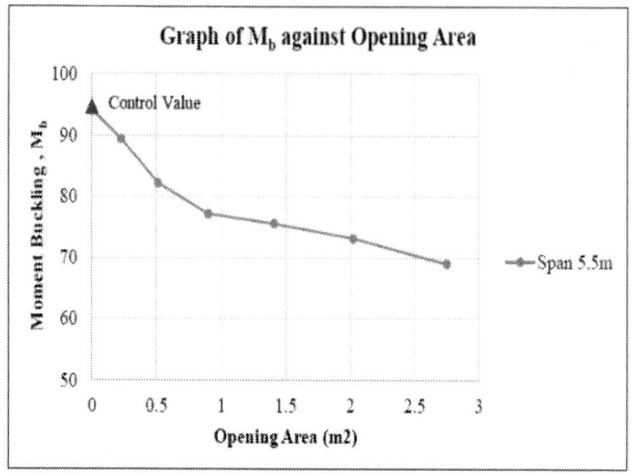

Figure 5. Graph of buckling moment, M_b against opening area for span 5.5 m.

The buckling moment capacity will increase with the decreasing number of opening and increasing thickness of web and flange of the section as shown in Figure 6 to Figure 9. Different length of span gives different results and rate of increases and decreases. Besides that, different opening arrangement was also give different buckling moment capacity. Meanwhile, Figure 10 to Figure 12 demonstrates the failure pattern of tapered steel section with perforation.

The best arrangement chosen was open-close-open (OCO) because it does not have too many or too few number of opening. It also has an average buckling moment compared to the other two arrangement which is open-open-open (OOO) and close-open-close (COC). Thickening the web simultaneously increase the depth and width. Not to mention thicker section should be considered for the optimum design in bending since the results show that the thicker the section, the higher the buckling moment capacity. The difference between non-perforated section with the recommended perforated section buckling moment is around 2% for 3.5m span and 4% for 5.5m span.

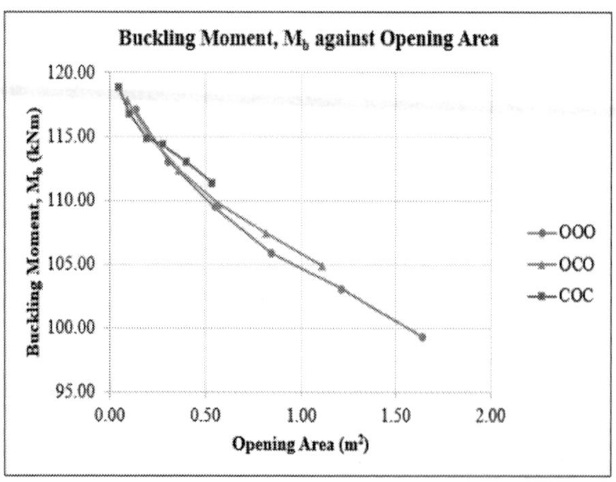

Figure 6. Graph of buckling moment, M_b against opening area for span 3.5 m span (effect of opening arrangement).

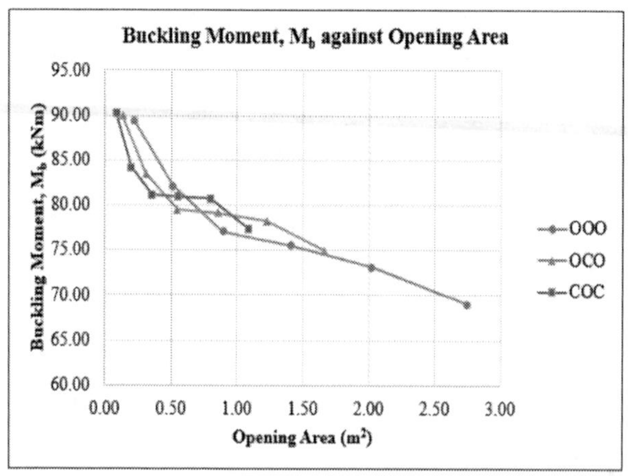

Figure 7. Graph of buckling moment, M_b against opening area for span 5.5 m span (effect of opening arrangement).

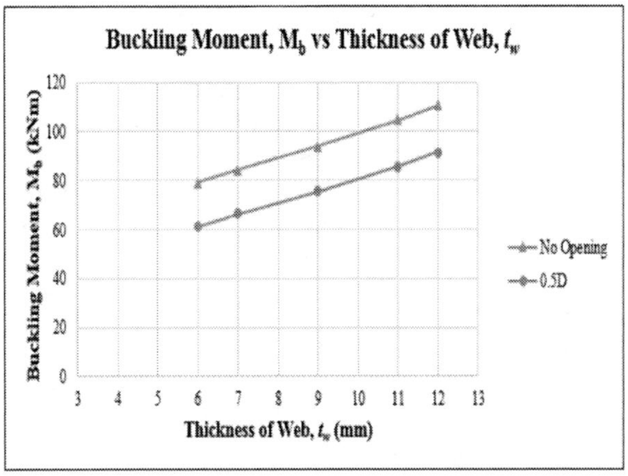

Figure 8. Graph of buckling moment, M_b against thickness of web, t_w.

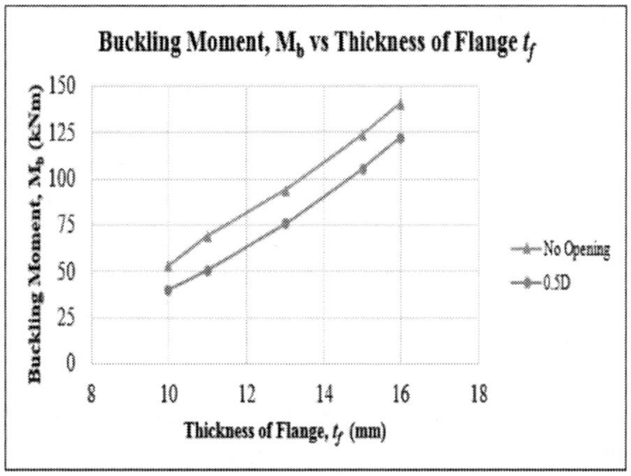

Figure 9. Graph of buckling moment, M_b against thickness of flange, t_f.

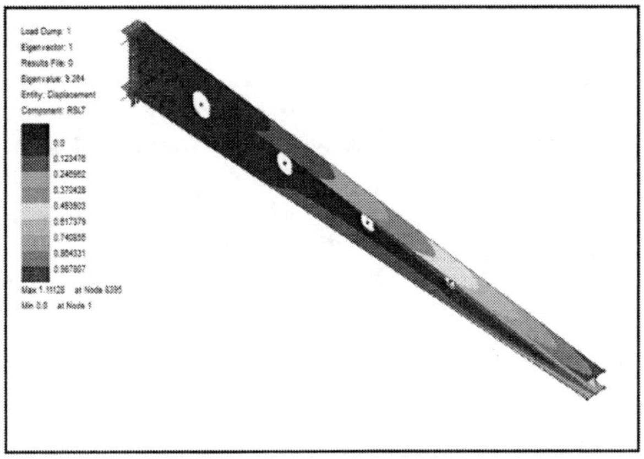

Figure 10. Contour displacement for tapered steel section with perforation with eigenvalue = 9.264.

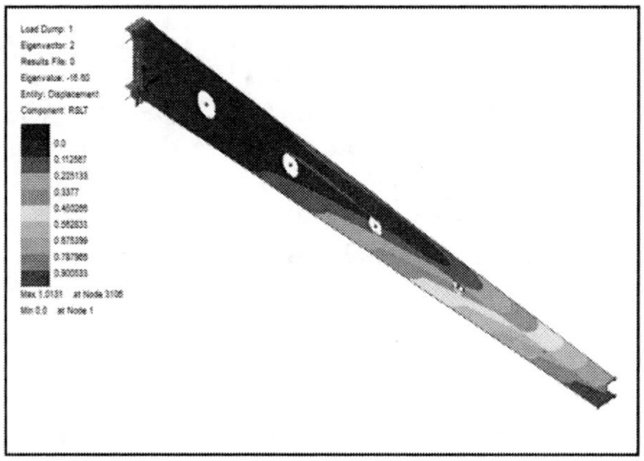

Figure 11. Contour displacement for tapered steel section with perforation with eigenvalue = 16.60.

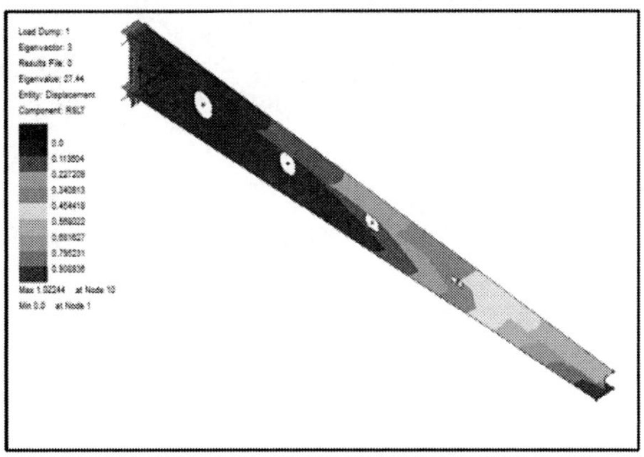

Figure 12. Contour displacement for tapered steel section with perforation with eigenvalue = 27.44.

4 Conclusion

The conclusions obtained from this study are the failure pattern for tapered steel section with and without perforation practically do not to have large dissimilarities. The buckling moment capacities of tapered steel section with perforation decreases when the ratio of tapering and opening size increases. Meanwhile, the buckling moment capacity will increase with decreasing number of opening and increasing the thickness of web and flange of the section. The buckling moment capacity for both tapered steel section with and without perforation increases for the increasing the thickness of flange and web. However, tapered steel section with perforation has lower buckling moment capacity compared to tapered steel section without perforation.

Acknowledgement

The authors wish to thank the School of Universiti Sains Malaysia for the financial support towards this research under Fundamental Research Grant Scheme (FRGS), account number: 203.PAWAM.6071239.

References

[1] B. Boley and V. Zimnoch, Lateral buckling of nonuniform beams, J. of the Aeronautical Sciences, **19**(8), 567-568, (1952).
[2] N. Trahair, Interaction buckling of tapered beams, Engineering Structures, **62**, 174-180, (2014).
[3] A. Andrade and D. Camotim, Lateral–torsional buckling of singly symmetric tapered beams: theory and applications, J. of Engineering Mechanics, **131**, 586-597, (2005).
[4] N.D. Lagaros, L.D. Psarras, M. Papadrakakis, and G. Panagiotou, Optimum design of steel structures with web openings, Engineering Structures, **30**, 2528-2537, (2008).
[5] M. Basher, N. Shanmugam and A. Khalim, Web openings in horizontally curved composite plate girders, J. of Constructional Steel Research, **65,** 1694-1704, (2009).
[6] A.Z. Abidin and B. Izzuddin, Meshless local buckling analysis of steel beams with irregular web openings, Engineering Structures, **50**, 197-206, (2013).
[7] T. Liu and K. Chung, Steel beams with large web openings of various shapes and sizes: finite element investigation, J. of Constructional Steel Research, **59**, 1159-1176, (2003).
[8] S.A. Ibrahim, Lateral torsional buckling strength of unsymmetrical plate girders with corrugated webs, Engineering Structures, **81**, 123-134, (2014).
[9] A.K. Sabat, Lateral-torsional buckling analysis of steel frames with corrugated webs, Master Thesis, Chalmers University of Technology, Sweden, (2009).
[10] J. Moon, J.W. Yi, B.H. Choi and H.E. Lee, Lateral–torsional buckling of I-girder with corrugated webs under uniform bending, Thin-Walled Structures, **47**, 21-30, (2009).

MATEC Web of Conferences **47**, 03001 (2016)
DOI: 10.1051/matecconf/20164703001

Modeling Zero – Inflated Regression of Road Accidents at Johor Federal Road F001

Joewono Prasetijo[1,a] and W. Zahidah Musa[1]

[1]*Faculty of Civil and Environmental Engineering, University Tun Hussein Onn Malaysia, 86400 Batu Pahat, Johor, Malaysia*

Abstract. This study focused on the Poisson regression with excess zero outcomes on the response variable. A generalized linear modelling technique such as Poisson regression model and Negative Binomial model was found to be insignificant in explaining and handle over dispersion which due to high amount of zeros thus Zero Inflated model was introduced to overcome the problem. The application work on the number of road accidents on F001 Jalan Jb – Air Hitam. Data on road accident were collected for five-year period from 2010 through 2014. The result from analysis show that ZINB model performed best, in terms of the comparative criteria based on the P value less than 0.05.

1 Introduction

Malaysia has experienced a remarkable period of economic expansion and growth in population, economy, industrialisation and motorisation. Since then, the increase in population and motorisation led to a consequent increase in the number of road traffic accidents [1]. According to the survey obtained by iRAP (International Road Assessment Programme) on Malaysian road networks in 2007, the casualties for traffic accident include over 6,000 fatalities in a year. Over 60% of these fatal accidents happen on rural roads. Nearly half of all fatalities occur on federal roads and over a quarter on state roads. It was estimated about 2% of Gross Domestic Product (GDP), approximately RM 9 billion, to Malaysia's economy. In addition, traffic accident has become a major social-economic problem in Malaysia, and it is considered as main cause of death. Due to road accidents, result in annual economic cost of 9 billion ringgit to the nation economy. Numbers of fatalities reach over 6,000 people per year in road accidents nationwide. The amount has been increasing since 2003 and reached 6,872 for year 2010 while 21,397 suffer injuries. Traditionally, the simple risk measures such as observed accident counts, rates, costs, and other similar methods were used to target locations with a number of accidents exceeding a chosen threshold as hot spots [2]. Unfortunately, these measures have several shortcomings as discussed in a number of studies [2]. Alternatively, statistical modelling approach relate to the traffic safety are used for identification of hazardous locations. This approach is extensively used in traffic safety studies for identifying major contributing factors to crashes and injuries, establishing proper relationships between crashes and explanatory variables, and predicting crash frequency and injury severity. A generalized Linear Modelling technique such as Poisson

[a] Corresponding author : joewono@uthm.edu.my

distribution was used and was found to be insignificant in explaining the effect of explanatory variables related to intersection geometry with accident count.

Generally Poisson regression analysis is the primary model which is use to find out the relation between independent variables and independent variable. The dependent variable must distribute as Poisson distribution. One of the assumptions of Poisson regression is the mean and variance must be equal, but actually most of the data will have the larger variance or over dispersion. The negative binomial regression is a method which can solve the over dispersion problem [3]. The problem outcome from the Poisson regression is when there are many zeros for dependent variable, the accuracy of the Poisson regression will be reduce, Zero-inflated Poisson regression (ZIP) is introduce to solve this problem [3]. The dependent variable in this study was accident data which occur to be excess zero. Thus, the zeros dependent variable depends on the other independent variables. Zeros dependent variable comes from the zero data that produce from data crashes. The motivation for the ZIP model is the model are comparable to those from fitting a NB regression model for general over-dispersion [4], the potential under-reporting of vehicle accidents especially minor injury and property damage accidents [5], and [6] proposed to identifying effective factors on crash occurrence probability and prove the efficiency of Zero-inflated regression models against traditional Poisson and Negative binomials regression models. At the end of likehood ratio test, it seems that the ZINBDR model gave better results than NBDR model [7].

2 Methodology

2.1 Poisson regression model, negative binomial model and zero inflation model

The methodology, on which this research proposal is based, firstly treats the mean accident frequency at a site as an unknown quantity. Due to a random, discrete, and positive nature of crash data, count data modelling techniques such as Poisson and negative binomial (NB) regression models are used in such cases. The Poisson regression model has been traditionally considered as the starting point in modelling crash data, with assumption of the mean of crash counts being equal to its variance (that is, equal-dispersion) [8]. In the Poisson regression model, the probability of n_i vehicle crashes occurring at a given road *section i*, $Pr(n_i|\mu_i)$, can be estimated by

$$Pr = \left(n_i|\mu_i \right) = \frac{\exp(-\mu_i)\mu_i^{n_i}}{n_i} \tag{1}$$

$$\mu_i = \exp\left(\sum_{j=1}^{M} \beta_j X_{ij} \right) \tag{2}$$

where μi is the expected number of vehicle crashes at the given section, X_j is a vector of covariates (road geometric, environmental and traffic information) and β_j is a vector of estimable regression coefficients.

Unfortunately, much of the cases, the crash data produce the variance which is greater than the mean, well known as over-dispersion. The over-dispersion is a result of extra variation in crash means across sites which can be caused by various factors such as model misspecification, omission of important covariates, and excess zero counts. In this case, applying a Poisson regression model for vehicles crash data would result in underestimation of standard error of the regression parameters. Therefore, the negative binomial model will be introduced with

$$\mu_i = \exp(\beta X_i + \varepsilon_i) \tag{3}$$

where the *exp* (ε_i) is gamma distributed with mean 1 and variance α (the *over dispersion* parameter). By taking the unobserved heterogeneity of the Poisson mean to be counted, the following variance differ from the mean is introduced.

$$Var[y_i] = E[y_i][1+\alpha E[y_i]] = E[y_i] + \alpha E[y_i]^2 \tag{4}$$

and the maximum likelihood function of the binomial regression model is shown as:

$$L(\mu_i) = \prod_i^N \frac{\Gamma(\theta + n_i)}{\Gamma(\theta . n_i)} \left[\frac{\theta}{\theta + n_i}\right]^\theta \left[\frac{\mu_i}{\theta + \mu_i}\right]^{n_i} \tag{5}$$

where θ is inverse dispersion parameter (1/α), Γ (.) is a value of gamma distribution, and N is the total number of road sections.

In some cases, excess zeros in crash data exist and considered as a result of over dispersion. In this case, the NB model cannot be used to handle the over dispersion which is due to high amount of zeros. To do this, zero-inflation (ZI) models including Zero Inflated Poisson (ZIP) and Zero Inflated Negative Binomial (ZINB) models can be alternatively used. Both the ZIP and ZINB models assume that all zeros count come from two different processes: the process generating excess zero count derived from a binary model, and the process generating non-negative counts for vehicle crashes including zero values. As description of ZIP model, let P_i as the probability of being excess zero for section *i*, and ($1 - P_i$) be the probability of crash counts derived from the Poisson distribution as it is given as

$$P(Y = y_i) = \begin{cases} P_i + (1-P_i)\theta^{\mu_i} & y_i = 0 \\ (1-P_i)\dfrac{\theta^{\mu_i} \mu_i^{y_i}}{y_i!} & y_i > 0 \end{cases} \tag{6}$$

where y is the number of vehicle crashes for section i and μ_i is the expected crash frequency in section i as a function of road section covariates, $\mu_i = exp\ (X_i\beta)$. The probability of being in the zero-crash-state, P_i, is often fitted using logistic regression model, as follows:

$$logit(P_i) = ln\left(\frac{P_i}{1-P_i}\right) = \gamma_0 + \gamma_i Z_i + \ldots + \gamma_N Z_N \tag{7}$$

where $Z = (Z_1, Z_2, \ldots, Z_N)$ is a function of some explanatory variables and $\gamma = (\gamma_1, \gamma_2, \ldots, \gamma_N)$ is the estimable coefficients.

Similar to ZIP model, the probability density function for the ZINB given mean μ_i, and dispersion parameter, α, is given by:

$$P(Y = y_i) = \begin{cases} P_i + (1-P_i)\dfrac{1}{(1+\alpha\mu_i)^{1/\alpha}} & y_i = 0 \\ (1-p)\dfrac{\Gamma\left(y_i + \frac{1}{\alpha}\right)(\alpha\mu_i)^{y_i}}{\Gamma(y_i+1)\Gamma\left(\frac{1}{\alpha}\right)(1+\alpha\mu_i)^{y_i + \frac{1}{\alpha}}} & y_i > 0 \end{cases} \tag{8}$$

The maximum likelihood method is used to estimate the parameters of the ZI models. In general, the probability of observing zero counts in ZI models is the sum of the probability of observing an excess zero in the first process and probability of observing a zero in the second process.

2.2 Study area and data collection

Study area of this research work was selected and considered from the south of Peninsular Malaysia after consulting with Malaysia road authorities such as Public Work Dept. (JKR), Malaysian Institute of Road Safety Research (MIROS), and Malaysia Royal Police. This study focused on F001 roadway which is Jalan Jb-Air Hitam. They will include those routes where high traffic volumes are present, large numbers of crashes occur and where a range of different road conditions are evident. The R software was used to analyze the data.

Table 1. A summary of some predictor variables.

Variable	Name	Variable	Name
Dependent	Accident Crashes	Independent	Median width
Independent	AADT	Independent	No. of lanes
Independent	Traffic flow	Independent	Land use
Independent	Lane width	Independent	Access density
Independent	Shoulder width	Independent	Horizontal alignment
Independent	Roadside hazard rating	Independent	Vertical alignment

3 Results and Discussion

The empirical distribution of crash counts is plotted in Figure 1. The figures indicate that a peak of extra zeros which representing crash free exists. It confirms the presence of excess zeros in road accident crashes. This pattern is typical for most type of vehicle crash data. Moreover, the ratio of the variance to mean for fatal is 4.2 indicating the crash data is over-dispersed implying the poisson model do not adequately fit the overdispersed crash data. Then as alternative we use the better regression model which is Zero-inflated poisson and Zero-inflated negative binomial regression model.

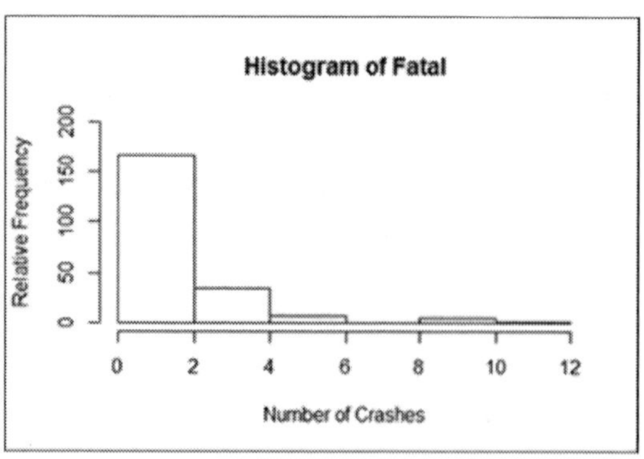

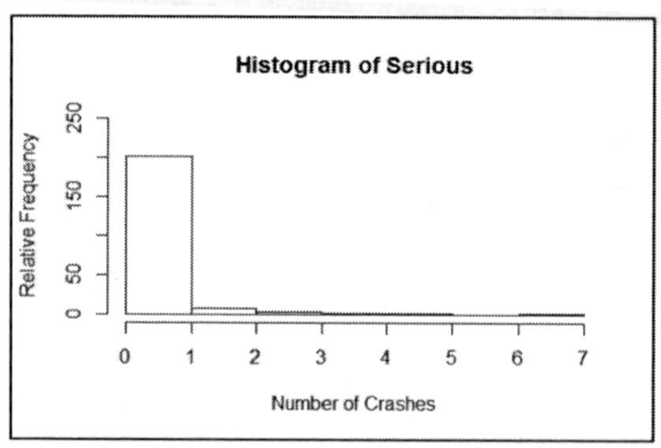

Figure 1. Histogram of observed accident crashes for F001 Jln Jb-Air Hitam during 2010-2014.

From Table 3, it shows the Poisson model coefficient. The less AIC (Akaike Information Criteria) the greater the model be more fitted. Hence from Table 3, AIC values for Poisson was 10154 which were high value compare to the AIC value for ZINB (727.2). The Poisson model assumes that the variance of the dependent variable is equal to mean but in many applications, count data were found to show the over-dispersion [9]. The Poisson regression model is fit but when consider the raw data, there are many zeros which the ZIP will be approach.

Table 2. Result for showing existence of over dispersion in crash data.

Variables	Standard deviation [SD]	Variance	Mean	Over dispersion= [var>mean]
Fatal	2.057139	4.23181917	1.526	Over dispersion
Serious	0.8486408	0.72019126	0.3116	Over dispersion
Minor	3.475564	12.079547924	1.558	Over dispersion
Wreckage	125.4524	15738.29633704	70.24	Over dispersion

Table 3. Poisson model coefficient.

	Poisson			
Pearson residuals:				
Min				
1Q		-13.2799		
Median		-5.8067		
3Q		-0.6374		
Max		5.1534		
AIC		9.4189		
		10154		
	Estimate	**Std. Error**	**Z value**	**Pr(>|z|)**
Intercept	4.904e+00	3.244e-02	151.160	< 2e-16
Fatal	-3.382e-02	3.685e-03	-9.178	< 2e-16
Serious	-7.509e-03	8.825e-03	-0.851	0.394826
Speed	-3.742e-03	5.447e-04	-6.870	6.41e-12
Volume	2.015e-04	5.288e-05	3.810	0.000139

Table 4. Zero inflated Poisson model coefficients.

	ZIP				ZINB			
Pearson residuals:								
Min		-1.1967				-0.8444		
1Q		-0.8433				-0.7490		
Median		-0.4249				-0.3374		
3Q		0.4513				0.3559		
Max		7.0431				5.9528		
Log-like hood:								
		-382.6				-359.6		
	Estimate	Std. Error	Z value	Pr(>|z|)	Estimate	Std. Error	Z value	Pr(>|z|)
Intercept	0.068389	1.988734	0.034	0.973	6.68847	0.00	0.00	0.00
Speed	- 0.024876	0.017027	- 1.461	0.144	-0.03952	0.00	0.00	0.00
Volume	0.001118	0.001710	0.654	0.513	0.01719	0.04404	0.39	0.696
SW	- 0.381369	0.655635	- 0.582	0.561	-11.47283	0.00	0.00	0.00
LW	0.271939	0.5223228	0.521	0.603	-0.34308	0.00	0.00	0.00

The absolute value of V is less than a threshold value such as 1.96 for 0.95 confidence level. From Table 5 the result for vuong was -1.93 which indicated that model 2 (ZINB) was preferred. Vuong test can indicate in any way if zero–inflated model is an improvement over standard a Negative Binomial regression [10]. It can compare between two model regressions. The vuong tests compare the zero-inflated model with an ordinary Negative Binomial model. In this analysis we can see that our test statistic is significant, indicating that zero-inflated model is superior to the standard Negative Binomial model. The less AIC (Akaike Information Criteria) the greater the model be more fitted. Hence in this analysis, Zero inflated Negative binomial was the most fitted model between the others due to the value of the AIC was 727.2. While for Zero inflated Poisson was 773.2, Negative Binomial was 734.16 and lastly Poisson was 10154. Good P-value for each independent variable was when P value was less than 0.05. The result from Table 4 show that most of the p value for ZINB was less than 0.05 which is 0.00.

Table 5. Vuong non-nested hypothesis Test-statistic.

	Vuong z-statistic	H_A	p-value
Raw	0.8474188	model1 > model2	0.198381
AIC-corrected	-1.9370687	model2 > model1	0.026368
BIC-corrected	-6.6298185	model2 > model1	1.6805e-11

4 Conclusion

This study is focusing on the Poisson regression and negative binomial model with many zero outcomes on the response variable. The zero-inflated Poisson regression model is more effective for many zero outcomes than Poisson regression. While, the zero-inflated negative binomial regression model is more effective for many zero outcomes than negative binomial regression. The application is work on the number of road accidents on federal road in the Johor, south of Malaysia.The zero-inflated negative binomial regression was used to find the relationship between dependent and independent variables when there are many zeros value in the dependent variable, where the relationship is the mixture between Negative binomial model and logistic model. This paper aimed to focus on developing a model which can properly handle stochastic nature of crash event. It also implies to find out the most relevant contributory factors at the hazardous sections and type of vehicle crashes. Lastly, this research can develop a statistical approach to estimate the safety performance of a road network in Johor federal roads. There are many further work to be done in this research (as the study is still ongoing) such as comparing the ZIP and ZINB model in terms of nested models, full probit and logit models, examine the coefficients of variables in different models and examine the diagnostic statistics of models to make it best performed to incorporate the models into traffic management systems.

References

[1] U. Radin, Updates of road safety status in Malaysia, IATSS Research, **29**, 78-80, (2005).
[2] K. Geurts and G. Wets, Black spot analysis method: literature review, **7**, 7-13, (2003).
[3] C. Pudprommarat, M. Khamkong and P. Bookkamana, Zero-inflated Poisson regression in road accidents on major road in the north of Thailand, *IRCMSA Proc.*, 323-330, (2005).
[4] A.H. Lee, M.R. Stevenson, K.Wang and K.K.W.Yau, Modeling young driver motor vehicle crashes: data with extra zeros, Accident Analysis and Prevention, **34,** 515-521, (2002).
[5] S.P. Miaou, The relationship between truck accidents and geometric design of road sections: Poisson versus negative binomial regressions, Accident Analysis and Prevention, **26,** 471-482, (1994).
[6] E. Ayati and E. Abbasi, Modeling accidents on mashad urban highways, Open J. of Safety Science and Technology, **4**, 22-35, (2014).
[7] M. Asrul and N.N. Naing, Analysis death rate of age model with excess zeros using zero-inflated negative binomial and negative binomial death rate: mortality AIDS co-infection patients, Kelantan, Malaysia, Procedia Economics and Finance, **2**, 275-283, (2012).

[8] M.H. Pour, J. Prasetijo, A.S. Yahaya and S.M.R. Ghadiri, Modeling vehicle-pedestrian crashes with excess zero along Malaysia federal roads, Procedia Social and Behavioral Sciences, 1218-1227, (2012).

[9] J. Garber, Stochastic Models Relating Crash Probabilities with geometric and corresponding traffic characteristics data, Research Report No. UVACTS-5-15-74, (2001).

[10] Q.H. Vuong, Likehood ratio test for model selection and non-nested hypotheses, Econometric, **57**, 307-333, (1989).

MATEC Web of Conferences **47**, 03002 (2016)
DOI: 10.1051/matecconf/20164703002

Structural Equation Modelling in Behavioral Intention to Use Safety Helmet Reminder System

Naida Rosli[1,a], Kamarudin Ambak[1], Basil David Daniel[1] and Mohd Erwan Sanik[1]

[1]*Smart Driving Research Centre, Universiti Tun Hussein Onn Malaysia, 86400 Parit Raja, Johor, Malaysia*

Abstract. Motorcycle is one of private transportation which has been widely used in many countries including Malaysia. However, motorcycles are the most dangerous form of motorized transport. Royal Malaysian Police (PDRM) statistics recorded that motorcycle is the highest vehicle (45.9%) involved in traffic accident compared to other vehicles. The potential cause of the death to the motorcyclist was due to the head injury. One of strategy to mitigate this problem is through proper usage of safety helmet. Therefore, this paper was introduce a new approach on motorcyclist safety by using the Technology Acceptance Model (TAM) with additional determinants that contribute to behavioral intention and to increase the proper usage of safety helmets among Malaysian motorcyclists. The Structural Equation Modelling (SEM) was used to test the structural TAM proposed. The evaluation for structural model showed the goodness of fit indices are excellent fit. This study found that perceived ease of use, perceived usefulness and social norm are significant towards behavioral intention to use Safety Helmet Reminder System (SHR).

1 Introduction

Currently, many people use motorcycles for their daily activities such as going to work or school. This is due to the fact that motorcycles are compact, agile, consume less fuel and affordable to buy and maintain than automobiles [1].Therefore, motorcycle is one of private transportation which has been widely used in many countries including Malaysia. Besides that, motorcycle is the dominating transport mode as a motorization is growing fast in Asian [2].

Despite the fact that motorcycles have more attraction and advantages than other vehicles do, they are still riskier than other transportation. Motorcycles are the most dangerous form of motorized transport [3]. Majority of road accidents fatalities involve motorcyclist, constituting more than 50% of the total number of fatalities [2]. Ambak et al. [4] also found that motorcycle is known as to be vulnerable road users in term safety-risk exposure and instability compared to other vehicles. Based on statistical report of road accidents in Malaysia 2012 by Royal Malaysia Police (PDRM) [5] shows that motorcycle is the highest vehicle (45.9%) involved in a traffic accident compared to car (30.1%), lorry (10.3%), bus (1.59%), van (1.93%) and others (10.35%).

In addition, it is reported that the potential cause of death to the motorcyclists was due to head injury. Pang et al. [6] highlighted that 62.9% of vital organ injury causing deaths among the motorcyclists are due to head injuries. Futhermore, Abdul Manan and Varhelyi [2] stated that injury

[a] Corresponding author : kamardin@uthm.edu.my

caused in motorcycle fatalities is mainly to the head. Therefore, Ambak et al. [4] suggested that one of strategy to mitigate this problem is through proper usage of safety helmet. Motorcycle helmet is the best shielding equipment that is possible to wear while riding a motorcycle, being most effective means of protection offered to motorcyclists [7]. However, Abdul Manan and Varhelyi [2] reported that 75% or motorcyclist involved in road accidents wore helmets. Jaafar et al. [8] which states that motorcyclists who wears safety helmets are the highest numbers of fatalities from the overall road accidents in Malaysia. There are several possible reasons why motorcyclists with their helmets on are still suffering head injuries, (1) Motorcyclists do not wear them properly either the helmets' sizes are too large or they are not fastened properly (2) use unstandardized motorcycle helmets (3) Motorcyclists wear helmets that have been used for too long (4) riding motorcycle in high speed. Jaafar [9] indicated that in motorcycle accidents, 23.3% involving the helmet coming off during the impact that causing 25.2% of motorcyclist suffering severe head injuries. The properly safety helmet use can reduce injuries by 25% [9].

Therefore, the aim of this study is to introduce a new approach on motorcyclist safety by using the Technology Acceptance Model (TAM) with additional determinants that contribute to behavioral intention and to increase the proper usage of safety helmets among Malaysian motorcyclists.

2 Reference Model and Hypothesis

2.1 Technology acceptance model

Technology Acceptance model was widely used in previous study to examine user perception and acceptance towards new system or technology. The Technology Acceptance Model was developed by Davis (1986). This model is grounded in both Theory Reasoned Action (TRA) and Theory Planned Behavior (TPB) [10,11]. The factor in the model, namely perceived usefulness (PU), perceived ease of use (PEOU) and attitudes toward usage (ATU), represents characteristic of the system [10]. While, the behavioral intention to use is an important factor that determines whether users will actually utilize the system [12].

Perceived Usefulness refer to the degree to which an individual believes that using a particular system would enhance his or her job performances" [10]. While peceived ease of use indicated that the degree to which an individual believes that using a particular system would be free of physical and mental effort" [10].

There have a two additional variables in this study to determine the behavioral intention towards using safety helmet reminder system. There are social influence and perceived safety. Social influence is defined as the degree which an individual believes that other people whose opinion is important to himself/herself think the same way about using a new system [13]. While perceived safety is defined as the degree to which an individual believes that using a system will affect his or her well-being [13]

2.2 Hypothesis

The construct in TAM was used to know the relationship between perceived usefulness, perceived ease of use, behavioral intention, social influence and perceived safety against exogenoues variable which is (behavioral intention to use safety helmet reminder system) that can affect the helmet use among motorcyclist. Therefore, the hypothesis was test to know the significance of this constructs. H1: Perceived usefulness significant positive impact on behavioral intention to use SHR among motorcyclist. H2: Perceived ease of use significant positive impact on perceived usefulness for using SHR among motorcyclist. H3: Behavioral intention to use SHR significant impact on helmet use among motorcyclist. H4: Social norm significant positive impact on behavioral intention to use a SHR among motorcyclist. H5: Social norm significant positive impact on perceived usefulness for using the SHR among motorcyclist. H6: Perceived safety significant positive impact on behavioral intention to use a SHR among motorcyclist.

3 Methodology

3.1 Instrument and measurement

This study utilized the survey questionnaire form as the main data instrument. The questionnaire was designed based on Technology Acceptance Model (TAM) and adapted from a few previous literature. The questionnaire was divided into 5 sections. A seven likert point scale was used to examine the agreement of respondents towards Safety Helmet Reminder System (SHR).

3. 2 Data collection

The quantitative data were collected by self-administered questionnaire form that distribute face to face among motorcyclists in Parit Raja, Batu Pahat, Johor. Only voluntarily respondents were asked to answer the questionnaire form based on their perception towards Safety Helmet Reminder System (SHR). Data collection took place from first May until end of July 2013.

3.3 Sampling method and sample size

This study used the stratified random sampling method approach to obtain the sample of respondents. The population was chosen among motorcyclist in Johor area and the strata in Parit Raja, Johor. The estimation of population in this research was taken from the total of the number motorcycle registered in Johor state. To determine the sample size for this study, the Formula of Cochran (1977) was used because this formula is suitable to determine the sample size with having the data categorical and continuous data. The total minimum of sample size required from the Cochran's formula calculation is 118. However, this study chosen 300 respondents as a sample size. This is because Weston and Goore [14] has stated that minimum sample size required for Structural Equation Modelling is 200. Then, the questionnaire were processed with the aid of SPSS version 22 while for the Structural Equation Modelling AMOS software was used to test the structural model.

4 Results and Discussion

4.1 Descriptive analysis

Table 1 present the descriptive statistics based on 300 respondents among motorcyclist. Most of the participant's aged varied between 21 to 29 years old (40.0%) with majority of them was male (72.3%) compared to female (27.7%). This have been indicated in previous study that singles and young males are the highest group in of motorcyclist in Malaysia [11]. We also noted that most of them attained education from higher institution (59.7%) and work in private sector (41.7%) with monthly income RM3001 and above (33.3%). Besides that, most of respondents possess full driving licenses (74.3%) and have riding experience 10 years and above (63.3%). Regarding experience with road accident, majority of respondents have an experience with road accident (69.3%) with frequency involvement 3 times and more (36.7%). The data clearly indicated that majority of respondents possessed full face helmet (283%). Ambak et al. [15] study also indicate that most of Malaysian motorcyclist owned full-shell helmets. While Ambak et al. [16] stated that majority of motorcyclist in Malaysia possessed helmets that have safety features and meet the standard requirements set by the authorities.

4.2 Reliability and correlation analysis

The reliability test was performed for all items selected in the proposed model by using Alpha Cronbach coefficient. Alpha Cronbach coefficient between .65 to .95 is considered satisfactory [17].

Table 1. Characteristic of the respondents (N=300).

Demographic	Frequency	(%)
Age (Years)		
20 and below	32	10.7
21 to 29	120	40.0
30 to 39	86	28.7
40 to 49	44	14.7
50 and above	18	6.0
Gender		
Male	217	72.3
Female	83	27.7
Marital Status		
Single	133	44.3
Married	167	55.7
Education Level		
Primary School	20	6.7
Secondary School	101	33.7
Higher Institution	179	59.7
Working Sector		
Government	50	16.7
Private	125	41.7
Self-Employed	65	21.7
Jobless	60	20.0
Monthly Income		
RM 1000 and below	12	4.0
RM 1001- RM 2000	72	24.0
RM 2001-RM 3000	55	18.3
RM 3001 and above	100	33.3
Not relevant	61	20.3
Type Of Motorcycle Licence		
Full Licence	223	74.3
Probation Licence (P)	30	10.0
Learning Licence (L)	1	0.3
No Licence	46	15.3
Riding Experience (Years)		
Less than 2 years	8	2.7
2-5 years	59	19.7
6-9 years	43	14.3
10 years and above	190	63.3
Experience Respondent Towards Road accident		
Yes	208	69.3
No	92	30.7
Frequency of Respondents Experiencing Road Accidents.		
1 times	52	17.3
2 times	46	15.3
3 times and more	110	36.7
Type of Helmet possesed		
Full Face	8	2.7
Full Shell	283	94.3
Half Shell	9	3.0

The result obtained is .795. Therefore, it is indicated that the data is reliable and valid enough. Table 2 shows the correlation between variables is correlated and significant for a few variables.

4.3 Structural equation modelling

The result of confirmatory factor analysis (CFA) for proposed model [χ^2 (N=300, df= 132) = 329.519, p<.05] is insignificant indicated that the model proposed is good fit for the data. The proposed model (TAM) showed that the goodness of fit indices NFI (.921), GFI (.901), IFI (.951), TLI (.936) and CF1 (.951) are greater than .90. While RMSEA value is .071. Chua [14] indicated that the goodness of fit indices must be greater than .90 while RMSEA value must lower than <1.0 for the excellent fit. These result verify the reliability of the model structure to predict behavioral intention to use SHR towards helmet use among motorcyclist. The Square multiple correlation (SMC) showed the highest variation percentage (72%) which indicate that the amount of construct variables was able to be explained in the model [11].

4.4 Hypothesis testing

Based on the analysis result that presented in figure 1, the full structural model was tested using the SEM. Perceived Usefulness (β=.50, p<0.001) and Social Norm (β=.20, p<0.001) are positively associated with behavioral intention to use SHR system. Thereby supporting H1 and H4. In addition, Social Norm (β=.20, p<0.001) is significant towards Perceived Usefulness and supporting H5. The path analysis between perceived usefulness and Perceived Ease of Use (β=.86, p<0.001) shows that the stronger and positive effect between these two variables. Therefore, H2 are accepted.

Table 2. Correlation between variables.

	PS	SN	PU	PEOU	BI	HU
PS	1					
SN	.426**	1				
PU	.633**	-.002	1			
PEOU	.610**	-.147*	.665**	1		
BI	.637**	.161**	.629**	.586**	1	
HU	.-0.86	.353**	.032	-.066	.059	1

PS: Perceived Safety, SN: Social Norm, PU: Perceived Usefulness,
PEOU: Perceived Ease of Use, BI: Behavioral Intention,
HU: Helmet Use. **Correlation significant at 0.01 level (2-tailed)

5 Conclusions

Safety helmet is the best protective and most effective for motorcyclist during riding. However, the function of safety helmet is meaningful if the motorcyclist use their safety helmet in a proper way. Therefore, a novel solution, Safety Helmet Reminder System (SHR) was introduced to mitigate this problem. Based on the result obtained, perceived usefulness, perceived ease of use and social norm are significant towards behavioral intention to use Safety helmet reminder system (SHR). This indicated that, most of respondent believes that SHR can enhances their performance and make them free from effort. Besides that, they believes that other people opinion is important for them to use the SHR system. However, based on the result obtained, perceived safety are insignificant towards behavioral intention to use SHR system which indicates that perceived safety does not affected towards behavioral intention to use SHR system among motorcyclist.

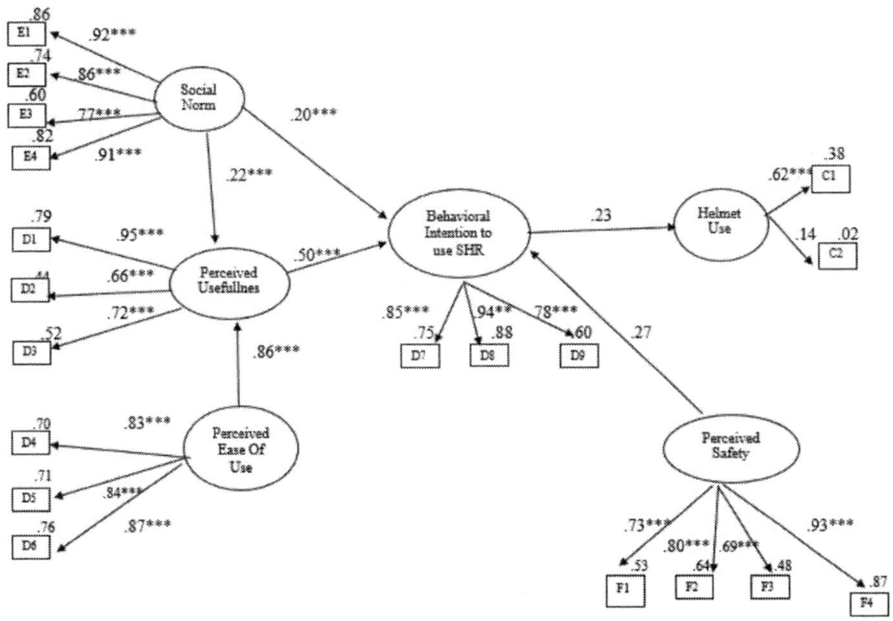

Figure 1. TAM model proposed to predict behavioral intention to use SHR (***p<0.001).

References

[1] S. Mc. Inally, A model for motorcycle rider skill development, L. Dorn (Ed), *Driver Behavior and Trainning*, Ashgate Publishing, Cranfield University, 35-43, (2003).

[2] M.M.A. Manan and A. Vrhelyi, Motorcycle fatalities in Malaysia, IATSS Research, **36**, 30-39, (2012).

[3] M.A. Akaateba, A.R. Gyimah and I. Yakubu, A cross-sectional observational study of helmet use among motorcyclists in Wa. Ghana, Accident Analysis and Prevention, **64**, 18-22, (2014).

[4] K. Ambak, A. Riza Atiq and I. Rozmi, Intelligent transport system for motorcycle safety and issues, European J. of Scientific Research, **28** (4), 600-611, (2009).

[5] Polis Diraja Malaysia, Laporan Perangkaan Malaysia 2010, Malaysia, (2012).

[6] T.Y. Pang, R.S. Radin Umar, A.A. Azhar, M.T. Mohd Nasir and H.H Hamdan, Injury characteristic of motorcyclist by Abbreaviated Injury Scale (AIS), Research Report 2/2000, National Road Safety Council Malaysia, Kuala Lumpur, (2000).

[7] F.A.O. Fernandez and R.J. Alves De Sousa, Motorcycle helmets- A state of the art review, Accident Analysis and Prevention, **56**, 1-21, (2013).

[8] T.R. Jaafar, M.F. Mustafa, S. kemin and R. Kasiran, Kemalangan jalan raya: Analisis data membabitkan pengguna motosikal, Jurnal Teknologi, **38**(B), 1-14, (2003).

[9] S.P. Baker, B. O'Neill, M.J. Ginsburg and L. Guohua, *The Injury Fact Book*, Oxford University Press, New York, (1992).

[10] F.D. Davis, Perceived usefulness, perceived ease of use, and user acceptance of information technology, MIS Quarterly, **13**, 319-340, (1989).

[11] K. Ambak, R. Ismail, R.A. Abdullah, A.A. Latif and M.E. Sanik, Application of technology acceptance model in predicting behavioral intention to use safety helmetreminder system, Research J. of Applied Science, Engineering and Technology, **5**(3), 881-888, (2013).

[12] R.H. Shroff, C.C. Deneen and E.M.W. Ng, Analysis of the technology acceptance model in examining students' behavioral intention to use an e-portfolio sytem. Autralasian, J. of Education Technology, **27**(4), 600-618, (2011).

[13] S. Osswald, D. Wurhofer and S. Trosterer, Predicting information technology usage in the car: Towards a car technology aceptance model, Automotive U1'12, Portsmouth, (2012).

[14] R. Weston and P.A. Jr. Gore, A brief guide to structural equation modelling, Couns. Psychol, **34**(5), 719-751, (2006).
[15] K. Ambak, R. Ismail, R.A. Abdullah, M.N. Borhan, Using structural equation modelling and the behavioral sciences theories in predicting helmet use, *Proceeding Of the Int. Conference on Andvances Sciences, Engineering and Information Technology*, Putrajaya, 639-645, (2011).
[16] K. Ambak, N. Rosli, B.D. Daniel and J. Prasetijo, Behavioral intention to use safety helmet reminder system using the extended technology acceptance model, *The 3rd Annual Int. Conference on Architecture and Civil Engineering*, Singapore, (2015).
[17] Y. P. Chua, *Kaedah dan Statistik, Uian Regresi, Analisis Faktor dan Analisis SEM*, McGraw Hill, Malaysia, (2014).

MATEC Web of Conferences **47**, 03003 (2016)
DOI: 10.1051/matecconf/20164703003

Gender Differences in Pedestrian Perception and Satisfaction on the Walkability of Kuala Lumpur City Center

Ahmad Kamil Arshad[1,a], Noor Iza Bahari[2], Wardati Hashim[2] and A.G. Abdul Halim[3]

[1]*Institute for Infrastructure Engineering and Sustainability Management, Universiti Teknologi Mara, 40450 Shah Alam, Malaysia*
[2]*Faculty of Civil Engineering, Universiti Teknologi Mara, 40450 Shah Alam, Malaysia*
[3]*Faculty of Civil and Environmental Engineering, Universiti Tun Hussein Onn Malaysia, 86400 Parit Raja, Johor, Malaysia*

Abstract. The quality of built environment usually influences the walkability of a city. This is because each pedestrian walk differently on different type of facilities or built environment provided to them. This paper aims to investigate whether gender differences influences the pedestrian perception and satisfaction level at three sidewalks located within the Kuala Lumpur City Center by means of questionnaire survey. A total of 317 pedestrians were involved in the questionnaire survey at the three different sidewalk locations. The result shows significant differences in mean satisfaction value between male and female pedestrians. Female pedestrian give lower satisfaction value for overall travel experience at Jalan Tuanku Abdul Rahman but higher satisfaction value at Petaling Street and Bukit Bintang compared to male pedestrian. Overall, the satisfaction level was rated between acceptable to satisfactory regardless of gender differences. Improvements should be made for the safety parameter because both male and female pedestrian gives lowest satisfaction level in that area.

1 Introduction

Walkability can be defined as to what extend does the build environment or the surrounding environment could encourage and support walking [1]. Walkability is usually linked to the quality of the built environment and connected by the quality of the pedestrian environment. The condition of the sidewalk plays an important role in increasing the level of walkability of the built environment. The sidewalk path context and quality of sidewalk network are adequate predictors to walkability that has the ability to affect the walking likelihood and increased walkability in a city [1-4]. Lack of walkable elements within the sidewalk can negatively affect pedestrian walking behavior [5]. Safe walking environment will also influenced pedestrian activities positively [6]. The tendency of people to walk more often and farther increased in an environment with high sidewalk qualities [2, 7].

Walking does not only give benefits in terms of personal health but also give benefits to the country through economic development in commercial areas with high pedestrian movement. A walkable city could act as an alternative solution in reducing traffic congestion with low impact towards the environment [8]. Due to these reasons, the Malaysia Government (through Dewan

[a] Corresponding author : drahmadkamil@salam.uitm.edu.my

Bandaraya Kuala Lumpur) and the private sector have cooperated to construct and upgrade the existing pedestrian network in Kuala Lumpur. This long term development includes a total of 45KM length of covered (including elevated) pedestrian network throughout Kuala Lumpur city [9]. In order to promote and facilitate pedestrianisation, enhancement projects such as sidewalk widening, provision of safer road crossing, landscape treatment and high kerb obstacle removal were initiated [10].

However, pedestrian perception and preference on the sidewalk facilities during their walking journey may also be closely related to socio-demographic characteristic and personal factors such as gender, age and transportation mode choice [11]. These characteristics differences give influence on the pedestrian travel behavior, thus influencing their perception and travel decision [12]. In order to create a desirable walking environment that promotes walking, pedestrian are the most appropriate group to be surveyed as they are the daily user of the facilities. This paper attempts to investigate whether gender differences influences the pedestrian perception and satisfaction level for sidewalks located within the Kuala Lumpur City Centre. The findings could assists the planer, local authorities and facilities provider in providing optimum pedestrian facilities that consider local pedestrian travel preferences and characteristics rather than using existing rules of thumb and research findings from other countries with different cultural background and characteristics.

2 Methodology

The major steps involved for this study include study area selection, paper-based on-street questionnaire survey and data analysis. Three study areas located within Kuala Lumpur City Center that have highest pedestrian volume of 100 pedestrian/hour or higher (determined by extrapolation of 15 minutes manual count) were selected for this study. Questionnaire surveys were then distributed to 100 or more respondents at each location to assess the pedestrian perception and satisfaction level during their journey at each sidewalk. Random sampling method was used in distributing the questionnaire survey to increase the generalisability [13]. The collected data were transformed from qualitative data to quantitative data during data analysis by giving weighted rating number to each level of satisfaction (qualitative data) given by the respondent to facilitate the understanding of the data obtained during the survey.

2.1 Study area

The Kuala Lumpur Structure Plan 2020 has identified City Center Commercial (CCC) zone that is located within the Kuala Lumpur City Centre. CCC is a commercial zone that promotes a broad range of commercial activities to be conducted within walking distance, this provide the highest potential in supporting Kuala Lumpur's economic growth [10]. CCC within Kuala Lumpur City Centre was divided into three different districts namely Tuanku Abdul Rahman (TAR), Golden Triangle and Old City Centre. Three locations were selected within each CCC district which is Jalan Tuanku Abdul Rahman (TAR), Jalan Bukit Bintang and Petaling Street as shown in Figure 1 and Figure 2. The facilities available in these areas include public transport system, shopping complexes, education centres and business centres within walking distance, thus offering a broad range of pedestrian activities including shopping, leisure, working and commuting transit.

2.2 On-street questionnaire survey

Paper-based on-street questionnaire survey method was carried out on each sidewalk. The survey was carried out during afternoon peak hour on working weekdays. Random sampling technique was used in selecting respondent. A total of at least 100 pedestrians using the sidewalk facilities were participated in the survey for each sidewalk location. The parameters surveyed include land use and accessibility, safety, path facilities / amenities, aesthetic / pleasurability and overall travel experience.

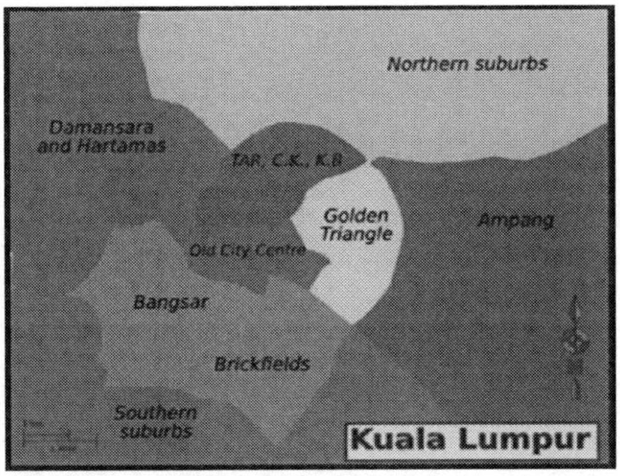

Figure 1. Kuala Lumpur District [14].

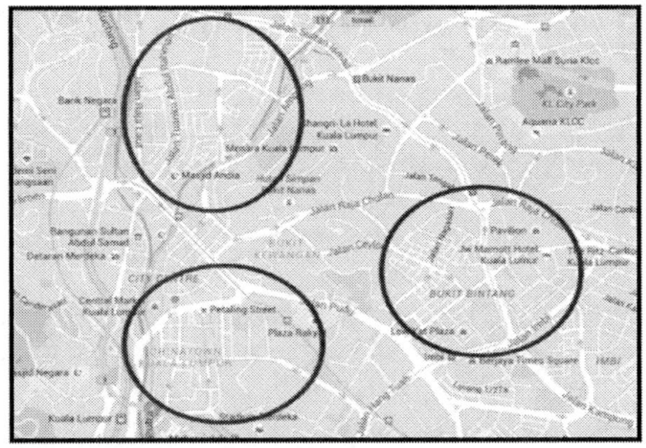

Figure 2. Selected study area in Kuala Lumpur [15].

2.3 Data analysis

To facilitate data analysis, the qualitative data collected was transformed into quantitative data using simplified weighted factor analysis method. Total Weighted Score (TWS) and Average Weighted Score (AWS) are calculated using Equation (1) and Equation (2) respectively. There are 10 scale of agreement involved in counting the weighted score as shown in Table 1. Weighted factor calculation example is also shown in Table 1. By setting the score to zero (0) as the borderline or acceptable satisfaction level, any positive (+ve) value calculated indicates that the respondents are satisfied, whereas negative (-ve) value calculated indicates that the respondents are dissatisfied with the existing sidewalk condition. The mean satisfaction rating was then determined using the same Likert scale of agreement in Table 1 to assess the pedestrian satisfaction level according to gender differences. Lastly, statistical analysis that involves reliability test was performed on the data set using SPSS Software to check the validity of the result.

$$\text{Total Weighted Score, TWS} = \sum (WF \times N)_1 + (WF \times N)_2 + \ldots + (WF \times N)_n \quad (1)$$

$$\text{Average Weighted Score, AWS} = \text{(Total Weighted Score)/(Number of Respondents, N)} \quad (2)$$

where: WF = Weighing Factor; N = Number of Respondents; n = Scale of Agreement; TWS = Total Weighted Score; AWS = Average Weighted Score.

Table 1. Weighing factor and weighted score calculation example.

Level of Satisfaction	Scale of Agreement, n	Weighing Factor ,WF	No. of Respondent, N	Weighted Score, WS
Not satisfied at all	1	-5	2	-10
Very dissatisfied	2	-4	4	-16
Moderately dissatisfied	3	-3	0	0
Slightly dissatisfied	4	-2	6	-12
Somewhat dissatisfied	5	-1	3	-3
Somewhat satisfied	6	1	6	6
Slightly satisfied	7	2	5	10
Moderately satisfied	8	3	0	0
Very Satisfied	9	4	3	12
Completely satisfied	10	5	1	5
Total Weighted Score			30	-8
Average Weighted Score				-0.267

3 Results and Discussion

The results are presented in two different sections. The first section used simplified average weighted factor analysis to assess pedestrian perception on the existing sidewalk condition. The second section used SPSS software to calculate mean satisfaction level according to gender.

3.1 Pedestrians' perception on the existing sidewalk condition

A total of 317 pedestrians were involved in the questionnaire survey at the three different sidewalk locations. Table 2 shows the results for pedestrian travel experience. Jalan TAR obtained the highest score (AWS =1.3379) which is higher than the average for all areas surveyed (AWS = 1.0387). This is followed by Bukit Bintang (AWS = 1.0612) and Petaling Street which was rated the lowest by the pedestrians (AWS = 0.6788). The AWS for other parameters on pedestrian perception is shown in Figure 3. Based on the overall results, the pedestrian rated the existing sidewalks in the range between somewhat satisfied to slightly satisfied based on the Likert Scale as shown in Table 1. However, all the pedestrians surveyed agreed that the safety parameter is slightly acceptable, as the score approached zero for all sidewalks especially at Petaling Street (0.259 AWS value).

Table 2. Average weighted score for overall travel experience by road.

Criteria	Jalan TAR		Bukit Bintang		Petaling Street		TOTAL	
	TWS	AWS	TWS	AWS	TWS	AWS	TWS	AWS
Overall Travel Experience	172	1.6863	158	1.3621	61	0.6559	**391**	**1.2572**
Pathway / Sidewalk Condition	144	1.3981	114	0.9828	64	0.7033	**322**	**1.0387**
Crossing	137	1.3301	115	1.0000	59	0.6413	**311**	**1.0032**
Sidewalk Amenities	127	1.2451	110	0.9565	60	0.6452	**297**	**0.9581**
Personal Safety	103	1.0000	39	0.3362	18	0.1957	**160**	**0.5145**
Adjacent Traffic/Driver's behavior	92	0.9020	62	0.5391	52	0.5591	**206**	**0.6645**
Aesthetics and Amenities	165	1.6176	192	1.6696	78	0.8478	**435**	**1.4078**
Accessibility	157	1.5243	189	1.6435	110	1.1828	**456**	**1.4662**
Mean AWS		1.3379		1.0612		0.6788		**1.0387**

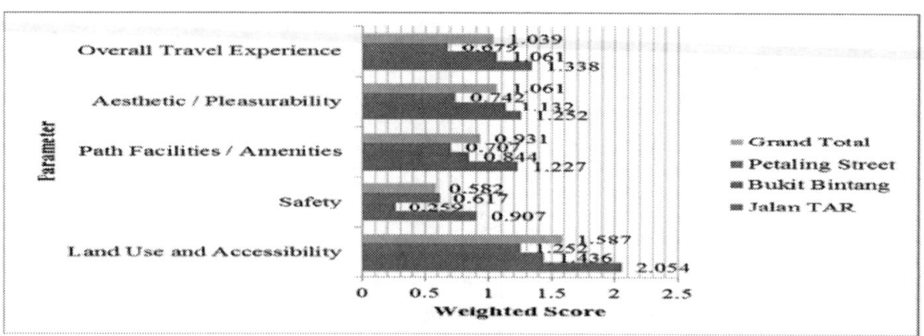

Figure 3. Mean of average weighted score for each parameter.

3.2 Pedestrian satisfaction level by gender difference

From the total of 317 respondents, 45% were male respondent and 55% were female respondent. The gender split for each sidewalk location is shown in Fig. 4. There is significance difference that can be observed in the male and female satisfaction level. Female pedestrian gave almost similar mean satisfaction level value for all sidewalks with Bukit Bintang receiving the highest value (6.53), followed by Petaling Street (6.36) and Jalan Tar (6.31). On the other hand, male pedestrian rated Jalan TAR as the most satisfied sidewalk and Petaling Street as the least satisfied with high difference in mean satisfaction value of 7.04 and 5.72 respectively. However, Bukit Bintang obtained almost similar mean satisfaction value for both male and female respondents. Both male and female pedestrian agreed that all sidewalks in the study location were lacking in terms of safety which can be seen by the lowest mean value of 5.89 and 5.98 obtained in this area for male and female pedestrians respectively. Overall, male pedestrian satisfaction level of sidewalk arrangement were in line or similar to the results calculated using average weighted score for overall travel experience where Jalan TAR is the most satisfied, followed by Bukit Bintang and Petaling Street.

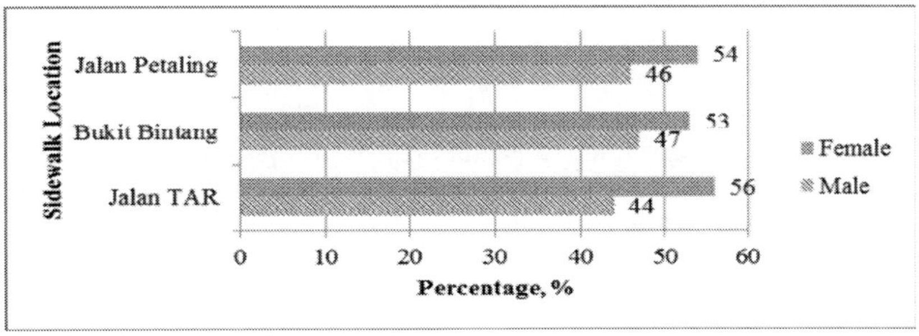

Figure 4. Gender split.

Table 3. Pedestrian mean satisfaction level by gender and road.

Criteria	Jalan TAR		Bukit Bintang		Petaling Street		Total All	
	Male	Female	Male	Female	Male	Female	Male	Female
Overall Travel Experience	7.24	6.70	6.50	6.76	5.67	6.40	6.49	6.63
Pathway / Sidewalk Condition	7.02	6.40	6.19	6.42	5.83	6.26	6.35	6.36
Crossing	7.09	6.28	6.02	6.56	5.70	6.35	6.26	6.40
Sidewalk Amenities	6.93	6.23	6.17	6.42	5.81	6.20	6.30	6.29
Personal Safety	6.71	6.05	5.63	5.95	5.35	5.94	5.89	5.98
Adjacent Traffic/Driver's behavior	6.91	5.79	5.85	6.10	5.70	6.18	6.14	6.02
Aesthetics and Amenities	7.11	6.66	6.70	7.11	5.86	6.54	6.58	6.79
Accessibility	7.31	6.38	6.77	6.94	5.81	7.04	6.65	6.78
Average	**7.04**	**6.31**	**6.23**	**6.53**	**5.72**	**6.36**	**6.33**	**6.41**

3.3 Statistical analysis

The reliability test was carried out using SPSS Software to calculate Cronbach's Alpha values. Table 4 shows that Cronbach's Alpha values obtained were between 0.812 and 0.954. The high value of Cronbach's Alpha coefficient indicates greater internal consistency of the item on the scale [16] and that the result is valid for analysis [17].

Table 4. Statistical analysis using reliability test.

No	Section	Reliability Test
		Cronbach's Alpha
1	Land Use and Accessibility	0.894
2	Safety	0.928
3	Path Facilities / Amenities	0.880
4	Aesthetic / Pleasurability	0.812
5	Overall Travel Experience	0.954

4 Conclusion

The respondents rated the existing sidewalk between acceptable to slightly satisfied for all sidewalk locations since there was no negative AWS value obtained for all the parameters. The safety parameter obtained the lowest AWS value for all sidewalks. Female pedestrian gave almost similar rating for all sidewalks. Male pedestrian gave same rating arrangement with AWS analysis arrangement. Improvements should be made for the safety parameter because both male and female pedestrian gives lowest satisfaction level in that area. The safety areas to be improved include safety while using the sidewalk, adjacent traffic, crossing safety, personal safety against crime and stray animals, safety while using the sidewalk during night and rainy days.

Acknowledgement

The author would like to express gratitude to Faculty of Civil Engineering, UiTM and to all volunteers in completing the survey for this study. This research is funded by Ministry of Science, Technology and Innovation (MOSTI) under eScience Fund.

References

[1] M. Southworth, Designing the walkable city, J. Urban Plan. Develop., **131**(4), 246-257, (2005).
[2] F. Jaskiewicz, Pedestrian level of service based on trip quality, Transportation Research Circular E-C019, National Research Council, Washington, (2000).
[3] K.R. Rahaman, J.M. Lourenco and J.M. Viegas, Perceptions of pedestrians and shopkeepers in European medium-sized cities: Study of Guimaraes, Portugal J. Urban Plan. Develop., **138**(1), 26-34, (2012).
[4] S. Lotfi and J. Koohsari, Neighborhood walkability in a city within a developing country, J. of Urban Plan. Develop., **137**(4), 402-408, (2011).
[5] S. Shamsuddin, N.R. Abu Hassan and S.F.I. Bilyamin, Walkable environment in increasing the liveability of a city, Procedia - Social and Behavioral Sciences, **50**, 167 – 178, (2012).
[6] J. Zakaria and N. Ujang, Comfort of walking in the city center of Kuala Lumpur, Procedia - Social and Behavioral Sciences, **170**, 642–652, (2015).
[7] C.E. Kelly, M.R. Tight, F.C. Hodgson and M.W. Page, A comparison of three methods for assessing the walkability of the pedestrian environment, J. Trans. Geo., **19**(6), 1500-1508, (2011).
[8] N.I. Bahari, A.K. Arshad and Z. Yahya, Assessing the pedestrians' perception of the sidewalk facilities based on pedestrian travel purpose, *IEEE 9th International Colloquium on Signal Processing and its Applications*, (2013).
[9] Information on http://app.kwpkb.gov.my/greaterklkv/entrypoint-project-pedestrian
[10] Information on http://www.dbkl.gov.my/pskl2020/ english/ urban_ design_and_landscape
[11] P.P. Koh and Y.D. Wong, Comparing pedestrians' needs and behaviours in different land use environments, J. Trans. Geo., **26**, 43-50, (2013).
[12] M.G. Boarnet and S. Sarmiento, Can land-use policy really affect travel behaviour?: A study of the link between non-work travel and land-use characteristics, Urban Studies, 35(7), 1155 – 1169, (1998).
[13] S.M. Tseng, Correlations between external knowledge and the knowledge chain as impacting service quality, Journal of Retailing and Consumer Services, **19**, 429-437, (2012).
[14] Information on https://en.wikivoyage.org/wiki/Kuala_Lumpur
[15] Information on https://www.google.com.my/maps/place/Kuala+Lumpur
[16] J.A Gliem, R.R. Gliem, Calculating, interpreting, and reporting cronbach's alpha reliability coefficient for likert-type scales, Midwest Research to Practice Conference in Adult, Continuing, and Community Education, **1992**, 82–88, (2003).
[17] D. George and P. Mallery, *SPSS for Windows Step by Step*, Allyn and Bacon, Boston, (2003).

MATEC Web of Conferences **47**, 03004 (2016)
DOI: 10.1051/matecconf/20164703004

Road Accident Analysis: A Case Study of Federal Route FT024 Yong Peng- Parit Sulong

Mohd Idrus Mohd Masirin[1], Walid A. Al-Bargi[1,a] and Joewono Prasetijo[1] and Basil David Daniel[1]

[1] *Faculty of Civil and Environmental Engineering, University Tun Hussein Onn Malaysia, 86400 Batu Pahat, Johor, Malaysia*

Abstract. Traffic accidents are considered as an unplanned and unfortunate event which is a serious concern to the community as well as the authority. An accident-counter measure can reduce the rate of road accidents by initially identifying critical locations. The total of road accidents along FT 024 between 2009 and 2012 is 907 cases. Road accidents during the same period range between 24 % and 26 % each year. These accidents killed 34 people and injuring another 101 people. This research aims to identify factors that may contribute to the cause of accidents and to study the effects of speed, volume and road geometry on road accidents. In this study, the locations labelled as *km 1, km 2, km 4, km 8, km 5* and *km 14* of Federal Route FT 024 Yong Peng – Parit Sulong were selected as the study-case sections based on Accident Point Weightage of ranking, in which each location has different road characteristics. Speed study was carried out at selected road sections to evaluate the influence of speed upon road accidents; and traffic volume count was conducted at the same selected road sections to determine the existing condition of the route. Besides, road geometry observations and measurements were also conducted at selected sections, they were also studied to evaluate influence of road designs upon road accidents. The extracted data were analyzed by using regression analysis on different variables to evaluate the relationship between accident Weightage point and other dependent variables that were considered to have considerable effects upon road accidents such as mean speed, volume, shoulder width, lane width and access point. P value below 0.05 was considered as statistically significant. After conducting data analysis, this study showed that the number of road accidents increases with the increment of speed and access point. On the other hand, volume has no strong relationship to road accidents which means that it may not have an effect on accidents. Moreover, based on the assumptions made by the researchers and project supervisors, road safety index development is important for the selected road sections along FT 024 to improve the road conditions in terms of safety and also to propose improvements at the selected locations in order to reduce accident rate.

1 Introduction

Road traffic accident is defined as "An accident that as accident which took place on the road between two or more objects; resulted in one or more persons being died or injured, and at least one moving vehicle was involved [1, 2]. Thus, RTA is collisions between road vehicles; between automobiles and pedestrians; between road vehicles and animals; or between road vehicles and geographical or

[a] Corresponding author : walidalbargi@gmail.com

architectural obstacles. Road traffic accidents (RTAs) have turned out to be a huge worldwide public health problem killing nearly 1.2 m people yearly and injuring about 4 times this number throughout the same period of time [3]. Almost 85% of all road accidents that have caused loss of lives in the world occur in developing countries [4]. Amongst the developed countries such as Britain, Japan and the United States of America, traffic related death rate (per 100,000 citizen per year) has reached 12.3, 3.85 and 3.59 respectively in the year 2009 [5, 6, 7].

One of the issues discussed on traffic in Malaysian highways is road accidents. The past thirty four years have seen a rapid increment of road crashes in Malaysia. Amidst a large volume of traffic on the road, accidents involving different kinds of vehicles inevitably happen. Globally, the number of people killed in road traffic crashes yearly is estimated at nearly 1.2 million. To top it all, the number of injuries could be as high as 50 million – a combined population of five large cities in the world [8].

Many research studies that incorporated different aspects of road traffic accidents have been performed by various workers at different durations across the world [9-16]. according to the official statistic of Federal Route 024; FT024 (Yong Peng-Muar/Yong Peng-Parit Sulong) experienced 1,033, 2,764 and 925 road accidents between the years 2007 to May 2010 respectively; killing 169 people and injuring 539 people accordingly. This study will be conducted to investigate and refresh the issues that cause traffic accidents along FT024. Till date there are a limited number of studies that have been conducted on Federal Route F024 Yong Peng – Parit Sulong. Therefore, realizing the need to create a baseline info on RTAs, the current research was conducted on the aforesaid reasons. This study will also be concentrating on the relationships between road condition, traffic speed, flow and accident rates to help in identifying hazardous locations, which needs treatments as well as a ranking on the black spot locations along FT024 Yong Peng – Parit Sulong.

2 Analysis of Accident Data

The data for this research were collected from Batu Pahat Traffic Police Station. The accident records were collected from the year 2009 to 2012 (4 Years). The purpose of the preliminary analysis of the traffic accident data is to determine the breadth of accident problem. At this stage, the general trends and statistics of accidents were established as shown in the Figure 1. The total number of road accidents occurred along FT024 between 2009 and 2012 is 907 cases. Figure 1 shows the highest number observed was in 2011 with 242 cases, followed by the year 2012 and 2010 with 226 and 220 accident cases respectively. The lowest number of accidents is in the year 2009 with 219 cases with a total of 34 persons lost their lives. From Figure 2, the year 2009 recorded the highest number of deceased road users (n=12), the second highest is in the year 2011 and 2010 with 11 and 6 deceased road users respectively. Generally, it could be concluded that there was a slight increase in the number of road accidents along Federal Route F024 Yong Peng – Parit Sulong from 2009 to 2011 and then, it dropped in 2012.

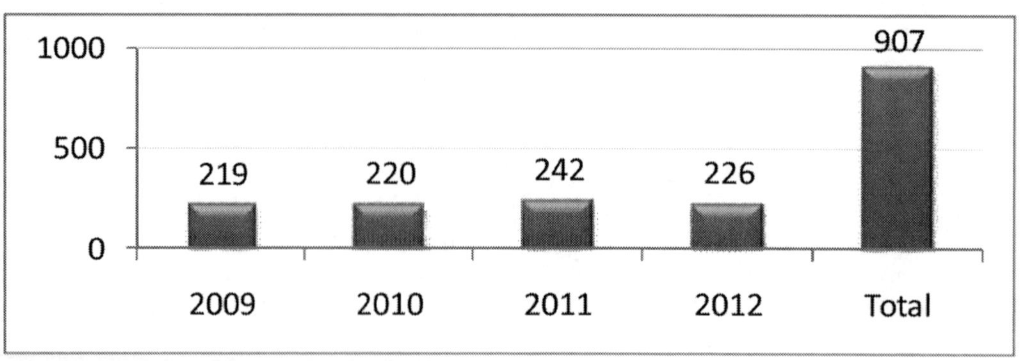

Figure 1. Accident frequency at FT 024, 2009-2012.

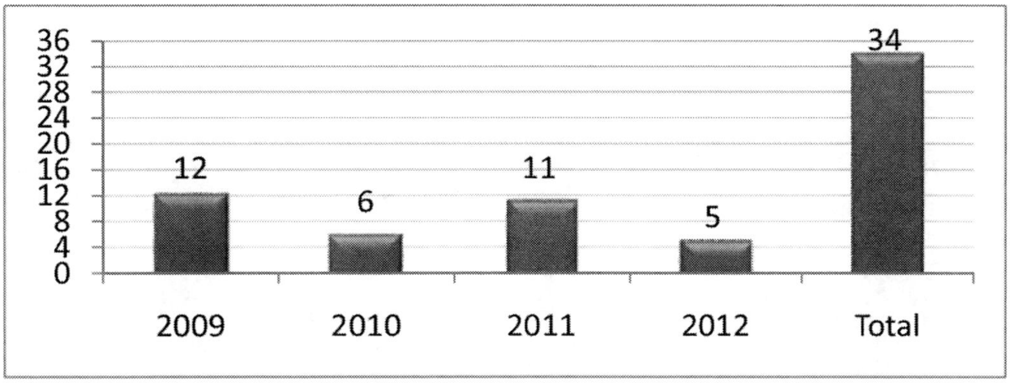

Figure 2. Fatal accidents at FT024, 2009-2012.

2.1 Percentage of road accidents according to seriousness

Figure 3 shows the breakdown of road accidents between 2009 and 2012 according to the seriousness of the accidents. Most accidents resulted in only slight damages, which accounts for 85.12% of the total number of accidents; that equals to 772 cases. Accidents that have caused serious injuries accounts for 0.33%, whilst those that have caused fatalities and non-serious injuries accounts for 3.75 % and 10.8 % respectively

2.2 Yearly road accidents frequency according to seriousness

From Figure 4, there is a huge difference in frequencies between slight-damage-only accidents and the other categories. Generally, fatal and serious injuries accidents reflect a dropping trend in terms of frequencies between 2009 and 2012. Moreover, there is a dramatic increase in the non-serious-injury cases in terms of frequencies. This is due to the increased frequency of non-serious-injury-only accidents from 12 to 41 cases between 2009 and 2012. Due to the increasing trend of non-serious-injury-only accidents, the overall frequency of all road accidents has increased between 2009 and 2011 and later, dropped in 2012.

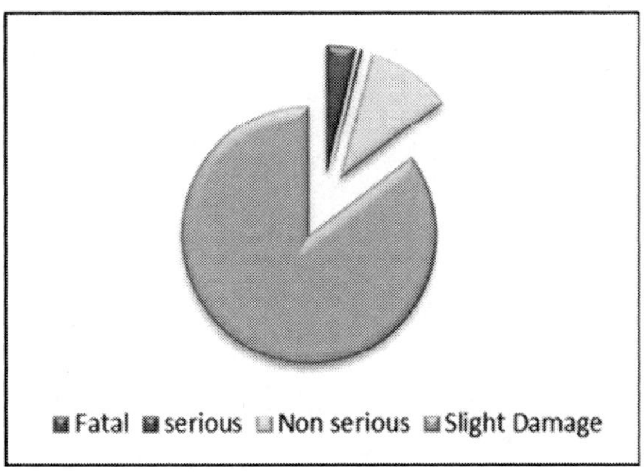

Figure 3. Percentage of road accidents according to seriousness at FT 024.

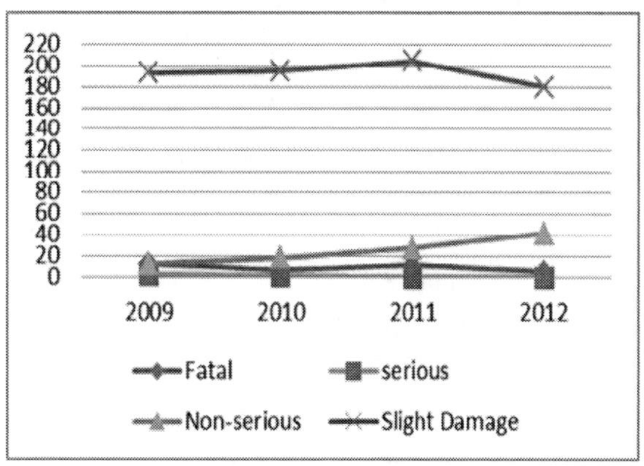

Figure 4. Breakdown of road accidents according to seriousness at FT 024 by year.

2.3 Ranking of the top ten accidents section

Table 1 shows the top ten worst accident sections at F024. The analysis was based on the data compiled over four years starting from 2009 to 2012 by using the accident-point weightage (APW) formula. Km 5 recorded the highest APW with 43.2 and the second highest was Km 2 with 39.6. These were followed by km 8 with 34 APW while the lowest was at km 9 and km 10 with 18 and 17.8 APW respectively. Generally, it can be concluded that Km 5 has highest number of fatalities.

Table 1. Analyses for Top Ten Accidents Ranking by Weightage (APW) at FT024, 2009-2012.

Sections	Type of road accident				Total	APW	rank
	Fatal	Serious	non-Serious	Sligth Damage			
KM 5	6	0	0	36	42	43.2	1
KM 2	2	0	9	102	113	39.6	2
KM 8	2	0	9	74	85	34	3
KM 14	3	0	5	33	41	28.6	4
KM 9	3	0	2	25	30	24.6	5
KM 4	2	0	2	53	57	24.2	6
KM 11	2	0	5	26	33	21.2	7
KM 7	1	1	5	31	38	19.2	8
KM 26	0	1	7	47	55	18	9
KM 13	2	0	1	25	28	17.8	10

2.4 Relationship amongst variables

Based on the accident point weightage (APW) calculated, and referring to the correlation coefficient obtained amongst the variables, P value below 0.05 was considered as statistically significant between APW and other variables. Figure 5 and 6 show that mean speed rate has the highest correlation with APW; 0.79, which means as the speed increases the APW increases this is followed by the access point; 0.77.

Regarding lane width, the negative sign in the correlation coefficient shows that the width of the lane relates to APW inversely as shown in Figure 7. This means that when lane width increases the APW reduces. Figure 8 shows that volume also has no strong coefficient with APW, this means it may not have directly affect the accident point weightage.

	APW	*Mean Speed*	*Volume*	*Lane width*	*Shoulder width*	*Access Point*
APW	1					
Mean Speed	0.793547162	1				
Volume	0.425012185	-0.080000331	1			
Lane width	-0.616954895	-0.966015654	0.227573899	1		
Shoulder width	-0.251779414	0.010374278	0.19034486	-0.187148979	1	
Access Point	0.775493866	0.374366415	0.891714163	-0.207514339	0.1124203	1

Figure 5. Correlation amongst variables

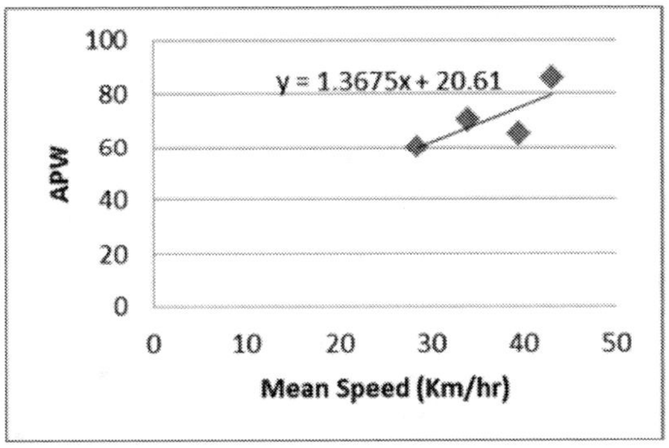

Figure 6. Relationship between APW and Mean Speed

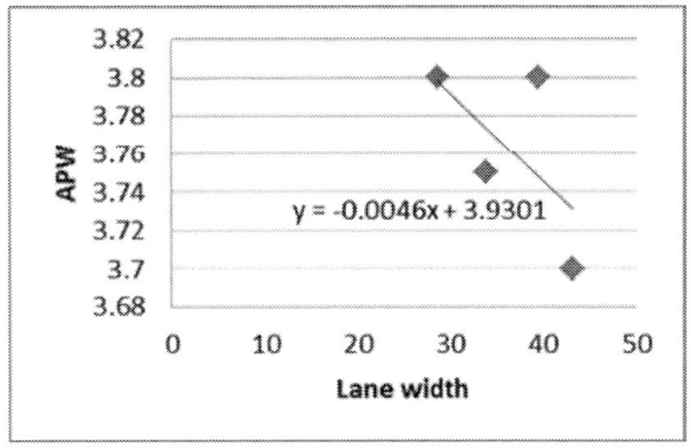

Figure 7. Relationship between APW and Lane width

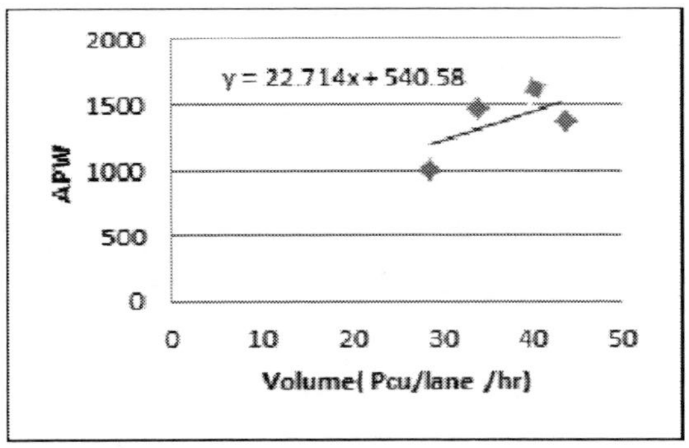

Figure 8. Relationship between APW and volume.

From the result of this study, it is noted that the factors which contribute to accidents at Federal Route 024 are: Vehicle speed and Number of access points. The effect of each contributory factor on the number of road accidents are as follows: (i) Speed increase contributes to accident increases, and (ii) A greater number of access points per length increases accidents.

3 Road Safety Development Index

The index covers all relevant parameters of road safety in selected sections at FT024 including infrastructure, which researchers and project supervisors have developed as a method to classify the Road Safety Index (RSDI) based on the observed data and analyzed parameters. These data were then given a range based on the assumptions made by the researchers and project supervisors. The following Table 2 demonstrates the ranges and assumptions for the Road Safety Index in accordance with "A Guide Geometric Design of Road "(REAM Manual, 2002)

Table 2. Index in relation to road conditions

No	RSDI range	Remarks for Section observation
1	Less than 20%	Inadequate
2	20% to 40%	Poor
3	40% to 60%	Average
4	60% to 80%	Good
5	More than 80%	Excellent

Speed study showed that the mean speed at Km 1, Km 2 and Km 8 were 60.04 , 61.18 and 70.21 respectively while Km 4 recorded a speed of 76.21 km/hr. Apart from that, as outlined in Arahan Teknik Jalan JKR (JKR Road Technical Specifications) the total capacity for a two-lane tow way is 2800 pcu/hr/lane. However, from the result gained in the study, the maximum peak hour at Km 1, Km 2, Km 4 and Km 8 hauled a traffic of 1851 pcu/lane/hr, 1788 pcu/lane /hr, 1441 pcu/lane/hr and 1477 pcu/lane/hr respectively.

Accordingly, referring to Table 2, we can evaluate the condition of the selected kilometers in their more specific aspects, as shown in Table 3. The road sections were labelled as poor, average and excellent in terms of speed, volume, lane width and shoulder width respectively. Based on these results, we can take the necessary measures to reduce speeding at the areas. Regarding the road shoulder in this area, it may need widening or redesigning in terms of their geometry

Table 3. Road safety development index.

Sections	**Rating parameters**			
	Speed	**Volume**	**Lane Width**	**Shoulder Width**
Km1	Poor	Average	Excellent	Average
Km2	Poor	Average	Excellent	Poor
Km4	Average	Good	Excellent	Good
Km 8	Poor	Good	Good	Good

4 Conclusions

This research is successful in achieving its aim to study the effects of speed, volume and road geometry on road accidents along Federal Route 24 (FT 024). Reviews of previous studies have revealed that most, if not all past researches have only involved a few sections of a location. This research, however, has established a relationship between road accidents with traffic volume and speed along selected sections of the route from Yong Peng to Parit Sulong.

In conclusion, this research has found that the peak hours and the normal hours of the sections mentioned above display differences in terms of traffic volume and speed. At peak hours the sections recorded a higher traffic volume compared to the normal hours due to the road activities; except at the Plaza Toll which recorded a higher traffic volume during peak hours. The summary of the study results shows that there is no direct relationship between accident rate and traffic volume. The initial results show that speed and access point have positive relationships with road accidents. The results have shown that the increase of speed and access point would also trigger an increase in the Frequency of road accidents.

Besides, the Road Safety index as discussed in this research has been found to be very effective as a tool to evaluate the condition of each road section in terms of road safety. This will help to improve on the results of the analysis, particularly for road geometry in future. It is hoped that further research to improve the existing Road Safety Index will evaluate better safety measures for road users.

References

[1] M.I.M. Masirin and M. Idris, Sustainable transportation system: Utilisation of energy resources, International Recognition Seminar, Unversiti Tun Hussein Onn Malaysia, (2013).

[2] N. Jha, D.K. Srinivasa, G. Roy and S. Jagdish, Epidemiological study of road traffic accident cases: A study from South India, Indian J. Community Med., **29**, 20-24 (2004).

[3] World Health Organization, *World Report on Road Traffic Injury Prevention: Summary*, Geneva, (2004).

[4] DA. Sleet and CM. Branche, Road safety: A new public health priority, Safety Science Monitor, **1**, 1-3, (2004).

[5] http://www.japantoday.com/category/national/view/traffic-accident-deaths-hit-57-yearlow-below-5000-in-2009

[6] IRTAD Database, Risk Indicators, OECD International Traffic Safety Data and Analysis, (2009). <http://www.internationaltransportforum.org/irtad/pdf/risk.pdf.>

[7] http://www.dft.gov.uk/pgr/statistics/datatablespublications/accidents/casualtiesmr/rrcgbmainresults2009

[8] L. Abdullah, Road accident models with two threshold levels of fuzzy linear regression, Journal of Emerging Trends in Computing and Information Sciences, (2012).

[9] P.L.L. Shrinivas, Studies undertaken to identify critical causes of accidents in the highways of Tamil Nadu, Indian Highways, **31**, 11-22, (2014).

[10] L. Sarangi, R.K. Parhi, Parida and P. Panda, A study on epidemiological factors associated with road traffic accidents presenting to the casualty of a private hospital in Bhubaneswar, J. of Community Medicine, **5**(2), (2009).

[11] N. Jha and C.S. Agrawal, Epidemiological study of road traffic accident cases: A study from Eastern Nepal, Regional Health Forum, **8**(1), 15–22, (2004).

[12] Q.S. Hossain, S.K. Adhikary, W.H.W. Ibrahim and R.B. Rezaur, Road traffic accident situation in Khulna City, Bangladesh, *Proceedings of the Eastern Asia Society for Transportation Studies*, **5**, 65–74, (2005).

[13] A. Persson, Road traffic accidents in Ethiopia: Magnitude, causes and possible interventions, Advances in Transportation Studies, **15**, 5-16, (2008).

[14] W. Odero, Road traffic accidents in Kenya: An epidemiological appraisal, East African J., **72**(5), 299–305, (1995).

[15] F. Mustakim, I. Rahman, B. David Daniel and K. Ambak, Accident investigation, blackspot treatment and modelling of federal route F050 Malaysia, *The 4th International Conference on Construction In The 21st Century*, Gold Coast, (2007).

[16] K. Ambak, K. Diman and B. David Daniel, A study on compliance of safety belt usage among car drivers in the district of Batu Pahat, *National Seminar on Civil Engineering Research*, Skudai, (2007).

MATEC Web of Conferences **47**, 03005 (2016)
DOI: 10.1051/matecconf/20164703005

Train Dwell Time Models for Rail Passenger Service

Hor Peay San[1,a] and Mohd Idrus Mohd Masirin[1]

[1]*Faculty of Civil and Environmental Engineering, University Tun Hussein Onn Malaysia, 86400 Parit Raja, Johor, Malaysia*

Abstract. In recent years, more studies had been conducted about train dwell time as it is a key parameter of rail system performance and reliability. This paper draws an overview of train dwell time models for rail passenger service from various continents, namely Asia, North America, Europe and Australia. The factors affecting train dwell time are identified and analysed across some rail network operators. The dwell time models developed by various researches are also discussed and reviewed. Finally, the contributions from the outcomes of these models are briefly addressed. In conclusion, this paper suggests that there is a need to further study the factors with strong influence upon dwell time to improve the quality of the train services.

1 Introduction

Rail transport is growing continuously and is becoming the preferred mode for land public transport in both developing and developed nations due to its sustainability [1]. As the facilities for railway transport continue to expand and improve, this has become a pulling force for the public to use the railway transport as the preferred choice for public transport. However, the quality of service, particularly its reliability is also a determining factor in the public's choice of mode. Train punctuality is a very important aspect of the reliability of the service. The perceived utility of the system will definitely decrease if trains are late or cancelled very often [2]. According to Puong [3], dwell time is a key parameter of system performance, service reliability and quality in any mode of public transportation. Over the years, more researches had been conducted about train dwell time. The following highlights the importance of dwell time studies: i) Dwell time studies gives insight into the travel time and headway variations and can produce effective timetables; ii) With known parameters affecting dwell time, efforts can be made at critical stations to reduce dwell time.

2 Factors Affecting Dwell Time

Train dwell time is the time a train stands at the platform usually for the purpose of allowing passengers to board or alight [4]. It is devoted to the loading and unloading processes of the train, along with door opening and closing processes. Thus, boarding and alighting at stations are likely the most significant factors causing dwell time variations.

[a] Corresponding author : hf130073@siswa.uthm.edu.my

Kraft, in his studies on street transit systems highlighted seven main categories of factors affecting dwell time: human, modal, operating policies, operating practices, mobility, climate and other system elements [5]. Each of these factors was further subdivided into selected areas for analysis and where possible, the effects of each of the factors on dwell time were quantified by the author. In a more recent report, Douglas categorised the factors affecting dwell time into five categories, which are passenger volume, passenger profile, train design, station design and operational factors [4].

Passenger volume, which refers to the number of passengers alighting and boarding a train, is considered as the main factor influencing its dwell time. The dwell time is determined by the number of passengers alighting and boarding and the speed at which they alight and board. Mixed flow of passengers tends to lengthen the time of dwell time as compared to uni-directional flow. Based on Weston's model, 35 s is needed for 40 passengers per door to alight or board (uni-directional) but the dwell time increases to 40 s for a mixed flow of 20 alighting and 20 boarding passengers [6]. Similarly, Douglas' model showed that a mix flow of 30 alighting and 30 boarding passengers requires 40 s compared to 60 alighting or boarding passengers which require between 25 to 35 s [4]. A high number of standing through-passengers is also likely to slow down the speed of boarding and alighting. Puong highlighted that for when there are 5 and above standing through-passengers (per door) on board, the marginal boarding time will increase significantly [3]. Douglas estimated that dwell time will increase by 9 s when there are 30 alighting and 30 boarding passengers with 20 standing through-passengers (per door) on board [4]. Passenger profile also has an effect on dwell time. Passengers with strollers or luggage and those who require special assistance (visually impaired and on wheelchairs) would extend the process of boarding and alighting [4] but it is uncertain as to how much the dwell time is affected. Regular passengers may alight slightly faster than those who are irregular and unfamiliar with the stations [4] but this is difficult to quantify as the distinction between regular and irregular travellers is not very obvious.

Train design also influences boarding and alighting times. The number of doors and their width will affect the alighting and boarding speed. Wider doors enable more passengers to board and alight, thus increase the rate of boarding and alighting process, though it may take slightly longer for the doors to open. Wider doors lead to 10% shorter times and narrower doors to 10% longer times than the mean value [7]. For trains with a minimum door width of about 80 to 90 cm, increasing the door width only results in a moderate increase in passenger flow rate [8]. This is because the passengers do not usually utilise the whole door width, especially if a queue is formed. A wider door only transfers the problem of congestion further into the train as narrow sections further into the train gradually becomes limiting factors for the passenger flow rate. The vertical distance between the platform of the station and the floor of the train heavily influences the dwell time. Heinz found out that there is no great difference between level entrance and 2-steps, particularly for passengers without any luggage but for passengers carrying luggage or with some form of impairment, the difference became apparent. For trains with 3 steps, the boarding and alighting times increased for all categories of passengers studied. He also found out that for horizontal distances, gaps of less than 5 cm, did not affect passengers while for gaps above 15 cm, passengers needed to adjust their step, and this cost a little more time [8].

Dwell times are also influenced by the design of stations. Narrow platforms become crowded easily, especially during peak hours [4]. The number, capacity and location of stairs, escalators and lifts influences the distribution of waiting passengers along the platform and hence the boarding time for trains. The timetable, which is an operational factor determine the length of dwell time, especially if the station is a timed stop. Trains arriving early will have longer dwell times so they don't depart before the scheduled time. At interchange stations, longer dwell times may be required to allow for trains to connection.

3 Previous Dwell Time Models

Various models had been developed by researchers to study the dwell times of trains of rail passenger services. One of the earliest train dwell time models was found to be a model developed by

Wirasinghe & Szplett (1984) for Calgary LRT [9]. Referring to Table 1, three separate equations were produced based on the percentage of boarders for a train. The coefficients for alighting passengers ranged from 0.4 s to 1.4 s per passenger, whilst the coefficients for boarding passengers ranged from 1.4 s to 2.4 s per passenger, where the lowest of each coefficient is found to be within the alighting and boarding category ($0.33 \leq \psi \leq 0.66$). Thus, the authors have proven that the number platform entrances and "friction" between boarding and alighting passengers have significant effect on dwell time and that dwell time should not be determined simply by using the average demand per door [9].

Table 1. Predictive equations for dwell time.

Group description (ψ = **fraction of boarders**)	**Predictive equation** $t = l + \lambda(a) + \mu(b)$*
Exclusive or dominant alighting $\psi \leq 0.32$	$t = 2 + 1.0(a) + 2.4(b)$
Alighting and boarding $0.33 \leq \psi \leq 0.66$	$t = 2 + 0.4(a) + 1.4(b)$
Exclusive or dominant boarding $\psi \geq 0.67$	$t = 2 + 1.4(a) + 1.4(b)$

* l = lost time (s)
λ = average time per passenger to alight (s)
μ = average time per passenger to board (s)
a = average demand per door for alighting
b = average demand per door for boarding

In 1989, Weston established a model for London Underground Ltd [8], which was later used as a main reference by many other researchers. He had incorporated a great number of variables into the model. As shown in Equation (1), the variables were number or boarding and alighting passengers, number of doors, peak door and average door factor, number of seats and the number of through passengers. Weston used a constant of 15 s for the process of door opening and closing, and for power to be applied for acceleration. However, the value should vary by rolling stock type and door type. The dwell time estimated increases with the number of passengers boarding and alighting but at a declining rate. This model therefore incorporates an 'alighting/boarding congestion effect'. The additional time needed in the event of a mixed flow (boarding and alighting at the same time) is also accounted for in the model.

$$SS = 15 + 1.4\left[1 + \frac{F}{35}\frac{(T-S)}{D}\right].\left[\left(\frac{FB}{D}\right)^{0.7} + \left(\frac{FA}{D}\right)^{0.7} + \left(0.027\left(\frac{FB}{D}\right)\left(\frac{FA}{D}\right)\right)\right] \quad (1)$$

where SS = station stop time (s), A = number of alighting passengers per train, B = number of boarding passengers per train, D = number of doors, F = peak door/average door factor, S = number of seats, and T = number of through passengers.

In 2007, Harris & Anderson applied the same model to estimate dwell times of metros and suburban railways around the world. They concluded that despite parameters had to be varied slightly, "Weston's formula for estimating dwell time appears to have validity around the world, and the overall structure of the approach appears sound" [10].

Lam et al. [11] conducted a study on train dwell time at the Hong Kong MTR system. They established regression models for dwelling delays due to congestion and developed a simulation model to assess the reliability of the estimated train dwell time. This model is a combination of three separate models developed based on data collected at three different stations. When comparing these three models, it was found that the constants and the coefficients for the number of alighting and boarding passengers were in the same order and of approximate value. Therefore, a generalised model that includes a fixed time for doors opening and closing, and the number of boarding and alighting passengers per train was produced (refer to Equation (2)). Lam et al. [11] also highlighted that dwell time does not increase infinitely with the increase of passengers as the train headway governs the maximum allowable dwell time of trains, which is around 3 minutes in Hong Kong.

$$DT = 10.5 + 0.021A + 0.016B \tag{2}$$

where DT = train dwell time (s), A = number of alighting passengers per train, and B = number of boarding passengers per train.

Puong [3] developed a model to analyse dwell times for the Massachusetts Bay Transport Authority (MBTA) in Boston based on an observation of 54 dwell times of trains with three or four single doors per carriage. All dwell times observed were below 90 s. He estimated that there was a constant of about 12 s for the process of door opening and closing and train starting. For each boarding and alighting passenger at a door, it was estimated that 2.27 s and 1.82 s was required respectively. The presence of standing through-passengers has no effect on the dwell time (specifically boarding time) if there were only 5 or below. But if there were more than 5, its effect is at cubed rate, as can be seen from Equation (3).

$$DT = 12.22 + 2.27B_d + 1.82A_d + 0.00062TS_d^3B_d \tag{3}$$

where DT = train dwell time (s), A_d = number of alighting passengers per door, B_d = number of boarding passengers per door, and TS_d = number of standing through-passengers per door.

Douglas was tasked by Transport for New South Wales (TfNSW) to review alternative ways of calculating train dwell time [4]. As a conclusion, Douglas established a model to predict train dwell time for RailCorp in Sydney. In his model (refer to Equation 4), the variables included were number of boarding and alighting passengers per door and the number of standing through passengers per door. He also estimated the function time to be 10 s. A power function of 0.7 gave a better fit than a linear function for both alighting and boarding. There was consideration for mixed flow of passengers and crowding caused by standing through passengers in the model. The best fit variable for standing through passengers was a linear function that multiplied standing passengers by the combined total of boarding and alighting passengers.

$$DT = 10 + 1.9A_d^{0.7} + 1.4B_d^{0.7} + 0.007(A_d + B_d)(Std_d) + 0.005(A_d.B_d) \tag{4}$$

where DT = train dwell time (s), A_d = number of alighting passengers per door, B_d = number of boarding passengers per door, and Std_d = estimated number of standing through-passengers per door.

4 Comparing Dwell Time Models

These five studies show how dwell time data can be used to develop a model to calculate dwell times according to passenger volumes. The model by Wirasinghe & Szplett was one of the earliest dwell time models established and proved that "friction" due to a mixed flow has a significant effect on dwell time and that dwell time should not be determined simply by using the average demand per door. This conclusion proved to be an important basis for models developed thereafter. When Weston came out with his model 5 years after Wirasinghe & Szplett, he managed to include the parameters found in Wirasinghe's model and further improved it by including a parameter for crowding effect.

Generally, all five studies agree that the increase of passenger volumes increase the dwell time of trains at stations, though at different rates due to factors such as door widths, platform gaps and movement of passengers. The model by Weston was non-linear and predicted that dwell times increased but at a decreasing rate as total of boarding and alighting increased. By contrast, the passenger dwell times in models by Lam, *et al.* and Puong increased proportionally. It is considered that the Weston formulation is more realistic especially for boarding time. Similarly, Douglas' model is also considered more realistic as he had adopted Weston's model.

The models by Weston, Puong and Douglas allowed for dwell times to increase with on-board train crowding. In Weston's model, a factor for excess demand compared to seats was applied that increased boarding and alighting times. Douglas's model also indicates that on-board train crowding increased the times for boarding and alighting. In comparison, the effect of on-board train crowding is only significant when there are more than 5 passengers per door in Puong's model. A summary of parameters included in the four models discussed is shown in Table 2.

Table 2. Parameters included train dwell time models.

	Parameters			
Model	**Passenger volume**	**Mixed flow effect**	**Crowding effect**	**Others**
Wirasinghe & Szplett Model (1984)	Yes	Yes	No	Yes
Weston Model (1989)	Yes	Yes	Yes	Yes
Lam et al. Model (1998)	Yes	No	No	No
Puong Model (2000)	Yes	No	Yes	No
Douglas Model (2012)	Yes	Yes	Yes	No

Referring to Table 2, it is clear that the models developed by Weston and Douglas are the best as they have included the three most influential parameters, namely passenger volume, mixed flow effect and crowding effect, though the model by Douglas is a simplified version of Weston's model. The Weston model can be considered as most relevant as it has been applied by Harris & Anderson on dwell time data for metros and suburban railways around the world.

5 Contributions of Dwell Time Models

The dwell time model established by Weston had later been tested by Harris & Anderson on over 30 railways around the world. Thus, the dwell time model could be applied to railways around the world but with adjustments to the parameters involved. Lam's model provides a reasonable estimate for the average train dwell time at MTR stations, and the reliability analysis can be used to give a reliable range for the estimated train dwell time for assessment.

The model by Puong uncovered the effect of on-board crowding on boarding times for the stations. This model enabled critical stations to be identified for maintaining high-frequency service during peak periods. The dwell time model by Douglas could be extended into a forecasting model for a particular line by inserting initial timetable data and patronage forecasts. It can also predict boarding, alighting and on-board passenger loads for each station along the route for an individual service.

6 Conclusions

Dwell time is a key parameter of system performance, service quality and reliability in the rail passenger service. With known factors affecting dwell time, the dwell time at critical stations could be minimised, thus reducing the headway variations and enhancing service quality and reliability. All five models studied in this paper agree that dwell time is very much influenced by the passenger volume, though at different rates due to factors such as door widths, platform gaps and movement of passengers. A mixed flow of passengers and on-board crowding are also expected to increase dwell time.

From this paper, the numbers of alighting, boarding and standing through-passengers were found to be most influential and are dominant parameters for developing dwell time models. It is recommended that further research be conducted to study on train on-board crowding and how the different types of passengers affect the boarding and alighting time, which affects the dwell time of a train.

References

[1] N.H. Nordin, M.I. Mohd Masirin, I. Ghazali and M.I. Azis, Appraisal on rail transit development : A review on train services and safety, *Int. Integrated Eng. Summit 2014*, Batu Pahat, (2014).
[2] Z. Alwadood, A. Shuib and N. Abd. Hamid, Rail passenger service delays: An overview, IEEE Business, Eng. Ind. Appl. Colloq., 449–454, (2012).
[3] A. Puong, Dwell time model and analysis for the MBTA red line, Massachusetts Inst. Technol. Res. Memo, 1–10, (2000).
[4] N. Douglas, Modelling train and passenger capacity, Douglas Econ., Sydney, (2012).
[5] W.H. Kraft, An analysis of the passenger vehicle interface of street transit systems with applications to design optimization, PhD Dissertation, New Jersey Inst. of Technology, Newark, (1975).
[6] J.G. Weston, Train service model - technical guide, London Underground Operational Research Note 89/18, (1989).
[7] P.B.L. Wiggenraad, Alighting and boarding times of passengers at Dutch railway stations, TRAIL Research Sch., Delft, (2001).
[8] W. Heinz, Passenger service times on trains, KTH Royal Institute of Technology, Stockohlm, (2003).
[9] D. Szplett and S.C. Wirasinghe, An investigation of passenger interchange and train standing time at LRT stations: (II) Estimation of standing time, J. Adv. Transp., **18**(1), 13–24, (1984).
[10] N.G. Harris and R.J. Anderson, An international comparison of urban rail boarding and alighting rates, *Proc. Inst. Mech. Eng. Part F J. Rail Rapid Transit*, **221**, 521–526, (2007).
[11] W.H.K. Lam, C.Y. Cheung and Y.F. Poon, A study of train dwelling time at the Hong Kong mass transit railway system, J. Adv. Transp., **32** (3), 285–295, (1998).

MATEC Web of Conferences **47**, 03006 (2016)
DOI: 10.1051/matecconf/20164703006

Pedestrian Footpath Level of Service (FOOT-LOS) Model for Johor Bahru

Basil David Daniel[1,a], Siti Naquiyah Mohamad Nor[1], Munzilah Md Rohani[1], Joewono Prasetijo[1], Mohamad Yusri Aman[1] and Kamarudin Ambak[2]

[1]*Faculty of Civil and Environmental Engineering, Universiti Tun Hussein Onn Malaysia, 86400 Parit Raja, Johor, Malaysia.*
[2]*Smart Driving Research Centre, Universiti Tun Hussein Onn Malaysia, 86400 Parit Raja, Johor, Malaysia*

Abstract. Pedestrian level of service (LOS) is an overall measure of walking conditions on a route, path, or facility. This is linked directly to factors that affect mobility, comfort and safety, reflecting pedestrian's perceptions of the degree to which the facility is pedestrian-friendly. A unique model called FOOT-LOS (short for pedestrian footpath level of service) was developed to facilitate the LOS measurement. The factors considered fall into three categories; physical, location and user attributes. Pedestrian conditions can be described through a grading system, i.e. from LOS A (ideal pedestrian condition) to LOS F (unsuitable pedestrian condition), based on an assessment of the factors affecting LOS. The assessment includes desktop and on-site assessment of LOS factors. The development of the model involved an iterative process that included testing and refinement. The research undertaken and the LOS model developed provide a sound basis for the measurement of LOS for pedestrian footpaths. The model not only provides a methodology for determining LOS, but also determines which factors predominantly contribute to the LOS of pedestrian footpaths.

1 Introduction

Walking is a means of active transportation that is healthy and sustainable. Residents in pedestrian-oriented cities were found to be healthier than private automobile-oriented cities [1], and areas with improved walkability rates resulted in better air quality [2]. Local authorities strive to create walkable cities by providing pedestrian facilities such as pedestrian footpaths, footbridges and crossings.

The level of service (LOS) for pedestrians can be defined as a measurement of the condition of walkway in a road and around public areas. This relates directly to factors that can affect the movement of pedestrians, and their perceptions on comfort and safety while walking. The purpose of this study was to identify the factors that can influence the perception of pedestrians in terms of safety and comfort, and to come up with a model that can be utilised in determining the LOS of a footpath.

To achieve the aforementioned objectives, 25 streets in Johor Bahru that have pedestrian footpaths were selected. These locations represent the diversity of characteristics of footpaths in Johor Bahru, which were rated from poor to excellent in terms of levels of comfort and safety by pedestrians in a survey that was conducted at the study sites.

[a] Corresponding author : basil@uthm.edu.my

2 Research Motivation

Pedestrianisation is a technique that aims to encourage walking among city folk by segregating pedestrians from vehicle traffic and providing idyllic walking environments. In Malaysia, pedestrianisation is not given enough attention since much of the city infrastructure is dedicated to road networks that predominantly cater to private automobiles. Local authorities are often in a bind when dealing with meeting demands for both vehicle traffic and pedestrians. Unsurprisingly, it is the infrastructure for vehicles that normally receives precedence over pedestrian needs [3].

Johor Bahru is a bustling city situated in the south of Peninsular Malaysia, boasting a population of nearly half a million people. While travelling by private automobiles is the foremost choice of transportation among city commuters, walking is also a mode of travel for those who cannot afford cars and those who opt to travel by foot to avoid traffic jams and crowded buses, for environmental reasons, and for travel over short distances. Most downtown streets in Johor Bahru have pedestrian footpaths, but not all are as idyllic as one would hope for (see Figure1). There is a need to regularly monitor the performance of pedestrian footpaths in order to ensure that pedestrians feel safe and comfortable when using these facilities. This in return would encourage more city folk to walk and reduce dependence on motorised travel. Hence, this study was conducted with the goal of producing a pedestrian level of service model, i.e. FOOT-LOS, for the evaluation of pedestrian LOS along footpaths.

Figure 1. Contrasting conditions of footpaths in Johor Bahru.

3 Factor Affecting Pedestrian Comfort and Safety

A study by Florida Department of Transportation managed to identify a number of factors that can affect pedestrian safety and comfort in a roadway. The factors included lateral clearance between pedestrians and traffic, traffic volume, vehicle speed and type of vehicle [4]. Vehicle-pedestrian interaction is an important factor to consider, given that motorist behaviour influences the risk of conflicts between vehicles and pedestrians [5]. Similarly for other pedestrian facilities such as crossings, it was found that pedestrian flow, surface condition and walkway width exerted some influence on pedestrian perception of comfort and safety [6].

In Malaysia, guidance for pedestrian facilities is provided by Jabatan Kerja Raya (JKR). Ideally, width of footpaths shall be 2.4 meters or greater, while the allowable minimum width is 0.9 meters. Footpaths shall have a height clearance of 2.0 meters or more, and lateral clearance of at least 1.0 meters from the traffic lane edge. The elimination of physical obstructions, minimising of changes in level, and surface skid resistance are also considered when providing pedestrian footpaths [7]. Good walkways usually have comfortable surfaces without cracks, depressions and bumps. As a result, pedestrians will have ease of travel when using walkways that have good surface quality [8].

4 Methods and Findings

4.1 Data collection

On-site measurements were conducted at 25 roads in the city of Johor Bahru, to obtain footpath and road widths. Manual counts were carried out to determine the volume of both pedestrians and vehicles, while a speed meter was used to measure vehicular speed. Pedestrian flows were determined from video recordings, while obstructions and surface condition of the footpaths were quantified by visual walkthrough surveys.

391 pedestrians between the ages of 18 and 50 years were randomly interviewed at the study locations and asked to rate the footpaths in terms of comfort and safety from the scale of 1 (poor) to 5 (excellent). This sample size was well sufficient as it exceeded the 370 sample size proposed by Krejcie & Morgan. 54% of the respondents used the footpaths regularly to get to work and college, while the remainder used the footpaths occasionally for shopping and recreational activities.

4.2 Model development

Initially, the FOOT-LOS model included seven contributory factors, i.e. pedestrian flow, traffic volume, footpath width, road width, vehicle speed, percentage of surface damage and number of obstructions. Multiple linear regression analysis was applied to relate pedestrian rating with these factors.

For this purpose five assumptions were made, i.e. the relationship between the independent and dependent variables were assumed to be linear, the variables were assumed to be normal, there was little or no multicollinearity in the data, there was little or no autocorrelation in the data and the error terms along the regression line were equal or homoscedastic. After numerous runs, it was found that by excluding vehicle speed (S) and pairing some of the factors, a significant model could be produced. Hence, the FOOT-LOS model took the form of:

$$FOOT - LOS = 0.7078(FW + RW) - 0.2138(SD + OBS) - 0.1909\left(\frac{P+V}{1000}\right)^2 \tag{1}$$

where: FOOT-LOS = pedestrian footpath level of service rating, FW = footpath width (meter), RW = road width (meter), SD = surface damage (% of area), OBS = number of obstructions (number per 100 meter), P = pedestrian flow (pedestrians/minute/meter), and V = traffic volume (vehicles/hour).

The FOOT-LOS model yielded a coefficient of determination (R^2) value of 0.97, which indicates that 97% of the variance in footpath level of service is predictable from the contributory factors. The F-test for significance of the regression parameters produced a Sig.F value of 0.000, showing that the FOOT-LOS model has statistically significant predictive capability. The parameter coefficients were tested for significance using t-tests and were found to be statistically significant, with the P-values below the significance level of 0.05 (see Table 1). This implies that the parameters are good predictors of the FOOT-LOS rating.

Table 1. Results of t-test for significance of FOOT-LOS model parameter coefficients.

Parameter	Coefficient	t-Statistic	P-value
FW, LW	0.7078	12.721	0.000
SD, OBS	-0.2138	-2.249	0.035
P, V	-0.1909	-2.096	0.005

The FOOT-LOS model can ultimately be used to determine the numeric rating of the level of service. Alternatively, the scores can be translated to letter grades ranging from A (excellent) to F (poor), as shown in Table 2.

Table 2. FOOT-LOS rating scale.

Level of Service	FOOT-LOS Score
A	$x > 8.5$
B	$7.0 < x \leq 8.5$
C	$6.0 < x \leq 7.0$
D	$5.0 < x \leq 6.0$
E	$4.0 < x \leq 5.0$
F	$x \leq 4.0$

4.3 Model validation

While the aforementioned statistical tests give credence to the predictive capability of the FOOT-LOS model, further validation is necessary. Footpaths on eight additional streets were randomly selected to test the accuracy between observed FOOT-LOS scores and predicted FOOT-LOS scores, which employed the model that was developed from this study (Equation (1)). As shown in Table 3, the predicted values were not far off the observed values, resulting in residuals that were relatively small.

Table 3. Observed and predicted FOOT-LOS score for model validation.

Road Name	FOOT-LOS Score		Residual
	Observed	Predicted	
Jalan Ibrahim Sultan	6.1	6.6	-0.5
Jalan Stulang Darat	7.8	7.8	0.0
Jalan Sawmill	6.4	6.9	-0.5
Jalan Tampoi	6.9	6.6	0.3
Jalan Datin Halimah	7.6	7.7	-0.1
Jalan Pasar	5.3	5.6	-0.3
Jalan Abdullah Ibrahim	7.5	7.5	0.0
Jalan Tan Hiok Nee	6.8	6.9	-0.1

5 Conclusion

The findings of this study show that pedestrian's perception of comfort and safety is significantly influenced not only by the walking environment and conditions, but also the adjacent road and traffic characteristics. Also, the FOOT-LOS model developed in this study was proven to be a reliable tool for quantifying the level of service of pedestrian footpaths and thus, the model can be used for performance analysis of such facilities.

Based on the FOOT-LOS model, a high level of service may be achieved if the width of pedestrian footpath is increased meet ideal specifications so that pedestrian flow is smoother and conflicts between pedestrians are minimised, the footpath is regularly maintained to provide a surface of acceptable walking quality, fixed elements that obstuct pedestrians removed from the footpath, and lateral separation between pedestrians and traffic is provided to reduce qualms regarding pedestrian safety.

References

[1] R. Ewing, T. Schmid, K. Killingsworth, A. Zlot and S. Raudenbush, Relationship between urban sprawl and physical activity, obesity and morbidity, American J. of Health Promotion, **18**(1), 47-57, (2003).

[2] L. Frank, S. Kavage and T. Litman, Promoting public health through smart growth: Building healthier communities through transportation and land use policies, Smart Growth British Columbia, (2006).
[3] K. Singaravelloo, Local Government and Intergovernmental Relations, in N.A. Siddiquee (Ed.), *Public Management and Governance in Malaysia: Trends and Transformations*, Routledge, (2012).
[4] B.W. Landis, V.R. Vattikuti, R.M. Ottenberg, D.S. McLeod and M. Guttenplan, Modeling the Roadside Walking Environment: A Pedestrian Level of Service, Transportation Research Record No. 1773, Part 2: Bicycle and Pedestrian Research, Transportation Research Board, Washington, (2001).
[5] I. Wahab, *Teknik Sistem Lalu Lintas and Pengangkutan*, Dewan Bahasa dan Pustaka, Kuala Lumpur, (1997).
[6] B.D. Daniel, S.L. Choong, K. Ambak and M.E. Abdullah, Pedestrian level of service for crosswalks at signalized intersections, *Proceedings of the National Seminar on Civil Engineering Research*, Universiti Teknologi Malaysia, Skudai, (2007).
[7] Jabatan Kerja Raya, A Guide on Geometric Design of Roads, Arahan Teknik (Jalan) 8/86, Kuala Lumpur, (1986).
[8] C.A. O'Flaherty, *Transport Planning and Traffic Engineering*, Butterworth Heinemann, (1997).

MATEC Web of Conferences **47**, 03007 (2016)
DOI: 10.1051/matecconf/20164703007

Systems Dynamic Approach in Development of Infrastructure in Entikong, Border Regions Indonesia – Malaysia

Ilyas Tommy[1], Dianariana Dwi[2], Suryani Fitri[2,a] and Henni[3]

[1]*Department of Civil Engineering, Faculty of Engineering, University of Indonesia, Depok 16424, Indonesia*
[2]*Department of Civil Engineering, Faculty of Engineering, University of Persada Indonesia, Jakarta, Indonesia*
[3]*Department of Industrial Engineering, Engineering Faculty, University of Persada Indonesia, Jakarta, Indonesia*

Abstract. National strategic area needs to be supported by the various sectors in order to spur growth in the surrounding areas so that the surrounding region can develop and reduced. Public works infrastructure that includes roads, infrastructure for housing, infrastructure, water resources, and energy becomes one of the important components in the economic development efforts of the region as well as the reduction of disparities between regions. Entikong, as a national strategic areas, border areas designated as a national strategic area that needs to be supported by the physical infrastructure of the structure-forming region in order to achieve existence to support the functioning of the economy, social and defense.

1 Introduction

Kalimantan border area has a very strategic location, because it is located in the midst of the territory of Southeast Asian countries, and is flanked by two international shipping lanes, the South China Sea and the Celebes Sea / Sulu. The strategic role of Kalimantan border area apart as a security belt, is also the gateway and the image of the nation. Moreover, with large forest resources within the scope of the global environment, this area is a buffer zone and the lungs of the world.

Kalimantan border area, is currently the border area whose interaction with the large neighboring countries, both in terms of trade, labor and cross social relationships and kinship. The strategic position held Kalimantan border area has not been well explored for public welfare and national income for his role as a security belt is very prominent which are necessary at that time. This resulted in the border area of Indonesia neglected and regarded as an area that should be kept secure. With the condition, remote and limited facilities and infrastructure while dealing with residential area which have good infrastructure and human resource capacity is high, then there is an unequal socioeconomic relations. This condition, if left unchecked, will cause adverse effects, particularly the image of Indonesia is getting worse in the international world.

Entikong is a district in West Kalimantan Province, Indonesia. Entikong has a land border with a country lane Malaysia, especially Sarawak so that the landline is often referred to as the Silk Road can pass directly by bus both from Indonesia and Malaysia

Based on data from Sanggau Planning and Regional Development Agency (BAPPEDA), livelihood Entikong districts by GDP per capita in 2011 Rp. 9,396,432.92 / year contribution of the

[a] Corresponding author : suryani.fitri21@yahoo.com

agricultural sector Rp. 46,009,480,000.00 is comprised of farmers 5,567 people (36%), traders 382 people (2.5%), has not been or is not working 3,465 people (23%). Entikong has a total population in 2013 is_15.056 people (men: 7682 inhabitants, women: 7374 inhabitants), a density of 30 inhabitants / km2, regional per capita income of Rp. 8,604,313.37 / year.

Frontier in the Sanggau district along ± 129.50 Km (15%) of the length of the border line of ± 966 Km West Kalimantan. Sanggau district strategic position is: in the middle of West Kalimantan Province, adjacent land with Sarawak (East Malaysia), located on the Trans Kalimantan Strip, located on the Trans Borneo Trails, located on track Kapuas River, and accessibility: ± 90 Km (2 hours) from / to Kuching, Sarawak, ± 119 Km from / to Pontianak, located in the Kapuas river as in Figure 1.

Figure 1. Strategic position of Sanggau District.

2 System and Dynamic Model Development Public Private Partnership in the Development of border Regions Indonesia-Malaysia

Dynamic system is a methodology to study the issues around who saw the problem an overall (holistic). This methodology is not like another methodology that examines the problems with the sort them into parts of smaller and mutually limiting. Primary concept of dynamic system is understanding of how all the objects in a system interacting with one another. Dynamic system according to the community dynamic system is a methodology to learn and managing complex feedback systems, such as those are commonly encountered in the business world and other social systems.

Based on the results of the first year of the study showed that there are four infrastructure sectors that need attention are: Transportation, Housing, Energy / Electricity and Water. The development of a dynamic system models of Public Private partnership in the border region Indonesia – Malaysia development based on infrastructure networking in the area Entikong Sanggau West Kalimantan was conducted in four infrastructure sectors, namely transport infrastructure, settlements infrastructure, energy infrastructure and water infrastructure, development models aim to get percentage of financing infrastructure development scheme for the Government and private sector [1, 2] such as public private partnerships scheme for funding infrastructure [3, 4]. The causal relationship diagram of the four infrastructure are as follows:

a) Transport infrastructure - sub-system of transport infrastructure is built to model the structure of relationships that make up the causal relationship between the various variables and parameters in the sub-system of transport infrastructure. A variety of variables and parameters as well as the causal structure of sub-systems that make up the transportation infrastructure. can be seen in the diagram the structure of a causal relationship (causal loop diagram / CLD) in Figure 2 below:

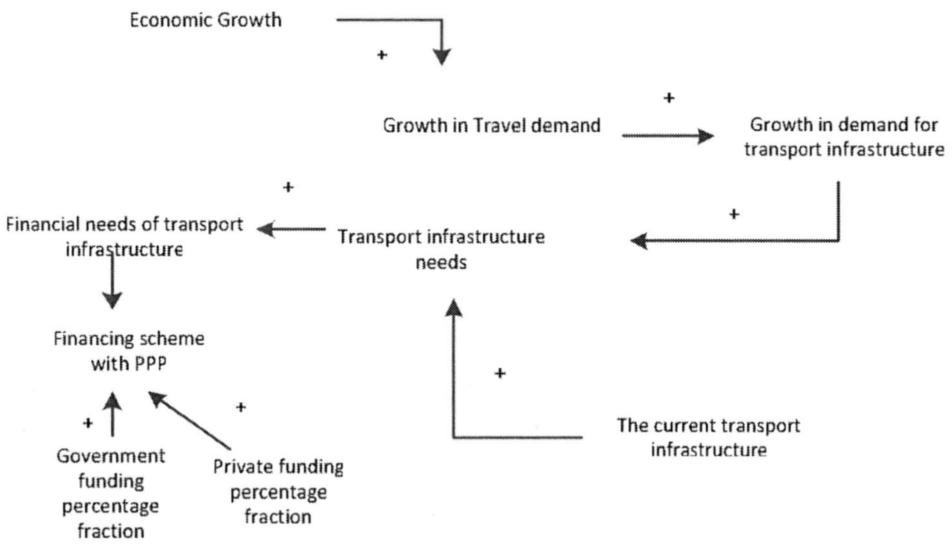

Figure 2. Causal loop diagram structure transportation infrastructure.

b) Settlements infrastructure - on sub system will illustrate the need for citizens Sanggau Settlements. Sub Settlements system infrastructure is built to model the structure of relationships that make up the causal relationship between the various variables and parameters in the sub-system of the settlements infrastructure. A variety of variables and parameters as well as the causal structure of sub-systems that make up the infrastructure of the Settlements can be seen in the diagram the structure of a causal relationship (causal loop diagram / CLD) in Figure 3.

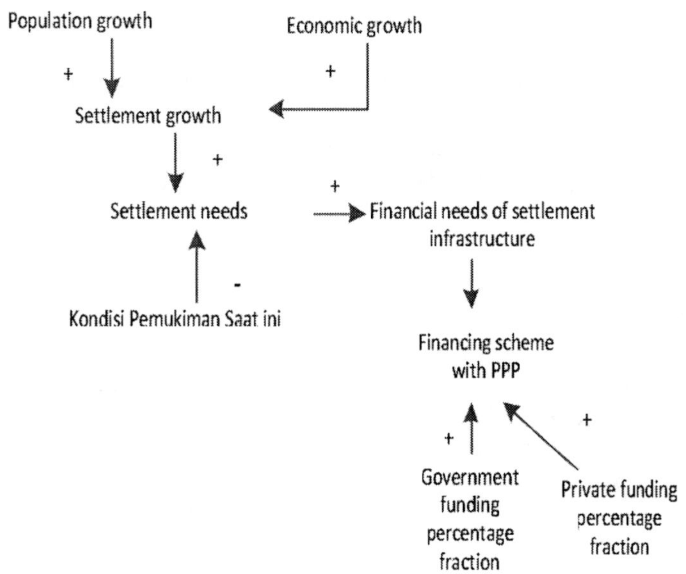

Figure 3. Causal loop diagram structure settlements infrastructure.

c) Energy/ electricity infrastructure - causal structure that forms the sub system intrastruktur energy/ electricity can be seen in the diagram causal relationship between the various variables and parameters in the sub-system energy infrastructure / electricity. A variety of variables and parameters as well as the causal structure of sub-systems that make up the infrastructure of energy/ electricity can be seen in the diagram the structure of a causal relationship (causal loop diagram/ CLD) in Figure 4.

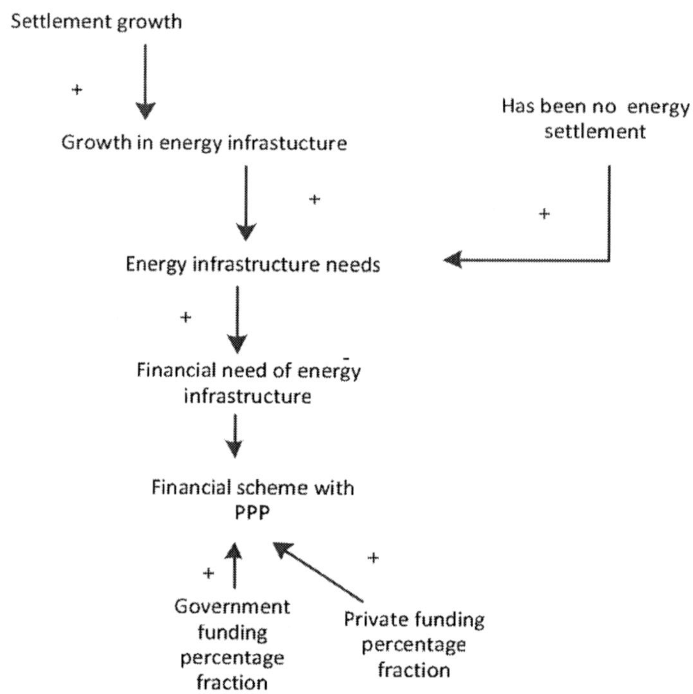

Figure 4. Causal loop diagram structure energi/ electricity infrastructure.

d) Clean water infrastructure - causal structure that forms the sub system intrastruktur clean water can be seen in the diagram causal relationship between the various variables and parameters in the sub-system of clean water infrastructure. A variety of variables and parameters as well as the causal structure of sub-systems that make up the clean water infrastructure can be seen in the diagram the structure of a causal relationship (causal loop diagram / CLD) in Figure 5.

Border regions needs to be addressed, because as the country face of entrance gate and then the border areas should be developed primarily related to Network of 4 its main infrastructure: Settlements, transportation, energy and water.

3 Conclusions

There are four infrastructure sectors that need attention are: Transportation, Housing, Energy / Electricity and Water. The development of a dynamic system models of Public Private partnership in the border region Indonesia – Malaysia development based on infrastructure networking in the area Entikong Sanggau West Kalimantan was conducted in four infrastructure sectors, namely transport infrastructure, Settlements Infrastructure, Energy Infrastructure and Water Infrastructure. Development in the border regions require approaches welfare and the environment, in addition to the security approach. For the example is constructions of a parallel road borders and development of plantations along the border line. The construction of a parallel road has a dual function, namely enhance border security and open borders inland isolation. Local economic development needs to be more emphasis on the spatial dimension. Regions need to combine sectoral cluster-based approach in which current business / sectors in the regions and the poor tend to cluster. There needs to be integration with the rural development strategies of urban development strategies. The village is generally still lagging behind in various types of infrastructure. This integration is expected to be developed rural-urban linkage and intercity networks (network Cities).

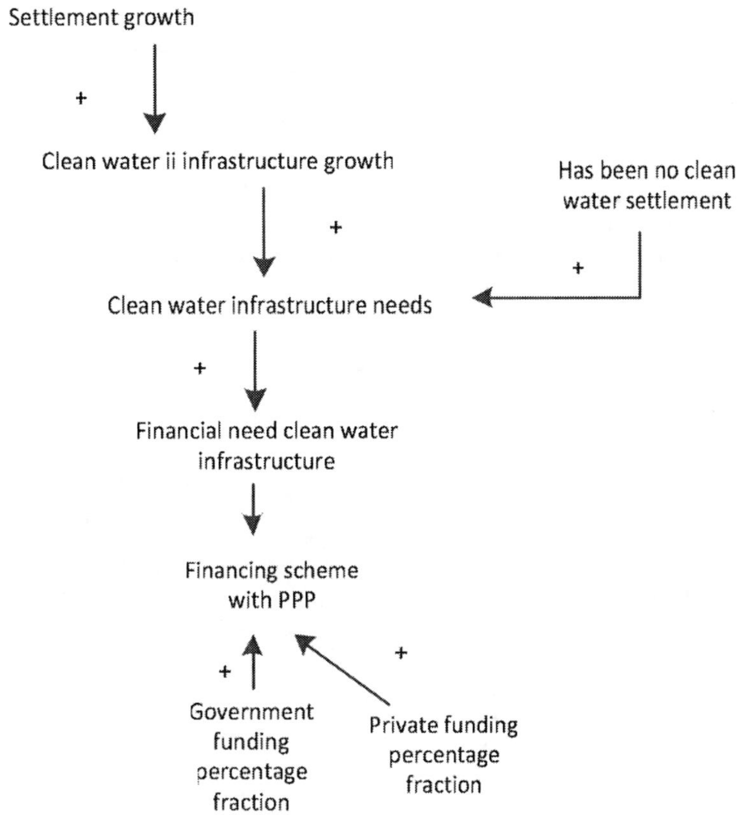

Figure 5. Causal loop diagram structure clean water infrastructure.

Acknowledgements

The main financial support from the Directorate General of Higher Education, Ministry of Technology Research and Higher Education Republic of Indonesia by National Strategic Grants 2015 is gratefully acknowledged.

References

[1] Bappenas, Indonesia Infrastructure Development, Ministry of National Development, Planning Agency of Indonesia, Jakarta, (2012).
[2] E. Heurkens, Changing Public Private Partnerships Roles in Urban Area Development in the Netherlands, Delft University of Technology, (2009).
[3] E.R. Yescombe, Public Private Partnerships Principles of Policy and Finance, Elsevier, (2007).
[4] R. Alasad, M. Ibrahim and O. Stephen, A system dynamics-based model for demand forecasting in PP infrastructure project – a case of toll road, Organization, Technology and Management in Construction International Journal, **5**, 791-798, (2013).

MATEC Web of Conferences **47**, 03008 (2016)
DOI: 10.1051/matecconf/20164703008

Behavioral Intention to Use Public Transport Based on Theory of Planned Behavior

Kamarudin Ambak[1,a], Kanesh Kumar Kasvar[1], Basil David Daniel[1], Joewono Prasetijo[1] and Ahmad Raqib Abd Ghani[1]

[1]*Smart Driving Research Center, Universiti Tun Hussein Onn Malaysia, 86400 Batu Pahat, Johor, Malaysia*

Abstract. An increase in population generates increasing in travel demand. In Malaysia, public transport become an important modes of transport that connection people. This paper presents behavioural intention to use public transport especially public bus based on Theory of Planned Behaviour (TPB). A questionnaire survey was conducted to identify factors that contribute and influence users to use public bus and to determine factor that most dominant using TPB model. A total of 282 questionnaires were distributed in selected area of Batu Pahat and Kluang. Correlation and regression analysis were used for this study. Results show that the Attitude toward public transport is the most dominant factor compared with Subjective Norm and Perceived Behavior Control that influencing users to use public bus. Majority respondents were agreed that they prefer to use public bus because it is cheap to travel and no other choices of other transfer modes. As for the recommendation, this study can be extended in future as part of strategic sustainable transportation system in Batu Pahat and Kluang areas.

1 Introduction

Malaysia is one of the substantial developing nations in South East Asia. Transportation plays vital role in Malaysia economic contribution. Modern day busy lifestyles have increased the value of time. Public transport interpreted as a shared passenger transport service that is applicable for public uses as the movement of people and good to meet the basic need of society that require mobility and access [1]. Rapid population growth and economic growth have been able to embark on extension and improvement of the current transportation system. Malaysia's population growth has been on a downward trend over the past 10 years in line with other urbanizing countries [2]. There are four type modes of public transport such as highway, railway, airway and water. The most common types of public transport that are widely used in Malaysia are buses, airlines, taxis, trains and rails. On the other hand, public transport in Malaysia becomes a very important travel mode on a daily basis in human society. The modes of public transport system that exist in the study areas are land transportation modes such as public buses and taxis. Numbers of people using route F050, Jalan Batu Pahat to Kluang rapidly growing due to the establishment of higher education institution and manufacturing factories [4].The important of public transport necessary in the study area due to increase in demand.

[a] Corresponding author : kamardin@uthm.edu.my

Generally, demand of public transportation increased rapidly from year to year. Mainly in urban areas, public transportation has become necessary for their lifestyle due to traffic congestion issues and more environments. Malaysian government highly recommends public to utilize public transport in order to decrease traffic flow, reduce accident rate and also initiative for more savings due to increase of fuel prices [19]. Here in Batu Pahat and Kluang area, most widely public transportation has become mobility and access for people to travel from one place to another place. Mostly, public here at the Batu Pahat and Kluang area, use public transport on a daily basis such as travel from home to school, workplace and also college or university. There are also most people using public transport to go bus stand and also purchase of groceries from the supermarket. Besides that, there are few people who prefer public transport as mode of transport as travelling to the malls for movies and shopping. Transportation quality is acknowledged to consist two parts: the objective aspect, such as performance measures and service measure such as how customer perceives the service [1, 2]. These two aspects are significant in influencing intention user to use public bus.

Therefore, this study are to identify factors that contribute and influence users to use public transport at Batu Pahat and Kluang area and to determine factor that most dominant based on Theory of Planned Behaviour (TPB). This study was conducted on public transport especially feeder buses at routes along Batu Pahat to Kluang such as Johore Motor and Causeway Link bus companies. This two bus service company had been major bus companies involved in stage buses along these routes for several years. Moreover, this study was conducted randomized in Batu Pahat and Kluang bus station. The chosen travel route is route F050 Jalan Batu Pahat to Kluang.

2 Theory of Planned Behaviour

The theory proposes a model which can measure how human actions are guided. It predicts the occurrence of a particular behaviour, provided that behaviour is intentional. The theory of planned behaviour (TPB) in Figure 1 was developed by Icek Ajzen in 1985 [5]. The theory of planned behaviour is an extension of the theory of reasoned action [6]. The model is out lined in the next figure and represents the three variables which the theory suggests will predict the intention to perform behaviour. The intentions are the precursors of behaviour. There are three main components of TPB namely subjective norms, attitude and perceived behavioural control.

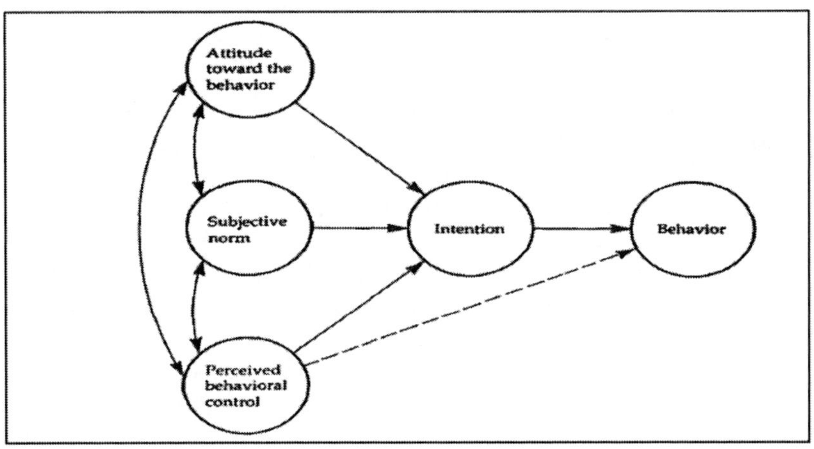

Figure 1. Theory of planned behaviour [5].

There are many studies that used TPB as model to study behaviour. The TPB has been applied successfully to predict and explain diverse behaviours, such as voting choices, weight loss, smoking cessation, social sciences, transportation and committing traffic violations [7, 18]. The effectiveness of U-pass programs has been examined in Canada using [8] TPB model and reduced car use among university students is generally found. A study by Wen and Ching [9] on Kaohsiung Mass Rapid Transit System, Taiwan shows public transit passengers' behaviours by constructing a comprehensive model considering public transit involvement, service quality, perceived value, satisfaction, and behavioural intentions at Taiwan,). In Vietnam, a study on the behavioural intention to use the bus while considering the perceived quality of bus service, problem awareness, and moral obligation [10]. Also, others related to traffic safety focus on motorcycle and helmet use in Malaysia [17].

3 Materials and Methods

This study chooses the questionnaire as research tool. Questionnaire developed based on Theory of Planned Behaviour (TPB). This questionnaire is divided into three parts, first part is the respondent's demographic analysis or information of the respondents, the second part is about perceived quality of public bus services and the third part is intention to use public bus based on Theory of Planned Behaviour (TPB). The five-point Likert scale used for the study and respondents answered the questions by lowest point as 1 (totally disagreed) and highest point as 5 (totally agreed).

The location of study that was determined at route F050, along Jalan Batu Pahat to Kluang. This route selected because the demand of public bus in this area are increasing due to rapid development of housing estate and office-shop building, also surrounded by educational institutes and factories. A pilot study of a total 20 respondents among public bus users were chosen to answer the questionnaire that was developed. Cronbach Alpha scores are used as a reference for analyzing the reliability of internal consistency questionnaire items. Cronbach alpha reliability coefficient that indicates the items connected to each other. The reliability of the questionnaire in the pilot study is to obtain Alpha value of 0.6 and above [12].

The questionnaires were distributed in selected bus terminal and bus stops along Jalan Batu Pahat and Kluang road. Respondents were randomly selected in term of gender, age, race, education level, purpose of using public bus and frequency of using public bus in a week. Respondents must be Malaysian citizens because the questionnaires were designed in "*Bahasa Malaysia*". According to [13] suggested that a minimum sample size of 200 is appropriate for any multivariate statistical analysis thus for this study total 282 respondents were chosen. The findings were analyzed using Statistical Package For Social Science (SPSS) version 22.0. Descriptive statistics, correlation and regression analysis were conducted to achieve study objectives and to determine relationship between variables.

4 Results and Discussion

Table 1 shows the descriptive analysis about respondent gender, age, level of education, occupation, race, income per month, and purpose of using public bus and frequent of using public bus per week. The descriptive statistics analysis and the findings shows that most of the public bus users are female ager ranged 20-30 years. It also shows that respondents were still pursuing in college or university and income less than RM1000.00. Meanwhile, most of respondents that using public bus was Malay. According to [14] majority populations in Batu Pahat and Kluang area were Malay. It is clear that, Batu Pahat and Kluang area is strategic location with many education institute established here. This shows public bus users mostly are students with less income and age not exceed 30 years. The study also shows education is main purpose of using public bus and frequency using public bus was 1 to 5 times per week.

Table 2 shows mean scores and standard deviations on quality of public bus services generally. Respondent's give their feeback on several aspects such as comfort, convenience, speed, punctuality, security, staff services, sufficient of seating places, cleanliness and breakdown experience. The respondents were influenced to use public bus because it is cheap to travel and cost-saving compare to private vehicles. Comprehensive improvements, such as Bus Rapid Transit systems, can provide large increases in transit use and attract large numbers of discretionary riders who would otherwise travel by automobile [14].

Table 1. Descriptive statistics.

Item		Frequency	Percentage (%)
Gender	Male	129	45.7
	Female	153	54.3
Age	Less Than 20	31	11.0
	Years	241	85.5
	20-30 Years	8	2.8
	30-40 Years	2	.7
	40 Years and Above		
Education Level		2	.7
	Primary School	26	9.2
	Secondary School	99	35.1
	Diploma/STPM	154	54.6
	Graduate		
Income		45	69.9
	Less Than 1000	31	16.0
	10001-2000	9	11.0
	2001-3000	27	3.2
	3001 and Above		
Purpose of Using Public Bus	Work	45	9.6
	Daily Routine	143	16.0
	Education	60	50.7
	Entertainment	7	21.3
Frequency of Using Public Bus	1-5 times	137	48.6
	6-10 times	73	25.9
	More than 11 times	72	25.5

Table 2. Mean score and standard deviation for quality of public bus services.

Item	Mean	Std. Deviation
Comfort of using public bus.	2.85	.79
Usage of public bus facilitates users.	3.41	.86
Bus driver obey to speed limit and drive with prudence.	2.90	.84
Public bus adheres of time punctuality.	2.46	.87
On-board security level when travelling in public bus.	2.89	.83
Level of services provided by bus driver and conductor is satisfactory.	3.21	.81
Adequacy of seating place in public bus.	2.81	.91
Cleanliness in public bus is satisfactory	2.65	.85
Experienced breakdown when travelling in public bus.	1.76	.44

From Table 3, there was a moderately positive relationship between intention and behavior with $r = 0.527$ and $p < 0.01$. While, attitude, subjective norm, perceived behavioral control and behavior demonstrate a weak positive relationship with $r = 0.272, 0.223, 0.270$ and $p < 0.01$. It means all components in TPB are found significantly contribute to behavior intention to use public bus. The relationship between intention and behaviour shows high significant value, but the main three components that focused on are attitude, subjective norms and perceived behavioural control. Similar to Health & Gifford [15] found that intention is strongly associated with behaviour compared with attitude linked with intention. Ambak et al. [16] were found a significant positive correlations with intentions and behavior toward helmet usage.

Table 3. Correlation analysis for TPB components.

	ATT	SN	PBC	INT	BEH
ATT	1				
SN	.291**	1			
PBC	.464**	.253**	1		
INT	.460**	.197**	.364**	1	
BEH	.272**	.223**	.270**	.527**	1

Notes**: Correlation is significant at the 0.01 level (2-tailed)
INT=Intention, ATT=Attitude, PBC=Perceived Behavioural Control,
SN=Subjective Norms, BEH=Behaviour

There are two level regression analyses were conducted from TPB model. Firstly, multiple regression analyses were conducted to examine the relationship between Intention and Attitude, Subjective Norms (SN) and Perceived Behavioural Control (PBC). Secondly, simple regression analysis was performed between Behaviour and Intention variables. Table 4 presents the coefficients value of Attitude, Subjective Norm and Perceived Behavioral Control on multiple regression analysis. Attitude and Perceived Behavioral Control components show the strong relationship (β_1=0.239 and β_2=0148, $p < 0.05$) with Intention. There are some studies show a relatively strong relationship between attitudes and behavioural intentions [15, 18].

Table 4. Coefficients of regression on attitude, subjective norm and perceived behavioural.

Coefficients					
	Unstandardized Coefficients		**Standardized Coefficients**		
Model	B	Std. Error	Beta	**t**	Sig.
1 (Constant)	1.388	.641		2.166	.031
ATT	.239	.040	.361	5.992	.000
SN	.043	.053	.044	.806	.421
PBC	.148	.047	.186	3.127	.002

a. Dependent Variable: Intention

Table 5 shows result on simple linear regression analysis. There is a strong relationship between Intention (β=0.441, $p<0.001$) and Behaviour variable in TPB model. In addition, the analysis shows that the coefficients of determination, $R^2 = 0.527$, which means 52.7% of variance in Intention can be explained in Behavior. A study [15] found that intention is strongly associated with behaviour compared with attitude linked with intention. Contrary with Sumaerdi et al. [17] mentiod that service quality and perceived value have direct and positive effect on behavior intention.

Table 5. Coefficients of regression on intention.

Coefficients					
	Unstandardized Coefficients		Standardized Coefficients		
Model	B	Std. Error	Beta	t	Sig.
1 (Constant)	3.694	.275		13.413	.000
INTENTION	.441	.042	.527	10.387	.000

a. Dependent Variable: Behaviour

5 Conclusion

The modes of public transport system that exist in the study areas are land transportation modes such as public buses and taxis. This study is focused on the behavioral intention to use public bus using the Theory of Planned Behavior (TPB). Descriptive statistics analysis of demographic, and experience, mean score and standard deviation are used to determine perceived quality of bus services The highest mean score that contributes and influence users to use public transport at Batu Pahat and Kluang area to use a public bus is factor on cheaper than private transportation. While, the lowest mean score that identified is factor on the use of public bus may reduce traffic congestion. Correlation and regression analysis were used to analysis TPB components. The results show that three significant factors in TPB that influence public bus user to use bus are Attitude, Perceived Behavioral Control and Intention. Factors that influence public bus user to use public bus are public bus cheap to travel and no choices of travel modes. It can be concluded that positive attitude toward saving cost and having no other travel modes will encourage public transport (bus) ridership.

Acknowledgment

The authors would like to thanks Universiti Tun Hussein Onn Malaysia (UTHM) for funding this study under the CoR Contract Research Grant Vot U395 and Smart Driving Research Center (SDRC), Faculty of Civil and Environmental Engineering for support and providing facilities to accomplish the study.

References

[1] L.A. Hoel and A.W. Sadek. *Transportation Infrastructure Engineering: A Multimodal Integration*, Cengage Learning, (2011).
[2] M.M. Rohani, D.C. Wijeyesekera and A.T.A. Karim, Bus operation, quality service and the role of bus provider and driver, Procedia Engineering, **53**,167-178, (2013).
[3] Land Public Transport Commission Malaysia, Annual Report, **1**, 26-27, (2012).
[4] A.A.S. Mohamad, Study towards service quality for Johore Motor bus at routes from Batu Pahat to Ayer Hitam, BEng Thesis, UniverstitiTun Hussein Onn Malaysia, Johor, (2007).
[5] I. Ajzen, From intentions to actions: A theory of planned Behaviour, In J. Kuhl and J.Beckmann (Eds.), Action-Control: From cognition to Behaviour, 11-39, (1985).
[6] I. Ajzen and M. Fishbein, Understanding attitudes and predicting social behaviour, Journal of Experimental Social Psychology, 45-76, (1980).
[7] I. Ajzen, Organizational behaviour and human decision processes, University of Massachusetts, (1991).
[8] J. Brown, D. Hess and D. Shoup, Unlimited access. Paper presented at the Conference of the Transit and University Communities, (1998).
[9] T.L. Wen and C.C. Fu, Behavioural intentions of public transit passengers: The roles of service quality, perceived value, satisfaction andInvolvement, Transport Policy Patent 18, 318-325, (2011),
[10] F. Satoshi and T.V. Hong, Psychological determinants of the intention to use bus in Ho Chi Minh City, Journal of Public Transportation, **12**, 1, (2009).
[11] A. Doug, Statistics in medicine, MRC Clinical Research Centre, Watford Road, United Kingdom, **1**, 59-71, (2006).
[12] J. Gliem and R. Gliem, Calculating, Interpreting, and reporting cronbach's alpha reliability coefficient for likert-type scales, *Midwest Research to Practice Conference in Adult, Continuing, and Community Education*, (2003).
[13] R. Weston and P.A. Gore, A brief guide to structural equation modeling, The Counseling Psychologist, **34**(5), 719 – 751, (2006).
[14] Nelson, Primitivo, Hussein and E. Crispin, An assesment of the van for hire as a local public transport mode in Bohol, Phillipines, University of the Philippines, (2005).
[15] Y. Heath and R. Gifford, Extending the theory of planned behaviour: predicting the use of public transportation, Journal of Applied Social Psychology, 30-78, (2002).

[16]K. Ambak, R. Ismail, R.A. Abdullah and M.N. Borhan, Prediction of helmet use among malaysian motorcyclist using structural equation modeling, Australian Journal of Basic and Applied Sciences **4**(10), 5263-5270, (2010).

[17]S. Sumaedi, G.M.Y. Bakti and M. Yarmen, The empirical study of public transport passengers' behavioral intentions: The roles of service quality, perceived sacrifice, perceived value, and satisfaction (Case study: Paratransit passengers in Jakarta, Indonesia), International Journal for Traffic and Transport Engineering, **2**(1), 83 – 97, (2012).

[18]M.B. Ahmed, K. Ambak, A. Raqib and N.S. Sukor, Helmet usage among adolescents in rural road from the extended theory of planned behaviour, Journal of Applied Sciences, **13**, 161-166, (2013).

[19]R. Ismail, M.H. Hafezi, R. Mohd Nor and K. Ambak, Passengers preference and satisfaction of public transport in Malaysia, Australian Journal of Basic and Applied Sciences, **6**(8), 410-416, (2012).

MATEC Web of Conferences **47**, 03009 (2016)
DOI: 10.1051/matecconf/20164703009

Pavement Response to Variable Tyre Pressure of Heavy Vehicles

Ahmad Kamil Arshad[1,a], Hairol Anuar Haron[2], Zanariah Abd Rahman[3] and A.G. Abdul Halim[4]

[1]*Institute for Infrastructure Engineering and Sustainable Management, Universiti Teknologi Mara, 40450 Shah Alam, Selangor, Malaysia*
[2]*Faculty of Civil Engineering, Universiti Teknologi Mara, 81750 Masai, Johor, Malaysia*
[3]*Faculty of Civil Engineering, Universiti Teknologi Mara, 40450 Shah Alam, Selangor, Malaysia*
[4]*Faculty of Civil and Environmental Engineering, Universiti Tun Hussein Onn Malaysia, 86400 Parit Raja, Johor, Malaysia*

Abstract. In recent years, the effect of overinflated tyre pressure and increased heavy vehicles' axle load on flexible pavements has become a subject of great concern because of the higher stress levels induced and damage caused to road pavements. This paper aims to evaluate the effect of variable tyre inflation pressures (using actual tyre contact/footprint area) to determine the responses of flexible pavement. A full scale experiment was conducted on a heavy vehicle with 1:1:2 axle configuration, 10 R 20 tyre size and attached trailer with constant axle load. Measurements were made for actual tyre-pavement contact area. KENPAVE linear elastic program was then used to analyse the effects of the measured actual tyre-pavement contact area and the results was compared using conventional circular tyre contact area. A comparative analysis was then made between the actual contact area and the conventional circular tyre contact area. It was found that high tyre inflation pressure produce smaller contact area, giving more detrimental effect on the flexible pavement. It was also found that the temperature of tyres when the heavy vehicles are operational give less significant impact on tyre inflation pressure for the Malaysian climate.

1 Introduction

The Government of Malaysia currently spends a substantial amount of money to maintain and rehabilitate road pavements. Heavy vehicles' overloading is one of the major factors contributing to pavement damage, in addition to other factors such as poor road construction and effect of the environment [1] [2]. As a consequence to the heavy vehicles' overloading, most of the truck operators increased the tyre inflation pressure to reduce tyre wear and tear thus increasing the life of the heavy vehicles` tyres. Therefore, a study on the detrimental effect of increasing tyre inflation pressure on the flexible pavement is important for designing flexible pavement structures.

A vehicle's loading is transferred to the pavement structure through the tyres. The interface between the tyre and pavement is generally assumed to be circular contact area and the relationship between tyre inflation pressure and contact area is given by following formula [3]:

[a] Corresponding author : drahmadkamil@salam.uitm.edu.my

$$p = \frac{P}{a^2 \pi} \tag{1}$$

where: p = tyre pressure in kPa, P = total load on the tyre in kN, and a = radius of contact in m.

Studies carried out on non-uniform tyre contact area using finite element and linear elastic programmes found significant effect on flexible pavement layers due to the increased in tensile and compressive strains [4-6]. According to Sebaaly, the magnitude and frequency of loading from the tyre to the pavement surface interface can influence the tyre contact area and tyre contact pressure [7]. In addition, Sebaaly and Tabatabaee found that tyre contact area depends on the tyre inflation pressure and tyre structure that influence the critical tensile strain at the bottom of the asphalt layer, surface deflection and compressive strength at the interface between the base and the asphaltic concrete [8].

This paper aims to evaluate the effect of variable tyre inflation pressures (using actual tyre footprint area) to determine the responses of flexible pavement. Analysis will be carried out to evaluate the impact of tyre inflation pressure on the allowable axle load repetitions for fatigue and permanent deformation based on Asphalt Institute's equations [9]. The analysis will determine the impact of overinflated tyre pressure on flexible pavements responses which has become a subject of great concern because of the higher stress level induced and causing increased damage to road pavements.

2 Methodology

A full scale experiment work was carried out to determine the effect of variable tyre inflation pressure on tyre contact area. The truck used in the experiment was a four-axle (single axle steer with single axle drive and dual tyre rear tandem) 32-tonne gross vehicle weight (GVW) and fitted with 10 R 20 tyres. Only the rear tandem axle was tested and the recommended tyre inflation pressure was chosen based on the tyre pressure data printed on the side wall of the tyre. Tyre inflation and deflation were carried out manually using a tyre inflation system with portable power source and an air compressor with a maximum working pressure of up to 1034 kPa (150 psi). The experiment was carried out on an existing flexible pavement at parking area in Universiti Teknologi MARA (UiTM), Shah Alam, Malaysia. Properties of existing pavement condition such as strength and pavement layer thickness was determined using falling weight deflectometer (FWD), coring test and dynamic cone penetrometer (DCP). The experimental works were conducted for two sessions, in the morning (0800hr-1000hr) and in the afternoon (1230hr-1430hr). The different sessions were conducted to measure the effects of temperature difference, particularly for the afternoon session, whereby the surface layer is at a higher temperature and therefore at a lower stiffness level. The lower stiffness value of the surface layer will reduce the load carrying capacity of the pavement. The temperature of the pavement surface and tyre were recorded during the experiment using non-contact infrared thermometer.

Tyre inflation pressure was set at the recommended tyre inflation pressure (676 kPa or 98 psi) printed at tyre side wall and then increased up to 50 % above the recommended tyre pressure at increments of 10%. The tyre was then decreased up to 50 % of the recommended tyre inflation pressure at increments of 10 %, (338 kPa or 49 psi - 676 kPa or 98 psi - 1014 kPa or 147 psi). Each inflation pressure was evaluated three times at separate sides of the tyre (right hand side outer tyre, right hand side inner tyre, left hand side outer tyre and left hand side inner tyre) for a total of 12 tyre footprint measurements for each inflation pressure.

Tyre footprint was obtained by free fall (static) measurement conducted for four days in the morning (0800hr-1000hr) and afternoon (1230hr-1430hr) using glossy black paint, stamped on an A3-size white paper underlain on a plastic cardboard of about 2 mm thick to avoid damaging the 80 gram A3 white paper due to the heavy vehicle loading. The tyre footprint was then scanned using a computer scanner.

Tyre footprint was calculated using computer software known as ImageJ v1.46 [9]. This software was designed to generate complicated non-uniform images. A scale was set in millimetres and calibrated with ordinary ruler measurement when scanning the tyre footprint images. After the scale was set, the image was then generated to calculate the non-uniform (actual) area for each region of the tyre contact area.

3 Results and Discussion

As the tyre pressure increases, the tyre-pavement contact area decreases for both sessions. The quadratic form models (for actual tyre contact area obtained versus tyre inflation pressure) during the morning session for both the right hand side (RHS) tyre and left hand side (LHS) tyre gave r^2 value of above 0.9 for all measurements of the tyre contact area when the tyre was deflated from 1014 kPa (147 psi) to 339 kPa (49 psi) (Figure 1). However, it was observed that there is a lower value of tyre contact area for the RHS outer tyre for both sessions. There is a large gap in value between those groups and the RHS outer tyre, possibly due to the problem of axle alignment or tyre tread condition. It can be seen that there is a big difference between conventional method and actual tyre contact area when the tyre inflation pressure starts to increase, however, for tyre pressure between 476 kPa (69 psi) to 607 kPa (88 psi), the difference is less than 10%.

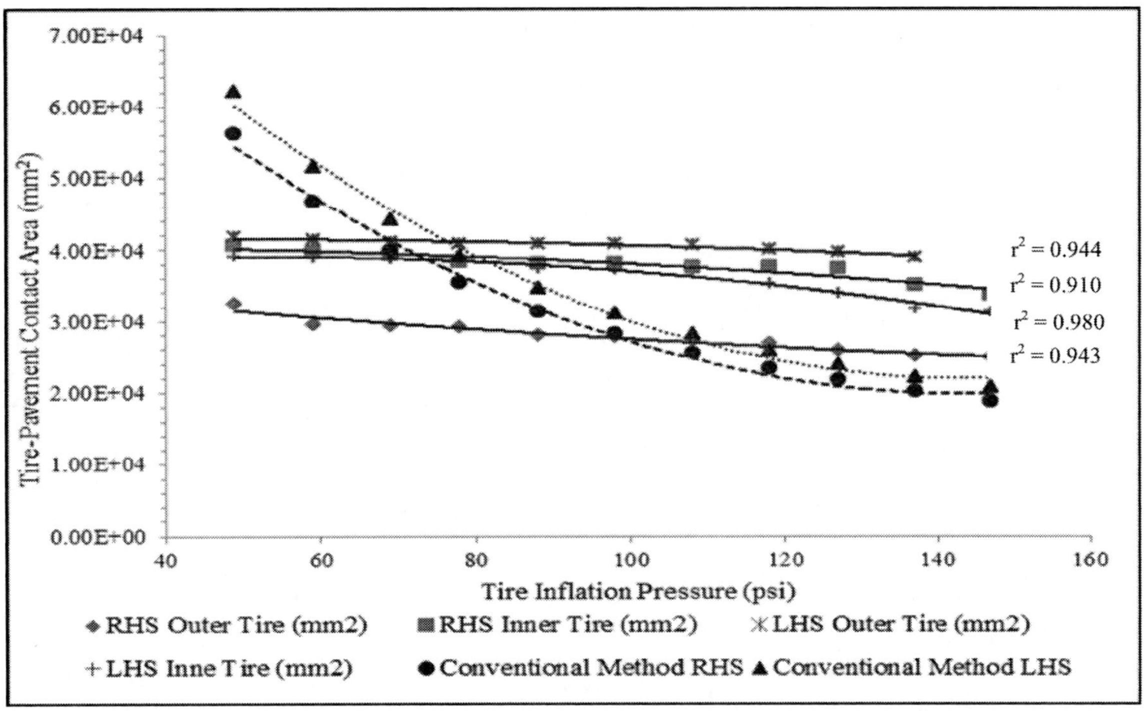

Figure 1. Correlation between tyre inflation pressure and tyre contact area for morning session.

For the afternoon session, the behaviour of tyre contact pressure act almost similar to the morning session. Figure 2 shows the quadratic form models for the RHS, LHS tyre axle and the conventional calculated circular area method. For tyre inflation pressure between 476 kPa (69 psi) to 538 kPa (78 psi), there is less than 10 % difference between conventional and actual non-uniform contact area. On RHS outer tyre, almost all tyre inflation pressure are similar except for 814 kPa (118 psi) and 876 kPa (127 psi) (over 10%) due to the effects of mechanical alignment problem or uneven tyre tread surface. However, it was observed that there is lower value of tyre contact area for the RHS outer tyre for both sessions.

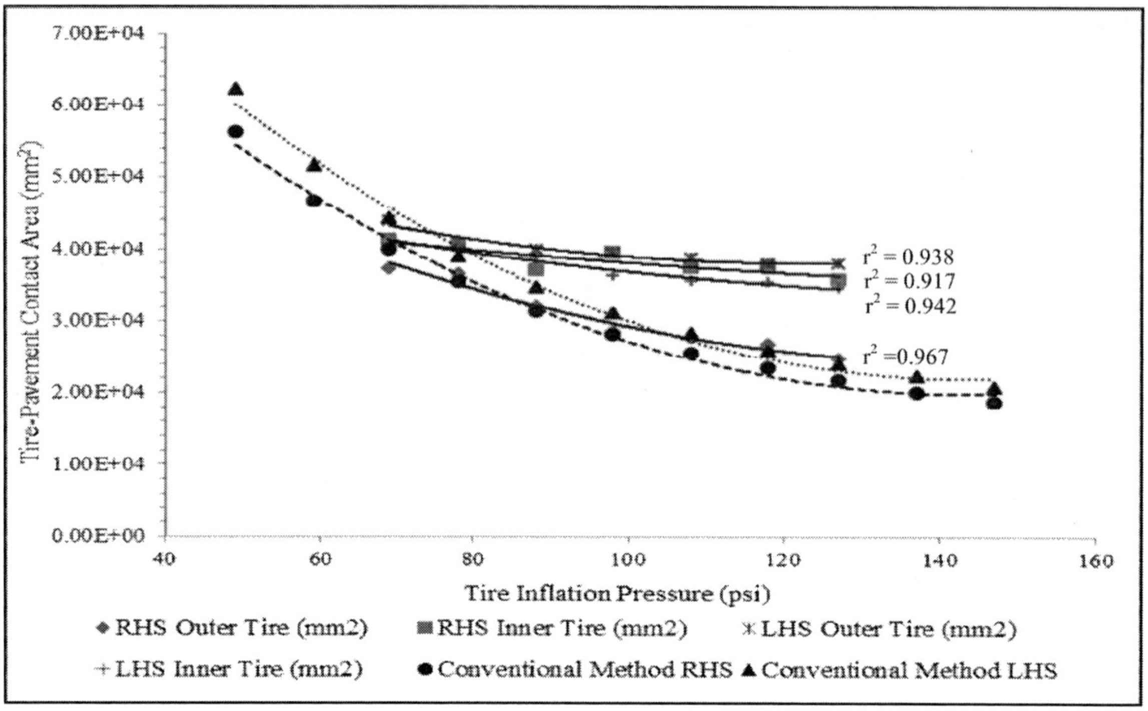

Figure 2. Correlation between tyre inflation pressure and tyre contact area for afternoon session.

4 Analysis of Pavement Response

Only the tyre inflation pressure range from 476 kPa (69 psi) to 876 kPa (127 psi) with constant axle load of 80 kN were analysed in order to evaluate the critical tensile strain at the bottom of asphalt layer, and critical vertical compressive strain at the top of subgrade in the middle of the tyre contact area. For the conventional circular area method, when the tyre inflation pressure increased (decrease of radius in tyre contact area), the tensile and compressive strains increased in value. This caused a decrease in the allowable axle load repetitions for both fatigue and permanent deformation, as shown in Table 1. However, the allowable load repetition for fatigue is lower than the allowable load repetition for permanent deformation in all cases.

Table 1. Pavement response for conventional circular area method.

Tyre Inflation Pressure (kPa)	Radius of Circular Contact Area by Conventional Method (cm)	Tensile Strain, ϵ_t (1×10^{-4})	Compressive Strain, ϵ_c (1×10^{-4})	Allowable Load Repetition for Fatigue Prediction (1×10^{4})	Allowable Load Repetition for Permanent Deformation Prediction (1×10^{4})
475.74	11.27	4.91	3.59	3.60	361
537.79	10.60	5.29	3.61	2.81	353
606.74	9.98	5.66	3.62	2.25	348
675.69	9.46	6.00	4.95	1.86	85.8
744.63	9.01	6.27	4.96	1.61	85.0
813.58	8.62	6.50	5.37	1.42	59.6
875.63	8.31	6.69	5.89	1.30	39.4

Table 2 shows the pavement response for the actual measured contact area in the morning session. As the tyre inflation pressure increased, the radius of the actual tyre contact area decreased. The actual tensile strain for the morning session increased when the actual contact area decreased and the difference was as high as 27% compared to the tensile strains calculated using the conventional circular area. For the compressive strain, the difference was only 12% but still above the conventional circular area method. As with the conventional circular area method, the allowable load repetition for fatigue is lower than the allowable load repetition for permanent deformation in all cases.

Pavement response for the afternoon session is shown in Table 3. The difference between conventional method and actual measured contact area is 25% for tensile strain and 12% for the compressive strain analysis. Similarly, as in the morning session, the allowable load repetition for fatigue is lower than the allowable load repetition for permanent deformation in all cases.

Table 2: Pavement response for morning session with actual tyre contact area.

Tyre Inflation Pressure (kPa)	**Radius of Actual Contact Area Morning (cm)**	**Tensile Strain, ϵ_t** (1×10^{-4})	**Compressive Strain, ϵ_c** (1×10^{-4})	**Allowable Load Repetition for Fatigue Prediction** (1×10^{4})	**Allowable Load Repetition for Permanent Deformation Prediction** (1×10^{4})
475.74	10.85	4.78	3.16	3.93	637
537.79	10.80	5.38	3.54	2.66	383
606.74	10.69	6.02	3.92	1.84	244
675.69	10.67	6.70	4.35	1.29	153
744.63	10.62	7.35	4.75	0.95	103
813.58	10.51	7.96	5.09	0.73	75.8
875.63	10.40	8.49	5.37	0.59	59.7

Table 3: Pavement response for afternoon session with actual tyre contact area.

Tyre Inflation Pressure (kPa)	**Radius of Actual Contact Area Afternoon (cm)**	**Tensile Strain, ϵ_t** (1×10^{-4})	**Compressive Strain, ϵ_c** (1×10^{-4})	**Allowable Load Repetition for Fatigue Prediction** (1×10^{4})	**Allowable Load Repetition for Permanent Deformation Prediction** (1×10^{4})
475.74	11.39	4.95	3.47	3.49	421
537.79	11.22	5.54	3.81	2.42	276
606.74	10.85	6.09	4.03	1.76	215
675.69	10.68	6.70	4.36	1.29	152
744.63	10.52	7.29	4.67	0.97	112
813.58	10.45	7.92	5.03	0.74	79.6
875.63	10.26	8.39	5.23	0.62	67.1

5 Conclusion

The actual measured contact area for the morning and afternoon session was measured and compared with the conventional calculated circular area. The results showed that there is a 10% difference between actual tyre contact area measured for tyre inflation pressure range of between 476 kPa (69 psi) to 607 kPa (88 psi) compared to the conventional circular area method. Above the 607 kPa (88 psi), the conventional circular area method gives much lower value for tyre contact area than the actual measured area, resulting in lower tensile strain value.

Using KENPAVE computer program, the pavement response showed a difference of 25% to 27% for tensile strains between actual measured contact area and conventional circular area method for both morning and afternoon sessions respectively. The compressive strain showed only 12 % difference between the measured and conventional circular area for both sessions. Similarly, as with the conventional circular area method and actual measured area, the allowable load repetition for fatigue is lower than the allowable load repetition for permanent deformation in all cases.

It is recommended that the study is expanded for other types of heavy vehicles on the full scale experimental approach to investigate the effects of tyre inflation pressure and temperature because of the difficulty to obtain the data when the heavy vehicles are in operation.

References

[1] Roadscanners Oy, Effect of axle and tyre configurations on pavement durability – a prestudy, The ROADEX Network, Finland, (2014).

[2] A.E. Behiry and A.Y. Beltagy, Mechanistic-empirical study of sensitivity of truck tire pressure to asphalt pavement thickness in Egypt, Int. J. of Eng. Res. and Apps., **3**(5), 1760-1771, (2013).

[3] E.J. Yoder and M.W. Witczak, *Principles of Pavement Design*, John Wiley and Sons, New York, (1975).

[4] D. Moazami, R. Muniandy, H. Hamid and Z.M. Yusof, Effect of tire footprint area in pavement response studies, Int. J. of Physical Sciences, **6**(21), 5040-5047, (2011).

[5] M.H. Nahi, A. Ismail and A.K. Arifin, Analysis of asphalt pavement under non-uniform tire-pavement contact stress using finite element method, J. of App.Sci., 11(14), 2562-2569, (2011).

[6] D.W. Park, A.E. Martin, J.H. Jeong,and S.T. Lee, Effect of tyre inflation pressure and load on predicted pavement strains, The Baltic J. of Road and Bridge Eng., 3(4), 181-186, (2008).

[7] P.E. Sebaaly, Pavement damage as related to tyres, pressure, axle loads, and configurations, ASTM STP 1164, 54-68, (1992).

[8] P.E. Sebaaly and N. Tabatabaee, Effect of tyre parameter on pavement damage and load-equivalency factors, ASCE J. of Trans. Eng., **118**(6), 805-819, (1992).

[9] Asphalt Institute, *Thickness Design Manual (MS-1),* Lexington, (1981).

[10] Information on http://rsb.info.nih.gov/ij/index.html

MATEC Web of Conferences **47**, 03010 (2016)
DOI: 10.1051/matecconf/20164703010

Assessment of Road Infrastructures Pertaining to Malaysian Experience

Norshakina Samsuddin[1,a] and Mohd Idrus Mohd Masirin[1]

[1]*Faculty of Civil and Environmental Engineering, Universiti Tun Hussein Onn Malaysia, 86400 Parit Raja, Johor, Malaysia*

Abstract. Road Infrastructures contribute towards many severe accidents and it needs supervision as to improve road safety levels. The numbers of fatalities have increased annually and road authority should seriously consider conducting programs or activities to periodically monitor, restore of improve road infrastructure. Implementation of road safety audits may reduce fatalities among road users and maintain road safety at acceptable standards. This paper is aimed to discuss the aspects of road infrastructure in Malaysia. The research signifies the impact of road hazards during the observations and the impact of road infrastructure types on road accidents. The F050 (Jalan Kluang-Batu Pahat) road case study showed that infrastructure risk is closely related with number of accident. As the infrastructure risk increase, the number of road accidents also increase. It was also found that different road zones along Jalan Kluang-Batu Pahat showed different level of intersection volume due to number of road intersection. Thus, it is hoped that by implementing continuous assessment on road infrastructures, it might be able to reduce road accidents and fatalities among drives and the community.

1 Introduction

Road infrastructure is the main element in road transport system in developing country like as Malaysia. It can be defined as the basic facilities, services and installation needed for the functioning of transport highway, roads and streets [7,8]. Implementation of road specific infrastructure can help to reduce road accident and improve the road safety level in all stage. In the world stage, about 1.3 million people die and as many as 50 million are injured or permanently disabled in road crashes [1]. The report has highlighted that almost half of those who die in road traffic crashes are pedestrian, cyclist and motorcyclist. It showed the vulnerable road user will offer to higher of casualties. They are generally the outcome of the interaction of many factors in chain in event involving the vehicle and the environment. The factor can be broadly categorized into human factor, road and environment. But, most of crashes are caused by human error but to fix the human mistake it's possible because in psychology finding human always make mistakes [2]. National safety study show investment in safer infrastructure is better to deliver twice the casualty saving provided by investment in either behavior or vehicle. In order to evaluate road infrastructure performance, auditing technique has been performed based on the Road Safety Audits Stage 5. Road infrastructure has been list according to the Guideline of Road Safety Audit Stage 5, Arahan Teknik Jalan (ATJ) and Guidebook for Traffic and

[a] Corresponding author : misshakina@gmail.com

Safety Audits in order to offer the safe road to road user. The components of the road infrastructure are road alignment, cross section, intersection (signalised and unsignalised), interchange, bridge, tunnel, vulnerable infrastructure, roadside safety and road surface. Poor design of the road infrastructure can be effect on the safe operation of the road. To fix the impact during the accident, road infrastructure can be used as pre-prevention techniques. Application of Road Safety Audits Stage 5 is a one of the method to ensure safe road infrastructure along the road in Malaysia.

2 Road Infrastructure Related Risk

Road infrastructure is one of the main factors determining the level of safety features in road transport system. In United State, road condition offer 53% factor on road crashes [6]. Poor road infrastructure can influence on the driver's compliance level and it will cause unsafe situation among the driver [9]. Serious consideration in road safety is the influence of the road geometry, roadside installation such as barriers, posts, road marking and etc. Related risk due to the bad infrastructure provided such as poor road side protection, poor road design, poor access management, work/construction zone safety, poor pavement management, poor traffic management and poor road hierarchy discipline or land use issue. Figure 1 shows the example of the infrastructure failure due to the road risk. It was also investigated that road infrastructures such as drains and other furniture may result in accident [3]. Mohd Idrus [4] also reported that undulating road surfaces may also contribute towards serious road accidents as studied in Batu Pahat, Malaysia.

Figure 1. Pavement defect and improper of ending point of barrier in Federal Route F001.

3 Safe Road Infrastructure

Safe road infrastructure may involve planning, design, implementation and maintenance in every level of activity. Therefore, proper management are needed to ensure all the element of road infrastructure was built and equipped according to the standard which able to function properly or not. Many lives could be saved and many accidents avoided if the existing road infrastructure was managed according to the best practice of safety engineering. The important to have a proper road infrastructure such as clear signs and road marking is essential if road users are to know what they are expected to do [10].

4 Applications of Road Safety Audits

Road safety audits have been declared as a proven method to prevent the road accident. RSA process aimed to identify of potential safety hazards on new road project at the appropriate stage, identify of hazardous features of an existing road so that it can be eliminated or treat before they become accident prone locations. It also to ensure the safety requirement of all road user are be consider in all stage and its aim to reduce the overall "through life" costs of a road project to the community. In Malaysia, they are five stages in RSA application which are: Stage 1 Audit: Feasibility and Planning Stage, Stage 2

Audit: Draft (Preliminary) Design Stage, Stage 3 Audit: Detailed Design Stage, Stage 4 Audit: The Pre-opening Stage and Stage 5 Audit: RSA of an Existing Road.

Figure 2. Sight distance and delinetion improvement [10].

4.1 Road safety audits stage 5

Application of the RSA in this stage will ensure safer road infrastructure operated on the existing road. It will help to identify safety problem which develop due to normal 'wear and tear' from traffic operation and identify hazards due to the way the landscaping matured such as trees, foliage obstructing traffic signs or distance impaired. Furthermore, at this stage audit will particularly look at the way each road user group 'sees' the road from their particular safety viewpoint. Road design standard made with regard to five main aspects namely functionality, economy, safety, comfort and aesthetic [8, 11]. The element list in this stage are vertical and horizontal alignment, cross section, intersection, expressway interchange and other grade separated intersections, traffic signal installation, traffic signing, roadside safety and landscaping, pavement marking, street lighting and general traffic management items. But in this study, its only focus on the road infrastructure consist of the road pavement conditions, road marking, cross section, road side safety and landscaping, road signing, street lighting, road barrier, traffic signal installation, road bridge and the number of intersection. All the items selected based on the road safety element that are related to the road environment and basic infrastructure needed along the road.

5 Method of Data Collection

Data collection on this study was conducted based on the research method according to the standard guideline on the locations of the studies. There two different data collection method, there are:

a) Road Infrastructure Data: Site observation was implemented to observe the state of weakness and lack of road infrastructure with reference guided manual of Road Safety Audits Stage 5, Arahan Teknik Jalan (ATJ) and Geometric Design Guide Road.The data analysed using HIRARC method where it helped to analyse the factors of the accident on the road based on the audits. The study areas in this stage only focus from KM 10 to KM 24. The area has been selected based on the data of the accident, zone area, and type of the infrastructure in that area.
b) Intersection Data: Preliminary works was carried in order to select the critical junction along the F050 road. Critical junctions was been selected based on the factor of level of users, location of the junction and strategic factor. Site observation was been conducted at the selected area (traffic volume count, junction audit) based on the JKR standards and RSA guidelines. The study areas only focus from KM 19 to KM 32. This is because the area must full fill the junctions characteristic such as main road of the junctions didn't have the road divider, unsignalised junctions and the junctions had been use for vehicles as the way to entering and exiting the house or the public utilities.

6 Results and Discussion

6.1 Road infrastructures analysis

From the data observation, HIRARC analysis was conducted to identify risk at every kilometer and every types of infrastructure. The risk assessed by considering the possibility danger at every type of infrastructure available along the road (KM 10 to KM 24). Figure shows the result of risk and analysis for each kilometer and percentage of risk analysis for each type of road infrastructure involved.

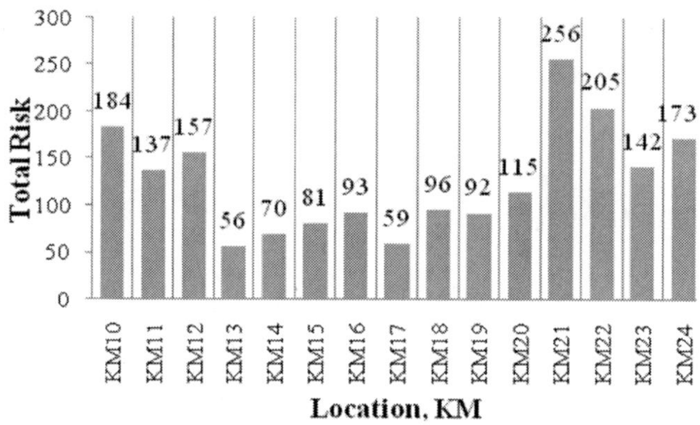

Figure 3. Total Risk based on the location.

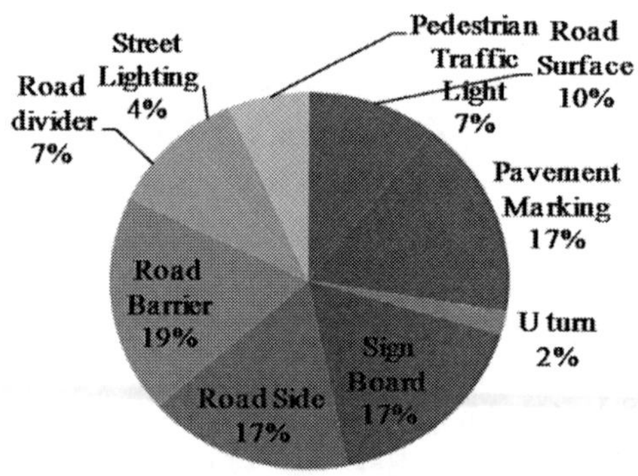

Figure 4. Percentage of Risk based on the types of infrastructure.

6.2 Road intersection analysis

From the data observation, critical junctions were selected based on the accident statistic along the F050 from KM 19 to KM 32. There for, 9 critical junctions were selected as follow in Table 1. Zone classifications are based on the number of the accident and number of the intersection in one kilometer area. There for, critical zone have been classified to the three phases, there are Less Critical Zone (1-19 accident), Semi Critical Zone (20-39 accident) and Critical Zone (40-59 accident).

Table 1. Critical junction choose and selection factor.

KM	No. of Junction	Junction Name	Junction Code	Selection Factor	Intersection Vol.
KM 21	1	Taman MajuBaru	TMB	• Shop house and residential areas. • Away from PintasPuding Traffic Light	2246
KM 22	3	Taman Senang	TS	• Linking many townships • Area near the city • Slopes intersection	2140
		Jalan Raja 1	JR1	• Intersection of shop house and housing • Urban areas and crowded	2354
		Taman BungaCempak aBaru	TBCB	• Shop houses and residential areas • Urban areas and crowded	2680
KM 23	1	Taman Manis 2	TM2	• Shop houses and residential areas • Linking many residential areas	1917
KM 24	1	Parit Haji Ali	PHA	• Intersection of big village	2032
KM 31	2	Taman UPC	UPC	• Near the corner of the road • The area of houses and shop house	1196
		Taman Ayer Hitam	TAH	• Near town and school • Slopes intersection	1367
KM 32	1	JalanSeroja (PejabatPos Air Hitam)	JS	• Urban area and crowded • Intersection of housing and Post Office • Near Traffic Light Ayer Hitam	1736

6.3 Related risk due to road infrastructure

Figure 5 shows the correlation between the infrastructure risks against accident. Based on the graph analysis against the accident rate, it can be conclude that the factors of road infrastructure can also cause of the accident.

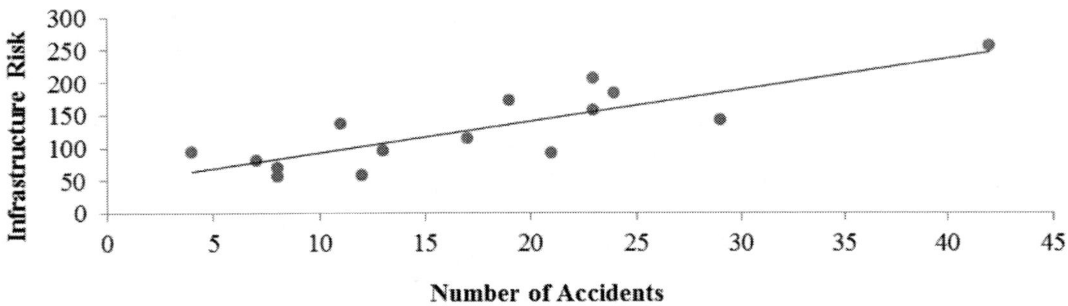

Figure 5. Graph of total infrastructure risk against accidents [4].

6.4 Related risk due to the access road

Figure 6 shows the correlation between volume of junction and accident data. Through the analysis it found that critical zone offer less of intersection with the high of accident rate and lowest to moderate of vehicle volume in each junction. It can be conclude that if the intersection has a high volume of vehicle, it was not guaranteed the accident always at high of the intersection. This is, probability of accident happen more to the location of the intersection and the layout and design of intersection. The dangerous location lead to serious accident occurs.

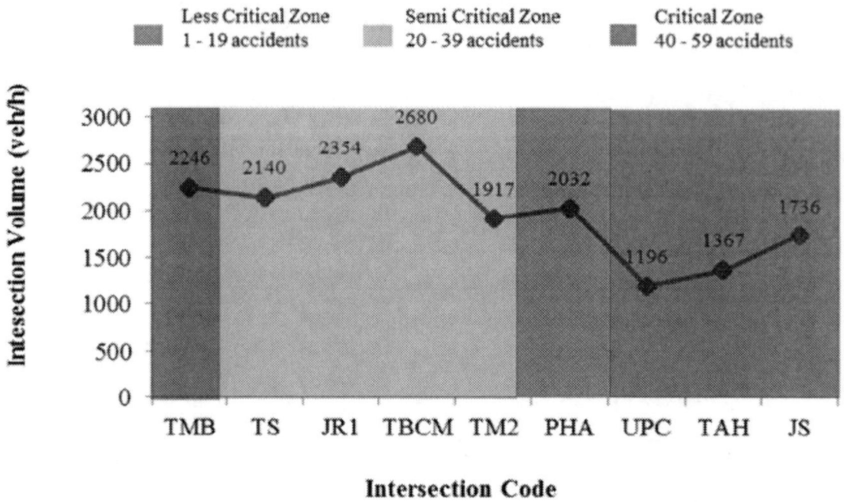

Figure 6. Relationship between the volumes of critical junction with accident zone.

7 Conclusion

Road safety infrastructure should be applied in all level of development, in order to maximize the impact due to engineering field and to solve the safety problem among the road user. By applying the good design principle, it will possible to avoid the safety problem occur. Through the implementation of road infrastructure audits and a set of data, we can define the hazard and problem occurs in that area. It will help the policy makers to make a decision considering the safety evaluation in the planning from the beginning as there is often when a need for extrasafety features. Installation of good and complete infrastructure components along the road would reduce the rate of accidents and it can make road user feel more comfortable driving on the roads.

References

[1] World Health Organization, Global Status Report On Road Safety 2013, Supporting A Decade of Action, Luxembourg, (2013).

[2] International Road Assessment Programme, iRAP Result, Malaysia, (2009).

[3] A.H.M. Nor, M.I.M. Masirin and M.E. Sanik, Site investigation of road drains for rural road on Batu Pahat soft clays (BPSC), J. Mech. and Civil Eng., IOSR, (2014).

[4] M.I.M Masirin, Performance Evaluation of Encapsulated Road Pavement on Different Ground Condition, PhD Thesis, British Library, United Kingdom, (2007).

[5] PDRM, Kemalangan Jalan Raya di Malaysia, Laporan Polis Diraja Malaysia 2013, Kuala Lumpur Malaysia, (2013).

[6] United Nation Road Safety Collaboration, A Policy Framework for Safe Infrastructure on Major Road Transport Networks, FIA Foundation Report, <available online http://www.irfnet.ch /files-upload/pdf.files/UNRSC%20Infrastructure%20Document%20lr%20Final.pdf>

[7] Inland Transportation Service, Road Infrastructure Safety Management on the Trans-European Networks, Consultation Paper TREN E3, European Commission, Brussels, (2006).

[8] M.I.M. Masirin, analysis of road infrastructural audit (RIS) along Jalan BatuPahat-Kluang Malaysia: A case study, *Proceeding of International Integrated Engineering Seminar*, Batu Pahat, (2014).

[9] D. Friday, Modification of road infrastructure to increase driver compliance in Uganda, International Journal of Advances in Management and Economics, **1**(3), 78-87, (2012).
[10] R.N. Zaman, Road Safety Audit Report MRSA 01/2012, Malaysian Institute of Road Safety Research, (2012).
[11] M. Sanusi, Audit Keselamatan Jalan Raya Dalam Kampus Universiti Teknologi Malaysia, Dissertation, Universiti Teknologi Malaysia, Skudai, (2006).
[12] Public Work Department, *Arahan Teknik Jalan*, Kuala Lumpur, (2006).

MATEC Web of Conferences **47**, 03011 (2016)
DOI: 10.1051/matecconf/20164703011

Passenger Rail Service Comfortability in Kuala Lumpur Urban Transit System

Noor Hafiza Nordin[1], Mohd Idrus Mohd Masirin[1,a], Mohd Imran Ghazali[2] and Muhammad Isom Azis[3]

[1]*Faculty of Civil and Environmental Engineering, Universiti Tun Hussien Onn Malaysia, 86400 Parit Raja, Johor, Malaysia*
[2]*Faculty of Mechanical and Manufacturing Engineering, Universiti Tun Hussien Onn Malaysia, 86400 Parit Raja, Johor, Malaysia*
[3]*Prasarana Negara Berhad, 59000 Bangsar, Kuala Lumpur, Malaysia*

Abstract. Rail transit transportation system is among the public transportation network in Kuala Lumpur City. Some important elements in establishing this system are ticket price, operation cost, maintenance implications, service quality and passenger's comfortability. The level of passenger's comfortability in the coach is important to be considered by the relevant authorities and system operators in order to provide comfort and safety to passengers. The objective this research is to study some parameters that impact the comfortability of passengers and to obtain feedbacks from passengers for different rail transit system. Site observations were conducted to obtain data such as noise, vibration, speed and coach layouts which will be verified by using the passenger feedback outcomes. The research will be focused in and around the Kuala Lumpur City for the duration of 10 months. Four rail transit systems were being considered, i.e. Train Type A (LRA), Train Type B (LRB), Train Type C (MRL) and Train Type D (CTR). Data parameters obtained from field observations were conducted in the rail coaches during actual operation using apparatus among others the sound level meter (SLM), vibration analyzer (VA) and the global positioning system (GPS). Questionnaires were prepared as passenger feedback instrument, focusing on the level of comfortability in rail coaches. The outcomes of these analyses showed that CTR was the best and most comfortable coach, followed by LRA, LRB and MRL. Using the passenger feedbacks (MBP) it was also found that the ranking values for the railway transit system using Rail Coach Comfortability (RCC) is RCC (MBP) were the same as the outcomes using the RCC (parameters). In conclusion, it was found that this research has successfully determined the level of comfortability as determined in its aim and objectives. From this research, it is hoped that the relevant authorities will continue to find ways to enhance the comfortability and safety of its train coaches especially its passengers and the community at large.

1 Introduction

Rail transportation service becomes the first choice mode of public transportation in city area. Rail service giving a better solution for traffic congestion and served passengers the quality of service and comfort [3]. It also provides a better quality of environment and reduces the air pollution cause by transportation used diesel as power of supply. Rail service use own rail tracks for transport passengers and freight, and its contrast to other public transportation where vehicle run on the road. Basic infrastructure requirements for developing railway system usually consists rail, track, rolling stock,

[a] Corresponding author: idrusmas@gmail.com

stations and signaling system [2]. A rail services is expected to connect effectively and efficiently the population in conducting their business and daily activities. It may be classified into different categories depending on its demographic nature or its types of service [5]. Demographically, it can be categories as urban and urban services whereas typically, it can be categories as cargo, high speed, light rail transit, mass rapid transit, monorail, passenger train aero-train, depending on its purpose and designation. In this paper, the focus will be mainly on the urban transit systems that exist in Kuala Lumpur city [1].

A rail transit system will require specific infrastructures to be conducted. It needs rail system with signaling and control systems, the stations as terminal stops with platforms and information system, and it also needs several sets of strains which will commute at designated frequencies to deliver its services effectively and efficiently [5, 6]. Thus, a rail transit system differs totally on its operation nature when compared with other transportation system. The most pertinent difference between train system and others are train will need a right of way system as it only operates on rails which does not same with other transportation system [7].

2 Urban Rail Transit Systems

Urban rail transit systems are mentioned earlies is getting more popular as it seems to be an effective alternative for urban transportation to reduce congestion. Many cities such as London, Paris, Seoul, Kuala Lumpur and China were successful in managing their traffic flow to reduce the travelling time of motorist on the roads [2]. It seems to successfully the numbers of vehicles on the roads. The needs of an effective and efficient urban transit system is more pertinent as the growth of population is getting more rapid due to the current economic needs of more human resources. Migration of the people from rural to urban areas due to availability of employment in the cities has increased the needs for an alternative transportation system [3].

As mentioned earlier, there are categories of urban transit systems and one of them is the rail transit system. In Malaysia, the most popular urban transit system are the bus and taxis, whereas rail transit system is only for long distance travel, even though currently in Kuala Lumpur it is popular as an urban transit system due to its punctuality and less hassle during peak hours. However, the Malaysian government is currently contemplating to encourage other cities to plan and built rail transit system into their urban transportation network [1]. This is again important to reduce congestion during peak hours, and to coupe with future population growth, to enhance and create the image and landscape of a modern city and to reduce air pollution due to vehicular carbon emission [5].

3 Impact of Passenger Comfortability on Rail Transit Service

Better riding quality, especially for passengers travelling long distance needs the high level of comfort. The key for sustainable of passengers using one of public transportation is when the riding comfort parameters such as noise, vibration, design layout and speed are acceptable to them. These parameters were studied by various parties such as rail operators, engineers, scientist and researchers [1, 11]. Most of them were concerned about punctuality, vibration and noise. Others are more concern on ways to attract passengers using the rail transit system [9]. The economics of establishing a rail transit system is always a concern, thus many authorities usually will give or stress more on ensuring that the ROI (Return of Investment) period will be considered thoroughly which depends on the population growth as the main demand indicator.

Comfortability in rail service depends on the perception of passengers whereas efficiencies and effectiveness will depend on the operation and delivery of the service [5, 11, 12]. If the service frequency in punctual and as expected, then the rating will be high while if it is not, then it will be considered as poor and not trustworthy. Comfortability of rail transit service will determine the numbers of commuters as the more popular it is. The more passengers it will be attracting. Thus, the ROI of a train system or operation will depend not only on its technical delivery but also on its marketing and ability to attract new passengers from time to time, while reducing the technical problems such as vibration and noise disturbances to passengers.

4 Results and Analysis

This is a research conducted in collaboration with Prasarana Negara Bhd. and Keretapi Tanah Melayu Berhad, two main railway operations in Malaysia. This study was carried out on rail transport system in the city of Kuala Lumpur. The results of the research conducted in the field, the data for the specified parameters have been obtained. The focus of the research is mainly on vibration and noise with another parameter is also discussed i.e. design layout which affects comfortability of passengers in railway coaches. All these parameters are analyzed to determine the level of passengers comfort and data from passenger feedback are used as the verification process. All of this data observation was carried out on four (4) types of selected rail transport system. The data obtained were used to determine the level of passenger comfort in coaches for every type of rail system under study. Selected routes with pre-determined number of stations were observed as samples of the research work. The data observed were based on field and passenger feedbacks.

Vibration parameters were analyzed for this study is the vibration displacement and velocity. It is considered important due to the potential cause discomfort while operating the passenger coaches. According to Table 1, it was found that the LRA system producing the least vibration and very similar to LRB system. This shows that light rail transport system, LRT, producing the least vibration level compared to other rail transport system. Whereas MRL system is the rail system with the highest income level of vibration. This may be due to system or a mono track has created a situation less stable over time. Therefore, the study of the real causes vibration viewed more than 5 times of that produced by the LRT system to be done to improve the rail transportation system based on vibration impact on passengers. CTR system, which is categories as sub-urban rail system, also viewed produce high vibration compared to the LRT system. But it's still an acceptable level of around 0.1μm for displacement and 0.1mm/s based on the specification operators and authorities rail transport system.

Table 1. Ranking of vibration parameter.

No.	Types of rail system	Displacement (μm)	Vibration Velocity (mm/s)	Ranking
1.	LRA	0.33553	0.03234	1
2.	LRB	0.14408	0.03400	2
3.	MRL	1.67692	0.42292	4
4.	CTR	0.85608	0.08517	3

*Ranking is based on displacement and vibration velocity

The data tabulated were obtained from concurrent observations with the vibration observations. According to collected data, Table 2 was produced and the comparability of the rail system is showed. The lowest average noise level for this study was developed by CTR system while the highest is MRL system. The range of sound, is the sound level of the highest and lowest is the biggest LRB rail systems while the smallest is MRL system. However, the observed range was almost in each other, a difference of 2 to 3 dBA only and is considered a small value. However, the level of sound emitted by the rail system being studied, it is still within the sound level required by the specifications of the original operator and the authorities are around 75dBA.

The layout was obtained from the operators which were approved for construction. In Table 3, the best rail transit system in terms of layout and the number of passengers is the CTR system. It's because this rail transit categories as sub-urban system, the better facilities provided to passengers because of long distance and travel of time. This resulted in CTR provides excellent facilities to give passengers the high level of comfort. While the lowest rail system is MRL system. However, this occurs because MRL system has limitations in terms of control and balance during operation. Its size is only built for 2 cars with the maximum number of passengers a total of 158 people led to poor levels of passenger comfort, especially at peak hours.

Table 2. Ranking for noise parameter.

No.	Types of Rail System	Standard (dBA)	Average Noise Level (dBA)	Difference Range in Noise Level (dBA)	Average Noise Level (dBA)	Ranking
1.	LRA	75	73.8 – 78.3	4.5	76.0	3
2.	LRB	75	69.0 – 75.9	6.9	72.0	2
3.	MRL	75	74.8 – 78.9	4.1	77.0	4
4.	CTR	75	68.2 – 74.0	5.8	71.2	1

*Ranking is based on average noise level generated

Table 3. Ranking for coach design layout.

No.	Types of Rail System	Number of Passengers (Specification)	Number of Passengers (Observation)	Floor Area Coaches (m^2)	Number of Doors	Number of OKU Seats	Ranking
1.	LRA	166	137	44.255	6	1	2
2.	LRB	156	192	40.152	6	-	3
3.	MRL	92	112	25.836	4	-	4
4.	CTR	259	171	62.700	6	2	1

*OKU – Disable person

Summary of field observation results based on all parameters that were observed, Table 4 showed the summary of the four types of rail systems in City of Kuala Lumpur. From the field data analysis, it shows that the level of comfort in the CTR system coach is the best, followed by LRB, LRA and MRL systems.

Table 4. Summary of field data analysis.

No.	Types of Rail System	Types of Parameter	Ranking (from field data analysis)	Total Point	Ranking
1.	LRA	Noise	3	10	2
		Vibration	1		
		Speed	4		
		Coach design layout	2		
2.	LRB	Noise	2	10	2
		Vibration	2		
		Speed	3		
		Coach design layout	3		
3.	MRL	Noise	4	14	3
		Vibration	4		
		Speed	2		
		Coach design layout	4		
4.	CTR	Noise	1	6	1
		Vibration	3		
		Speed	1		
		Coach design layout	1		

Note: LRA, LRB, MRL and CTR denote the different types of rail system in Kuala Lumpur. LR means the light rail, MRL means the monorail and CTR means the Commuter system.

MRL system is the worst because of the effects of noise and vibration in the coach is the lowest such an arrangement coaches. If the ranking was made without involving the CTR system as it is categorized as sub-urban rail transit, the LRB system is the best as it categorized as urban rail transit, followed by the LRA system and MRL system. However LRB system and LRA system is distinguished only by the speed compliance during operation. This system does not use weighted method but rather direct observation values based on ranking made by respondent's feedbacks and field observations value.

Summary of passenger feedbacks were obtained through questionnaires conducted during the research study duration. It took ten months for the researchers to gather enough feedbacks from the passengers as many were reluctant to sit and answer the questionnaires due to their busy schedules. The followings are the results obtained as shown in Table 5 and Table 6.

Table 5. Ranking from RCC (parameters) and RCC (passenger feedbacks).

No.	Types of rail system	RCC (Parameter)	RCC (MBP)
1	LRA	2	3
2	LRB	2	2
3	MRL	3	4
4	CTR	1	1

The results and analyses of the research case study showed that different train systems will generate different comfortability levels. The research was able to rank the different train systems which operate in Kuala Lumpur City according to field data observations and passenger feedback analysis. From these analyses, as shown in Table 6, it was found that the commuter system is ranked the best whereas light rail transit is ranked second and third. However, trailing close is the monorail which is considered a new addition to the urban rail transit in Kuala Lumpur. The commuter system is classified as a sub-urban train service and it is expected that it will give a better service compared to the light rail and the monorail. By disregarding the commuter system, it leaves the light rail as the best urban rail transit option over the monorail and the setbacks of the service found during the research can be improved for better future rail transit service. The ranking showed in Table 7 were based on respondent feedbacks which gave their opinions as whether the rail coach condition and environment are best or worst. From Table 5, CTR was voted the best coach in all questions posed to passengers whereas MRL gets 7 worst votes from 7 questions. LRB and LRA are ranked second and third best if compared to MRL.

Table 6. Respondent rating (best or worst) through questionnaire for different rail coaches.

Question		A1	A2	B	CI	C2	D	E1	E2	E3	E4
Rating	Best	CTR	CTR	CTR	CTR	CTR	CTR	CTR	CTR	CTR	CTR
	Worst	MRL	MRL	MRL	LRA	LRA	MRL	MRL	MRL	LRB	MRL

Table 7. Overall performance ranking of different rail transit services in Kuala Lumpur.

Overall Feedback Ranking Analysis			
No.	**Nos of Best Ranking (Based on Parameters and Feedbacks)**	**Nos of Lowest Ranking (based on Parameters and Feedbacks)**	**Ranking**
1.	LRA = 0	LRA = 2	**3**
2.	LRB = 0	LRB = 1	**2**
3.	MRL = 0	MRL = 7	**4**
4.	CTR = 10	CTR = 0	**1**

5 Conclusions

As a conclusion, it may be summarized that passenger comfortability is important as it may be the pulling factor for the public to commute to and from their offices or for leisure, using the urban rail transit system. The case study conducted by the researcher is a continuing research project. Thus, determining the comfortability level or range will somehow enhance credibility of the train services and enhance the traffic mobility in a city such as Kuala Lumpur. In many metropolitan cities such as London, Paris, New York, Seoul, Shanghai and Tokyo, traffic congestions are tremendously organized and controlled during peak hours. However, in cities such as Jakarta, Bangkok, New Delhi and several cities without rail transit systems, traffic mobility and congestion are still experienced by the local residents. Thus, this has an impact to the well-being of the city population and the economy of the country.

References

[1] N.H. Nordin, Kajian Tahap Keselesaan di dalam Koc bagi Sistem Transit Rel yang Berbeza, Master Thesis, Universiti Tun Hussein Onn Malaysia, Johor, (2013).

[2] N.F. Johari, Comparative Analysis of Train Coach Noise and Vibration for LRT (Light Rail Transit) and Monorail in the City of Kuala Lumpur, Bachelor Thesis, Universiti Tun Hussein Onn Malaysia, Johor, (2014).

[3] Z. Zakaria, Z. Hussin, M. F. Abdul Batau and Z. Zakaria, Service Quality of Malaysian Public Transports, Cross-Cultural Communication, **6**(2), 84 – 90, (2010).

[4] S. Bachok, M.M. Osman, M. Murad and M. Ibrahim, An Assessment Of Commuter's Perception Of Safety And Comfort Levels Of 'Women-Only Coach': The case study of KTM Komuter Malaysia, *The 4th International Conference on Sustainable Future for Human Security, Sustain*, (2013).

[5] M.I.M. Masirin and M. Idris, Urban transportation system: Utilisation of energy resources, *International Recognition Seminar 2013*, Batu Pahat, (2013).

[6] S. Ponnuswamy, *Railway Transportation Engineering, Operation and Management*, Alpha Science, Oxford, (2012).

[7] V. Vuchic, *Urban Transit: Operations, Planning and Economics*, John Wiley Publication, New Jersey, (2005).

[8] CPCS Transcom Limited, The state of Rail Safety in Canada, (2007).

[9] N.H. Nordin, M.I. Mohd Masirin, M.I. Ghazali and M.I. Azis, Appraisal on rail transit development: Train services and safety, *International Integrated Engineering Summit*, (2014).

[10] F. Akabal, M.I. Mohd Masirin, Z.A. Akasah and M.M. Rohani, Review on selection and suitability of rail transit station design pertaining to public safety, *International Integrated Engineering Summit*, Batu Pahat, (2014).

[11] M.I. Mohd Masirin, N.F. Johari, N.H. Nordin and A.H. Abdullah, A field study on urban transits in city of Kuala Lumpur: Passenger views on train noise and vibration, Journal for Applied Mechanics and Materials, **773-774**, 839-844, (2015).

[12] M.I. Mohd Masirin, A.M. Salin, A. Zainorabidin, D. Martin and N. Samsuddin, Appraisal on Malaysian rural rail transit operation and management: Issues and solution in integration, *International Integrated Engineering Summit*, Batu Pahat, (2014).

MATEC Web of Conferences **47**, 03012 (2016)
DOI: 10.1051/matecconf/20164703012

Railway Station Facilities in Rural and Urban Services Based on Observation and User Feedback

Mohd Idrus Mohd Masirin[1,a], Ili Izzati Ahmad Fuad[2] and Noor Hafiza Nordin[2]

[1]*Smart Driving Research Center, Universiti Tun Hussein Onn Malaysia, 86400 Parit Raja, Johor, Malaysia*
[2]*Faculty of Civil and Environmental Engineering, Universiti Tun Hussein Onn Malaysia, 86400 Parit Raja, Johor, Malaysia*

Abstract. Public transport is one of the facilities available in Malaysia. Overwhelming demand from the community has demanded public transport as an alternative for some people to do their daily activities. The increasing of Malaysian population brought many problems to the environment especially traffic congestion problem. The Government spends a lot of money to overcome this problem [10]. Public transport has been upgraded to ensure user comfort. Most people in the city are more interested in using railway transport as traffic congestion and travel time accuracy is better than other public transport [1, 5]. To influence public demand, the facilities available at the train station should be complete [3, 5]. The choice of location for the study are Batu Gajah station (station 1) and Kluang station (station 2) for rural services while Bandar Tasik Selatan station (station 3) and KL Central station (station 4) for urban services. The results of this study showed that, based on data and information obtained from field survey work and the feedback questionnaires from users, basic facilities at railway stations surveyed for urban and rural services can be identified and listed. An analysis of the status of the facilities at railway stations can be established through the safety of passengers and observations in the field. The facilities available at the urban service station are more complete than in rural service station. Facilities at railway station should be improved to attract people using the facilities.

1 Introduction

In Malaysia, public transport such as trains becomes an alternative transport to travel from one destination to another. There are many advantages by using public transport which is can reduce traffic congestion and increase the country's economy. Referring to the statistical population of Malaysia, in 1980s the number of population is about 18 million, and on 2013 the number of population has reached to 28.3 million. This number will continue to increase and it will cause many problems if there is no good traffic management and transport facilities [1, 7]. Traffic congestion is considered one of the problems, especially in urban areas. Public transport such as train is able to create sustainable urban development where it can reduce environmental pollution [2]. This public transport can reduce carbon emissions and toxic gases. Tourist also use train as their transport due to the convenient and less expansive compared to other public transport. Safety of user should be emphasized by management to attract people to use train [9, 11].

[a] Corresponding author : idrusmas@gmail.com

Railway facilities are important elements to attract commuters to use the rail services [6]. Passengers perceived that rail services are considered safe if the facilities are adequate and easy to access. Facilities in trains and stations are necessary to assist operators in providing good services. Elements that may affect quality train services are in-train facilities, station facilities, operation, maintenance and policies. In-train facilities will contribute towards passenger comfortability such as the provision of information, facilities in the coaches, safety, thermal comfort and technical elements during operation [5]. These factors are similar in urban and rural areas but with different approach and outlooks [4, 8, 12]. A good rail transit station should be able to make passengers feel comfortable and safe when they are at the station awaiting the train. Thus, it is important to provide and ensure that the railway station facilities are always audited, maintained and upgraded from time to time. However, in planning, it is easy to identify and suggest an ideal railway station facility and design but, several constrains and challengers such as financial, policies and procedures may dampen the realization of a good proposal for construction of a railway station [13].

2 The Research Methods

The research methodology is the most important aspect in a study. The results of the correct selection method will facilitate the search for information to carry out this study. There are many approaches that should be taken. Apart from literature review, on-site monitoring is important to get a more accurate analysis results. Additional information is also needed, such as journals, reference books, the internet, newspapers and others to provide more approaches to the study. Literature review is used to increasing understanding of the comparative facilities of the in urban and rural services. Site observation method and questionnaire method is used to achieve the objectives.

Several problems can be identified from site observation method. Through these problems, researchers can get a solution to overcome the problems. The researchers could identify the type of facilities available at the train station for rural services and urban services. 4 stations are selected for this study, where two stations for urban service and two stations for rural service. Station for rural services are Batu Gajah station (station 1) and Kluang station (station 2) while station for urban services are KL Central station (station 4) and Bandar Tasik Selatan station (station 3). Each station has different facilities. 6 elements needed to assess the facilities at train station which are information, safety, service, emergency, comfort and ventilation system. The objective of this study is to identify the basic facilities at the train station for urban services and rural services. To achieve this objective, observation form is needed. The purpose of this form is to facilitate the process of analyzing the study.

The questionnaire method was conducted in order to obtain information from users of railway services [3]. The questionnaire was based on objective of the research. Several questions which is related to the facilities at the railway station need to be answered by the respondents. Each respondent was asked the same question. Therefore, the comparison can be made between a respondent to another respondent. This is an effective method because the data could be obtained quickly. Questionnaires were distributed to 100 sets at the train station in the rural areas. The questionnaire is divided into seven sections. Part 1 is consisting of respondent information. Meanwhile, part 2 consists of four questions that involve information related question. Part 3 contains three questions that involve security related questions. Service related questions were asked in section 4 which containing 7 questions. 4 questions were asked in section 5 which are emergencies related question. 6 questions were asked in section 6 which are comfort related question. Section 7 contains two questions related to the ventilation system.

3 Results and Discussion

Observations work was done by researcher using observation form. It is to determine the facilities based on users safety and comfort at train station. 6 elements needed to assess the facilities at train station which are ventilation system, information, safety, emergency, comfort and service. The numbers of respondent for the elements were then converted to percentages to identify the composition of the feedbacks during the observations. It is considered easier to conduct analysis and to determine the majority opinions of the respondent.

Figure 1 showed percentages of signage at stations 1 and 2 are 25%, while at stations 3 and 4 are much higher which are 35% and 80%. Percentage of the board was higher at station 3 and 4 which are 50% and 80% compared to station 1 and 2 which are 50% and 40%. The percentage of signboards also shows that the rate of rural services are low which 40% at stations 1 and 30% at station 2 compared to urban service station which is 50% at station 3 and 90% at station 4. Percentage of visual notice board also show high percentage at station 3 and 4 which are 50% and 80% compared to the percentage of stations 1 and 2 which is 40% and no visual notice board found in station 2.

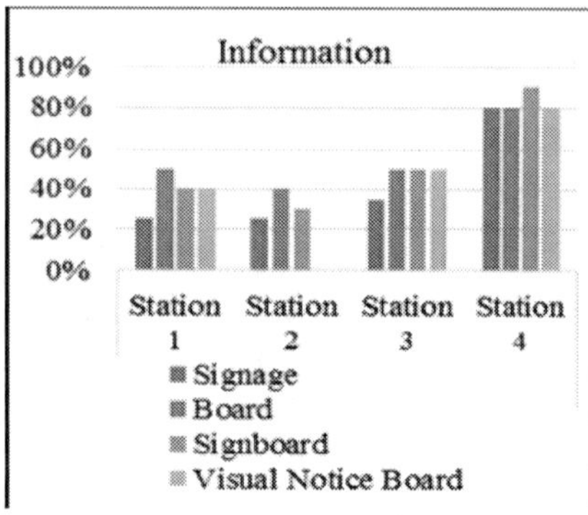

Figure 1. Graph percentages of information facilities at stations.

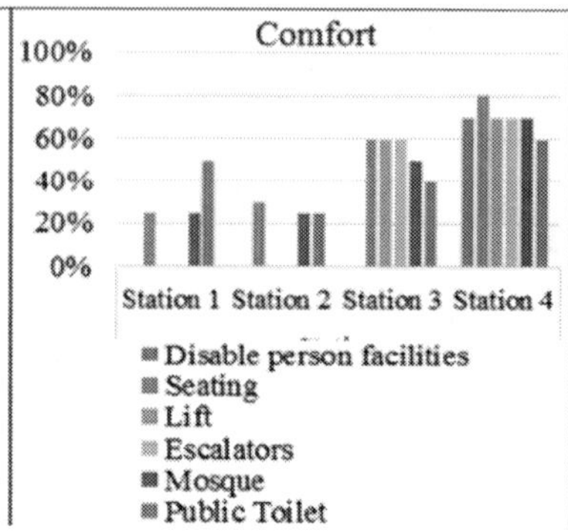

Figure 2. Graph percentages of comfort facilities at stations.

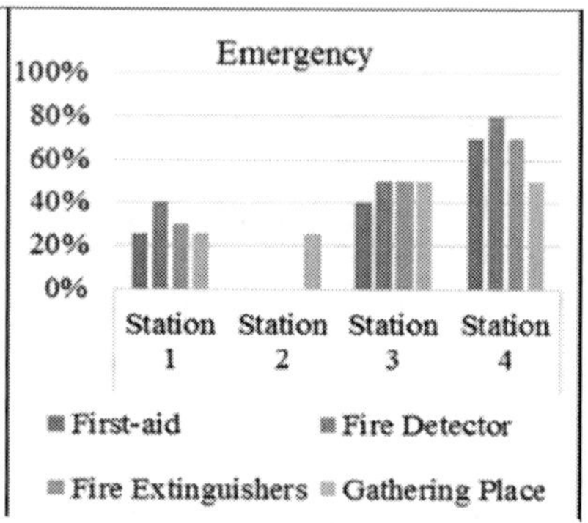

Figure 3. Graph percentages of emergency facilities at stations.

Figure 2 showed facilities for disable person only available at station 4 with percentage 70%. Seating showed low percentage at stations 1 and 2 which are 25% and 30% compared to station 3 and 4 which are 60% and 80%. Lifts and escalators only allocated at station 3 and 4, which are 60% and 70%. In addition, the percentage of mosque facilities at stations 1 and 2 is also lower than the percentage at stations 3 and 4. The percentage of the public toilet facilities at stations 1 and 2 are 50% and 25% compared to the percentage of stations 3 and 4, which give a high percentage. In addition, the percentage of parking is also high at station 3 and 4 exceed 50% compared to the two other stations. This because this station (3 & 4) are the interchange station for 2 lane which are LRT, KTM. The percentage of first-aid facilities in Figure 3 showed a high rate at station 3 and 4 which are 40% and 70% compared to station 1 which is 25% and no first aid facilities provided at stations 2. The percentage at station 1 and 2 for fire detector is lower than the percentage at station 3 and 4 which are 50% and 80%. The percentage of fire extinguishers at station 1 and 2 are also lower which are 30% and it was not provided at the station 2 while at station 3 and 4 give a high percentage which are 50% and 70%. Percentage of gathering place was also high at station 3 and 4, with a percentage of 50% compared to 25% of the station 1 and 2. The percentage of closed-circuit camera at station 1 and 2 in Figure 4 showed a low rate of 25% at station 1 and there is no closed circuit cameras at the station 2 were found showed in Table 6. Station 3 and 4 are higher which are 40% and 80%. The police officer was found at station 1 and 2 are 25% for both stations compared to station 3 and 4, 25% at station 3 and 80% at station 4. Figure 5 showed the percentage of ticket counter facilities give a high percentage at station 3 and 4 which are 60% and 80% compared to station 1 and 2 which are 40% and 20%. Customer service also give low percentage at station 1 and 2 which are 50% compared to station 3 and 4 which are 50% and 75%. In addition, the automated teller machines also available at station 4, with 75% percentage. Kiosk and business facilities only available at station 1 with a percentage of 25%. For urban services, station 3 and 4 give a high percentage of 50% and 80%. Many bins available at urban services with 60% percentage at station 3 and 80% at station 4 compared to rural services with 30% percentage at station 1 and 35% at station 2. Locker and ticket vending machine are available at station 4 with 90% and 75% percentage. Figure 6 showed the percentage of fan facility at station 1 and 2 are low which are 40% and 25% compared to station 3 and 4, which showed a high percentage rate which is 50% and 80%. In addition, air conditioner is only available at station 4.

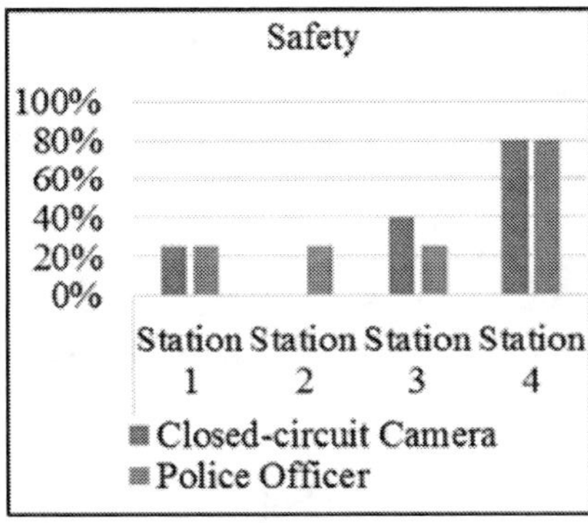

Figure 4. Graph percentages of safety facilities at stations.

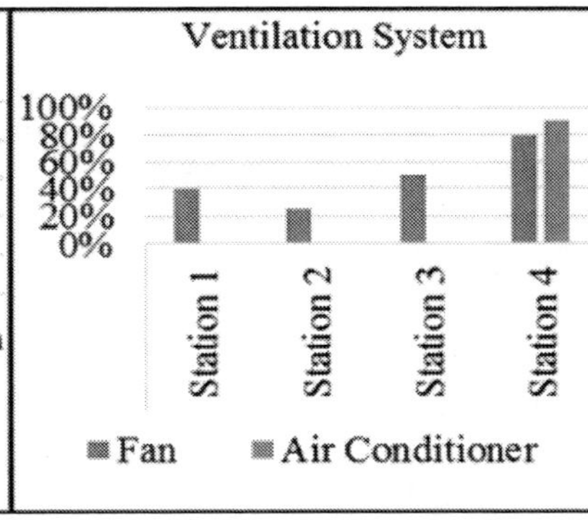

Figure 5. Graph percentages of services facilities at stations.

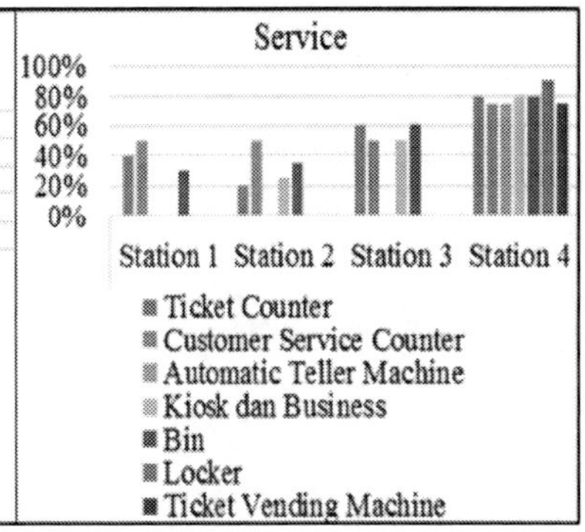

Figure 6. Graph percentages of ventilation system facilities at stations.

User Feedback on urban services demonstrates a better percentage than rural services. Many of the respondent from urban services state that the information at the station is easy to understand, easy to get the information, the location of the information facilities are at the right place and the information given is correct as showed in Figure 7. From the graph in Figure 8, an urban service is more comfort than rural services. This is because urban services station has enough seating, more facilities for disable people, provide more facilities for toilet and mosque.

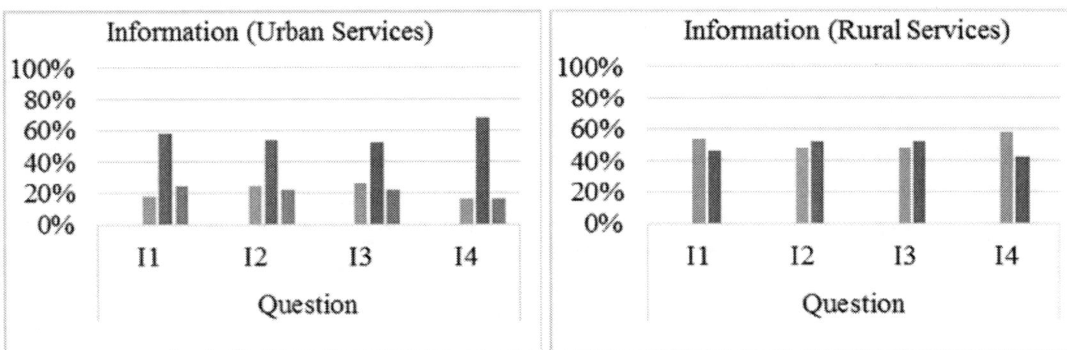

Figure 7. Comparison of information facilities in urban and rural area.

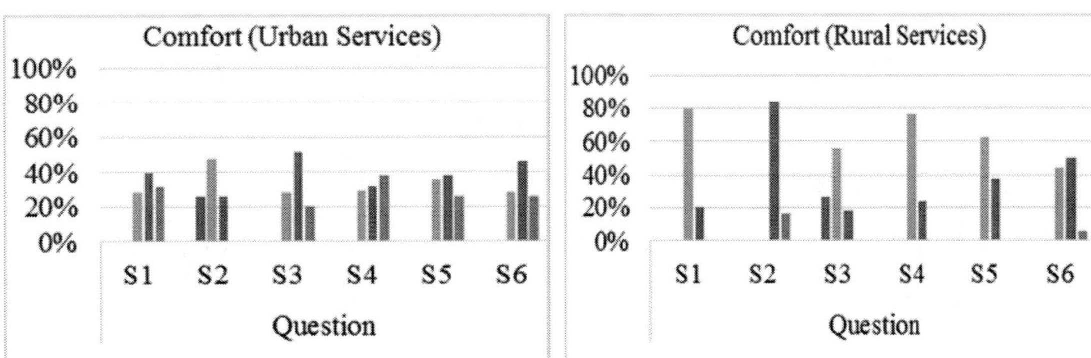

Figure 8. Comparison of comfort facilities in urban and rural area.

Facilities for emergency are better at urban services station as showed in Figure 9. Some of the respondent disagrees about the gathering place at rural services station. This is because the location is not suitable.

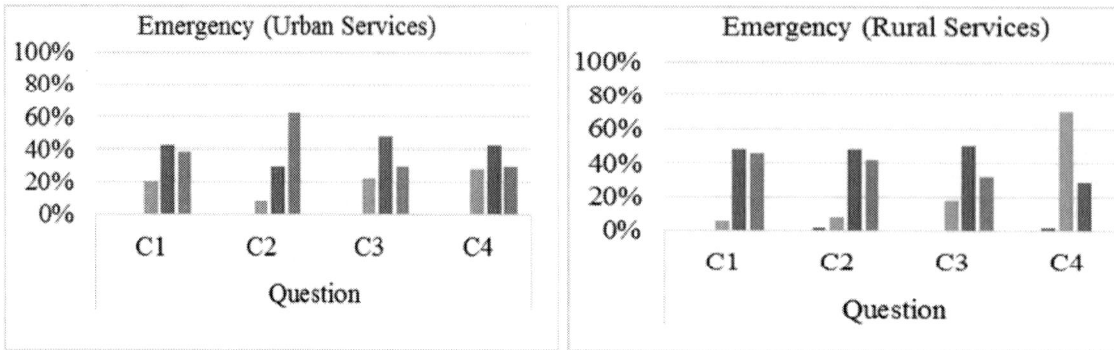

Figure 9. Comparison of emergency facilities in urban and rural area.

From Figure 10, it can conclude that the safety at urban services station is much better than in rural services station. Many facilities are available at station for urban services such as closed-circuit camera, and police officer. Through the questions contained in the service element, clearly showed in Figure 11 that station for urban services is much better than rural services. The analysis from Figure 12 showed that the ventilation system at station for urban services is much better than station for rural services. This is evidenced by the percentage of respondents who rate the highest ranking for urban services. Many respondents are not satisfied with the ventilation system and the respondent feel uncomfortable at station for rural services.

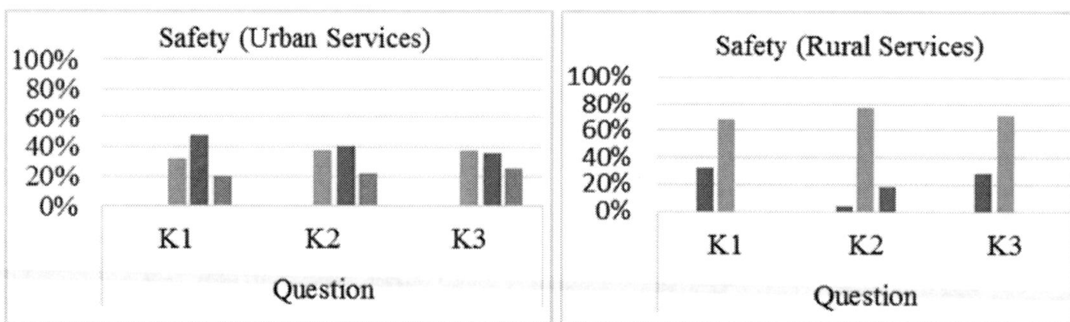

Figure 10. Comparison of safety facilities in urban and rural area.

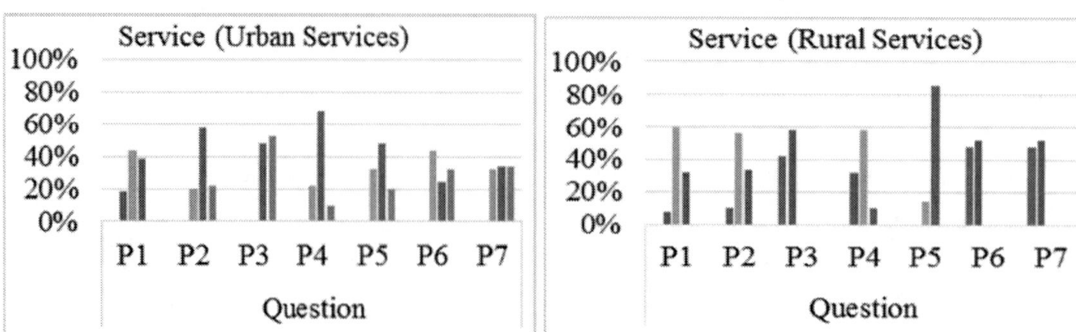

Figure 11. Comparison of service facilities in urban and rural area.

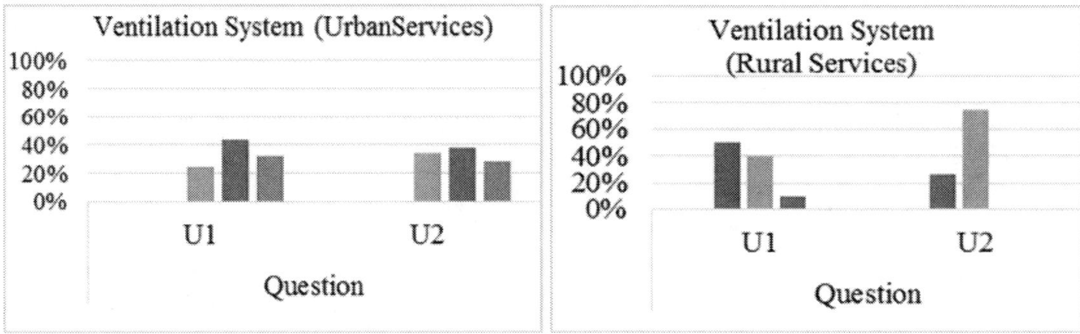

Figure 12. Comparison of ventilation system facilities in urban and rural area.

4 Conclusions

This study can provide a lot of advantages and benefits to the community. It was found that facilities available at the train station were not adequate and some users were uncomfortable with the conditions. However, the facilities at the railway station can be improved and maintained. Improvement of facilities at the railway station can be done if the operation management focuses on safety and comfort of users. Train services will improve thus can attract people to choose train as their daily transport. Through the data analysis, it was proven that facilities at the urban and rural services station have different facilities. Researchers also found that the train stations for urban services have more facilities train station for rural services. These differences were found due to the demand levels of the service, the location of the stations and the landscape around the stations. Thus, when constructing a train station considerations were not focused only on budget and design but also on the surrounding landscape of the stations.

References

[1] Ibrahim, Penduduk Malaysia Kini 28.3 Juta, Berita Harian, 29 Julai, (2011).

[2] M.S. Mohd Zakaria, Kajian Perbandingan Prestasi Operasi Monorel dan Keretapi Aliran Ringan (LRT) Bagi Penilaian Kualiti Perkhidmatan (QOS) Transit, Degree Thesis, Universiti Tun Hussein Onn Malaysia, Johor, (2014).

[3] M.I. Mohd Masirin, N.F. Johari, N.H. Nordin and A.H. Abdullah, A field study on urban transits in City of Kuala Lumpur: Passenger views on train noise and vibration, Journal for Applied Mechanics and Materials, **773-774**, (2015).

[4] F. Akabal, M.I. Mohd Masirin, Z.A. Akasah and M.M. Rohani, Review on selection and suitability of rail transit station design pertaining to public safety, *International Integrated Engineering Seminar*, Batu Pahat, (2014).

[5] N.H. Nordin, M.I. Mohd Masirin, M.I. Ghazali and M.I. Azis, Appraisal on rail transit development: Train services and safety, *International Integrated Engineering Seminar*, Batu Pahat, (2014).

[6] V. Vuchic, *Urban Transit: Operation, Planning and Economics*, John Wiley, New Jersey, (2005).

[7] G. Li and C. Toda, Discussions on the local rail transit system in the urbanization, *The 9th International Conference on Traffic and Transportation Studies*, 193-198, (2014).

[8] Black, Rural communities and sustainability, in Cocklin, C. and Dibden J. (Eds) *Sustainability and Changes in Rural Australia*, University of NSW Press, Sidney, (2005).

[9] W. Village, Train station as places for community wellbeing, (2006).

[10] C.F Huang and Y. Xia, Research on the role of urban rail transit in promoting economic development, *International Conference on Green Buildings and Sustainable Cities*, 520-525, (2011).

[11] Victorian Rail Industry Operators Group Standards, Railway Station Design Standard and Guidelines: Revision, (2011).
[12] M. Goggln, Design Guidelines: Guide to Station Planning and Design, Network Rail, (2011).
[13] N. Mohajeri, G.R. Amin, Railway Station Site Selection using Analytical Hierarchy Process and Data Envelopment Analysis, 107-114, (2010).

MATEC Web of Conferences **47**, 03013 (2016)
DOI: 10.1051/matecconf/20164703013

Construction of Buildings on Peat: Case Studies and Lessons Learned

Ali Abdul-Wadoud Mahmod[1,a], Sabarudin Mohd[1], Mohd Idrus Mohd Masirin[2], Saiful Azhar Ahmad Tajudin[2], Ismail Bakar[2], Adnan Zainorabidin[2], Azrul Zulwali Kifli[1] and Ling Jen Hua[1]

[1]*Peat Technology Research Institute, University College of Technology Sarawak, 96000 Sibu, Sarawak, Malaysia*
[2]*Research Center for Soft Soil, Universiti Tun Hussein Onn Malaysia, 86400 Parit Raja,Johor, Malaysia*

Abstract. Building construction on soft soils including on peat has many challenges and difficulties. The failed and deteriorated buildings have a big impact on the community. The IConCEES International Workshop 2015 which was conducted on October 2015 convened as a joint venture between Universiti Tun Hussein Onn Malaysia (UTHM) and the University College of Technology Sarawak (UCTS). The aim was to invite regional experts from academia and the industry to formally present and discuss the various construction problems encountered when working with peat. The discussions were divided into two divisions; infrastructure and building construction. This paper discusses the outcomes of the workshop and focuses on the factors and relevant challenges when constructing buildings on peat. The experts have discussed regulatory and construction issues including: drainage issues, site investigation practices, monitoring and construction guidelines. A few suggestions were outlined as a remedy to these problems and to better assist the peat practitioner at work.

1 Introduction

Peat is that separated soil material consisting mainly of plant remains in various degrees of decomposition; fibric, sapric and hemic. Peat in characterized by its very high moisture content, high compressibility and very low shear strength. Consequently, building on peat entails a huge amount of uncertainty, as all textbook bearing capacity, consolidation and shear strength theories are based on inorganic soils. Hence it is vital that special work and more research be conducted on peat to overcome this difficulty and to further understand its behavior.

Peatland is widely scattered all over Southeast Asia and Malaysia. Figure 1 shows the distribution of Peatland in Southeast Asia, of which Malaysia is an integral part. Knowing this it becomes vital for construction engineers to study the impact of construction on peat in its various forms of decomposition.

As such, a workshop co-organized by Universiti Tun Hussein Onn Malaysia (UTHM) and The University College of Technology Sarawak (UCTS) was held on the grounds of UCTS to discuss the various construction issues on Peat. Experts and panelists from all over the Peat engineering industry spectrum discussed in two days all the relevant construction issues and problems attributed to working on Peat subsoil. The workshop was divided into two separate sessions: infrastructure and

[a] Corresponding author : ali.mahmood@ucts.edu.my

superstructure/buildings. While this paper is devoted to the discussion on the superstructure/building presentations and session, another will present the results and discussion on the infrastructure session. The superstructure/building session was further subdivided into two parts, the presentations and the plenary session.

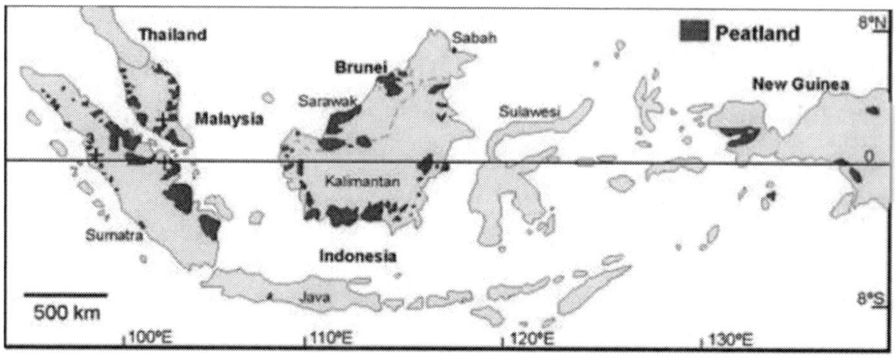

Figure 1. The distribution of peatland in Southeast Asia [4].

2 Sharing of Experiences and Knowledge

2.1 Challenges and reducing building failures

Mohd Masirin [1] discussed an overview on the construction of buildings and the effect due to the presence of Peat on the structural integrity and livelihood of buildings. Continuous monitoring for buildings on Peat is required and appropriate design considerations and materials are essential in order to minimize construction defects. Implications of tipping, differential and uniform settlement on buildings were described, as well, with Sibu, Sarawak, as a case study. It is common knowledge that foundation makes the cost of constructing a building on Peat relatively high due to the nature of the underground formation. Preconstruction difficulties, post construction failures, cost for construction, maintenance issues, short and long term impacts for construction on peat were discussed. Seven recommendations were outlined in order to avoid construction problems on peat, they are, among others, introduction of appropriate and suitable soil improvement methods, identification of suitable building designs, introduction of a technique to safeguard the foundation against dewatering effects, using lightweight materials for buildings and strengthening the foundation with appropriate methods for long term resilience against consolidation. Construction technologies on peat, currently employed by engineers, among others, are: piled foundation, raft foundation, light weight concrete, stabilized soil on Peat and geogrids. Continuous research on construction methods and materials is demanded, to determine an economic and effective solution. However, standards and guidelines need to be established in order that JKR and other local agencies are able to better evaluate developers' project proposals before any endorsement are made.

2.2 Case studies of building construction failures on peat

Phienwej [2] described the various structural and physical problems associated with construction on Peat and suggested some remedies. There are four case studies highlighted; (i) the subsidence in Ramkhanhaeng University area, Bangkok, (ii) the campus of the Asian Institute of Technology, (iii) the Nakhon Luang Aquifer, and (iv) Bangkok International Airport. Prefabricated vertical drains (PVD) were implemented as a remedy to the airport subsidence problem. It was, also, suggested that several techniques could be applied to other cases to strengthen the soil such as: vacuum consolidation and lightweight materials and soil cement columns for deep mixing methods. Figure 2 shows some foundation subsidence in the Asian Institute of Technology [2].

Figure 2. Some foundation subsidence in the Asian Institute of Technology in Thailand [2].

2.3 Case study on bridge failure

Ong [3] described the construction of a 13.5 km single-carriageway linking the existing Igan Bridge to a proposed university site in Sibu, Sarawak, East Malaysia. The soft soils of maximum 30 m thick, which were found below the 5-m thick Peat layer. They were treated with pre-fabricated vertical drains to reduce post-construction settlements so that design requirements were met. After site investigation and soil characterization tests, it was successfully constructed using hydraulically placed sand-fill to replace the underlying fibrous peat. Settlement plates were installed to monitor the settlement of the sand layer-PVD system. It was shown that the rate of settlement increased markedly when PVD was installed. Field instrumentation results cross-checked with Asoaka observational method to measure settlements was successfully implemented to provide a good understanding of ground treatment of fibrous peat and soft soil. Figure 3 shows the installed PVD and the settlement plates [3].

Figure 3. a) (left) PVD installed after sand-fill reached existing ground level and b) (right) condition of settlement plates when site was flooded [3].

2.4 Physio-chemical behaviour of peat

Nazir [4] emphasized that the construction and design guide for the various types of Peats, which are: Hemic, Sapric and Fibric, should be outlined for engineers. Full scale site studies need to be conducted to determine the most suitable Peat stabilization and reinforcement methods. Proper site investigation needs to be thoroughly conducted to determine the various heterogeneous types of Peats present. The physio-chemical behavior of tropical peat and the most effective economical methods need to be understood and further investigated.

2.5 Case study on housing project construction

Jafaar [5] investigated a structural building failure on soft soil. A case study of a housing units in Shah Alam, Selangor was discussed. It was revealed that irregular settlement of foundation structure has occurred due to foundation settlement, which was caused by lack of consultant supervision during construction. It was also revealed that no problem on the design and structural plans existed, which could have contributed to the failure. So it was suggested that settlement should be monitored for the affected building. In the future, proper consultant supervision should be performed on all construction projects.

3 The Plenary Session

3.1 Drainage problems

These were discussed with two case studies: the Medan mall and bus station in Sibu, Sarawak. Time and cost are the critical issues discussed and the suggested solution was to have a working platform; not temporary solution (like bamboo and geotextile mattress, which is only considered for temporary access). It was emphasized that ground treatment should not be overlooked; surrounding areas should be treated, as well. Ground water management was, also, one of the issues discussed since its drainage affects the peat foundation to a great extent. Replacement of the Peat layer to a certain depth (by sand for example) was considered as the best method. It was mentioned that building failures in Sibu were, partly, because of the type of pile (Bakau pile) used and the change in water table under drainage. Hence, it was emphasized that design of foundations has to account for lateral movement within the Peat layer. Also, as far as long term monitoring is concerned, installation of piezometers and settlement plates is mandatory.

3.2 Site investigation practices

These were also discussed, and it was stated that site investigation is very important before any construction work could take place. Building construction guidelines on Peat, especially for pile use in foundation engineering, need to be included in the existing Malaysian construction codes and practices. High strength concrete is required for structural use, however, lightweight concrete (replacement for brick wall or non-load bearing walls, for example) is recommended for non-structural members.

3.3 Monitoring

Workshop panelists stated, that settlement is the main problem with peat soils, hence, expert opinion is required for validation on peat research in this regard. While, guidelines do not state the preferred method; the expert has to decide on it based on field and laboratory data. Current guidelines only state the two available options: replacement or reinforcement (stabilization). And, in essence, financial issues are the key. Suitable design for buildings on peat was also discussed as it relates to settlement, environmental problems and structural failures.

3.4 Guidelines

Guidelines for specifications on building construction on peat testing is needed, as far as pre-construction and post construction practices. While some were specified in the Guidelines presented by CREAM (Construction Research Institute of Malaysia), additional testing schemes suitable for building construction and monitoring are needed. In that regard, it is recommended that developers, consultants and contractors, as a mandatory practice, should install piezometers and settlement plates for short and long term monitoring of buildings constructed on Peat. Table 1 illustrates the building issues encountered and the proposed solutions discussed by the panelists.

Table 1. Building issues encountered and the proposed solutions.

Issues encountered on site	Proposed solutions
Drainage problems	• Ground treatment • Ground water management • Peat replacement • Design of foundations • Long term monitoring
Site investigation practices	• Pile use as part of building construction guidelines • High strength concrete for structural members • Lightweight concrete for nonstructural members
Monitoring	• Suitable design for building on Peat • Expert opinion for research validation
Construction guidelines	• Guidelines for specifications on building construction on peat • Testing needed: pre-construction and post construction • Piezometers and settlement plates are mandatory for installation in building on peat

4 Conclusions

Experts from academia and the industry presented and debated for two days in the IConCEES International Workshop 2015 in UCTS the impact of building construction on peat. Various construction problems and issues were thoroughly discussed including drainage issues, site investigation practices, monitoring and construction guidelines. The panelists suggested solutions to overcome these construction problems. However, more work needs to be done in terms of both regulation and research before building construction on peat can be fully implemented with the minimum amount of settlement and drainage problems.

Acknowledgment

The authors would like to acknowledge with gratitude the priceless support and expert opinion rendered by the keynote and expert speakers during IConCEES International Workshop 2015. This includes YB Datuk Dr. Abu Bakar Mohamad Diah, Prof. Ir. Dr. Zuhairi Abdul Hamid, Assoc. Prof. Dr. Noppadol Phien-wej, Assoc. Prof. Ir. Dr. Naser Abdul Ghani, Prof. Ir. Dr. Ramli Nazir, Prof. Dr. Fauziah Ahmad, Ir. Liew Shaw Shong, Dr. Ir. Slamet Widodo, Dr. Ir. Dominic Ong Ek Leong, Ir. Ghazali bin Abd Aziz, Assoc. Prof. Ir. Dr. Abd Halim and Ir. Muhd Salmizi Ja'afar.

References

[1] M.I. Mohd Masirin, Peat and building construction, *IConCEES International Workshop 2015*, Sibu, Sarawak, (2015).
[2] N. Phienwej, Thailand experiences in geotechnical difficulties in road and embankment construction in soft soils, *IConCEES International Workshop 2015*, Sibu, Sarawak, (2015).
[3] D.E.L Ong, Alternative method of road embankment construction on fibrous peat using hydraulic sand-fill, *IConCEES International Workshop 2015*, Sibu, Sarawak, (2015).
[4] R. Nazir, A review on Malaysia peat soil properties for construction purposes, *IConCEES International Workshop 2015*, Sibu, Sarawak, (2015).
[5] M.S. Jafaar, Construction on peat: Case study on soft soil building construction, *IConCEES International Workshop* 2015, Sibu, Sarawak, (2015).

MATEC Web of Conferences **47**, 03014 (2016)
DOI: 10.1051/matecconf/20164703014

Construction of Infrastructure on Peat: Case Studies and Lessons Learned

Ling Jen Hua[1, a], Sabarudin Mohd[1], Saiful Azhar Ahmad Tajudin[2], Syazie Nordzaima Ali Mohamad[1], Ismail Bakar[2], Mohd Idrus Mohd Masirin[2], Adnan Zainorabidin[2] and Ali Abdul-Wadoud Mahmood[1]

[1]*Peat Technology Research Institute, University College of Technology Sarawak, 96000 Sibu, Sarawak, Malaysia*
[2]*Research Center for Soft Soil, Universiti Tun Hussein Onn Malaysia, 86400 Parit Raja, Johor, Malaysia*

Abstract. Construction of infrastructures on peat land is a very challenging task due to its properties of low shear strength, high compressibility and high water content. This paper summarizes various solutions which could be adopted for the construction of infrastructure on peat, as reviewed by the experts and panels during IConCEES International Workshop 2015. Engineers could (a) avoid peat, such as to transfer the load to the hard layers through end bearing piles or to replace the peat with the other soils, or (b) construct on peat with special precautions, such as by reducing the weight of the construction materials and dewatering the peat to improve the engineering properties. This paper serves to generate new ideas and give insights of the problems commonly encountered by the industry. Some of the proposed solutions might never be tested on peat. This would rely on the researchers to take up the challenge to further investigate and address the technical issues outlined in this paper.

1 Introduction

Peat is brownish-black in dye which consists of decayed organic and mineral substance (Figure 1). It is a naturally occurring highly organic substance derived primarily from plant material [1]. It is formed when organic matter accumulates more quickly than it decays. This usually happens when the organic matters are preserved below high water table like in swamps or wetland.

In Malaysia, approximately 8% of the total land area (32,975,800 ha) is covered with peat. Sarawak is the state having the largest area of peat, which is approximately 69% of the total peat land area. Although the utilization of peat land is quite low, the demand has been rapidly increasing due to the population growth and various economic reasons. Normally, engineers prefer to avoid having constructions on peat. However, this option might not always be feasible.

Peat is often considered problematic for the construction due to its low shear strength, high compressibility and high water content. These characteristics lead to the problems and difficulties in the construction in various aspects such as pre-construction difficulty, post construction failures, cost of construction, maintenance issues, as well as short and long term impacts. The challenges of constructing infrastructures on peat are to adopt a sustainable design, select suitable materials, reduce the cost of construction, prolong the service duration, and minimize the necessity of maintenance.

[a] Corresponding author : lingjenhua@ucts.edu.my

For this, IConCEES International Workshop 2015 was organised on 12-13 October. This workshop invited the 12 keynote and expert speakers from various countries, such as Malaysia, Thailand and Indonesia, and assembles about 47 panels from the sectors of academic, industry, governments and R&D to study the problems and propose solutions for the construction on peat. The expert speakers and the panels are divided into two groups, discussing on two main themes, which are buildings and infrastructures. This paper reports the outcomes obtained from the workshops, focusing mainly on infrastructures.

Figure 1. Large amount of organic substance and water content in peat [2].

Table 1. The area (ha) of peat in Malaysia [3].

State	Area (ha)	Percentage (%)
Sarawak	1,697,847	69.08
Selangor	164,708	6.70
Pahang	164,113	6.68
Johor	143,974	5.86
Sabah	116,965	4.76
Terengganu	84,693	3.45
Perak	69,597	2.83
Kelantan	9,146	0.37
Negeri Sembilan	6,245	0.25
Federal Territory	381	0.02
Total	2,457,669	100

2 Proposed Solutions

The most common problems related to the construction of infrastructures can be categorised as (a) settlement of peat, (b) low bearing capacity of peat, (c) poor stability of peat during excavation and (d) high ground water table. Various solutions were proposed to resolve these problems.

Settlements could occur at various modes, such as tipping, differential and uniform settlements, depending on the conditions of peat. These settlements, in excessive, affect the durability and smooth riding of roads or highways. To eliminate the settlement of roads, pile foundation with suspended slabs can be used.

The arching effects of aggregate could be used to substitute the suspended slabs to control the settlement. Aggregates can be arranged in the form of arches, and being supported by the pile foundation. The loads of the traffic and road embankment are supported by the aggregate arch and to be transferred to the hard layer underneath the ground through the piles. Geosynthetics could be used depending on the thickness of the embankment and the span-height ratio of the aggregate arch. In the case that the embankment thickness is insufficient where the span-height ratio is relatively high, it behaves as a partial arching and the aggregate arch might not be effective. Thus, the geotextile is required between the pile caps and the aggregate arch to provide restraint to the arch (Figure 2).

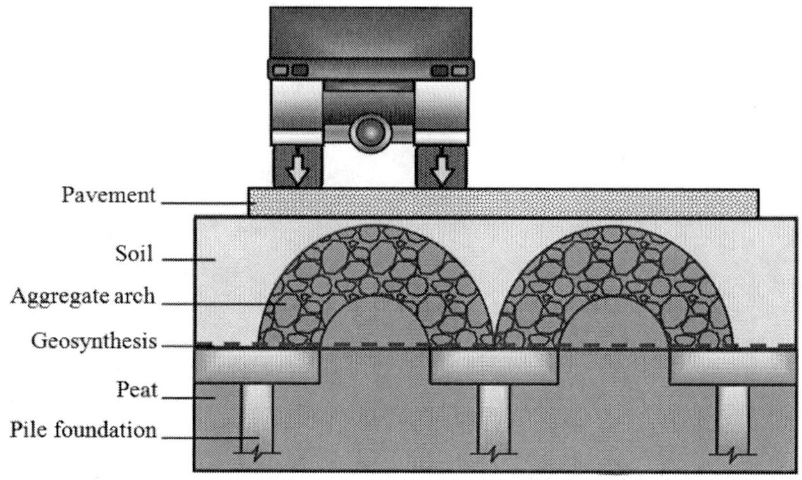

Figure 2. Aggregate arches with pile foundations.

This system, however, inherits certain limitations. It probably requires a huge amount of end bearing piles and thick embankment might be required to ensure a full arching effect. Investigation should be carried out to (a) determine the feasibility, and (b) model the behaviour of the system, to study the effects of the span-height ratio, distance between the piles and the thickness of the embankment. Researchers also need to work out a suitable construction method and the design guide, with and without Geosynthetics, for this system.

The lightweight Geo-materials, such as EPS, foam concrete and tire shreds, could be used as the materials for the construction of roads and highways. In principle, this method minimizes the weight imposed onto the road embankment, and thus, reduces settlement. On the peat with high ground water table, the buoyancy force of the water could be utilized to float these lightweight Geo-materials. For this, it is essential to ensure the weight of road pavement is equal to the weight of excavated peat in order to prevent settlement.

There are several technical issues that need to be resolved while adopting the system. For example, the water level fluctuation, the phenomena of ground water drawdown during the construction, the vertical and lateral restrains of the platform, undesired stresses incurred due to differential settlement of the peat and others. Studies should also be carried out to investigate the strength and the load resisting parameters, to develop a workable system and to acquire the design guide for the system.

For the construction of pipelines, differential settlement at the supports could cause undesired stresses and damages of the pipes. The challenge is to minimize the differential settlement throughout the entire pipeline. Lightweight materials could be used to encase the pipelines to be supported by the low bearing capacity of the peat. For this, the feasibility of the system, the mechanism of supporting the pipelines, the methods of construction or installation, and the connections of the pipeline need to be studied.

Due to the low bearing capacity of peat, it is sometime difficult to access the site during the construction stage (Figure 3). Lightweight materials in the forms of mattresses could be laid and floated on the peat as the temporary platform and access for the construction. For the research, the scopes that could be studied involve (a) determination of suitable materials with the desired engineering and physical properties, (b) the stability of floating mattress platform under various loading conditions, (c) the prolonged life span and the service duration, (d) the mechanism of installation and the performance of joints of the floating platform.

Figure 3. Severe settlement of the tractor due to poor bearing capacity of peat [4].

Deep excavation on peat with high ground water table could lead to instability of soil or slope. The dramatic drawdown of the ground water table as a result of excavation leads to high mobility of the seepage flow and this causes the failure of the slope. To resolve this, a groundwater barrier is required to cutoff or control the flow in order to ensure the stability of the slope. This could be done through various mechanisms, such as gravity mass, diaphragm wall, sheet piles, jet grouting, stone columns and soil mixing piles. These methods are relatively costly compared with the dewatering scheme.

Dewatering the construction site on peat is known to be able to improve the engineering properties of the peat in order to ensure the stability of the slopes during excavation. However, excessive extraction of the water leads to significant settlement of the ground and, in most of the time, a large region is affected. This could affect the existing structures nearby, which are beyond the construction site. For this, a recharge pond lining system could be used to maintain the ground water table at the regions where excavation works are not required in order to minimize the affected region. Investigation is required to acquire the response of dewatering and its affected region and to model the rate of recharging process. Studies could be extended to determine the setback distance or area and mechanisms to minimize the required setback so that it is applicable for small construction sites.

In addition, the centralized sewerage network system could be strategized to be used as a mechanism of dewatering during the construction stage prior to its actual usage after the construction. Similarly the combination of subsoil and open channel drainage systems could be strategized and used to control the ground water table and to maintain the engineering properties of the peat. In normal days, the subsoil drain can be used to retain water and slowly discharge the water. Under the overflow condition, the open channel can be used to rapidly discharge the water. The feasibility of these methods, in a separate or combined system, needs to be studied.

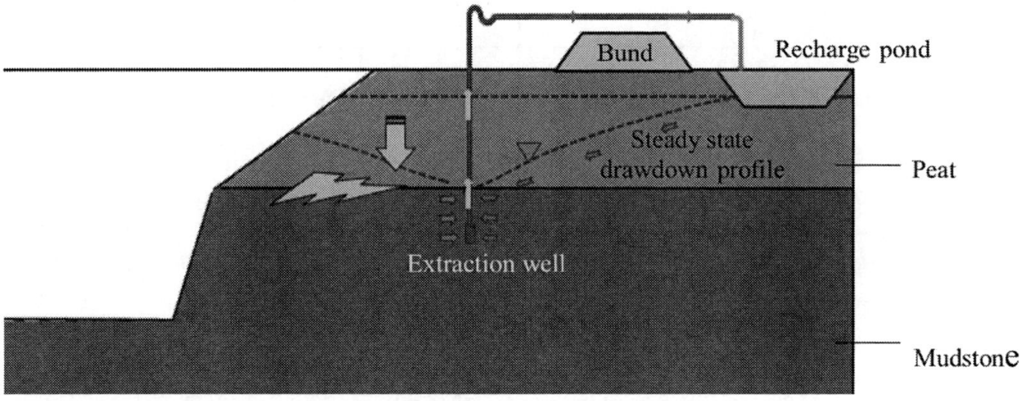

Figure 4. Dewatering scheme using extraction well and recharge pond [2].

Peat could be removed to expose a firm layer of sufficient bearing capacity and replaced with suitable materials with the desired engineering properties, e.g. road embankment. This method is economically viable for the depth of up to 4.0 m.

Earth filling at a deep peat area causes lateral movement of the soil. This causes the undesired lateral load to the pile foundations of the nearby structures, and this could jeopardize the stability and the integrity of the structures. For this, suspended slabs are usually adopted with a minimum amount of earth filling. The void underneath the suspended slab is recommended to be filled and compacted in order to prevent accumulation of the hazardous gases.

3 Discussions

These methods function on the basis of (a) avoiding the use of the peat by transferring load to the hard layers underneath it, (b) reducing the weight of materials to be lighter than water to "float" the structures on the peat, (c) controlling the seepage flow to avoid the instability of the slope during excavations, (d) improving the properties of the peat through dewatering, and (e) replacing the peat with the soils with better engineering properties (Table 2).

Table 2. Classifications of the approaches used by various method.

Methods	Approaches				
	Load Transfer	**Reduce Weight**	**Seepage Flow Control**	**Dewatering**	**Peat Replacement**
Pile foundation • with suspended slab • with aggregate arch	√				
Lightweight Geo-materials • EPS, Foam concrete, tire shreds		√			
Groundwater Barriers • gravity mass • diaphragm wall • sheet piles • jet grouting • stone columns • soil mixing piles			√		
Dewatering schemes • Extraction well and recharging pond • Sewerage system • Subsoil drainage				√	
Excavation and earth filling					√

Out of the methods proposed, not all have successfully been applied on peat at a full scale. Some of these methods are at the conceptual stage, at least for the time being. Some have been tested on soft soils, but not on peat. It is still unsure of the feasibility and practicality of these methods. For this, simulation models are required, experimentally, numerically or analytically, to determine the feasibility and to acquire the response of the peat prior to adoption of the stated systems. These include the aggregate arches, dewatering schemes by using sewerage and drainage network systems.

4 Conclusions

IConCEES workshop provides a platform for the experts and panels from the sectors of academic, industry, government and R&D gather thoughts and propose various solutions to the construction of infrastructure on peat. The aim was to generate new ideas and give insights on the problems encountered by the industry for the researchers to investigate and to resolve. At this stage, we are yet to acquire a perfect solution. Different method inherits its own limitations and there are still many technical issues to be addressed. These are the challenges to be undertaken by the researchers all around the world. By having those limitations and technical issues resolved, we are a step closer to a better solution.

Acknowledgment

The authors thank the selfless sharing of knowledge and expertise by the keynote and expert speakers during IConCEES Workshop 2015. These speakers include YB Datuk Dr. Abu Bakar Mohamad Diah, Prof. Ir. Dr. Zuhairi Abdul Hamid, Assoc. Prof. Dr. Noppadol Phien-Wej, Assoc. Prof. Ir. Dr. Naser Abdul Ghani, Prof. Ir. Dr. Ramli Nazir, Prof. Dr. Fauziah Ahmad, Ir. Liew Shaw Shong, Dr. Ir. Slamet Widodo, Dr. Ir. Dominic Ong Ek Leong, Ir. Ghazali bin Abd Aziz, Assoc. Prof. Ir. Dr. Abd Halim and Ir. Muhd Salmizi Ja'afar.

References

[1] Construction Research Institute of Malaysia, Guidelines for Construction on Peat and Organic Soils in Malaysia, 129, (2015).
[2] L.S. Shong, Problems and solutions of peat soils in engineering applications, presentation slides, *IConCEES international workshop 2015*, Sibu, Sarawak, (2015).
[3] Wetlands International Malaysia, A quick scan of peatlands in Malaysia, Petaling Jaya, Malaysia, 74, (2010).
[4] A.L.L Kong, Jurutera, Institution of Engineers Malaysia, June, No.6, (2005).

MATEC Web of Conferences **47**, 03015 (2016)
DOI: 10.1051/matecconf/20164703015

Relationship of Physical Parameters in Pb-Contaminated by Stabilization/Solidification Method

Saiful Azhar Ahmad Tajudin [1,a], Mohamad Azim Mohammad Azmi[1], Shahiron Shahidan[1], Mohd Hazreek Zainal Abidin[1] and Aziman Madun[1]

[1]*Faculty of Civil and Environmental Engineering, University Tun Hussein Onn Malaysia, 86400 Parit Raja, Johor, Malaysia*

Abstract. This research was performed to investigate the relationship between compressive strength, density and water absorption of Pb-contaminated soil treated by Ordinary Portland cement (OPC) incorporated with sugarcane bagasse (SCB) using the Stablisation/Solidification (S/S) method. The SCB and OPC varying from 5% to 15% was added to stabilize and solidify the Pb-contaminated soil. The cylindrical samples were prepared and cured at room temperature for 28 days. Unconfined compression test, density test and water absorption test were conducted on these samples. The relationship between these variables were determined using simple statistical method by Analysis of Variance (ANOVA). Results indicate that there is a strong relationship between compressive strength and density with a regression coefficient of 77.4%. There is also a significant relationship between density and water absorption of approximately 70.7%. In addition, compressive strength and water absorption produced a strong relationship which is 73.6% of variability in strength development. In a nutshell, the strong relationship between these three variables represent the actual physical characteristics in S/S application.

1 Introduction

The stabilization/solidification (S/S) method has been widely accepted and has become an important part of environment technology worldwide. Stabilization refers to techniques which reduce the hazard potential of waste by converting the contaminants into their least soluble, mobile or toxic form [1]. The physical nature characteristics of waste are not necessarily changed by stabilization. On the other hand, solidification refers to a technique that encapsulates waste in a monolithic solid of high structural integrity. Generally, S/S method involves mixing of waste with binder to improve the physical and chemical properties of soil. Typically, there are many binders which have been developed for S/S method and the most commonly used is Portland cement [2]. The S/S method with cement is relatively common due to universal availability, desirable hydration properties and suitability for heavy metal immobilization [3-6]. This process is particularly suitable for heavy metal remediation and has been widely applied over several decades, especially in the U.S.

[a] Corresponding author : saifulaz@uthm.edu.my

The addition of cement to soil is mainly to modify its physical characteristics, increase strength, stiffness and durability while reducing water permeability through a chemical hardening reaction [7]. Basically, the physical characteristics involve several tests such as unconfined compression strength test (UCS), density and water absorption test. These tests are important to measure the effectiveness and efficiency of soil remediation in the S/S method. Malviya and Chaundhary [8] stated that these tests produce significant relationship during the S/S process that affects strength development. Therefore, this study was undertaken to find out the relationship between UCS, density and water absorption of S/S samples containing cement incorporated with sugarcane bagasse.

2 Materials and Methods

2.1 Contamination of soil

Clay soil samples were collected at a location near to the Research Center for Soft Soil (RECESS), University Tun Hussein Onn Malaysia (UTHM). The top of the soil to a depth of 3 meters was withdrawn in order to avoid taking the humus and plant sources [9]. The soil sample was oven-dried at 105oC for 24 hours. Dried soil was crushed into fine powders prior to storage in a polyethylene plastic bag. The initial concentration of Pb is 500 ppm. Treatment of the lead contaminated soil was conducted using Type II ordinary Portland cement (OPC) as the main binder and supplemented by treated and untreated sugarcane bagasse (SCB). The SCB was collected from hawkers around Parit Raja, Johor and divided into two parts: the first part was untreated and the second part was treated. The untreated SCB was raw as collected but the treated SCB was exposed under acidic conditions to eliminate the possibility of cellulose fibers and to liberate the lignin. For the treated part, the SCB was boiled with 0.1 MHCl for approximately 45 minutes, with the residue washed free of sugars and hydrolysis products. This procedure was repeated 3 to 4 times until the filtrate was virtually colorless [10]. Both types of SCB were dried in the oven under a temperature of 105 °C for 24 hours in order to eliminate the moisture content before being grinded into a particle size of 90μm.

2.2 Production of S/S sample

The binder (OPC with untreated and treated sugarcane bagasse) was added to the contaminated soil at binder-to-dry soil (B/Sd)(w/w%) ratio. Table 1 summarizes the mix design of sample. The S/S sample was then compacted into a split mould to form a sample of 38 mm in diameter and 76 mm in height. A specially designed miniature hand compacting tool was used to compact the mixture in 4 layers, 50 blows each and cured at a controlled condition for 28 days (temperature = 25±2 oC, humidity > 70%)[11].

Table 1. Mix design of sample.

Mixing Type	Sample	Percentage of Binder	Total Weight of Sample (g)
(Clay + Pb) + OPC (control)	A	5%OPC	160
	B	10%OPC	160
	C	15%OPC	160
(Clay + Pb) + OPC + SCB (untreated)	D	2.5%OPC2.5%SCB	160
	E	5%OPC5%SCB	160
	F	7.5%OPC7.5%SCB	160
(Clay + Pb) + OPC + SCB (treated)	G	2.5%OPC2.5%SCB	160
	H	5%OPC5%SCB	160
	I	7.5%OPC7.5%SCB	160
Clay + Pb	J	-	160

2.3 Unconfined compressive strength (UCS)

The UCS test was carried out using the GeocompLoadTrac II system available in RECESS UTHM. Testing was conducted according to BS1377-7:1990. The samples were extruded from the moulds, levelled, measured for length and diameter, weighed and subjected to uniaxial compression test (UCT) at a constant rate of strain of 1% per minute. Before testing, the loading device was adjusted carefully to ensure the upper platen made contact with the sample. Extra care was taken to ensure that both ends of the sample were as flat as possible to minimize bedding error during the test, especially for the stiffer samples.

2.4 Density test

The density test is an easy test which is carried out in order to determine the tapped bulk density of samples. In this test, the bulk density is measured by calculating the volume of the soil and weight. The testing involved the measurement of height and diameter of each sample. Finally, the weight of soil is divided by the total soil volume.

2.5 Water absorption test

Water Absorption is one of the important tests for the S/S method, with which infiltration into the disposed solidified matrices can be detected [12]. The water adsorption test was conducted according to ASTM C 140. Principally, the test consists of two major factors which are saturating the specimen followed by drying. The specimens will be immersed in water until a change in mass takes place during 24 hours. Then, the obtained saturated mass of soil will be weighed and dried in the oven at a temperature of 105°C ± 5°C. Thereafter, the water absorption by immersion will be estimated using the saturated weight minus dry weight and divided by the dry weight [13].

2.6 Relationship analysis

The results of UCS, density and water absorption of the sample were then analyzed using simple statistical analysis which is analysis of variants (ANOVA) to determine the relationship between the variables.

3 Results and Discussion

3.1 Relationship between strength and density

Generally, there is a linear relationship between strength and density of sample. Figure 1 presents a compressive strength against density at 28 days. The influence of strength on the density values is showed using a regression coefficient. Distribution of data shows a linear relationship between the compressive strength and density as follows.

$$\sigma = 1389.4 + 0.24\,\alpha \qquad (1)$$

with σ being the compressive strength (kPa) and α is the density (Kg/m^3) of the sample.

Statistical analysis showed that there was a significant relationship between two variables as P-value in the analysis of variance (ANOVA) indicated that $P = 0.006$, which is lower than 0.05. Linear Equation (1) described that one density unit of 0.24 increased the strength by approximately 1389.4. In addition, the positive value indicates that the strength increases with the density. Apart from that, the strong regression coefficient of 77.4% shows the variability in compressive strength.

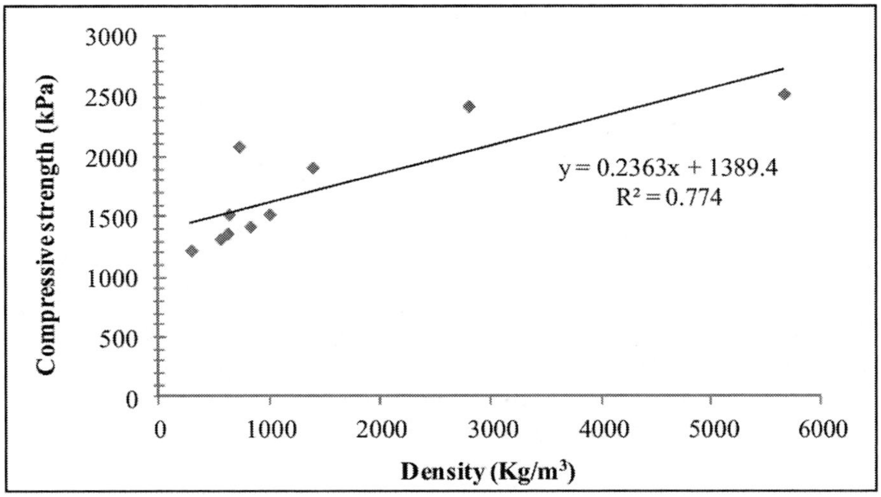

Figure 1. Relationship between compressive strength and density at 28 days.

3.2 Relationship between strength and water absorption

Figure 2 shows that the compressive strength of samples decreases with increasing water absorption. Through observation of the data, the linear model was used to describe the relationship between the strength and absorption of water. This linear equation expresses the relationship as follows:

$$\sigma = 4054.9 - 77.17\ \alpha \tag{2}$$

where σ is the compressive strength (kPa) and α is the water absorption (%) of the sample. Statistical analysis showed that there were significant correlations at the 95% confidence level between two variables. This is because the P-value in the analysis of variance is 0.007, which is lower than 0.05. However, regression analysis showed that this model explains 73.6% of variability in strength. This means that there is a strong relationship between these two variables. In addition, the linear Equation (2) indicates that one water absorption unit of 77.17 may increase the sample strength by approximately 4054.9.

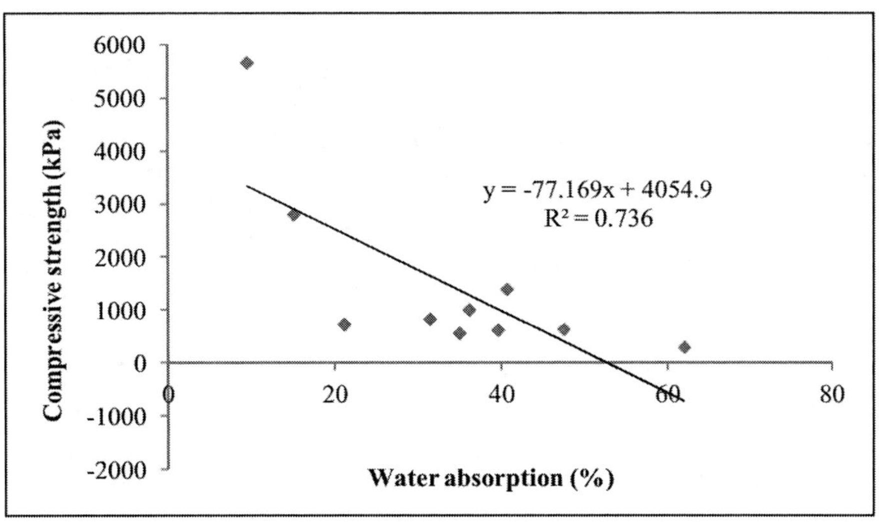

Figure 2. Relationship between compressive strength and water absorption at 28 days.

3.3 Relationship between density and water absorption

In general, the low density of the material will reduce the absorption of water[14]. Therefore, based on the Figure 3, a linear relationship between density and water absorption can be obtained as follows:

$$\rho = 2583.7 - 25.23\ \alpha \tag{3}$$

with ρ as the density (Kg/m^3) and α as the water absorption (%) of the sample.

Statistical analysis showed that there were significant correlations at the 95% confidence level between density and water absorption. It is proven by the P-value obtained in analysis of variance (ANOVA) which is 0.007, greater than 0.05. Additionally, analysis of regression coefficients (R^2) showed that this model can explain 70.7% of the variability in density where it indirectly describes the strong relationship between these two variables. On the other hand, the linear Equation 3 showed that one water absorption unit of 25.23 may affect the density at 2583.7.

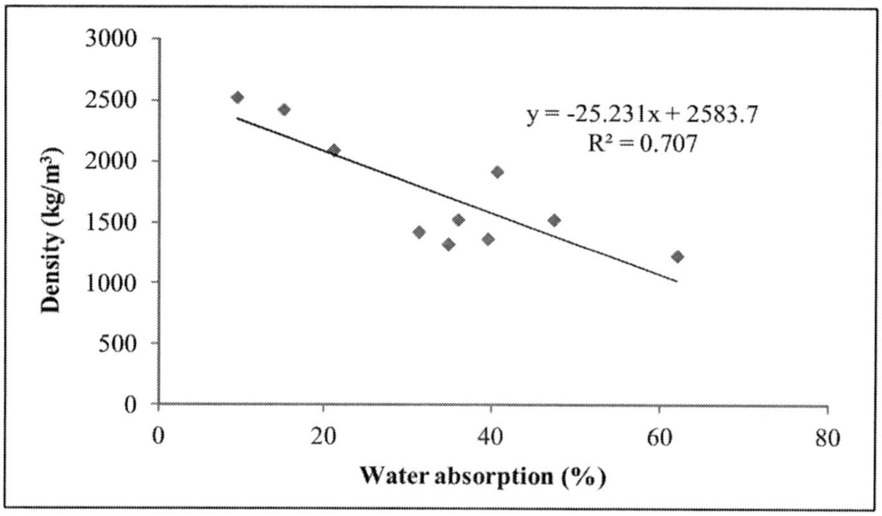

Figure 3. Relationship between density and water absorption at 28 days.

4 Conclusions

Analysis of variance (ANOVA) showed that there is a strong relationship between cement content and bagasse with compressive strength, density and water absorption. However, there are significant weaknesses in this relationship as the negative value of density occured as a result of high percentage of water absorption which is over 10%. Similarly, the relationship between strength and water absorption is also negative due to the high percentage of water absorption. In a nutshell, this study concludes that there is significant relationship at the 95% confidence level between the strength-density, strength-water absorption and density-water absorption.

References

[1] M. Nur, S. Mohamed, S. Azhar, A. Tajudin, A.A. Kadir, A. Madun, M. Azim, M. Azmi and N.S. Nordin, Leachate characteristics of contaminated soil containing lead by stabilisation/ solidification technique, J. of Applied Mechanics and Materials, **774**, 1443–1447, (2015).

[2] M. Azim, M. Azmi, S.A. Tajudin and A.A. Kadir, Remediation of zinc contaminated soils by stabilization / solidification technique, Int. J. of Zero Waste Generation, **1**(2), 1–4, (2013).

[3] M.J. Harbottle, A. Al-Tabbaa and C.W. Evans, A comparison of the technical sustainability of in situ stabilisation/solidification with disposal to landfill, J. Hazardous Material, **141**(2), 430–440, (2007).
[4] Q.Y. Chen, M. Tyrer, C.D. Hills, X.M. Yang and P. Carey, Immobilisation of heavy metal in cement-based solidification/stabilisation: A review, Waste Management, **29**(1), 390–403, (2009).
[5] R.B. Kogbara, A. Al-Tabbaa, Y. Yi and J.A. Stegemann, pH-dependent leaching behaviour and other performance properties of cement-treated mixed contaminated soil, J. Environmental Sciences, **24**(9), 1630–1638, (2012).
[6] Y.J. Du, M.L. Wei, K.R. Reddy, Z.P. Liu and F. Jin, Effect of acid rain pH on leaching behavior of cement stabilized lead-contaminated soil, J. Hazardous Material, **271**(2014), 131–40, (2014).
[7] A. Tarmizi, A. Karim, I.A. Rahman, A.A.A. Latiff and M.R. Taha, The effects of strength in lead contaminated residual soil solidified/ stabilized with cement, Int. J. of Civil and Environmental Engineering, **12**(6), 8-12, (2012).
[8] R. Malviya and R. Chaudhary, Factors affecting hazardous waste solidification/stabilization: A review, J. Hazardous Material, **137**(1), 267–276, (2006).
[9] M. Azim, S.A.A. Tajudin, A.A. Kadir, M.N.S. Mohamed, N.S. Nordin and M.S. Adnan, Leaching behavior of lead contaminated soil sample by using sugarcane bagasse in stabilization/ solidification method, J. of Applied Mechanics and Materials, **774**, 1481–1485, (2015).
[10] M.A. Janusa, C.A. Champagne, J.C. Fanguy, G.E. Heard, P.L. Laine and A.A. Landry, Solidification/stabilization of lead with the aid of bagasse as an additive to Portland cement, Microchem. J., **65**(3), 255–259, (2000).
[11] M. Ho, A. Tarmizi, C. Chan and I. Bakar, Leachability and strength of kaolin stabilized with cement and rubber, Int. J. of Sustainable Construction Engineering and Technology, **2**(1), 89–104, (2011).
[12] F.M. Khalaf and A.S. DeVenny, New tests for porosity and water absorption of fired clay bricks, J. Materials Civil Engineering, **14**, 334–337, (2002).
[13] F. Wang, H. Wang and A. Al-Tabbaa, Leachability and heavy metal speciation of 17-year old stabilised/solidified contaminated site soils, J. Hazardous Materials, **278**, 144–51, (2014).
[14] R. Nemes and Z. Jozsa, Strength of lightweight glass aggregate concrete, J. Material Civil Engineering, **18**, 710–714, (2006).

MATEC Web of Conferences **47**, 03016 (2016)
DOI: 10.1051/matecconf/20164703016

Evaluation of the Permanent Deformations and Aging Conditions of Batu Pahat Soft Clay-Modified Asphalt Mixture by Using a Dynamic Creep Test

A. M. Al Allam[1,a], M. I. M. Masirin[1], M. E. Abdullah[1] and A. S. Bader[1]

[1]*Facultyof Civil and Environmental Engineering, Universiti Tun Hussein Onn Malaysia, 86400 Parit Raja, Johor, Malaysia*

Abstract. This study aimed to evaluate the permanent deformation and aging conditions of BatuPahat soft clay–modified asphalt mixture, also called BatuPahat soft clay (BPSC) particles; these particles are used in powder form as an additive to hot-mix asphalt mixture. In this experiment, five percentage compositions of BPSC (0%, 2%, 4%, 6%, and 8%) by weight of bitumen were used. A novel design was established to modify the hot-mix asphalt by using the Superpave method for each additive ratio. Several laboratory tests evaluating different properties, such as indirect tensile strength, resilient stiffness modulus, and dynamic creep, was conducted to assess the performance of the samples mixed through the Superpave method. In the resilient modulus test, fatigue and rutting resistance were reduced by the BPSC particles. The added BPSC particles increased the indirect tensile strength. Among the mixtures, 4% BPSC particles yielded the highest performance. In the dynamic creep test, 4% BPSC particles added to the unaged and short-term aged specimens also showed the highest performance. Based on these results, our conclusion is that the BPSC particles can alleviate the permanent deformation (rutting) of roads.

1 Introduction

With the increase in the number of vehicles and the amount of load on road pavement, numerous casualties related to pavement deformation have occurred [1]. Truck tire pressure has also increased; as such, permanent deformation (rutting) has been considered as one of the most important types of asphalt pavement failure [2]. The rutting of asphalt pavements is also regarded as the main problem of flexible pavements [3]. In severe cases, permanent deformation gradually occurs through small distortions caused by repeated shear deformations under applied wheel loads; rutting or permanent deformation may be attributed to the plastic properties of an asphalt mixture under repeated loading; however, this phenomenon is unlikely related to the elastic properties of asphalt mixture [4]. Furthermore, rutting may be caused by mechanical deformation. Therefore, the mechanism by which permanent deformation occurs should be elucidated to enhance the performance of pavement design [5]. The asphalt binder aging of a hot-mix asphalt (HMA) mixture should also be understood to implement reasonable pavement designs and to apply these designs in construction [6]. Various

[a] Corresponding author : alallm84@yahoo.com

additives for modified asphalt mixtures have been used to change the formation stage and to improve the engineering properties of asphalt mixtures [7].

Oxidative aging is the reaction of oxygen with an asphalt binder; the rate of this reaction depends on the characteristics of the asphalt binder and temperature [8]. Moreover, oxidation and volatilization cause asphalt pavements to retreat by increasing the stiffness of the binder, by enhancing the sensitivity to cracking, and by adversely affecting the structural performance of pavements [9]. As such, aging HMA pavements are evaluated to facilitate the prediction of the remaining fatigue on the aged HMA surfaces [10]. Furthermore, the manner by which asphalt undergoes aging differs on the basis of various aging processes; for instance, various asphalt types may age differently in the same environment. The severity of aging among different asphalts can also evaluated using different parameters [11]. For example, unmodified binders in pavements commonly harden and oxidize to a grade that exceeds the generally acceptable specifications of pavement aging [12]. Binder aging also influences the mechanical behavior of asphalt. Therefore, the stiffness of the aged binder increases as aging temperature increases [13]. Moreover, the effect of long-term aging on the pavement depends largely on the properties of the materials constituting the asphalt mixture [14]. Therefore, pavement aging should be investigated to help enhance the design of durable flexible pavements on the basis of mechanistic–empirical principles [6].

HMA mixtures are viscous materials generally used in pavement construction. Therefore, the techniques necessary to develop HMA on the basis of complicated behaviors should be understood to enhance the mechanical properties of the asphalt [15]. [16] used an additive called Nanoclay to improve the mechanical properties of the HMA; the performance of the Nanoclay-modified HMA is more efficient than that of the unmodified bitumen under dynamic creep. Clay nanoparticles are common materials used in asphalt construction. For instance, Nanoclay is used as an additive to reduce the degree of the permanent deformation and fatigue of asphalt mixtures [17]. Similar to Nanoclay, bentonite is used as an additive in a HMA mixture to enhance the mechanical properties of the asphalt mixture [18]. Likewise, hydrated lime can be added to HMA to improve the behaviors of resilient modulus and dynamic modulus [19]. Furthermore, kaolinite clay combined with polymer-modified bitumen has been investigated to enhance storage stability; the use of the expanded clay in bitumen also improves the layers of wearing, as indicated by the volume of the mixture determined via a novel calculation method [20]. Recycled brick powder can also be used as an additive to asphalt mixture to enhance the mechanical performance of the asphalt mixture; the mixture modified with the recycled brick powder is more efficient than the mixture modified with limestone as a filler [21].

2 Experimental Testing Method

2.1 Asphalt binder

Malaysia experiences a tropical climate with high temperature and high humidity all throughout the year. As such, the production of performance-grade asphalt binder in this country depends on high temperature. The current study did not include low-temperature environments, because the temperature in this country hardly reach above 30 °C during daytime and the temperature rarely reach below 20 °C during nighttime. Dense-grade asphalt mixtures were prepared using 80/100 penetration-grade bitumen; the specific gravity of the asphalt binder is 1.03.

2.2 Batu Pahat soft clay (BPSC) particles as an additive

BPSC particles were used as an additive to a HMA mixture; bitumen was modified with various ratios of BPSC (0%, 2%, 4%, 6%, and 8%) by weight of bitumen. Bitumen was heated in a draft oven at 150 °C for 1 h; afterward, bitumen was mixed with the corresponding percentages of the BPSC particles for 15 min at 150 °C. The modified bitumen was prepared using a high shear mixer at 3000 rpm; the modified BPSC was prepared using aggregates through a wet process. Table 1 presents the properties of the BPSC particles used in this research.

Table 1. Properties of BatuPahat soft clay.

Parameters	Results
Bulk Density (mg/m^3)	1.36
Specific Gravity	2.66
Plastic Limit (%)	29.70%
Liquid Limit (%)	37.55%
Plasticity Index (%)	7.85%
Moisture Content (%)	29.35%

2.3 Aggregate properties and structure

The allocation of the particle sizes of Superpave aggregate was determined through a design aggregate process. The aggregate gradation curve is shown in Figure 1. The aggregate gradations in this study were selected on the basis of ASTM D 3515-96 (D-4). Table 3 explains the gradation of the Superpave aggregate mix design with specific gravity.

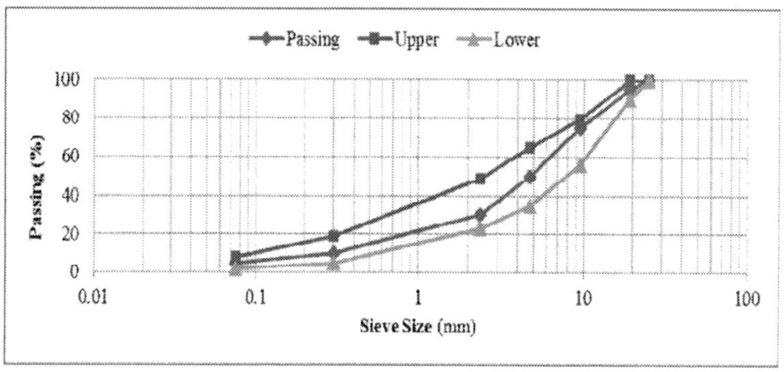

Figure. 1 Gradation curve of the aggregate mix design.

2.4 Specimen preparation and compaction

The HMA test specimens were prepared using a Superpave gyratory compactor. The Superpave samples were prepared in accordance with AASHTO T312 and PP-28-200. To enhance the Superpave HMA mixture, we mixed the components at optimum binder content with an acceptable volumetric property of 4% air voids on the basis of the Superpave standard of the design number of gyrations. The projected traffic load chosen in this study was <30 million ESALs, whereas a design number of gyrations (N_{des}) of 125 was selected. The compatibility of the asphalt mixture was assessed in terms of the initial number of gyrations (N_{ini}) of nine gyrations and at the maximum number of gyrations (N_{max}) of 205 gyrations.

2.5 Aging procedures

The properties and performance of HMA can be accurately predicted by using aged test samples. All of the specimens were subjected to mixture conditioning. The mixture conditioning of the volumetric mixture design procedure was applied to laboratory-prepared loose mixtures only. The aggregate was placed in a forced-draft oven for 2 h to 4 h to equilibrate at the selected temperature. The asphalt binder was heated to 155 °C for 2 h. The mixture and the pan were placed in a forced-draft oven for 2 h ± 5 min at 135 °C for the unaged condition and at compaction temperature. The mixture was stirred after 60 ± 5 min to preserve regular conditioning. The short-term aging procedure was designed to simulate the aging process during mixing and construction.

Table 2. Gradation of the aggregate superpave mix design with specific gravity.

Sieve Size	Percent Passing (%)		
	Upper Limit	Lower Limit	Design
19 mm	100	90	95
9.5 mm	80	56	75
4.75 mm	65	35	50
2.36 mm	49	23	30
0.300 µm	19	5	10
0.075 µm	8	2	5

3 Test methods

3.1 Resilient modulus test

In contrast to the stiffness modulus, the resilient modulus has not been commonly used to represent the HMA stiffness because more emphasis has been placed on the dynamic modulus [22]. The resilient modulus test was conducted at 25 and 40 °C. The resilient modulus at 25 °C was based on the resistance of the mixture to fatigue; the resilient modulus at 40 °C corresponds to the resistance of the mixture to rutting. These specifications were in accordance with the test standard procedure of ASTM D4123. In general, the stiffness modulus of asphalt mixtures is considerably reduced when temperature is increased. Strain was measured with a linear vertical differential transducer (LVDT); strain and applied stress were used to compute the resilient modulus of the design procedures. The specimens were kept in a universal testing machine (UTM) at 25 ± 0.5 °C for least 4 h before the test was performed. In the stiffness modulus tests, the samples were subjected to conditioning loading pulses. The applied force was automatically adjusted to determine the requested horizontal strain or deformation value [23].

3.2 Dynamic creep test

A dynamic creep test can be conducted by establishing repeated axial stress loading cycles and by measuring the resultant axial strain. In this study, the test was performed to assess the permanent deformation (rutting) potential of the asphalt mixtures; the test was also performed using the same specimens for the resilient modulus test. The concepts involved in the dynamic creep test are similar to those in the repeated shear test at a constant height. The test was also carried out using a UTM to apply the repeated axial stress pulses; vertical deformation was determined using LVDTs [24]. At high temperatures, an asphalt binder becomes less viscous. This decreased viscosity produces less stiff pavement that can be susceptible to creep caused by traffic loads. Hence, creep resistance is a mixture property desirable at high temperatures. In this study, the specimens with the a diameter of 100 mm and a loading stress of 100 kPa were subjected to 3,600 loading cycles for approximately 1 h. The test was conducted at 40 °C, and all of the samples were placed in a controlled temperature chamber for at least 1 h to reach a uniform mixture temperature [25].

4 Results and Discussion

4.1 Resilient modulus test

The resilient modulus test is a repeated indirect load tension test. The test method is described in ASTM D 4123. In the indirect tensile modulus test, a non-destructive process is applied to observe the effect of temperature and loading rate [23]. All of the samples were tested at 25 and 40 °C. The resilient modulus at 25 °C corresponds to the resistance of the mixture to fatigue; by contrast, the resilient modulus at 40 °C displays the resistance of the mixture to rutting. The loading pulse width and the pulse repetition period used in this study were set to 100 and 1000 ms, respectively. The stiffness modulus of the asphalt mixtures is considerably reduced when the temperature was increased. The results of the resilient modulus tests are presented in Figures 2 and 3 for unaged and short-term aged specimens, respectively. For unaged specimens, the least susceptibility to fatigue deformation with the highest resilient modulus of 4944 MPa was observed in 4% BPSC particles, followed by 2% BPSC particles with the resilient modulus of 3993 MPa, 6% BPSC particles with the resilient modulus of 4895 MPa, and 8% BPSC particles with the resilient modulus of 4232 MPa; these particles were compared with the control sample with the resilient modulus of 4187 MPa. Similar trends were observed in the samples exposed to short-term aging.

As temperature increased to 40 °C, the addition of 4% BPSC particles yielded a higher resilient modulus than the control sample (Figure 3). This finding indicated that 4% BPSC particles were less susceptible to rutting than the control mix with a resilient modulus of 762 MPa; likewise, 2% BPSC particles were less susceptible to rutting than the control sample. Table 3 shows the aging index that represents the difference between unaged and aged samples at 25 and 40 °C. To calculate the aging index, we divided the resilient modulus of the aged samples by the resilient modulus of the unaged samples. The results revealed that the short-term aged samples exhibited a higher stiffness modulus than the unaged specimens. Therefore, the highest value was determined at 4% BPSC particles during short-term aging. The aging indices (Table 3) showed that the susceptibility to oxidative aging was significantly reduced when the percentage by weight of the BPSC particles increased, especially in short-term aging.

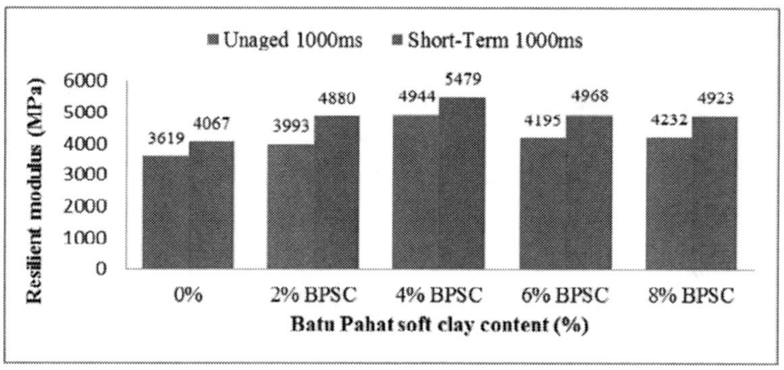

Figure 2. Resilient modulus test at 25 °C (unaged and short-term aged specimens).

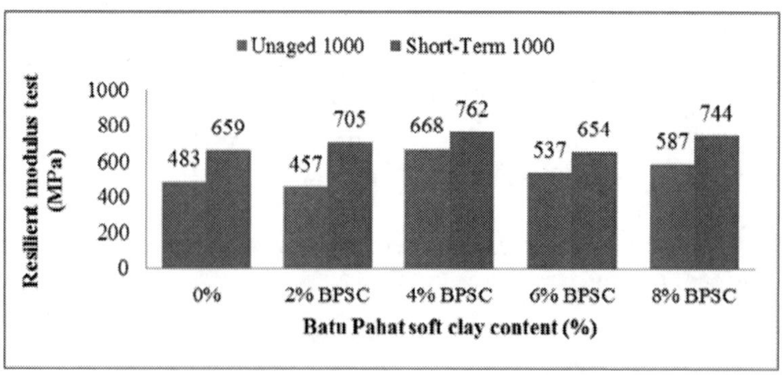

Figure 3. Resilient modulus test at 40 °C (unaged and short-term aged specimens).

4.2 Dynamic creep test

Dynamic creep test is commonly used to determine the resistance of HMAs against permanent deformation [26]. This test is usually conducted by applying a static load to an asphalt mix sample, and the resultant rutting is observed through time. After the load is applied to the surface, the asphalt pavement deforms, despite the plurality of the deformation recapture after the load is extracted. This deformation is greatly affected by the number of cyclic loads and the environmental temperature. Thus, the dynamic creep test was conducted to describe the permanent deformation (rutting) performance of the control mixture and the BPSC-modified asphalt mixtures. During the loading cycles, a repetition time of 1000ms (pulse periods of 100ms with the rest periods of 900ms) was applied. Before the test was performed, all of the specimens were placed in a controlled temperature chamber for 3 h to reach the desired temperature of 40 °C. The dynamic creep test was conducted for 3,600 cycles.

Table 3. Aging indices for the resilient modulus test at 25 and 40 °C.

Samples	Temperature (°C)	Aging Index	
		Unaged	Short-Term Aged
Control	25	1.00	1.12
2% BPSC	25	1.00	1.22
4% BPSC	25	1.00	1.10
6% BPSC	25	1.00	1.18
8% BPSC	25	1.00	1.16
Control	40	1.00	1.36
2% BPSC	40	1.00	1.54
4% BPSC	40	1.00	1.14
6% BPSC	40	1.00	1.21
8% BPSC	40	1.00	1.26

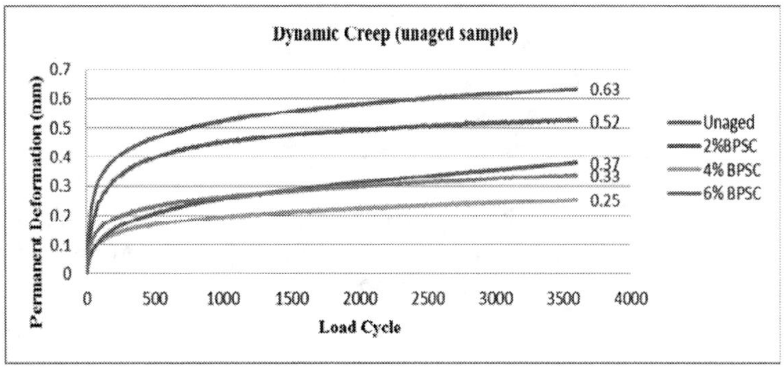

Figure 4. Dynamic creep test for the unaged specimens.

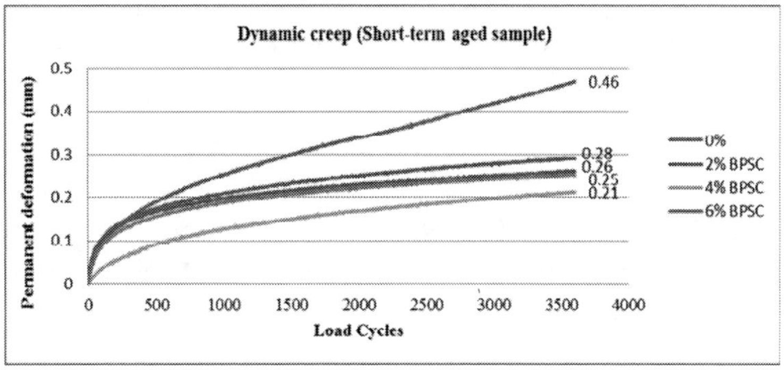

Figure 5. Dynamic creep test for the short-term aged specimens.

In Figure 4, the lowest permanent deformation was observed at 0.25 mm and a load cycle of 3,600 in 4% unaged BPSC particles, followed by 2%, 6%, and 8% of BPSC particles by weight and the control sample. Therefore, the BPSC particles improved the resistance to permanent deformation compared with the control mixes. Figure 5 also shows that 4% short-term aged BPSC particles showed a lower permanent deformation value than the unaged specimens at 0.21 mm and a load cycle of 3600. Hence, the lowest permanent deformation value was observed in 4% short-term aged BPSC particles, followed by 2% BPSC particles at 0.26 mm, 6% BPSC particles at 0.28 mm, and 8% BPSC particles at 0.25 mm compared with the control mix at 0.46 mm. Similar to the resilient modulus test, the dynamic creep test was conducted to calculate the aging index; in this procedure, the aged dynamic creep was divided by the unaged dynamic creep. The results showed that 4% BPSC exhibited the highest performance among the mixes. Thus, permanent deformation likely decreases after the short-term aging occurred because aging conditions increased the stiffness of the asphalt mixtures. Table 4 shows the aging indices calculated through the dynamic creep test.

Table 4: Aging indices from the dynamic creep test at 40 °C.

Samples	**Temperature (°C)**	**Aging Index**	
		Unaged	**Short-Term Aged**
Control	40	1.00	0.73
2% BPSC	40	1.00	0.50
4% BPSC	40	1.00	0.84
6% BPSC	40	1.00	0.76
8% BPSC	40	1.00	0.75

5 Conclusions

On the basis of the limited laboratory work performed in this study, we concluded that 4% BPSC particles were the optimal additive to enhance the performance characteristics of the HMA mixture. The moisture susceptibility of the mixes also improved, and the strength of the asphalt mixtures increased when the BPSC particles were added. Moreover, the resilient modulus of the mixtures prepared using the modified BPSC bitumen was higher than that of the control mixture. Stiffness significantly increased at 25 °C but decreased at 40 °C. This research demonstrated that BPSC can decrease the susceptibility to permanent deformation (rutting) and can improve the resistance to fatigue in asphalt mixtures. The dynamic creep test was also performed at 40 °C and at different stress levels. The rutting properties of the BPSC-modified asphalt were evidently enhanced compared with those of the control mixture; among the mixtures, 4% BPSC showed the most efficient performance in terms of good resistance against permanent deformation.

Acknowledgment

The authors would like to acknowledge the contributions and the financial support by Office for Research, Innovation, Commercialization and Consultancy Management (ORICC) Found No (U033). The mixture testing work was completed in the University Tun Hussein Onn Malaysia.

References

[1] T.B. Moghaddam, M. Soltani and M.R. Karim, Experimental characterization of rutting performance of Polyethylene Terephthalate modified asphalt mixtures under static and dynamic loads, Constr. Build. Mater., 487-494, (2014).
[2] R. Garba, Permanent deformation properties of asphalt concrete mixtures, (2002).
[3] A. Golalipour, E. Jamshidi, Y. Niazi, Z. Afsharikia and M Khadem, Effect of aggregate gradation on rutting of asphalt pavements, Procedia-Social and Behavioral Sciences, 440-449, (2012).

[4] G.P. He and W.G. Wong, Laboratory study on permanent deformation of foamed asphalt mix incorporating reclaimed asphalt pavement materials, Constr. Build. Mater., **8**,1809-1819 (2007).
[5] B.A Feyissa, Analysis and modeling of rutting for long life asphalt concrete pavement, TU Darmstadt, (2009).
[6] P.E.Y. Wang, Y. Wen, K. Zhao, D. Chong and S.T.A. Wong, Evolution and locational variation of asphalt binder aging in long-life hot-mix asphalt pavements, Constr. Build. Mater., **68**,172-182, (2014).
[7] H. Chen and Q. Xu, Experimental study of fibers in stabilizing and reinforcing asphalt binder, Fuel, **7**, 1616-1622, (2010).
[8] J. Reed, Evaluation of the effects of aging on asphalt rubber pavements, Arizona State University, (2010).
[9] R. Lolly, Evaluation of short term aging effect of hot mix asphalt due to elevated, Arizona State University, (2013).
[10] H.P Bell, Fatigue evaluation criteria for aged hot-mix asphalt surfaces, Mississippi State University, (2009).
[11] C.T Chiu., Investigation of laboratory asphalt aging processes for development of an effective procedure to characterize asphalt durability, University of Florida, (1994).
[12] S.H Jung, The effects of asphalt binder oxidation on hot mix asphalt concrete mixture rheology and fatigue performance, Texas A&M University, (2006).
[13] S. Saoula, K. Soudani, S. Haddadi, M. E. Munoz and A. Santamaria, Analysis of the rheological behavior of aging bitumen and predicting the risk of permanent deformation of asphalt, (2013).
[14] H.S Manual, American Association of state highway and Transportation Officials (AASHTO). Washington, (2010).
[15] Y. Xiao, Evaluation of engineering properties of hot mix asphalt concrete for the mechanistic-empirical pavement design, (2009).
[16] S. Ro and S.R.T. Mi, Nanoclay-modified asphalt mixtures for eco-efficient construction. Nanotechnology in Eco-Efficient Construction: Materials, Processes and Applications, 108, (2013).
[17] J. Yang and S. Tighe, A Review of advances of nanotechnology in asphalt mixtures, Procedia-Social and Behavioral Sciences, 1269-1276, (2013).
[18] H. Ziari, Evaluation of fatigue behavior of hot mix asphalt mixtures prepared by bentonite modified bitumen, Constr. Build. Mater., 685-691, (2014).
[19] I. Hafeez, M. A. Kamal, M. R Ahadi, and M. J Msawil, Effects of hydrated lime on fatigue and rutting behavior of HMA mixtures in dynamic modulus testing, Transportation Research, **1**, 23, (2011).
[20] C. Ouyang, S. Wang, Y. Zhang and Y. Zhang, Preparation and properties of styrene–butadiene–styrene copolymer/kaolinite clay compound and asphalt modified with the compound, Polymer Degradation and Stability, **2**, 309-317, (2005).
[21] M. Z. Chen, J. T. Lin, S. P. Wu and C. H. Liu, Utilization of recycled brick powder as alternative filler in asphalt mixture, Constr. Build. Mater., **4**,1532-1536, (2010).
[22] F. M. Nejad, A. R. Azarhoosh, G. H. Hamedi and M.J. Azarhoosh, Influence of using nonmaterial to reduce the moisture susceptibility of hot mix asphalt, Constr. Build. Mater., 384-388, (2012).
[23] T. Alataş and M. Yilmaz, Effects of different polymers on mechanical properties of bituminous binders and hot mixtures, Constr. Build. Mater., 161-167, (2013).
[24] A. Khodaii and A. Mehrara, Evaluation of permanent deformation of unmodified and SBS modified asphalt mixtures using dynamic creep test, Constr. Build. Mater, **7**, 2586-2592 (2009).
[25] M. Dehnad, A. Khodaii and F.M. Nejad, Moisture sensitivity of asphalt mixtures under different load frequencies and temperatures, Constr. Build. Mater., 700-707, (2013).
[26] T. Alataş, M. Yılmaz, B. V. Kök and A. fatih Koral, Comparison of permanent deformation and fatigue resistance of hot mix asphalts prepared with the same performance grade binders, Constr. Build. Mater., 66-72, (2012).

MATEC Web of Conferences **47**, 03017 (2016)
DOI: 10.1051/matecconf/20164703017

Estimate of Small Stiffness and Damping Ratio in Residual Soil Using Spectral Analysis of Surface Wave Method

Nor Faizah Bawadi[1,2,a] , Khairul Anuar Mohd Nayan[2], Mohd Raihan Taha[2] and Noor Adila Omar[2]

[1]*School of Engineering and Environment, Universiti Malaysia Perlis, 02600 Arau, Perlis, Malaysia*
[2]*Department of Civil and Structural Engineering, Universiti Kebangsaan Malaysia, 43600 Bangi, Selangor, Malaysia*

Abstract. Research in the important parameters for modeling the dynamic behavior of soils has led to rapid development of the small strain stiffness and damping ratio for use in the seismic method. It is because, the experimental determination of the damping ratio is problematic, especially for hard soils sample. Many researchers have proved that the surface wave method is a reliable tool to determine shear wave velocity and damping ratio profiles at a site with very small strains level. Surface wave methods based on Rayleigh waves propagation and the resulting attenuation curve can become erroneous when higher modes contribute to the soil's response. In this study, two approaches has been used to determine the shear strain amplitude and damping ratio of residual soils at small strain level using Spectral Analysis of Surface Wave (SASW) method. One is to derive shear strain amplitude from the frequency-response curve and the other is to derive damping ratio from travel-time data. Then, the results are compared to the conventional method.

1 Introduction

In many geotechnical engineering applications, the level of strains in the soil is small [9, 11], especially when dealing with normal functionality of the structures. Within the range of small strain (10-4% or less), [17], the mechanical behavior of the ground can be properly modeled by means of viscoelasticity [3]. Depending on the history of loading, soil may compact or dilate, its stiffness may depend on the magnitude of stress levels, soil deformations are time-dependent, etc. In fact, soil behavior is considered to be truly elastic in the range of small strain.

Specifically, the shear modulus and damping ratio profiles at a site can be determined by means of Rayleigh waves propagation [14], by assuming a layered half-space with horizontal, mono-phase, homogeneous and isotropic layers. Rayleigh waves propagate on the free surface of a medium and during propagation they are subjected to dispersion and attenuation. Generally, viscoelasticity dispersion and attenuation phenomena are correlated [8], so that shear modulus and damping ratio are coupled dynamic properties of the soil. Nevertheless, at very small deformations uncoupling dispersion and attenuation phenomena may represent a good approximation of soil behavior.

[a] Corresponding author : faizahbawadi@yahoo.com

2 Basic Theory of Seismic Method

In current practice of geotechnical testing, the measurement of low-strain dynamic properties of soils can be performed with a variety of techniques using in-situ or laboratory tests. Surface wave measurements has been used to investigate the attenuation and damping ratio of a layered soil deposit [13]. In their studies, an attenuation curve was constructed from the observed spatial attenuation of Rayleigh wave amplitudes and then was inverted to obtain the material shear damping ratio. The strain levels can be categorized into three groups: the very small strain level, where the stiffness modulus is constant in the elastic range; the small strain level, where the stiffness modulus varies non-linearly with the strain; and large strain level, where the soil is close to failure and the soil stiffness is relatively small [2, 5, 10, 15]. The significance of small strain non-linear behavior of soils in deep excavations also was examined [7].

2.1 Strain amplitudes of soil deposit

A review of literature revealed that there are a number of factors affecting the estimation of soil stiffness at small strain level. These factors are not only related to natural and soil inherent structures but also to its geological history. Initial shear modulus can be defined as shear modulus at infinitely small strain amplitude. It is therefore necessary to obtain strain amplitude in soil deposits when the SASW test is to be conducted. The strain amplitude, γ is defined as the ratio of peak particle velocity, υ, to shear wave velocity, V_s:

$$\gamma = (\upsilon/V_s) \tag{1}$$

At small strain conditions, particle motion resulting from propagation of shear waves is non-destructive. As γ increases, the shear modulus will decrease from the maximum small strain value, G_0. In-situ test have commonly been assumed to be small strain and the measurement of V_s will be directly related to the G0 [16]. The appropriate strain must be known, as well as the appropriate modulus reduction with increasing strain.

2.2 Damping ratio and shear stress of soil

The G_0 is an important soil parameter related to the predictions of the ground movements and field data interpretations. The increase of the internal disorder in the material and hence increase of the dissipative properties [9], in terms of material damping. In soil dynamics and earthquake engineering, the G_{max} and the damping, D ratio are important parameters in soil characterization. Through the relationship of damping and shear strain [1, 12], the characteristic of normal or shear strain can be determined by using following equation:

$$D = (D_{max}/1 + \ln(1 + (e-1)(\gamma_r/2\gamma))) \tag{2}$$

where D_{max} is maximum damping, 33%, D is damping at corresponding strain level and γ is the shear strain at measured damping. An elastic model is not able to describe energy dissipation and hence a viscoelastic model must be used. The dissipative phenomena take place at very low cyclic strain in soils and must be taken into consideration. Soil dynamics tests under cyclic loading, both in normal triaxial or shear test have shown common hysteretic behavior. The skeleton or the backbone curve of the hysteresis is known to follow the form of curve [4]. This curve can be expressed in the form of hyperbolic stress-strain soil model which the shear stress, τ of soil can be defined using the following equation:

$$\tau = (G_0\gamma/(1 + (\gamma/\gamma_r))) \tag{3}$$

where G_0 is shear modulus at small strain (less than about 10^{-4}%), γ is shear strain under loading and γ_r is the characteristics of shear strain.

3 Research Method

In this study, determination of importance parameters of residuals soils was performed by using the following procedure: (1) measuring V_s at the field and (2) estimating the variation of G_0 and D of soil corresponding γ level, based on interpretation from WinSASW 3.2.12 software.

3.1 SASW field testing

The field measurements were conducted using a seismograph and three vertical geophones having a natural frequency of 1Hz and geophone calibration factor is 400V/m/s. The test site was a residual soil with meta-sedimentary origin where 27 stories commercial building is to be build located at Damansara, Selangor. The configuration of field measurements was set up using Common Array Profiling (CAP) [6]. Four different seismic impact sources were used to generate energy over a broad frequency range. The location of Spectral Analysis and Surface Wave (SASW) test performed at the nearest point of Standard Penetration Test (SPT) and Instrumented Pile Load (IPLT) has shown in Figure 1.

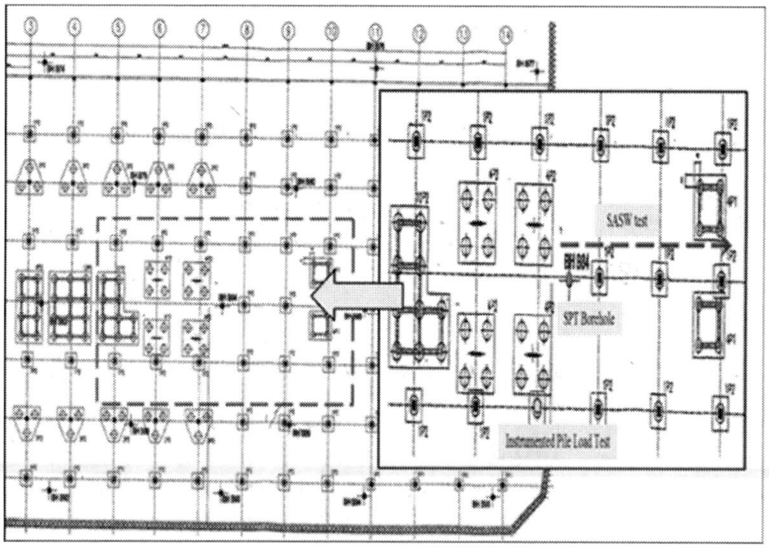

Figure 1. Site map with marked testing locations (N.T.S drawing).

3.2 Data analysis and interpretation

The SASW data was obtained using the National Instrument USB6289 data acquisition system with the WinSASW 3.2.12 and analysis software to determine V_s. This paper summarized a procedure for estimating the shear strain amplitude and damping ratio of soils at small-strain level from SASW method. The shear strain amplitude estimated from travel-time data as shown in Figure 2(a) and damping of soil measured from frequency-response curve as shown in Figure 2(b). In addition, the result obtained at a site with the two approaches are compared to the conventional method.

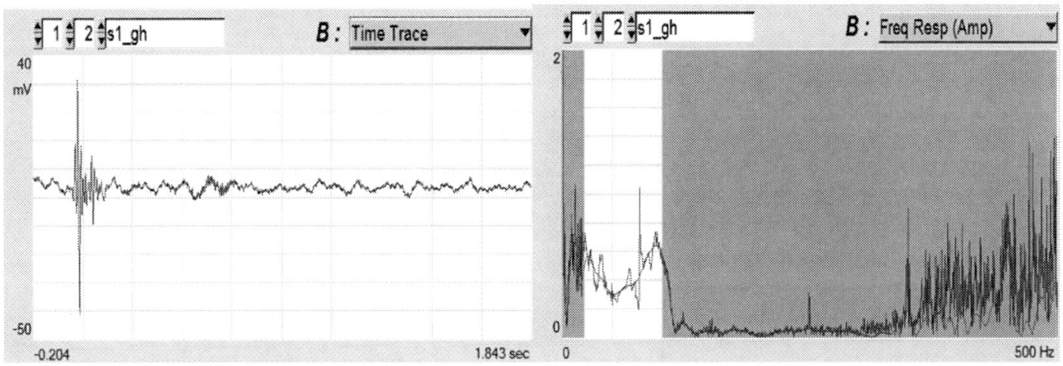

Figure 2. Data analysis for (a) travel-time curve and (b) frequency-response curve in determining strain amplitude and damping of soil.

4 Results and Discussion

The SPT and SASW test results identified numerous soil layers at the site indicate over 20.30 meters of medium dense silty sand, stiff sandy clay and silt which is a residual soil and has underlain by weathered rock deposit of granite as provided in Figure 3. Direct comparison of the shear wave velocity, V_s and normalized shear wave velocity, V_{s1} profiles is illustrated in Figure 4(a). The V_s profile is roughly divided into three sections, V_s is about 218m/s, 257m/s and 516m/s at 0~6.45m, 6.45~7.60m and below 7.6m which encountered the rock formation respectively. The V_{s1} performed after consider the overburden pressure of the soil layer.

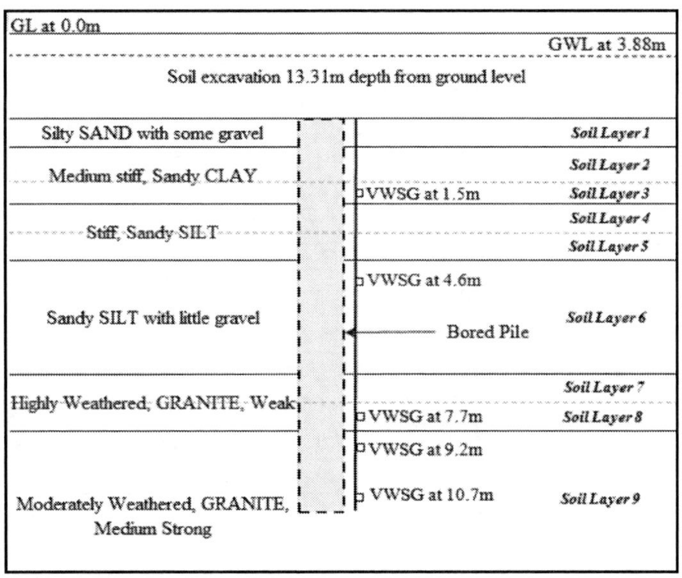

Figure 3. The soil profile of the test site.

The dynamic strain level caused by the shear waves can be estimated based on attenuation response on the soil by conducting SASW testing. In this study, the peak particle velocity can be defined by the interpretation of travel-time data that obtained from WinSASW software. Furthermore, the results of shear strain amplitudes for all sources can be determined using Equation (1) as presented in Figure 4(b). Meanwhile, in order to measure the damping from the signal spectrum recorded from SASW test, the half-power bandwidth method has been used for the interpretation of frequency-response curve using Equation (2). The results of damping based on this interpretation has shown in Figure 4(c).

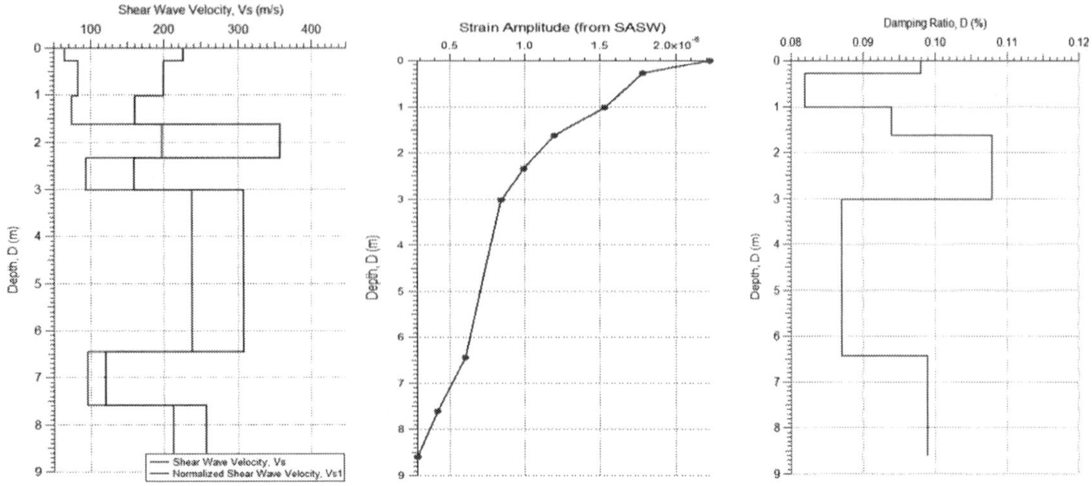

Figure 4. Profile of (a) shear wave velocity, (b) strain amplitude and (c) damping for the test site using SASW method.

In general, prediction on characteristic of strain amplitudes and damping in SASW method will decreases with depth of soil. In this study, analysis of the performed SASW field data shows that the strain amplitudes caused by several assorted hammer as seismic source are generally less than 10^{-4}%. Summary of the shear wave velocity, shear strain amplitudes and damping as tabulated in the Table 1 below.

Table 1. Summary of the result for the test site.

Layer	Depth, z (m)	Shear Wave Velocity, V_s (m/s)	Shear Modulus, G_0 (kN/m^2)	Shear Strain Amplitudes, γ		Damping, D (%)	Shear Stress of Soil, τ (kN)
				SASW Test (x10^{-6})	IPLT Test (x10^{-4})		
1	0	64.48	6494.2	3.5	4.6	9.8	242.46
2	0.28	83.33	5373.9	2.4	4.6	8.2	275.29
3	1.08	74.68	3368.1	2.06	4.52	9.4	88.15
4	1.63	197.09	21516.7	1.56	4.46	10.8	258.84
5	2.34	93.04	3536.4	1.29	4.41	10.8	35.31
6	3.02	237.99	16672.2	1.10	4.3	8.7	313.93
7	6.45	95.46	2019.6	0.78	3.77	9.9	17.02
8	7.6	212.22	11323.9	0.54	3.36	9.5	66.69
9	8.57	647.32	71235.8	0.34	2.7	9.5	261.96

From this study, the result shows that as shear strain increases, the shear modulus will decreased. The appropriate shear strain must be known, as well as the appropriate modulus reduction with increasing strain. Shear stress, τ of the soil have been determined using Equation (3) and the result has been tabulated in Table 1 above. The shear stress for a different types of soil affected by the dynamics properties of soils such as shear wave velocity, shear modulus, strain amplitudes and damping.

5 Conclusions

As a conclusion, it can be summarized that the effects of shear strain amplitudes on shear modulus and damping can significantly influenced the predicted shear stress of soil. Thus, shear strain amplitudes should be considered when selecting shear modulus and damping for geotechnical engineering design.

It is also shown that the characterization of the physical properties of the soil structure in terms of the shear strain amplitudes and the damping of soil layer can be satisfactory obtained using frequency-response and travel-time curve from SASW data analysis. In addition, this method has the advantage of being fast, practical and non-destructive tool for engineers in measuring shear modulus and damping at very small strain conditions compared to the conventional method. Implication of the results and future research directions are also presented.

References

[1] C.P. Abbiss, Calculation of elasticities and settlements for long periods of time and high strains from seismic measurements. Geotechnique, **33**(4), 397-405, (1983).

[2] J. H. Atkinson and G. Sallfors, Experimental determination of stress-strain-time characteristics in laboratory and in situ tests, *Proc. of the 10th European Conference on Soil Mechanics and Foundation Engineering*, Florence, Italy, 915-956, (1991).

[3] R.M. Christensen, Theory of Viscoelasticity - An Introduction, Ed. Academic Press, 245, (1971).

[4] B.O. Hardin and V.P. Drnevich, Shear modulus and damping in soils, J. of the Soil Mechanics and Foundation Division ASCE, **98**(SM7), 667-692, (1972).

[5] K. Ishihara, Soil Behavior in Earthquake Geotechnics, Oxford University Press, New York, (1996).

[6] Joh Sung-Ho, Jang Dae-Woo, Kang Tae-Ho and Lee Il-Wha, Evaluation of stiffness profile for a subgrade cross-section by the CAP(Common-Array-Profiling)-SASW technique, J. of the Korean Geotechnical Society, Korean Geotechnical Society, **21**(4), 71-81, (2005).

[7] G.T.C Kung, C.Y. Ou and C.H. Juang, Modelling small-strain behavior of Taipei clays for finite element analysis of braced excavations, Computer and Geotechnics, **36**(1), 304–319, (2009)

[8] C.G. Lai, Simultaneous inversion of Rayleigh phase velocity and attenuation for near-surface site characterization, PhD Dissertation, Georgia Institute of Technology, Atlanta, (1998).

[9] R. Lancellotta, Geotechnical Engineering, second ed., Taylor and Francis, New York, (2009).

[10] R.J .Mair, Developments in geotechnical engineering research: application to tunnels and deep excavations, *Proc. of the Institution of Civil Engineers-Civil Engineering*, Unwin Memorial Lecture 1992, **97**(1), 27-41, (1993).

[11] C. Mancuso, R. Vassallo and A. d'Onofrio, Small strain behavior of a silty sand in controlled-suction resonant column-torsional shear tests, Canadian Geotechnical J., **39**(1), 22-31, (2002).

[12] K.A.M. Nayan, M.R. Taha, N.A. Omar, N.F. Bawadi, S.H. Joh and M.N. Omar, Determination of ultimate pile bearing capacity from a seismic method of shear wave velocity in comparison with conventional methods, J. of Technology (Sciences and Engineering), **74**(3), 99-102, (2015).

[13] G.J. Rix, C.G. Lai and A.W. Spang Jr., In-situ measurement of damping ratio using surface waves, J. of Geotechnical and Geo-environmental Engineering, *Proc. of the ASCE*, **126**(5), 472–480, (2000).

[14] G.J. Rix, C.G. Lai and S. Foti, Simultaneous measurement of surface wave dispersion and attenuation curves, Geotechnical Testing J., **24**(4), 350 – 358, (2001).

[15] A. Sawangsuriya, P. J. Bosscher and T. B. Edil, Alternative testing techniques for modulus of pavement bases and subgrades, *Proc. of the 13th Annual Great Lakes Geotechnical and Geoenvironmental Engineering Conference,* Geotechnical Applications for Transportation Infrastructure, ASCE, Geotechnical Practice Publication, 108-121, (2005).

[16] S. S. Tezcan, Z. Ozdemir and A. Keceli, Allowable bearing capacity of shallow foundations based on shear wave velocity, J. of Geotechnical and Geological Engineering, **24**, 203-218, (2006).

[17] M. Vucetic, Cyclic threshold shear strains in soils, J. of Geotechnical Engineering, **120**(12), 2208–2228, (1994).

MATEC Web of Conferences **47**, 03018 (2016)
DOI: 10.1051/matecconf/20164703018

Influence of Agricultural Wastes on Shear Strength Properties of Soil

Nik Daud Nik Norsyahariati[1,a], Abdul Jalil Falil Nisa[1], Muhammed Abubakar Sadiq[1] and Ghafar Azmi Juliana[1]

[1]*Department of Civil Engineering, Faculty of Engineering, Universiti Putra Malaysia, 43400, Serdang, Selangor, Malaysia*

Abstract. Fiber reinforced soil is an effective and reliable technique for improving strength and stability of subgrade soil. Many studies have been conducted to investigate the influence of randomly distribution fiber inclusions in soil. Agricultural waste fiber was chosen as it is cheap, locally available, bio-degradable and eco-friendly. For this study, coir fiber and palm oil fiber will be implemented. Three types of soil with different properties were used to be mixed with both fibers. The present work will be focused on the influence of 1% waste materials into soils in terms of physical, chemical, morphological and mineralogical properties of waste-soil mixture and its relationship to the shear strength properties. Test result indicates that with addition of both fibers, the optimum moisture content (OMC) increased and the maximum dry density (MDD) decreased. Fibers inclusion can significantly decrease and increased the cohesion and internal angle of friction of soils, respectively.

1 Introduction

Agricultural waste materials are too abundant nowadays due to high demand in food industry and increasing of housing area that lead to clearance of plantation area. Due to this condition, the oil palm and coconut waste will be generated more. If those waste did not have a proper management, it will brings negative impact to the environment. Process of altering some soil properties by different methods, either mechanical or chemical can be defined as soil reinforcement. This is due to produce an improved soil material which has all desired engineering properties. Soil generally been reinforced to increase their strength and durability or to prevent erosion and dust formation in soils. The main purpose of this soil reinforced is to hold under the design use conditions and for the design life of one project. The properties of soil reinforcement totally depend on soil testing.

Addition of fiber may affect the engineering properties of soil in few aspects. Fiber reinforced soil behaves as a composite material in which fibers of relatively high tensile strength are embedded in a matrix of soil. Shear stresses in the soil mobilize tensile resistance in the fibers, which in turn imparts greater strength to the soil. Fibers from agricultural waste materials (coconut coir and palm oil) are chose to be implemented in this study. Discrete fibers are simply added and mixed randomly with soil, like in mixing soil with cement, lime, or other additives [1]. It is more economical both in terms of cost and energy to increase the bearing capacity of the soil rather than going for deep foundation or

[a] Corresponding author : niknor@upm.edu.my

raft foundation. Inclusion of randomly distributed fiber makes the soil mass homogenous and maintains the isotropy in strength [2]. According to [3] and [4], the used of coir fiber is very cost effective compared to other material in construction area such example in slopes and bunds. In addition, inclusion of fiber only changes the physical properties of soil and has no impact on the environment. The introduction of reinforcing elements in a soil mass increases the shear resistance of the medium [5]. Soil stabilization improves the workability and the durability of the soil. The objectives of this study are i) to determine the properties of soils and agricultural waste materials, ii) to investigate and analyses the influence of agricultural waste on soil shear strength properties.

2 Materials and Methods

2.1 Materials

There are three soil samples were used in this study. Two samples (Soil A and Soil B) were collected at Universiti Putra Malaysia (UPM), Serdang, Selangor area and another sample (Soil C) was taken from Klang, Selangor area. These samples were collected based on the surrounding and the land use. Disturbed sample was used in this study and those sample were collected by using shovel and keep it in plastic for avoiding losses of moisture content. All three samples are being air-dried before mixing with agricultural waste fibers (Figure 2) in order to study the influence of fibers in soil.

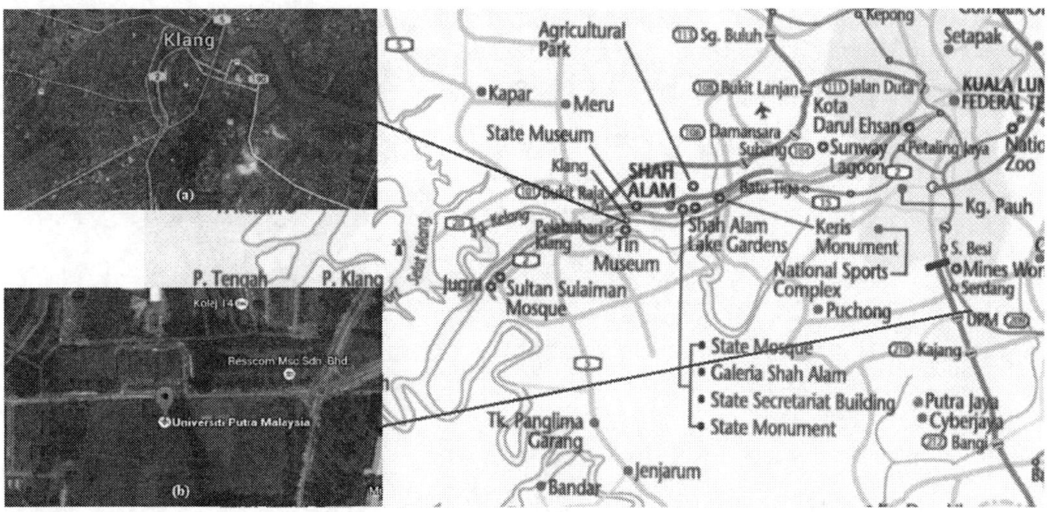

Figure 1. Sample location for soils used in this study (a) Klang site and (b) UPM site.

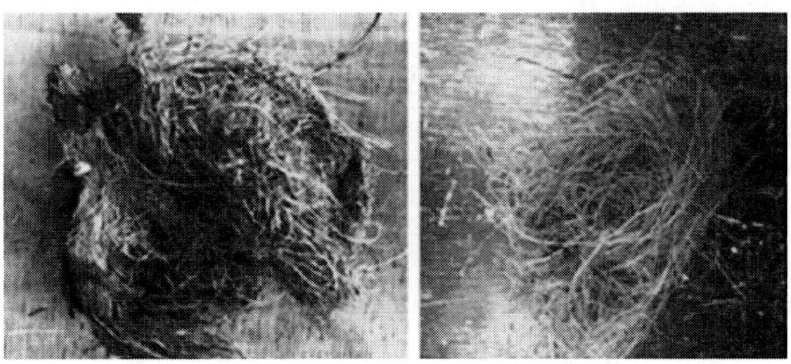

(a) (b)

Figure 2. Agricultural waste fiber (a) palm oil fiber and (b) coir fiber.

2.2 Methods

The reinforced soil specimens were prepared manually by mixing the dry soil, water and fibers. During the mixing process, the required amount of water was added to the soil prior to adding the fibers. Fibers were mixed manually with the wet soil at small increments. Particular care was taken to achieve satisfactory uniform mixtures. Soil sample with mix proportion of 0% and 1% of fibers by weight was prepared by referring to the previous study [6-8]. Then a few tests were conducted to obtain the properties of each waste-soil mixtures such as physical properties tests, chemical properties tests, structural and morphological tests and mechanical tests. Direct Shear Box test is used to determine the shear strength parameters of all soil samples used in this study with and without mixing fiber under undrained condition. All tests basically conducted accordance to [9].

3 Results and Discussion

3.1 Physical properties

Coir-stabilized lateritic soils, the maximum dry density of the soil decreases with the addition of coir and the value for optimum moisture content of the soil increases with an increase in percentage of coir [10]. With increasing the water content to the fiber reinforced, value of cohesion and suction decreased. Water plays an important role in lubricating layer on the interface of fiber and soil [1]. The increase of water content may also weaken the interfacial mechanical interactions between fiber and soil matrix and fiber will not capable to bear the tensile load. Thus, the increase of water content can induce a decrease of interfacial friction and cohesion.

Table 1. Basic properties for natural soils.

Properties	Soil A 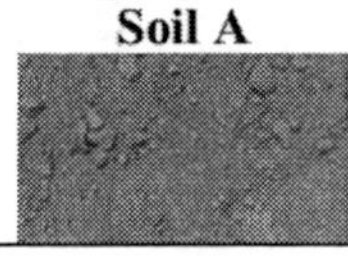	Soil B 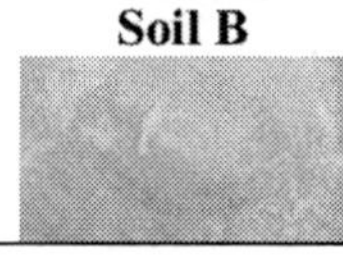	Soil C 
Moisture Content (%)	31.33	28.05	239.00
Specific Gravity	2.19	1.77	1.29
Grain size analysis			
• Gravel	33.2	20.6	59
• Sand	62.4	66.8	38.2
• Silt and clay	4.4	12.6	2.8
• USCS	SW	SC	Pt
Consistency limit			
• Liquid Limit (%)	47.6	61	120.4
• Plastic Limit (%)	36.84	34.01	Non plastic
• Plasticity Index (%)	10.76	26.99	-
Compaction study			
• MDD (Mg/m^3)	1.38	1.46	0.72
• OMC (%)	23.6	21.5	36.2
pH	2.73	5.65	3.32
OC (%)	12	4.8	84.3
Specific Surface Area (m^2/g)	33.168	16.513	1.947

Soil A can be classified as gravelly sands for having 33.2% of gravel, 62.4% of sand and 4.4% of silt and clay. This soil is red in color and wide range in grain size with substantial amounts of intermediate particle sizes. Atterberg consistency limit tests carried out on the soil sample show that

Soil A has liquid limit of 47.6%, plastic limit of 36.84% and plasticity index of 11%. The plots of plasticity index against liquid limit on the plasticity chart in Figure 3 showed Atterberg limits plot under "A" and it is in the zone of inorganic silts and very fine sand, clayey silts with slight plasticity. The soil is very firm and dry in place. For Soil B, grain size distribution shows it is clayey sands, sand-clay mixtures with 12.6% silt and clay, 66.8% sand and 20.6% gravel compositions. According to Unified Soil Classification System (USCS), this type of soil is medium in dry strength. Based on Figure 3, Soil B is in the zone of inorganic silts having liquid limit of 61%, plasticity index of 27% and very low percentage of organic content with the value of 4.8%. Mainly, consistency limits depends on the type and amount of clay fraction [11]. Soil C is readily identified by the color and odor that it is peat soil. Soil C is black in color and contains high organic matter content. The percentage of organic content showed that this soil has higher amount of organic matter which is 84%. Peat normally formed from decomposed vegetation and well known for having difficulty on constructing road. Soil C is very high in moisture content that will lead to stability problem.

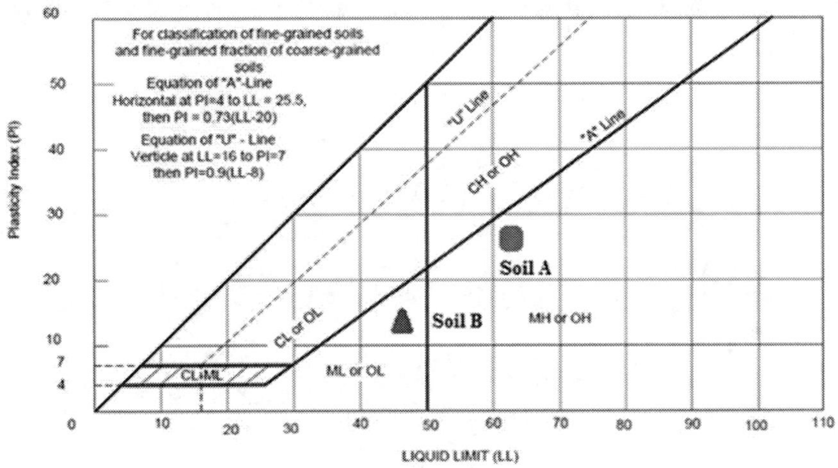

Figure 3. Particle size distribution of soils using in this study.

3.2 Chemical properties

The chemical compound in 1 gram of each type of soils and fibers are tested by using XRF equipment. The value of aluminium (Al_2O_3) in soil A is 30% and 11% of iron (Fe_2O_3). It is also contains large amount of silica (SiO_2). For soil B, it showed that the highest element in this soil is silica about 46% and aluminium 24%. Based on grain analysis, soil A has a high percentage of sand with silt and clay. Soil C contains 7% of silica and 2% of aluminium, which is very small amount compared to soil A and B. However, soil C showed a high organic content among others. For fiber, both coir and palm oil fibre consists of aluminium, silica and calcium oxide compound. Coir fiber seems to have greater amount of silica about 45% compared to palm oil which having 7% silica.

3.3 Morphological and mineralogical properties

The mineralogical properties for each sample were determined by using X-ray diffractometer (XRD) equipment. Soil A contained kaolinite, gibbsite and quartz, Soil B contained quartz and muscovite and Soil C contained quartz and calcium strontium. The SEM images of the structure and morphology of the different samples are presented in Figure 4. These show the rough surface morphologies of the ground specimens. The SEM provided micrographs that show up bond formation, surface texture, mineral structure and geometry of the specimens. For laterite, it can be seen that the soil particles is attached to each other and rounded in shape. Flaky arrangements of clay can be seen as matrix between the fine grains. Soil is formed in stack of several layers in basic sheet units. These are generally flat and smaller in size, thus their surface areas per weight are large. From Figure 4, sample of peat have shown that the surface have porous granules of organic material. SEM analysis examined the surface topology of untreated fibers were shown in Figure 5. Coir fiber and palm oil fiber surface is covered with protrusions and with voided area left by detached protruded material. This porous surface morphology is useful for better mechanical interlocking of soil with the fiber.

(a) (b) (c)

Figure 4. SEM images of natural soil (a) gravelly sand, (b) clayey sand and (c) peat.

(a)

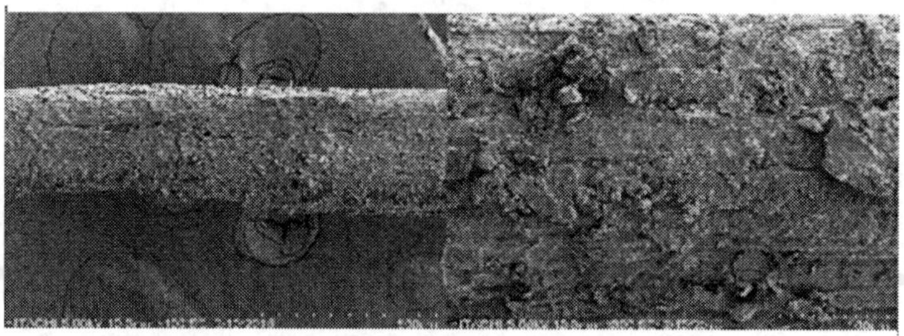

(b)

Figure 5. SEM images of natural fiber (a) coir fiber and (b) palm oil fiber.

3.4 Shear strength properties

The soil's shear strength parameters in terms of the internal friction angle and the cohesion value are evaluated and compared for the unreinforced and reinforced conditions of the soil sample. Table 2 shows the value of cohesion and angle of friction for each unreinforced and reinforced soil samples. It can be clearly seen that the inclusion of fibres increased the friction angle of soil. Coir-soil mixtures values for friction angle increased about 10%, 36% and 43% for laterite, clay and peat, respectively. Findings show that coir fibre exhibits greater enhancements in the strength of soil by having higher coefficient of friction [12]. Coir retains much of its tensile strength when wet and it has low tenacity but the elongation is much higher [13] and as for palm fiber, a constant palm fiber length, with increase in fiber inclusion, the maximum and residual strengths were increased [14]. It is observed that the palm fibers act to interlock particles and group of particles in a unitary coherent matrix thus the strength properties of soil can be increased [15]. In terms of cohesion, addition of 1% of both fibers into soil samples decreased the cohesion value and caused the soils less cohesion. It is believed that this happens due to the fibers separating soil particles. Singh and Mittal [7] conducted an experiment on the subgrade soil added with different ratios of coconut coir and found out that there is considerable improvement in compressive strength in case of all the soils on account of treatment with coconut coir fibre. Values of unconfined compressive strength are increase as more percentage of soil is replaced with coir fibre. It is noted that the compressive strength of soil increases when treated with coir fibre up to 1%. It might be due to the increase in shear parameter. Fiber reinforcement benefit on the mechanical properties is governed by the interfacial friction and cohesion [16].

Table 2. Shear strength parameters.

Type of Soil	Fibre (%)	Shear strength parameters	
		c (kPa)	ø°
Laterite	0%	48.33	20.6
	CF 1%	34.00	22.8
	POF 1%	38.66	24.9
Clay	0%	19.33	22.1
	CF 1%	15.66	30.0
	POF 1%	18.66	29.0
Peat	0%	52.00	14.6
	CF 1%	36.33	20.6
	POF 1%	50.33	15.6

4 Conclusions

Throughout the study, the used of fibres to treat or stabilize subgrade soils has been recognized, discussed and analyzed. The study was focused on determining the properties for waste-soil mixtures and investigating the influence of agricultural waste materials into soils for the purpose of subgrade works. These objectives were determined by conducting various laboratory tests on soil specimens with 1% of coir fibre and palm oil fibre. It can be summarized and concluded as follow: i) The optimum moisture content (OMC) for each soils increased with the inclusion of fibres due to the higher absorption ability of fibres. The maximum dry density (MDD) decreased when 1% fibres mixed with soils due to the soil being heavy in weight compared to fibre and was replaced by the light of weight fibre, and ii) The cohesion and angle of friction decreased and increased respectively with the inclusion of both fibres for all soil samples. The cohesion decreased due to the fibres acting as a separator for soil particles and causing less cohesion.

In conclusion, the properties for all waste-soil mixtures were determined and it showed that both fibres influenced the properties of each soils in terms of strength.

Acknowledgement

This study has been funded by Fundamental Research Grant Scheme-FRGS (No. 03-01-13-1172FR) granted by Ministry of Education (MOE), Malaysia. The help and assistance various governmental and non-governmental organizations in this study are gratefully acknowledged.

References

[1] J. Li, C. Tang, D. Wang, X. Pei and B. Shi, Effect of discrete fiber reinforcement on soil strength, J. of Rock Mechanics and Geotechnical Engineering, **6**, 133-137, (2014).

[2] R.R. Singh and E.S. Mittal, Improvement of local subgrade soil for road construction by the use of coconut coir fiber, Int. J. of Research in Engineering and Technology, 3(5), 707-711, (2014).

[3] B.M. Lekha, S. Goutham and A.U.R. Shankar, Evaluation of lateritic soil stabilized with arecanut coir for low volume pavements, Transportation Geotechnics, **2**, 20–29, (2015).
[4] S. Vishnudas, H.H.G. Savenije, P.V.D. Zaag, K.R. Anil and K. Balan, The protective and attractive covering of a vegetated embankment using coir geotextiles, Hydrology Earth Syst. Sci., **10**, 565-574, (2006).
[5] S. Akbulut, S. Arasan and E. Kalkan, Modification of clayey soils using scrap tire rubber and synthetic fibres, App Clay Sci, **38**, 23-32, (2007).
[6] H.P. Singh and M. Bagra, Improvement in CBR value of soil reinforced with jute fiber, Int. J. of Innovative Research in Science, Engineering and Technology, **2**, 3447-3452, (2013).
[7] R.R. Singh and E.S. Mittal, Improvement of local subgrade soil for road construction by the use of coconut coir fiber, Int. J. of Research in Engineering and Technology, **3** (5), 707 – 711, (2014).
[8] H.P. Singh, Effects of coir fiber on CBR value of itnagar soil, Int. J. of Current Engineering and Technology, 1283-1286, (2013).
[9] British Standard Institution, BS 1377 Methods of Test for Soils for Civil Engineering Purposes, London, (1990).
[10] U. Ravishankar and S. Raghavan, Coir stabilised lateritic soil for pavements, *Indian Geotech. Conf.*, Ahmedabad, (2004).
[11] Y.A. Yeliz and K. Abidin, Method dependency of relationships between specific surface area and soil physicochemical properties, Applied Clay Science, **50**, 182–190, (2010).
[12] R. Ayyar, R. Krishnaswamy and S. Vismanadham, Geosynthetics for foundations on swelling clay, *Int. Work on Geotex*, Bangalore, (1989).
[13] S. Babu and K. Vasudevan, Strength and stiffness response of coir fiber-reinforced tropical soil, J. Mater. Civil Eng., **20**, 571–7, (2008).
[14] M. Marandi, H. Bagheripour, R. Rahgozar and H. Zare, Strength and ductility of randomly distributed palm fibers reinforced silty- sand soils, Am. J. Appl. Sci., **5**, 209–20, (2008).
[15] Z. Jamellodin, Z. Talib, R. Kolop and N. Noor, The effect of oil palm fibre on strength behaviour of soil, *The 3rd SANREM Conf*, Kota Kinabalu, (2010).
[16] F. Ahmad, F. Bateni and M. Azmi, Performance evaluation of silty sand reinforced with fibers, Geotext Geomembr, **28**, 93–9, (2010).

MATEC Web of Conferences **47**, 03019 (2016)
DOI: 10.1051/matecconf/20164703019

A Coupling Simulation Between Soil Scour and Seepage Flow by Using a Stabilized ISPH Method

Tomotaka Nogami[1,a], Mitsuteru Asai[1], Kiyonobu Kasama[1] and Taro Arikawa[2]

[1]*Department of Civil Engineering, Graduate School of Engineering, Kyushu University, Nishi-ku, Fukuoka 819-0395, Japan*
[2]*Department of Science and Engineering, Chuo University, Bunkyo-ku, Tokyo 112-8551, Japan*

Abstract. In 2011, the example that breakwaters collapsed because of the basic ground's destabilization was reported by Tohoku-Kanto earthquake tsunami. Fluid-Structure-Soil coupling simulation is desired for a systematic comprehension of the breakwater collapse mechanism, and it may help to develop next disaster prevention method. In this study, A particle simulation tool based on the SPH has been modified and improved to analyze seepage flow and soil scouring. In seepage flow analysis, as a first step, this simulation treat the surface flow and seepage flow interactions by using governing equation. In the scouring analysis, soil scour is judged by an empirical criteria based on quicksand quantity formula.

1 Introduction

In 2011, Tohoku-Kanto earthquake, tsunami caused serious damage to the port structures such as a breakwater and seawalls. Damage mechanisms of these structures have been studied in the past, and there are mainly three causes; I. Horizontal force due to the water level difference between the front and rear breakwater, II. Soil scour and erosion behind the seawall during overflow and III. Piping destruction associated with the decline of the bearing capacity by seepage flow. Fluid-Structure-Soil coupling simulation is desired for a systematic comprehension of the seawall collapse mechanism, and it may help to develop next disaster prevention method. In this study, a particle simulation tool based on the SPH has been developed to solve seepage failure problem and scouring phenomenon. In this study, as a first step, surface flow and seepage flow are described by the same governing equations, and an SPH formulation has been developed. After that, efficiency and adequacy of the proposed simulation technique has been validated through an application to one test. On the other hand, by using the empirical judging equation, we implemented the numerical analysis to replicate phenomenon of scouring in the back of breakwater.

2 Seepage Flow Analysis

In this part, as a first step, surface flow and seepage flow are described by the same governing equations, and an SPH formulation has been developed.

[a] Corresponding author : t_nogami@doc.kyushu-u.ac.jp

2.1 Governing equation

Fluid flowing outside the soil is generally modeled as Newtonian fluid, and described Navier-Stokes equation as:

$$\frac{Dv}{Dt} = \frac{1}{\rho^w}\left(-\nabla P + \rho^w g\right) + \left(v_w + v_T\right)\nabla^2 v \tag{1}$$

In addition to the above equation, the mass conservation law was also used as governing equation.

$$\frac{D\rho}{Dt} + \rho\nabla \cdot v = 0 \tag{2}$$

Here, v, P, g is water's velocity, pressure, gravitational acceleration. vw is the kinematic viscosity coefficient of water and ρw is specific density. vT is the kinematic eddy viscosity coefficient of water. On the other hand, the soil is generally described Darcy's law as:

$$\frac{\mu}{K} v = -\nabla P + \rho^w g \tag{3}$$

Here, K is a specific permeability coefficient in soil mound and μ is viscosity. In a field of the hydrology, the Darcy's law has been used as an equation of motion widely. However, it is reported that Darcy's law is not established in a flow field with a low Reynolds number, and then the non-Darcy flow rule is adapted as a model of seepage flow in this study. The traditional Extended Darcy's law suggested by Forchheimer is given by:

$$\frac{\mu}{K} v + \frac{\rho^w C}{\sqrt{K}} v|v| = -\nabla P + \rho^w g \tag{4}$$

The above equation is not enough to simulate collapse behavior of seawall and breakwater structure, and it is necessary to unify the governing equations both for the free surface flow and seepage flow. Akbari equation [1, 2] is introduced as an unified governing equation as follows:

$$\frac{C_r(\varepsilon)}{\varepsilon}\frac{Dv_D}{Dt} = -\frac{1}{\rho^0}\nabla P + v_E(\varepsilon)\nabla^2 v_D - a(\varepsilon)v_D - b(\varepsilon)v_D|v_D| + g \tag{5}$$

Here, ^vD is Darcian velocity or average velocity, and it has a relationship with the regular velocity v as $^vD^{=\varepsilon v}$. Meanwhile, ε is porosity in the ground. Various coefficients in the Akbari's equation are given as a function of the porosity and their expressions of relations are as follows.

$$C_r(\varepsilon) = 1 + 0.34\frac{1-\varepsilon}{\varepsilon} \quad \text{: inertia coefficient} \tag{6}$$

$$v_E(\varepsilon) = \frac{v_w + v_T}{\varepsilon} \quad \text{: effective viscosity coefficient} \tag{7}$$

$$a(\varepsilon) = \alpha_c \frac{v_w(1-\varepsilon)^2}{\varepsilon^3 D_{50}^{\ 2}} \quad \text{: linear coefficient} \tag{8}$$

$$b(\varepsilon) = \beta_c \frac{(1-\varepsilon)}{\varepsilon^3 D_{50}} \quad \text{: nonlinear coefficient} \tag{9}$$

D_{50} is expressed the average particle size of ground materials and the value of α_c, β_c set $\alpha_c = 1000$ and $\beta_c = 1.1$.

2.2 ISPH method

In this paper, smoothed particle hydrodynamics (SPH) method was adopted to solve the Akbari's unified equation for free surface and seepage flow. The basic concept in SPH method is that any function ϕ attached to particle "i" at a position $\boldsymbol{r}_i$ is written as a summation of contributions from neighbor particles. Note that, the triangle bracket $< \phi_i >$ means SPH approximation of a function ϕ.

$$\varphi(r_i) \approx \langle \varphi_i \rangle = \sum_j \frac{m_j}{\rho_j} \varphi_j W(r_{ij}, h) \tag{10}$$

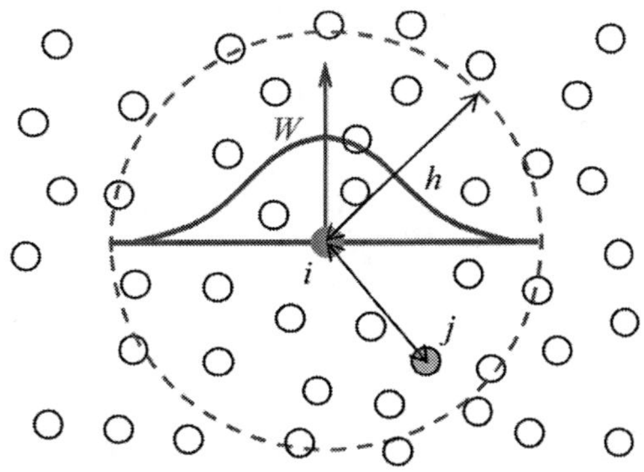

Figure 1. Particle placement and influence radius in the SPH method.

The Laplacian of the function can be assumed by using the above defined SPH approximation as follows:

$$\langle \nabla^2 \varphi_i \rangle = \sum_j m_j \left(\frac{\rho_i + \rho_j}{\rho_i \rho_j} \frac{\boldsymbol{r}_{ij} \cdot \nabla W(r_{ij}, h)}{\boldsymbol{r}_{ij}^2 + \eta^2} \right) (\varphi_i - \phi_j) \tag{11}$$

The main concept in an incompressible SPH method [3] for Navier-Stokes equation is to separate the governing equations of the incompressible fluid by using projection method. In this method, the pressure is calculated implicitly and the velocity fields are updated explicitly. In this study, the same idea of ISPH for the Navier-Stoke equation is applied to solve the unified equation as follows:

$$\frac{D\bar{\rho}}{Dt} + \bar{\rho} \nabla \cdot \frac{\boldsymbol{v}_D}{\varepsilon} = 0 \tag{12}$$

$$\frac{C_r(\varepsilon)}{\varepsilon} \frac{D\boldsymbol{v}_D}{Dt} = -\frac{1}{\bar{\rho}} \nabla P + \frac{\boldsymbol{g}}{\varepsilon} + \nu_E(\varepsilon) \nabla^2 \boldsymbol{v}_D - a(\varepsilon)\boldsymbol{v}_D - b(\varepsilon)\boldsymbol{v}_D |\boldsymbol{v}_D| \tag{13}$$

Here, the density can be assumed as a constant value because of the incompressible assumption. Equation (12) can change as follows:

$$\nabla \cdot \boldsymbol{v}_D = 0 \tag{14}$$

In the incompressible SPH method and treatment, the final pressure Poisson equation is given by:

$$\langle \nabla^2 P_i^{n+1} \rangle = \frac{C_r(\varepsilon_i)}{\varepsilon_i} \frac{\rho_i^0}{\Delta t} \langle \nabla \cdot \boldsymbol{v}_{Di}^* \rangle \tag{15}$$

During numerical simulation, the 'particle' density may change slightly from the initial value because the particle density is strongly dependent on particle locations in the SPH method. If the particle distribution can keep almost uniformity, the difference between 'physical' and 'particle' density may be vanishingly small. In other words, accurate SPH results in incompressible flow need to keep the uniform particle distribution. For this purpose, the different source term in the pressure Poisson equation can be derived using the 'particle' density. In stabilized ISPH method [4], the pressure Poisson Equation (15) reformulated as:

$$\left\langle \nabla^2 P_i^{n+1} \right\rangle \approx \frac{C_r(\varepsilon_i)}{\varepsilon_i}\left(\frac{\rho_i^{\,0}}{\Delta t}\left\langle \nabla \cdot \boldsymbol{v}_{Di}^{\ *} \right\rangle + \alpha\varepsilon_i \frac{\rho_i^{\,0} - \left\langle \rho_i^{\,n} \right\rangle}{\Delta t^2} \right) \tag{16}$$

where α is relaxation coefficient, $\boldsymbol{u}_i$* is temporal velocity and triangle bracket < > means SPH approximation. Note that this relaxation coefficient is dependent on the time increment and the particle resolution. Then, the reasonable value can be estimated by the simple hydrostatic pressure test using the same settings on its time increment and the resolution. In this study, α=0.01.

2.3 Validation test

In validation, the experimental test ralated to seepage problem in breakwater structure is simulated. This experiment was conducted by Kasama et al. [5]. The penetration collapse of the breakwater in case given water level difference was observed. The water level differenceΔh is selected as 40mm, 80mm, 120mm, 145mm, and only in case of 145mm the mound collapsed. Experimental outline and analysis model is shown in Figure 2 and Figure 3. In Figure 2, blue points are water pressure gauge and Figure 4 shows their number. In this analysis, the particle distance d_0 = 1cm, time increment Δt = 0.001s and the total numbers of particles are about 8 hundred thousand. Piezo water head shown below was used for a comparison between experimental and analysis. In analysis, Δh is selected 145mm as the biggest water level. The analysis result of comparison of piezo water head in case of 145mm is shown in Figure 5. From this comparison, the value of the analysis is slightly different.

$$Piezo = \frac{P}{\rho g} + \Delta z \tag{17}$$

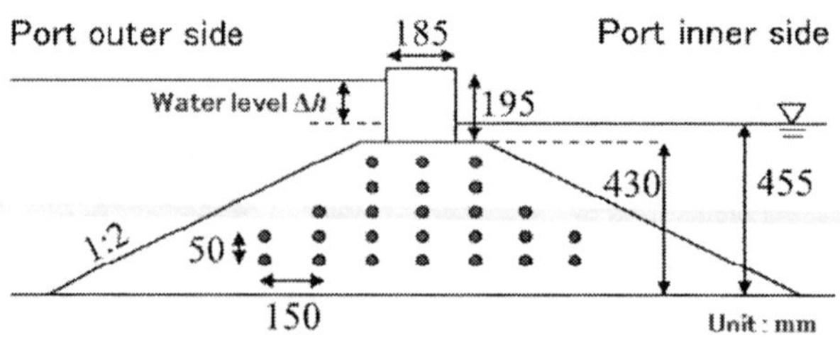

Figure 2. Experimental outline.

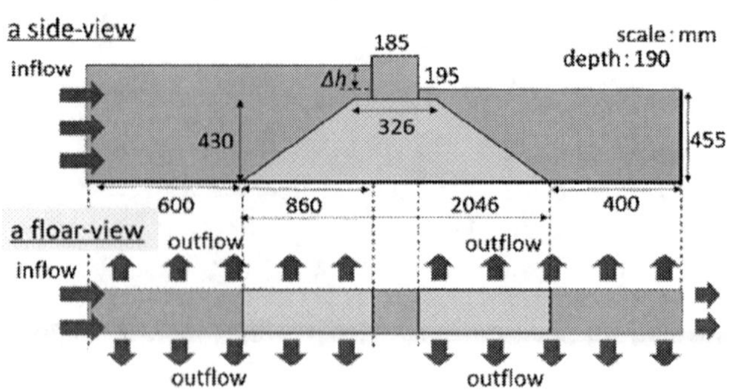

Figure 3. Analysis result.

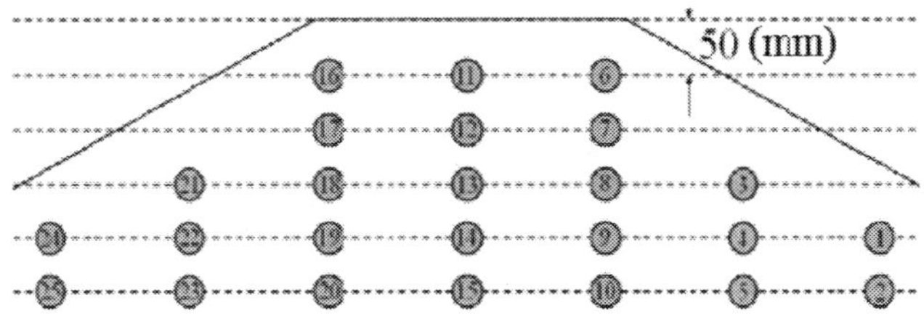

Figure 4. Water pressure gauge number.

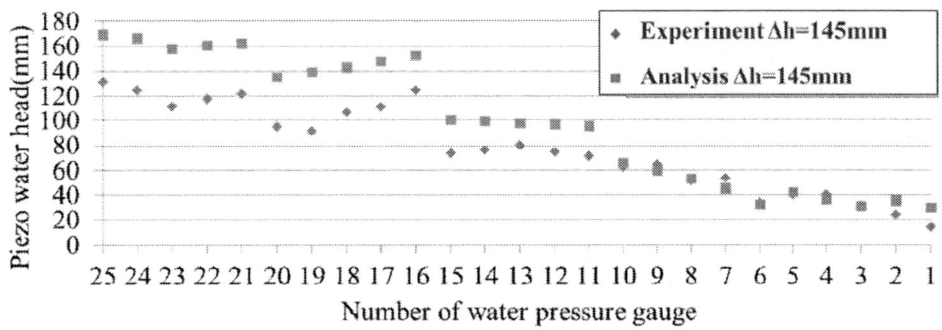

Figure 5. Analysis result (Piezo water head).

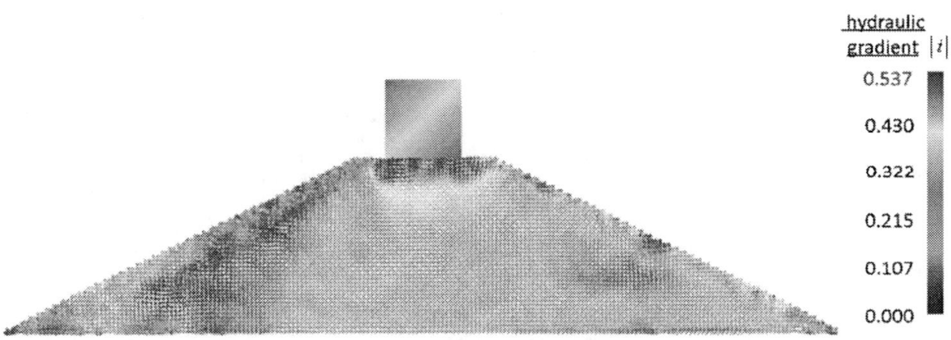

Figure 6. Analysis result (hydraulic gradient).

However, an analysis result behavior shows a good agreement with the hydraulic experiment test. Finally, the seepage failure can be estimated by using our simulated hydroulic gradient by refering to its limitation as shown in Figure 6. In the figure, red color region shows the possible domain for failure by seepage flow.

3 Seepage Flow Analysis

3.1 Scouring judging formula

Soil scour may depend on shear velocity of fluid flow and particle size on the see floor. In this study, soil scour is judged by an empirical criteria based on quicksand quantity formula [6]. The velocity of threshold of sediment movement is shown as follows.

$$u_{limit} = \sqrt{\frac{2A_3(\mu_f \cos\theta - \sin\theta)\left(\rho_{sp}/\rho_l - 1\right)gd}{(C_D + \mu_f C_L)A_2}} \quad \text{: water flow slope downward} \tag{18}$$

$$u_{limit} = \sqrt{\frac{2A_3(\mu_f \cos\theta + \sin\theta)\left(\rho_{sp}/\rho_l - 1\right)gd}{(C_D + \mu_f C_L)A_2}} \quad \text{: water flow slope upward} \tag{19}$$

Here, A_3 is three-dimensional shape coefficient of the soil particle, μ_f is coefficient of static friction, θ is slope inclination, ρ_{sp} is density of soil particle, C_D is drag coefficient, C_L is lift coefficient, A_2 is two-dimensional shape coefficient of the soil particle. Once soil particle meets the above condition, the fixed soil particle is shited as a moving particle based on fluild flow with a density as the soil material.

3.2 Validation test

By using the empirical equation based on the correlation between shear velocity and sediment discharge, we implemented the numerical analysis to replicate phenomenon of scouring in the back of breakwater. As analysis model, the experiment that was conducted by Arikawa [7] is adapted. In this experiment, overflow of tsunami is reproduced by hydraulic test. Water level difference 20cm is selected and validated in this study. Experimental outlines and Analysis models are shown in Figure 7 and Figure 8. The analysis result of comparison of each time and scouring shape in 15s is shown in Figure 9 and Figure 10. From this comparison, our model can estimate tendency and the depth of the soil souring quantitatively, but it still has some difference with experimental result especially in the shape of soil souring. In this study, soil particle's sedimentation after soil scour occured doesn't be considered yet, then the shape after the soil scouring couldn't be reproduced. In the future work, this analysis would be expanded to be able to consider soil sedimentation.

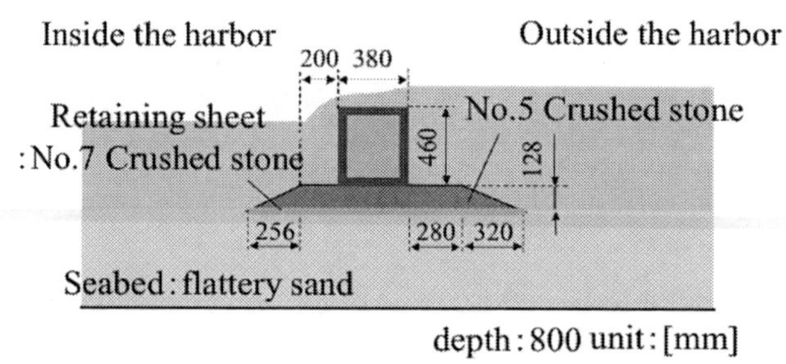

Figure 7. Experimental outline.

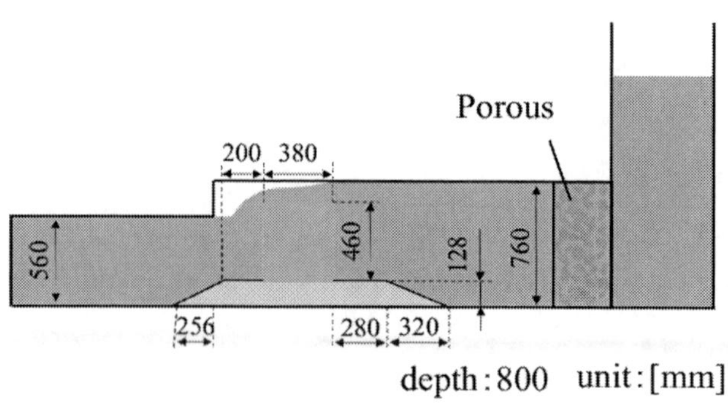

Figure 8. Analysis model.

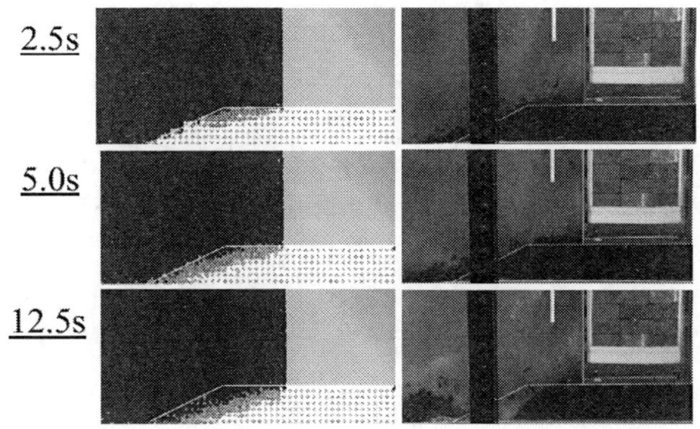

Figure 9. Analysis result.

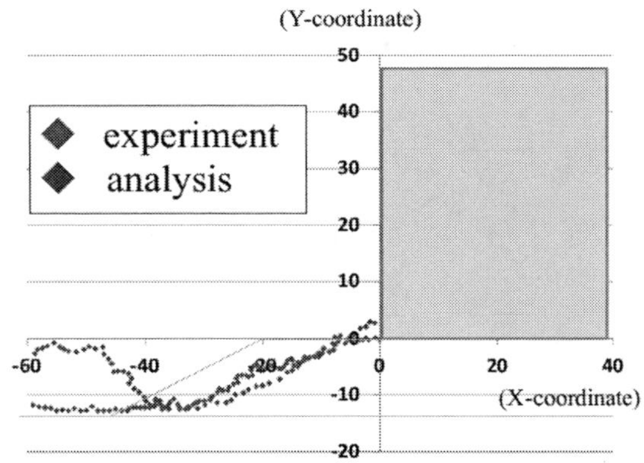

Figure 10. Scouring shape (15s).

4 Summary

In this study, a particle simulation tool based on the SPH has been developed to solve both of the seepage failure problem and the soil scour problem. In seepage flow analysis, a unified government equation which can model both of free surface flow and seepage flow is applied to a particle simulation. From the verification and validation test, the effect of porosity and reasonable surface flow and seepage flow behavior has been given. In scouring analysis, scouring phenomenon was reproduced by using the scouring judging formula. In the future works, this analysis method would be expanded to soil scour and seepage flow coupling simulator.

Acknowledgement

This work was supported by JSPS KAKENHI Grant Numbers 26282106, 15K12484.

References

[1] H. Akbari, Modified moving particle method for modeling wave interaction with multi layered porous structures, Coast. Eng., **89**, 1-19, (2014).
[2] H. Akbari and M.M. Namin, Moving particle method for modeling wave interaction with porous structures, Coast. Eng., **74**, 59-73, (2013).
[3] M.A.A.A. Abdelraheem, An improved incompressible smoothed particle hydrodynamics to simulate fluid-soil-structure interactions, Kyushu University, (2012).

[4] M. Asai, A.M. Aly, Y. Sonoda and Y. Sakai, A stabilized incompressible SPH method by relaxing the density invariance condition, Int. J. for Applied Mathematics, **2012**, (2012).
[5] K. Kasama, K. Zen and Y. Kasugai, Model experiment for the instability of caisson-type composite breakwater under tsunami condition, *Proceedings of Coastal Engineering JSCE,* (2013).
[6] H. Gotoh, *Computational Mechanics of Sediment Transport*, Morikita Shuppan Co. Ltd., (2004).
[7] T. Arikawa, S. Oikawa and S. Moriyasu, Stability of the Breakwater with Steel Pipe Piles under Tsunami Overflow, Independent Administrative Institution, Port and Airport Research Institute, Japan, (2015).

MATEC Web of Conferences **47**, 03020 (2016)
DOI: 10.1051/matecconf/20164703020

Soil Plasticity Model for Analysis of Collapse Load on Layers Soil

Masyitah Md Nujid[1,a] and Mohd Raihan Taha[1]

[1]*Department of Civil Engineering and Structural Engineering, Universiti Kebangsaan Malaysia, 43650 Bangi, Selangor, Malaysia*

Abstract. Natural soil consist of soil deposits which is a soil layer overlying a thick stratum of another soil. The bearing capacity of layered soil studies have been conducted using different approach whether theoretical, experimental and combination of both. Numerical method in computer programme has become a powerful tool in solving complex geotechnical problems. Thus in numerical modelling, stress-strain soil behaviour is well predicted, design and interpreted using appropriate soil model. It is also important to identify parameters and soil model involve in prediction real soil problem. The sand layer overlaid clay layer soil is modelled with Mohr-Coulomb and Drucker-Prager criterion. The bearing capacity in load-displacement analysis from COMSOL Multiphysics is obtained and presented. In addition the stress distribution and evolution of plastic strain for each thickness ratio below centre of footing are investigated. The results indicate the linear relation on load-displacement which have similar trend for both soil models while stress and plastic strain increase as thickness ratio increase.

1 Introduction

Layered soil may consist of soil deposits which is a soil layer on top or called surface layer of finite thickness overlying a thick stratum of another soil. A homogeneous soil layer is considered if depth from footing base on top layer to lower layer H_o approximately in Equation (1).

$$H_o = 0.5B \tan (45+\phi/2) \quad (1)$$

where B is a width of shallow footing and ϕ is internal friction angle. In practice it is usually assumed that H = 2B The ultimate bearing capacity equation in Equation (2) for a strip footing in a uniform soil states that,

$$q_u = cNc + \gamma DNq + \frac{1}{2} B\gamma N\gamma \quad (2)$$

where q_u = ultimate bearing capacity (kN/m^2); c = cohesion; γ = effective unit weight (kN/m^3); B = width of the footing (m); D = depth of embedment (m); and N_c, N_γ and N_q = bearing capacity factors,

[a] Corresponding author : masyitahnujid@gmail.com

which are a function of the effective stress friction angle ϕ. In Equation (2) indicates that q_u increases linearly with B or D. Also few assumptions are taken such that ϕ and γ are constant with depth and the soil is dry. Thus for effective soil which shear strength increase linearly with depth at failure is expressed in Equation (3).

$$\tau_f = c + \sigma'_n \tan\phi' \tag{3}$$

However the bearing capacity of layered soils cannot be treated anymore as single soil since the soil strength and stiffness parameters are differ or increase linearly with depth. In present day numerical analysis such as finite element difference and finite element method using appropriate constitutive equation become a powerful technique in geotechnical problems to analyze complex behavior of stress and strain due to external loading and geometrical such as multiple layers of soil [1, 5, 7]. Several studies on load-deformation relation of shallow foundation by numerical analysis were conducted using von Mises, Mohr Coulomb (MC) and Drucker-Prager (DP) criterion due to some limitations in finite element package. However the constitutive soil models (von Mises, MC and DP) incapable to model certain cases of problem such as stress path dependency and coupling of volumetric and shear response [2, 4, 6].

In present study, this paper is attempted to implement soil plasticity models which are Mohr-Coulomb and Drucker-Prager models for bearing capacity of strip footing on layered soil. The bearing capacity is presented in load-displacement analysis which is obtained from COMSOL Multiphysics finite element package. The behavior of layered soil under strip footing is also being analyzed. In addition this paper will look into factor of thickness ratio which govern the development of stress behavior of layered soil under strip footing.

2 COMSOL Multiphysics

The COMSOL Multiphysics finite element software offers real simulation problems with partial differential equation applied in the software with elements of multiphysics such as heat, particle tracing and water flow. In this study there are two branches have been selected: 1) Solid Mechanics; Geomechanic and 2) Mathematics; ODE and DAE. Normally the processes of numerical simulation involve with assigning geometry, material, physic branch, mesh, analyze and viewing result [8]. The bearing capacity is obtained from load-displacement curve and similar geotechnical problem using this software i.e [9]. The study analysis is done in function of thickness ratio and properties of soils are remain constant throughout the analysis.

3 Constitutive Soil Model

By assuming the footing problem in plane strain condition and the foundation is treated as an elastic-perfectly plastic material. Before yielding within the elastic range, the stress-strain relationship of the soil is described by Hooke's law. Beyond the elastic range, the soil is assumed to behave as perfectly plastic in accordance with Mohr-Coulomb and Drucker-Prager (extended Von Mises) yield criterion. In COMSOL Multiphysics (v4.4) programme these two models were chosen to be used in analysis.

3.1 Mohr-Coulomb model

The simplest and controversial constitutive soil model is Mohr-Coulomb which accepted in geotechnical problems due to simplicity to define soil parameters. The strength of soil mass using the simple elastic-perfectly plastic model requires cohesion (c) and internal angle of friction (ϕ).

For simple elasto-plastic soil plasticity obey Mohr-Columb yield criterion in plane strain deformation,

$$f(\sigma) = 0 \tag{4}$$

where Equation (4) in principal stresses space,

$$f(\sigma) = \frac{1}{2}(\sigma_{max} - \sigma_{min}) + \frac{1}{2}(\sigma_{max} + \sigma_{min})\sin\phi - c\cos\phi = 0 \tag{5}$$

where σ_{max} and σ_{min} are the biggest and smallest principal stresses. The Mohr-Coulomb criterion defines an irregular hexagon pyramid in the prinsipal stress space which give rise to singularities in the derivatives of the yield function.

3.2 Drucker-Prager model

The Drucker-Prager model approximate the Mohr-Coulomb criterion by a circle on the cross-sectional shape of the yield surface by a smooth function based on the invariant I_1 and J_2 together with two material constants. The yield surface for the Drucker-Prager criterion is given by,

$$f = \alpha I_1 + \sqrt{J_2} = k = 0 \tag{6}$$

In which I_1 and J_2 = the first stress invariant and the second stress invariant respectively; and α and k = constants which are functions of strength parameters of the test soil and has two material constants for the plane strain condition can be expressed as follows,

$$\alpha = \frac{\tan\phi}{(9 + 12\tan\phi)^{1/2}} \tag{7}$$

and

$$k = \frac{3c}{(9 + 12\tan\phi)^{1/2}} \tag{8}$$

The flow rule relation between the plastic strain increment in a given direction and the current level of stress in the same direction reads as,

$$d\varepsilon_{ij} = d\lambda \frac{\partial Q}{\partial \sigma_{ij}} \tag{9}$$

Where $d\lambda$ is the plastic multiplier and Q is the plastic potential. If the yield surface f and Q are identical then it is called an associated flow rule and in oppose is called non-associated flow rule.

4 Geomatry of Layered Soils, Soils Properties, Boundary Condition and Mesh Discretization

Geometry of layered soils in numerical modelling was modelled in two dimensional (2D), plane strain problem with z-axis and footing was founded on surface of elastic half space soil (infinite element). In this analysis a flexible and smooth strip footing carried a vertical load on sand (Layer 1) overlaid clay (Layer 2) with static response. An infinite layer was drawn on the right side of the model to indicate flexible layer. Figure 1 shows geometry of layered soils in modelling where width of footing B = 1.0m. In order to consider effect of thickness ratio between top and bottom layers in layered soil, thickness of top layer, h play a vital role to indicate whether soil is homogeneous or inhomogeneous. If h < B the soil is inhomogeneous and h > B the soil is homogeneous. The physical and mechanical parameters of layered soils (i.e γ,ν,E,ϕ,Ψ) are given in Table 1. The properties of soils are remain constant throughout the analysis.

The finite element solutions were affected by selection boundary conditions and mesh refinement and size. Since the layered soil was symmetrical in geometry thus only half of it was modelled and assigned a prescribed displacement at nodes to the right of centerline. The bottom of the geometry model was assigned by a fixed end boundary condition. The right vertical boundary was perfectly smooth and a roller was assigned on right of the geometry problem (Figure 1). For non-uniform mesh discretization, coarse element mesh would be assigned at bottom, left and right sides of soil geometry model. Under footing base, the soil element would be discretized into medium fine and near at edge of footing, it was very fine mesh because of significant displacement change in this region. The element sizes affect the finite element solution where divergence could occur at very small size but require larger computational time [5]. The finite element mesh has 3337 degrees of freedom solved.

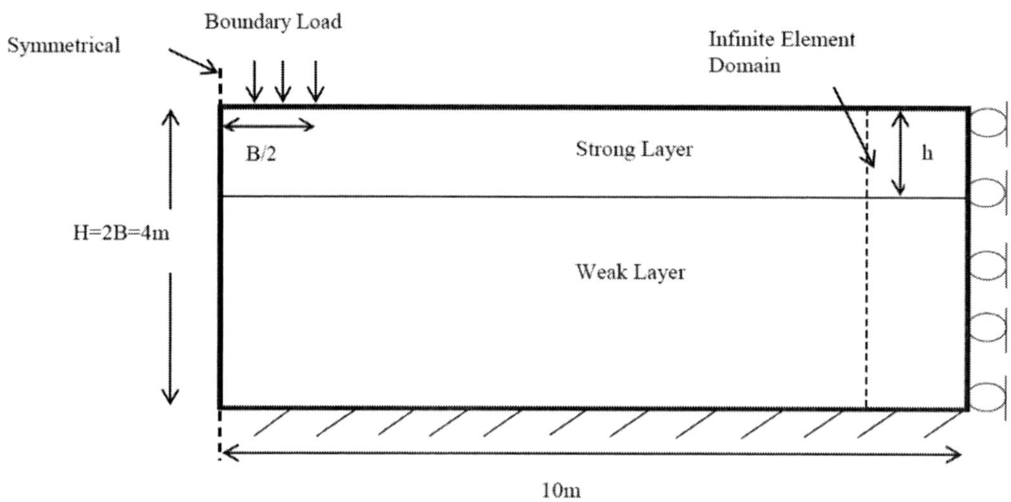

Figure 1. Geometry of layered soils model.

5 Results and Discussion

In Figure 2 show the load-displacement graphs for both Mohr-Coulomb and Drucker-Prager criteria. The graphs show that applied footing pressure versus the centerline displacement (directly beneath the footing's center) in the y direction. These graphs indicate similarity trend of footing pressure and displacement at beneath of footing's center for Mohr-Coulomb and Drucker-Prager criteria. By assuming that critical depth at h = 2.0 the footing pressure reached the collapse load at approximately 1.1MPa. Figure 3(a) shows a linear relation between applied load and the displacement where an algebraic equation that controls the applied pressure so that the models reach the desired displacement increments. The desired vertical displacement in the analysis was -0.05m in Figure 3(b).

5.1 Stress distribution

At the desired vertical displacement (0.05m) the von Mises stresses distribution show the value for each increment of thickness ratio (h/B) from 0.5 to 2.0 as in Table 2. In the table listed the stress distribution started at 0.561MPa (h/B = 0.5), then dropped to 0.546MPa (h/B = 1.0), increased to 0.559MPa (h/B = 1.5) and finally at h/B = 2.0 the stress distribution was 0.643MPa.

5.2 Effect of thickness ratio on the analysis

The load-displacement relation shows that the footing pressure increase linearly with thickness ratio from 0.5 to 2.0 with maximum footing pressure at beneath footing center was approximate 1.1MPa. This was happened due to availability of ODE and DAE branch in COMSOL Multiphysics reached to desired vertical displacement at 0.05m at maximum collapse loading. The soil plasticity models used indicate similarity trend for both Mohr-Coulomb and Drucker-Prager criterion which lied on single line. The capability offered in the Drucker-Prager model matched to Mohr-Columb model with existing two constant material parameters which related to strength of soil. This model overcome the limitation of Mohr-Coulomb model in yield criteria (Table 1). Meanwhile the stress distribution in von Mises showed the sudden dropped at thickness ratio of 1.0. However the difference still acceptable in range

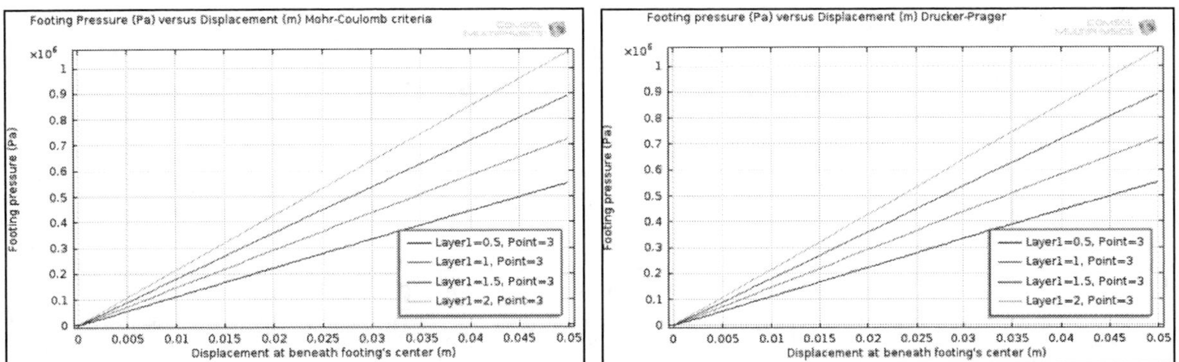

(a) (b)

Figure 2. Load-displacement curve, (a) Mohr-Coulomb criteria and (b) Drucker-Prager criteria.

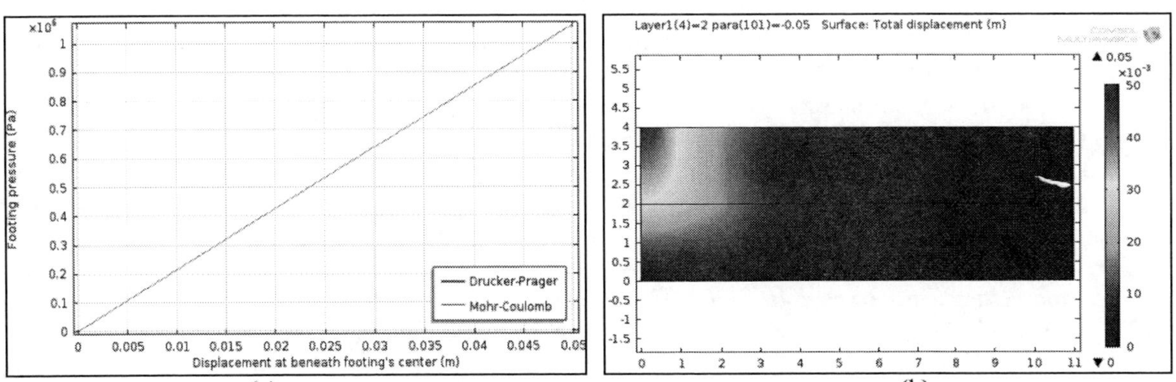

(a) (b)

Figure 3. (a) Linear relation load-displacement, (b) total vertical displacement.

Table 1. Von Mises stress distribution at different thickness of layer 1.

Thickness Ratio (h/B)	Stress Distribution (MPa)
0.5	0.561
1.0	0.546
1.5	0.559
2.0	0.643

6 Summary

The paper summarized and concluded effect of thickness ratio with implemented soil plasticity soil model on bearing capacity of strip footing on layered soil. Load-displacement relation showed linearly with desired vertical displacement for both Mohr-Coulomb and Drucker-Prager criterion on single line. The footing pressure increase as thickness ratio increase however stress distribution slightly drop at h/B = 1.0 then continued to increment. It is important for geotechnical analysis using finite difference and finite element method for simulation in prediction real soil behavior on field with certain assumption in analysis for examples are geometry, boundary condition, discretization element and size. Besides choosing right solution of constitutive relation of stress and strain in soil model, it also could predict the real soil behavior itself and interaction with structures.

Acknowledgement

The first author would like to express gratefully acknowledgement to Ministry of Higher Education, Malaysia and Universiti Teknologi MARA for given a great opportunity in sponsoring the research study in the Geotechnical Engineering field.

References

[1] O.A. Abdulhafiz and A.A. Awad, Derivation on bearing capacity equation for a two layered system of weak clay layer overlaid by dense sand layer, Pertanika J. Sci. and Technol., **13**(2), 213-235, (2005).
[2] A.H. Akhaveiss, Implementation of generalized plasticity in load- deformation behavior of foundation with emphasis on localization problem, World Academy of Science, Engineering and Technology, **39**, 4-9, (2010).
[3] COMSOL Multiphysics v4.4, (2013).
[4] C.S.Desai, H.V. Phan and S. Sture, Procedure, selection and application of plasticity models for a soil, Int. J. Numer. Anal. Meth, Geomech., **5**, 295-311, (1981).
[5] D.V. Griffiths, Computation of bearing capacity factors using finite elements, Geotechnique, **32**(3), 195-202, (1982).
[6] J. Lee and R. Salgado, Estimation of bearing capacity of circular footings on sands based on cone penetration test, Journal of Geotechnical and Geoenvironmental Engineering, **131**(4), 442-452, (2005).
[7] N. Maniharan and SP. Dasgupta, Bearing capacity of surface footings by finite elements, Computers and Structures, **54**(4), 563-586, (1995).
[8] J. Mwebesa, D. Kalumba and R. Kulabako, Stimulating bearing capacity failure of surface loading on sand using COMSOL, *Proceedings of the Second International Conference on Advances in Engineering and Technology*, Entebbe, Uganda, (2011).
[9] M.M. Nujid and M.R. Taha, A review of bearing capacity of shallow foundation on clay layered soils using numerical method, Electronic Journal Geotechnical Engineering, **19**, 811-825, (2014).

MATEC Web of Conferences 47, 03021 (2016)
DOI: 10.1051/matecconf/20164703021

The Study of Soil-Roots Strength Performance by Using Pennisetum Setaceum Grass

Nor Azizi Yusoff[1] , Mohamad Fazrin Mohamad Fauzi[1] and Firdaurs Abdullah[1,a]

[1]*Faculty of Civil and Environmental Engineering, Universiti Tun Hussein Onn Malaysia, 86400 Parit Raja, Johor, Malaysia*

Abstract. Vegetation is regarded as a new engineering material contributes to slope stability and reduces superficial soil erosion, both by soil shear reinforcement and influencing soil water content. Yet, it is aware that roots growth occurred below the soil surface and it is considered as dynamic organs that respond to physical, chemical and hydrological induced by surrounding substrate. This study was carried out to investigate the soil shear strength performance by using Pennisetum Setaceum grass at the different planting period up to three months. The grass was planted in a 150 mm x 200 mm x 700 mm test plot with a suitable soil medium. Pull-out strength test was conducted for each plot to determine the soil pull-out strength according to different planting period. The results show that there is an increment in the pullout strength of rooted-soil sample over different planting period with average results of 6.7 kPa, 11.4 kPa and 25.8 kPa for month 1, 2 and 3 respectively. Based on this study, the soil pull-out strength developments for vegetated soil were observed. The higher pull-out strength was recorded for different growth period. Moreover, root weight was increased for different growth period. Based on the tensile test, the diameter of the root affects the rate of resistance against the tensile forces. This indicated that the roots structure growth affects the soil shear strength.

1 Introduction

Slope stabilization is the process of creating and implementing measures resistance to erosion and slope failure. The use of ground cover plants to overcome the slope failure was widely practiced throughout the world. Grass planting is the most effective method to stabilize the banks and slopes. The function of the grassroots is for gripping the particles of land slope and prevent it from slipping. It will be a strong vegetative barriers that slow the flow of water and filter and trap sediment in runoff [1]. In the event of rainfall on sloping surfaces, infiltration of rainwater into the slope will cause the pore water pressure increases, particularly near the surface of the slope. The increase in pore water pressure can make a decrease of shear strength, this facilitates the occurrence of failure. Through this study, infiltration of rainwater that weaken the strength of the slope near the surface will be examined in depth the effects of root growth on the slope in the context of slope stabilization. Interactions between plants and soil texture when planted with plants is carried out from time to time. In addition the strength of the performance of different construction periods has been investigated using the direct shear test machine and pull-out strength test. In addition, the strength test of the roots as incurred a

[a] Corresponding author : gf130019@siswa.uthm.edu.my

tensile force was also studied in different diameters. Therefore, in this study, interaction between plants and soil as a result of roots growth in the context of the slope stabilization is studied. It was aimed to determine the pull-out strength performance of soil and Pennisetum Setaceum grass roots at different growth period of up to 3 months. Besides, the relationship between pull-out strength and tensile roots strength for vegetated soil at different growth period was examined. Knowing the influence of root tensile strengths towards the mechanical reinforcing effect, a tensile test has been conducted on the roots of Pennisetum setaceum grass.

2 Materials and Methods

2.1 Plot preparation

Four plots with dimension of 150 mm x 200 mm x 700 mm had been considered (Figure 1). All plots were planted with Pennisetum setaceum as shown in Figure 2(a). The grass was obtained from nearest nursery (Figure 2 (b)).

2.2 Plot maintenance

2.2.1 Plant watering activities

The grass uniformly was watered to ensure that the planted grass gets enough water to grow. However, the frequency of the plants watering normally affected by several factors such as the weather conditions, the maturity of the plant and plant species and soil types.

Figure 1. A wooden plot.

Figure 2. Guinea grass; (a) planted in plots and (b) nursery grown.

2.2.2 Pull-out strength test procedure

The selected plants that reached the required growth period were identified and marked using a tagging stick. Nine samples were selected for each of the same species (3 grasses for every month). Before setting up the pull-out equipment (Figure 3 (a)), the surface of the soil in the plot is cleared. Prior to the pull-out strength test, several factors that will affect the results must be carefully taken into consideration. For example, the equipment to be used must be in good condition, well-calibrated especially the portable scale. Pennisetum Steaceum grass clumps should be tied firmly using an adhesive tape and placed closer to portable scale that clung onto that grass clump (Figure 3 (c)).

2.3 Root screening process

Once pull-out test is completed, the rooted sample is washed to remove soil from the clod leaving only the root specimens (Figure 4). This screening process enabled the root growth to be measured in term of mass at different planting period up to 3 months.

Figure 3. Pull-out strength test; (a) the pull-out machine, (b) test set-up, (c) portable scale being attached to grass clump and (d) Grass being completely pull-out from the soil.

Figure 4. Root screening process, leaving only roots out of soil clod.

2.4 Tensile test procedure

Sample preparation is done by grouping the grass roots according to diameter size. Every root specimen is cut at the same length of 10 mm (Figure 5 (b)). This is done to ensure that the roots are in the uniform span to avoid error. To measure the root diameter, TMI Precision Micrometer (Figure 5(c)) had been used because it's capable to read up to 4 decimal places of diameter accuracy compared with common vernier calipers. Tensile strength of roots was determined using testing machine Testometric Micro 350 (Figure 5 (d)).

3 Results and Discussion

3.1 Pull-out test

Figure 6(a) shows the maximum pull-out strength for of *Pennisetum Setaceum* grass samples in the first month, which is 6.500 kPa, and 6.911 kPa. Average maximum pull-out strength of these two data is represented by a curve in Figure 6(b). There are only two data recorded for that month due to another sample indeliberately destroyed. For the second month planting period, the maximum pull-out strength recorded are 11.470 kPa, 11.263 kPa and 11.592 kPa (Figure 6(c)). The average maximum pull-out strength of month 2 is shown in Figure 6(d). Lastly, on the third month, results recorded on maximum pull-out strength for three sample of grass are 20.590 kPa, 28.680 kPa and 28.080 kPa (Figure 6(e)). Theirs average is shown in Figure 6(f).

Figure 7 shows the increment trend of the average pull-out strength from first month to third month of Pennisetum setaceum. The data shows that there has been an increase of pull-out strength from the first to the third month which is 6.705 kpa, 11.442 kpa and 25.783 kpa for month 1, 2 and 3 respectively.

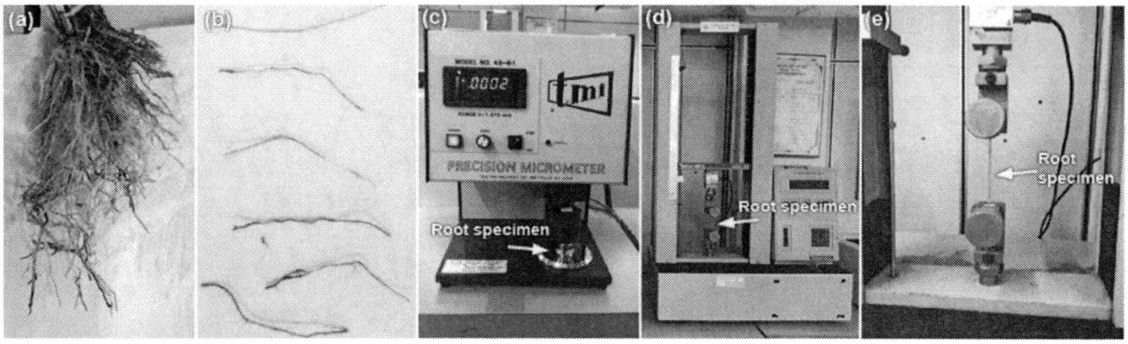

Figure 5. Root tensile test; (a) roots after being washed, (b) roots were cut into same length, (c) root diameter is measured using TMI Precision Micrometer, (d) Testometric Micro 350 and (e) root specimen is being clamped by a pair of wedge grip.

3.2 Pull-out strength comparison over time

The comparisons performed on the results of pull-out tests for different planting period showed a linear increment for the average pull-out strength encompassed different planting period (Figure 8).

3.3 Effect of root weight on pull-out strength

As expected at the beginning of this study, soil pull-out strength increased with the growth of roots [2]. However, based on Figure 9, this study showed mass of roots only increased drastically on month 2 compared to month 1. It is noticed very small increase in mass of root for month 3 compared to month 2. The same pattern of increment also observed in the graph of root mass against average pullout strength (Figure 10).

3.4 Tensile strength

Result of the maximum tensile resistance of roots in diameter ranged from 0.28 to 1.36 mm shows that Pennisetum grass exhibited the maximum tensile force of 22 N (Figure 11). Hence, based on the observation of graph maximum tensile resistance versus diameter, an increase in diameter would increase the tensile resistance. Similar trend was also observed by [2, 3]. Finally, the maximum tensile strength (force/root area) of Pennisetum grass was 279 MPa (Figure 12). From the regression analysis between tensile strength and root diameter, it can be observed that the tensile strength decreased with increasing root diameter which is in line with the previous root studies conducted by [2, 4]. According to [5], during soil shearing, the finer and thinner roots that have high tensile strength would slip while the thicker roots with lower tensile strength would break.

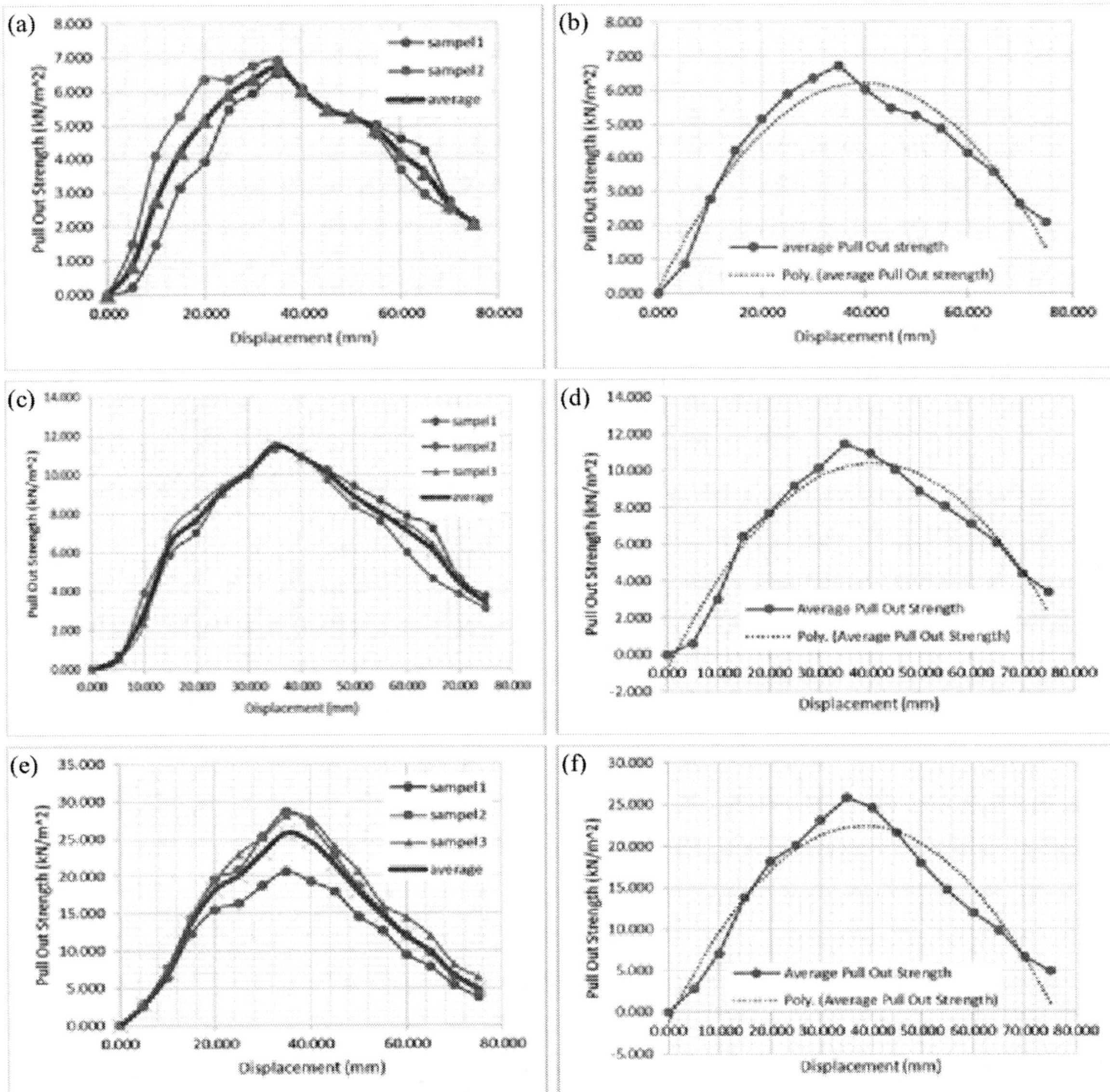

Figure 6. Maximum pull-out strength against vertical displacement curve for three replicates Pennisetum setaceum grass of planting period; (a) 1st month, (c) 2nd month and (e) 3rd month.

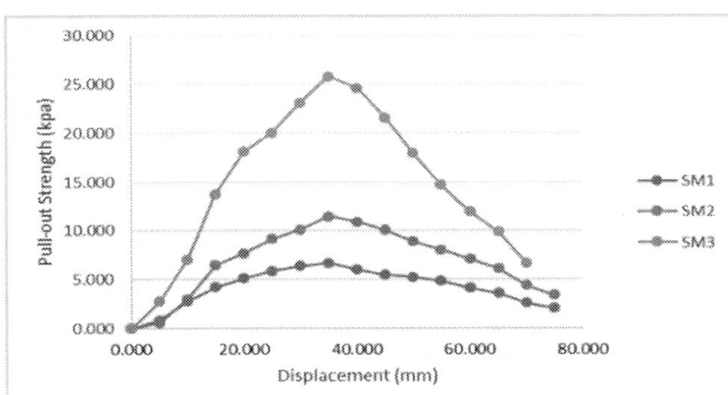

Figure 7. Average pull-out strength versus displacement for month 1, 2 and 3.

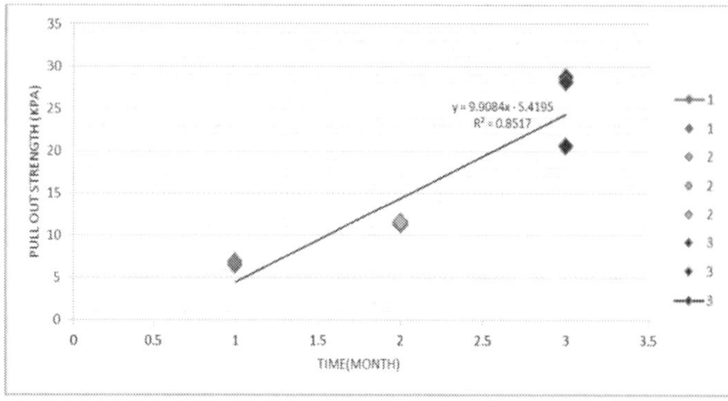

Figure 8. Max. pull-out strength versus growth period.

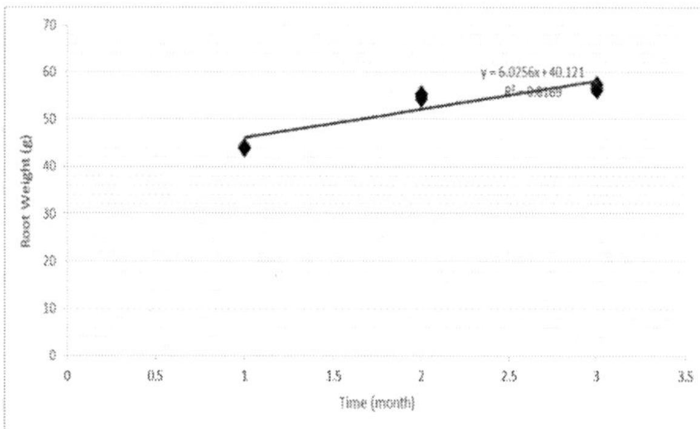

Figure 9. Root weight versus time.

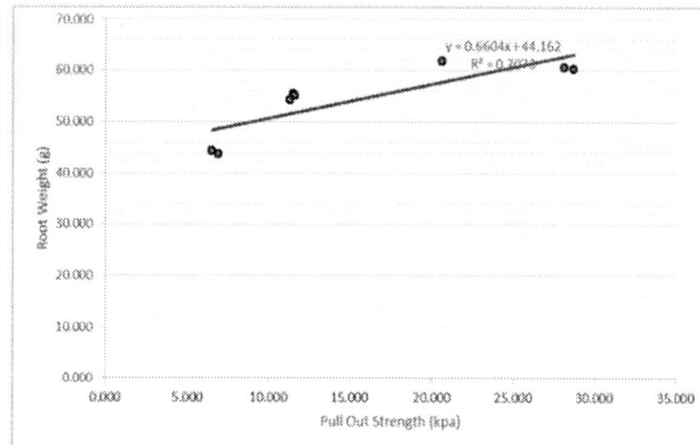

Figure 10. Root weight versus average pull-out strength.

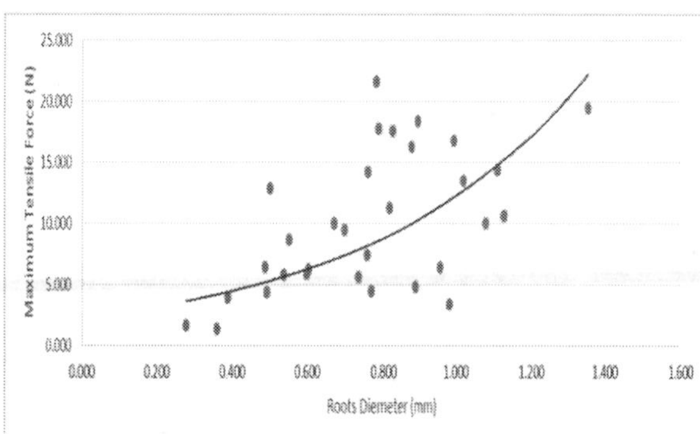

Figure 11. Max. tensile force versus root diameter.

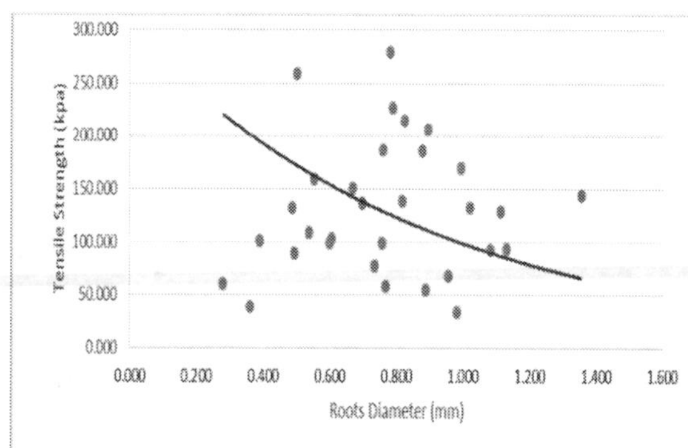

Figure 12. Tensile stress versus root diameter.

4 Conclusion

It can be concluded that the pull-out strength of Pennisetum increasing over the time. This result was clearly demonstrated by series of pull-out tests that has been conducted in this study. Moreover, soil pull-out strength showed an increment due to plant growth in the root structure from time to time. It was due to the roots grip into the ground, giving some mechanical reinforcement effect resulted in strengthened of soil. As for the tensile strength, the value decreases with increasing root diameter.

Acknowledgement

This research publication was funded by Multi-Disciplinary Research Grant (MDR), Vote no. U088. The researchers would like to thank staff on duty at Packaging Laboratory, Faculty of Mechanical and Manufacturing Engineering, Universiti Tun Hussein Onn Malaysia for the full permission given on the use of tensile strength equipment.

References

[1] M.S. Zainordin, N.A. Yusoff, T.N.T. Chik, M.A.H. Manap, Z. Sayuti, M.F. Abdullah, The study of soil-roots strength performance of soil slope by using guinea grass. J. Applied Mechanics and Materials, **802**, 10-15, (2015).

[2] N.Osman, M.N. Abdullah, C.H. Abdullah, Pull-out and tensile strength properties of two selected tropical trees. Sains Malaysiana, **40**(6), 577-585, (2011).

[3] M. Tosi, Root tensile strength relationships and their slope stability implications of three shrubs species in the Northern Apennines (Italy). Geomorphology, **87**(4), 268-283, (2007).

[4] N. Pollen-Bankhead, A. Simon, and R.E. Thomas, The reinforcement of soil by roots: recent advances and directions for future research. Treatise on Geomorphology, **12**(8), 107-124, (2013).

[5] S.B. Mickovski, L.P.H. van Beek, Root morphology and effects on soil reinforcement and slope stability of young vetiver (Vetiveria zizanioides) plants grown in semi-arid climate, Plant Soil, **324** (1-2), 43-56, (2009)

MATEC Web of Conferences **47**, 03022 (2016)
DOI: 10.1051/matecconf/20164703022

The Effect of Soil Particle Arrangement on Shear Strength Behavior of Silty Sand

Nik Daud Nik Norsyahariati[1,a], Kam Ru Hui[1] and Abd Ghafar Azmi Juliana[1]

[1]*Department of Civil Engineering, Faculty of Engineering, Universiti Putra Malaysia, 43400 Serdang, Selangor, Malaysia*

Abstract. The presence of fines in sandy soil is recognized as a problem in geotechnical engineering. It is often assumed that the strength and liquefaction potential of a soil decreases with increasing fines content. Sand with a significant amount of fines is always encountered in geotechnical engineering projects. The main interest of this paper is to study the effect of particles arrangement and stress behaviour on sand and silty sand sample. The objectives are to determine the basic properties as well as the morphological and mineralogical properties and the relationship of those properties to the stress behaviour of the soils. Sand-silt mixtures with fines contents of 20% and 40% were prepared and a series of testing was carried out to determine their basic properties, morphological and thin section properties, and mechanical properties by using direct shear box equipment. Results show that the basic properties and morphological properties of soil affect the mechanical behaviour of soil. Shape of the soil particle can influence the packing of the soil, hence altering the mechanical behaviour of the soil. From the thin section, the soil is observed to be well graded and have a dense packing while the minerals observed were mainly quartz and rutile.

1 Introduction

Most of the natural and man-made soil deposits contain a certain amount of silt and/or clay. Due to increasing of shortage of clean sands near project locations, the use of mixed sand, which includes silty sand, clayey sand, sandy clay and sandy silt, will be practical option in reclamation works. Fines may also be produced during dredging process itself, causing the clean sands to contain a certain amount of fines.

The presence of fines in sandy soil is recognized as a problem in geotechnical engineering. It is often assumed that the strength of a soil decreases with increasing fines content. However, regarding to the liquefaction potential it is not a drawback, indeed it is advantage in geotechnical application. Thus, the maximum fines content is usually limited to 10% only in most geotechnical projects. It has been recognized that the availability of soils which have small amount of fines is very rare. In most of projects soil with maximum amount of fines above 10% are encountered.

The shape of soil grains affects the engineering behaviour of the soils. Round grains are more likely to slip and roll than angular fragments. Natural sand and crushed sand has difference shape. Natural sand has the roundness and roughness. The shape of natural sand can be altered during

[a] Corresponding author : niknor@upm.edu.my

transportation or through weathering. On the other hand, crushed quartz sand tends to be equidimensional or elongated in shape. This is because quartz has no cleavage. Particle rotation is a basic component of deformation during compressibility of granular materials [1], which is affected by the shape of the grains. As the angularity increases, the ratio of rolling to sliding contacts decreases, which leads to greater shearing resistance [2]. Dense packing enhances the effect of angularity. The total number of contacts between particles increases and cause the rotation resistance increases. In shear behaviour, the friction angle can be affected by the gradation of sand such as sands with the lowest friction angle tend to be medium-fine, well-rounded, and poorly graded sands and Sands with the highest friction angle tend to be coarser grained, well-graded, and/or angular sands [3]. The addition of non-plastic silt to the host sand increases both the peak and critical-state friction angles of the soil [4]. It appears that the more angular non-plastic silt particles tend to interact with irregularities on the surface of sand particles, increasing the "jamming" effect during shearing.

The aim of this study is to study the stress behaviour on silty sand with the relation toward the particle arrangement. Specific objectives are (i) To determine the basic properties of sand with varying silt content, (ii) To determine the morphological and thin section properties of the soils and (iii) To analyze the effect of silt content on the strength behavior of the soil.

2 Materials and Methods

2.1 Materials

Silty sand is collected from Tasik Angsana, Sungai Pelik, Sepang, Selangor (GPS coordinate: 2°37'21.9"N 101°42'00.1"E) as indicated with a red pointer in Figure 1. There was no vegetation on the surface of the soil, so the top soil layer was scraped away and the soil sample was collected from the sub soil layer using a shovel.

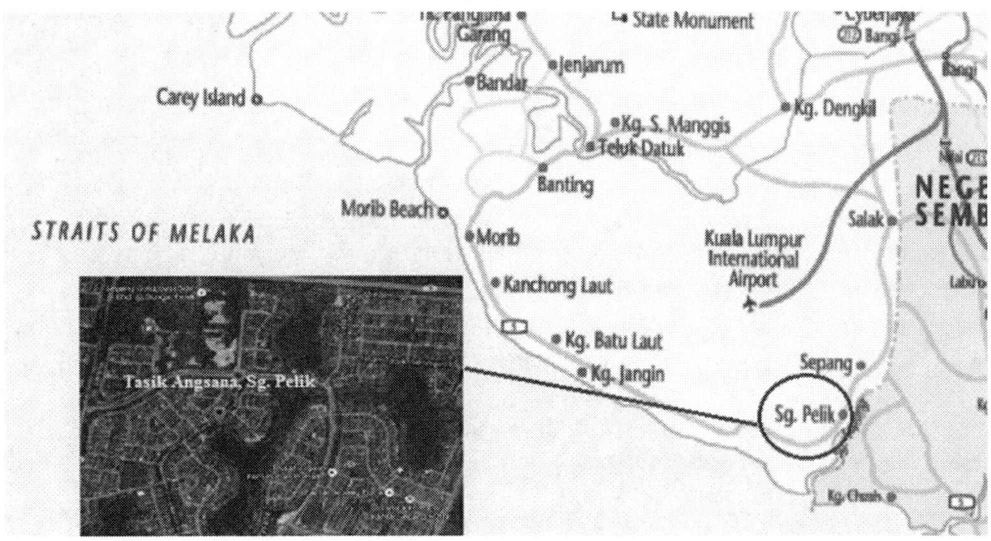

Figure 1. Soil sampling location (Source: Google Maps).

2.2 Methods

The soil is oven-dried at an average 105°C and then sieved in order to obtain sand (particle retained 0.063mm sieve) and silt (particle passed 0.063mm sieve) portion. The fines and the coarse fraction were reconstituted in varying proportions of fines to coarse ratios to study the effects of silt content on the stress behaviour of sands, from 20% silt; 80% sand and 40% silt; 60% sand. Since the natural soil has 20% silt and 80% sand, the soil is not altered or remixed. Sample of 40% silt 60% sand was produced by mixing the clean sand and pure silt by weight.

A complete basic properties testing program was performed on all the sand-silt mixtures based on standard [5]. Basic properties including grain size distribution, specific gravity, linear shrinkage, liquid limit and plastic limit were determined for the each sand-silt mixture. Morphological and mineralogical properties testing program was performed on the natural soil sample, which is 20% silt and 80% sand. Morphological properites of the samples were determined by using Scanning Electron Microscope – SEM (Hitachi S-3400N Variable Pressure) and mineralogical properties were determined by using and thin section procedure. The thin sections were produced according to well established methods [6, 7]; although modified as necessary to accommodate the delicate nature of the oil sands with a view to preserving its integrity and minimizing alterations of the oil hosted in the quartz rich samples during the course of thin section production [8]. Thin sections were done by mixing the oven dried soil sample at an average 105°C with epoxy in a plastic mould and left 24 hours at room temperature to harden as shown in Figure 2(a). The hardened block is then cut into a smaller block then glued to a glass slide with epoxy and left to cure for 24 hours at room temperature as shown in. The block of sample on the glass slide is then filed down to 1 mm thick as shown in Figure 2(b). The thin section is then viewed under a polarized light microscope at different magnifications to observe the soil structure, soil matrix and minerals present in the soil.

Figure 2. (a) Hardened sample of the natural soil sample (20% silt, 80% sand) mixed with epoxy, (b) Thin section slide of the soil sample.

A series of direct shear box test (TKA-DDS-5C Microprocessor-controlled Direct Shear Apparatus) were carried out on each of the sand-silt mixture to determine the stress behavior of the soils. The procedures of the test were referred to BS 1377.

3 Results and Discussion

The findings on the basic properties will be discussed in this section, followed by the morphological and mineralogical properties, and finally detailed analysis and discussion on the stress behaviour of the soils.

3.1 Basic properties

Several basic properties testing including grain size distribution, specific gravity, linear shrinkage, liquid limit, plastic limit and compaction test were carried out. The initial moisture content of natural soil sample collected (20% silt, 80% sand) was 21.16%. Table 1 shows as the silt content in sand increases, the liquid limit, the maximum dry density and optimum moisture content increases. It is shows that fines percentages gave a significant effect to the related properties. The detailed results of all basic properties are tabulated in Table 1.

Table 1. Index properties of sand- silt mixture samples.

Silt content in sand (%)	20	40
USCS classification	SM	SM
Coefficient of uniformity, C_u	2.000	3.385
Coefficient of curvature, C_c	28.000	43.706
Specific gravity, G_s	2.70	2.71
Liquid limit (%)	24.40	25.32
Plastic limit (%)	23.96	24.78
Plasticity index (%)	0.44	0.54
Maximum dry density (Mg/m^3)	1.62	1.69
Optimum moisture content (%)	10.5	10.8

*USCS - Unified Soil Classification System; SM – silty sand

The specific surface area value for the natural soil sample (20% silt, 80% sand) is 1.986 m^2/g. The relatively low value indicates that the soil mostly consist of sand and has a relatively high permeability and is slightly plastic.

3.2 Morphological and mineralogical properties

From the Scanning Electron Microscope (SEM) image in Figure 3 and thin section image (Figure 4), silt and sand particles shape are observed and compared to the designated chart [9] in order to determine the roundness and sphericity of a soil particle.

Figure 3. SEM image of the natural soil sample oven-dried at an average 105°C at 300× magnification.

It can be observed that the particle is basically sub-angular and has a medium sphericity. The surface of the soil particles generally has a semi-rough texture. Since the soil particles are sub-angular, the ratio of rolling to sliding contacts is decreased leading to greater shearing resistance. The effect of angularity is enhanced for dense packing. The total number of contacts between particles increases and as a result, the rotation resistance increases, leading to the rotational frustration.

(a) (b)

Figure 4. Random sampling on the thin section at 400× magnification of the natural soil sample.

The soil structure is generally comprised of sand and silt indicated with yellow arrows in the figure. The shapes of the sand grains are mostly sub-angular and medium sphericity, similar to the observations from the SEM photograph. The variation of sand and silt particles cause the arrangement of soil to have a fairly dense packing and this contributes to the stress behaviour of the soil [10]. The sample is also randomly sampled at different spot on the slide to determine the minerals present in the soil (Figure 4).

By referring to the mineral identification table [11], minerals observed in the thin section are quartz, rutile, spinel, garnet, augite and biotite. Using an X-ray diffractometer equipment, the dominant mineral identified in the sample is quartz. According to the minerals present in the sample, most of the minerals also appears in granite rock such as quartz and biotite [12] and it can be summarized that samples used for this study are originate from an igneous parent rock.

3.3 Shear strength properties

The results from the direct shear tests were analysed and the shear strength parameters were determined by using Mohr's circles (Figure 5) for each sand-silt mixture at different stresses (100 and 300 kPa). The cohesion, friction angle and shear strength and normal stress at failure point of each sample are determined and summarised in Table 2.

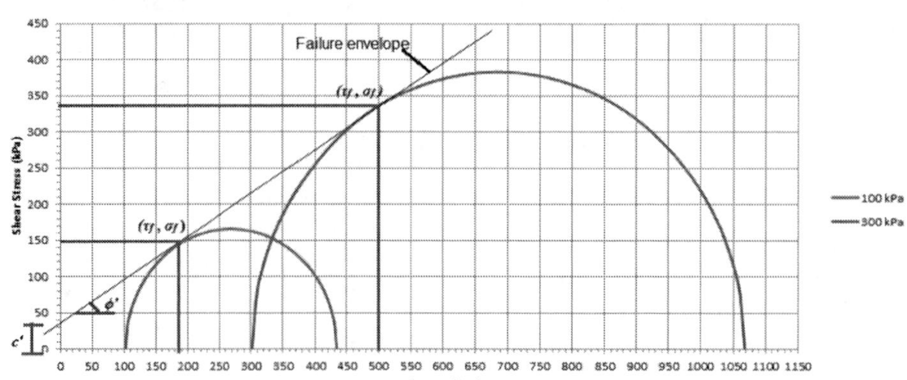

Figure 5. Stress conditions at failure the natural soil sample (20% silt, 80 % sand).

Table 2. Shear strength parameters for silty sand samples.

Silt Content	Cohesion, c' (kPa)	Friction angle, ϕ' (°)	Shear strength, τ_f (kPa)		Normal stress, σ_f (kPa)	
			100	300	100	300
20%	38.407	31.479°	150	350	180	500
40%	60.235	27.686°	180	380	210	530

It is observed that the shear stress is higher in the sample with 40% silt content. As the silt content in sand increases, the cohesion of the soil increase, while the friction angle decreases. This is due to the gradation of the soil samples and the shape of soil particles [13]. Since the shape of the soil particles are the same in all the samples, the gradation will be compared. In sample of 20% silt content, there were sufficient fines to support the sand particles to prevent sliding and rolling. However, at 40% silt content, the silt particles are more dominant in contributing to the stress behaviour. Therefore, the friction angle is lower. Cohesion of the 20% silt content sample is much lower compared to the 40% silt content sample. Obviously, this is because there are more fine particles with higher surface area.

4 Conclusions

The purpose of this study is to find out the stress behaviour of silty sand. A series of experiment has been carried out and analysed. The following conclusions can be obtained from the results of the experimental work.

a) The basic properties of soil affect the mechanical behaviour of soil. The gradation of the soils affects the packing of the soils, causing them to exhibit different strengths.
b) The morphological and mineralogical properties of soil particles affect the mechanical behaviour of soil. Shape of the soil particle can influence the packing of the soil, hence altering the mechanical behaviour of the soil. The minerals available in the soil may also contribute to the strength of the soil.
c) Direct shear box test results shows that the friction angle of soil increases then decreases as the fine content increases. Cohesion of soil also increases then decreases as fine contents increases. This is mainly due to the packing of the soil particles.

Acknowledgement

This study has been funded by Fundamental Research Grant Scheme-FRGS (No.: 03-01-13-1172FR) granted by Ministry of Education (MOE), Malaysia. The help and assistance various governmental and non-governmental organizations in this study are gratefully acknowledged.

References

[1] G. Mesri and B. Vardhanabhuti, Compression of granular materials, Canadian Geotechnical, 369-392, (2009).

[2] K. Mair, K.M. Frye and C. Maronez, Influence of grain characteristics on the friction of granular shear zones, Journal of Geophysical Research, **107**(10), 41 – 49, (2002).

[3] C.A. Bareither, T.B. Edil, C.H. Benson and D.M. Mickelson, Geological and physical factors affecting the friction angle of compacted sands, Journal of Geotechnical and Geoenvironmental Engineering, **134**(10), 1476-1489, (2008).

[4] J.A. Carraro, M. Prezzi and R. Salgado, Shear strength and stiffness of sands containing plastic or nonplastic fines, Journal of Geotechnical and Geoenvironmental Engineering, **135**(9), 1167-1178, (2009).

[5] BS 1377, *Methods of Test for Soils for Civil Engineering Purposes*, British Standard Institution, London, (1990).

[6] A. Jongerius and G. Heintzberger, Methods in soil micromorphology, a technique for the preparation of large thin sections, Soil Survey Institute, Netherlands, (1975).

[7] E.A. Fitzpatrick, *Micromorphology of Soils*, Chapman and Hall, London, (1984).

[8] J.D. Bell, O. Eruteya and O. Oono, Application of petrographic image analysis and multivariate statistical techniques for textural studies of oil sand samples, *Annual Convention and Exhibition*, Long Beach, 2012).

[9] N.A. Miller and J.J. Henderson, Correlating particle shape parameters to bulk properties and load stress at two water contents, Agronomy Journal, **103**(5), 1514-1523, (2011).

[10] C.A. Hecht, Geomechanical models for clastic grain packing, Pure Appl. Geophysic, **161**, 331–349, (2004).

[11] J. Faithfull, Identification tables for common minerals in thin section, (1998). <retrieved from http://www.hmag.gla.ac.uk/john/teaching/mintable.htm.> (accessed on June, 2015).

[12] G. Vahid, O. Husaini, M.Y. Zainuddin, K.H. Bujang, M. Ratnasamy and N.A. Mohd, A new model of microcracks propagation in granite rock, Australian Journal of Basic and Applied Sciences, (2010).

[13] A.B. Christopher, B.E. Tuncer, H.B. Craig and M.M. David, Geological and physical factors affecting the friction angle of compacted sands, Journal of Geotechnical and Geoenvironmental Engineering, **134**(10), 1476 – 1489, (2008).

MATEC Web of Conferences **47**, 03023 (2016)
DOI: 10.1051/matecconf/20164703023

Investigation of Light Non-Aqueous Phase Liquid Migration Single and Double-Porosity Soil Using Light Transmission Visualization Method (LTV)

Motasem Y. D. Alaziaza[1,a], Su Kong Ngien[1], Mustafa M. Bob[2], Samira A. Kamaruddin[3] and Wan Mohd Faizal Ishak[4]

[1]*Faculty of Civil Engineering and Earth Resources, Universiti Malaysia Pahang, 26300 Gambang, Kuantan, Malaysia*
[2]*Department of Civil Engineering, College of Engineering, Taibah University, Madinah, Saudi Arabia*
[3]*Razak School of Engineering and Advanced Technology, Universiti Teknologi Malaysia, 54000 Kuala Lumpur, Malaysia*
[4]*Centre of Earth Resources Research and Management, Universiti Malaysia Pahang, 26300, Gambang, Kuantan, Malaysia*

Abstract. Two experiments were conducted to investigate the migration of light non-aqueous phase liquid (LNAPL) in two different porous media using LTV technique. The first media was fine sand as a single-porosity and the second media was S300 kaolin as a double-porosity. The two media were packed separately in the flow chamber and then LNAPL was injected from the top of the model and the migration of LNAPL (Toluene) was observed using a digital camera connected to a laptop and controlled using special software. The images were captured according to a predetermined time interval set by the software. The results show a significant difference in the migration of LNAPL in the two porous media, where the migration in the double-porosity media was much faster than the migration in the single-porosity. This finding is due to the occurrence of inter-aggregate pores in the double-porosity soil structure. This paper shows that more attention should be given towards the pollution of the groundwater sources that are located within or near the double-porosity media especially the agricultural soil.

1 Introduction

Contamination from leakage of underground storage tanks and huge pipelines is the main pollution hazard to the groundwater compared to other sources. These contaminants which are known as non-aqueous phase liquid (NAPL)have a low solubility in water and occur in the subsurface as a separate phase and are considered as one of the most spread hazardous chemical [1]. Based on the density of liquids, NAPLs are classified into two types: the first type is light non-aqueous phase liquid (LNAPL) which has density less than water and the second type which is termed dense non-aqueous phase liquid (DNAPL) is denser than water. LNAPL include many hydrocarbon fuel components, for instance, toluene, benzene, xylenes (BTEX) and ethyl benzene. Tetrachloroethylene (PCE) and trichloroethylene (TCE) are examples of DNAPL materials [1]. The migration of LNAPL in

[a] Corresponding author : my.azaiza@gmail.com

subsurface system depends on its relative density. When LNAPL enter the subsurface system, it will pass through unsaturated soil and migrate downward due to gravity and float on the surface of water table[2] resulting in deterioration of groundwater quality. On the other hand, DNAPL will pass through unsaturated zone and continue its downward migration under the effect of gravity until it reaches the saturated zone [3]. Experimental technologies such as non-invasive imaging techniques have gained more attention during the last decades which make the characterization and understanding of the multiphase system more accurate [4]. A number of image analysis techniques have been used previously to investigate and measure multiphase fluid contents in laboratory experiments [5]. Accordingly, LTV technique based on image analysis is one of the most important and promising techniques.The theory of this method depends on the passing of electromagnetic energy through the tested media where the distribution of liquid saturation is measured as variation in light intensity field. There is a linear relationship between saturation and light intensity due to closer matching between refraction index of the matrix and water relative to the matrix and air [6]. TheLTV experiments are usually conducted using a thin flow chamber connected with a fan for cooling the chamber and a built in lighting source behind the flow chamber. LTV is considered a cheap technique and requires only limited equipment [7]. This paper investigates the migration of LNAPL in two different types of porous media, the first media is single-porosity soil and the second media is double-porosity soil using LTV technique. The materials that used arenatural fine sand as a single-porosity and S300 kaolin as a double-porosity. The migration of LNAPL was monitored using digital camera and capturing of the photos was carried out according to specific time intervals.

2 Materials and Methods

2.1 Experimental setup

A 2-D flow chamber made of acrylic was used in our experiments. The dimensions of the flow chamber were 45 cm × 30 cm × 1 cm. The light box contains light source consisted of three 18-W fluorescent tubes (OSRAM, L18W/780 SMARTLUX, Sky White, Malaysia). The flow chamber was packed with the porous media and fixed inside steel frame in front of the light source. A digital cameraNikon D7100(Nikon SDN, BHD, Malaysia) was fixed on the same position during the experiments at a distance of 1 m from the light box. The camera captured images that has 24 mega pixels (6000 × 4000 pixels), and a 14 bit dynamic range that results in 16,384 grey levels. The camera was connected to a laptop and the capturing of the images was controlled by Nikon camera control pro 2 software. The setup of the experiment is shown in Figure 1.

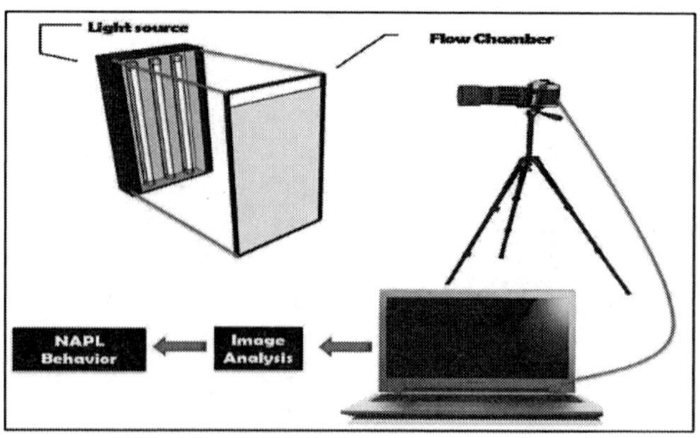

Figure 1. Experimental setup.

2.2 Preparion of porous media and LNAPL

The first media was natural fine sand which retained on sieve 212 μm. The sand was washed with water and dried in the oven for 48 hours at 45°C. After drying, the sand was packed by pouring layer by layer into the flow chamber through a funnel, and then tapped the outer frame of the model using a plastic hammer to achieve a dense pack. The other media was S300 kaolin which presents the double-porosity soil structure. To create the double-porosity structure, the procedure in previous study[8] was followed. The powdered kaolin was weighted and mixed with determined water quantity with hand to get a water content of 30%. This percentage was chosed because kaolin starts to disintegrate at 35% water content [9]. After mixing the kaolin with water, it was placed in a sealed plastic bag for 24 h. Then, the mix was broken by hand and passed through a 2.36 mm sieve to obtain the granules structure of the kaolin. The same procedure was repeated to pack the model with kaolin, but the difference here was the compaction process was using a rectangular piece made of acrylic with the same width of the model. Each experiment used 50 ml toluene as a LNAPL. Since the LNAPL is colorless, 0.1 g of Oil Red O was used to dye the LNAPL for better visualization. Kechavarzi et al.[10] has stated that mixing a small amount of dye is enough. This concentration is deemed to be sufficient to facilitate good visual observation of the LNAPL movement through the acrylic wall of the flow chamber. The dye was weighted and then the toluene was added and mixed with the dye using spatula. The LNAPL was injected using injection syringe needle from the top of the model. The needle was pushed to penetrate the soil to about 2 cm below the top of the model before LNAPL was released to allow the movement of it downward directly and avoid the movement of the LNAPL on the top of the soil.

2.3 Image acquisition

After the injection of the LNAPL, the digital camera starts to capture images automatically based on predetermined time intervals. Table 1 lists the time intervals for image acquisition for the two porous media.

Table1. Frequency of image acquisition.

Period (min)	Frequency Interval	
	Single-porosity (sec)	Double-porosity (sec)
0 – 6	-	5
6 - 9	-	10
9 – 12	-	30
After12	-	60
0 – 50	15	-
50 – 110	30	-
110 – 212	60	-
212 – 270	180	-
After 270	Randomly according to the change	-

3 Results and Discussion

3.1 Single-porosity experiment

According to Figure 2, the migration of the LNAPL was very slow and the change in the behavior and movement of the LNAPL is very small. It was also observed that the migration of the LNAPL in the sand was fast during the first hour, and then the migration decreased slowly. This behavior is the same in the case of horizontal direction. It can be concluded that rabid toluene migration occurred during the first 1 hr, and later the migration rate becomes slower.

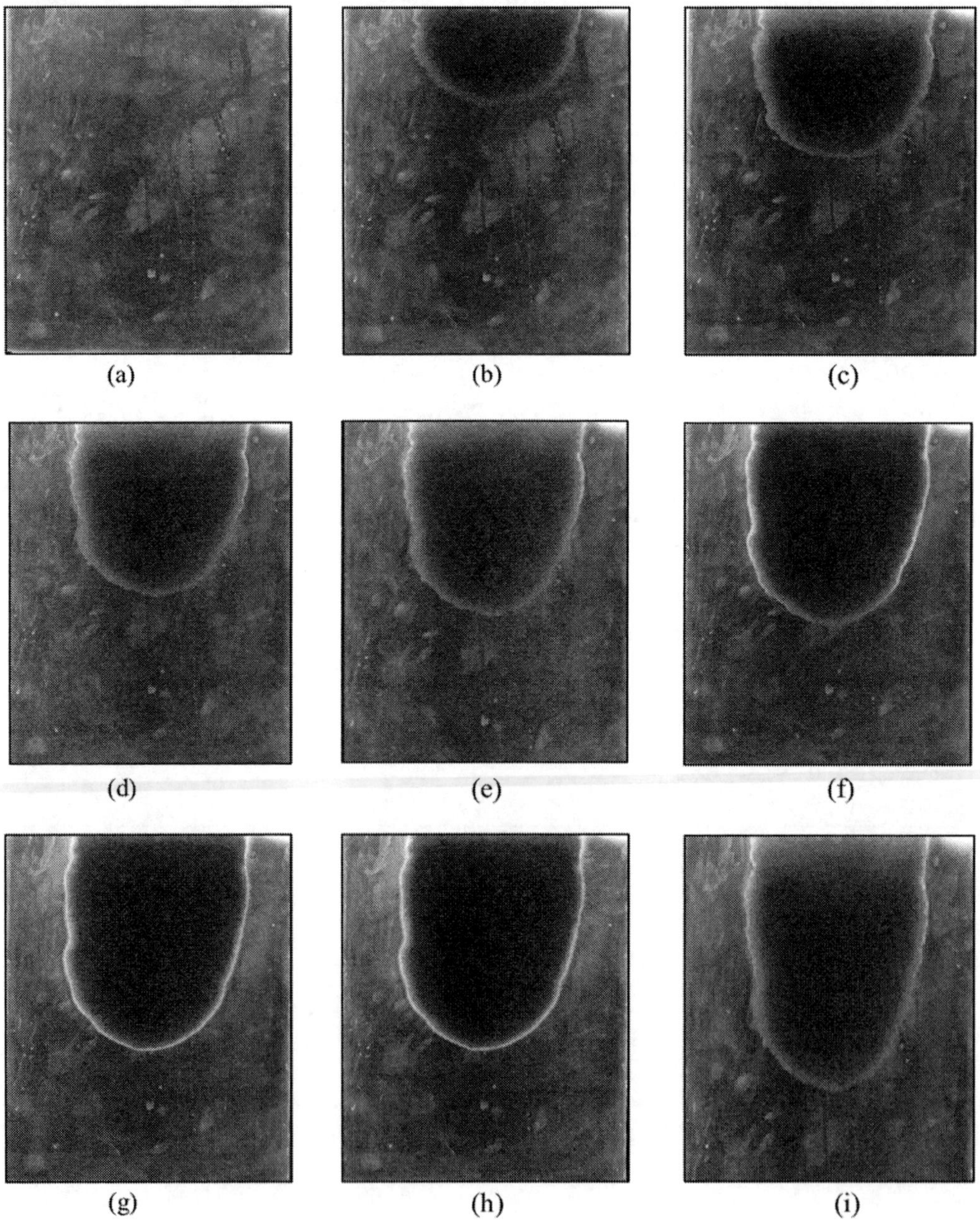

Figure 2. Toluene migration in single porosity soil after (a) 0 sec (b) 1min (c) 3min (d) 10 min (e) 30 min (f) 1 hr (g) 5 hr (h) 1 day (i) 4 days.

3.2 Double-porosity experiment

Based on the captured images shown in Figure 3, a significant difference in the velocity of migration of the LNAPL occurs in double-porosity kaolin media.LNAPL reached the bottom of the model in 27 minutes and stopped the migration in 44 minutes. This is due to the occurrence of inter-aggregatepores (secondary porosity) between the kaolin granules.This result was observed previously as mentioned by Ngien et al. [11] when they compared the migration of LNAPL between single and double-porosity soil using light reflection method. Furthermore, it was observed that the migration of LNAPL in double-porosity soil is fast in the horizontal direction.

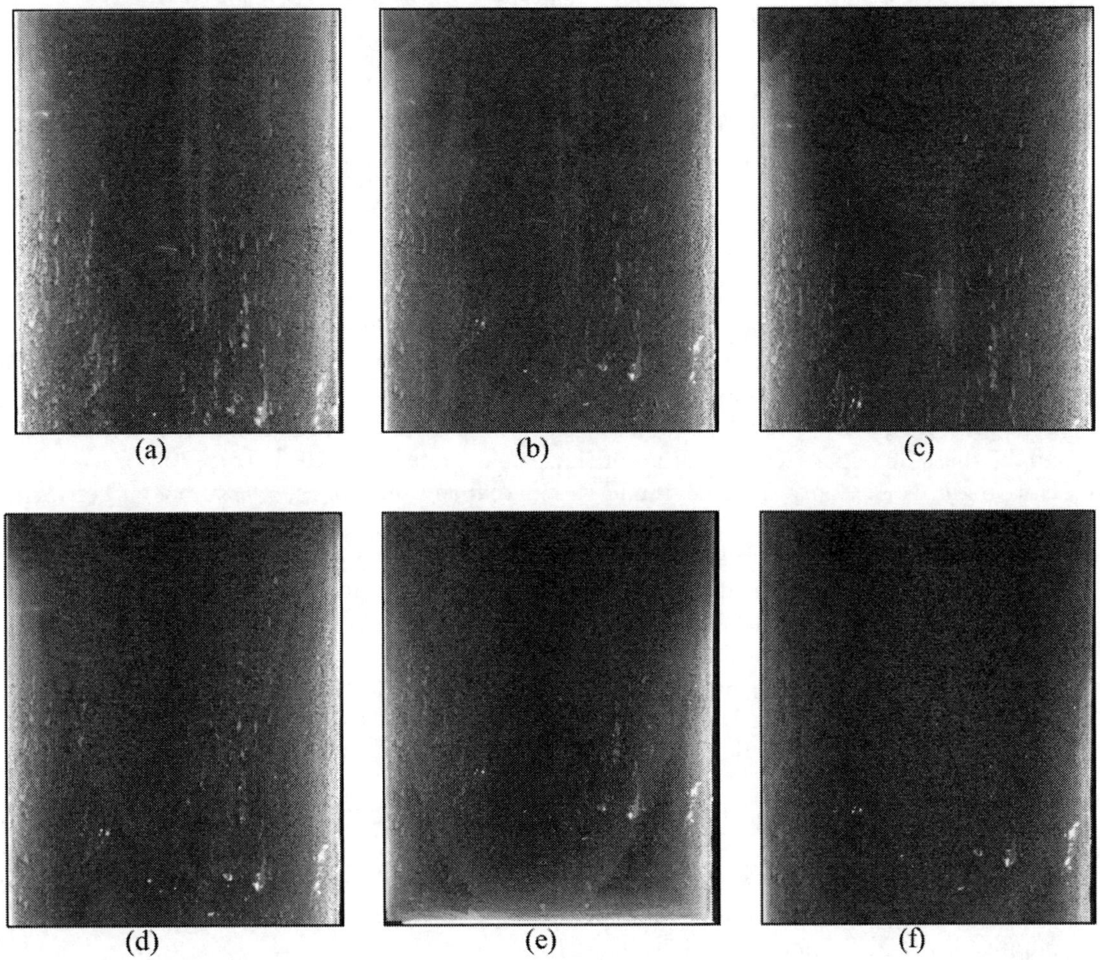

Figure 3. Toluene migration in double-porosity soil after (a) 0 sec (b) 5 sec (c) 30 sec (d) 1 min (e) 27 min (f) 44 min.

Based on the observations, it was noticed that at the first moments of LNAPL injection, the migration of the LNAPL was very fast during the first minutes. This is obvious in Fig. 3 (c) and (d), where LNAPL had migrated about half the model depth during the first 30 seconds and about three quarter in 1 minute. However, after 1 minute, the migration was decreasing slowly and this is clear in Fig. 3 (e), where LNAPL take about 26 minute to migrate from the three quarter the model depth to reach the bottom of the model. After LNAPL reaches the bottom of the model, its migration continue but in the horizontal direction as shown in Fig. 3 (f). This may be due to the higher compaction in the bottom of the model than of the top of the model. The more compaction in the bottom of the model resulted in smaller inter-aggregate pores which led to slower migration than the top of the model.

4 Conclusions

Laboratory experiments using LTV technique were conducted to investigate the migration of the LNAPL in single-porosity and double-porosity soil structures. The results show a significant difference in the migration movement of the LNAPL through the two media. The migration of the LNAPL in double-porosity soil was much faster than the migration in single-porosity. This is most likely due to the occurrence of the inter-aggregate pores in the double-porosity soil structure. This finding depicts the risk of the pollution in the groundwater sources that are located in the region of double-porosity soil especially agricultural soil through which the contaminants will diffuse very fast compared with other types of soil. Further studies are needed in this area of research using LTV especially the migration of LNAPL under different conditions for double-porosity soil such as fully and partially saturated.

References

[1] C.J. Newell, S.D. Acree, R. Ross and S. Huling, Ground water issue: Light nonaqueous phase liquids, U.S. Environmental ProtectionAgency. Groundwater Services, Houston, (1995).

[2] B.L. Morris, A.R. Lawrence, P. Chilton, B. Adams, R.C. Calow and B.A. Klinck, Groundwater and its susceptibility to degradation: A global assessment of the problem and options for management, J. Early Warning and Assessment Report Series, RS. **03**-3. United Nations Environment Programme, Nairobi, Kenya. (2003).

[3] K. Soga, J. Page and T.A. Illangasekare, Review of NAPL source zone remediation efficiency and the mass flux approach, J. Journal of Hazardous Materials, **110**, 13-27, (2004).

[4] B. Agaoglu, N.K. Copty, T. Scheytt and R. Hinkelmann, Interphase mass transfer between fluids in subsurface formations: A review, J. Advances in Water Resources, **79**, 162-194, (2015).

[5] M. Oostrom, J. Dane and T.W.A. Wietsma, Review of multidimensional, multifluid, intermediate-scale experiments: Flow behavior, saturation imaging, and tracer detection and quantification, J. Vadose Zone Journal, **6**, 610-637, (2007).

[6] R.J. Glass, T.S. Steenhuis and J.Y. Parlange, Mechanism for finger persistence in homogeneous, unsaturated, porous media: Theory and verification, J. Soil Science, **148**, 60-70, (1989).

[7] H. Wang, X. Chen and J.W. Jawitz, Locally-calibrated light transmission visualization methods to quantify nonaqueous phase liquid mass in porous media, J. of Contaminant Hydrology, **102** , 29-38, (2008).

[8] A. Bagherieh, N. Khalili, G. Habibagahi and A. Ghahramani, Drying response and effective stress in a double porosity aggregated soil, J. Engineering Geology, **105**, 44-50, (2009).

[9] R. Sa'ari, N. Rahman, Z. Yusof, S. Ngien, S. Kamaruddin and M. Mustaffar, Application of digital image processing technique in monitoring LNAPL migration in double porosity soil column, J. Jurnal Teknologi, **72**, 23-29, (2015).

[10] C. Kechavarzi, K. Soga and P. Wiart, Multispectral image analysis method to determine dynamic fluid saturation distribution in two-dimensional three-fluid phase flow laboratory experiments, J. Journal of Contaminant Hydrology, **46**, 265-293, (2000).

[11] S.K. Ngien, N.A. Rahman, M.M. Bob, K. Ahmad, R. Sa'ari and R. W. Lewis, Observation of light non-aqueous phase liquid migration in aggregated soil using image analysis, J. Transport in Porous Media, **92**, 83-100, (2012).

MATEC Web of Conferences **47**, 03024 (2016)
DOI: 10.1051/matecconf/20164703024

Development of Continuous Speed Profile Using GPS at Johor Federal Roads F0050

Joewono Prasetijo[1,a] and Zaffan Farhana Zainal[1]

[1]*Faculty of Civil and Environmental Engineering, Universiti Tun Hussein Onn Malaysia, 86400 Parit Raja, Johor, Malaysia*

Abstract. Road accidents are one of the most relevant issues in today's society. It causes hundreds of accidents every year from over the world. Every year 1.2 million of people are killed and between 20 and 50 million people are offended due to the road accidents. Three main types of accidents in Malaysia is collision with passenger cars, collisions with other motorcycles and single-motorcycle accidents. F0050 is ranked the district with the highest road fatalities in Johor for five consecutive years. Motorcyclists and their pillion riders were the highest contributors – 60% or 648 fatalities, followed by car drivers and passengers numbering 266 fatalities. One reason of accident occurrence can be lack of road design consistency which most drivers make fewer errors in the vicinity of geometric features. Geometric design consistency is emerging as an important component in highway design relate to the safety performance. The result shows the continuous speed profiles along F0050. Since motorcyclist have a higher fatality per distance traveled, this study will develop the potential relationship between design consistency which is represented by continuous speed profiles by using Global Positioning System (GPS).

1 Introduction

Nowdays, road accidents are one of the most relevant issues in society. It causes hundreds of accidents every year from over the world. Every year 1.2 million of people are killed and between 20 and 50 million people are offended due to the road accidents. If these trends continue, road traffic accidents are will be predicted to be the third important to contributed to the global burden of disease and injury by 2020. Among ASEAN countries [1, 2], Malaysia has the highest road fatality and more than 50% of the road accident fatalities involved motorcyclist that represent more than two thirds of road accident victims. About 300 000 accidents reported and 1% of it is fatal accidents in Malaysia for every year. Motorcyclist is the main vehicle in Malaysia that contributes about 55% of all registered vehicle in Malaysia. The potential to exposure in road accident are highest among the user of motorcycle because it directly exposed and come in direct contact with the impacting vehicles or obstacle during a collision and resulting in severe injuries and fatality [3]. Based on previous study, accidents that occur either car driver or motorcyclist was not injured but the injured parties is either passengers or the occupants of the other vehicle around 97% of motorcyclist were injured or killed in these collisions compared with 50.5% of car drivers. It also stated that at higher speed motorcycles

[a] Corresponding author : joewono@uthm.edu.my

and car collision than car and car collisions motorcyclist involved were 95.4% likely to be injured while car drivers were 0.9% likely to be injured [4]. Mainly, motorcyclist is 17 times more dangerous than passengers car and also the motorcyclist will be in front of higher risk of injury or fatality because the level of exposure to injury is higher if compared to other vehicle such a passenger car in accident [5]. The reason why road accident keep rising in Malaysia because one of the lack of road geometric design consistency where the drivers make a fewer errors due to the road geometric features. If the consistency are poor, the potential for accident happen are highest [6, 7]. Geometric design consistency is emerging as an important component in highway design relate to the safety performance. Geometric design is influenced by the vehicle, drivers and traffic characteristics. Geometric design elements that influence traffic operations include number and width of lanes, the presence and widths of shoulders and highway medians, and the horizontal and vertical alignment of the highway.

The importance of design consistency and its contribution to road safety can be justified based on interaction between driver, vehicle and roadway interaction. Most consistency concepts today deal with the variation of vehicles speed but less variation types (heterogeneous) on road section and this speed variation is affected mainly by horizontal alignment and vertical alignment of roads [2, 6, 7]. Design consistency is measured by estimating the speed variation along successive geometric elements and evaluation of the speed deviation from the average speed [8]. A road that has a good level of consistency can affect the driver to drive in a stable and careful. While for the poor level of consistency, can affect high speed along a different path among different drivers and can cause accidents. Operating speeds profile is used to assess the design consistency by identifying locations with speed variability between geometric design/features and crash rate. Some previous studies estimated consistency of horizontal alignment.

2 Method

In many years, the use of Global Positioning System (GPS) technologies has used to perform traffic data collection for transportation studies. In transportation engineering, GPS devices has been frequently used in study of travel time, route choice, car following and behaviors' of driver speed. By previous studies, most of them measure speed using automated traffic counters or radar speed gun measurement at specific points along the segment by assuming it representative of the speeds along the segment. But this assumption not is able when the road has different geometric elements that can affect vehicle speed. The data collected with the GPS device contains highly accurate measurements of the test vehicle's position and speed at each second in time. GPS data can also be utilized to calculate the values of other traffic measures related to work zone traffic conditions, including average travel speeds, deceleration and acceleration rates, traffic delays, and vehicle queue lengths [9]. The GPS device was configured to record the vehicle positions and speeds at one-second time intervals. Two persons, a driver and a GPS recorder, were needed to perform the data collection. The average vehicle travel speed along distance traveled are be measured by GPS at every segment. Based on previous study [10], they suggested at least 300 m from the study site to any intersection or other element or stop sign that may affecting operating speeds.

According to [8], continuous speed profiles are used to determine the global speed variation along a road segment and determining a single consistency value for the whole road segment. Parameters for this model are the bounded area between the profile and the average speed along a two-lane highway segment. As a design consistency increased, crash rates will shown to decrease significantly. Refer Equations (1): Ra is defining as the normalized relative area (per unit length) bounded among the speed profile and the average speed line. It can be practical use to individual speed profile or to the operating speed profile. First, calculate the average speed of the speed profile along the segment of road. Next, the areas bounded between the speed profile and the average speed lines were calculated (a_i). The consistency measure is the sum of the areas divided by the length of the segment (L).

$$\mathrm{Ra} = \frac{\left(\sum a_i\right)}{L} \quad (1)$$

where, Ra = relative area (m/sec) measure consistency, Σa_i = sum of i areas bounded between the speed profile and the average speed (m^2/sec), and L = entire segment length (m).

3 Site Selection

The study was conducted at Johor Federal Roads. The roads consist of two lanes rural which is the road at F0050 (Batu Pahat-Kluang).The road Batu Pahat-Kluang (F0050) is the main road connecting two cities in Johor. In this study, Figure 1 shows the area of study, the data are record starting intersection at KM5 until intersection Ayer Itam at KM32. According to the Star, reported on October 2, 2013 [11], based on Road Safety Department, F0050 is ranked the district with the highest road fatalities in Johor for five consecutive years. Motorcyclists and their pillion riders were the highest contributors – 60% or 648 fatalities, followed by car drivers and passengers numbering 266 fatalities.

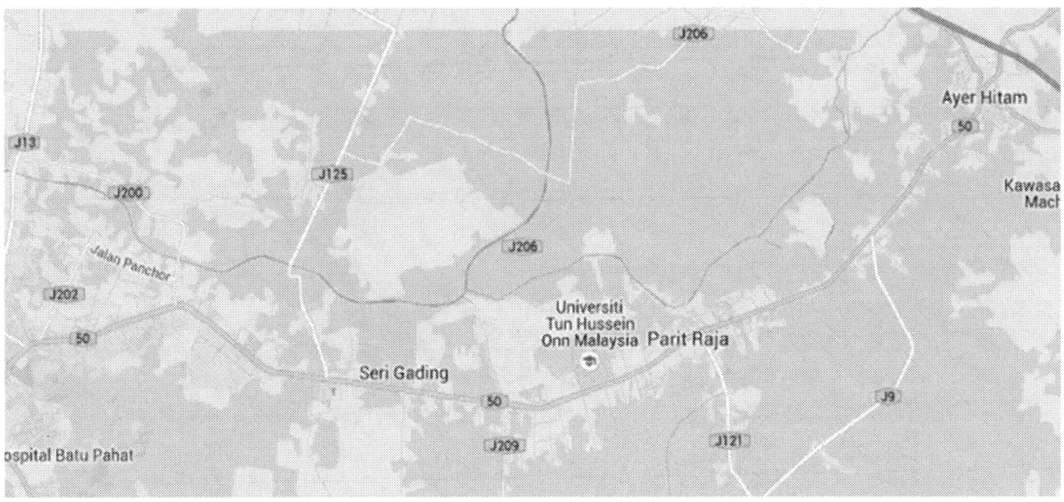

Figure 1. Area study.

4 Results and Discussion

Speed data were collected in the traffic stream under free - flow conditions along the study area using GPS which is primary data. The consideration in selection it should be free to avoid zero speed. The data should consider every elements of road design (geometric); lane width, shoulder width, access point/density access point which may not the same along the segment. The constantly geometric features will change along the segment, according to environmental conditions and it will cause changes in vehicles speed. Table 1 shows a summary of secondary data that included of road geometric, traffic information and accident information.

Data was collected on weekdays during off-peak hours where it more free flow traffic and the duration is selected to avoid from peak hour in morning and the evening. The test drivers were instructed to drive the equipment; GPS along the segment. Figure 2 shows the average speed of vehicle along the segment. The average speed was calculated as the sum of all speeds that divided by the number of speed observation. In this case, the average speeds of vehicles based on the data taken by GPS. The average length is 25 km. The GPS was used to obtain the continuous speed vehicles. Figure 3[12] shows the example of speed profile and consistency measured. Operating speed profile was plotted for each travel direction. The starting point of the analysis was the prediction of speeds that depending on the geometric characteristics along the segment. The independent measures was proposed to evaluate consistency; the relative area bounded by the speed profile.

Table 1. A summary of secondary data.

Variable		Type	Description
Dependent	Accident Crashes	Count	Type of accident crashes (Fatal, Serious, Minor, Wreckage)
Independent	AADT	Continuous	Average annual daily traffic per km
Independent	Traffic flow	Continuous	Average daily traffic including all motorized forms of traffic (i.e. motorcycles, cars, bus, trucks)
Independent	Lane width	Continuous	Lane width (ranging from 3.5m to 3.7m)
Independent	Shoulder width	Continuous	Road shoulder width (ranging from 0 to 2.4m)
Independent	Roadside hazard rating	Qualitative	Scale ranging from (1 (best) to 7(worst))
Independent	Median width	Continuous	Median width (ranging from 3.5 to 9.0m)
Independent	No. of lanes	Count	Total no. of lanes in the direction of travel
Independent	Land use	Qualitative	Level of activity along roadway (1 for no activity level, 2 for low activity level (e.g. school, factory), 3 for high activity level (e.g., residential or commercial)
Independent	Access density	Count	Number of intersection and minor access points per km along roadway
Independent	Horizontal alignment	Continuous	Horizontal curvature (1/km)
Independent	Vertical alignment	Qualitative	Indicator of vertical slope along road's length (1 for flat, 2 for rolling/undulating)

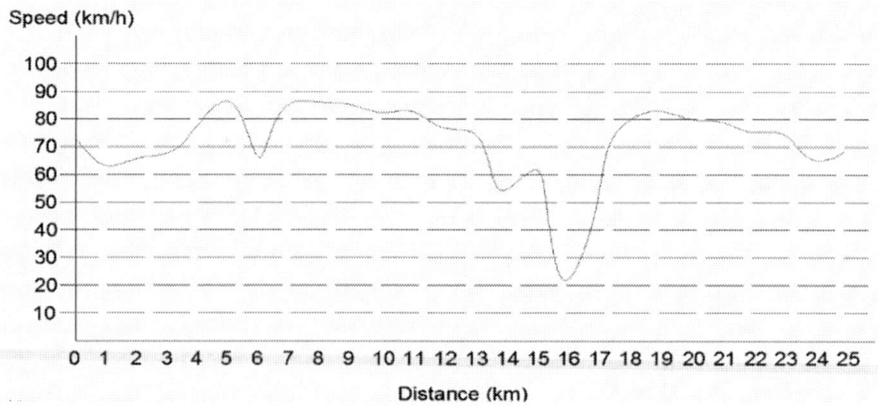

Figure 2. Average speed profile.

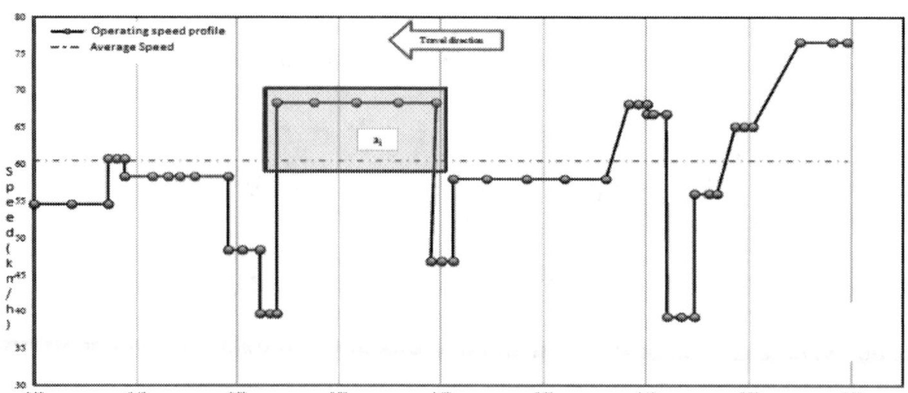

Figure 3. Example of speed profile and consistency measures.

Figure 4 shows the linear speed profile of vehicle along the road segment. The design consistency parameter will based on continuous operating speed profile models. The Ra was calculated based on speed profile along the road. Analysis of road design consistency is a good way to reduce road accident. The design consistency parameter will based on continuous operating speed profile models. The advantages of operating speed profile are based on continuous data but not based on speed-spot data collection. A plot of the 85th percentile operating speeds along the segment provided the "speed profile" under free-flow condition. Plot the 85th percentile speed based on average speed along the road. The 85th percentile speed was predicted based on such independent parameters as the degree of curve, length of curve, and the deflection angle. The concern of this study is the horizontal alignments which consist of the straight line. The 85th percentile values were obtained from real speed for each data through analysis. The profiles were plotted for observed 85th percentile values for real speed in each through along the segment. Speed prediction model was utilized in the present study for the development of the speed profile.

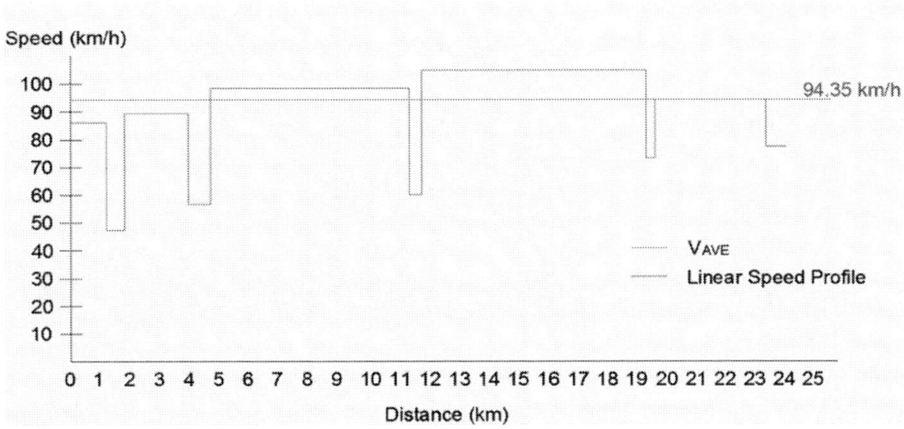

Figure 4. Linear speed profile.

5 Conclusions

Road accidents cause a thousand of victims in every year on the road. The outcomes from this study are valuable and can be used to implement and can be important role in designing or redesigning to reduce the road accident among motorcyclist and cars that occur in Johor Federal Roads especially F0050 (BatuPahat – Kluang). By using GPS devices, vehicles are recorded for every second over an entire trip at any time of day and any under weather condition. It does not need to assume that vehicle speeds at one location are representative of speed along the entire segment and also using GPS devices, able generate continuous speed profiles. After collecting data and road segmentation, the next step is to develop 85th percentile speed profile which is required for developing the consistency model. An average of 85th percentile speed will be presented at every road segment. Estimation of geometry for consistency in design is approach for enhancing safety in Johor Federal Roads.

6 Future Work

The study is still ongoing. There are some parts of the analysis have not been obtained. After the speed profile along the segment, area boundary between motorcycles and cars will be calculate to find the design consistency of the road. Combining the results on the road consistency with accidents frequency along the sections, it is expecting to produce the relationships between accidents frequency and road design consistency. Therefore, applying Poisson distribution in modeling is to overcome the shortcoming of conventional regression model. The last findings of the study that model will illustrate relationship between design consistencies and expected number of motorcycles and other vehicle (cars) crashes along the road over the years. The findings in identification of hazardous location for motorcycles will become the effective countermeasures to reduce crash frequency and severity of motorcyclist, in term of road design and traffic behavior.

Acknowledgement

The authors would like to thank all individuals and organization that have made this study possible. Thank you is also extended to the authorities of Faculty of Civil Engineering and Environment, Universiti Tun Hussein Onn Malaysia.

References

[1] M.M.A. Manan and A. Várhelyi, Motorcycle fatalities in Malaysia, IATSS Research, **36**(1), 30-39, (2012).
[2] F.J.C. Torregrosa, P.Z. Ana Maria and G.G. Alfredo, New geometric design consistency model based on operating speed profiles for road safety evaluation, Interm Report, (2011).
[3] S.M.R. Sharifah Allyana, H.Z. Zarir, A.M. Abdul Rahmat, M.F. Siti Atiqah, F.P. Noor, S.V. Wong and M.J. Jamilah, Recent trend of fatal motorcycle crashes in Malaysia, *Proc. of the 8th International Forum of Automotive Traffic Safety*, (2010).
[4] D. Hung, M. Stevenson and R. Ivers, Prevelance of helmet use among motorcycle riders in Vietnam, Injury Prevention, 409-413, (2006).
[5] U. Radin, Updates of road safety status in Malaysia, IATSS Research, **29**, 78-80, (2005).
[6] R. Lamm, B. Psarianos and T. Mailaender, *Highway Design and Traffic Safety Engineering Handbook*, (1999).
[7] J. De Oña and L. Garach, Accidents prediction model based on speed reduction on spanish two-lane rural highways, Procedia-Social and Behavioral Sciences, **53**, 1010-1018, (2012).
[8] C. Mattar-Habib, A. Polus and H. Farah, Further evaluation of the relationship between enhanced consistency model and safety of two-lane rural roads in Israel and Germany, European J. of Transport and Infrastructure Research, **4**(8), (2008).
[9] Y. Jiang and S. Li, Measuring and analyzing vehicle position and speed data at work zones using global positioning systems, ITE J., **72**(3), 48-53, (2002).
[10] K. Schurr, P. McCoy, G. Pesti and R. Huff, Relationship of design, operating, and posted speeds on horizontal curves of rural two-lane highways in Nebraska, Transportation Research Record: J. of the Transportation Research Board, **1796**, 60-71, (2002).
[11] C. Tan, Batu Pahat ranked highest in road fatalities, The Star, (October 2, 2013). <retrieved from http://www.thestar.com.my> (assessed on July 27, 2015).
[12] F. Russo, R. Mauro and G. Dell'Acqua, Rural highway design consistency evaluation model, Procedia-Social and Behavioral Sciences, **53**, 952-960, (2012).

MATEC Web of Conferences **47**, 04001 (2016)
DOI: 10.1051/matecconf/20164704001

Assessment Framework of Building Facade in Optimizing Indoor Thermal Comfort of Green Building Index (GBI) Certified Office Building

Asniza Hamimi Abdul Tharim[1] and Muna Hanim Abdul Samad[2,a]

[1]*Faculty of Architecture, Planning and Surveying, Universiti Teknologi Mara (Perak), Malaysia*
[2]*Architectural Programmed, Universiti Sains Malaysia, Penang, Malaysia*

Abstract. During the past decade, the construction industry has seen a new trend in the development of "green" or "sustainable" construction concept around the world with vast support from prominent organization, together with the introduction of sustainable building codes. The establishment of green building certification systems worldwide is seen as one of the most efforts in the emerging green building movement. In order to support the development of the "green" and "sustainable" concept in Malaysia, Green Building Index (GBI) was launched by the government on 21 May 2009 that created to promote sustainability in the built environment and raise awareness of environmental issues. However, the construction industry seems to have focused only on findings the "right mechanism" for an environmentally sustainable "final result" in order for the building to be certified as green with the lacking of continuous assessment on the building performance after the certifications. This study is purposely conducted to investigate the performance of various rated Green Building Index (GBI) Non-Residential New Construction office buildings and the influence on Indoor Thermal Comfort (ITC) of the selected buildings. The aim is to develop an assessment framework for optimum green building architectural façade to be used for office buildings in Malaysia as well as to analyse the occupants' perception, satisfaction and performance in the selected Green Building Index (GBI) rated office indoor environment. This research is still in its infancy; therefore the paper is focused on research aims, research scope and methodology, and expected deliverables for the proposed research.

1 Introduction

Sustainability is fast becoming a global phenomenon with a wide range of professional attentiveness and perspectives. It is believed that the idea of sustainability first emerged 40 years ago [1]. Since then, expansion of the sustainability worldwide continuous to accelerates with the supports from the industries and society [2]. Over the last few decades, a common definition of sustainability has been developed. According to the "United Nations- Our Common Future" [3] sustainability is a concept that incorporates a wide range of social, economic and environmental issues with the idea of responsible stewardship that integrate systems in order for them to meet the needs of present without compromising the future needs. Sustainability comprises of doing good things for environment and

[a] Corresponding author : mhanim@gmail.com

societies that subsidise to the improvement in energy management and conservation, human being social life over and above the nation's economic development. Hence, sustainable development is a dynamic process that incorporates a diversity of levels and balancing scales vacillating from economic and agricultural development to the management of human defrayals, economics, societies, ecological systems and building practices that link to the global concern [4-6]. One of the most significant aspects in nation sustainability in most countries is the construction industry that comprises the sustainable development and construction that also known as the 'green' construction. The sustainable development of the construction industry has been a worldwide crusade that has been growing for the last two decades [7]. Sustainable development has a complex relationship with growth; subsequently the population and economic growth increase the pressure on the environment. Nevertheless, in order to achieve sustainability, the excessive economic growth is involved to attain an adequate level of social development with the quality of life. Therefore, it is crucial to understand the various aspects of sustainable development and how it affects the economics growth by understanding the proper use of environment resources that comprises two main components that are the human physical and nature.

2 Research Background

Sustainable in the construction industry is defined as the degree of growth and expansion that fulfil the need of the current state of development without compromising the need for future generations [8]. Hence, the construction industry often being accused of contributing to the negative impact towards the world environment in a most of the aspects, ranging from excessive consumption of resources to direct environment pollution [9] include climate change, ozone exhaustion, forfeiture of biodiversity and many others [7]. Therefore, implementation of sustainable development in the construction industry is seen as the best way to address, reduce and eliminate these interrelated problems in order to conserve the environment for the sake of future generation and planet [10]. However, [11] raise concern about the sustainable and buildability of the construction industry are key elements in building design that are typically governed by laws and regulations that may differ from country to country thus creating barriers to complete and standard implementation of sustainable construction in worldwide. Previous studies by various researchers believed that the market for sustainable buildings is increasing as the people in the construction industry has acknowledged that sustainable in construction industry may ease the negative impacts on the environment and bring significant improvement to the social and environmental benefits of all nations worldwide [12-14]. Therefore, as more owners and clients seek to develop sustainable buildings, the construction industry is adapting to new requirements in order to meet owners' concerns. Recently, a significant change in the construction industry has been the increased interest in green design and sustainable materials. Sustainable materials are the potential resource to mitigate the impact on the environment and bring significantly and holistic benefits [15].

Similarly, the development and introduction of sustainable building codes is taking place around the world with vast support from prominent organization [16]. The ICC define building codes as the code that "*provide reasonable controls for the design, construction, use, occupancy and maintenance of buildings and their facilities and various components*" [17]. Thus, the establishment of green building certification systems worldwide be seen as one of the significant efforts among the numerous efforts in the emerging green building movement, it is also believed to be one of the most prominent and guarantees systematic approach to continuing countries effort toward promoting sustainable environment [18]. The first green building standard was introduced in United Kingdom (UK) known as the Building Research Establishment Environmental Assessment Method (BREEAM). Since then several country have developed their own green standard [19], including the Leadership in Energy and Environmental Design (LEED) of the United State of America (USA), Comprehensive Assessment System for Building Environmental Efficiency (CASBEE) Japan and Green Star Australia [20]. Consequently, as an acceleration of the Malaysian sustainable construction industry and as a manifesto of the government's seriousness in implementing "green" initiatives for the country, The National Green Technology Policy was launched by the Former Prime Minister of Malaysia Tun Dr

Mahathir Mohamed in July 2009 with the aim to be a driver to support the national economy and promote sustainable development in Malaysia. In order to expand the use of green technology in Malaysia, Green Building Index (GBI) was launched by the government on 21 May 2009 and it is one of the incentives announced in Budget 2010 under the heading of *"Promoting Construction of Green Building"*. The Green Building Index (GBI) is a green rating index on environmentally friendly building with the ability to save utility costs and preserve the quality of the external and internal environment. Therefore, it is crucial for the sustainable construction not only focused on environmental sustainability but also integrations of all aspects that can contribute to improved health, satisfaction and wellbeing amongst building users [21]. Some research suggests that occupants in environmentally sustainable buildings feel better psychologically, although this area of research is in its early stages. The construction of green and sustainable buildings has increased due to increasing level of awareness on the sustainability issues around the worlds with the aims to reduce the negative environmental impact to building indoor space and increase occupant's satisfaction [22, 23].

3 The Issue of Indoor Thermal Comfort and Green Rated Building

Malaysia is classified to have tropical climate condition that are mainly hot and humid throughout the year. Normally the first six months will be sunny months and the second half of the year will be the wet months. Szokolay [24] described the tropical climate of hot and humid as one of the extreme and hardest climate to compromise in term of the building design. The after effect of global warming that the world is facing today is the increase in outdoor and indoor temperatures. Although climates, living conditions and cultures differ widely throughout the world, the temperature that people choose for comfort under similar conditions of clothing, activity, humidity and air movement has been found to be very similar. To date, however the construction industry seems to have focused only on finding the "right mechanism" for an environmentally sustainable "final result" such as energy efficiency or water conservation in order for the building to be certified as green with the lacking of continuous assessment on the building performance after the certifications [25].

The primary objective of a building is to provide shelter, space and comfort for the people to live, work and interact with them [26]. In this respect, it is worthwhile to consider and take into account the role of building façade as not only being a wrapper of a building but also a barrier with various vital functions that influence the indoor environmental quality and comfort [27]. Often the achievement of Indoor Thermal Comfort (ITC) in buildings is determined by the thermal performance of façade [28 and 29], to a large extent ranking second under local climate characteristic. The question raised is how far the building façade affects indoor thermal performance is conducted for the mechanical ventilation buildings in hot and humid country of Malaysia? Research on this matter has been conducted by various researchers in different aspects. Earlier research conducted by [30] found out that thermal comfort is affected by the thermal interaction between the building and external environment. Researcher [31] had conducted research on the impacts of architectural components on energy consumption in mechanically ventilated building as well as its health related effects on the performance of building occupants. Although there were many research conducted on the effects of external building envelope and its influence to the indoor air quality, intelligent façade design solutions in hot and humid climate have been barely studied compared with those in cold climate [32].Nevertheless, a variety of green building certification systems had been introduced worldwide but one question remains to be clarified: *are the scheme currently applied in green building certification fostering buildings of Indoor Environmental Quality satisfying their occupants?* [33]. Gou et al. [34] suggested the need of research in the involvement of additional factors influencing the indoor air quality satisfaction of occupants to be conducted in tropical countries. It is vastly believed that the green building can provide a better indoor environment that contributes to health, wellbeing and user's productivity and performance. If these green building are designed well, it can increase comfort and create healthier working conditions.

4 The Proposed Research

The main aim of this proposed research by the author is to develop assessment framework on the selection of architectural building facade to be used as a construction façade guideline for office buildings. This study also intended to examine the objective and subjective measurements of overall post- occupancy thermal comfort in a Non-Residential New Construction (NRNC) Green Building Index (GBI) [35] rated office building with different types of building façade in hot and humid country of Malaysia to achieve overall thermal comfort of the internal environment as well as to identify on the correlation of these architectural building façade heat radiation with the satisfaction level of occupants in post occupied green office building in Malaysia. However this research will not touch on the aspects of the energy efficiency or the Indoor Environment Quality (IAQ) of the selected green rated office buildings, but will focus more the correlation between the objective and subjective measurement of the Indoor Thermal Comfort (ITC) of the working space. Extension empirical research on indoor Thermal Comfort in hot and humid climate in various green aspects is not new in this country, thus the study on the impacts of façade to building indoor thermal comfort remains relatively new in Malaysia. Therefore this study intended to:

i. Measure both on objective and subjective parameters of rated Green Building Index (GBI) office building Indoor Thermal Comfort (ITC)
ii. Study on Non- Residential New Construction Green Building Index (GBI) rated office building in Malaysia
iii. Covering various types of green building façade

As for the hypothesis, it is assumed that the Green Building Façade (GBI) certified buildings will be able to reflect the ultimate thermal comfort in the building indoor space compared to the uncertified Green Building Index (GBI) buildings in Malaysia. It is also predicted that the occupants (employees) in the certified office buildings should be in the most satisfaction conditions while working in the buildings. The scope of this study focuses on the measurement of Thermal Comfort in a working space of a Green Building Index (GBI) rated office building in Malaysia. Thermal comfort measurement will cover both the objective and subjective parameters. Objective measurement will be done by using the fieldwork equipment measurement of the indoor temperature, relative humidity, air velocity and heat radiation, while the subjective measurement will be conducted by questionnaire survey to the selected building occupants that will covers the aspects of user's perception and satisfaction in five offices building with different architectural façade. The questionnaire survey covers the demographic profile of building users and their perception and satisfaction on four main aspects of the Indoor Thermal Comfort (ITC). The proposed research project is purposely to be conducted in the state of Kuala Lumpur. The study will focus on various Green Building Index (GBI) rated Non- Residential New Construction Office Building for comparative studies on the best selection of green building architectural facade. There are two types of sampling used for this proposed research project; Random Sampling for the subjective measurements (questionnaire survey) and Selected Sampling for the objective measurements.

5 Significant of the Proposed Study

This study adds to the existing body of knowledge by examining the contribution of different building façade radiation transmitted to the indoor space of building. In additions this study will be the benefits reference for architect (designers) and developer in constructing good indoor thermal comfort building through the implementation of suitable building façade for hot and humid country like Malaysia that helps to ease the process of obtaining green building certifications in this country.

References

[1] W.M. Adams, The future of sustainability: re-thinking environment and development in the twenty-first century, report of the IUCN renowned thinkers meeting, (2006). <Retrieved from http:// cmsdata.iucn.org/downloads/iucn_future_of_sustanability.pdf>

[2] C.J Kibert and M.E. Rinker, The next generation of sustainable construction, Building Research and Information, **35**(6), 595-601, (2007).

[3] United Nations, Report of the world commission on environment and development: Our common future, (1987). <Retrieved from http://www.un-documents.net/wced-ocf.htm>

[4] B. Givoni, *Climate Consideration in Building and Urban Design*, Van Nostrand Reinhold, New York, (1998).

[5] P. Bansal, Evolving sustainably: a longitudinal study of corporate sustainable development, Strategic Management Journal, **26**(3), (2005).

[6] P.R. Berke and M.M. Conroy, Are we planning for sustainable development, Journal of the American Planning Association, **66**(1), (2000).

[7] C.J. Kibert, *Sustainable Construction: Green Building Design and Delivery*, John Wiley and Sons, New Jersey, (2008).

[8] F. Duchin, *Global Scenarios about Lifestyle and Technology, the Sustainable Future of the Global System*, United Nations University, Tokyo, (1995).

[9] G.K.C Ding, Sustainable construction – the role of environmental assessment tools, Journal of Environmental Management, **86**(3), 451-64, (2008).

[10] T. Waas, A. Verbruggen and T. Wright, University research for sustainable development: definition and characteristics explored, Journal of Cleaner Production, **18**(7), 629-636, (2010).

[11] M. Duris, I. Nemec and M. Polak, Buildability and sustainability – a question of professional responsibility, International Congress – Global Construction: Ultimate Concrete Opportunities, Dundee, (2005).

[12] W. Wang, R. Zmeureanua and H. Rivard, Applying multi-objective genetic algorithms in green building design optimization, Building and Environment, **40**(11), 1512-1525, (2005).

[13] R. Ries, M. Bilec, N.M. Gokhan and K.L Needy, The economic benefits of green buildings: A comprehensive case study, The Engineering Economist, **51**(3), 259-295, (2006).

[14] C. Thormark, The effect of material choice on the total energy need and recycling potential of a building, Building and Environment, **41**(8), 1019-1026, (2006).

[15] L. Florez, D. Castro and J. Irizarry, Measuring sustainability perceptions of construction materials, Constr. Innovation, **2**(4), 217–234, (2013).

[16] G.B. Hanna, Energy efficiency building codes for Egypt, Journal of Energy and Power Engineering, **5**(12), 1134-1141, (2011).

[17] International Code Council, *Building Department Administration*, ICC Publications, Country Club Hills, (2007).

[18] H.H. Liang, C.P. Chen, R.L. Hwang, W.M. Shih and S.C. Lo, Satisfaction of occupants toward indoor environment quality of certified green office buildings in Taiwan, Building and Environment, **72**,232-242, (2014).

[19] V. Potbhare, M. Syal, M. Arif, M.M.A. Khalfan and C. Egbu, Emergence of green building guidelines in developed countries and their impact on India, Journal of Engineering, Design and Technology, **7**(1), 99-121, (2009).

[20] A. Zimmerman and C.J. Kibert, Forum informing LEED's next generation with the natural step, Building Research and Information, **35**(6), 681-689, (2007).

[21] A. Smith and M. Pitt, Sustainable workplaces and building user comfort and satisfaction, (2011).

[22] S. Korkmaz, M. Horman, K. Molenaar and D.Gransberg, Influence of project delivery methods on achieving sustainable high performance buildings, Research Sponsored by the Charles Pankow Foundation, DBIA, (2010). <available at: www.dbia.org/NR/rdonlyres/ F15664D5-B617-487C-9C95-EE4F5E453A9D/0/CPF_ThrustII_05212010_Final.pdf.>

[23] Green Building Council of Australia, Green star overview, (2011). <available at: www.gbca.org.au/ green-star/green-star-overview/>

[24] S.V. Szokolay, *Introduction to Architectural Science: The Basis of Sustainable Design*, Elsevier, (2008).
[25] J. Yang, Editorial: promoting integrated development for smart and sustainable built environment, Smart Sustain. Built Environment, **1**(1), 4–13, (2012).
[26] M. Bessoudo, A. Tzempelikos, A.K. Athienitis and R. Zmeureanu, Indoor Thermal Environmental Conditions near Glazed Facades with Shading Devices, Part 1: Experiments and Building Thermal Model, Building and Environment, **45**, 2506-2516, (2010).
[27] S. Drake, *The Third Skin: Architecture, Technology and Environment*, UNSW Press, Sydney, (2007).
[28] E. Gratia and A. De Herde, Is day natural ventilation still possible in office buildings with a double-skin façade?, Building and Environment, **39**(4), 399–409, (2004).
[29] W. Liping and W.N. Hien, The impacts of ventilation strategies and facade on indoor thermal environment for naturally ventilated residential buildings in Singapore, Build. Environment, **42**(12), 4006–4015, (2007).
[30] A.P. Gagge, A.P. Fobeles and L.G. Berglund, A standard predictive index and human response to the thermal environment, ASHREA Transactions, **92**(2), 709-731, (1986).
[31] C.K. Cheung, R.J. Fuller and M.B. Luther, Energy efficient envelope design for high rise apartments, Energy and Buildings, **37**(1), 37-48, (2005).
[32] C.E. Ochoa and I.G. Capeluto, Strategic decision- making for intelligent buildings: comparative impact of passive design strategies and active features in hot climate, Building and Environment, **43**, 1829-1839, (2008).
[33] H.L. Han, P.C. Chen, L.H. Ruey, M.S. Wen, C.L. Shih and Y.L. Huey, satisfaction of occupants towards indoor environment quality of certified office building in Taiwan, Building and Environment Journal, **72**, 232-242, (2013).
[34] Z. Gou, D. Prasada and S.Y. Lau, Are green buildings more satisfactory and comfortable?, Habitat Int., **39**, 156-161, (2013).
[35] Buildings, GBI, (2013).

MATEC Web of Conferences **47**, 04002 (2016)
DOI: 10.1051/matecconf/20164704002

Rethinking Affordable Housing Delivery: An Analytical Insight

Abdullateef Olanrewaju[1,a], Seong Yeow[1] and Lee Lim[1]

[1]*Department of Construction Management, Faculty of Engineering and Green Technology, Universiti Tunku Abdul Rahman, 31900 Kampar, Perak, Malaysia*

Abstract. Malaysia currently has a housing shortage of 12 million units. Towards the year 2020, this would require an annual supply of a minimum of 2 million homes. With the current production rate, the deficits will remain over the next 100 years. The crisis will lead to poor standards of living, un-affordable rental price, high mortgage payment, abandonment, and dilapidation of the existing housing stock. Lack of affordable housing is already a source of concern among many Malaysians. The middle and lower income earners spend more than 30% of their income on rent or for mortgage payment. Housing rent or mortgage is a basic need like foods, education, clothing and health. The government has introduced various measures to address the housing deficit. Despite these measures, the housing supply and distribution gaps continue to grow. To better understand the issues, there is a need to examine how the procurement planning in the affordable housing delivery supply chain and market is formulated, explore how cost of affordable housing would be reduced, identify the user value systems in affordable housing, and identify the criteria in the current regulatory framework? For these purposes, this paper reviews the relevant literature to reach preliminary findings on the stated issues. Among the factors found accounting for housing shortages, the overwhelming factors were that of poor policies, strategies, practice, management, and finance. The findings are meaningful in framing an affordable housing delivery model. It could also be useful to stakeholders involved in affordable housing delivery in Malaysia and elsewhere.

1 Introduction

This paper reports initial findings of an ongoing large research project aimed at developing an affordable housing model for an emerging economy. This study addresses the issues obstructing the supply of affordable housing based on a review of existing literature. To clarify, 'affordability' is an elastic word. At one end, a product is very affordable yet at the other end, it is completely unaffordable to the same group of buyers. This makes the issue of affordable housing somewhat complex to understand and explain. Most authors and governments used the terms incorrectly [1] resulting in different definitions of 'affordable housing'. Often, the term is used quite loosely to represent all types of housing developed by governments, its agencies, or in partnership with the developers or social landlords. It would seem that the meaning of affordable housing is more easily defined than understood. The question then arises; at what point does a house that was once

[a] Corresponding author : abdullateef.olanrewaju@ymail.com

unaffordable become affordable? This underscores the importance of household income. The demand for houses is income elastic, because its elasticity is greater than 1. In other words, the percentage change in the quantity of homes demanded is more than the corresponding changes in price. Housing is affordable if a rental cost or mortgage repayment does not exceed 30% of household income for households in the lowest 40% of the income distribution range [2]. For rental purposes, the 30% would include utilities bills, including for electricity, water, gas, sewage and garbage collection. In cases of a mortgage, the amount includes the actual consideration, tax, insurance, utilities, and maintenance. This definition is referred to as the 30/40 rule. Affordable housing is also defined in terms of 'median multiplier' [3]. The Demographia compares the median house prices to the median household income to measure affordable housing based on annual international Housing Affordability Survey. Based on either of these definitions, the cost of housing in all major cities in Malaysia is 'severely unaffordable' to both the middle-income and low-income earners. For instance, monthly expenditure for the rental is 26.7% of the household income [4] however, for the low and middle-income earners is 35% [5]. 2014 inflation in the housing, water, electricity, gas and other fuels category increased to 3.4% (2013: 1.7%), reflecting the upward adjustment in electricity tariffs and a broad-based increase in rentals across different types of residential property [6]. In terms of the income groups, the mean income is RM 6,141. However, the median income which is the acceptable criteria is RM4, 585.

"One of Malaysia's longstanding development objectives is the provision of affordable housing for Malaysians in both rural and urban areas, with a focus on lower-income groups" [7]. Many schemes were put in place both at the state and federal levels toward meeting the housing requirements in Malaysia. At the federal level, some of the schemes include the Housing Loan Scheme, Housing Maintenance Program, People's Housing Program, MyHome Scheme, Transit House Program 1Malaysia Civil Servants Housing Program (PPA1M) PR1MA, Government Housing Loan, and the Affordable Housing Policy Wilayah Persekutuan (RUMAWIP). While most of the schemes are under the Ministry of Urban Wellbeing, Housing and Local Government, the others are either under the Prime Minister's Department, Ministry of Federal Territories, or Ministry of Finance. 80% of the households belong to the low and median income groups [8]. If the housing needs of these groups are not met; it will imply low quality of life and consequently lead to low national productivity. Lack of affordable housing would lead to other social problems including traffic congestions, air pollution, crime and urban sprawl. Unaffordable housing could be the result of many factors, notably increases in house prices or decreases in household income or a combination of both. Methods to increase household income fall outside the scope of this research.

2 Background and Theoretical Framework

The provision of affordable and liveable social housing is critical particularly as the government is committed to encourage greater home ownership among the bottom 40% households. The intense pressure on the provision and maintenance of affordable housing are extreme. By 2020, over 70% of Malaysians are expected to reside in urban areas. The greater Kuala Lumpur alone is expected to accommodate an additional one million residents by 2020 [9]. Housing requirements are expected to increase remarkably due to the rapidly growth in population, increase in foreigners (expatriate, students and tourists), migration, changing economic status of the citizens, change in tastes, and dilapidation of the existing stock. In 2020, the population is estimated to reach 32.4 million. While the requirements for housing for the poor and low cost will remain, emphasis will be shifted to delivering vibrant housing that is commensurate with the country's status. Table 1 presents allocations to the housing sectors for twenty years.

Table 1. Federal government allocation to the housing sector [7, 10-12].

Malaysian Plan	6MP	7MP	8MP	9MP
Amount (RM Million)	3,850	6,125	6,972	9,647

The total allocation for the housing sector under the 9MP was RM11,512 million. Approximately 111,000 units of public affordable housing were built during the Ninth Plan period [7]. There is more than 150% increase in the allocations to the housing sector over the period. Under the Tenth Malaysia Plan (10MP) it was stipulated that housing development will focus on the provision of adequate housing, as well as ensuring a safe, healthy, harmonious and secure living environment equipped with complete necessary public amenities and quality recreational facilities require in the improvement of standard of living. During the 10MP, a total of 78,000 units of new affordable public housing were slated for construction and to reduce the burden on the bottom 40%, low-cost public housing units are provided to household income with less than RM2,500 per month [7]. Subsidises of between 30% and 75% of total construction costs for public housing units is provided. Added to these is the establishment of Housing Maintenance Fund with an initial funding of RM500 million to assist the residents of both public and private low-cost housing units. In continuation of government initiatives towards housing supply, under the 11MP a total of 606,000 houses will be built for those in low and middle-income groups this is in addition to 47,0000 to be built or repaired specifically for the poor. To interpret, during the Plan, approximately 130,600 public homes will be built annually.

However, there is high level of insufficiency in homes supply. Lack of affordable house is now a source of concern among Malaysians. To illustrate, available statistics show that Malaysia currently has 4,831,791 representing an increase of 2.2% compare to 2013 (i.e. 4,725,109 actually) residences and of the total properties transacted 65% is residential and represents 50.4% in terms of value [13]. That is, home production is approximately 107,000 per annum. The plausible implication of this statistics is that there are approximately 6six persons per home. However, this is inadequate to achieve the above set objectives, countries like Australia and UK have on average 2.5 persons per home. Malaysia's growing population is increasing at a rate of about 1.7% a year. This means that in a year, approximately 500,000 persons are added. To interpret, in one year, about 400,000 persons will not have homes to live in. This will lead to increase in prices of homes, or many will have to pay unaffordable prices to own or rent a home. Secondly, there will be an increase in maintenance demand. A major factor that determines maintenance demand is the size of the households [14]. Hence, the existing housing stock would not be part of national gross fixed capital formation that is required in a country to stimulate and support national development. Policy makers need to take the supply-side imbalance into consideration to provide accommodation that is affordable in all respects. With increased median incomes from RM2, 841 in 2009, RM3, 626 in 2012, to RM4, 585 in 2014 [15], and with the Gini coefficient sliding down from 0.431 in 2012 to 0.401 in 2014 [15], household income is improving. The mean monthly house income of the top 20%, middle, 40% and bottom 40% of households are RM9, 796; 4,372, and RM1, 852 [15]. Table 2 contains the incomes distribution by household in 2012.

Table 2. Percentage distribution of households and income share by monthly household income class [15].

Income class (RM)	Household	Income
999 and below	5.0	0.8
1,000 – 1,999	17.6	5.4
2,000 – 2,999	15.9	7.8
3,000 – 3,999	16.7	11.5
4,000 – 4,999	11.1	9.9
5,000 – 5,999	7.8	8.6
6,000 – 6,999	6.0	7.7
7,000 – 7,999	4.5	6.7
8,000 – 9,999	5.7	10.1
10,000 and above	9.7	31.5

From Table 2, more than 60% of Malaysians are either lower income earners or middle income earners, and as such their housing needs are not being met, leading to a housing crisis. Household expenditure for housing, water, electricity, furnishing, household equipment and building maintenance is the highest among the various house classes [5]. Furthermore, in 2012, 23% of households earned less than RM2, 000 monthly, 55% earned less than RM4, 000 monthly and 74% less than RM6, 000 per month [16]. Out of the 6,659,036 active EPF members, 8% earned between RM4, 001 and RM6, 000, 24% earned between RM2001 and RM4000, while 60% earned less than RM2000 monthly [17]. For 2013 and 2014, while the total percentage change in the consumer price index was 2.7% for housing, water, electricity, gas and house maintenance collectively was 4% [18]. Based on the recent information on the consumer price, the low and middle-income earners spend about 35% of their monthly incomes on housing, utilities bills, and house routine maintenance [19]. From 2000 to 2014, there is more than 100% increase in the price of houses from the period quoted (Table 3), and this will be the case in the years to come particularly with the increase in the cost of living and the uncertainties that surround the implementation of the Goods and Services Tax. While it is probably a touch early to speculate on the extent of increase in cost caused by the GST, observers in the industry hypothesised that the likely increase would be around 4%.

Table 3. Annual House Price Index and house price1999-2013 [13].

Year	Index 2000=100	1-Yr % Change	House price (RM)
2000	100	-	138,712
2001	101.1	6	140,507
2002	103.6	1.1	144,830
2003	107.7	2.5	150,705
2004	112.9	4	156,549
2005	115.6	4.8	160,654
2006	117.8	2.4	165,111
2007	124	1.9	172,464
2008	129.8	5.3	179,359
2009	131.8	4.7	184,002
2010	140.7	1.5	196,720
2011	154.6	6.7	216,720
2012	172.8	9.9	241,591
2013	192.9	11.8	266,304
2014	213.6	11.6	295,345

While the growth rate for the Housing Price Index was 11.6 in 2014, the corresponding growth rate for the Consumer Price Index was 3.4. In addition, the total approved loan to the housing sector increased to RM129,051 million in 2014 as compared to RM124,230 million in 2013 but RM426,800 million housing loan is still outstanding[6].

3 Summary and Further Research

While there are commitments to provide more affordable housing, the housing shortage is on the increase. Hence there is a need to rethink the affordable home delivery. Firstly, incentives to prospective homebuyers provide opportunities to have access to buy a house, but do not translate to lowering the cost of houses. This could be explained by the inadequacies in the formulation of the initiatives. For illustration, while the government allow applicants to withdraw from the EPF to finance the 10% down payment for the PR1MA scheme, a major problem with this initiative is that about 68% of EPF contributors aged 54 have less than RM50, 000 in their accounts [17] and more than 60% of the contributors actually earned less than RM2,000 monthly. For applicants to withdraw RM30, 000 for a property that costs RM300, 000 will be very unlikely. Having only about RM20, 000 left in their account at the age of 55 would lead to members retiring into poverty. Out of EPF's 14,192,832 members, only 6,659,036 are active or those who have contributed at least once within 12-month [17]. Based on the My First Home / Skim Rumah Pertamaku (SRP), affordable housing is those that cost between RM90,000 and RM170,000, while this is affordable, to be eligible for this scheme now, homeowners are required to purchase a property that cost RM400,000. Providing subsidy of up to RM30, 000 per unit to developers would lead to ineffective over production apart from the fact that it is limited to 10,000 units. Essentially, the subsidy could only reduce costs paid by homeowners but will not necessarily increase home production. While there are some schemes, e.g. Skim Rumah Pertamaku, (SRP) Perumahan Rakyat 1Malaysia (PR1MA), Rumah Mesra Rakyat (RMR1M), Program Rumah Mampu Milik (RMM), that help to reduce the burden on the entry costs, the house price itself is unaffordable. To date, only 560 units of 158,807 of the PR1MA housing have been completed, though 41,187 are under construction [20, 21], yet a total of 1,02 million have registered for the scheme [20].

Secondly, homes in Malaysia are 'severely unaffordable' or very expensive to own or rent [2]. Therefore, the question to ask is why the houses are so expensive and how can costs be reduced? Both in absolute and relative terms, subsidises such allowing prospective homebuyers to utilise part of the EPF to finance the down payment do not have significant impact to prompt the homebuyers to own a home because these do not contribute to housing economics. However, in real time, what is required is to explore how the cost of construction could be reduced. This is critical to any strategies that will be beneficial to prospective homebuyers and the governments. Hence, issues of procurement planning, regulatory and polices framework, method of financing and availability of basic materials and products are elementary problems that will result in affordable housing delivery. Other fundamental aspects that are often over looked by either the government or developers are the size, locations, functionality, employment, transport, access to services. In other words, the homes must be adequate all in respects. Most of the MyHome housing is not located in the cities or close to other infrastructures. But residing in affordable housing in the suburb is no longer cost-saving. Increases in transportation costs, access to children school among other is wiping saving accrued from paying reduced rent or mortgage on affordable housing due to location factor. Specifically, the current initiatives are inadequate to offer solutions to the housing requirements because the fundamental aspects that would lead to price reduction have not been evaluated and disregarded for homeowners or home users' value systems.

Thirdly, the existing measures on affordable housing do not connect affordable housing to sustainability. However, as government aims to provides housing that will contribute to sustainability requirements, it becomes imperative to integrate the supply of affordable housing with sustainable development. Evidently, the affordable housing delivery needs to be integrated with sustainability. The requirements of sustainability on the housing delivery will lead to significant energy optimisation, enhancement of durability of housing, waste minimisation, improvements on social impacts, good indoor environment, reduces pollution, consideration of project life cycle-cost, and active user-friendliness, and user comfort and total satisfaction. Therefore, the possibilities of matching sustainability requirements with affordable housing require exploration. Since research on this aspect is grossly inadequate, it could not be established if the two are compatible and to what extent. This study shows that the 11MP is ambitious in increasing the access to affordable housing. However, what is still not clear is if there are definitive interpretations and understandings of the concept of affordable housing. Some unanswered questions have also been exposed, for instance, as the 11MP also hinges considering sustainable issue on affordable housing, on what basis is this initiative based?

To conclude, this study begins to reveal the benefits and shortcomings of the existing housing programmes and financing methods in the affordable housing procurement processes. The housing deficits, which are expected to rise, are a clear indication that the existing housing policy is not formulated and implemented as expected and hence not operational as planned. In order to develop a systemic framework for the supply of affordable housing, fundamental concepts, strategies and methods need to be evaluated. These include the procurement planning in affordable housing delivery, behaviours credentials of home owners and users, and users' value systems and sustainability criteria in affordable housing delivery. All these ought to precede any decisions on the development of affordable housing. This research is concerned with investigating and evaluating these aspects through exploratory methods. The supply of affordable housing requires different funding sources and systemic delivery models that involve the various stakeholders in the supply chain and marketplace.

References

[1] D.M. Jewkes and M.L. Delgadillo, Weaknesses of housing affordability indices used by practitioners, Journal of Financial Counseling and Planning, **21**(1), 43-52, (2010).

[2] U.S. Department of Housing and Urban Development, *Housing Impact Analysis*, Government Printing Office, Washington, (2006).

[3] Demographia, The 11th annual international housing affordability survey: 2015 ratings for metropolitan market, (2015). <Available at: www.demographia.com/dhi.pdf>

[4] Department of Statistics, *The Malaysian economy in Brief*, (2014).

[5] Department of Statistics, Malaysia report on household expenditure survey, (2011). <available at: http://www.statistics.gov.my/portal/download_household/files/Expenditure/Laporan_PIR_Report_on_HES_2009-10.pdf>

[6] Bank Negara, Annual Report 2014, (2015).

[7] Government of Malaysia, Tenth Malaysian Plan 2011-2015, Economic Planning Unit Prime Minister's Department, Putrajaya, (2010).

[8] Department of Statistics, Report on characteristics of household 2010: Population and housing census of Malaysia, (2014).

[9] National Transformation Programme (2012) Annual Report.

[10] Government of Malaysia Seventh Malaysian Plan 1996-2000, Economic Planning Unit Prime Minister's Department, Putrajaya, (1996).

[11] Government of Malaysia, Eighth Malaysian Plan 2001-2005, Economic Planning Unit Prime Minister's Department, Putrajaya, (2001).

[12] Government of Malaysia Ninth Malaysian Plan 2006-2010, Economic Planning Unit Prime Minister's Department, Putrajaya, (2006).

[13] National Property Information Centre (NAPIC), Key Data 2014: Property Market Report, (2014). <available at: http://napic.jpph.gov.my/portal>

[14] A.A. Olanrewaju, Green maintenance management initiatives for university buildings, Built Environment Journal, **8**(1), 17-24, (2011).

[15] Department of Statistics, Household Income and Basic Amenities Survey Report 2012, (2012).

[16] Khazanah Research Institute (KRI), The State of Household. Kuala Lumpur, (2014).

[17] Employees Provident Fund (EPF), Annual report: living our values, (2014), <available at:http://www.kwsp.gov.my/portal/documents/10180/5382689/1._OUR_ORGANISATION.pdf>

[18] Department of Statistics, Quarterly Construction Statistics, (2014).

[19] Department of Statistics, Consumer Price Index, States Indices, (2014).

[20] New Straits Times (NST), RM30 billion loan facility for PR1MA projects, (Saturday 22August, 2015).

[21] The Star, Rent-to-own your home: scheme will help PR1MA applicants who cannot get a loan, (Saturday 22 August, 2015).

MATEC Web of Conferences **47**, 04003 (2016)
DOI: 10.1051/matecconf/20164704003

Maintenance of Heritage Building: A Case Study from Ipoh, Malaysia

Seong Yeow Tan[1,a], Abdullateef Olanrewaju[1] and Lim Tat Lee[1]

[1]*Department of Construction Management, Faculty of Engineering and Green Technology, Universiti Tunku Abdul Rahman, 31900 Kampar, Malaysia*

Abstract. Heritage buildings represent the tangible cultural heritage of a community. However, many of the heritage buildings have being left neglected. Ipoh as a city rich in heritage, has many dilapidated heritage buildings which are experiencing a resurging interest. However, the problems faced by many owners are the lack of technical information of such buildings, leading to premature abandonment and demolition. Therefore, the objectives of this study are to determine the types and extent of defects present in the building studied. To achieve these objectives, a case study of a century old heritage property in Ipoh, Perak was conducted. The study aims to provide reference to owners and those responsible for the conservation of heritage buildings with similar conditions to identify and prioritize critical defects in relation to the building life span to determine its condition. The findings determined the probable causes of defects such as settlement and façade cracks, which are over 30 years old, were attributed to leaking plumbing pipes, rainwater ingress and the construction of an adjacent 20 story apartment building. The major issues to address were stabilizing the foundation through cement grouting, reinforcing the existing structural systems and roof systems as well as arresting the decay of timber floor structure. In conclusion, major maintenance guidelines are need to address structural issues and weather tightness of the building envelope, especially its roof and drainage systems.

1 Introduction

Cultural heritage constitutes the practices and objects passed on from previous generations. Building as an object of built heritage represents the historical evidences and values of the local community [1]. Malaysia is endowed with a number of heritage buildings in many state capitals that serve as sources of livelihood and prestige to the owners and nation. While some of the heritage buildings are declared 'national heritage', a number are owned by private individuals or associations which are yet to be registered but contribute no less to the heritage and tourism landscape [2]. These notable building typologies include the shop houses of Southeast Asia and were used for various purposes including education, residences, offices, community halls, businesses and warehouses [3]. However, many of these heritage shop-houses have being left neglected, leading to scores of dilapidations and abandonment. This represents an economic loss to various stakeholders [4]. Ipoh proliferates with many heritage sites as it was a tin mining center of the world and capital of Perak state, besides being

[a] Corresponding author : tsyeow@utar.edu.my

the fourth largest city in Malaysia. Since the great fire of 1892, the original part of Old Ipoh city was rebuilt with brick buildings and clay roof tiles. The Chinatown part of Old Ipoh was concentrated on Leech Street [5]. However, Ipoh is experiencing a resurgent interest in heritage buildings and as part of a private led conservation effort; a case study on one of the century old heritage buildings along the intersection of Leech Street and Patrick Street was commissioned. The problem faced by many owners was a lack of technical based studies conducted to understand the types of physical defects present in such buildings and to conclude with appropriate solutions. The objective of this study is to identify the types and extent of defects as well as appraise the causes of these defects. The general significance of the study is that, by understanding the causes of defects through the life cycle of the building, a general guideline for heritage buildings maintenance can be framed for these buildings to be usable instead of being demolished for redevelopment. This is significant to provide reference to neighboring owners and those responsible for the conservation of heritage buildings with similar conditions to identify and prioritize critical defects in relation to the building life span to determine its condition.

2 Literature Review

The awareness to conserve Ipoh's heritage is increasing due to its rich mix of colonial era heritage buildings comparable with the straits settlements of Penang, Melaka and Singapore [5]. Old Ipoh town is currently experiencing a resurgence in interest due to the influx of outstation tourist as Perak is targeting to attract 5 million tourist in 2015 [6] due to its unique 'sense of place'. The preservation of the colonial era buildings were mainly due to the Rent Control Act that discouraged the redevelopment of the buildings from 1966 until its repeal in 1997 [7]. While the city council has guidelines on the conservation of these buildings, many building owners have minimal understanding of maintaining the authenticity of the building fabric and practically demolished the entire building except the front façade wall. The lack of specialized training and knowledgeable consultants and contractors probably contributed to the 'demolition approach' as they are more familiar with conventional work methods. However, the definition of heritage building is to retain its authentic character as evidence to a site's history and the minimal intervention approach is desired [4].

However, due to the inability of building owners to extensively repair a building while it is in use, as this would disrupt business operations, many such heritage buildings' ground floor would be functional while the first floor would be left abandoned once it was deemed unsafe to use. Old Ipoh mainly suffers from poor building conditions, high vacancy rate and low building utilization, as well accessibility. This depresses the rental rates, property prices and the variance may depend on different streets [5, 7]. This raises the question of finding equilibrium between maintaining the authenticity of the materials and introducing new renovations in order to allow the continual use of the building as a building left vacant would invite much more problems. Generally, local heritage conservation lacks a holistic heritage maintenance planning [8].

In Malaysia, the thrust of heritage conservation of many of the nation's landmark buildings were undertaken by the public sector. However, the funds are limited for preventive measures to maintain all relevant buildings [9] as well as the lack of public awareness of internationally recognized guidelines such as the Burra Charter [10]. The public is now starting to be aware of the value of heritage buildings as a limited resource that forms part of the urban scenery as well as a tourism landmark and attraction [11]. This is where the private owners come in but they require the right guidance and approach to their property conservation independent of the oversight of the National Heritage Department [10]. The introduction of the National Heritage Act 2005 is a step in the right direction though it remains to be seen if building owners perceive it to be so. The potential is the private owners unlocking their financial investment in a manner that is sustainable and relieves the dependency of public funding. This requires an understanding of how the building fabric that is well conserved can impart its value in attracting businesses and customers and to reinvigorate the urban life. In Malaysia the model closely resembles the BCO (Build-Conserve-Operate) model [12]. However, most conservation management lacks the skill to interrogate the building's history due to the lack of archival materials. Most of the heritage conservation literature focus on the organizational management, programmatic and awareness approach but the technical application and building survey approach is lacking [13]. Therefore, this study is an initial investigation towards closing this gap.

3 Research Method

The study includes desktop sourcing of literature and interview with the neighboring shop owners who are in their 80s. The research method for the study includes a building survey approach: (a) taking dimensional scaled photographs of all building elevations, (b) to record building dimensions, (c) observation survey of ground level condition and estimate the extent of settlement, (d) dilapidation survey by visual observation, (e) determine building elements that have been renovated, and (f) observation of site context that may attribute to building condition. The case study process also requires an understanding of, (i) age of the building, (ii) the construction of historical buildings and the materials used, (iii) interpretation of the architectural style, (iv) historical record on the urban development, (v) introduction of modern utilities, (vi) history of defects appearing, and (vii) history of its functional use. This building survey study will serve as a guideline to manage the building maintenance for modern use.

4 Findings and Discussion

The building studied (Figure 1) is at the intersection of Patrick and Leech Street which faces the Kinta Height apartment block (Figure 2) that used to be the town market before it was relocated. The building is classifiable as Edwardian architecture as it was constructed between 1900 and 1910.

Figure 1. Case study building at Patrick Street.

Figure 2. Kinta Heights apartment.

However, the original owner of the building as well as its original function could not be verified from historical records. From the interview, the building was used as a furniture shop before being converted to a school and later for retail of cooking wares with the first floor being used as store for about 40 years until 2014 where it was vacated. It was a period where plumbing technology was introduced but before connected sewer was established as the architecture utilized the 'bucket outhouse' system. Its original function was most probably designed for warehousing due to its internal riveted structural steel system, span and height as well as rear courtyard which is a different typology from the conventional shop houses. Table 1 shows the estimated period of major failures experienced by the building. The clay roofing was possibly replaced in the 1970s with asbestos roofing sheets. Building settlement were reported during the 1980s when the internal plumbing had to be replaced and the construction of Kinta Heights began. The building also has systemic cracks on its arches and ground floor slabs which had a 'saucer' shaped settlement and which is part of a larger eccentric settlement that most likely occurred during the construction of the 20 storey Kinta Heights. The building cracks are dormant but the overall crack pattern on the wall, edges of openings and the floor seem to indicate a horizontal soil movement occurring on the façade side facing Kinta Heights. Other major defects include the failure of 1st floor timber balcony due to exposure to moisture and subsequent decay. The floor slab in the corridor has sunken between 75-80mm (Figure 3) on the exterior and interior (Figure 4) of the building and this leads to a visible 'bow' of the building façade (Figure 5).

Table 1. Estimated period of element failure.

1903-1970	Likely failure of terracotta barrel roof tiles, replaced with asbestos roof. The ground floor fenestration renovated to folding door for commercial use.
1903-1980	Likely failure of plumbing with leaks causing floor settlement. Replaced with exposed piping.
1982-2014	Likely crack defects start on walls due to settlement from the construction of the adjacent Kinta Heights apartment block. Excessive loading and rain water ingress contributed to decay of timber elements.

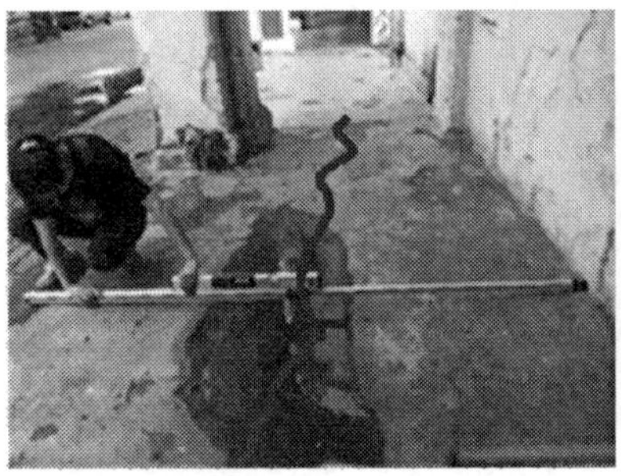

Figure 3. Settlement and water ponding

Figure 4. Interior floor crack.

Figure 5. Façade bowing effect.

The areas with most settlement also coincide with the ingress of the plumbing pipe from the public main into the building. This pipe has since been replaced with and aboveground piping route, indicating that the previous pipe had leaked and this can cause soil erosion. The building's adjacent location to Sungai Kinta (150m) may also contribute to the issue due to the fluctuation of water table levels during monsoon seasons which leads to annual flash flood. The primary concern is to ensure that the foundation is stabilized and further settlements are prevented and any causes of soil erosion are addressed. The crack patterns found on the arches and piers are part of the pattern of building envelope stress that forms, especially on a brick load bearing construction that does not cater to tensile stresses due to eccentric settlement. Other defects relate to moisture ingress through rising damp (failure of water barrier) and falling damp (failure of roof drainage). The building is deemed in good condition as most of the the lime plaster is not delaminating, nor the majority of the timber structure decaying or attacked by termites, indicating overall water tightness for the preservation of the building elements as well as effective ventilated drying of moisture.

4.1 Proposed solutions

Soil samples of the site are to be taken to determine the suitable of using cement grouting for foundation remediation. The benefits of this method are to avoid the need for excavation, cement the non-cohesive soil particles to prevent erosion and to improve the bearing capacity of the soil as well as to seal the adjacent public drain from future leaks. To further ensure that future problems are avoided, all plumbing pipes should be installed above ground and visible for maintenance. As the building is to retain its structural integrity, it is recommended that minimal destruction of its original envelope be permitted as it acts as a complete load bearing system. Instead, it requires future spatial needs to be adapted to its layout and this requires that the floor system must be upgraded and reinforced for modern fire compartmentalization standards and loadings. The introduction of new walls on the first floor is to be avoided unless they are lightweight partitions. Of concern are the balcony concrete slabs exposed to rain. A major failure for these floor systems over a century is that the timber battens that support the clay tiles and concrete topping above will sag over time due to water induced decay. This is a result of water seeping through the cracked concrete topping slab. Overtime, the previous owner abandons any repairs and the moisture seeps to the timber joists which in turn rots, loses strength and further sags, pulling at the adjacent brick walls, causing cracks. In such cases, the retention or demolition of such slabs to be replaced with reinforced concrete slabs must be considered in the context of future loading, water tightness and the necessity of historical authenticity.

At the ground floor of the building party wall, rising damp was detected along the entirety of the envelope. A chemical injection method is recommended for sealing off the ground water rising through capillary action. An alternative method is to allow for the rising damp to naturally ventilate and dry up to the average height is 4'(1200mm). Any repairs to the plaster and paint are to use compatible technologies such as lime plaster and lime paint to ensure the 'breathability' of the finishes. As the original buildings do not consider the integrated design of plumbing and drainage, it is recommended that all future supply of water and drainage be kept to the rear perimeter of the building to avoid introducing any moisture to the interior. All piping ingress and egress to utilize existing openings to avoid damage to existing strip foundation. The final source of damp is the zinc roof gutter laid parallel to the party wall which had, over a century, rusted with rainwater seeping to the interior, causing a portion of the first floor and ceiling timber to rot over time. The roof is currently using asbestos roofing sheets as the roof was replaced in the early 1970s and poses a health hazard. It is recommended to be replaced with clay roof tiles that has good insulation properties and allows the roof system to ventilate while retaining the originality of the architecture.

5 Conclusion and Recommendation

This study has revealed the nature and extent of building defects which have profound implications on its choice of maintenance solutions. In order to reduce the cost of maintenance, other buildings of a similar nature must consider these guidelines: (a) the foundation and structural issues needs to be addressed sensitively by remediation and reinforcements methods (b) the maintenance and upkeep of the main roof system and especially the roof gutter, sealing off any exposed balcony against further infiltration of rainwater by introducing a layer of fenestration, (c) the compatibility of modern construction technologies with the existing building, (d) avoid covering up by ceilings or over-cladding which will prevent visual monitoring especially on the connecting joints of different building components, (e) prevent embedded plumbing and electrical utility, (f) maintenance of drainage system, (g) prevent major hacking of load bearing sections which results in stressing the building envelope, (h) for newly introduced structures to avoid point loading design but to spread the load over a larger area and (i) upgrading of the 1st floor timber floor as well as the design of the staircase suitable for modern fire prevention and escape requirements. A suitable policy by the fire department is required to allow for heritage buildings to retain their authenticity while mitigating fire risk.

References

[1] Y. Ahmad, The scope and definitions of heritage: from tangible to intangible, International J. of Heritage Studies, **12**(3), 292-300, (2006).

[2] C.L. Cartier, Conserving the built environment and generating heritage tourism in Peninsular Malaysia, Tourism Recreation Research, **21**(1), 45-53, (1996).

[3] W.H Wan Ismail and S. Shamsuddin, The old shophouses as part of Malaysian urban heritage: The current dilemma, *Proc. of the 8th International Conference of the Asian Planning Schools Association*, (2005).

[4] S.N. Harun, Heritage building conservation in Malaysia: experiences and challenges, Procedia Engineering, **20**, 41-53, (2011).

[5] W.W. Lim, Hew-David, Y.K. Tong-Gerald and G.G. Goh, Revitalisation of the old township of Ipoh, Malaysia, J. of Place Management and Development, **7**(1), 57-73, (2014).
[6] Ipohecho, Perak Targets Five Million Tourists, (2015), <available at: http://www.ipohecho.com.my/v3/article/2015/02/01/perak-targets-five-million-tourists>
[7] S. Atumi, The repeal of rent control in Malaysia, Cornell Real Estate Review, **2**, 29-38, (2003).
[8] A.F. Mohd-Isa, Z. Zainal-Abidin and A.E. Hashim, Built heritage maintenance: A Malaysian perspective, Procedia Engineering, **20**, 213-221, (2011).
[9] R. Talib, A.G. Ahmad, N. Zakaria and M.Z. Sulieman, Assessment of factors affecting building maintenance and defects of public buildings in Penang, Malaysia, Architectural Research, **4**(2), 48-53, (2014).
[10] R. Abdul Rashid and A.G. Ahmad, The implementation of maintenance works for historical buildings- a review of the current scenario, Procedia Engineering, **20**, 415-424, (2011).
[11] N.F. Nik-Azhari and E.Mohamed, Public perception: Heritage building conservation in Kuala Lumpur, Procedia Social and Behavorial Sciences, **50**, 271-279, (2012).
[12] S. Macdonald, Leveraging heritage: Public-private and third sector partnership for the conservation of the historic urban environment, *ICOMOS Conference*, Paris (2011).
[13] R. Abdul Rashid and A.G. Ahmad, Overview of maintenance approaches of historical buildings in Kuala Lumpur- a current practice, Procedia Engineering, **20**, 425-434, (2011).

MATEC Web of Conferences **47**, 04004 (2016)
DOI: 10.1051/matecconf/20164704004

Accidents Preventive Practice for High-Rise Construction

Kai Chen Goh[1,a], Hui Hwang Goh[2], Mohd Faizal Omar[3], Tien Choon Toh[4] and Abdullah Asuhaimi Mohd Zin[5]

[1]*Faculty of Technology Management and Business, Universiti Tun Hussein Onn Malaysia, 86400 Batu Pahat, Johor, Malaysia*
[2]*Faculty of Electrical and Electronic Engineering, Universiti Tun Hussein Onn Malaysia, 86400 Batu Pahat, Johor, Malaysia*
[3]*School of Quantitative Sciences, Universiti Utara Malaysia, 06010 Sintok, Kedah, Malaysia*
[4]*Faculty of Engineering and Science, Universiti Tunku Abdul Rahman, 53300 Kuala Lumpur, Malaysia*
[5]*Faculty of Electrical Engineering, Universiti Teknologi Malaysia, 81310 Skudai, Johor, Malaysia*

Abstract. The demand of high-rise projects continues to grow due to the reducing of usable land area in Klang Valley, Malaysia. The rapidly development of high-rise projects has leaded to the rise of fatalities and accidents. An accident that happened in a construction site can cause serious physical injury. The accidents such as people falling from height and struck by falling object were the most frequent accidents happened in Malaysian construction industry. The continuous growth of high-rise buildings indicates that there is a need of an effective safety and health management. Hence, this research aims to identify the causes of accidents and the ways to prevent accidents that occur at high-rise building construction site. Qualitative method was employed in this research. Interview surveying with safety officers who are involved in high-rise building project in Kuala Lumpur were conducted in this research. Accidents were caused by man-made factors, environment factors or machinery factors. The accidents prevention methods were provide sufficient Personal Protective Equipment (PPE), have a good housekeeping, execute safety inspection, provide safety training and execute accidents investigation. In the meanwhile, interviewees have suggested the new prevention methods that were develop a proper site layout planning and de-merit and merit system among sub-contractors, suppliers and even employees regarding safety at workplace matters. This research helps in explaining the causes of accidents and identifying area where prevention action should be implemented, so that workers and top management will increase awareness in preventing site accidents.

1 Introduction

Construction site is a work place which exist a lot of hazardous activities and accidents [1]. The safety and health issues in construction industry need to be pay more attention due to the frequently development of high rise building in the country. Every year, a high number of accidents were happen at building construction site especially at the high rise building construction site [2]. Many of the unpredictable hazards and accidents are available at the construction site and thousands of workers were injured and killed in every year [3]. The most common accidents at high-rise building

[a] Corresponding author : kaichen@uthm.edu.my

construction site are workers fall from height and injuries caused by fallen objects [3]. According to the National Fire Incident Reporting System (NFIRS), high rise building can be classified into four ranges which are 7-12 stories 13-24 stories, 25-49 stories, and 50 stories or more [4]. In general, 75 feet (23 meter) or seven stories will be the cut-off point for the most purpose [4]. In year 2011, 4,937 cases of accidents were reported in the construction industry and it was the second highest record of accidents reported among all the industries [5]. In the same year, there are total 9,057 cases of people falling from height accidents and 4,689 cases of struck by falling object accidents occurred [5]. The figures, which are based on the statistics of Social Security Organization, are considered high when compared to the amount of the existing workforce in Malaysia. This study is focus to identify the causes of accidents at high-rise building construction site and to identify the preventive measures for accidents at high-rise building construction site. Besides that, this research helps in explaining the causes of accidents and identifying area where prevention action should be implemented, so that workers and top management will increase awareness in preventing site accidents.

2 Literature Review

There are four types of high-rise accidents, which are by scaffolding, people fall from height, struck by falling object and plant and machinery. The workers are always exposed to the risk of collapse of the scaffolding. Major of the scaffold accidents occurred were due to the use of defective materials for scaffolding and coupled with the unskilled and careless workmanship in erection of scaffolds [6]. In construction industry, falls is not only the most fatal accidents but it also is the most frequent occurred accidents [7, 8]. Everybody in the construction site has the risk to expose to fall in anywhere and anytime especially at the higher level [9]. In general, lack of the safety measure at the construction sites is one of the causes the occurrence of fall accidents [2]. Struck by falling object accidents can be defined as people who work at construction site is struck by the equipment, private vehicles, falling materials, vertically hoisted materials and horizontally transported materials [7]. The improper rigging method had caused the accident happen [10]. Overloading is one of the factors that will cause the cranes collapse in the high-rise building construction. During the operation, the amount of allowable handling load by the crane is always not proper control by the supervisor [11]

Accident is an event of unpredictable and it may occur due to the following causes, lack of training, improper equipment and working platform, wrong safety attitude, inadequate housekeeping, failure to use personal protective equipment (PPE), and problem procurement method and sub-contracting method. The lack of training can caused accidents at construction site [12]. Because of the lack of training in safety and technology knowledge, workers are haven't ability and sufficient knowledge to predicts the potential risk and the way to avoid the accidents [13]. The use of unsafe working platforms also may put workers at risk when the equipment is not properly used, maintained or stored [14]. Construction worker's safety attitude is influence by their understanding and realizing of risk, management, safety rules and the working procedures [15]. The unsafe actions are include do not follow the standard safety procedures, constructing barbarously and deciding to proceed work in an unsafe conditions. [13]. The poor housekeeping in the workplace can be considered as a risk factor for occupational injuries [16]. Working without wearing any PPE may highly increase the probability for occurrence of any undesired accident. The various reason of workers refuse to wear PPE during working are such as feel uncomfortable with the gears while performing their job at site and consider it as an disturbing item to their work output. According to Ali, et al. [17], the International Labour Organization has revealed that some of the workers felt uncomfortable while wearing any types of PPE and it indirectly decreases their work performance. Sub-contractors usually have poor safety awareness at the construction site. Poor coordination, lack of proper instructions and misunderstanding between working trades all can lead to construction accidents [13].

According to Occupational Safety and Health Act 1994, section 24 [18], the employees are responsible to wear or use at all the times of any protective equipment or clothing which provided by the employer. Good housekeeping is the best defenses against accidents during construction [19]. Therefore, good housekeeping is an important part in any construction site which help to reduce occurrence of accidents and improve the overall safety performance [20]. Inspection is one of the common forms of analysis to prevent any accidents. . Inspections are part of a preventive or proactive to accident prevention [19]. Training is a necessary and important part of accident prevention policy [21]. Training is a necessary and important part of accident prevention policy [21]. Each of the company should ensure that the training program is an integral part of the instruction given to all men in construction work and make sure them particular attention towards to the safety issues [19]. The nature of the construction industry's rapidly changing conditions, associated work hazards, and the characteristics of construction organizations further aggravate the situation [22]. Therefore, there are necessary to implement the appropriate prevention action to prevent the occurrence accidents effectively.

3 Methodology

This research is conducted by qualitative methods only. Data which collect from the primary source which are semi-interview were analyzed with the exploratory data analysis method According to Silverman [23], the important most things in the method of qualitative analysis are focused in the documents, transcripts, diagrams and text that make up the data. This research is choosing to adopt qualitative content analysis method. Content analysis involves establishing categories and then counting the number of instance when those categories are used in the particular item of text. Matrix table is used to assists in analyzing the interview information data. The collected data were analyzed at next stage to rearrange the data in form of easy to understand and clear to read. Discussion has been done to determine the research objectives successfully achieved. At the end of research, recommendation and conclusion were developed.

4 Data Analysis and Discussion

The findings presented in the research provide basic and justification for the conclusion to be drawn in the study. Therefore the findings have to be meticulously analyzed and explained in a logical organized manner. Using tables for the data display analyze the obtained data. The data were divided into three sections. Table 1 is about the perception of interviewees on the types of at high rise building construction site, Table 2 is about the perception of interviewees on the causes of accidents at high rise building construction site and Table 3 is about ways to prevent accidents at high rise building construction site. Semi-structured interviews were carried out with safety officers who were involved in different types of high-rise building projects in Malaysia. There are 7 interviewees who come from 3 companies involved in this semi-structured interview. They are involved in different types of high-rise buildings, which consist of services apartments, office blocks and condominiums.

Table 1. Perception of types of accidents at high-rise building construction site.

Perception	Annotations
Falling from height	• *"Workers are less awareness that they are working at high condition and common Personal Protective Equipment (PPE) not sufficient provided or equipped properly. Besides that, unsafe workplace condition that caused by poor housekeeping will cause workers slip and falling from height. In addition, poor housekeeping also will cause workers hit by falling working tools." (R1, R2, R3 and R5)*
Struck by falling object/working tool	• *"There is not enough guidance for workers regarding the safety matters at construction site." (R4)* • *"Government rules and regulations are not enough strong to control the safety and health issues at construction site. Besides that, it should be executed seriously to avoiding the unsafe action at construction site." (R6 and R7)*

From the results on the collected data from interview with 7 safety officers who conducted different projects on high rise building in Kuala Lumpur, all the interviewees have the same perception that the most frequently happened accidents at high rise building construction site were people falling from height accidents and accidents due to struck by falling object or working tools. According to the above annotation the highest accidents that occur at construction site in Malaysia are workers falling from height followed by accidents due to struck by object falling from height. Therefore, according to the past report and the existing interview collected data prove that the most frequently occurred accidents at high rise building construction site in Malaysia is people falling from height followed by accidents due to struck by object falling from height.

Table 2. Perceptions of major causes of accidents at high rise building construction site.

Question	Perception	R1	R2	R3	R4	R5	R6	R7
Major causes of accidents	Unsafe acts (workers' attitude)	√		√			√	
	Unsafe condition (poor housekeeping)	√					√	
	Unskilled workers		√		√	√		
	Lack of training		√		√		√	√
	Poor site safety management		√					√
	Cost saving			√	√			
	Less awareness of hazardous activity					√		√

Different perception of interviewees regard the major causes of accident were due to the unsafe acts at construction site. Worker's unsafe acts are always due to operating machinery without qualification or authorization, horseplay, operating equipment at unsafe speed and lack of or improper use of Personal Protective Equipment (PPE). According to R1, R2 and R6, they believe that worker's unsafe acts caused unsafe condition exists at construction site. Generally, unsafe condition is a condition in the workplace that is likely to cause property damage or injury. Unsafe condition is normally due to poor housekeeping, defective tools and equipment and congestion in the workplace. "Most of the occurred accidents were due to workers who lack training. Their safety knowledge and working skill are not sufficient in handling their task. (R2, R4, R6, R7). According to the experiences of the interviewees, most of them agree that worker who lacks training is one of the major causes of accidents which occurred at high rise building construction site. Workers who lack training or unskilled workers had not been enough qualification to operate certain machinery or working tools, therefore unsafe condition workplace do exists. Besides that R2 and R7 stated that not only worker who lack training is the major cause but also the poor site safety management. Safety officer and project team who do not have safety awareness will cause poor site safety management available whereas R5 and R7 agree that both employees and employer are less awareness of hazardous activities which exist anytime at the construction site. "....and some contractors are trying to save project cost in providing sufficient PPE and training for the workers." (R3 and R4). Other than that, R3 and R4 do experience that one of the major causes of accidents at high rise building construction site is cost saving by employer. The cost for preparing safety training and PPE are extremely expensive, therefore safety training and PPE might insufficient provided. The causes of accident at high rise building construction site not only due to lack of training, improper equipment and working platform, wrong safety attitude and inadequate housekeeping but also caused by cost saving by contractor, poor site safety management and less awareness of hazardous activities workers and employers.

All the interviewees have been implemented several alike of accidents prevention methods that were provided regular training to workers regarding safety awareness at construction site, accidents investigation and safety inspection. Besides that, interviewees from company A's also implemented accidents prevention method by giving instrument for new workers, daily toolbox, safety introduction awareness campaign and having site monthly meeting to follow up the safety matters at construction site. According to R5, they not only implemented the above prevention method but also implemented by providing regular training to management level and supervision team regarding safety awareness at construction site. Besides that, R7 also have been implement site monthly meeting to prevent any unsafe act and unsafe condition at construction site.

Table 3. Type of accidents prevention methods used by company.

Type of used prevention method	Annotations
Regular training to workers regarding safety awareness at construction site Accidents investigation Safety Inspection	• *"Basically we are providing regular training to workers regarding safety awareness at construction site, execute accident investigation and implement safety inspection." (All interviewees)*
Instrument for new workers Daily toolbox Regular training to management level and supervision team regarding safety awareness at construction site Safety introduction awareness and campaign Site monthly meeting	• *"In addition, our company also provide instrument for new workers, implement daily toolbox checking, regular training to management level and supervision team regarding safety awareness at construction site, safety introduction awareness campaign site monthly meeting." (R1, R2 and R3)* • *"Not only training for workers but also to management level and supervision team." (R5)* • *"In controlling the safety matters, we are having meeting every end of the month."*

5 Conclusion

This study has provided an overview on the causes of accidents and preventive measures at high-rise building construction site. The causes of accidents at high rise building were due to workers lack of training, unskilled workers, workers unsafe acts, unsafe condition, poor site safety management and less awareness of hazardous activities at work site. In addition, implement de-merit and merit system among sub-contractors, suppliers and even employees regarding safety at workplace matters and develop a proper site layout planning as one of the prevention method at high rise building construction site. As a summary, although some preventive measures have been result in the reduction of the number of accidents, but there is still no room for satisfying in view of high fatality. In addition further research should focus on the audition and inspection of safety at construction site to ensure the effectiveness of accidents prevention methods during project operation. Therefore, it is important that all parties to cooperate to implement the preventive measures in dealing with safety matters at high-rise building construction site.

Acknowledgment

The authors express gratefully acknowledgment to the industry stakeholders for their valuable contributions towards the success of this research. This research is fully funded by Universiti Tun Hussein Onn Malaysia (UTHM) and Ministry of Higher Education (MOHE) under Vot: 1521, FRGS on Flood Disaster Management 2015 and Vot 1445, Research Acculturation Collaborative Effort (RACE) Phase 2/2013.

References

[1] T. Dėjus, Accidents on construction sites and their reasons, *Proc. of the 9th International Conference: Modern Building Materials, Structures and Techniques*, 241-247, (2008).

[2] F.K. Wong, A.P. Chan, M.C. Yam, E.Y. Wong, T. Kenny and K.K. Yip, Findings from a research study of construction safety in Hong Kong: Accidents related to fall of person from height, Journal of Engineering, Design and Technology, **7**, 130-142, (2009).
[3] O. Murty, B. Chung, L. Yin and T. Loo, Pattern of injuries in fatal accidents of construction workers: A retrospective study of 10 years (1996-2005), The Malaysian Journal of Forensic Pathology and Science, 44, (2006).
[4] J. R. Hall, *High-Rise Building Fires*, One-Stop Data Shop, Quincy, Massachusetts, (2005).
[5] Social Security Organisation, Anual Report 2011, Social Security Organisation Malaysia, (2011).
[6] J.C. Rubio-Romero, M. Rubio and C. García-Hernández, Analysis of construction equipment safety in temporary work at height, Journal of Construction Engineering and Management, (2012).
[7] J. Hinze, X. Huang and L. Terry, The nature of struck-by accidents, Journal of Construction Engineering and Management, **131**, 262-268, (2005).
[8] D. Zhou and Q. Pang, Study on prevention measures of falling accidents during construction, *ICPTT 2012*, 1494-1501, (2012).
[9] Y.H. Lin and C.Y. Chen, Statistical analysis of fatal occupational falls in the Taiwan construction industry from 1996-2007, *Proc. of the 40th International Conference on Computers and Industrial Engineering*, 1-4, (2010).
[10] W. Wu, H. Yang, Q. Li and D. Chew, An integrated information management model for proactive prevention of struck-by-falling-object accidents on construction sites, Automation in Construction, **34**, 67-74, (2013).
[11] D.J. Edwards and G.D. Holt, Case study analysis of construction excavator H&S overturn incidents, Engineering, Construction and Architectural Management, **17**, 493-511, (2010).
[12] A. Gibb, S. Hide, R. Haslam, S. Hastings, A. Suraji and A. Duff, Identifying root causes of construction accidents, Journal of Construction Engineering and Management, **127**, 348-349, (2001).
[13] Z. Chen and Y. Wu, Explaining the causes of construction accidents and recommended solutions, *Proc. of the International Conference on Management and Service Science*, 1-5, (2010).
[14] S. Chi, S. Han, and D. Kim, The relationship between unsafe working conditions and workers' behavior and their impacts on injury severity in the U.S. Construction Industry, Journal of Construction Engineering and Management, (2012).
[15] S. Mohamed, T.H. Ali and W.Y.V. Tam, National culture and safe work behaviour of construction workers in Pakistan, Safety Science, **47**, 29-35, (2009).
[16] V.M. Dufort and C. Infante-Rivard, Housekeeping and safety: An epidemiological review, Safety Science, **28**, 127-138, (2001).
[17] A. Ali, S. Kamaruzzaman and G. Sing, A Study on causes of accident and prevention in Malaysian construction industry, Journal of Design + Built, **3**, 95-104, (2010).
[18] Occupational Safety and Health Act 1994, 514, (2002).
[19] P. Hughes and E. Fereett, *Introduction to Health and Safety in Construction*, Elsevier, United Kingdom, (2007).
[20] N.V. Schwatka, L.M. Butler and J.R. Rosecrance, An aging workforce and injury in the construction industry, Epidemiologic Reviews, **4**, 156-167, (2012).
[21] B. Hare and I. Cameron, Site manager safety training, Engineering, Construction and Architectural Management, **18**, 568-578, (2011).
[22] R.M. Choudhry and D. Fang, Why operatives engage in unsafe work behavior: Investigating factors on construction sites, Safety Science, **46**, 566-584, (2008).
[23] D. Silverman, *David Silverman Interpreting Qualitative Data*, SAGE, India, (2011).

MATEC Web of Conferences **47**, 04005 (2016)
DOI: 10.1051/matecconf/20164704005

Assessment of Outdoor Thermal Comfort and Wind Characteristics at Three Different Locations in Peninsular Malaysia

Mohd Hafizal Hanipah[1,a], Abd Halid Abdullah[1], Nor Azwadi Che Sidik[2], Riduan Yunus[1], Mohd Nor Azam Yasin[1] and Muhammad Noor Afiq Witri Muhammad Yazid[2]

[1]*Faculty of Civil and Environmental Engineering, Universiti Tun Hussein Onn Malaysia, 86400 Batu Pahat, Johor Malaysia*
[2]*Faculty of Mechanical Engineering, Universiti Teknologi Malaysia, 81310 Skudai, Johor, Malaysia*

Abstract. Urbanization and rapid growth in construction have led to the problems of global warming and urban heat island throughout the world. In order to reduce these problems particularly in hot and humid climatic region, a research on current level of outdoor thermal comfort and wind characteristics based on the local weather conditions around Malaysia should be conducted. This paper reports on the analysis of outdoor thermal comfort level at hottest temperatures and wind characteristics at three locations in Peninsular Malaysia by using hourly climatic data recorded by Malaysian Meteorological Department (MetMalaysia). The level of outdoor thermal comfort was assessed based on the Universal Thermal Climate Index (UTCI). The results showed extreme heat stress conditions have occurred at Alor Setar, Kuantan, and Subang with UTCI values of 51.2°C, 49.7°C, and 49.0°C respectively taking into account only temperature data from the year 2012 to 2014. However, for 20 years data from 1994 to 2014, the calculated UTCI also showed extreme heat stress conditions with their respective values of 49.6°C, 43.8°C, and 49.7°C for Alor Setar, Kuantan, and Subang respectively. Meanwhile, the hourly mean wind speed for three years data at Alor Setar, Kuantan, and Subang, were 1.70m/s, 1.69m/s, and 1.63m/s respectively. The highest mean wind speed of 11.6m/s was observed at Subang, while no wind movement (i.e. 0m/s) was considered to be the lowest hourly wind speed for all three locations. The observed prevailing wind direction for all the three locations was from the north (0°). It can be concluded that Peninsular Malaysia is generally facing extreme heat stress problem due to unfavourable climatic conditions.

1 Introduction

Global warming and urban heat island have become major concerns in developed and developing countries throughout the world [1-3]. Inevitably, Malaysia has also been facing this heat stress and thermal comfort problems [4-6]. The yearly mean temperature for Central Peninsular Malaysia for the period of 1969-2014 shows an increment of 0.35°C per decade [7]. The warmer urban climate may have some negative impacts towards people outdoor thermal comfort in Malaysia. Major problem such as urban heat island can directly affect the outdoor human thermal comfort level. Currently in

[a] Corresponding author : hafizal@uthm.edu.my

Malaysia, there is very limited research being conducted on these issues with regard to building construction and urban development. It is even more critical as the building arrangement parameters such as packing density (λ_p), frontal area density (λ_f), height-to-width ratio (λ_s) etc [2,8-10] have not been considered during earlier stage of urban design. Therefore, the primary aim of this research study is to determine current extreme outdoor thermal comfort index and wind characteristics. For future studies, factors affecting outdoor thermal comfort in urban areas, and dominant building arrangement parameters in hot humid climate will be examined.

2 Study Area

In general, Malaysia is a South East Asian country which is located in the equatorial zone. The geographic coordinate is 2°30' in the north latitude and 112°30' in the east longitude. Throughout the year, Malaysia experiences a wet and humid condition with daily temperature ranging from about 24°C to 38°C [7]. Wind speed and wind direction across the Peninsular Malaysia are influenced by the monsoon seasons, namely southwest monsoon, northeast monsoon and two short inter-monsoons.

Figure 1. Map of Peninsular Malaysia showing the case study stations.

The wind speed range is below 8m/s during southwest monsoon season, and 5m/s to 10m/s for northeast monsoon season. While during inter-monsoons, the wind is generally light and variable. Malaysia has high humidity where the monthly mean relative humidity is ranging from 70% to 90% throughout the year varying from different places and times. In addition, Malaysia receives about 6 hours of sunshine per day on average [7]. For this study, three different locations or cities around Peninsular Malaysia were chosen; which include Alor Setar (i.e. Northern region), Kuantan (i.e. East Coast region facing the South China Sea), and Subang (i.e. Central region) as shown in Figure 1 and stated in Table 1. The weather data representing the hot and humid tropical climate of Malaysia from three principal weather stations namely Alor Star, Kuantan, and Subang were utilized to investigate outdoor thermal comfort conditions and wind characteristics of the three respective locations mentioned earlier.

Table 1. Location of wind speed stations used in the study.

Station	Latitude	Longitude	Height above sea level (m)
Alor Setar	6° 12' N	100° 24' E	3.90
Kuantan	3° 46' N	103° 13' E	15.20
Subang	3° 08' N	101° 33' E	16.60

3 Methods

The hourly data for wind speed and direction within the three years period of 2012-2014 were obtained from the MetMalaysia. The hottest temperature of the day for the past 20 years from 1994-2014 were also obtained and analyzed. At all the selected weather stations, wind speed and direction were measured using the anemometer and wind vane which were installed at the height of 10 m. The instrument used for measuring the surface wind speed is rotating cup type anemometer where the cups are mounted symmetrically at right angle to a vertical shaft. Installing the anemometer at 10m height is in accordance to ref. [11], which is basically for climatological and practical reason. It has been agreed that this elevation should be the standard meteorological reference level. The data on wind speed and direction were taken at 1 hour interval and stored in a data logger. The hourly wind speed and direction data for three years were analyzed in order to investigate the wind characteristics. The hottest ambient temperatures were analyzed and calculated to get UTCI values. In this study, UTCI was calculated by a program based on regression equation which is devised by International Society of Biometeorology Commission 6 by COST (European Cooperation in Science and Technology) Action 730 under the umbrella of the WMO Commission on Climatology [12]. The weather parameters, which include value of ambient temperature (°C), solar radiation (W/m^2), relative humidity (%), and wind speed (m/s) are required for the calculation of UTCI (°C) [13].

4 Universal Thermal Climate Index (UTCI)

The UTCI is divided into 10 different stress levels which comprise no thermal stress, 4 heat stress levels, and 5 cold stress levels [14]. UTCI is stated as the equivalent ambient temperature of a reference environment that causes the same physiological response of a reference person as would the actual environment [15]. It is based on Fiala multi-node model [16] of human thermal regulation in combination with an adaptive clothing model. UTCI is appropriate for thermal assessments in all climates and seasons, and on any scale [12].

5 Results and Discussion

The determination of outdoor thermal comfort and wind characteristics were made by conducting detailed analysis on weather data such as ambient temperature, relative humidity, sun radiation, min-max wind speed, mean wind speed, wind direction and its frequency. Heat stress is one of the major concerns of this study. Figure 2 shows the monthly mean wind speed in three years. From the observation, the windiest months were in January and February which recorded approximately 1.84m/s and 1.87m/s respectively. The minimum mean wind speed indicated in the month of November was just below 1.40m/s. On the following month, the mean wind speed tremendously increased up until the month of January before it gradually decreased until the month of April which might be due to northeast monsoon season. Between the months of April to November, the trend of mean wind speed showed gradual increase and reach a peak at about 1.73m/s before decreasing quite steeply due to southwest monsoon season.

Furthermore, the yearly mean wind speed for three years data analysis at Alor Setar, Kuantan, and Subang were at 1.70m/s, 1.69m/s, and 1.6m/s respectively. The yearly mean wind speeds trend at all locations show only slight differences. The mode of wind speed at Kuantan expressed no wind movement at all (i.e. 0m/s) as the dominant readings at 13.94% of frequency. Meanwhile, the remaining locations demonstrated calmness and no wind with the range of frequency from 10% to 13%. On average, the maximum wind speed at all locations was about 10m/s except for Subang with the maximum wind speed of 11.6m/s. The variations of yearly mean wind speed are also constant as the display values of standard deviation are in the range of 1.19 to 1.39.

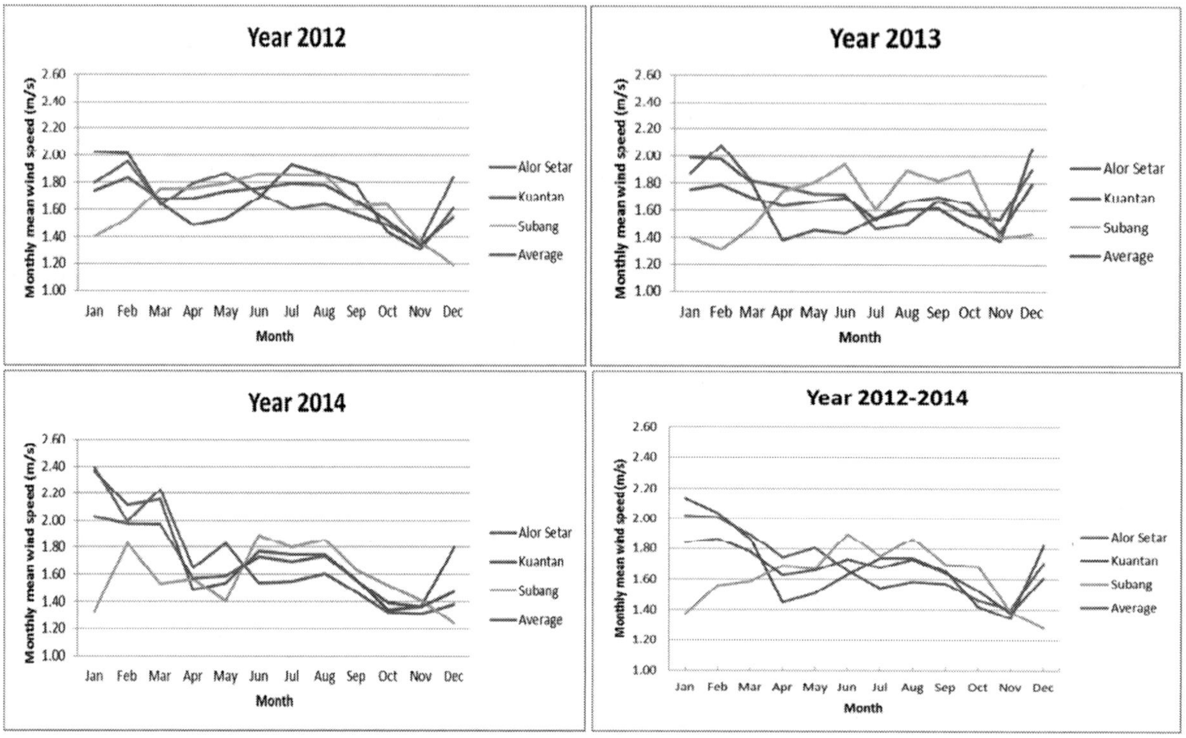

Figure 2. Monthly mean wind speeds at 10m height for the selected sites.

Besides the value and trend of wind speed, the direction of the wind is the most significant in evaluating the possibilities of outdoor thermal comfort. The urban planner, architect and engineer can strategize the orientation of the building in order to increase building permeability and indirectly, will increase the induced wind as a matter of reducing urban heat island. In general, the annual trend of prevailing wind as shown in Table 2 in all locations sticks at the same North direction (0°) with average frequency of about 12.21%. It indicates that Kuantan has the highest frequency which is at 13.94%. In contrast, Alor Setar illustrates the less dominant wind direction with frequency at about 10.21%.

Three years weather data were analyzed in order to obtain UTCI values at the worst heat stress cases. In general, it can be seen from Table 3 that during hottest temperatures, all the 3 locations experience extreme heat stress conditions, except for Kuantan which is slightly better with sometimes experiencing only very strong heat stress conditions. In the year 2012, Subang has the highest UTCI value of 47.6°C. For the following year, the calculated UTCI value of 51.2°C for Alor Setar is the highest highlighting the most extreme heat stress condition as compared to the other locations. Meanwhile, in the next two years, Subang station once again experienced extreme heat stress condition with the highest calculated UTCI value of 49.7°C. Based on Table 3, the range of maximum heat stress in three years data are in between 42°C to 52°C at all locations which indicate occurrence of very strong heat stress and extreme heat stress conditions. Hence, it is worth to mention that heat stress would be a common problem in Peninsular Malaysia which would enhance people's level of discomfort especially to whom living in urban areas. The problem of urban heat island is crucial nowadays which might be due to the variation of solar radiation and inadequate wind speed to wipe out the heat entrapped around the building.

Table 2. Annual wind rose distribution and frequency in the study.

Location/ Year	2012	2013	2014	(2012-2014)
Alor Setar	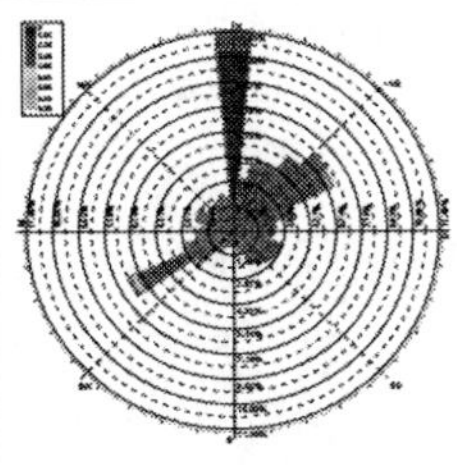	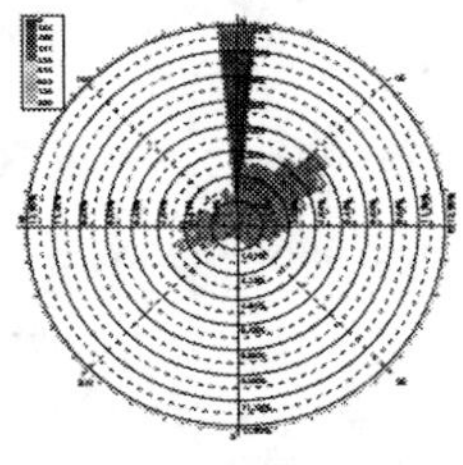	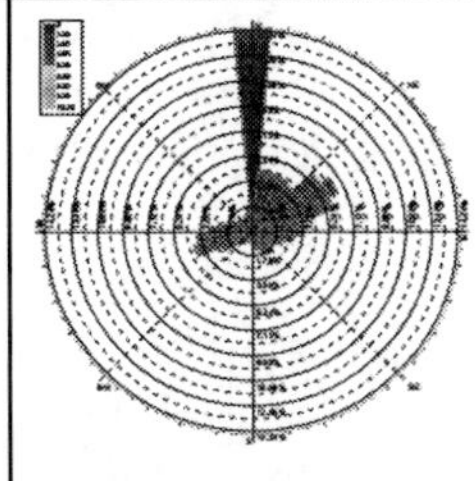	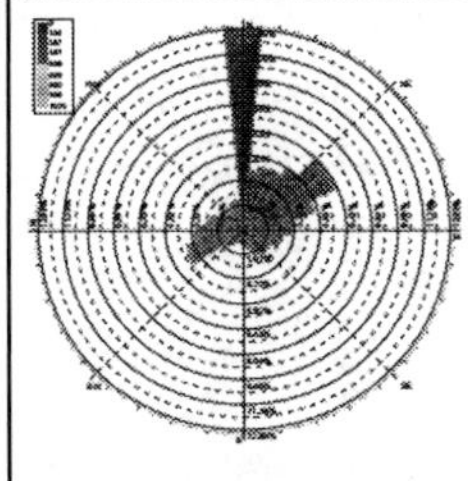
Kuantan	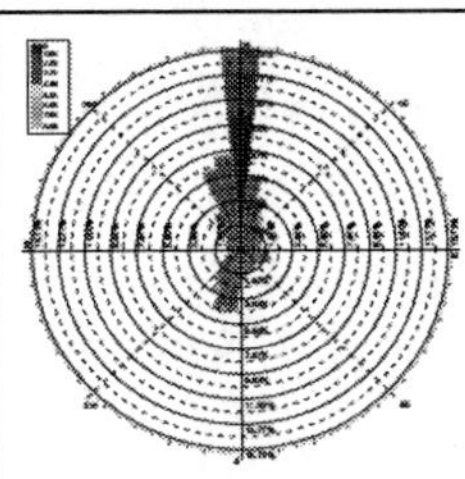	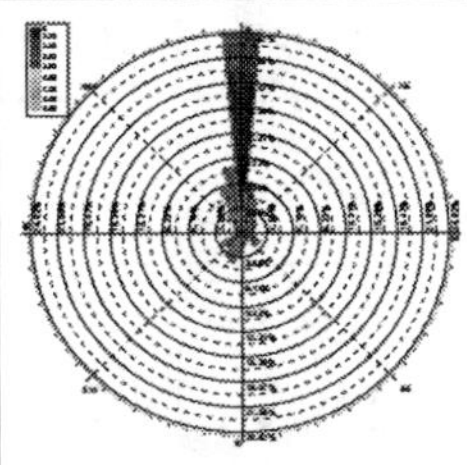	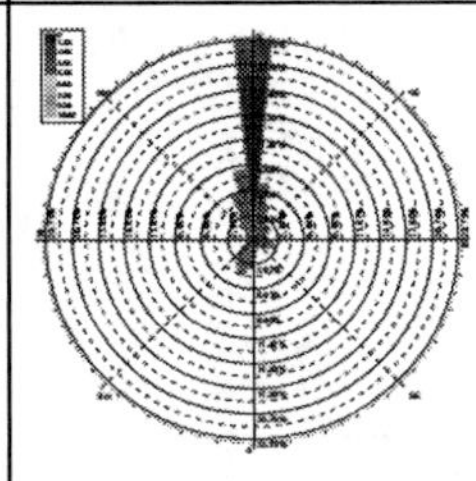	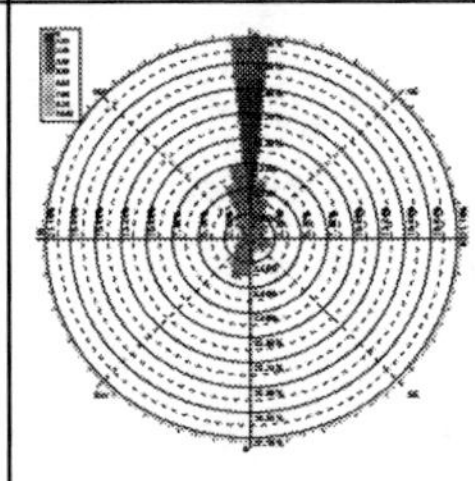
Subang	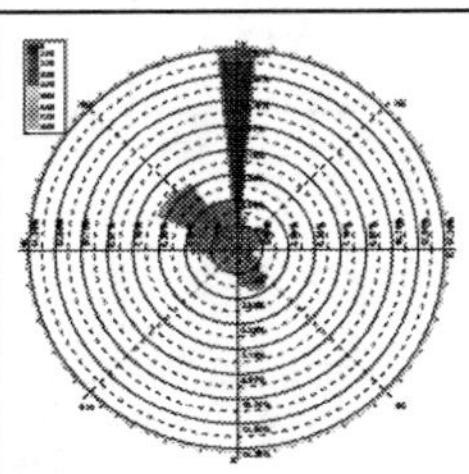	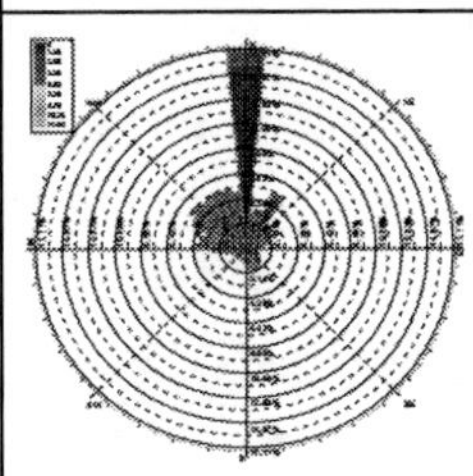	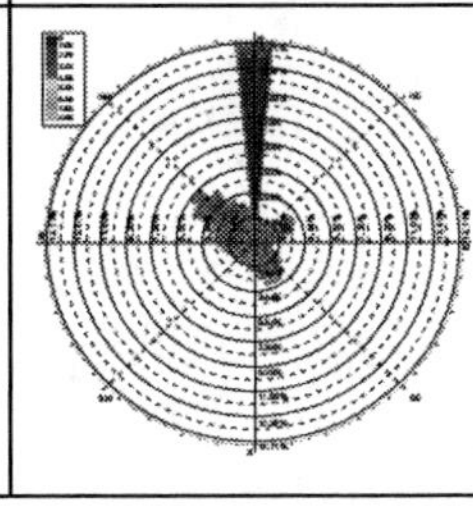	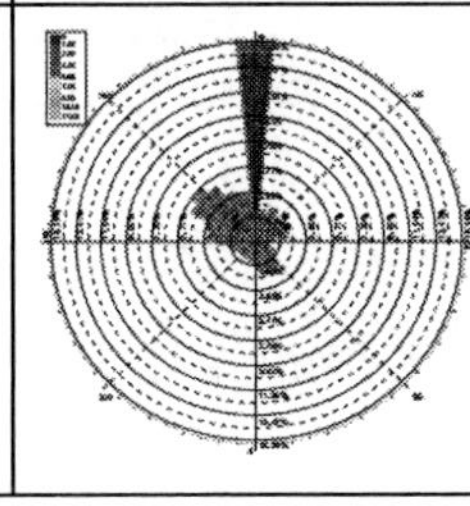

Table 3. Annual UTCI in the study at hottest temperature in 2012-2014 (3 years).

Location	Date	Time	Tem-perature (°C)	Relative Humidity (%)	Wind Speed (m/s)	Solar Radiation (W/m²)	UT-CI (°C)	Stress Category
Alor Setar	7.2.2012	1600	35.4	41	1.90	855.56	46.7	extreme heat stress
	1.4.2013	1500	37.2	42	1.70	1025.00	51.2	extreme heat stress
	28.3.2014	1700	37.5	37	2.10	644.44	46.2	extreme heat stress
Kuantan	3.10.2012	1600	34.5	52	4.00	702.78	43.2	very strong heat stress
	19.6.2013	1600	36.0	48	4.80	758.33	45.1	very strong heat stress
	24.4.2014	1400	35.3	55	1.60	922.22	49.0	extreme heat stress
Subang	5.6.2012	1500	36.4	49	6.7	938.89	47.6	extreme heat stress
	18.6.2013	1400	36.0	39	6.0	933.33	46.2	extreme heat stress
	5.3.2014	1500	37.8	17	3.4	1097.22	49.7	extreme heat stress

By analyzing the 20 years weather data (1994-2014), the hottest days during that period for all the three locations were determined and together with other weather parameters their UTCI values were calculated as detailed in Table 4. Based on UTCI calculation, it was shown that the extreme heat stress conditions would occur at Alor Setar and Subang. Table 4 also shows that Subang has not only the highest UTCI value of 49.7°C but also the highest solar radiation intensity of 1097.22 W/m^2. Although Subang's solar radiation is 30% higher than that of Alor Setar, their UTCI values differ only slightly at 0.01°C. This might be due to the value of wind speed at Subang which is 47% higher at about 3.4m/s as compared to that of Alor Setar. Again, it is crucial to mention that wind speed variation plays an important factor affecting heat stress performance.

Table 4. UTCI index at the hottest temperature in 1994 to 2014 (20 years).

Location	Date	Time	Temp-erature (°C)	Relative Humidity (%)	Wind Speed (m/s)	Solar Radia-tion (W/m^2)	UT-CI (°C)	Stress Category
Alor setar	27.3.1998	1600	39.0	36	1.8	769.44	49.6	extreme heat stress
Kuantan	26.5.1998	1500	36.6	50	4.0	536.11	43.8	very strong heat stress
Subang	5.3.2014	1500	37.8	17	3.4	1097.22	49.7	extreme heat stress

6 Conclusion

To recapitulate, there are several major findings from this study. The findings are the mean wind speed for three years data at Alor Setar, Kuantan, and Subang, were found to be 1.70m/s, 1.69m/s, and 1.63m/s respectively. The highest hourly wind speed was observed to be 11.6m/s at Subang on 13rd July 2013 at 4.00p.m. It was also observed that no wind movement (i.e. 0 m/s) could be considered as the lowest hourly wind speed for all 3 locations, and Kuantan showed the highest frequency of calmness at13.94%. Alor Setar, Kuantan, and Subang demonstrated the prevailing wind from North direction (0°) with about 10% of frequencies. The extreme heat stress conditions were calculated to be at Alor Setar and Subang as the latest trend of UTCI in year 2012 to 2014. Meanwhile, Kuantan demonstrated very strong heat stress conditions. The 20 years weather data from 1994 to 2014 were analyzed where the values of UTCI at Alor Setar, Kuantan, and Subang were 49.6°C, 43.8°C, and 49.7°C respectively. Here, it should be noted that the current work is just a preliminary study in estimating the prevailing wind and wind speed in order to have a comprehensive wind data base for obtaining a good prediction of outdoor thermal comfort. Apart from that, this study can also contribute towards the awareness on the current level of heat stress problem in Peninsular Malaysia. Ultimately, it is hope that the overall findings from this study can be beneficial to researchers, state's authorities and country's policy makers for achieving a sustainable environment.

Acknowledgements

The author would like to express gratitude to Ministry of Education Malaysia and Universiti Tun Hussein Onn Malaysia for funding this study, Malaysian Meteorological Department for providing weather data, and all individuals for their help throughout the course of this research study.

References

[1] A. Mochida and I.Y.F. Lun, Prediction of wind environment and thermal comfort at pedestrian level in urban area, J. Wind Eng. Ind. Aerodyn., **96**(10-11), 1498–1527, (2008).
[2] A. Abd Razak, A. Hagishima, N. Ikegaya and J. Tanimoto, Analysis of airflow over building arrays for assessment of urban wind environment, Build Environ., **59**, 56–65, (2013).
[3] T, Asawa, A. Hoyano, and K. Nakaohkubo, Thermal design tool for outdoor spaces based on heat balance simulation using a 3D-CAD system, Build Environ., **43**(12), 2112–2123, (2008).
[4] A.H. Abdullah and F. Wang, Design and low energy ventilation solutions for atria in the tropics, Sustain Cities Soc., **2**(1), 8-28, (2012).
[5] F. Wang and A.H. Abdullah, Investigating thermal conditions in a tropic atrium employing CFD and DTM techniques, Int. J. Low-Carbon Technol., **6**, 171-186, (2011).
[6] N. Makaremi, E. Salleh, M.Z. Jaafar and A. Ghaffarian Hoseini, Thermal comfort conditions of shaded outdoor spaces in hot and humid climate of Malaysia, Build Environ., **48**, 7-14, (2012).
[7] Malaysian Meteorological Department, Malaysia Weatherwise Malaysia. <available at: http://www.met.gov.my>
[8] J. Hang, Y. Li, M. Sandberg, R. Buccolieri and S. Di Sabatino, The influence of building height variability on pollutant dispersion and pedestrian ventilation in idealized high-rise urban areas, Build Environ., **56**, 346-360, (2012).
[9] J. Hang, Y. Li and M. Sandberg, Experimental and numerical studies of flows through and within high-rise building arrays and their link to ventilation strategy, J. Wind Eng. Ind. Aerodyn., **99**(10), 1036-1055, (2011).
[10] T. Kubota, M. Miura, Y. Tominaga and A. Mochida, Wind tunnel tests on the relationship between building density and pedestrian-level wind velocity: Development of guidelines for realizing acceptable wind environment in residential neighbourhoods, Build Environ., **243**(10), 1699-1708, (2008).
[11] B.D. Katsoulis, *Theoretical and Applied Climatology*, (1993).
[12] G. Jendritzky, R. de Dear and G. Havenith, UTCI-Why another thermal index?, Int. J. Biometeorol., **56**, 421-428, (2012).
[13] Climate Chip, Climate change health impact and prevention. <available at: www.climatechip.org>
[14] Glossary of Terms for Thermal Physiology, Journal of Thermal Biology, **28**, 75-106, (2003).
[15] K. Blazejczyk, Y. Epstein, G. Jendritzky, H. Staiger and B. Tinz, Comparison of UTCI to selected thermal indices, Int. J. Biometeorol., **56**, 515-535, (2012).
[16] D. Fiala, K.J. Lomas and M. Stohrer, A computer model of human thermoregulation for a wide range of environmental conditions: The passive system, J. Appl. Physiol., **87**, (1999).

MATEC Web of Conferences **47**, 04006 (2016)
DOI: 10.1051/matecconf/20164704006

Leadership Quality for Malaysia Construction Leader to Steer a Success Construction Project

Nur Ain Ngah Nasaruddin[1,a] and Ismail Abdul Rahman[1]

[1]*Faculty of Civil and Environmental Engineering, Universiti Tun Hussein Onn Malaysia, 86400 Parit Raja, Johor, Malaysia*

Abstract. Issues that dominate the construction industry for the long time such as time overrun, cost overrun, quality of construction and also safety during the construction process are difficult to resolve because of several factors involved which amongst them are leadership quality in the industry. Solving these issues can make the industry more sustainable in nature. This paper presents the need of leadership quality among construction parties involved to steer a construction project successfully. Literature work found that 11 leadership qualities to ensure construction project success. From the frequency analysis, five most dominant leadership qualities are team building, communication skill, visioning, planning & strategy skill, and decision making. Beside the literature, this study also seeks an opinion from a senior leader of a successful construction company regarding the need of these qualities amongst parties to resolve the construction issues. The results from an interview session indicate that he fully agreed with the need of all the qualities in the construction industry. This shows that there is a strong relationship between solving the issues with the leadership quality in ensuring construction project success. These qualities should be further investigated to acquire a better representation of leadership quality in resolving construction issues for Malaysia construction industry.

1 Introduction

In nature, construction industry is a labor intensive, needs skills & commitment, and effective organization of manpower [1]. It is considered as one of the world's largest industries that holding 18% of gross domestic product (GDP) [2]. In the construction project implementation process, it involves activities such as planning, designing, and executing of the projects from beginning to completion for controlling time, scope, cost and quality [3]. Thus, the implementation of the project requires construction leaders who can play an important role in ensuring that the processes are undertaken smoothly.

Construction leaders from management team may vary either they are from contractors, site engineers, project managers and project site supervisors. Basically, the qualities of leadership during project execution affect project performance either positively or negatively [2]. According to [4], leadership is the ability to guide the project team in achieving project objectives and balancing project constraints. Therefore, good construction leaders need to be an effective and successful strategist [5].

[a] Corresponding author : nurain_ngah@yahoo.com

An effective leader is built from inference, experience and instinct [6]. Whereas, strategic leaders generate a viable organization that will be capable of working in all circumstances [7].

2 Construction Project Failure

A construction project is considered fail if it faced cost and time overruns, not meeting the required quality as stipulated in the contract document and faced with safety issue during the construction.

2.1 Issues on time overrun

Time overrun can be defined as the non-completion of the project within the original or agreed contract period [8]. It can be caused by several factors including the lack of building materials, changes to the original designs, and new requirements being added to the original plan, etc [9]. However, in real life situations, many projects have been delayed due to changing of main contractor after work had started [9]. Statistics show that only 20.5% and 33.35% of public sector and private sector projects respectively are completed within the time specified in the contracts in Malaysia [10]. Therefore, there is a need to have good construction leaders who are able to control the construction process by avoiding or minimizing time overrun.

2.2 Issues on cost overrun

According to [11] defined construction cost as the degree to which the general conditions promote the completion of a project within the estimated budget. However for cost overrun occur when the final cost of the project exceeds the original estimates [12]. Significantly, construction projects facing cost overrun due to fluctuation in price of raw materials, lack of experience, unforeseen site conditions, high cost of machineries, and inaccurate or poor estimation of original cost [13]. In Malaysia only 46.8% and 37.2% of public sector and private sector projects respectively are completed within the budget [10]. Construction leader who able to manage effectively and strategically the project will keep the cost within the budget.

2.3 Issues on quality

Quality is defined as meeting the customer's expectations, or compliance with customer's specification according to appearance, performances, and reliability of the project for a given price range [14]. However, quality may sometimes be ignored in construction industry in order to reduce the costs or to shorten the project time [15]. A surveys conducted by [16] identified several unethical conducts and ethical dilemmas in the construction industry such as corruption, negligence, conflict of interest, bid cutting, under bidding, collusive tendering, cover pricing, frontloading, bid shopping, withdrawal of tender, and payment game. Significantly, a study by [17] show that majority of 72.7% construction leader agreed that unethical conduct can be the main cause for poor quality project.

2.4 Issues on safety

Safety has been a major issue and concern in the construction industry when it comes to work-related accidents [18]. According to [19], in 2014, Malaysia construction accident show that 72 workers die at the job site as a result of work, 6 workers suffering permanent disability, and 94 workers suffering non-permanent disability. From previous studies by [20], among the factors that lead to the construction accidents are the workers' negligence, failure of workers to obey work procedures, work at high elevation, operating equipment without safety devices, poor site management, harsh work operation, lack of knowledge, low skill level of workers, failure to use personal protective equipment's, and poor workers attitude about safety.

3 Importance of Construction's Leadership Quality

Like the other effort, the importance of leadership quality to construction project success are to ensure the smooth running of projects [21], to satisfy conflicting requirements in support of organizational success [2], to increase the company's profitability [22], to create new opportunities for companies to apply its skills [22], and to develop leader to professional ladders [2].

With good quality of construction's leader a project will run smoothly with less disruption due to project constraints; such as tight schedule and budget, limited resource, risky and others [4]. Good construction leader should be able to recognize project strength, weakness, opportunity, and threat in order to complete construction project successfully. These will increase company profit because less or no repetition of works and also delay. Furthermore, proactive leaders tend to create new opportunities rather merely completing the projects. Also, good leaders with ethical behavior will lift their images in their profession.

4 Research Methodology

This study was involved in several lists of step as below:

i. Identified Relevant Reference. Identification of relevant references should be geared to the quality of leadership of the construction practitioners. The reference is covering from books and journal articles.
ii. Able to Identify and Extract Important Fact or Point Related to Study. A thorough literature work of research articles was conducted to identify the leadership quality in the construction industry.
iii. Organized the Facts into Systematic Manner. For instance, listing the leadership quality and estimating frequency for all the quality. Descriptive statistic using frequency was used in order to analyze and interpret the result into a meaningful form.
iv. Similarity Checking. The process involves checking the fact obtained. The purposes of these approaches are to avoid duplication of points or facts, generalization of several points or facts, and rephrase or reword the facts according to the study.
v. Verification from the Expert. An interview session with senior successful construction leaders in Malaysia was carried out to strengthen the information obtain from the literature review. Besides, an interview session was carried out to give an insight view of Malaysia construction regarding the leadership quality required to steer a successful construction project.

5 Leadership Quality to Tackle Issues

One of the main components in ensuring a success of a construction project is the leadership quality amongst construction leader. A good leader will be able to manage the project effectively and also be able to strategize construction activities based on the risk management approach. With proper leadership skill among the construction parties will enable to navigate the construction process in a sustainable manner. It is time for Malaysia to have a sound leader quality among the construction parties to make sure our construction industry attain world standard by 2020 as envision being developed country. Features of leadership quality are extracted from research articles produced by researchers who explored the leadership quality amongst the construction parties as in Table 1.

Table 1 indicates that there are 11 quality features that are being noted by researchers in the field of construction. Out of these 11 features the 5 most highlighted by the researchers are team building, communication skill, visioning, planning & strategy skill and decision making. From the literature analysis, an interview session with construction leader was carried out to give an insight view of Malaysia construction regarding the leadership quality required to steer a successful construction project.

Table 1: Leadership quality for construction project success

No.	Leadership Quality	Construction Industry Related Sources							Frequency
		A	B	C	D	E	F	G	
1	Communication skill	√	√	√		√			4
2	Team building	√	√		√	√		√	5
3	Decision making	√				√		√	3
4	Sense of responsibility			√			√		2
5	Visioning		√		√		√	√	4
6	Planning & strategy skill		√	√	√			√	4
7	Relationship management skill					√		√	2
8	Empowerment	√	√						2
9	Ethics	√							1
10	Influencing		√	√					2
11	Integrity					√	√		2

(Sources: A [23], B [24], C [25], D [26], E [27], F [28], G [29])

6 Interview with Construction Leader

An interview was carried out with one of the prominent and successful construction leader in the northern part of peninsular Malaysia regarding his opinion on the importance of leadership quality to steer success of his construction company. He is managing director from grade G7 contractor company with more than 30 years of experiences in construction industry. In the interview session, he mentioned that to be a good leader in construction industry, one must possess qualities such as honesty and trust with the employees, teamwork, planning, time management, decision making, empowerment, and good communication.

He emphasized that as leaders in implementing construction project, they have to be knowledgeable and skillful in handling their worker or might lose them. For example, issue like a wage dispute between workers is quite famous where the workers tend to demand more. With strong relationship quality, leaders will be able to treat all employees fairly and equitably, and caring for the welfare of workers. As leaders they should be able to manage risk such as price of construction material increase due to economic fluctuation and also late payment from the client. With good quality leaders, they will able to anticipate possible kinds of risk in the implementation of the project. Improper handling of these issues, my result to cost and time overrun which lead to project failure, thus it there is a strong correlation between the leadership with the project failure.

7 Conclusion

This study has provided an insight issues in construction project implementation which needs a strong leadership quality. It can be stated that there is no single leadership quality is suitable in all project circumstances. As construction leaders, they must use their knowledge and skills to adopt or adapt any specific leadership quality based on work that need to be done. If the issues are tackled with strong leadership quality, the leaders can become as role models to their employee.

References

[1] R. Morton and A. Ross, *Construction UK: Introduction to the Industry*, Oxford, (2008).

[2] R. Udhayakumar and P. Karthikeyan, Expected leadership qualities for a project manager to manage construction projects, Int. J. of Innovative Research and Development, **3**, 57-61, (2014).

[3] J.J. McKeon, *Becoming a Construction Manager*, New Jersey, (2011).

[4] PMBOK Guide, *A Guide to the Project Management Body of Knowledge*, Project Management Institute, United States, (2008).

[5] http://www.stuba.sk/new/docs/stu/ustavy/ustav_manazmentu/NAB2012-2/ clanok9.pdf.

[6] K. Gharehbaghi and K. McManus, The construction manager as a leader, Leadership and Management in Engineering, **3**(1), 56-58, (2003).

[7] http://www.alnmag.com/articles/2011/06/three-levels-leadership.

[8] D.A.R. Dolage and D.L.G. Rathnamali, Causes of time overrun in construction phase of building projects: A Case Study on Department of Engineering Services of Sabaragamuwa Provincial Council, The Institution of Engineers, Sri Lanka, **66**(3), 9-18, (2013).

[9] http://www.thestar.com.my/Opinion/Letters/2015/04/28/Minimising-project-delays/.

[10] I.R. Endut, A. Akintoye and J. Kelly, Cost and time overruns of projects in Malaysia, *Proc. of the 2nd Scottish Conference for Postgraduate Researcher of the Built and Natural Environment*, 243-252, (2005).

[11] E.W.M. Lam, A.P.C. Chan and D.W.M. Chan, Benchmarking success of building maintenance projects, Emerald Insight, **28**(5/6), 290-305, (2010).

[12] N. Azhar, R.U. Farooqui and S.M. Ahmed, Cost overrun factors in construction industry of Pakistan, *Proc. of the First International Conference on Construction in Developing Countries*, 499-508, (2008).

[13] A.S. Ali and S.N. Kamaruzzaman, Cost performance for building construction projects in Klang Valley, Journal of Building Performance, **1**(1), 110-118, (2010).

[14] K.N. Jha and K.C. Iyer, Critical factors affecting quality performance in construction projects, Total Quality Management, 1155-1170, (2006).

[15] A.M. Turk, ISO 9000 in construction: An examination of its application in Turkey, Building and Environment, 501-511, (2006).

[16] C. Vee and M. Skitmore, Professional ethics in the construction industry, Engineering, Construction and Architectural Management, **10**(2), 117-127, (2003).

[17] H. Abdul-Rahman, C. Wang and X.W. Yap, How professional ethics impact construction quality: Perception and evidence in a fast developing economy, Scientific Research and Essays, **5**(23), 3742-3749, (2010).

[18] M. Alhajeri, Health and Safety in the Construction Industry: Challenges and Solutions in the UAE, PhD Thesis, Coventry University, (2014).

[19] http://www.dosh.gov.my/index.php?option=com_content&view=article&id=1225:occupational-accidents-statistics-by-sector-2&catid=502&Itemid=1280&lang=en.

[20] A.R.A. Hamid, M.Z.A. Majid and B. Singh, Causes of accidents at construction sites, Malaysian Journal of Civil Engineering, **20**(2), 242 – 259, (2008).

[21] M. R. Senam, K. Abdul Rashid, A. A. Sarkawi and R. M. Zaini, construction project leadership from the perspective of Islam, International Journal of Islamic Thought, **6**, 46-56, (2014).

[22] D.B. Russell, D.D. Dunne, How Healthy is your company's leadership development culture?, Construction Management Association of America, 1-13, (2004).

[23] J.V. Farr, S.G. Walesh and G.B. Forsythe, Leadership development for engineering managers, Journal of Management in Engineering, **13**(4), 38-41, (1997).

[24] V.K. Verma and R.M. Wideman, Project manager to project leader? and the rocky road between, *Project Management Institute's 25th Annual Seminar and Symposium*, 1-9, (1994).

[25] B. Benator and A. Thumann, *Project Management and Leadership Skills for Engineering and Construction Projects*, Unites States, (2003).

[26] http://www.maconline.org/tech/ estimating/newsletter.pdf.

[27] A.R.J. Dainty, M. Cheng and D.R. Moore, Competency-based model for predicting construction project managers' performance, Journal of Management in Engineering, **21**(1), 2-9, (2005).

[28] http://www.rhgraham.org/RHG/Recent_publications_files/ELE%20White%20 Paper-102109.pdf

[29] P.P. Bhangale and R. Devalkar, Study the importance of leadership in construction projects, International Journal of Latest Trends in Engineering and Technology, **2**(3), 312-318, (2013).

MATEC Web of Conferences 47, 04007 (2016)
DOI: 10.1051/matecconf/20164704007

Influential Factors in Estimating and Tendering for Construction Work

Toh Tien Choon[1,a], Lim Cheng Sim[1], Ting Connie[2], Ali Kherun Nita[3], Aliagha Godwin Uche[4] and Goh Kai Chen[5]

[1]Department of Surveying, Lee Kong Chian Faculty of Engineering and Science, Universiti Tunku Abdul Rahman, 43000 Kajang, Selangor, Malaysia
[2]Asia Pacific Logistics Center, Eppendorf Asia Pacific Sdn Bhd, 47600 Subang Jaya, Selangor, Malaysia
[3]Department of Quantity Surveying, Faculty of Built Environment, Universiti Teknologi Malaysia, 81310 Skudai, Johor, Malaysia
[4]Department of Real Estate, Faculty of Geoinformation and Real Estate, Universiti Teknologi Malaysia, 81310 Skudai, Johor, Malaysia
[5]Department of Construction Management, Faculty of Technology Management and Business, Universiti Tun Hussein Onn Malaysia, 86400 Parit Raja, Johor, Malaysia

Abstract. Cost estimation, bid/no bid decision, estimating accuracy, and cost overrun are the four main cost estimating issues (MCEI) investigated by many previous researchers from different parts of the world for decades. Factors relating to client characteristics, consultant and design parameters, contractor attributes, project characteristics, contract procedures and procurement methods, and external market conditions [1] are examined for their influences on those four MCEI separately. However, very little attention is paid on the comparison of factors rankings across those four MCEI. Results from comparing those factors rankings improve the understandings on cost estimating issues. Thus, this study aims to compare the factors rankings across the four MCEI and evaluate the degree of agreement between the four sets of ranks for their influential factors. Four previous studies, each examining factors affecting each of the four MCEI are selected through literature survey for analysis. Influential factors from each previous study are assigned to their related categories: client, consultant, contractor, project, contract, and environment. Kendall's coefficient of concordance is used to assess interjudge reliability. The computed correlation indicates a moderate degree of association between the four sets of ranks. It is found that the four MCEI are influenced by different factors with different ranks and hence should be treated as different MCEI and be managed differently.

1 Research Background and Motivation

Factors with respect to client characteristics, consultant and design parameters, contractor attributes, project characteristics, contract procedures and procurement methods, and external market conditions [1] are often examined for their influences on the four MCEI separately. For instance, factors related to contractual arrangements [2,5], contractor attributes [2], market conditions [5] are found to be the main causes of cost overrun; factors related to client characteristics, contractor attributes are found to be the major determinants for bid/no bid decision [3]; and factors related to market conditions and

[a] Corresponding author : tohtc@utar.edu.my

contractual arrangements are found to be the critical determinants of estimating accuracy [4]. But studies on comparing factors rankings across those four MCEI are rare. Results from this kind of study are useful in two ways. First, if there is a high correlation among the factors rankings, a list of factors can be produced in assisting practitioners to manage the four different MCEI concurrently. Second, if there is no correlation among the factors rankings; it simply provides a clear conclusion on what and when a factor is impactful on those four MCEI. Therefore, this study aims to compare the factors rankings across the four MCEI and evaluate the degree of agreement between the four sets of ranks for their influential factors.

2 Research Methodology

Unlike many other studies in the field of estimating and tendering for construction work that collected and analysed primary data, this study analyses secondary data collected from high impact journal articles. The data used in this study are hence reliable. Because there is a lack of studies in this nature, it is reminded that this study is conducted with an exploratory purpose. Therefore, at the first stage, 4 previous studies examining factors affecting the four MCEI are selected through literature survey for analysis. Each of these 4 previous studies is examining factors affecting only one of the four MCEI. To ensure currency of the results from this study, the 4 previous studies selected are all published in 2014 and 2015. Studies on factors affecting the four MCEI received very little attention in 2015. They are found in 4 high impact journals relating to the field of construction management and economics. Cheng's study [2]: 'An exploration into cost-influencing factors on construction projects' is published in *International Journal of Project Management*; Jarkas *et al.*'s study [3]: 'Critical factors determining bid/no bid decisions of contractors in Qatar' is published in *Journal of Management in Engineering*; Mahamid's study [4]: 'Factors affecting cost estimate accuracy: Evidence from Palestinian construction projects' is published in *International Journal of Management Science and Engineering Management*; and Rosenfeld's study [5]: 'Root-cause analysis of construction-cost overruns' is published in *Journal of Construction Engineering and Management.*

At the second stage, a content analysis is used to analyse textual information in all the four journal articles and systematically identify influential factors from each journal article. A conceptual analysis is carried out to analyse and interpret the influential factors by coding them into the six categories of factors: client, consultant, contractor, project, contract, and environment [1]. These six categories are employed because they are broad enough to include all possible factors affecting the four MCEI. At the third stage, Kendall's coefficient of concordance is used to assess interjudge reliability. Interjudge reliability is the degree to which two or more judges are in agreement with one another. It examines the degree of agreement between the four sets of ranks for their influential factors assigned to their related categories. The value of Kendall's coefficient of concordance (W) ranges from 0 to +1, with '0' indicating no agreement and '1' indicating complete agreement. Other details about the technique including the conduct of the test can be found in Sheskin [6]. In this study, the null hypothesis: there is no agreement between the four sets of ranks for the six categories of factors (H_0: $W = 0$); and that the alternative hypothesis: there is a significant agreement between the four sets of ranks for the six categories of factors (H_1: $W \neq 0$) are to be tested.

3 Results and Discussions

Table 1 to Table 4 show the results from the conceptual analysis done in this study. Based on severity indices by descending order, factors influencing construction projects cost [2] are coded into the categories of factors of Contract, Contractor, Environment, Project, Consultant, and Client, respectively, Table 1. Based on relative importance indices by descending order, factors determining bid/no bid decisions of contractors [3] are coded into the categories of factors of Client, Contractor, Project, Environment, Contract, and Consultant, respectively, Table 2. Based on importance indices by descending order, factors affecting cost estimating accuracy in construction projects [4] are coded into the categories of factors of Environment, Contract, Client, Contractor, Consultant, and Project, respectively, Table 3.

Table 1. Ranking of cost-influencing factors on construction projects [2].

No.	Factor	Severity Index	Rank	Related Category
1.	Clearly define the scope of project in the contract	94.78	1.5	Contract
2.	Cost control	94.78	1.5	Contractor
3.	High fluctuation in commodity	89.57	3.5	Environment
4.	The gap between the construction plan and the reality is too great	89.57	3.5	Project
5.	Project team (coordination capability and the understanding of operational procedure)	86.96	5	Consultant
6.	Project valuation does not match the collected payment	86.09	6	Client

Table 2. Ranking of factors determining bid/no bid decisions of contractors [3].

No.	Factor	Relative Importance Index	Rank	Related Category
1.	Previous experience of contractor with employer	87.39	1	Client
2.	Need for work	86.30	2	Contractor
3.	Size of project	84.13	3	Project
4.	Availability of other projects	81.52	4	Environment
5.	Tender documents quality level	80.00	5	Contract
6.	Identity of designer	57.39	6	Consultant

Table 3. Ranking of factors affecting cost estimating accuracy in construction projects [4].

No.	Factor	Importance Index	Rank	Related Category
1.	Fluctuation in the currency exchange rate	86.29	1	Environment
2.	Contract management	85.48	2	Contract
3.	Financial status of the owner	83.87	3	Client
4.	Previous experience of the contractor	79.84	4	Contractor
5.	Lack of coordination between designers and contractors	69.35	5	Consultant
6.	Project location	58.87	6	Project

Table 4. Ranking of universal root causes for construction-cost overruns [5].

No.	Factor	Percentage	Rank	Related Category
1.	Premature tender documents (drawings, bill of quantities, specifications, contracts and legal documents)	86.70	1	Contract
2.	Tender-winning prices are unrealistically low (suicide tendering)	65.10	2	Environment
3.	Insufficient, unstandardized owner's brief	35.90	3	Client
4.	Insufficient information about ground conditions	28.70	4	Project
5.	Shortage in high-quality management personnel	27.90	5	Contractor
6.	Lack of standard requirements from designers and poorly enforced professional liability of designers	16.90	6	Consultant

Based on percentages by descending order, universal root causes for construction-cost overruns [5] are coded into the categories of factors of Contract, Environment, Client, Project, Contractor, and Consultant, respectively, Table 4. Contractor and Contract are found to be the two most important categories affecting construction projects cost, Client is found to be the most important category affecting bid/no bid decisions of contractors, Environment is found to be the most important category affecting cost estimating accuracy in construction projects, and Contract is found to be the most important category affecting construction-cost overruns, Table 5. The computed correlation $W_c = 0.38$ indicates a moderate degree of association between the four sets of ranks. The tabled critical 0.05 and 0.01 values for $m = 4$ and $n = 6$ are $W_{0.05} = 0.512$ and $W_{0.01} = 0.629$. Since $W_c = 0.38$ is less than $W_{0.05} = 0.512$, the null hypothesis H_0: $W = 0$ cannot be rejected.

Table 5. Summary for the rankings of related categories of factors for cost estimation, bid/no bid decision, estimating accuracy, and cost overrun.

	Related Category						
MCEI	**Client**	**Consultant**	**Contractor**	**Project**	**Contract**	**Environment**	**Totals**
Cost estimation	6	5	1.5	3.5	1.5	3.5	
Bid/no bid decision	1	6	2	3	5	4	
Estimating accuracy	3	5	4	6	2	1	
Cost overrun	3	6	5	4	1	2	
$\sum R_j$	13	22	12.5	16.5	9.5	10.5	$T = 84$
$(\sum R_j)^2$	169	484	156.25	272.25	90.25	110.25	$U = 1282$

Apparently, the rankings of the same set of categories of factors for the four MCEI are not similar. The contractor is the party who is responsible to estimate the construction costs and he or she should refer to the contract-related documents when performing the construction cost estimation tasks. This is because the contractor is the one who constructs the project and therefore only he or she will know the exact costs of the construction. An experienced contractor will be wary when deciding whether to bid or not to bid for a project. He or she will definitely stay away from a client who is known to be a bad paymaster or difficult to deal with. The environment which is beyond the control of the parties involved in construction has a huge impact on estimating accuracy especially the market conditions. During a poor market condition, construction prices will go up and as a result affect the accuracy of the estimate. Changes either addition or omission made to the contract during the construction stage will change the scope of work and subsequently increase the project costs.

4 Conclusions and Recommendations

It is found that the four MCEI are influenced by different factors with different ranks and hence should be treated as different MCEI and be managed differently. Nonetheless, it is worth noting that as contractual arrangements are impactful on construction costs, contract management staffs should always deliver high quality of work to minimise project costs and prevent cost overruns. Another point to note is that the consultant has very little impact on all the four MCEI and therefore allowing the client to engage him or her without much hurdles. Seeing that this study is exploratory in nature, replications and further studies on this topic can test and confirm the finding from this study.

References

[1] T.M.S.Elhag, A.H. Boussabaine and T.M.A. Ballal, Critical determinants of construction tendering costs: Quantity surveyors' standpoint, International Journal of Project Management, **23**(7), 538-545, (2005).

[2] Y.M. Cheng, An exploration into cost-influencing factors on construction projects, International Journal of Project Management, **32**(5), 850-860, (2014).

[3] A.M. Jarkas, S.A. Mubarak and C.Y. Kadri, Critical factors determining bid/no bid decisions of contractors in Qatar, Journal of Management in Engineering, **30**(4), (2014).
[4] I. Mahamid, Factors affecting cost estimate accuracy: Evidence from Palestinian construction projects, International Journal of Management Science and Engineering Management, **10**(2), 117-125, (2015).
[5] Y. Rosenfeld, Root-cause analysis of construction-cost overruns, Journal of Construction Engineering and Management, **140**(1), (2014).
[6] D.J. Sheskin, *Handbook of Parametric and Nonparametric Statistical Procedures*, Chapman and Hall, Boca Raton, (2011).

MATEC Web of Conferences **47**, 04008 (2016)
DOI: 10.1051/matecconf/20164704008

Prediction of Low Cost Housing Demand in Malaysia Using ARIMA Model

N. Y. Zainun[1,a], F. E. Mohamed Ghazali[2] and M. S. Mohd Sallehudin[1]

[1]*Faculty of Civil and Environmental Engineering, University Tun Hussein Onn Malaysia,86400 Parit Raja, Johor, Malaysia*
[2]*School of Civil Engineering, University Sains Malaysia, 14300 Nibong Tebal, Pulau Pinang, Malaysia*

Abstract. Among the key challenges in construction industry sector faces are matching supply of and demand for affordable housing. It is very crucial to predict low-cost housing demand to match the demand and supply so that the government can plan the allocation of low cost housing based on the demand. In Johor, housing provision is very crucial due to urbanization. The supply of houses seems to be swamping the demand for luxury condos and houses especially in Johor Bharu. Thus the aim of this study is to predict low-cost housing demand in Johor, Malaysia using ARIMA model. Time series data on low-cost housing demand have been converted to Ln before develop the model. The actual data and forecasted data will be compared and validate using Mean Absolute Percentage Error (MAPE). After that, the results using ARIMA method will be compared with ANN method. The results show that MAPE analysis for ARIMA is 15.39% while ANN is 18.27%. It can be conclude that ARIMA model can forecast low cost housing demand in Johor quite good.

1 Introduction

One of Malaysia's longstanding development objectives is the provision of affordable housing for Malaysian, with a focus on lower-income groups [1]. Low cost housing can be defined as a development projects sold at the price set by the government that is between RM25, 000 to RM42, 000 [2]. Low cost housing built is intended to provide housing that is affordable for low earners in rural and suburban areas. The target groups for this project are households with monthly income of between RM500 to RM750 [3].

To provide adequate housing and affordable for Malaysians, especially for those with low incomes has become the main agenda through Malaysia plans prior to now [4]. However, there is mismatched data between the supply and demand for low-cost housing in Malaysia [5]. In some places, the supplies of low cost housing are exceeding compare to demand and lead to wastage of construction and of course has an impact on the cost and economic aspects. While in other areas the demand is exceeding supply provided, which supplies low-cost houses are insufficient, especially in urban areas [6]. Therefore, an alternative approaches need to be done to resolve these issues.

There are many series of forecasting methods can be used to predict the housing demand such as Artificial Neural Network (ANN), Autoregressive Integrated Moving Average (ARIMA), Power

[a] Corresponding author : nryasmin@uthm.edu.my

Model and Multiple Log linear Regression [7]. In this study ARIMA model known as the Box-Jenkins time series is used because it has good accuracy for the short term forecasting.

2 Scope and Limitation of Study

This study will focuses on forecasting low cost housing demand in Johor, Malaysia only. Previous time series data from [6] will be used to forecast low-cost housing demand in Johor using ARIMA model.

3 Methodology

The time series data were changed to Ln then analyzed using ARIMA software adopted from SPSS 20.0. Results were validated using MAPE where actual and forecasted data were compared to determine the accuracy of the model. Finally, the MAPE value will be compared to establish the performance of model.

4 Results and Discussion

Table 1 shows the monthly time series data on low cost housing demand in Johor from January 2000 to January 2007.

Table 1. Time series data on low cost housing demand.

Month / Year	Jan	Feb	Mac	Apr	May	Jun	Jul	Aug	Sept	Oct	Nov	Dec	Total
2000	90	272	360	325	360	405	497	488	334	474	472	530	4607
2001	419	485	572	661	1005	544	544	434	318	550	555	356	6443
2002	614	343	546	416	566	529	445	544	398	415	482	209	5507
2003	404	477	424	312	299	414	365	498	280	269	211	264	4217
2004	262	299	283	230	235	303	533	233	215	227	246	303	3369
2005	243	182	230	205	232	180	188	166	209	196	121	210	2362
2006	170	140	223	174	190	199	275	308	259	134	197	197	2466
2007	157												157
Total	2359	2198	2638	2323	2887	2574	2847	2671	2013	2265	2284	2069	29128

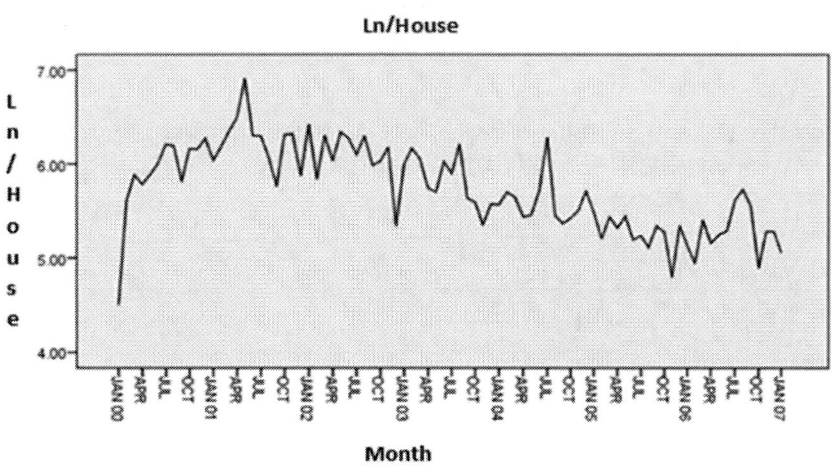

Figure 1. Housing demand (Ln).

Figure 1 are time series housing demand in Ln. Data were change to Ln to get idea for p, d, q value for a non seasonal ARIMA model ARIMA (p, d, q), where p is the number of autoregressive terms, d is the number of nonseasonal differences needed for stationarity, and q is the number of lagged forecast errors in the prediction equation. p and d were determined using coefficient Auto Correlation (AC) and Partial Auto Correlation (PAC) where AR(p) and MA(q) are the components for the time series.

Figure 2 views the ACF and PACF of housing demand using ARIMA (0,0,0). It can be seen that there are a few extrusions and improvements were done to create stationary data. From the p,d,q value, significant calculations for each parameters were done to determine the best model. In this study, three model were used; ARIMA (1,0,1); ARIMA (1,0,0) and ARIMA (2,0,0). ARIMA model produced the lowest value Akaike Information Criterion (AIC) and Schwarz Criterion (SC) is the best model.

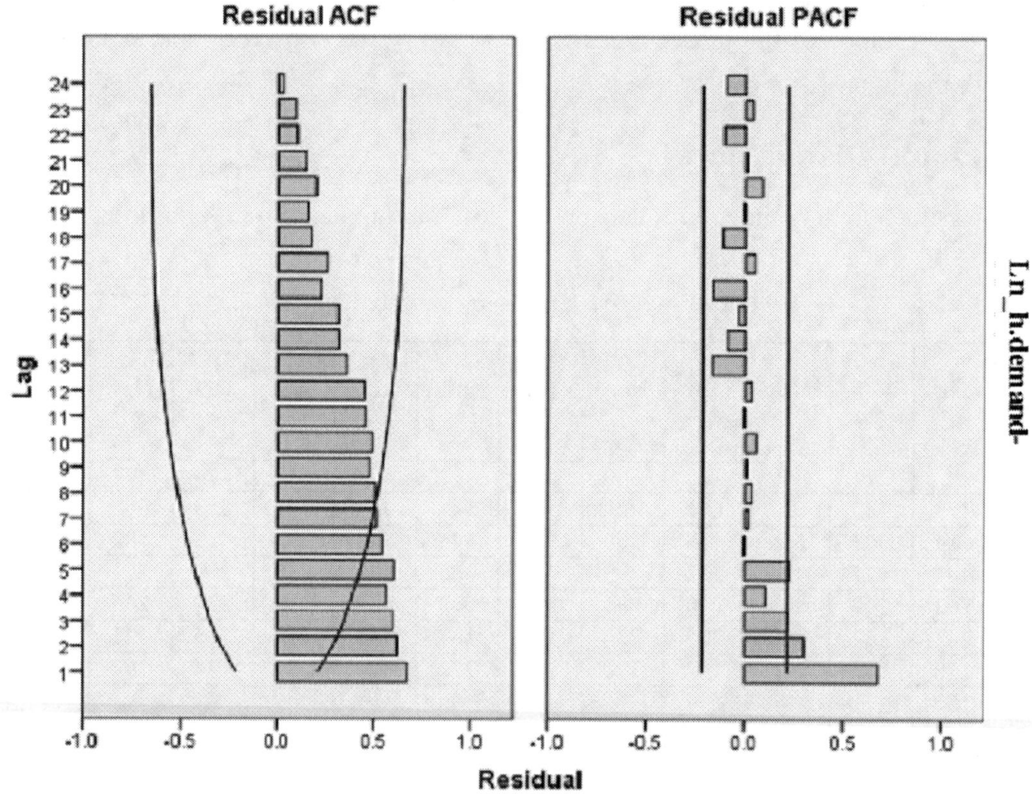

Figure 2. ACF and PACF of housing demand using ARIMA (0,0,0).

Table 2 views the AIC and SC value for each ARIMA model. It can be seen that model ARIMA (1,0,1) is the best compare to ARIMA (1,0,0) and ARIMA (2,0,0).

Table 2. Value for AIC and SC for each model.

Model	AIC	SC
ARIMA (1,0,1)	0.0178	0.2288
ARIMA (1,0,0)	0.0370	0.2499
ARIMA (2,0,0)	0.0656	0.2239

Figure 3 plots the improve ACF and PACF using ARIMA (1,0,1) and Table 3 shows the analysis value from model ARIMA (1,0,1).

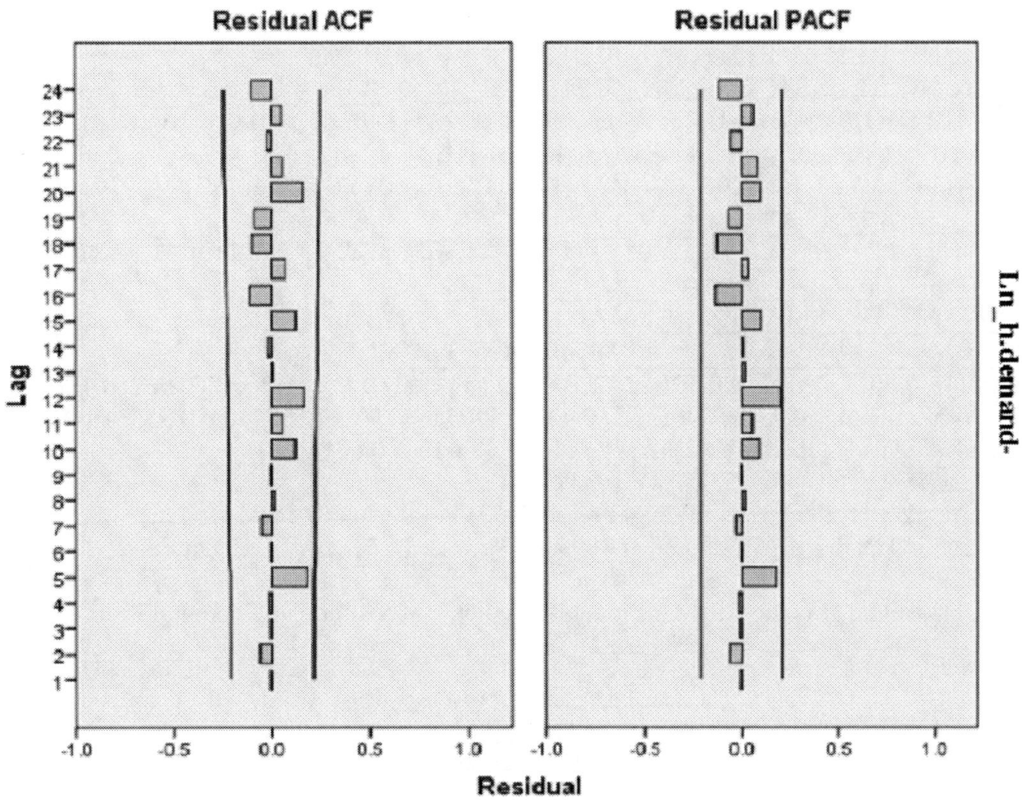

Figure 3. ACF and PACF of housing demand using ARIMA (1,0,1).

Table 3. Analysis value for model ARIMA (1,0,1).

	Intersection	AR(1)	Variable (1)	MA(1)
Coefficients	10.2767	0.6014	-0.0002	-0.4317
Standard Error	1.8472	0.0767	0.0000	0.1283
t-Statistic	5.5634	7.8451	-5.4500	-3.3625
p-Value	0.0000	0.0000		

From Table 3, |tcalc|> ttable (= 2.25) and p less than alpha. Therefore, null hypothesis can be rejected and can be concluded that the coefficient is different from zero.

Table 4. Comparison forecasting value between tentative models.

Month	Actual Data	ARIMA (1,0,1)	ARIMA (1,0,0)	ARIMA (2,0,0)
September 2006	5.56	5.54	5.68	5.71
Oktober 2006	4.90	5.54	5.62	5.58
November 2006	5.28	5.31	5.23	5.09
Disember 2006	5.28	5.31	5.23	5.38
Januari 2007	5.06	5.31	5.23	5.39
MAPE	-	3.90	4.42	5.72

Table 4 is the comparison forecasting value between tentative models ARIMA. It shows that ARIMA (1,0,1) have the lowest MAPE with 3.9%.

Therefore, the regression model equation was rewritten:

(1 - φ1B)(1-B)(1-B12)Yt = (1- θ1B)(1-
θ2B12)at..Equation (1)
· (1 - φ1B)(1-B12-B+B13)Yt = (1- θ2B12- θ1B + θ1θ2B13)at
· (1 - φ1B)(1-B12-B+B13)Yt = (1- θ2B12- θ1B + θ1θ2B13)at 48
· (1-B12-B+B13- φ1B+ φ1B13+ φ1B2- φ1B14) Yt = (1- θ2B12- θ1B + θ1θ2B13)at
· (1 - B12 – (1+ φ1)B + (1+ φ1)B13 + φ1B2 - φ1B14) Yt = (1- θ1B - θ2B12 + θ1θ2B13)at
· Yt – (1+ φ1)Yt-1 + φ1Yt-2 – Yt-12 + (1+ φ1)Yt-13 - φ1Yt-14 = at - θ1at-1 – θ2at-12 + θ1θ2at-13
· Yt = (1+ φ1)Yt-1 - φ1Yt-2 + Yt-12 - (1+ φ1)Yt-13 + φ1Yt-14 + at - θ1at-1 – θ2at-12 + θ1θ2at-13

AR(1)	SMA(1)	MA(1)
0.6014	0.9553	-0.431655405

Yt = (1+ 0.6014) Yt-1 – 0.0.6014Yt-2 + Yt-12 - (1+ 0.6014) Yt-13 + 0.6014Yt-14 + 0.6014Yt-14+ at + 0.431655405at-1+0.431655405at-12+(0.431655405x0.9553)at-13
Yt = 1.289 Yt-1 – 0.6014Yt-2 + Yt-12 - 1.2894Yt-13 + 0.6014Yt-14 + at +0.431655405at-1 – 0.9553at-12 + 0.8395at-13
Yt = Yt-12 + [1.2894 Yt-1 - 1.2894Yt-13 - 0.6014Yt-2 + 0.6014Yt-14] +[at + 0.431655405at-1 – 0.9553at-12 + 0.8395at-13]........................Equation (2)

Therefore, Equation (2) is the ARIMA model to predict low-cost housing demand in Johor.

Table 5. The actual data and forecast data using ARIMA (1,0,1) method.

Time series	Actual	Forecasted	|Actual – Forecasted|	Percentage Error
June 2006	199	191	8	4.02
July 2006	275	196	79	28.73
August 06	308	124	84	27.27
September 06	259	255	4	1.54
MAPE				15.39

Table 5 shows the actual and forecasted data from June 2000 to September 2006 using ARIMA (1,0,1) model. From the calculations, MAPE value obtained was 15.39%. Predictive ability is very good if the MAPE is less than 10% while MAPE less than 20% is good [8]. The results show that MAPE value for ARIMA less than 20%.

5 Conclusion

Since the MAPE value is less than 20%, it can be conclude that ARIMA model can predict low-cost housing demand in Johor quite good. It is recommend further study should be done to reduce the error of performance since the results generated are able to assist the construction of low-cost housing scheme in terms of the accuracy of necessity based on actual demand. Subsequently there would be a minimal possibility of the procurement of either under-construction or over construction of low cost houses particularly in the state of Johor.

References

[1] The Tenth Malaysia Plan, (2011-2015), Economic Planning Unit, Prime Minister's Department, Putrajaya, (2010).

[2] Ministry of Housing and Local Government Malaysia, Guidelines for the Implementation of the New Price of Low Cost Housing (Amendment) Act 2002, (2002).
[3] R. Mahamud and K. Hussein, Studies on potential medium income group in own homes in area Johor Bahru, University Technology Malaysia, Vote Upp, 71, 693, (2002).
[4] The Ninth Malaysia Plan (2006-2010), Chapter 21 Providing Housing and Urban Services Quality, Prime Minister's Department, Putrajaya, (2006).
[5] S. Shuid, Low medium cost housing in Malaysia: Issues and challenges, International Islamic University Malaysia, (2004).
[6] N.Y. Zainun, Computerized Forecasting Model to Forecast Low-Cost Housing Demand in Urban Area in Malaysia using Artificial Neural Networks (ANN), PhD Thesis, Loughborough University, (2012).
[7] N.Y. Zainun and M.Z. Abd Majid, Evolution on various forecasting models using artificial neural networks (ANN), Proc. *of the 2nd International Conference on Innovation in Architecture, Engineering and Construction*, United Kingdom, (2003).
[8] S. Harun, Forecasting and Simulation of Net Inflows for Reservoir Operation and Management, PhD Thesis, University Technology Malaysia, Johor, (1999).

MATEC Web of Conferences **47**, 04009 (2016)
DOI: 10.1051/matecconf/20164704009

Assessing the Use of Failure Case Studies in Civil Engineering Education

Wael Elleithy[1], Choong Wee Kang[1] and Lau Teck Leong[1,a]

[1]*Department of Civil Engineering, University of Nottingham Malaysia Campus, Semenyih 43500, Selangor, Malaysia*

Abstract. Civil engineering failures, unfortunately, occur from time to time leading to serious consequences. It is of importance that civil engineering undergraduate students are made aware of cases of failure in the past, and hence, to enhance their ability of identification of potential deficiencies. Investigating past cases of civil engineering failures has been carried out in UNMC for the last five years. The students are required to research, investigate and write essays reporting on these failures, detailing the causes and lessons learned. This study was conducted in order to assess the effectiveness of such approach. The outcomes of this study indicate that the use of civil engineering failure as case studies enhances the students' educational experiences. In addition to the technical learning benefits from past civil engineering failure case studies, there are also impacts on students' attitudes which has significant implications in their expected profession.

1 Introduction

Many civil engineering educators integrate case-based teaching as an inductive method of teaching, see e.g. [1-11] and the references cited therein. As reported in [9], case studies have potential for positive impacts in the affective domain as well as in the cognitive domain. The study of past cases of failure offers the civil engineering students valuable insights into technical, ethical, and professional issues in civil engineering [1, 2, 4, 6]. Approaches for accomplishing this, and the benefits of integration of cases of civil engineering failures are subject of professional discussions [1-9]. A first approach for inclusion of cases of failure into a civil engineering curriculum is to introduce a core or an upper-level module in failure analysis and forensic engineering [2]. However, adding a core module to an already existing full curriculum may lack practicality. A second approach is the inclusion of past civil engineering cases of failure into existing modules [2, 5].

In order to enhance the undergraduate students' understanding and learning experience at UNMC, an additional module was incorporated for the past five years that is substantially devoted to investigate cases of major constructed civil engineering projects [10], case studies of major civil engineering projects employing sustainable agenda [11], and past cases of civil engineering failures. The students are free to choose case studies and are required to find, sort, synthesize and present information relating to the main issues surrounding cases of construction, sustainability of major civil engineering projects, and cases of civil engineering failure. Finally, the students report/present their findings in the form of several essays. The outcomes of [10] indicate that the use of past civil

[a] Corresponding author : teckleong.lau@nottingham.edu.my

engineering projects as case studies enhances the students' educational experiences in higher education in the context of quality cognitive learning. A great majority of the students reported that the case studies are necessary and beneficial to their educational experience, and should be integrated into further civil engineering modules. The outcomes of [11] indicate that the use of major civil engineering projects employing sustainable agenda as case studies develops the students' appreciation of the contemporary engineered environment and sustainability in civil engineering. A common view by the lecturers and students, as indicated in [11], is that the degree course needs to further cover the key issues of sustainable construction.

This paper reports on the results and assesses the use of failure case studies in civil engineering education. The impacts of the use of failure case studies on students' learning and educational experiences were assessed through surveys and focus group discussions. An outline of the paper is as follows. The employed methodology is presented in the succeeding chapter. The students' and lecturers' feedback are presented and analysed in the Survey Questionnaires chapter.

2 Methodology

Surveys were conducted on three different groups of Year 2 and Year 3 students in addition to twelve of civil engineering academics with five to twenty years of teaching experience (Table 1). The questionnaire used in this survey was designed to determine how influential is the civil engineering failure case studies on students' aspiration for an ethical career. Furthermore, the lecturers' and students' views on the extent of coverage and the method of inclusion of cases of civil engineering failures as part of the civil engineering curriculum are evaluated.

Fully completed surveys on the use of past civil engineering failures as case studies were analysed. Simple descriptive statistics was mostly used in summarizing students' and lecturers' responses. When appropriate, differences between different groups' responses are highlighted. The results and analysis are presented in the succeeding chapter.

Table 1. Participating groups of students and lecturers.

Group	Description	Year of study	Year of survey	Size
S1	Students	2011-12 (Yr 2)	2012-13 (Yr 3)	43
S2	Students	2012-13 (Yr 2)	2012-13 (Yr 2)	61
S3	Students	2013-14 (Yr 2)	2013-14 (Yr 2)	43
L	Lecturers		2013-14	12

3 Survey Questionnaires

3.1 Students' feedback

The survey instrument is shown in Table 2(a) and (b). In what follows, the analysis of the students' responses is presented.

Table 2(a). Use of failure case studies in civil engineering - students' survey instrument.

Q1.a	"Have the investigated cases of failure led to major research studies?"
Q1.b	"Have the civil engineering cases of failure contributed to change in/improvement of the engineering practice?"
Q2.a	"Have the civil engineers who were involved in the design/construction faced any legal/ professional consequences?"
Q2.b	"How do the consequences/liabilities that civil engineers face affect your motivations in becoming a practicing civil engineer?"
Q3	"Has your study/investigation of past civil engineering failure cases motivated you for a better quality self-learning?"

Table 2(b). Use of failure case studies in civil engineering - students' survey instrument.

Q4	"Has your study/investigation of past civil engineering failure cases motivated you to practice ethically in your future career?"
Q5	"In which way should the case studies be included in civil engineering modules?"
Q6	"Should the civil engineering failure cases be included in further civil engineering modules?"

Q1.a: 78% of the students indicated that the investigated cases of failure have led to major research in civil engineering (Figure 1). 59%, 85%, and 78% were S1, S2 and S3 students, respectively. The response of group S1 students' with regards to project led research was expected as the survey was conducted while this group of students is in their third year.

Q1.b: The results (Figure 2) indicate that the students have developed an increased awareness of lessons learned from failure cases and the consequent improvement of the engineering practice.

Q2.a: 77% of the students indicated that the engineers involved in the investigated civil engineering cases of failure faced legal/professional consequences (Figure 3). 78%, 77%, and 76% were S1, S2 and S3 students, respectively. Among the three different groups, the reported levels of awareness of consequences faced by engineers were nearly identical.

Q2.b: 63% of the students were affected positively (Figure 4). 70%, 54%, and 68% were S1, S2 and S3 students, respectively. An increase in taking the challenges positively was observed was observed for S1 students.

Q3: Among the three different groups of students, a nearly identical level of motivation for better self-learning was reported (Figure 5).

Q4: 92% of the students have been motivated for ethical practice (Figure 6). 98%, 90%, and 92% were S1, S2 and S3 students, respectively. An increased motivation was observed for S1 students.

Q5: Analysis of results (Figure 7) indicates an increased preference of group/discussion based inclusion for S1 students.

Q6: 85% of the students prefer inclusion of case studies in further modules (Figure 8). A slight decrease in desire for inclusion in further modules was noted for S1 students.

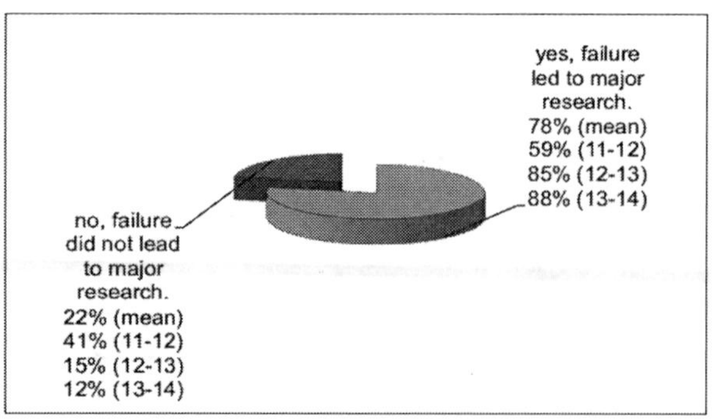

Figure 1. Leading to major research.

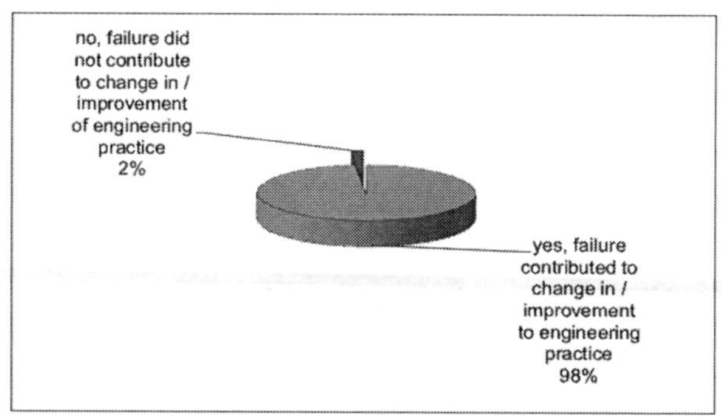

Figure 2. Contribution to changes in practice.

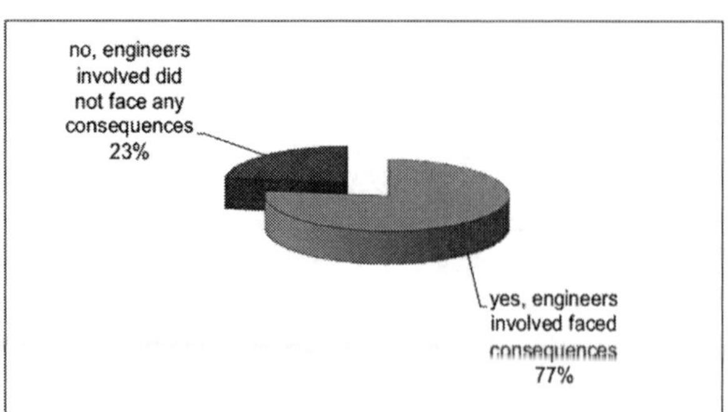

Figure 3. Consequences faced.

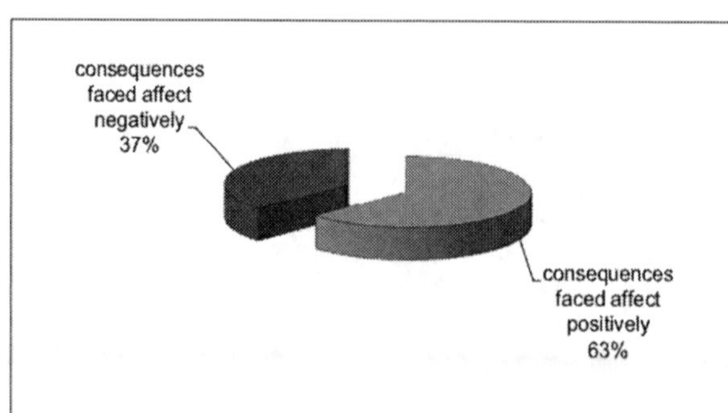

Figure 4. Motivation - practicing civil engineers.

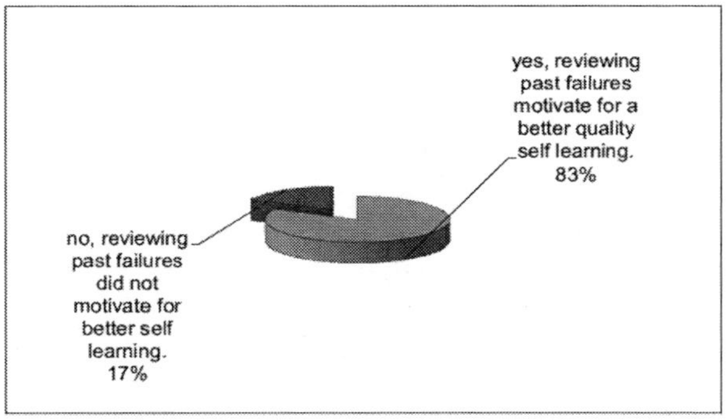

Figure 5. Motivation for self-learning.

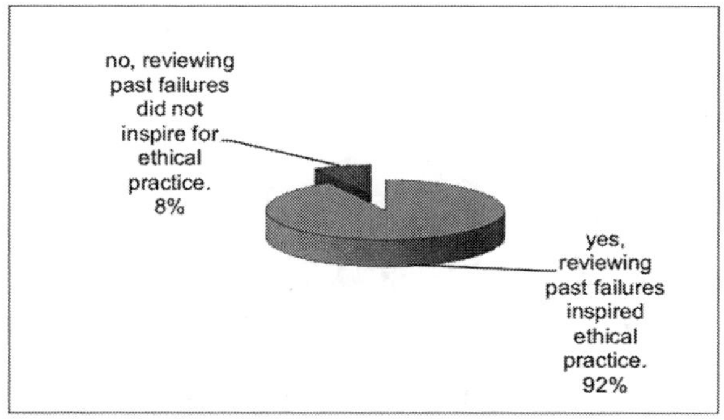

Figure 6. Motivation for ethical practice.

It is noted that the trend of the students' responses in this section is comparable to those reported in [10] and [11], on the use of case studies in construction of major civil engineering projects and sustainability of major civil engineering projects, respectively.

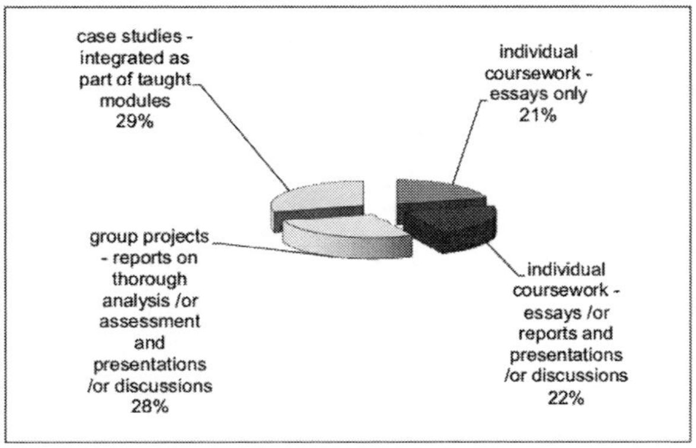

Figure 7. Preferred method of inclusion.

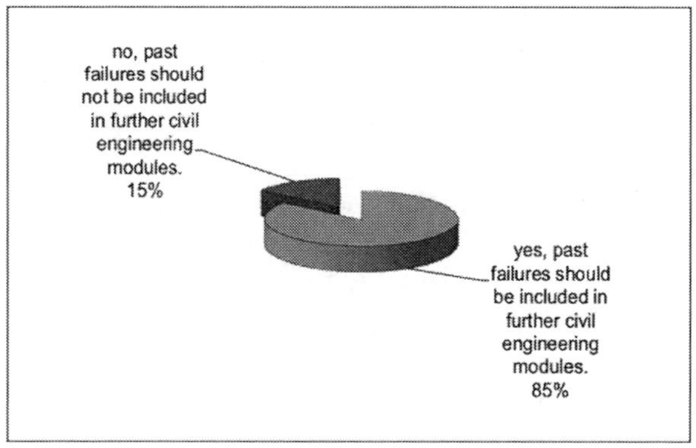

Figure 8. Inclusion in further modules.

3.2 Lecturers' feedback

A questionnaire comprising 5 questions was designed to investigate the lecturers' views on use of past civil engineering failures as case studies (Table 3).

Table 3. Use of failure case studies in civil engineering – lecturers' survey instrument.

Q1	"How would the consequences/liabilities that are faced by civil engineers involved in the design/construction of the studied/investigated past civil engineering failure cases affect the students' motivation in becoming civil engineers?"
Q2	"Would the study/investigation of past civil engineering failure cases motivate the students for a better quality self-learning?"
Q3	"Would the study/investigation of past civil engineering failure cases motivate students to practice ethically in their future career?"
Q4	"Would the study/investigation into critical information, such as: whether there were public enquiries held and the causes of failure of the studied/investigated cases lead to a better students' understanding in Civil Engineering?"
Q5.a	"Should past civil engineering failure cases be included in further civil engineering modules?"
Q5.b	"Indicate how would be the most appropriate methods to be included in the civil engineering modules."

In what follows, the analysis of the lecturers' responses is presented:

Q1: The lecturers' expectations (Figure 9) exceeded those of S1, S2 and S3 students by 13%, 29% and 15%, respectively.

Q2: The students' expectations (Figure 5) exceeded the reported lecturers' ones (Figure 10).

Q3: As in Q2, the students' expectations (Figure 6) exceeded those of the surveyed lecturers (Figure 11). The reported results of Q2 and Q3 may be attributed to students' enthusiasm.

Q4: The results (Figure 12) indicate that 75% of the lecturers believe that investigation into critical information leads to a better students' understanding in civil engineering.

Q5.a: 83% of the lecturers suggested that civil engineering failure cases should be included in further civil engineering modules (Figure 13).

Q5.b: The results (Figure 14) indicate that 39% and 28% of the lecturers recommend having the case studies as group and individual projects coupled with presentations/discussions, respectively. 28% of lecturers recommend having the case studies as parts of the taught modules. 5% of the lecturers recommend maintaining the case studies in essays form.

It is noted that the trend of the lecturers' responses in this section is comparable to those reported on the use of case studies in sustainability of major civil engineering projects [11].

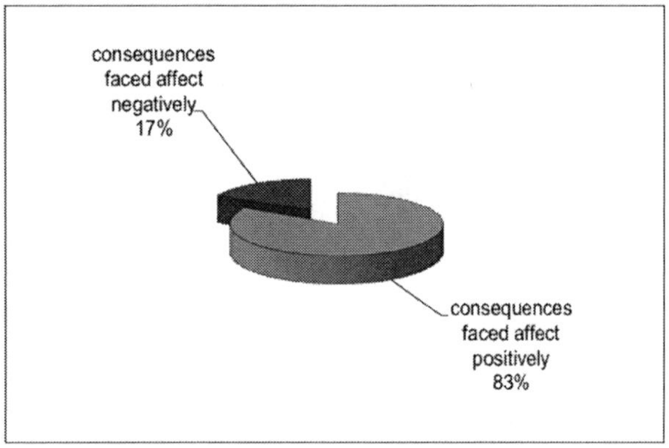

Figure 9. Motivation - practicing civil engineers.

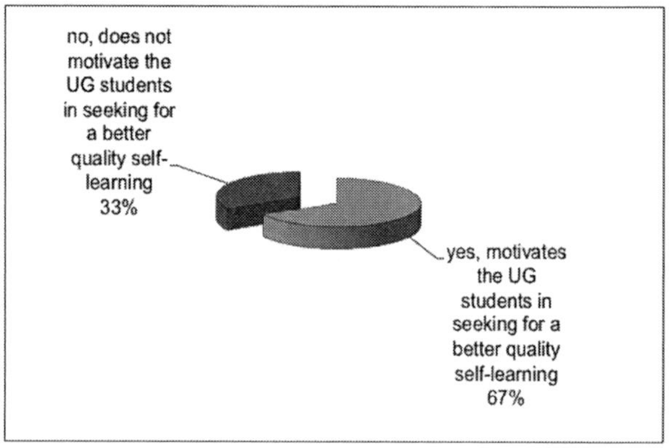

Figure 10. Motivation - self-learning.

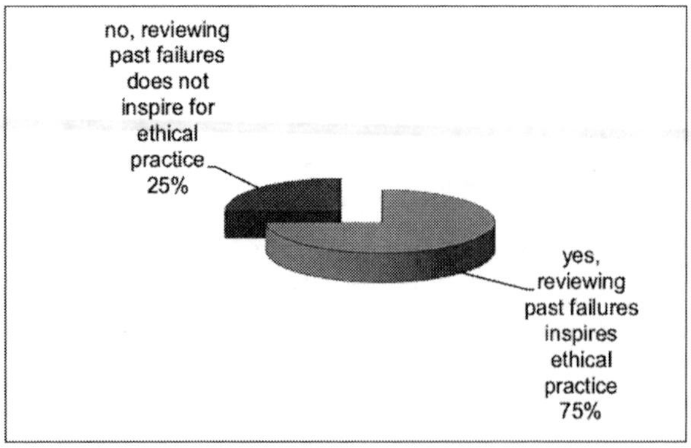

Figure 11. Motivation - ethical practice.

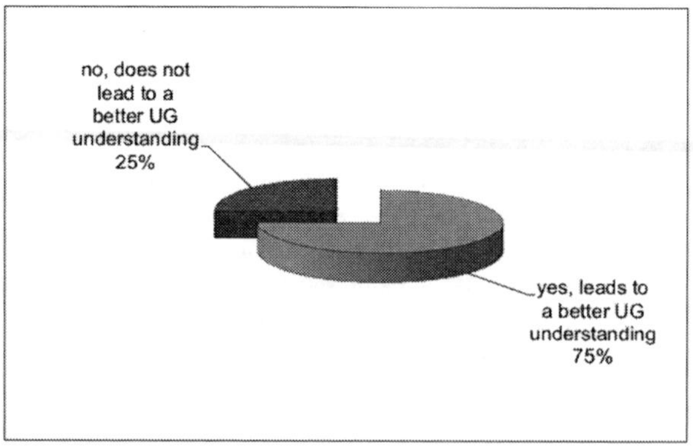

Figure 12. Investigation - critical information.

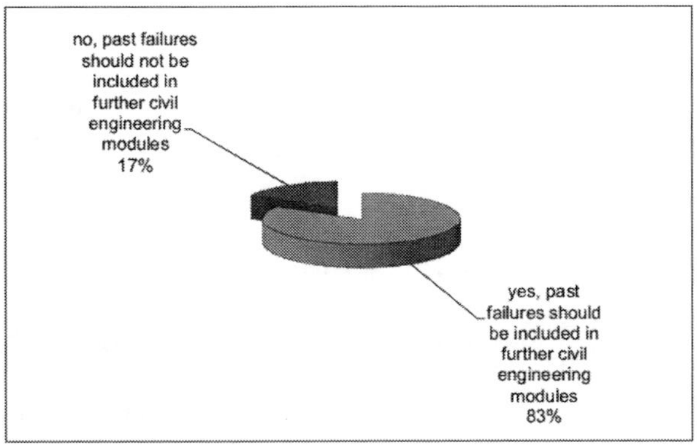

Figure 13. Inclusion of cases of failure.

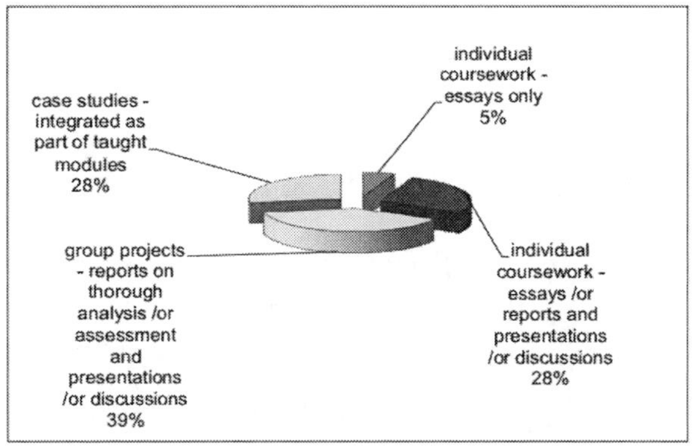

Figure 14. Integration of cases of failure.

4 Summary

Lessons learned from case studies have substantially affected the practice of civil engineering. According to Delatte et al. [9] "the history of the development of practice in many engineering disciplines is, in large part, the story of failures, both imminent and actual, and of the changes to designs, standards and procedures made as the result of timely interventions or forensic analyses".

It is of importance that civil engineering undergraduate students are made aware of cases of failure in the past. In this study, data was collected from academics involved in supervising students carrying out case studies of civil engineering failures and different groups of civil engineering students about their perceptions of this approach in improving student technical abilities and professional and ethical responsibilities. The study provided some insights into the value that civil engineering academics and students place on the use of failure cases in civil engineering.

The authors believe that civil engineering failure cases should be spread through the course modules, rather than developing and delivering specific modules focusing in failure analysis and forensic engineering. The authors, furthermore, believe that civil engineering failure cases make students aware of their professional/ethical responsibilities, and increase students' ability to apply knowledge of civil engineering to real life problems. This confirms with the surveyed students' and lecturers' responses, and the results presented in [9].

The presented results and analysis give only a brief of the lecturers' and different groups of civil engineering students' views at UNMC. A comparative investigation is underway of the appreciation, attitudes and perceptions regarding civil engineering failure cases in civil engineering involving civil engineering academics and students at three different campuses of the University of Nottingham. Comparison between students in successive years of the program and in different campuses would be strengthened by calculation of statistical significance for the given populations, and sample sizes needed to obtain sufficient statistical power.

Acknowledgement

The support of the University of Nottingham Malaysia Campus is acknowledged.

References

[1] N.J. Delatte, Using failure case studies in civil engineering education, Forensic Engineering, K.L. Rens, O. Rendon-Herrero, P.A. Bosela, eds., ASCE, 430-440, (2000).

[2] N.J. Delatte, K.L. Rens, Forensics and case studies in civil engineering education: State of the art, Journal of Performance of Constructed Facilities, **16**, 98-109, (2002).

[3] N.J. Delatte, Using failure case studies in civil engineering courses, *Proc. of the 2003 American Society for Engineering Education Annual Conference and Exposition*, Nashville, (2003).

[4] N.J. Delatte, Case studies for civil engineering educators, *Proc. of the 2005 Forensic Engineering Symposium/Structures Congress*, New York, 1539-1555, (2005).

[5] A. Adekoya and J. Patel, The use of failure case studies to enhance students' understanding of structural behavior and ethics, *International Conference on Engineering Education*,15-20, (2006).

[6] N. Delatte, *Beyond Failure: Forensic Case Studies for Civil Engineers*, ASCE Publications, (2009).

[7] N. Delatte, Failure literacy in structural engineering, Engineering Structures, **32**, 1952-1954, (2010).

[8] T.A. Newson and N.J. Delatte, Case methods in civil engineering teaching, Canadian Journal of Civil Engineering, **38**, 1016-1030, (2011).

[9] N. Delatte, P. Bosela J. Bagakas, Implementation and assessment of failure case studies in the engineering curriculum, Forensic Engineering, 458-466, (2012).

[10] T.L. Lau and W. Elleithy, The use of past projects as means to enhance the education experience of civil engineering students, Applied Mechanics and Materials, **357-360**, 2634-2639, (2013).

[11] W. Elleithy and T.L. Lau, Sustainable construction - the use of case studies in civil engineering education, *International Conference on Learning and Teaching in Computing and Engineering*, Kuching, 151-157, (2014).

MATEC Web of Conferences 47, 04010 (2016)
DOI: 10.1051/matecconf/20164704010

Impact of Past Project on the Educational Experience of Civil Engineering Students

Lau Teck Leong[1,a] and Wael Elleithy[1]

[1]*Department of Civil Engineering, University of Nottingham Malaysia Campus, 43500 Semenyih, Selangor, Malaysia*

Abstract. Civil engineering lecturers need to develop effective early emphasis and educational challenges in order to prepare the current students with in-depth fundamental principles, applicable technical knowledge, and realistic appreciation on challenges faced by practicing civil engineers. In view of its importance, the use of past civil engineering projects as case studies have been integrated into the existing civil engineering curriculum at the University of Nottingham Malaysia Campus. In order to enhance our undergraduate students' understanding, cognitive learning desire, and overall learning experience, an additional module has been incorporated in the past five years to investigate major past civil engineering projects. Students are free to choose any of past projects as their case studies and are required to find, sort, synthesize, and present critical elements that contributed to the overall success of the past projects in the form of discussion essays. In this paper, the universal consensus between the lecturers and the students on the effectiveness of past civil engineering projects as case studies to enhance students' lifelong learning desire and aspiration towards ethical practice are investigated. The lecturers' and students' views on the extent of coverage of the key elements and the preferred disciplines from the past projects are furthermore evaluated. The outcomes indicate that the use of past projects as case studies enhanced students' educational experiences that conduce for cognitive learning desires (73% of lecturers & 81% of students), motivated for future ethical practice (82% of lecturers & 86% of students), and reassured aspiration (100% of lecturers & 97% of students) in becoming practicing civil engineers upon graduation. A common view derives from this study is that the curriculum needs to further its extent of coverage on the industry practices.

1 Introduction

In recent years, increasing complexities of civil engineering projects in the context of technical, organizational, regulatory, and economical requirements have been noted [1]. Therefore, civil engineering lecturers need to develop effective educational approaches that embed sufficient and appropriate challenges in order to prepare the current civil engineering students with in-depth fundamental principles, applicable technical knowledge, and appreciation towards realistic technical challenges faced by practicing civil engineers. These have become increasingly important as the rate of participation in worldwide universities increases [2], which in turn, increases the rate of diversity among undergraduate (UG) students. The diversity is not limited to their academic quality, but will be

[a] Corresponding author : teckleong.lau@nottingham.edu.my

brought forwards with them into their professions [3]. In the context of civil engineering practice, these diversities can lead to significant adverse effects to the safety of the society.

In view of the importance of the quality of graduates and as a great challenge faced in engineering profession is engineering ethics [4], an early emphasis via means of specific module has been carried out in the University of Nottingham Malaysia Campus (UNMC) for the last five years to stimulate cognitive learning and motivate for future ethical practice via investigation into past civil engineering projects in the second year of its undergraduate studies [6]. Such practice is to ensure for the delivery of the paramount importance in undergraduate education [5], that is, the students' ability of identification of potential deficiencies through fundamental principles learnt during their year one, and with the aim to motivate for lifelong learning desire. Furthermore, this practice aims to make the students aware of the challenges faced and the advancements derived from past projects. Students are free to choose any past projects as their case studies and are required to find, sort, synthesize, and present critical elements that contributed to the overall success of the past projects in the form of discussional essays.

In this paper, the universal consensus between the lecturers and the students on the effectiveness of past civil engineering projects as case studies to enhance students' lifelong learning desire and aspiration towards ethical practice are investigated. The lecturers' and students' views on the extent of coverage of the key elements and the preferred disciplines based on the findings from the investigation into past civil engineering projects are furthermore evaluated.

2 Methodology

Surveys in the form of questionnaire were conducted on eleven civil engineering academics, with five to twenty years of teaching experience, who have supervised UG students (Year 2) carrying out case studies in the form of essays into past civil engineering projects. The questionnaire used in this survey was designed to determine how influential is case studies on students' aspiration for an ethical career, continual lifelong learning desire, and educational experiences. Furthermore, the questionnaire was designed to evaluate lecturers' and students' views on the extent and the preferred disciplines for further integration. The survey results are summarized and analyzed in the following section in the form of pie charts and are compared with the previous results [6] from thirty-nine Year 2 and forty Year 3 students. Discussions of the results are, furthermore, presented. furthermore evaluated.

3 Analysis of Results and Discussions

Question (1): How do challenges faced by engineers involved affect students' aspirations?

The results (Figure. 1a) indicate that 100% of the lecturers believe the students would appreciate the challenges faced by engineers involved. Whilst, previous results [6] (Figure. 1b) on students' view indicate that 97% of the students take the challenges positively, while 3% take it negatively.

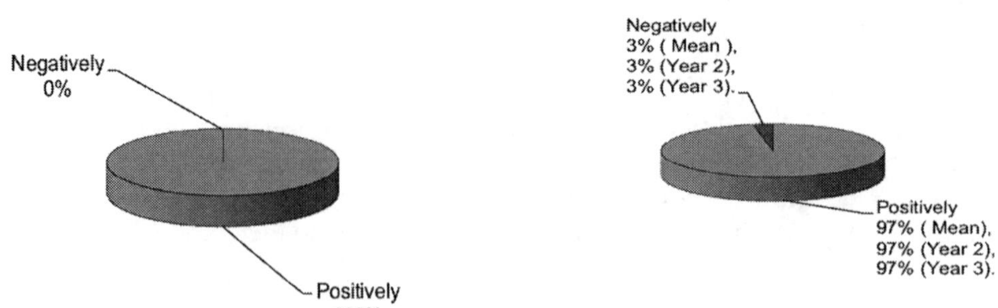

(a) Lecturer's view (b) Students' view [6]

Figure 1. Lecturers' and students' views on students' aspiration in becoming practicing civil engineers.

Although there are 3% of the students did not take challenges positively, majority of the students took the challenges faced positively as expected by their lecturers. Such agreements between lecturers and students confirms the use of past projects motivate students' aspiration in becoming a practicing civil engineer.

Question (2): Does study /or investigation of past civil engineering projects motivate students in seeking for a better quality self-learning?

The results (Figure. 2a) indicate that 73% of the lecturers believe the students after investigation of past project would be motivated for a better quality self-learning, while 27% of the lecturers do not think so. Whilst, previous results [6] (Figure. 2b) indicate that 81% (74% of Year 2, 88% of Year 3) of the students were motivated for a better quality self-learning, while 19% (26% of Year 2, 12% of Year 3) of the students were not motivated.

Comparing with the lecturers' views, the students exceeded (8%) the lecturers' expectations on self-motivation for quality self-learning. Further analysis into these results indicates an increase in motivation for better self-learning as these students progress into their year 3. Such increase is believed to be attributed to the continual development of the students' technical competencies, and hence, the ability to recognize their limitations thus desire for quality self-learning.

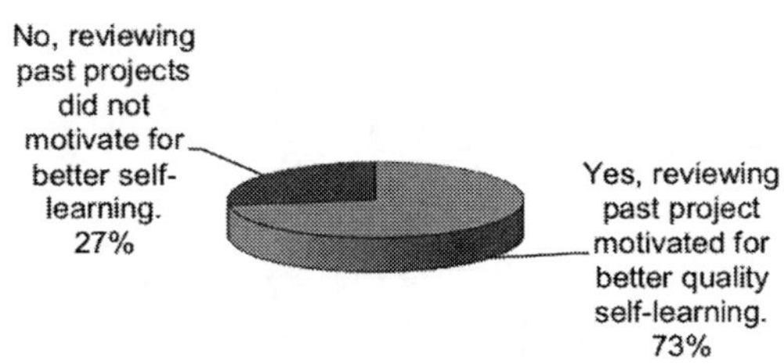

(a) Lecturer's view

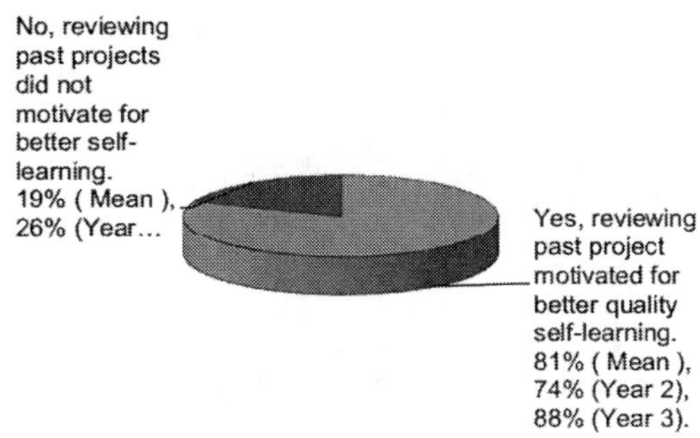

(b) Students' view [6]

Figure 2. Lecturers' and students' views on students' motivation for better quality self-learning.

Question (3): Does study /or investigation of past civil engineering projects inspire students to practice ethically in their future career?

The results (Figure. 3a) indicate that 82% of the lecturers believe the students would be inspired for ethical practice, while 18% do not think so. Whilst, previous result [10] (Figure. 3b) indicates that 86% (85% of Year 2, 87% of Year 3) of the students were inspired for ethical practice, while 14% (15% of Year 2, 13% of Year 3) of the students were not.

Comparing with the lecturers' views, the students exceeded (4%) the lecturers' expectations on self-inspiration for ethical practice. Further analysis into the results indicates a gradual increase in inspiration as these students progress into their year 3. Such increase is believed to be attributed to the students' increasing exposure to engineering ethical issues during their industrial placements and those that have been embedded in the civil engineering program.

Question (4): In your opinion, does study /or investigation of past civil engineering projects lead to a better students' understanding in Civil Engineering?

The results (Figure. 4a) indicate that 73% of the lecturers believe the students would be led to a better self-understanding, while 17% do not think so. Whilst, previous results [6] (Figure. 4b) indicates that 90% (86% of Year 2, 94% of Year 3) of the students believe reviewing the past project have led to a better self-understanding, while 10% (14% of Year 2, 6% of Year 3) of the students did not think so.

Comparing the lecturers' view, the students exceeded the lecturers' expectation (17%) in self aspiration for better understanding. Further analysis into these results indicates an increase as these students progress into their year 3. Such increase is as expected, as it is attributed to the continual development of students' technical competencies and the ability to recognize the importance of quality understanding.

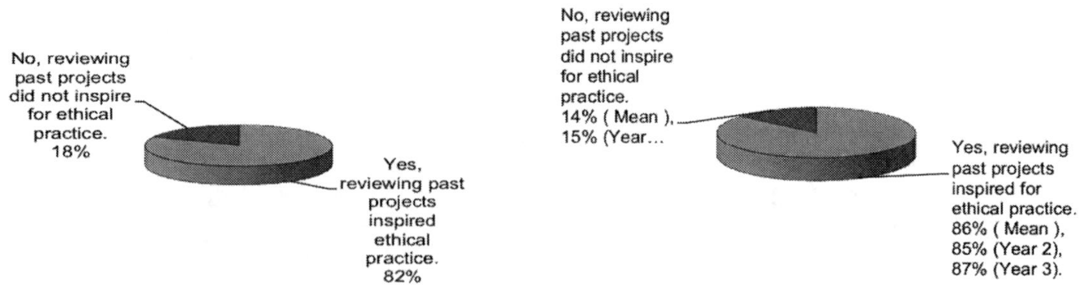

(a) Lecturer's view (b) Students' view [6]

Figure 3. Lecturers' and students'views on students'students' inspiration for ethical practice in future career.

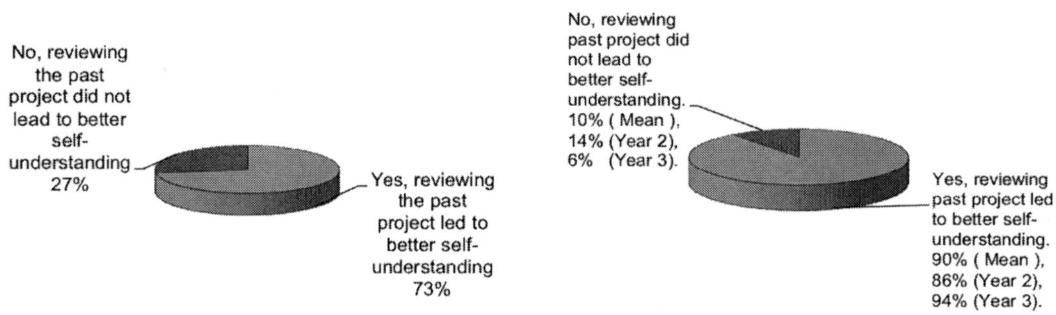

(a) Lecturer's view (b) Students' view [6]

Figure 4. Lecturers' and students'views on students'students' desire for better understanding.

Question (5): Indicate your preference in which way should the case studies be included in Civil Engineering modules/courses?

The results (Figure. 5a) indicate that 38% of the lecturers preferred to have the case studies as group projects, 32% as part of the taught modules, 27% as individual coursework in the form of report coupled with presentations /or discussions, and the remaining 11% of the students preferred to maintain the case studies in essays form. Whilst, previous results [6] (Figure. 5b) indicate 31% of the students preferred to be as group projects, 27% as part of the taught modules, 23% as individual coursework, and the remaining 19% preferred to maintain in essays form.

Despite some differences noted in the percentage distributions, the overall ranking of preferences are essentially identical between the lecturers and the students. That is, group projects (first preference), part of taught module (second preference), individual report coupled with presentation /or discussion, and lastly, the individual essay. However, further analysis of results indicate an increase in preference for group or discussion based integration with respect to coursework or taught based integration, as students progress into their year 3.

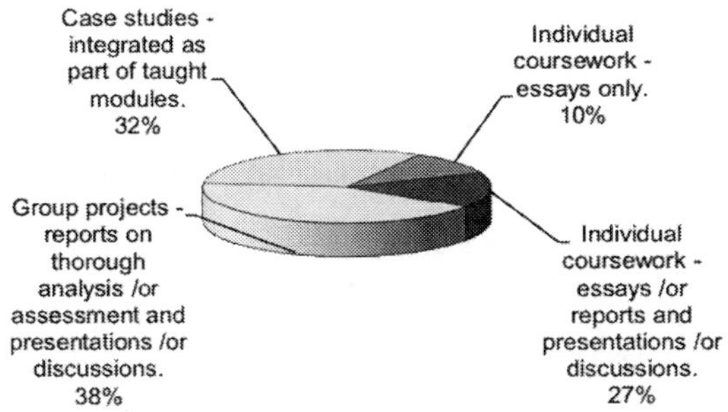

(a) Lecturer's preference

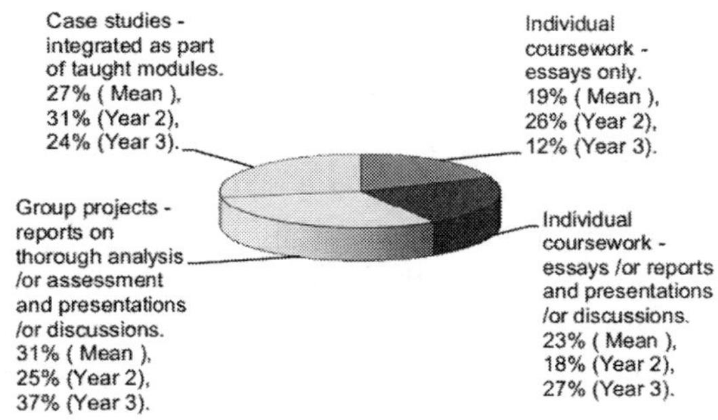

(b) Students' preference [6]

Figure 5. Lecturers' and students' preference on integration of past projects.

Question (6a): Should the past projects be included in further Civil Engineering modules/courses?

The results (Figure. 6a) indicate that 100% of the lecturers preferred to have the case studies to be integrated into further modules. Whilst, previous results [6] (Figure. 6b) indicate that only 83% of the students preferred to have the case studies integrate into further modules, while 17% preferred not to. However, further analysis into these results indicates an increase in students' desire for integration as these students process into their year 3.

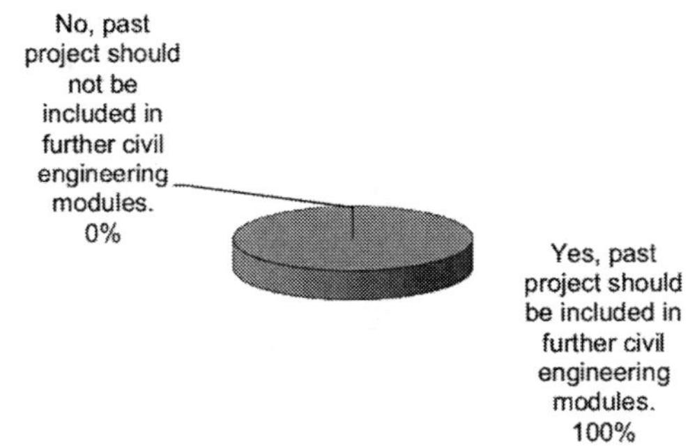

(a) Lecturer's preference

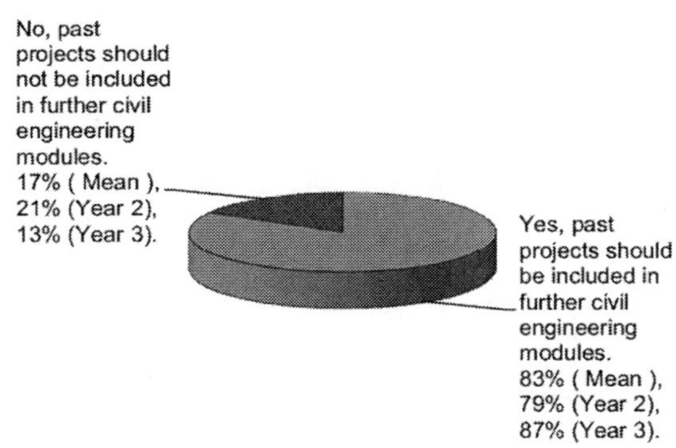

(b) Students' preference [6]

Figure 6: Lecturers' and students' preference on on inclusion of past projects.

Question (6b): If yes, please indicate the preferred courses/disciplines.

The result (Figure. 7a) indicates that 29% of the lecturers preferred to have the past project included in structural modules, 28% in hydraulic modules, 26% in geotechnical modules, 17% in transportation modules, and 0% in management. Whilst, previous results [10] (Figure. 7b) indicates that 42% of the students preferred structures, 28% management, 18% geotechnics, and 8% hydraulic modules, and 4% transportation modules. Comparing the lecturers' view with respect to that of the students indicates major differences in preferences, that is, hydraulic has risen from the forth to the second place, transportation has risen from the last to the forth place, management has dropped from the second to the last place, while, structures & geotechnics remains un-changed.

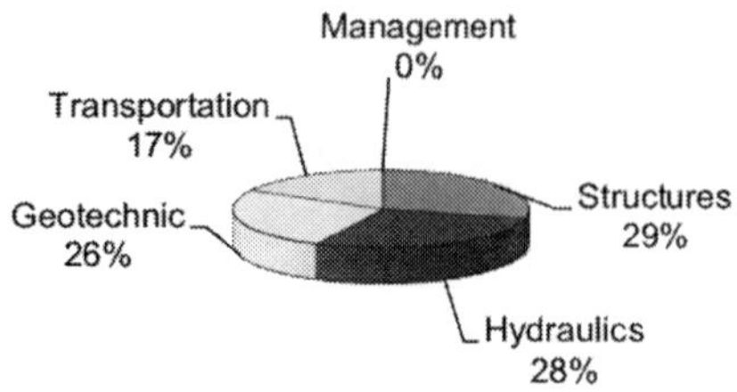

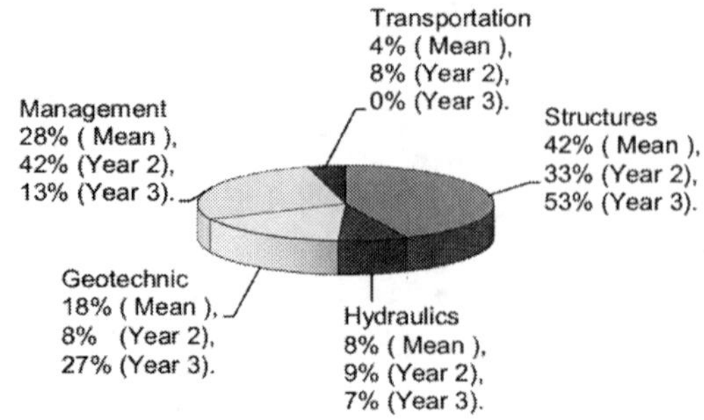

(a) Lecturer's preference (b) Students' preference [6]

Figure 7. Lecturers' and students' preference among civil engineering disciplines.

4 Conclusions

The outcomes of this study indicate that the lecturers' views agree with the students' views at UNMC. The outcomes indicated that he use of past civil engineering projects enhances the students' educational experiences in higher education in the context of quality cognitive learning, and in their understanding/appreciation of civil engineering practice. Challenges faced by engineers involved influence our students, in a positive manner, and instill the level of moral for future ethical practice. All academics and majority of our students felt that the case studies are necessary and beneficial to their educational experience, and should be integrated into further civil engineering modules. Our students appreciate the importance of quality self-learning during their higher education. The use of past civil engineering projects reassures their aspiration in becoming civil engineers upon graduation.

Acknowledgements

The support of the University of Nottingham Malaysia Campus is acknowledged.

References

[1] R. Pietroforte, Civil engineering education through case studies of failures, Journal of performance of constructed facilities, **12**, 51-55, (1998).
[2] P.G. Altbach, L. Reisberg and L.E. Rumbley, Trends in Global Higher Education: Tracking an Academic Revolution, Report for UNESCO 2009 World Conference on Higher Education, (2009).
[3] P. Aruna, The problem with fresh grads, (2011). <available from: http://www.thestar.com.my/story/?file=%2f2011%2f7%2f17%2fnation%2f9103031 &sec=nation> (accessed on 30th July 2015).
[4] W.A. Wulf, Great achievements and grand challenges, The Bridge, (2000). <available from: http://www.asce.org/communty/environmental/nae_gage.cfm.> (accessed on 15th July 2015).
[5] F.C. Hadipriono, Analysis of events in recent structural failures, Journal of Structural Engineering, **111**, 1468-1481, (1985).
[6] T.L. Lau and W. Elleithy, The use of past projects as means to enhance the education experience of civil engineering students, Applied Mechanics and Materials, **357-360**, 2634-2639, (2013).

MATEC Web of Conferences **47**, 04011 (2016)
DOI: 10.1051/matecconf/20164704011

A Conceptual Cost Benefit Analysis of Tailings Matrices Use in Construction Applications

Ali A. Mahmood[1,a] and Maria Elektorowicz[2]

[1]*Department of Civil Engineering, University College of Technology Sarawak, 96000 Sibu, Sarawak, Malaysia*
[2]*Department of Building Civil and Environmental Engineering, Concordia University, Montreal, Quebec, H3G 1M8, Canada*

Abstract. As part of a comprehensive research program, new tailings matrices are formulated of combinations of tailings and binder materials. The research program encompasses experimental and numerical analysis of the tailings matrices to investigate the feasibility of using them as construction materials in cold climates. This paper discusses a conceptual cost benefit analysis for the use of these new materials. It is shown here that the financial benefits of using the proposed new tailings matrices in terms of environmental sustainability are much higher when compared to normal sand matrices.

1 Introduction

Mine tailings are the leftovers from mining activity and they contain all the hazardous materials used in the milling process with the exception of the ore. Tailings are traditionally stored in ponds not far away from the mine they originated from. Mine tailings have been the subject of several environmental catastrophes such as the Merriespruit dam tailings failure that resulted in the deaths of 17 people and extensive damages to a residential township in South Africa in 1994 (Fourie and Papageorgiou [1] and Fourie et al. [2]). Another disaster took place in the Philippines in 2002 were tailings spilled into Mapanuepe Lake and eventually into the Saint Tomas river. Low lying villages were flooded with mine waste. At least 250 families were evacuated from that area [3].

Researchers around the world, having realized the potential dangers involved in the traditional method of storage of mine tailings in ponds, took up the task of attempting to beneficially re-use these tailings or at least modify them in a way less harmful to the environment. Researchers like Demers and Haile [4], Zou and Sahito [5] Celik et al. [6] Swami et al. [7] and Roy et al. [8] attempted to investigate the possibility of using mine tailings as filler or construction materials with the addition of binders that include Portland cement. Other researchers like Fall et al. [9], Fall et al. [10], Benzaazoua et al. [11], Ercikdi et al. [12], Fall and Pokharel [13], Ercikdi et al. [14] and Helinski et al. [15] investigated the use of mine tailings in the most traditional manner by using them as cemented paste backfill. Portland cement, however, is relatively expensive, hence the need to use additives other than cement to reduce cost without compromising effectiveness. This effectiveness is specifically important in cold climates such as that of Canada.

[a] Corresponding author : ali.mahmood@ucts.edu.my

Calsifrit is a special cementitious material that was developed recently by NovaFrit International [16]. It is a totally amorphous material, a matrix of calcium and sodium fluoro-aluminosilicate. It is a homogenous solid substance possessing high reactivity potential and shows cementitious properties when finely ground. Calsifrit will be used in conjunction with slag and fly ash as Portland cement replacements mixed with mine tailings. These combinations will form the mine tailings matrices under consideration herein. No research to date has been published on the conceptual cost benefit analysis of using mine tailings-Calsifrit matrices in construction.

This paper discusses a conceptual cost benefit analysis of re-using these mine tailings matrices in the construction of road bases. This analysis is part of a broader and more comprehensive experimental and numerical research program into the re-use of mine tailings matrices as construction materials in cold regions [17].

2 Conceptual Framework

A cost benefit analysis example on the use of these tailing matrices in construction is given here. This cost benefit analysis is implemented to ascertain the benefits involved in the use of these tailing matrices in the road construction industry. This analysis is primarily based on the environmental sustainability that the use of these matrices carries in comparison with the traditional, soil based methods.

This section attempts to quantitatively outline some of the benefits involved in the use of tailings matrices in construction, specifically road construction. In the current study, it is thought that the following approach is the best for this study and will be adopted by the identification of materials and costs and benefits, quantification of costs and benefits and then, that will be followed by the discussion and conclusion.

The two types of materials involved that make up the cost benefit analysis comparison are the following:

a. The tailing matrices including Ordinary Portland Cement (OPC), Calsifrit, fly ash and slag
b. Granular material normally used as sub-base in roads [18]

It is assumed in this analysis that two typical road sub-base sections are built that incorporate either one of these materials. The following is the identification of costs and benefits:

Benefits:

a. Tailings dusting is fully eliminated using this new technique. This dusting is considered a serious environmental incident prompting Hudson's Bay Mining and Smelting Company Limited to issue a control guidelines report for their tailings impoundment systems [19]
b. Failure of tailings dams with all the imposed hazards due to the traditional storage of tailings in ponds. A study by Rico et al. [20] reports 147 worldwide tailings dam disasters. Once the solidified/stabilized tailings matrices are in-place, this hazard is eliminated, thereby saving lives and property that are otherwise prone to tailings dangers
c. Heavy metal seepage to the ground water table is eliminated or much reduced by this approach of stabilizing/solidifying the tailings material
d. Freeing up the space previously occupied by the tailings dam
e. Freeing up resources and personnel connected with the supervision on the tailings dam
f. Using up industrial wastes such as fly ash and slag
g. Using Calsifrit, identified above

Costs:

a. Trucks to transfer to the site the tailings from local mines in the area
b. Machinery that spread and compact the tailings in the base course layer to the required thickness and density
c. Specialized workforce
d. Importation of OPC

3 Discussion

Cost benefit analysis attempts to put a monetary value on all items included in the analysis. However, this section is more focused on a conceptual quantitative analysis related to the environmental sustainability of the tailings matrices and their effects. Hence a procedure explained by Shutt [21] and used and cited by Amjad [22] will be used here whereby symbols will be given based on the importance priority of each cost or benefit item.

Amjad [22] performed a cost benefit analysis for three potential alternatives for a road construction project through an industrial-residential town. He was able to demonstrate through the use of a combined conceptual-monetary approach the financial implication of each alternative and the best solution for the problem.

When analyzing environmental issues, it is often difficult to assign explicit monetary values to the gains associated with sustainability schemes. Therefore, based on that, and in order to quantify the un-quantifiable, it is decided here to assign an abbreviation for each of the following basic values: M for monetary benefits, S for increased safety, I for intangibles and T for time saving. These abbreviations denote annual flow of cost or benefit. Also, a plus sign (+) will be added to denote benefits and a minus (-) to denote costs. A duplication of the sign will indicate added importance. Tables 1 and 2 show the benefits and costs associated with this approach for the tailings and sand matrices respectively.

Table 1. Cost benefit analysis of the use of tailings matrices.

Factor	Benefits	Costs	Balance
Tailings dusting	S+ I+		S+ I+
Tailings dam failures	S++ M+ T+		S++ M+ T+
Heavy metal seepage to the ground water	S+ M+ T+		S+ M+ T+
Freeing up the space previously occupied by the tailings dam	M+		M+
Freeing up resources and personnel connected with the supervision on the tailings dam	M+ T+ I+		M+ T+ I+
Fly ash	S+	M-	S+ M-
Calsifrit	S+	M-	S+ M-
Slag	S+	M-	S+ M-
Trucks		M-	M-
Machinery		M-	M-
Workforce		M-	M-
On Site Delivery	M+		M+
Importing OPC		M-	M-

M: monetary benefits, S: increased safety, T: time saving, I: intangibles

Table 2. Cost Benefit Analysis of the use of sand matrice

Factor	Benefits	Costs	Balance
Fly ash	S+	M-	S+ M-
Calsifrit	S+	M-	S+ M-
Slag	S+	M-	S+ M-
Trucks		M-	M-
Machinery		M-	M-
Workforce		M-	M-
On Site Delivery		M-	M-
Importing OPC		M-	M-

M: monetary benefits, S: increased safety, T: time saving, I: intangibles

It is seen through Table 3 that the financial implications of using the sand matrices are higher than those of the tailings matrices.

Table 3. Summary of findings and order of preference.

Factor	Tailings Matrices	Sand Matrices	Order of Preference	
			Tailings Matrices	Sand Matrices
Tailings dusting	S+ I+	-	1	2
Tailings dam failures	S++ M+ T+	-	1	2
Heavy metal seepage to the ground water	S+ M+ T+	-	1	2
Freeing up the space previously occupied by the tailings dam	M+	-	1	2
Freeing up resources and personnel connected with the supervision on the tailings dam	M+ T+ I+	-	1	2
Fly ash	S+ M-	S+ M-	1	1
Calsifrit	S+ M-	S+ M-	1	1
Slag	S+ M-	S+ M-	1	1
Trucks	M-	M-	1	1
Machinery	M-	M-	1	1
Workforce	M-	M-	1	1
On Site Delivery	M+	M-	1	2
Importing OPC	M-	M-	1	1
Preference Totals			13	19
Overall Order of Preference			1	2

M: monetary benefits, S: increased safety, T: time saving, I: intangibles

4 Conclusion

It can be seen that financial benefits in terms of environmental sustainability are much higher for the tailings matrices in comparison to normal sand matrices. Consequently, it can be concluded that tailings matrices present a higher sustainability priority material for use in road sub-bases.

References

[1] A.B. Fourie and G. Papageorgiou, Defining an appropriate steady state line for Merriespruit gold tailings, Canadian Geotechnical Journal, **38**(4), 695-706, (2001).

[2] A.B. Fourie, G.E Blight and G. Papageorgiou, Static liquefaction as a possible explanation for the Merriespruit tailings dam failure, Canadian Geotechnical Journal, **38**(4), 707-719, (2001).

[3] Wise, World Information Service on Energy-Uranium Project, (2012). <available at: http://www.wise-uranium.org/index.html>

[4] B. Demers and G. Haile, Management of tailings stabilized by lime and cement at Canadian electrolytic zinc, Valleyfield, Quebec, *Proc. of Sudbury 2003 Mining and the Environment*, Sudbury, Ontario, (2003).

[5] D.H. Zou and W. Sahito, Suitability of mine tailings for shotcrete as a ground support, Canadian Journal of Civil Engineering, **31**(4), 632–636, (2004).

[6] O. Celik, I.Y. Elbeyli and S. Piskin, Utilization of gold tailings as an additive in Portland cement, Waste Management and Research, **24**(3), 215-224, (2006).

[7] R.K. Swami, N.K.S. Pundhir and S. Mathur, Kimberlite tailings a road construction material, Transportation Research Record: Journal of the Transportation Research Board, **2**(1989), 131–134, (2007).

[8] S. Roy, G.R. Adhikari and R.N. Gupta, Use of gold mill tailings in making bricks, a feasibility study, Waste Management and Research, **25**(5), 475–482, (2007),

[9] M. Fall, M. Benzaazoua and S. Ouellet, Experimental characterization of the influence of tailings fineness and density on the quality of cemented paste backfill, Minerals Engineering, **18**(1), 41–44, (2005).

[10] M. Fall, M. Benzaazoua and E.G. Saa, Mix proportioning of underground cemented tailings backfill, Tunnelling and Underground Space Technology, **23**(1), 80–90, (2008).

[11] M. Benzaazoua, B. Bussiere, I. Demers, M. Aubertin, E. Fried and A. Blier, Integrated mine tailings management by combining environmental desulphurization and cemented paste backfill, Application to mine Doyon, Quebec, Canada, Minerals Engineering, **21**(4), 330–340, (2008).

[12] B. Ercikdi, A. Kesimal, F. Cihangir, H. Deveci and I. Alp, Cemented paste backfill of sulphide-rich tailings, importance of binder type and dosage, Cement and Concrete Composites, **31**(4), 268–274, (2009).

[13] M. Fall and M. Pokharel, Coupled effects of sulphate and temperature on the strength development of cemented tailings backfills, Portland cement-paste backfill, Cement and Concrete Composites, **32**(10), 819–828, (2010).

[14] B. Ercikdi, F. Cihangir, A. Kesimal, H. Deveci and I. Alp, Effect of natural pozzolans as mineral admixture on the performance of cemented-paste backfill of sulphide-rich tailings, Waste Management and Research, **28**(5), 430–435, (2010).

[15] M. Helinski, M. Fahey and A. Fourie, Coupled two-dimensional finite element modeling of mine backfilling with cemented tailings, Canadian Geotechnical Journal, **47**(11), 1187-1200, (2010).

[16] NovaFrit International, 1200 Garnier Street, Ville Ste-Catherine, Quebec, J0L 1E0, (2010).

[17] A.A. Mahmood, Experimental and computational assessment of tailings binder matrices for construction purposes in cold regions, PhD Thesis, Concordia University, Montreal, (2012).

[18] Foth and Van Dyke Associates, All season road construction cost analysis-Clark County, Wisconsin, Technical Report, Wisconsin, (2003).

[19] Hudson Bay Mining and Smelting Co. Limited, FFTIS Dust Control Guidelines, Document No. LAI-612, Flin Flon, Manitoba, (2007).

[20] M. Rico, G. Benito, A.R. Salgueiro, A. Díez-Herrero and H.G. Pereira, Reported tailings dam failures, a review of the European incidents in the worldwide context, Journal of Hazardous Materials, **152**(2), 846-852, (2008).

[21] R.C. Shutt, Economics for the construction industry, Addison Wesley Long Man Limited, Harlow, Essex, (1997).

[22] A.A. Amjad, Cost benefit analysis for construction projects, IEP-SAC Journal, 85-90, (2005).

MATEC Web of Conferences **47**, 04012 (2016)
DOI: 10.1051/matecconf/20164704012

Quantity Surveying Undergraduates' Awareness on Cost Significant of High-Rise Condominium Projects in Malaysia: The Case of a Private University in Malaysia

Lee Wah Peng[1,a], Ng See Seng[1], Toh Tien Choon[1], Lim Cheng Sim[1], Yong Ching Khian[2] and Goh Kai Chen[3,f]

[1]*Department of Surveying, Lee Kong Chian Faculty of Engineering and Science, Universiti Tunku Abdul Rahman, 43000 Kajang, Selangor, Malaysia*
[2]*Department of Mathematical and Actuarial Sciences, Lee Kong Chian Faculty of Engineering and Science, Universiti Tunku Abdul Rahman, 43000 Kajang, Selangor, Malaysia*
[3]*Department of Construction Management, Faculty of Technology Management and Business, Universiti Tun Hussein Onn Malaysia, 86400 Parit Raja, Johor, Malaysia*

Abstract. Recent developments in quantity surveying (QS) emphasised the importance of identifying cost significant elements (CSE). The knowledge on CSE of high-rise condominium projects (HRCP) is essential as high-rise residential multi-unit projects are the next option in building construction due to limited land areas in urban areas. This study aims to determine the levels of awareness among QS undergraduates of a private university in Malaysia on CSE of HRCP in Malaysia. The respondents' knowledge on CSE has not achieved a satisfactory level. Both male and female respondents have the same levels of awareness on CSE. Remedial strategies to improve this situation are recommended.

1 Introduction

Cost significant technique is a cost estimating method. Poh and Horner [1] demonstrated the way the cost-significance theory can create a simple, well-structured, and sufficiently accurate measurement method by deriving a cost-significant model for student hostels in Singapore. Using a cost significant technique, Tas and Yaman [2] developed a generic computer-based cost estimating model for Turkish public building construction projects for use in their detailed design phase. These studies produced their cost significant estimating models based on cost significant items or work packages. However, recent developments in the area of QS underscored the importance of identifying CSE. Cartlidge [3] explained that for elemental cost estimating for building works, cost checking on the CSE must be conducted against every pre-established cost target from the Formal Cost Plan 2 Stage. Obviously, different types of buildings will have different CSE. The one and only comprehensive theory on CSE and their effects can be found in Smith and Jaggar [4]. They provided a good explanation on CSE and their effects for high-density residential multi-unit projects.

HRCP can be seen easily in urban or city areas because housing developments are moving towards constructing vertically rather than horizontally due to limited land areas. The cost planner should be able to make all project team members aware of the implications of any decision that they make on the

[a] Corresponding author : leewp@utar.edu.my

building's design. Hence, this requires the cost planner to be equipped with a sound knowledge on CSE. This knowledge can be learned from courses during the study in the university or from working experiences. In the HRCP, the CSE are: Frame (Columns, Upper Floors), Finishes (Wall, Ceiling and Floor), Envelope (External Walls and Windows), Services, Internal Subdivision (Internal Walls, Screens and Doors), Preliminaries, and Fittings whereas the non CSE are: External Doors, Stairs, Roof, External Works, and Substructure [4]. Studies on CSE are vital, new, and so far, only received little attention both in Malaysia and overseas. So, this study aims to determine the levels of awareness among QS undergraduates of a private university in Malaysia on CSE of HRCP in Malaysia.

2 Method

The quantitative approach, specifically, a questionnaire survey methodology is adopted in this study in determining the levels of awareness among QS undergraduates on CSE of HRCP. Questionnaire responses based on a Likert scale (1 = strongly disagree, 2 = disagree, 3 = neither agree nor disagree, 4 = agree, 5 = strongly agree) can be utilised to effectively and efficiently test their understandings on the knowledge about CSE compared with the qualitative approach where the data are to be collected in the form of words or texts. In the January 2015 trimester, the total of 219 QS undergraduates who are in their third year out of the four-year bachelor degree programme, have been invited to take part in the questionnaire survey. These students became the targeted respondents because they have been exposed to subjects such as construction technologies, construction measurements and estimating, and construction economics. This means that they should already have the knowledge necessary for identifying CSE of a building. The final year QS undergraduates are not being targeted in this study because it would be difficult to implement any remedial strategies if later on it was to be found that their levels of awareness on CSE was actually low, seeing that they have graduated by then.

An online version of the survey questionnaire has been developed using the systems provided by SurveyMonkey.com. Before the full-fledged survey is carried out, a pilot study is conducted with 3 academic staffs and 7 respondents in identifying the weaknesses in the survey questionnaire. It is found that all of them are very satisfied with the survey questionnaire and thus the first draft of the survey questionnaire does not need any modification and is then considered final. The web address of the online survey questionnaire has been posted in the university's web-based learning environment (WBLE) in the form of a hyperlink. All the targeted respondents who are already the subscribers of the WBLE can access to the online survey questionnaire by clicking that hyperlink. A total of twelve statements about CSE and a question on gender are included in the survey questionnaire. The Likert scale (1 = strongly disagree, 2 = disagree, 3 = neither agree nor disagree, 4 = agree, 5 = strongly agree) is applied to each of the statements. Thus, the potential respondents will need to indicate their levels of agreement or disagreement with the statements about CSE. All questions require only an answer for each question. The respondents should answer all questions. One of the researchers has briefed the respondents about the questionnaire survey before they started to respond.

The questionnaire survey is conducted in the 2nd week of March 2015. A gentle reminder is sent out after a week since the data collection process started. All the 212 survey questionnaires, as the other 7 respondents who took part in the pilot study are excluded from the full-fledged survey, are collected in 10 days time. All the respondents have responded to the survey questionnaire and that all the questions have been answered. The response rate is 100%. The 212 (males = 107; females = 105) fully completed survey questionnaires collected are analysed statistically. In defining the levels of awareness explicitly, four levels of awareness according to the calculated mean for each statement, specifically, (1 – 2 = slightly aware, 2.01 – 3 = somewhat aware, 3.01 – 4 = moderately aware, and 4.01 – 5 = extremely aware) are derived. In addition, it is also very interesting to find out if there is a difference in terms of the levels of awareness on CSE between the male and female respondents. Thus, the null hypothesis: there is no difference in terms of the levels of awareness on CSE between the male and female respondents; and that the alternative hypothesis: there is a significant difference in terms of the levels of awareness on CSE between the male and female respondents are to be tested.

3 Results

Table 1 shows the ranking of overall respondents' levels of awareness on CSE of HRCP in Malaysia. Frame (columns, upper floors) is ranked 1st with a mean of 3.93. Finishes (wall, ceiling and floor) are ranked 2nd with a mean of 3.59. External doors are ranked 3rd with a mean of 3.49. Envelope (external walls and windows) is ranked 4th with a mean of 3.42. Services are ranked 5th with a mean of 3.35. Internal subdivision (internal walls, screens and doors) is ranked 6th with a mean of 3.31. Preliminaries are ranked 7th with a mean of 3.21. Fittings are ranked 8th with a mean of 3.11. Stairs are ranked 9th with a mean of 2.98. Roof is ranked 10th with a mean of 2.77. External works are ranked 11th with a mean of 2.61. And lastly, substructure is ranked 12th with a mean of 2.13. Table 2 indicates the respondents' levels of awareness on CSE of HRCP in Malaysia based on gender. Apart from the differences in the ranking of the statements, both the male and female respondents perceived that frame (columns, upper floors) is the most significant building element affecting the HRCP cost in Malaysia, and they thought that roof, external works, and substructure are building elements that are of cost significance in the HRCP in Malaysia. Table 3 shows the results of the Mann-Whitney U test comparing the male and female respondents' levels of awareness on CSE of HRCP in Malaysia. For each of the 10 out of the total of 12 statements, namely, S02 to S04 and S06 to S12, the probability value (p) is not less than or equal to 0.05, so the result is not significant. For each of the remaining 2 out of the total of 12 statements, namely, S01 and S05, the p-value is less than 0.05, so the result is significant. This means that, generally; there is no statistically significant difference in the levels of awareness on CSE between the male and female respondents. Thus, the null hypothesis is accepted at a 95% confidence level and that the alternative hypothesis is rejected (p-value > 0.05).

Table 1. Ranking of overall respondents' levels of awareness on CSE of HRCP in Malaysia.

Statement	Mean	Standard Deviation	Rank
S02 – Frame (Columns, Upper Floors) is a cost significant element.	3.93	0.818	1
S08 – Finishes (Wall, Ceiling and Floor) are a cost significant element.	3.59	0.952	2
S06 – External Doors are NOT a cost significant element.	3.49	0.990	3
S04 – Envelope (External Walls and Windows) is a cost significant element.	3.42	0.983	4
S10 – Services are a cost significant element.	3.35	0.984	5
S07 – Internal Subdivision (Internal Walls, Screens and Doors) is a cost significant element.	3.31	0.953	6
S12 – Preliminaries are a cost significant element.	3.21	1.041	7
S09 – Fittings are a cost significant element.	3.11	1.033	8
S03 – Stairs are NOT a cost significant element.	2.98	0.976	9
S05 – Roof is NOT a cost significant element.	2.77	1.000	10
S11 – External Works are NOT a cost significant element.	2.61	1.013	11
S01 – Substructure is NOT a cost significant element.	2.13	0.866	12

4 Discussion

The results reveal that both the male and female students have the same levels of awareness on CSE of HRCP in Malaysia. It therefore proves that gender is not a barrier in understanding the knowledge of construction particularly about CSE in this context. It also shows that the female students can also master the knowledge about construction as good as their male counterparts in an area which is male dominant from project conception to project completion. Based on this evidence, it has been expected that both the male and female students can equally provide advices on matters related to CSE.

Table 2. Respondents' levels of awareness on CSE of HRCP in Malaysia based on gender.

Statement	Gender					
	Male		Female		Total	
	Mean	Standard Deviation	Mean	Standard Deviation	Mean	Standard Deviation
S01	2.26	0.904	2.00	0.809	2.13	0.866
S02	3.92	0.859	3.95	0.777	3.93	0.818
S03	3.04	0.980	2.91	0.972	2.98	0.976
S04	3.37	0.957	3.48	1.010	3.42	0.983
S05	2.93	1.016	2.62	0.965	2.77	1.000
S06	3.60	0.999	3.37	0.973	3.49	0.990
S07	3.25	0.982	3.37	0.923	3.31	0.953
S08	3.55	0.954	3.63	0.953	3.59	0.952
S09	3.00	1.037	3.23	1.021	3.11	1.033
S10	3.41	1.027	3.30	0.940	3.35	0.984
S11	2.64	1.039	2.57	0.989	2.61	1.013
S12	3.19	1.100	3.23	0.983	3.21	1.041

Table 3: Results of the Mann-Whitney U test comparing the male and female respondents' levels of awareness on CSE of HRCP in Malaysia

Statement	Mann-Whitney U	Wilcoxon W	Z	Asymptotic Significance (2-tailed)
S01	4718.500	10283.500	-2.219	0.026 *
S02	5589.500	11367.500	-0.072	0.943
S03	5123.500	10688.500	-1.157	0.247
S04	5137.500	10915.500	-1.155	0.248
S05	4626.000	10191.000	-2.353	0.019 *
S06	4883.000	10448.000	-1.728	0.084
S07	5166.000	10944.000	-1.073	0.283
S08	5306.500	11084.500	-0.766	0.444
S09	4887.500	10665.500	-1.710	0.087
S10	5284.000	10849.000	-0.792	0.428
S11	5353.500	10918.500	-0.634	0.526
S12	5521.500	11299.500	-0.224	0.823

a. Grouping Variable: Gender
* Significant as p-value < 0.05

On the one hand, the students are moderately aware that frame (columns, upper floors) is of cost significance. This may be because they understand that the frame is the building element that forms the building's skeleton [5] and it is usually made of reinforced concrete that is expensive. However, based on these very obvious reasons, it was expected that the students would be able to confidently tell or be extremely aware that the frame is indeed a cost significant element. This result signifies that there are some students who are still not very sure about it and therefore will affect their capabilities to advise others regarding the CSE of a project.

Finishes (wall, ceiling and floor) are another building element that has been perceived to be of cost significance. This could be because condominiums are luxurious residential units that will have large quantities of higher quality of internal finishes such as odourless paints for the wall finishes, plastered ceilings for the ceiling finishes, and homogeneous tiles for the floor finishes. The effects of the uses of different finishes on construction costs which have been stressed in the subjects like construction technologies are accounted for the students' understanding on this matter. Similarly, however, it was expected that they would be extremely aware that the finishes are in fact of cost significance. This result also shows that some students might not understand it as well as others. Nonetheless, the ability of the students in ranking the finishes as the next most important cost significant element after the frame worth complimenting as this is probably the case from what the researchers observed.

On the other hand, external doors are a building element that is not of cost significance seeing that they are usually less expensive than other building elements such as the frame and the finishes and are only in small quantities. Again, the students are only moderately aware of it. Indeed, this is the only non cost significant element which they are able to recognise. While recognising the CSE of HRCP is important, recognising those that are not of cost significance is equally important. So, when teaching the students, emphasis must be put in explaining the reasons why some building elements are CSE and the reasons why others are not. The students seem to be lack of confidence in this matter.

This scenario is evident as the students could not demonstrate that the building elements of stairs, roof, external works, and substructure are in fact not CSE for the HRCP. On top of that, the absence of statements about CSE that they are extremely aware of is worrying. This implies that the students are not performing well in this particular area. The overall mean for all the 12 statements on CSE is equal to 3.16, indicating that in general the students are merely moderately aware of the CSE of HRCP. This further reveals that their knowledge on CSE is rather weak and therefore improvements are necessary in order to enable them to be able to advise the project team members particularly the client or the developer professionally as early as during the internship when they are in their senior years of study or as soon as completing their study from the university.

While the learning outcomes related to the CSE are not provided explicitly in the syllabuses of the subjects of construction technologies, construction measurements and estimating, and construction economics, the student after studying these subjects should already gain some theoretical knowledge on CSE for a given project. This may be a new challenge for educators to instill relevant knowledge to their students but the desired result is achievable. Hence, it is recommended that the contents of CSE should be emphasised more in the relevant subjects and that their learning outcomes are to be written out clearly in the syllabuses. The students' improvements in this particular area will no doubt increase the competitiveness and add value to the profession. Besides that, workshops, seminars, and talks on CSE can also be organised for the sake of the students in the effort to increase the levels of awareness on CSE among the QS undergraduates.

References

[1] P.S.H. Poh and R.M.W. Horner, Cost-significant modelling – its potential for use in south-east Asia, Engineering, Construction and Architectural Management, **2**(2), 121-139, (1995).

[2] E. Tas and H. Yaman, A building cost estimation model based on cost significant work packages, Engineering, Construction and Architectural Management, **12**(3), 251-263, (2005).

[3] D. Cartlidge, Estimator's *Pocket Book*, Routledge, New York, (2013).

[4] J. Smith and D. Jaggar, *Building Cost Planning for the Design Team*, Butterworth-Heinemann, Burlington, (2007).

[5] M. Mehta, W. Scarborough and D. Armpriest, *Building construction: Principles, Materials, and Systems*, Pearson Education, New Jersey, (2003).

MATEC Web of Conferences **47**, 04013 (2016)
DOI: 10.1051/matecconf/20164704013

The Insight of Industrialised Building System (IBS) By Bumiputera Construction Players

W. M. N. W. Muhammad[1,a], M. A. Azman[1], M. K. F. Othman[1], N. A. Hadi[1], S. N. S. Sahimi[1] and M. F. Mohammad[2]

[1]*Faculty of Architecture, Planning and Surveying, Universiti Teknologi Mara, 94300 Kota Samarahan, Sarawak, Malaysia*
[2]*Faculty of Architecture, Planning and Surveying, Universiti Teknologi Mara, 40450 Shah Alam, Selangor, Malaysia*

Abstract. The term Industrialised Building System (IBS) is widely recognised by construction players in Malaysia since its first implementation since 1960s for IBS pilot projects of Pekeliling Flats and Rifle Range Flats. The aim for the implementation is to promote better system in delivering construction end-products which offers efficiency and effectiveness. At the same time, Bumiputera construction players are the majority of parties in construction industry especially in Small and Medium Enterprises (SMEs). Therefore, the purpose of this paper is to identify the perception and awareness of IBS implementation among Bumiputera players. It is found that, Bumiputera construction players have sound knowledge on the IBS and optimistic towards further implementation for future projects. Nonetheless, issues on payment methods and coordination of IBS project delivery are found to be the negative perceptions which are considered as hindrance for Bumiputera construction players' involvement in IBS project.

1 Introduction

The Malaysian Construction Industry has recently been geared by the government with the term of Industrialised Building System (IBS) to represent the adoption of the construction industrialised and the use of prefabrication of components in building construction. To make the interpretation clearer, IBS is defined as a construction technique in which components are manufactured in a controlled environment (on or off site), transported, positioned and assembled into a structure with minimal additional site work [1, 2, 3]. To support this definition, an Industrialised Building System may be defined in which all building components such as wall, floor slab, beam, column and staircase are mass produced either in factory or at site under strict quality control and minimal on site activities [4]. This profitable system can produce high quality product in future, minimise the resource wastage and enhance the value of end products. This is further supported with the argument that an industrialisation process is an investment in equipment, facilities, and technology with the objective of maximising production output, minimising labour resource, and improving quality [4]. Construction Industry Development Board (CIDB) has classified the system into 6 categories which are precast concrete framed system, formwork system, steel framing system, prefabricated timber framing

[a] Corresponding author : neodanish@gmail.com

system, block work system and innovative system [5]. The classification is being classified according to the structural aspects for each component.

IBS has been successfully implemented in Japan, United Kingdom, Sweden and also Netherlands [5]. Apart from that, this IBS system has also being applied in Hong Kong and Singapore [6]. It has been well establish in many part of the world. The implementation of IBS in Malaysia is not at infancy stage. IBS has been already being introduced in early 1960s. This was when Public Work Department (PWD) and Ministry of Housing and Local Government (MHLG) of Malaysia visited several European countries and evaluate their housing development program. IBS in Malaysia was first initiated in two pilot projects in Jalan Pekeliling, Kuala Lumpur and Rifle Range, Penang with a series of 17 and 18-storey flats. Throughout the period, IBS was anticipated to change and to be used in a broader scale [6].

Key players who are directly connected to IBS construction are client, designer, contractor and specialist manufacturer [6]. Bumiputera contracting firms only started to get in the industry after the application of the New Economic Policy (NEP) in 1970 [7]. The government has implemented the NEP with the purpose to attract more Bumiputera businesses to get involved in economic activities. Construction industry was one of the main sectors that recorded high involvement of Bumiputera entrepreneurs [7]. To improve this, the government plays an important role in promoting the Malay or Bumiputera entrepreneur's development in the industry. In order to operate any construction, contracting firms need to get registered with two government agencies which are Pusat Khidmat Kontraktor (PKK) and also Construction Industrial Development Board (CIDB). PKK has been establishing in year of 1981 where their responsibilities are to register, lead and train especially the Bumiputera entrepreneurs that involves in the industry. Meanwhile, CIDB was only established 13 years after that which is in 1994. The objectives of the establishment are to control all activities in the industry especially on the contractor's registration, worker training, site safety, material controlling and also on the foreign workers. With the significant involvement of Bumiputera in Malaysia construction industry, it is important to gain the Bumiputera construction players' insight towards the adaptation and implementation of IBS among them. This is indispensable in order to sustain the involvement of Bumiputera in construction industry as well as to keep abreast with the current needs of IBS application by construction players. Several initiatives are already being conducted by Government and CIDB in order to promote IBS among Bumiputera construction players such as tax reduction incentives for Bumiputera components manufactures since 2005 and offer training scheme and financial loans for Bumiputera manufacturer and installer since 2004 [8]. However, the participation from Bumiputera being part of the construction players as to be a contractor neither to be a manufacturer was insufficient in IBS practice. This is highlighted by the failure in achieving Key Performance Index (KPI) for the Bumiputera Development category in IBS Roadmap 2003-2010 [9]. Hence, it is important to focus on the participation by Bumiputera since previous literature has shown that one of the barriers of the implementation of IBS is due to lack of involvement by Bumiputera as the contractors, erectors and manufacturers [3].

2 Methodology and Findings

The focus of this study is on getting the insight of the Bumiputera construction players towards IBS implementation in Malaysia. The quantitative research has been carried out with the respondents for this study comprise of Bumiputera constructors as the party that directly involved in the utilisation of IBS in construction and Bumiputera client as the owner as well as the end user of the building. The total of 250 questionnaires were prepared and distributed to both parties in gaining the respondent's feedback and 29.2% (73) of questionnaires distributed were returned. The returned questionnaire were being analyse using Statistical Package for Social Science (SPSS) version 16.0. The number of questionnaires received is considered acceptable enough by considering the population studied. The non-parametric statistical methods such as Mann-Whitney U was selected due to the nature of the instrument designed through the use of questionnaire which was assumed to be skewed. The results have been tabulated in sections. Section 1 presents respondents' background such as designation, experience in the construction industry and IBS project. All these data are nominal scales. Other sections (2 and 3) are as regards to the understanding of IBS and the perception and readiness of IBS usage in construction projects.

2.1 Section 1: Respondents' background

Table 1 depicts the majority of respondents are constructor which comprise of Bumiputera contractors and supplier with 82.19% of total respondents. In terms of their background in construction industry, majority of respondents are considered at senior level since 46.58% of respondents are having more than 10 years' experience. On the other hand, their experience in IBS projects are not abreast with their construction industry experience where majority of the respondents having less than 5 years' experience in IBS project.

Table 1. Respondent's designation and experience.

Category	Classification	Frequency	Percentage (%)
Designation	Client	13	17.81
	Constructor	60	82.19
Respondents' experience in the construction projects	< 5 years	14	19.18
	5-10 years	25	34.25
	>10 years	34	46.58
Respondents' experience in the IBS projects	<1 years	15	20.00
	< 5 years	22	69.33
	5-10 years	5	6.67
	>10 years	3	4.00

2.2 Section 2: Understanding IBS

Realibility Statistics on Section 2 is shown in the Table 2. The alpha value shows that the reliability of scale is considered acceptable which closed to 0.6 (alpha is 0.557) [10]. One sample t-test shows that all factors are statistically significant ($p<0.05$, 2-tailed).

Table 2. Reliability Statistics on Section 2.

Cronbach's Alpha	Cronbach's Alpha Based on Standardized Items	N of Items
0.557	0.563	7

Table 3 shows the mean score for respondent's understanding on IBS. This is crucial to gather this information in order to identify level of understanding and knowledge of IBS among Bumiputera construction players. There are 7 relevant variables have been asked to the respondents in getting the feedback. Generally, all respondents rated mean above 3.00, which ranging from 3.3919 to 4.1892. Out of all variables, it is found that most of respondents agree on the argument that IBS manage to improve and modernise the conventional or traditional construction method (mean = 4.1892). This is then followed by the awareness among the respondents on the government roles on the enforcement to the construction players in adopting IBS in construction projects. Nevertheless, it is indispensable to note the lowest rank variable by the respondents which is towards negative variable. Respondents found that IBS in Malaysia is not well accepted by construction players in the supply chain since the IBS operation payment is difficult. This is due to the condition where the payment method being used for IBS project is identical to payment method for conventional or traditional project. This has discouraged construction players in engaging to IBS. Therefore, the payment method and mechanism is required in bettering the situation since IBS adopters require reliable and safer payment and procurement [5].

Table 3. Respondent's understanding on IBS.

Respondent's Understanding on IBS				
Variables	**N**	**Mean**	**Std. Deviation**	**Ranking**
Familiarity with the IBS construction methods	75	3.4400	1.01662	6
Government enforcement in adapting IBS in construction projects	75	3.8533	0.84938	2
IBS usage gives the opportunity to all construction players to participate in IBS projects	74	3.6216	0.85533	4
IBS can improve and modernize the conventional construction method	74	4.1892	0.77062	1
The price of IBS components is controlled by manufacturers/suppliers	75	3.7467	1.05386	3
IBS in Malaysia is not well accepted by construction players in the supply chain because the IBS operation and payments are difficult	74	3.3919	0.88855	7
Lack of support and slow adoption of IBS from the private sector	73	3.5205	0.91454	5

2.3 Section 3: Perception and readiness of IBS usage in construction projects

Realibility Statistics on Section 3 is shown in the Table 4. The alpha value indicates that the reliability of scale is acceptable (alpha is 0.60). The original value is 0.477 if all questions remained in the analysis. Therefore, one factor was deleted to improve the reliability. One sample t-test shows that all factors are statistically significant ($p<0.05$, 2-tailed).

Table 4. Reliability statistics on section 3.

Cronbach's Alpha	Cronbach's Alpha Based on Standardized Items	N of Items
0.477	0.812	12

Table 5 shows the perception and readiness by the respondents towards the usage of IBS in construction project. It is found that the majority of respondents are in favour that it is not difficult in learning and being skilful in adapting IBS in the construction project (mean score = 4.5068). This indicates that Bumiputera construction players are keen and interested in the application of IBS or their projects. This is supported with the awareness on the need to train conventional small and medium scale contractors on the application of IBS which promotes and adding value for them and widening the opportunity in construction industry. In addition, installing IBS components require precision and stability for instance, the installation for suspended beam and slab should be perfectly levelled and positioned as per stipulated in the design [11]. Besides, specialized and additional engineering knowledge are required in designing, manufacturing and constructing a functional IBS components. It is necessary for all parties involved from designers to erectors to have sufficient and decent knowledge on IBS practice [12]. In order to keep abreast with the current needs and scenario of Malaysia construction industry, Bumiputera construction players are required to equip themselves with the IBS knowledge besides encouraging more participation of the IBS projects by the government. However, respondents' perception towards the issue of storage and transportation of the IBS components to be manageable are rated as the least preferable by the respondents (mean score = 3.5205). This is due to the current situation of IBS application which requires high efficiency and accuracy in coordination in order to avoid problems of storage and transporting the IBS components to the site as well as the installation. Failure in coordination and lack of competency in handling IBS components might result in serious problem such as delay in project delivery, difficulty in managing payments to the IBS suppliers and contractors and wastage of materials in producing IBS components.

Table 5. Perception and readiness of IBS usage in construction projects.

Perception and readiness of IBS usage in construction projects				
Variables	**N**	**Mean**	**Std. Deviation**	**Ranking**
Learning the concept of IBS is easy for the company	73	3.8767	0.84894	8
It is easy for the company to get info on IBS	73	3.6575	0.88536	10
The transportation & storage of IBS component to be manageable	73	3.5205	1.01529	11
The installation of IBS components will required fewer workforce	73	3.8219	0.83902	9
It is easy for the company to become skilful at using IBS	73	4.5068	6.06704	1
Using IBS would improve overall project performance and quality	72	4.0139	0.83063	3
Using IBS would enhance project delivery efficiency	73	3.9315	0.87128	6
Using IBS would make it easier to complete a project	73	3.9863	0.84151	4
IBS is useful in delivering construction projects	73	3.8904	0.87499	7
Conventional SMEs (G3-G5) contractors should be trained to become IBS contractors	73	4.1370	0.73248	2
Local SMEs (G3-G5) contractors are still lacking in respondent's area	73	3.9726	0.89710	5

3 Summary

The research manages to reveals that Bumiputera construction players are aware with implementation of IBS and already equipped with the knowledge and skill of IBS. Level of understanding is at satisfactorily level and depicts the positive indication on the adoption of IBS among them. Their optimistic view on the perception towards IBS indicates the keenness of IBS usage in future projects. Nonetheless, issues on method of payment for IBS project and coordination on transportation and storage of IBS components need to be further discussed in improving IBS project delivery. Further research are required in order to identify the critical success factors and determinants for the implementation of IBS among Bumiputera in Malaysia construction industy.

Acknowledgements

These researchers would like to express gratitude for the financial support from Kementerian Pendidikan Malaysia (KPM) under purview of Research Acculturation Grant Scheme (RAGS) 600-RMI/RAGS 5/3 (237/2014) and Research Management Institute (RMI) Universiti Teknologi MARA (UiTM).

References

[1] K.A.M. Kamar, Z.A. Hamid and N. Dzulkalnine, Industrialised building system (IBS) construction, measuring the perception of contractors in Malaysia, *IEEE Business, Engineering and Industrial Applications Colloquium*, (2012).

[2] Construction Industry Development Board, *IBS Roadmap*, Kuala Lumpur, (2003).
[3] K.A.M. Kamar, M. Alshawi and Z.H. Hamid, Barriers to industrialised building system (IBS), The case of Malaysia, *Proc. of International Postgraduate Research Conference*, (2009).
[4] W.A. Thanoon, L.W. Peng, M.R. Abdul Kadir, M.S. Jaafar and M.S. Salit, The essential characteristics of industrialised building system, *Proc. of International Conference on Industrialised Building Systems*, Kuala Lumpur, 283-292, (2003).
[5] M.F. Musa, M.F. Mohammad, M.R. Yusof and R. Mahbub, The way forward for industrialised building system (IBS) in Malaysia, *Proc. of the International Civil and Infrastructure Engineering Conference*, 163-175, (2015).
[6] A.S. Abd Shukor, M.F. Mohammad, R. Mahbub and F. Ismail, Supply chain integration in industrialised building system in the Malaysian construction industry, The Built and Human Environment Review, **4**(1), (2011). <retrieved from http://www.tbher.org/index.php /tbher/article/view/48>
[7] M. Jaafar, T. Ramayah and A.H.M. Salleh, Adaptation strategies of contracting firms: An analysis by firm size, *Proc. International Conference in the Built Environment in the 21st Century*, Kuala Lumpur, 387-394, (2006).
[8] M.Q. Oleiwi, *Industrialized Building System: A Malaysian Approach*, Filspay Academy, (2015).
[9] Construction Industry Development Board, Transformation of construction industry through industrialized building system (IBS) roadmap, (2011).
[10] S. Moss, H. Prosser, H. Costello, N. Simpson, P. Patel, S. Rowe, S. Turner and C. Hatton, Reliability and validity of the PAS-ADD checklist for detecting psychiatric disorders in adults with intellectual disability, Journal of Intellectual Disability Research, **42**,173-83, (1998).
[11] A.M. Abdullah and S.A. Malik, Key indicators of SWOT analysis between bumiputera and non-bumiputera industrialized building system (IBS) contractors in Selangor, *Proc. of International Conference on Statistics in Science, Business and Engineering*, (2012).
[12] A.B. Rahman and W. Omar, Issues and challenge in the implementation of IBS in Malaysia, *Proc. of the 6th Asia Pacific Structural Engineering and Construction Conference*, Kuala Lumpur, (2006).

MATEC Web of Conferences **47**, 04014 (2016)
DOI: 10.1051/matecconf/20164704014

The Relationship of Stressors and Stress on Injury Incident of Construction Workers in Penang

Khairul Ammar Muhd Ali[1,a], Muhamad Arif Arshad[2], Amalina Amirah Abu Bakar[1] and Nor Janna Tammy[1]

[1]*Fakulti Kejuruteraan Awam, Universiti Teknologi Mara, 13500 Permatang Pauh, Pulau Pinang, Malaysia*
[2]*Gaintrend Sdn Bhd, Taman Perindustrian Beringin, 14000 Bukit Tengah, Pulau Pinang, Malaysia*

Abstract. Construction Workers (CWs) are the main source of manpower that is necessary to every construction project. Non conducive and hazardous working environment at construction site will affect the physiological health of the construction labour. This study is conducted to explore the impact of job stress and emotional stress to the CWs that potentially lead to injuries incident in Penang. Twelve stressors were identified through factor analysis. Then, the stressors are classified into five main categories. Questionnaires were developed according to the stress-stressor relationship. The correlation between factors of injury incident (stressor) and stress shows that lack of autonomy and inappropriate safety equipment lead to the emotional stress among the CWs with 0.287 and 0.204 respectively. In addition, poor physical environment causes the job stress among CWs with the correlation of 0.270.

1 Introduction

Standard Operation Procedure (SOP) of work is a guideline used to control quality of product or services given by the workers. Department of Occupational Safety and Health Malaysia (DOSH) investigates that accidental events in construction sector are the third highest compared to factory and farming sector for the total accident in Malaysia. In 2013, the construction industry is the highest number of death among the other industries which contributes about 37.5% or 69 deaths from total death of 185 deaths [1]. This study investigated the relationship type of stress and injuries incident (stressor) among construction workers in Penang. Therefore, by identifying which stressor gave high impact to the type of stress, the superiors of the construction project especially those who are working directly with the construction labour could help them to improve the working environment.

2 Construction Accident

According to Occupational Safety and Health Association (OSHA) Europe, construction workers are easier to get involved in an accident two times and more potential to face death three times compared to workers in other industries [2]. In Malaysia, DOSH categorised the construction accident due to the effects after the accident happened which are death, non-permanent disability and permanent disability. From the same source, the death category is the most happened in Malaysia construction

[a] Corresponding author : khairulammar.ali@ppinang.uitm.edu.my

industries [1]. The most construction of fatal accidents in 2007 and 2008 were fall down from high level, slip and stuck by objects [4].

2.1 Physiological stress contribute to construction accident

Physiological stress positively was a factor to an accident among Hong Kong construction workers [12]. Construction workers are more 1.7 times more susceptible to have psychological health problems like psychiatric disorders included stress-related problems than other industries workers [5]. Too much work load, period duties, unfair reward, way of communication between labour and foreman and lack of skills in completing task are the causes of stress problems to construction workers [6]. Stresses problems among workers will affect the awareness and compliance of safety and health issues and will be one of the factors contributing to accidents in construction site.

2.2 Effects of construction accident

The operations of construction project need to be stopped immediately to give space for the authorities to do the investigation on site, and as a result, the project will be delayed from the time line. Since the project is delayed, the contractor needs to pay penalty to the developer as a compensation of delayed time. Secondly, it will affect the reputation of a company. Since the accident was recorded by the authority especially related to fatal tragedies, the company will get bad reputation, perception and trust from people and workers for the next project. Thirdly, the effect of accident will impact labours'psychology. This problem will affect to either the labour, victim of the accidental event or others around him/her. Labours will face trauma and stress to do work in construction site, thus will produce awful quality to the production. Hence, accident in construction site also gives negative impact to employer, employee and client [7].

2.3 Construction workers in Malaysia

According to Department of Statistic Malaysia (DOSM), amount of labour in Malaysia construction sector is divided into two groups which are the citizens (93759 people) and non-citizens (181,730 people) [8]. Malaysian Investment Development Authority (MIDA) stated that only 12 countries are allowed to work in Malaysia construction industries which are known as Indonesia, Cambodia, Nepal, Myanmar, Laos, Vietnam, Philippines (male only), Pakistan, Sri Lanka, Thailand, Turkmenistan, Uzbekistan and Kazakhstan [9]. Most non-citizen construction labours have different cultures and languages to communicate compared to Malaysian. While having difficulties to understand between site supervisor and general workers, this will cause error information as well as in action. Thus, the situation can contribute stress among workers and stress may cause an accident during the construction.

2.4 Construction in Penang

According to CIDB Malaysia, from 2012 to 2014, Penang ranked at number 5 out of 14 states with the highest numbers of construction after Johor, Selangor, Wilayah Persekutuan and Perak [10]. DOSH recorded that from 2012 until February 2014, the total number of constructions in Penang was 1138 with 376 numbers of accidents [11]. The percentage of number of accident to number of construction in Penang is 33.04 %. 6.17% for Perak where the number of constructions were 3016 and number of accidents are 186. Selangor had two times higher number of constructions compared to Penang, but the percentage of construction accident was low. Hence, there is a need to study the reasons (stressor) that contributed to a high number of construction accidents in Penang.

2.5 Stressor

Stressor are factors why physiology stress happened among human. The factors of stressor might differ depending on types of industries as well as the difficulties of a job, number of task, environment and many more. Table 1 shows five groups of stressor in construction industries stated by Leung, 2010 [12].

Table 1. Group of stressor and factor contribute to stressor.

Group of stressor	Factors of stressor
1) Task Stressor	1. Work overload
	2. Role Ambiguity
	3. Lack of autonomy
2) Organizational Stressor	4. Unfair reward & treatment
	5. Inappropriate safety equipment
3) Personal Stressor	6. Type A behaviour
	7. Optimism
4) Interpersonal Stressor	8. Inter-role conflict
	9. Poor workgroup relationship
	10. Lack of feedback
5) Physical Stressor.	11. Poor physical environment
	12. Unsafe environment

2.6 Stress

Stress is a force or pressure to something in a society and stress can lead to lack of performance [13]. Today, people have started to become aware of "human stress" especially the stress experienced by construction industry personnel. Stress can be divided into two types which are job stress and emotional stress [12].

3 Methodology

3.1 Stressor identification

The identification of factors that contributed to stress among construction labours is grouped into five main categories as shown in Figure 1.

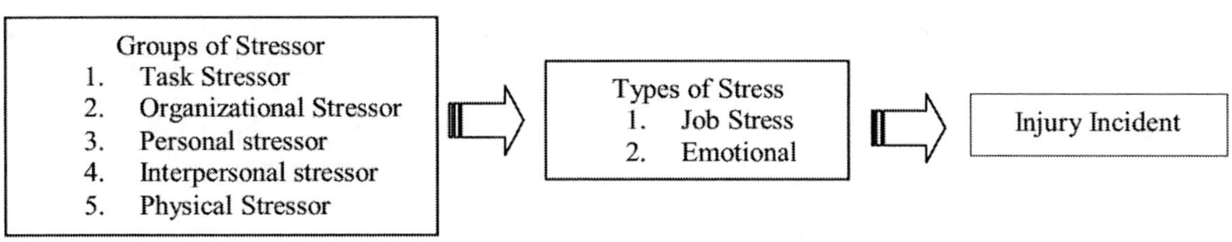

Figure 1. Process of stress-stressor relationship.

An injury incident may happen due to stress either job stress or emotional stress. The factors which are causing stress can be developed by these five groups of stressor as shown in Figure 1. This process of stress-stressor relationship is adopted form Leung, 2010 which previously studied on construction workers'accidents in Hong Kong [12]. Emotional scale is a compilation of 13 questions used to measure the level stress of construction labour due to emotional problems [14]. Out of 13 questions, only 4 were used in this study. Job stress scale developed by Gmelch in 1982 was used to measure the actual and expected abilities work need to be completed by labour. However, this study adopted the improvised job stress scale which in line with construction work scope produced by Leung, 2010 [12]. For the information of injuries, the numbers of accident happened in the same particular project was collected and the data was used to project the potential future incident. The period of accidents occurred in the particular project is collected between the previous two years of time since it is considered still as a short reference period [15].

3.2 Questionnaire

The questionnaires were developed according to the stress-stressor relationship and the scales mentioned earlier. The respondents of this questionnaire were only non-executive workers since these people conquered the construction sites in Malaysia as labours and are able to be involved in construction accidents [4]. The seven points Likert scale were used as a scale agreement which ranging from 1 (strongly disagree) to 7 (strongly agree). The questionnaires were given to CWs in Penang construction areas.

3.3 Data analysis

Analyses involved in this study are reliability analysis, correlation analysis and multiple linear regressions. Reliability analysis is used to find either the question is related to the objectives or not. This method is a common method used as an indicator to internal consistency. Correlation analysis is a statistical method used to determine whether a linear or curvilinear relationship between variance exists [16]. This study used this equation to determine linear relationship between of stressor, stress and injury incident. Multiple Linear Regressions analysis is applied to find the relationship between two variables that continue from correlation analysis process. The possibilities can be positive linear relationship, negative linear relationship, curvilinear relationship or no discernible relationship [16]. In this study, the relationships of stressor-stress and stress-injury incident were revealed.

4 Discussion

The alpha Cronbach's value gained from Reliability analysis shows that the job stress and emotional stress is 0.65 and 0.68 respectively. According to Bluman, 2005, the acceptable value is more than 0.6 [16]. Hence, the questionnaires are valid to be used in this research. The Pearson's correlation analysis was used to determine linear relationship between of stressor, stress and injury incident. The result indicates that the highest correlation between various stressor and job stress is poor physical environment with correlation value of 0.270. The highest correlation between various stressor and emotional stress is lack of autonomy with correlation value of 0.287. Moreover, inappropriate safety equipment is the second highest stressor with value of correlation recorded as 0.204. The Multiple Regression Analysis method is used to discover the interdependent relationship between two type of stress, various stressor and injuries incident based on previous correlation analysis. In addition, to prove that the job stress has no relationship to the cause of accident in construction site, the curvilinear interdependent relationship between construction workers injuries incident and two types of stresses were developed. The curvilinear results are shown in Figure 2.

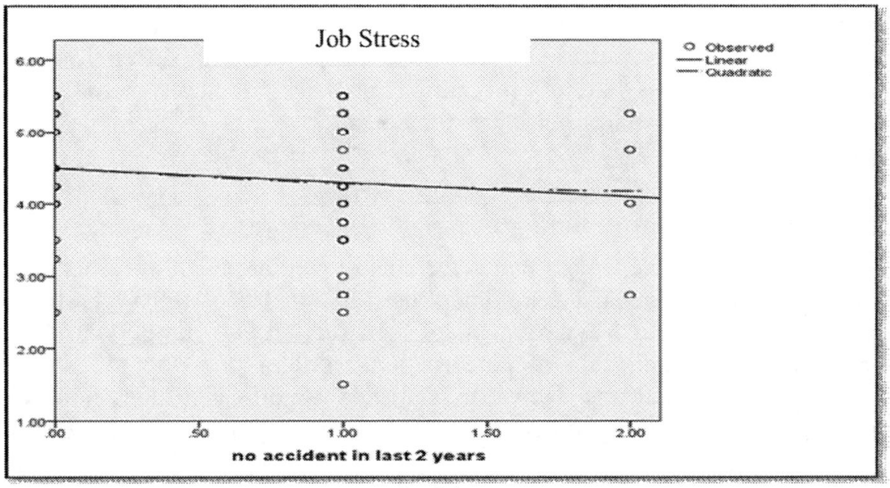

(a)

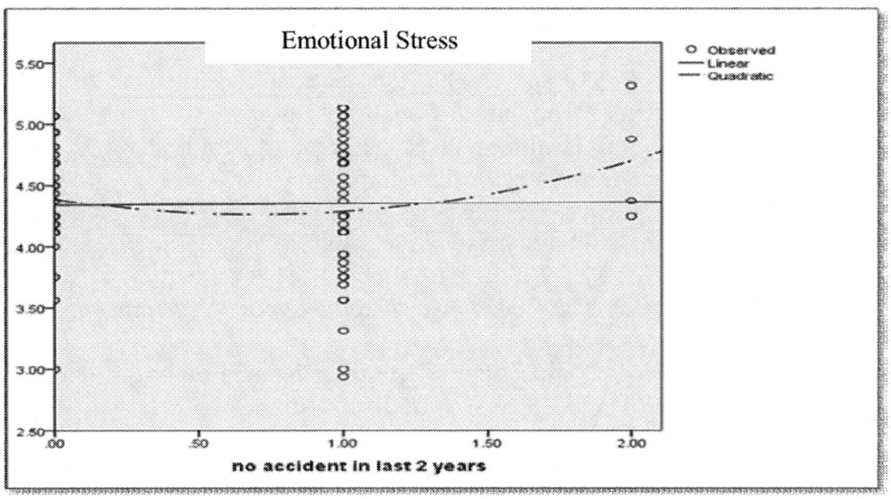

(b)

Figure 2. Curvilinear relationship between number of accident and (a) job and (b) emotional stress.

These curvilinear shows that the job stress resulted in negative curvilinear relationship compared to emotional stress that resulted in positive curvilinear relationship. A positive curvilinear shows that emotional stress could lead to potential of injuries incident to the construction workers. Based on the analysis, the relationship of stressor-stress-injuries incident is simplified statistically as shown in Figure 3.

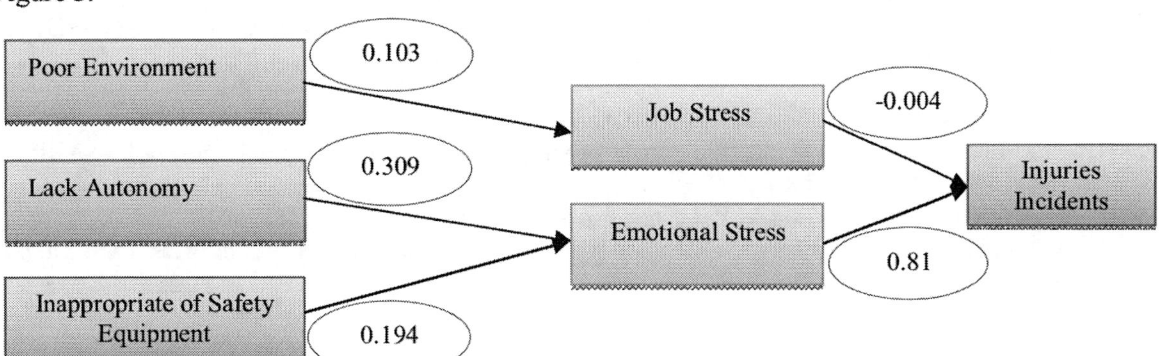

Figure 3. Integrated stressor-stress-injury incident statistical model for construction workers.

Since it is identified that the emotional stress could affect the construction workers; it is a need to reduce numbers of accident. Murray et al, 1997 stated that people who suffer from emotional stress may act to poor safety behaviour like less precautions and focus to safety compliance [17]. Hence, those people may face high potential to get injury incidents.

5 Conclusion

In conclusion, from this study, the factors that contributed to emotional stress among labours are lack of autonomy and inappropriate of safety equipment used in construction project while the factor that affect job stress is a poor physical environment of construction site. Emotional stress has positive relationship with the injury incidents of construction workers. For the job stress, there is no relationship with the injury incidents. However, statistical significance does not necessarily mean practical significance and lack of significance does not necessarily means that the null hypothesis is true [16].

References

[1] Department of Occupational Safety and Health, Occupational accidents statistics by sector until December 2013, (2013). <retrieve from http://www.dosh.gov.my>
[2] European Agency for Safety and Health at Work, Safety and health at work: How dangerous is construction, (2014). <retrieve from https://osha.europa.eu>
[3] S. Hide, S. Atkinson, T.C. Pavitt, R. Haslam, A.G. Gibb and D.E. Gyi, Causal factors in construction accidents, Health and Safety Executive, (2003).
[4] D.N.M.A Abdullah and G.C.M. Wern, An analysis of accidents statistics in Malaysian construction sector. *International Proceedings of Economics Development and Research*, 3, (2011).
[5] J.S Peterson and C. Zwerling, Comparison of health outcomes among older construction and blue-collar employees in the United States, American Journal of Industrial Medicine, **34**, 280–287, (1998).
[6] R. Choudhry, D. Fang and S. Ahmed, Safety management in construction: Best Practices in Hong Kong, J. Prof. Issues Eng. Educ. Pract., **134**(1), 20–32, (2008).
[7] S.X. Zeng, H.C. Wang and C.M. Tam, A survey of construction site safety in China, 243-248, (2001).
[8] Department of Statistic Malaysia, Population and vital statistics, (2012). <retrieve from http://www.statistics.gov.my>
[9] Malaysian Investment Development Authority, Employment of foreign workers, (2014). <retrieve from http://www.mida.gov.my>
[10] Construction Industrial Develepment Board Malaysia, Buletin 2014, (2014). <retrieve from https://www.cidb.gov.my>
[11] Department of Occupational Safety and Health, Occupational accidents statistics by state until December 2013, (2013). <retrieve from http://www.dosh.gov.my>
[12] M. Leung, Y. Chan and K. Yuen, Impacts of stressors and stress on the injury incidents of construction workers in Hong Kong, (2010).
[13] U. Lundberg, Stress, subjective and objective health, International Journal of Social Welfare, **15**(1), 41-48, (2006).
[14] C. Maslach, S.E. Jackson and M.P. Leiter, *Maslach Burnout Inventory*, (1986).
[15] D.D. Landen and S. Hendricks, Effect of recall on reporting of at work injuries, Public health Reports, **110**(3), 350, (1995).
[16] A.G. Bluman, Elementary Statistics, *A Step by Step Approach*, McGraw-Hill, (2004).
[17] C.J.L Murray and A. Lopez, *Global Health Statistics*, Harvard University Press, Cambridge, (1996).

MATEC Web of Conferences **47**, 04015 (2016)
DOI: 10.1051/matecconf/20164704015

Knowledge-Based Energy Damage Model for Evaluating Industrialised Building Systems (IBS) Occupational Health and Safety (OHS) Risk

Nor Haslinda Abas[1,a], Nick Blismas[2] and Helen Lingard[2]

[1]*Faculty of Civil and Environmental Engineering, Universiti Tun Hussein Onn Malaysia, 86400 Parit Raja, Johor, Malaysia*
[2]*School of Property, Construction and Project Management, RMIT University, Melbourne VIC 3001, Australia*

Abstract. Malaysia's construction industry has been long considered hazardous, owing to its poor health and safety record. It is proposed that one of the ways to improve safety and health in the construction industry is through the implementation of 'off-site' systems, commonly termed 'industrialised building systems (IBS)' in Malaysia. This is deemed safer based on the risk concept of reduced exposure, brought about by the reduction in onsite workers; however, no method yet exists for determining the relative safety of various construction methods, including IBS. This study presents a comparative evaluation of the occupational health and safety (OHS) risk presented by different construction approaches, namely IBS and traditional methods. The evaluation involved developing a model based on the concept of 'argumentation theory', which helps construction designers integrate the management of OHS risk into the design process. In addition, an 'energy damage model' was used as an underpinning framework. Development of the model was achieved through three phases, namely Phase I – knowledge acquisitaion, Phase II – argument trees mapping, and Phase III – validation of the model. The research revealed that different approaches/methods of construction projects carried a different level of energy damage, depending on how the activities were carried out. A study of the way in which the risks change from one construction process to another shows that there is a difference in the profile of OHS risk between IBS construction and traditional methods. Therefore, whether the option is an IBS or traditional approach, the fundamental idea of the model is to motivate construction designers or decision-makers to address safety in the design process and encourage them to examine carefully the probable OHS risk variables surrounding an action, thus preventing accidents in construction.

1 Introduction

The Malaysian construction industry plays a significant role in the development and growth of the country's domestic economy, generating further demands for construction activities [1-3]. However, the industry has earned the reputation of being a highly hazardous industry due to its high rates of accidents and fatalities [1-7]. There is therefore an urgent need to improve health and safety performance of Malaysia's construction industry.

---

[a] Corresponding author : nhaslin@uthm.edu.my

One initiative of the Malaysian government to address occupational health and safety (OHS) in construction is the widespread adoption of Industrialized Building Systems (IBS), commonly termed 'offsite' construction. An IBS approach is commonly deemed to be safer than conventional construction because it changes the nature of the construction process, and requires fewer workers onsite. Claims have been made that IBS, or more specifically offsite construction, can reduce site accidents [8-11]. However, the extent of this impact on safety and health in construction is still unclear, as there are no current systems to comparatively assess OHS risks in different construction processes. McKay [10] has identified the OHS risks of both onsite and the offsite processes, but presents a static assessment that lists hazards of specific processes, rather than comparing the extent of the risks of the various processes. There is therefore a need for a robust dynamic method for comparing different construction processes, such as IBS and traditional approaches, to determine the relative safety and health performance of these processes.This study explores the impact of IBS on construction workers' safety and health by providing a thorough investigation of the activities and associated risks in IBS and traditional construction, with the aim of enabling a comparative evaluation of the OHS risks involved in different construction approaches.

In order to evaluate and compare the OHS risks throughout traditional and IBS construction processes, a structured method incorporating specialist OHS knowledge and guidance is required. An 'argumentation theory model' [12]; as cited in Yearwood and Stranieri [13] building on the work of Cooke et al. [14] is proposed as a method to integrate the management of occupational health and safety risk into the design process. Cooke et al.'s work was developed from structured knowledge in the context of uncertainty and discretionary decision making, involving expert reasoning regarding design impacts on OHS risk represented by 'argument trees' [14]. Their model explored the use of argumentation theory in product design (of buildings), which focused on the implementation of physical aspects of the design. Cooke et al. did not evaluate the design or health and safety characteristics of the construction process itself. This study presents the development of a process-centric model that consists of a series of argument trees for best practice reasoning that can be used by designers or decision makers when examining the OHS risks posed in different construction processes. The argument trees consist of structured knowledge which was developed by focusing on the process involved to build a product in various construction processes. The model provides consideration of product and process design concurrently, thus contributing to the body of knowledge. In addition to Cooke et al.'s model, an 'energy damage model' [15] is used as an underpinning framework for developing the present model. The development of this model suggests options for the decisions that can be made by product and process designers, in such a way as to assess the extent to which their design decisions mitigate the OHS risks associated with various types of damaging energies in construction, and thereby offering a more rigorous relative comparison of OHS risks between IBS and traditional approaches.

This paper serves to outline the development of a knowledge-based energy damage model to assess IBS OHS risk in construction processes at the design stage. This paper expands on an earlier working papers by [16, 17], in which the main background of this research is described.

2 Development Process of the Model

The risk assessment model developed in this study applied an argumentation framework [14] with an energy damage model [18] as an underpinning framework for developing the model. The concept of argumentation was chosen since it provides the framework for modelling best practice reasoning that can be made by product and process designers, in such a way as to assess the extent to which their design decisions mitigate the OHS risk in construction. Meanwhile, the concept of the energy damage model was chosen due to: i) its unique characteristic which views hazards as a source of potentially damaging energy and risk; and ii), its ability to provide strategies to prevent injury which does not depend on human or behaviour changes. The combination of these concepts thereby offers a more rigorous relative comparison of OHS risks between IBS and traditional approaches. The development process of the model consists of three phases, which are: Phase I – data collection or knowledge acquisition; Phase II – development of the argument tree (model); and Phase III – validation of the model.

2.1 Phase I knowledge acquisition

Knowledge acquisition involves collecting the data that will underpin the model, which are the construction process, the associated risks and the design aspects which have potential to impact construction risks. Data was collected on on-going construction projects, which used the selected construction approaches in order to observe the process involved and to identify the associated OHS risks. This involved a number of site visits to the IBS manufacturing factory and construction sites. Five sites were visited, of which four were considered to represent IBS/offsite and one represented traditional construction. The detailed studies involved talking to the managerial staff, operatives and supervisors, and also observing and documenting operations. Document analysis techniques were also used to obtain the information regarding 'safe design' aspects generally associated with construction processes. This data source was extracted from OHS guidance material, industrial standards and codes. These were used to provide the design related information during the development of the argument trees.

2.2 Phase II knowledge processing (argument tree mapping)

The knowledge acquired from Phase I was compared according to the process involved in selected construction approaches. Within this phase, the risks/hazards associated with the activity were categorized into several damaging energies. Next, the information related to design aspects which have potential to impact upon the damaging energies gathered in secondary sources were used to develop an initial representation of the relevant knowledge in the argument tree.

In developing a model that represents the reasoning for decisions around different construction processes, the same method of modelling as the expert reasoning system in ToolSHeD, was used. Unlike ToolSHeD, the argument trees developed in this study differ in two fundamental ways: i) they seek to model process risks as opposed to product risks, and ii) they are undepinned by an 'energy damage model' to assess OHS risks in the design.

The design factors act as the barriers to prevent such energies from causing injury or damage in energy transfers. The extent of damage depends on the amount of energy deflected by the 'barriers'. The use of 'argument trees' for modeling expert reasoning is capable of providing several OHS risk levels for the activities being carried out, and the OHS risk levels were dependent on the design factors/features selected by designers when they plan to execute a construction project using one particular construction approach. Figure 1 depicts a typical design OHS argument tree for the potential gravitational damaging energy from lifting concrete panels. For example, in the activity of 'loading component onto truck', one of the design factors is 'the weight of the element'. Within this factor, there are several options of the component weight that the designer can choose, and the corresponding values of the options. If the weight of the component (beam) were 0.5 tons, then the relative value would be 3. This value, together with other values pertinent to the design factors/features within the argument tree (depending on designers' decisions) are then inferred at the end of the argument tree (refer to 'D' in Figure 1), which eventually gives a risk level of the activity, according to the design decisions opted by the designers.

As mentioned in the previous paragraph, the design factors act as the barriers that prevent such energies from causing injury or damage in energy transfers. If the activity of lifting the component is to be lifted up to more than 3 meters, the potential damage of gravitational energy, that is, falls of object (component), is likely to be high. In other words, the impact of the damage is higher when compared to the component being lifted to a lower height. This results in the weakening of the barriers. However, the combination of other design factors could balance the magnitude of the OHS risk level, thus improving efficiency of the barriers. The use of argument trees for modelling expert reasoning is also better suited to represent the level of 'how safe is safe enough' regarding the designer's decision about the 'barriers' to be used to counter damaging energies during the construction process. This level of decision will determine the value of a risk rating at the 'root' node of an argument tree.

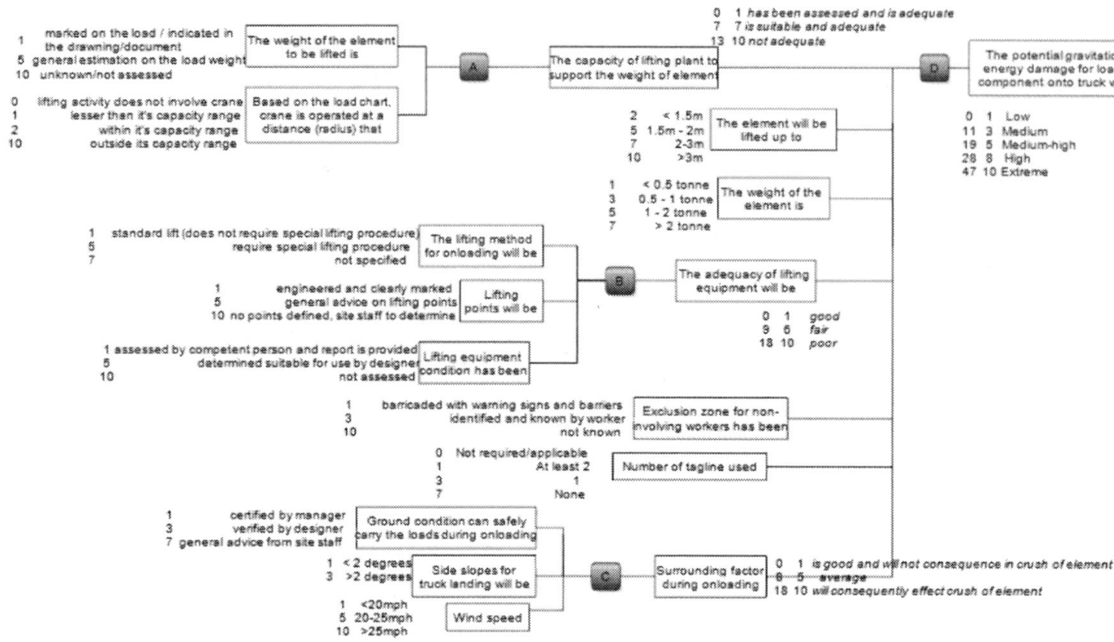

Figure 1. Example of argument tree showing the inference procedure.

As shown in Figure 1 above, the argument tree consists of a set of linguistic values with numerical values, that are assigned to each node of the argument tree, regardless of its position. These values are relevant to the design options available to a designer when making judgement upon aspects of a design, pertinent to the risk of wall cladding construction. The information about the values was sourced from a comprehensive document analysis, such as guidelines and method statements. It can be seen that the tree has linguistic values with corresponding numerical values in the child nodes inferring values to the parent nodes. The numerical values in the tree were initially assigned by the author, which were then modified according to an expert panel's judgement in the validation process. The inference procedure, denoted by A, B, C, and D, continues until ultimately inferred at the root node – the final risk rating. The risk rating at the root node indicates either 'extreme', 'high', 'medium' or 'low'. It is measured by calculating the potential of damaging energies which are contingent upon the values decided by the designer at every child node.

2.3 Phase III validation of the model

In order to make the model applicable and reliable, this study used scenario testing to validate the model. From the created tool (i.e. the argument tree model), a set of five scenario cases containing knowledge from the trees representing different damaging energies in different construction processes were developed. Each case scenario was assigned with two different scenarios. An expert panel's judgement was obtained to see whether or not their ratings on the scenario case given matches the tool-generated score. In this research, 12 experts were invited to participate. The experts were from various backgrounds and were chosen based on one of the following categories: (i) have experience in contruction projects, or (ii) are responsible for safety in their organisation. They were experienced and credible in their positions and were willing to give commitment to the research project.

The validation questions consisted of five cases in which each case was assigned with two different scenarios. Given some photographs (for each case), the description of the activities carried out related to the photos and also the scenario information listed within the survey, the respondents were asked to rate the risk of these activities based on the scenarios given. In this close-ended question, respondents were given a Likert scale ranging from 1 (if the respondents think the risk is low) to 10 (if the respondents think the risk is extremely high). The open-ended questions provided an opportunity for respondents to give a reason for their choices using their own words. The comments were useful for the researcher to obtain additional information from respondents' insights that may be overlooked during the development of the argument trees.Once the participants responded to the survey, the responses were analyzed and compared with the results (rating) generated from the tool.

The expert panel's judgement results were compared with those of the tool to determine the consistency of risk ratings between both approaches. Such consistency provides evidence that the knowledge contained in the model is valid. Further, the expert panels' insights on the risk provided valuable information that was adopted in amending the linguistic and numerical values in the argument trees. From the validation process, it was found that the knowledge within the validated argument trees was valid and reliable, involving only minor changes being made in the argument trees.

3 Demonstration of the Application of the Model

In order to demonstrate the application of the model, a 'mock trial study' was done, which provides a realistic opportunity to field test the model, and thus visualize the workability of the model. The 'mock' process was done on the existing cases utilised during the model developement. It is acceptable to use the existing cases because these cases were previously used to collect data and information about the process and the risks associated with the process - but not to rate the risk. The information from the case studies was used as if it were the designer's choice of process design arising from discussions made in the 'design' stage. Figure 2 depicts an example of the potential gravitational energy risk profile generated by the tool for the steps of the process involved in the selected construction approaches. The x-axis presents the process steps involved, whereas the y-axis presents the risk magnitude. The lines which plotted the graph represent the selected construction approaches used for this study.

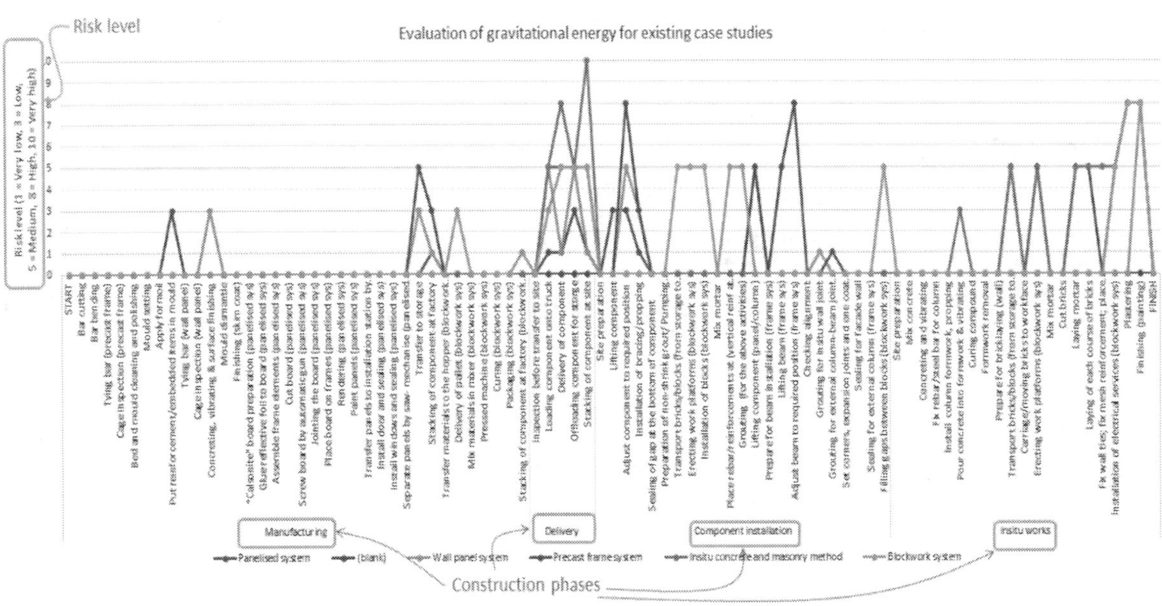

Figure 2. Example of the evaluation of the model for gravitational energy.

Particulaly noticable are changes to riskover the different steps of the construction processes. This provides evidence that there is a difference in the profile of OHS risk between IBS construction and the traditional method and that these change as work progresses through its component activities. The magnitude of OHS risk for each activity from the damaging energy perspective is also estimated within the tool. The development of this tool demonstrates that it is possible to pick up differences in the risk magnitudes of particular activities between various construction processes. The issue is not that construction approach A is better than B, but rather 'where is the risk?', 'what type of risk is it?', and 'how bad is the risk?'. The tool provides a link between the design and construction phases with regards to construction safety.

4 Conclusion

The development process of the argument trees model has been discussed above. The model developed in this study differs from the existing tool in that it is a process-driven model rather than a product-driven model. This is important because product risk is mainly associated with user/operator safety, and not the builder's safety. Process risk however concentrates on the constructor risk, as the constructor is responsible for planning, managing and monitoring the construction process. Using the model, constructors and designers can evaluate the risks in the construction process required for their specified design prior to the construction stage. This could help all parties to formulate a plan for workers' OHS while balancing projects within the planned cost, time and scope.

The use of this model can provide a mechanism for the comparison of hazards of different construction approaches. The tool enables a risk profile to be generated across different phases of a process, thereby providing powerful assistance to designers and contractors. Currently, OHS is considered post-design, and therefore options for hazard elimination are largely removed. This tool offers scenario testing in earlier design phases, so that OHS can inform design, rather than the current approach in which OHS bears little influence over design. Therefore, whether the option is an IBS or traditional approach, the fundamental idea of the model is to motivate construction designers or decision-makers to address safety in the design process and encourage them to examine carefully the probable OHS risk variables surrounding an action, thus preventing accidents in construction.

It is anticipated that the argument tree model will contribute by providing a way for designers to integrate construction process knowledge into design to eliminate or reduce hazards during construction. The fundamental idea will encourage construction designers or decision-makers to address safety in process design and also encourage them to carefully examine the probable OHS risk variables surrounding the construction process.

References

[1] A.R.A. Hamid, M.Z.A. Majid and B. Singh, An Overview of construction accidents in malaysia," in in issues in construction industry, M.I. Mohamad, Ed., Penerbit Universiti Teknologi Malaysia, Johor Bahru, 110–123, (2008).

[2] S.S. Hosseinian and Z.J. Torghabeh, Major theories of construction accident causation models: A literature review, Int. J. Adv. Eng. Technol., **4**(2), 53–66, (2012).

[3] D.N.M.A. Abdullah and G.C.M. Wern, An analysis of accidents statistics in Malaysian construction sector, *Proc. of International Conference on E-business, Management and Economics*, 3, (2011).

[4] C.L. Foo, Budgeting for occupational safety and health management and its implementation, *MBAM Annual Safety Conference*, (2005).

[5] SOCSO, Annual Report for 2009, Kuala Lumpur, (2009).

[6] SOCSO, Annual Report for 2000, Kuala Lumpur, (2000).

[7] DOSH, Accident Statistic, Kuala Lumpur, (2013).

[8] A.G.F. Gibb, *Off-site Fabrication: Prefabrication, Pre-assembly and Modularisation*, Whittles Publishing Services, United Kingdom, (1999).

[9] T. M. Toole and J. Gambatese, "The trajectories of Prevention through Design in construction.," J. Safety Res., vol. 39, no. 2, pp. 225–30, Jan. 2008.

[10] L. J. McKay, "The Effect of Offsite Construction on Occupational Health and Safety," Loughborough University, 2010.

[11] M. Gangolells, M. Casals, N. Forcada, X. Roca, and A. Fuertes, "Mitigating Construction Safety Risks Using Prevention Through Design.," J. Safety Res., vol. 41, pp. 107–122, 2010.

[12] S. Toulmin, The Uses of Argument, vol. 70. 1958, p. 262.

[13] J. L. Yearwood and A. Stranieri, "The Generic/Actual Argument Model of Practical Reasoning," Decis. Support Syst., vol. 41, no. 2, pp. 358–379, Jan. 2006.

[14] T. Cooke, H. Lingard, N. Blismas, and A. Stranieri, "ToolSHeD TM The Development and Evaluation of a Decision Support Tool for Health and Safety in Construction Design ToolSHeD," Eng. Constr. Archit. Manag., vol. 15, no. 4, pp. 336–351, 2008.

[15] D. Viner, Accident Analysis and Risk Control. New Delhi, India: Viner Pty Ltd., 1991.

[16] N. H. Abas, N. Blismas, and H. Lingard, "Development of a Knowledge-Based Energy Damage Model to Assess Occupational Health and Safety (OHS) Construction Risks in Malaysia," in CIBW099 Prevention: Means to the End of Construction Injuries, illnesses, and Fatalities, 2011.

[17] N. H. Abas, N. Blismas, and H. Lingard, "Knowledge-based energy damage model for evaluating Occupational Health and Safety (OHS) Construction Risks in Malaysia," in HKU-HKHA International Conference 2013 - Construction and Housing in the 21st Century, 2013, pp. 408–415.

[18] D. Viner, Accident analysis and risk control. Melbourne: Derek Viner Pty Ltd., 1996.

MATEC Web of Conferences **47**, 04016 (2016)
DOI: 10.1051/matecconf/20164704016

Drivers to Enhance ICT Uptake in IBS Management Processes in the Construction Industry

Peniel Ang Soon Ern[1,a], Narimah Kasim[2] and Goh Kai Chen[2]

[1]*Department of Civil Engineering Technology, Faculty of Engineering Technology, Universiti Tun Hussein Onn Malaysia, 86400 Parit Raja, Johor, Malaysia*
[2]*Department of Construction Management, Faculty of Technology Management and Business, Universiti Tun Hussein Onn Malaysia, 86400 Parit Raja, Johor, Malaysia*

Abstract. In Malaysia, as a result of the government's initiative and future outlook in advancing the use of innovative technologies, the Industrialised Building System (IBS) approach is actively promoted through several strategies and incentives as an alternative to conventional building methods. Extensive uptakes of modern Information Communication Technology (ICT) applications are able to support the different IBS processes for effective production. However, it is argued that ICT uptake at the organisational level is still in its infancy. This raises the importance to identify critical drivers for effective uptake of ICT in the IBS production management process. Critical drivers to ICT uptake were identified through questionnaire survey with industry stakeholders which was analysed using mean index and critical t-value with the use of SPSS software. The top ten priority ranked drivers reflect the People, Technology and Policy elements to ICT uptake. These identified critical drivers are able enhance ICT uptake in IBS management process in the construction industry.

1 Introduction

Industrialised Building System (IBS) is defined as a construction technique in which components are manufactured in a controlled environment (on or off site), transported, positioned and assembled into a structure with minimal additional site works [1]. The benefits of IBS, also called offsite construction have been widely studied and include reductions in time, defects, health and safety risks, environmental impact, and whole-life cost with a consequent increase in predictability, productivity, whole-life performance and profitability [2-6].

Appropriate implementation of Information Communication Technology (ICT) will facilitate more effective and productive materials management processes. Thus, widespread use of modern Information Technology (IT) tools will support the different IBS processes by enabling more accurate documents and good conditions for an effective production. Besides that, errors are discovered early and problems in the manufacturing and assembly phases can be avoided [7].

However, [8] indicated low IT adoption and limited technology availability to have generally discouraged IBS uptake in Malaysia. In point of fact, IT has been identified as one of the Critical Success Elements (CSF) or enablers for IBS adoption [9]. Hence, this paper aims to highlighting the critical drivers to enhance ICT uptake in the IBS management process.

[a] Corresponding author : peniel@uthm.edu.my

2 Uptake in IBS Management Process

More sophisticated solutions of ICT based technologies in IBS processes are emerging, such as wireless communication, bar-coding and Radio Frequency Identification (RFID) for tagging technologies. A report by Construction Research Institute of Malaysia [10] pointed technology as a main support to impact IBS in the future. 'Building Information Modeling (BIM)' and 'new technology for IBS invented' were classified as the highest ranking scenario of positive impacts on high-high analysis (high impact-high possibilities). In addition, it was stated that further improvement of manufacturing technology is a strong solution to tackle the weakness and threats of IBS adoption.

However, the current practice of ICT uptake in the tracking of IBS components in Malaysia is rather uncommon. There is a lack of effective logistics management in the Malaysian construction industry, where technology intervention plays a very critical role [11]. IBS production process needs a sustained improvement and management in order to ensure proper components are developed and handed over at the right time, in the proper order and without damages.

Recent findings on ICT uptake in IBS management process indicate that most of the problems or issues faced were normally solved with human solution or by technical menial methods [12]. ICT applications were not regarded as the best practices in overcoming the IBS components management problems. Few problems that cause the limited benefits of ICT uptake in the construction project level are lack of software integration (lack of a standardised platform for information exchange [15] and lack of standard processes for construction project management [16].

The shortcomings of ICT uptake in the construction industry bring forth a need to review possible driver elements for improving ICT uptake in the industry. Several research publications [17-21] indicate that people, process and technology are the three key elements that need to be considered for successful uptake of technologies. Together, these three elements create business value [21]. However, he further states that "the people, processes, and technology need a leader", just as "an orchestra needs a conductor". The same analogy can be applied to the adoption and uptake of new and innovative technologies within construction companies. The "conductor" in this case is the policy making. To successfully implement and use any new technology it requires policy makers to plan and drive policies and strategies. Policies are "written principles or rules to guide decision-making" [22]. All four elements are considered important for an organisation to effectively employ ICT. A company cannot employ ICT if it satisfies the requirements of just one element and not the others. Hence, all four elements - policy, process, people and technology - need to work hand-in-hand and symbiotically (refer Figure 1).

Figure 1. Driver elements for ICT uptake.

Taking this into account, the ICT uptake model has been so structured that for an organisation to employ ICT it must have: (a) Policy Makers that believes in the technology and takes strategic measures to drive its adoption, uptake and usage in order to mobilise the industry to implement and benefit from it, (b) Processes that enable and support the successful adoption of the technology, (c)

People who have adequate skills, understanding of, and belief in, the technology; and finally (d) Technology applications and infrastructures which are adequate. The four elements to ICT uptake will form the theoretical basis to identify critical drivers for the enhancement of ICT uptake in IBS management process.

3 Methodology

This paper applies the method of purposive sampling, where the samples are selected based on the knowledge of a population and the purpose of the study [23]. This is a form of non-probability sampling in which it includes specialist knowledge of the research issue. Apart from the government IBS policy makers, IBS precast manufacturers are the selected respondents which form the biggest population among the other four IBS systems (e.g. IBS formwork, IBS steel systems) in Malaysia (refer Table 1).

Table 1. IBS manufacturers in Malaysia.

Types of IBS Components	Total Manufacturer
Pre-cast concrete framing, panel and box system	30
Steel framework system	24
Steel framing system	25
Prefabricated timber framing system	4
Block work system	25
Others (on-site)	10
Grand total	118

Out of 41 questionnaires sent, a total of 31 was returned, giving a 76% high response rate from the construction industry, which in general stands at only 25% to 35% [24]. The questionnaire was sent to 33 organisastions from the listed companies, whom the majority are manufacturers under the Registered IBS System Provider (RISP) list in Malaysia. In addition, 8 questionnaires were distributed to the government policy makers who are direct decision makers to the issues pertaining to IBS. The data collection focused on the IBS precast manufacturing process, where ICT applications are more significantly used.

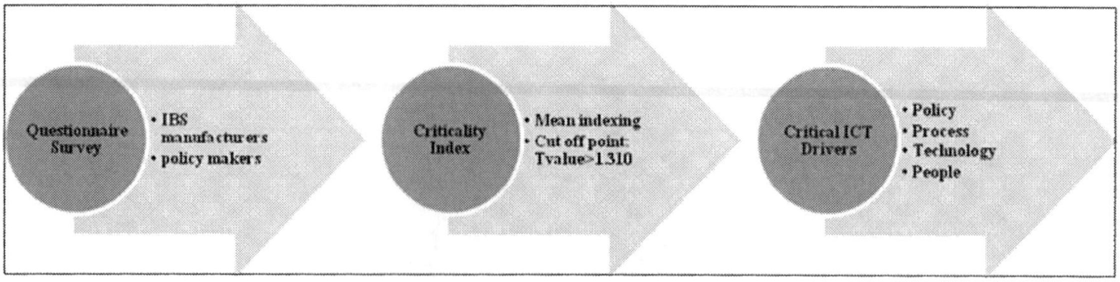

Figure 2. Flowchart of research methodology.

Referring to Figure 2, the IBS manufacturers and government policy makers were asked to rate the drivers for increased ICT uptake in IBS process management from Insignificant to Very Significant range. In order to identify the critical rating from all the sub-components of the ICT-barriers, the mean is fixed at scale '3.5' since ratings above '3' represent 'moderate significant', '4' represent 'significant' and '5' represent 'very significant'. The rankings for the ICT uptake drivers can be obtained by calculating the critical value of the variables. The critical t-value, $t_{(30,\ 0.10)} = 1.310$ with confidence interval 90% ($\alpha = 0.10$) was set in this study. T-test analysis was employed to rule out the ICT driver components which carry the mean value lesser than 3.5. ICT uptake drivers with t-values that are greater than the critical t-value of 1.310 will be shortlisted as the critical drivers to ICT uptake. Similar studies which applied mean indexing and t-tests are [25] which rated the criticality index for the importance of the sustainability-related cost components and [26] using mean indexing to discuss the importance of evaluation elements for customer satisfaction in project management. Apart from that, t-test analysis has been used in past research in identifying the relative important indicators [27, 28]. The critical drivers to ICT uptake are identified from the four driver elements (as shown in Figure 1).

4 Data Analysis

Critical drivers function as key enablers to specifically catalyse the proposed solutions to improve ICT uptake amongst the stakeholders. Hence, Table 2 summarises the top ten critical drivers to ICT uptake with t-values greater than 1.310. Ten of these critical ICT uptake drivers are under the ICT uptake elements of People, Technology and Policy, where Process is no longer within the top ten drivers.

Table 2. Critical drivers to ICT uptake.

Rank	Driver Elements	Drivers	T-Value (cut off point>1.310)
1	People	Top management support and commitment	8.469
2	Technology	Improve availability of the technologies	8.088
3	Policy	More cooperation between government and IBS players to promote and implement ICT	7.478
4	People	Devise better training procedures	7.045
5	Technology	Provide adequate ICT support (in-house or external)	6.225
6	Policy	Refine IBS Roadmap to encourage greater ICT uptake	5.903
7	Technology	Allocate adequate resources to implement and upgrade ICT tools	5.491
8	Technology	Consistent commitment and support to use BIM as a primary process and technology	5.479
9	Policy	More ICT course syllabus to start from tertiary level	4.507
10	Policy	Revision of reward systems	4.452

There are two critical drivers under the *People* element, namely *top management support and commitment* and *devise better training procedures.* [29] refers *People* to adequate skills, understanding of and belief in the technology. Thus, top down commitment is crucial in every organization, so to ensure all decision making are able to be passed down. Strong commitment from top management, including the senior leadership of an organization, is critical to ensuring technological acceptance by creating a supportive context for IT within the organization [30, 31].

The second most critical ICT uptake driver element is *Technology*. In the latest study technology plays a moderating role on the relationship between government and a construction organisation readiness level in implementing e-procurement [32]. In order to increase the adoption of ICT in IBS management process, there is a need to first *improve the availability of technologies*. Studies showed that appropriate training increases the likelihood that IT will be accepted, since training tends to bring about better understanding and more favorable attitudes toward the technology and to result in more frequent use of it [33, 34]. Adequate resources are to be rightfully allocated to implement and upgrade ICT tools whilst consistent commitment to support the use of BIM in the IBS processes. Commitment focuses on end-users who devote themselves to using ICT whereas organisational commitment focuses on top managers who support end-users to use IT/ICT [35]. An organisation's top management also needs to be committed to support and allocate adequate resources for technology investment such as ICT. Commitment is essential to avoid IT project failure [36, 37].

Lastly, *Policy* is also found to be a critical driver element to greater ICT uptake improvement. The role of government has a stronger influence on a construction organisation readiness level in implementing ICT [38]. IBS Manufacturers and the Government policy makers agreed that *more cooperation is needed between the government and IBS players* in order to promote and implement ICT. Other than that, the *IBS Roadmap has to be refined* and *more ICT course syllabus* to touch the grass root education. One recurring ICT driver mentioned throughout this study is to *revise the reward system* [38].

5 Conclusion and the Way Forward

The result from the criticality index provide the rankings for the ten most critical ICT uptake drivers. It is a comprehensive and detailed account of different critical drivers to ICT uptake, enabling decision makers and organisational strategists a platform to work on further development required in adopting ICT. Further research can be executed to incorporate strategies to eliminate barriers to ICT uptake.

Acknowledgement

This work was supported by RAGS Universiti Tun Hussein Onn Malaysia (UTHM).

References

[1] IBS Roadmap 2003-2010, Construction Industry Development Board , Kuala Lumpur, (2003).
[2] A. Gibb and Pendlebury, Buildoffsite: Glossary of Term Buildoffsite UK, London, (2005).
[3] T. Venables and R. Courtney, Modern methods of construction in Germany - playing the off-site rule, Rep. of a Department of Trade and Industry (DTI) Global Watch Mission, London, (2004).
[4] W. Pan, A.G.F. Gibb and A.R.J Dainty, Perspectives of housebuilders on the use of offsite modern methods of construction, Constr. Manage. Econ., **25**(2), 183–194, (2007).
[5] V.W.Y. Tam, C.M. Tam, S.X. Zeng and W.C.Y Ng, Towards adoption of prefabrication in construction, Building and Environment, **42**, 3642–3654, (2007).
[6] C.M. Eastman and R. Sacks, Relative productivity in the AEC industries in the United States for on-site and off-site activities, Journal of Construction Engineering and Management, **July**, 517-526, (2008).
[7] J. Lessing, A. Ekholm and L. Stehn, Industrialized housing – definition and categorization of the concept, *The 13th International Group for Lean Construction*, Sydney (2005).
[8] Z. Hamid, K.A.M. Kamar, M. Zain, K. Ghani and K. Rahim, Industrialized building system (IBS) in Malaysia: The current state and R&D initiatives, Malaysia Construction Research Journal, **2**(1), 1-13, (2008).
[9] K.A.M. Kamar, *Industrialised Building System (IBS): Chronology of Adoption in Malaysia*, International Islamic University Malaysia Press, (2011).
[10] CREAM, The 7th Industrialised Building System (IBS) Roundtable Workshop. Technology Foresight Report, Construction Research Institute of Malaysia, 1-27, (2011).
[11] K.A.M. Kamar and Z.IHamid, Modernising the Malaysian construction industry through the adoption of industrialised building system (IBS), *The Sixth International Conference on Multi-National Joint Ventures for Construction Works*, Kyoto, 1-14, (2010).
[12] P.S.E. Ang. Barrier-driver model for enhancement of ICT implementation in IBS production management process in construction industry, PhD Thesis, Universiti Tun Hussein Onn Malaysia, Johor, (2015).
[13] M. Alshawi, Future IT challenges in project management. In project management and IT, *The 7th Annual Conference of the Project Management Institute: Gulf Branch*, Bahrain, 1-19 (2000).
[14] L. Faraj and M. Alshawi. SPACE: Simultaneous prototyping for an integrated construction environment, INCITE, 466-484, (2000).
[15] J. Underwood and M. Alshawi, Data and process models for the integration of estimating and valuation, Computer-Aided Civil and Infrastructure Engineering, **12**(5), 369-381, (1997).

[16] M.A. Hunt, Strategic implementation of information technology within the construction industry, Doctoral Dissertation, University of Salford, (1995).
[17] S. Basu and P. Deshpande, Wipro's people processes: A framework for excellence, (2004). <available from http://www.wipro.com/insights/wipropeopleprocesses.htm>
[18] C. Goolsby, Integrated people processes tools = Best-of-breed service delivery, (2001). <available from http: //itpapers.news.com/>
[19] F. Xerox, Aligning people processes and technology, (2003). <available from http: //www.fujixerox.com.au>
[20] B. Larkin, Aligning people process and technology, (2003). <available from http: //www.paperlesspay.org/articles/Technology.pdf>
[21] B. Emmett, IT service management: People process technology = business value, The IT Journal, Third Quarter, (2002). <available from http://www.hp.com/execcomm/itjournal/third_qtr_02/ article2a.html>
[22] Clemson, Definition of Policy, Clemson University, (2007). <available from http //www.clemson.edu/research/orcSite/orcIRB_DefsP.htm.>
[23] A.R. Md Nor, *Statistical Methods in Reserch*, Pearson, Malaysia, (2009).
[24] R. Fellows and A. Liu, *Research Methods for Construction*, Wiley-Blackwell, West Sussex, (2008).
[25] K.C. Goh and J. Yang, Importance of sustainability-related cost components in highway infrastructure: perspective of stakeholders in Australia, Journal of Infrastructure Systems, **20**(1), (2013).
[26] J.B. Yang and S.C. Peng, Development of a customer satisfaction evaluation model for construction project management, Building and Environment, **43**(4), 458-468, (2008).
[27] J. Wong and H. Li, Development of a conceptual model for the selection of intelligent building systems, Building and Environment, **41**(8), 1106-1123, (2006).
[28] Z. Shehu and A. Akintoye, Major challenges to the successful implementation and practice of programme management in the construction environment: A critical analysis, International Journal of Project Management, **28**(1), 26-39, (2010).
[29] K. Ruikar, C.J. Anumba and P.M. Carrillo. Verdict-An e-readiness assessment application for construction companies, Automation in construction, **15**(1), 98-110, (2006).
[30] P. Weill, The relationship between investment in information technology and firm performance: a study of the valve manufacturing sector, Information Systems Research, **3**(4), 307-333, (1992).
[31] D.S. Bajwa, A. Rai and I. Brennan, Key antecedents of executive information system success: A path analytic approach, Decision Support Systems, **22**(1), 31-43, (1998).
[32] Q. Tran, D. Huang, B. Liu and H.M. Ekram, A construction enterprise's readiness level in implementing eprocurement: a system engineering assessment model, Systems Engineering Procedia, 131-141, (2011).
[33] L. Raymond, The impact of computer training on the attitudes and usage behavior of small business managers, Journal of Small Business Management, **26**(3), (1988).
[34] V. Venkatesh, Determinants of perceived ease of use: Integrating control, intrinsic motivation, and emotion into the technology acceptance model, Information Systems Research, **11**(4), 342-365, (2000).
[35] V. Peansupap, An exploratory approach to the diffucion of ICT in a project environment, Phd Thesis, (2004).
[36] D. Leonard-Barton and J.J. Sviokla, Putting expert systems to work, Harvard Business Review, 91-98, (1988).
[37] R.L. Thompson, C.A. Higgins and J.M. Howell, Personal computing: Toward a conceptual model of utilization, MIS Quarterly, **15**(1), 125-143, (1991).
[38] R. Mahbub, Readiness of a developing nation in implementing automation and robotics technologies in construction: A case study of Malaysia, *The 6th International Conference on Construction in the 21st Century*, Kuala Lumpur, (2011).

MATEC Web of Conferences **47**, 04017 (2016)
DOI: 10.1051/matecconf/20164704017

Motivational Factors on Adopting Modular Coordination Concept in Industrialized Building System (IBS)

Riduan Yunus[1, a], Azeanita Suratkon[1], Mia Wimala[2], Hazrul Abdul Hamid[1] and Siti Rahimah Mohd Noor[1]

[1]*Jamilus Research Centre, Universiti Tun Hussein Onn Malaysia, 83000 Parit Raja, Johor, Malaysia*
[2]*Institut Teknologi Nasional Bandung, 40124 Bandung, Indonesia*

Abstract. Modular coordination (MC) is recognized as a tool towards rationalization and industrialization. The implementation of MC concept in the design stage may improve the constructability and construction time. However, the implementation of MC in Industrialized Building System (IBS) implementation is still low compare to other developed countries such as the United Kingdom, Sweden and Japan. This paper examined the interrelationship between motivational factors of stakeholders in adopting MC concept using Interpretative Structural Modeling (ISM). Questionnaire survey was engaged in this study to identify significant motivational factors. Then, semi-structure interviews were used to collect qualitative data. ISM was adopted to build relationship between factors and develop an initial model to promote the adoption of MC in IBS construction. Seven (7) significant motivational factors were identified in this research namely 1) 'stakeholder's commitment', 2) 'reduce site disruption', 3) 'increase productivity', 4) 'high skilled workers', 5) 'site sustainability (environment, economy and social benefits)' 6) 'standardization' and 7) 'enabling 'open building' concept'. The result using *Matrice d'Impacts Croises Multiplication Applique an Clasment* (MICMAC) shows that there are three factors can be categorized as Independent / Driving Factors namely 'stakeholder's commitment', 'standardization' and 'enabling "open building" concept'. These factors should be explored in details to enhance the adoption of IBS in Malaysia. The findings provide a very good platform for a further research in formulating an efficient solution to promote MC concept adoption among the stakeholders. This scenario will improve the deliverables of IBS construction and eliminate negative perception in its implementation.

1 Introduction

The development of construction industry is rapid. Advanced technology and methods need to be integrated swiftly in order to compete in this industry. Malaysia as a developing country laid a comprehensive blue print in order to ensure the industry is competitive [1]. One of the strategies is to improve the adoption of Industrialized Building System (IBS). IBS have a potential to improve sustainability and keep the GDP of the country at the promising state [2-5]. However, the adoption of IBS in Malaysia is still low due to resistance towards change and lack of information to support the transformation [5]. Therefore, it is vital to identify the motivational factors to enhance the adoption of IBS.

[a] Corresponding author : riduan@uthm.edu.my

Modular Coordination (MC) is a standard of measurement for IBS elements to coordinate the dimension and spaces. It is where a building and its components are being dimensioned and positioned in a basic unit or module known as 1 M which is equivalent to 100 mm. The implementation of MC concept towards design prefabricated components may enhance the total constructability in the construction project [6]. The modular is an intermediate stage towards an eventual evolvement of open system which allows for easy adaptation of prefabricated components to any layout and their interchangeability within the building. Malaysian Standard MS 1064 was introduced by the local Malaysia government to catalyze the development of IBS implementation.

Malaysian Standard MS 1064 was prefaced by Malaysia government which obligates architects, suppliers, and contractors to follow the standard of specification established for the size of fundamental modular unit [7]. However, the adoption of this standard is still low due to misconception and lack of motivation to use MC.

2 IBS and Modular Coordination

IBS is a construction system with a combination of components that promote simplification and minimise on-site work. IBS components are manufactured in a factory either on or off site. The components are then positioned and assembled into structures. The production in a controlled environment able to reduce the numbers of workers involved, shorten the construction period, increase quality of buildings, reduce cost and enhance occupational health and safety [8-10]. With the coordination for measurement system and open building concepts, this construction technique will improve the project deliverables.

Open building concept permits intercrossed applications. It is adaptable towards MC and standardization. Open building system enables the production of building components produce by various different factories. As the results, the products are compatible for any IBS's buildings even the components are supplied from different IBS's suppliers [8]. In order to objectify the implementation of open building system, the adoption of MC is vital as it is an international system of dimensional standardization in building.

MC usage is capable to be applied into a wide range of building technologies, ranging from component building through partial prefabrication to rationalized traditional building method. In additions, components which are coordination on a modular basic might be used in renovation programmers as well [11]. The coordination in components' dimension will allow the components to be used for any projects. However, open building system will also provide a high degree of design flexibility but required a maximum coordination between the designer and manufacturer [12]. Higher commitment is required in early involvement of project between key stakeholders [10].

3 Research Method

A combination of questionnaire survey and semi-structured interviews were adopted in this research. First, the questionnaire survey helps to identify significant motivational factors to adopt MC as a measurement concept. Then, semi-structured interviews assist researcher to understand the interrelationships between identified factors. Interpretive Structural Modeling (ISM) technique was used to build relationship between and develop model around the element of a system [13]. The ISM method was applied to identify interrelationships between identified significant factors. The interviewees provide information required by sharing their industrial experiences, appropriate judgments and suggestions. Semi-structured interviews are fit in exploring stakeholders' perception regarding industrial issues enabling information delivered to interviewer to develop a final model based findings.

4 Motivational Factors

The survey investigated 18 motivational factors which were identified from literature review. The respondents were asked to rate the level of significance of each of these factors based on their judgment and experience. The respondents were divided into contractors, designers and manufacturers. The data was then analyzed by comparison and synthesis among others.

Table 1 summarizes the descriptive and inferential statistics for the "more" significant factors contributing to adoption of MC in IBS construction. Standard deviations show uniformity with all below 1, thus representing data accuracy in this research. The cut-off value in selecting the significant factors is 4.50 which indicate the factors are very significant. As a result, there are seven 'very significant' motivational factors, namely 1) Reduce site disruption, 2) Enabling "open building" concept, 3) Increase productivity, 4) High skilled workers, 5) Site sustainability (environment, economy and social benefits), 6) Standardization and 7) Stakeholder's commitment.

Table 1. Ranking of motivational factors to adopt modular coordination.

No.	Motivational Factors	N	Minimum	Maximum	Mean	Std. Dev
1	Reduce site disruption	41	3	5	4.61	0.586
2	Enabling "Open building" concept	41	3	5	4.59	0.547
3	Increase productivity	41	3	5	4.59	0.631
4	High skilled workers	41	3	5	4.59	0.591
5	Site sustainability (environment, economy and social benefits)	41	3	5	4.56	0.634
6	Standardization	41	3	5	4.54	0.552
7	Stakeholder's commitment	41	3	5	4.51	0.597
8	Government support and policy	41	3	5	4.49	0.553
9	Adaptability and buildability	41	3	5	4.46	0.636
10	Open systems	41	3	5	4.46	0.636
11	Local materials and availability	41	2	5	4.46	0.711
12	Supplier competitiveness	41	3	5	4.44	0.634
13	Efficiency in assembly and installation works	41	3	5	4.44	0.634
14	Sustainable procurement (e.g design charrette, post-costruction or demolition consideration)	41	2	5	4.41	0.741
15	Restricted on for IBS construction	41	2	5	4.41	0.805
16	Eliminated additional works and reduction of waste generation	41	3	5	4.39	0.666
17	Aesthetical value	41	1	5	3.66	1.063
18	New innovation and creativity	41	1	5	3.56	0.950

5 ISM Analysis and Results

Analysis was conducted by using ISM method leads to several level of findings. The first analysis takes place in ISM is developing Structural self-interactive matrix (SSIM). The entire interviews were conducted by implementing face to face interview. Table 2 shows on how SSIM matrixes were developed by assigning value V, A, X or O.

Table 2. The SSIM indicators relationship between variables.

SSIM	(i, j)	Reachability Matrix
V	Factor i will lead to Factor j	$e_{ij} = 1$ while $e_{ji} = 0$
A	Factor i will be achieved by Factor j	$e_{ij} = 0$ while $e_{ji} = 1$
X	Factor i and factor j will help achieve each other	$e_{ij} = 1$ while $e_{ji} = 1$
O	Factor I and factor j are not related	$e_{ij} = 0$ while $e_{ji} = 0$

By understanding the guidelines in Table 2, the SSIM matrix was developed as shown in Table 3. This matrix then was converted into a binary matrix which then called as an initial reachability matrix (Table 4). The transitivity of the matrix was verified to establish the final reachability matrix (Table 5).

Table 3. Structural self-interactive matrix.

No.	Factors	7	6	5	4	3	2
1	Reduce site disruption	V	V	X	A	X	A
2	Enabling "Open building" concept	A	X	V	O	V	
3	Increase productivity	A	A	X	A		
4	High skilled workers	O	O	X			
5	Site sustainability (environment, economy and social benefits)	A	A				
6	Standardization	A					
7	Stakeholder's commitment						

Table 4. Structural self-interactive matrix.

No.	Factors	1	2	3	4	5	6	7
1	Reduce site disruption	1	0	1	0	1	0	0
2	Enabling "Open building" concept	1	1	1	0	1	1	0
3	Increase productivity	1	0	1	0	1	0	0
4	High skilled workers	1	0	1	1	1	0	0
5	Site sustainability (environment, economy and social benefits)	1	0	1	1	1	0	0
6	Standardization	1	1	1	0	1	1	0
7	Stakeholder's commitment	1	1	1	0	1	1	1

Table 5. Final reachability matrix.

No.	Factors	1	2	3	4	5	6	7	Driving power
1	Reduce site disruption	1	0	1	1	1	0	0	4
2	Enabling "Open building" concept	1	1	1	1	1	1	0	6
3	Increase productivity	1	0	1	1	1	0	0	4
4	High skilled workers	1	0	1	1	1	0	0	4
5	Site sustainability (environment, economy and social benefits)	1	0	1	1	1	0	0	4
6	Standardization	1	1	1	1	1	1	0	6
7	Stakeholder's commitment	1	1	1	1	1	1	1	7
	Dependence	7	3	7	7	7	3	1	

The factors with the same reachability set and intersection set are located at the same level of the ISM hierarchy. The sets with the same reachability and intersection are given the top-level position as they will not help achieve any other variable above their own level [14]. Four iterations were conducted to allocate all factors in 3 levels (Table 6).

Table 6. Iteration for level partitions

No.	Factors	Reachability set	Antecedent set	Intersection set	Level
1	Reduce site disruption	1 3 4 5	1 2 3 4 5 6 7	1 3 4 5	1
2	Enabling "Open building" concept	1 2 3 4 5 6	2 6 7	2 6	2
3	Increase productivity	1 3 4 5	1 2 3 4 5 6 7	1 3 4 5	1
4	High skilled workers	1 3 4 5	1 2 3 4 5 6 7	1 3 4 5	1
5	Site sustainability (environment, economy and social benefits)	1 3 4 5	1 2 3 4 5 6 7	1 3 4 5	1
6	Standardization	1 2 3 4 5 6 7	2 6 7	2 6 7	2
7	Stakeholder's commitment	1 2 3 4 5 6 7	7	7	3

Based on the level partitions of the factors and the final reachability matrix, the final ISM based model was established (Figure 1).

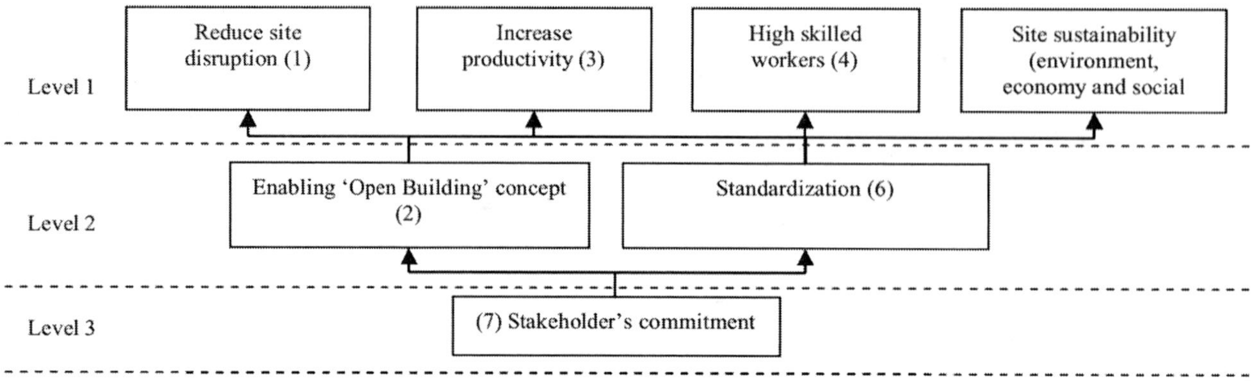

Figure 1. ISM-based model of interrelations among significant motivational factors of adopting MC in IBS implementation.

The ISM model provides a visualized the motivational factors in adopting MC concepts in IBS implementation in a structured and systematic format. The results depict the interrelationships between the factors and their influence level in the system. It can be noted that the factors that related to benefits of MC concepts are located on the first level of the ISM model. These benefits only can be achieved if there are commitments from the stakeholders which relate to standardization and enabling the 'Open Building' concept.

6 *Matrice d'Impacts Croisés Multiplication Appliquée an Classment (MICMAC)* Analysis

MICMAC analysis is shown in Figure 2. The analysis helps to categorize the factors into a four different cluster, namely autonomous, dependent, linkage and independent.

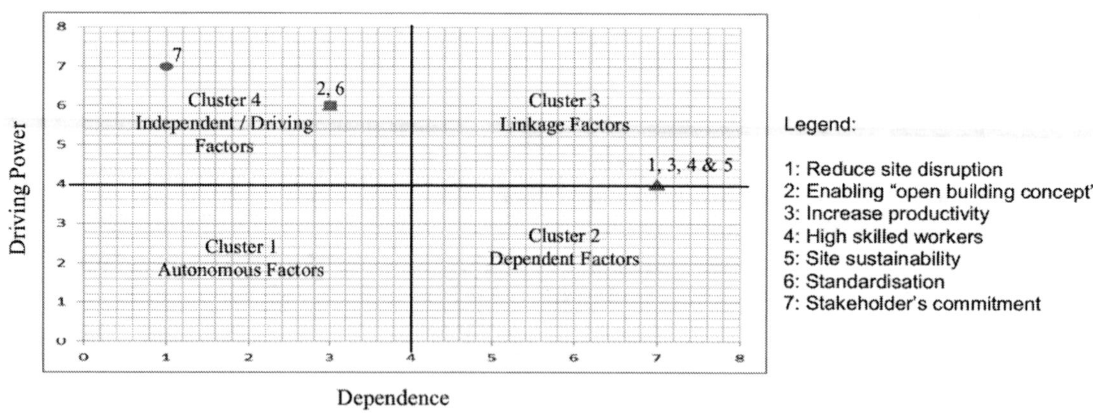

Figure 2. Driving power and dependence diagram.

The results confirmed that there are three factors can be clustered as independent or driving factors. The factors are 1) Stakeholder's commitment, 2) Enabling "Open building" concept and 3) Standardization. The analysis shows that these three factors are vital to ensure the successful of MC adoption. Thus, stakeholders are required to give more attention in getting all players participate in promoting MC. Local authorities should provide a strong fundamental to stakeholders by developing effective policies and legislative in implementing MC for IBS construction. On the other hand, 4 factors are clustered in the middle of 2 (linkage factors) and 3 (dependent factors). The factors are 1) Reduce site disruption, 2) Increase productivity, 3) High skilled workers and 4) Site sustainability (environment, economy and social benefits). These factors are depending to the driving factors and linked to each other.

7 Conclusions

All stakeholders play important roles in promoting MC on the implementation of IBS in Malaysia. Several measures that should be taken include the establishment of policies and legislative associated to the promotion of MC, development of standard for MC, and empowerment of open building concept. In addition to the participation of stakeholders, there are some dependent factors that also need to be considered as motivational factors for adopting MC such as to reduce site disruption, increase the productivity, engage the high skilled workers, and support the site sustainability.

References

[1] Construction Industry Development Board Malaysia, Construction Industry Master Plan 2006-2015, Kuala Lumpur, (2006).

[2] Z. Abd Hamid, Industrialised building system (IBS) in Malaysia: The current state and R&D initiatives, Malaysian Construction Research Journal, **1**(2), 1-11, (2008).

[3] K.A.M. Kamar, Collaboration initiative on green construction and sustainability through industrialized buildings systems (IBS) in the Malaysian construction industry, International Journal of Sustainable Construction Engineering and Technology, **1**(1), (2010).

[4] M.N.M. Nawi, A. Lee and K.M. Nor, Barriers to implementation of the industrialised building system (IBS) in Malaysia, The Built and Human Environment Review, **4**, 22-35, (2011).

[5] R. Yunus and J. Yang, Improving ecological performance of industrialized building systems in Malaysia, Construction Management and Economics, **32**(1-2),183-195, (2014).

[6] N.R.B. Zainol, A. Al-Mamun and P.Y. Permarupan, Overview of Malaysian modularity manufacturing practices, American Journal of Industrial and Business Management, **3**(7), (2013).

[7] MS 1064, Guide to Modular Coordination in Building, Construction Industry Development Board Malaysia, (2001).

[8] N. Blismas and R. Wakefield, Drivers, constraints and the future of offsite manufacture in Australia, Construction Innovation: Information, Process, Management, **9**(1), 72-83, (2009).

[9] L. Jaillon and C.S. Poon, Sustainable construction aspects of using prefabrication in dense urban environment: A Hong Kong case study, Construction Management and Economics, **26**(9), 953 - 966, (2008).

[10] R. Yunus and J. Yang, Critical sustainability factors in industrialised building systems, Construction Innovation: Information, Process, Management, **12**(4), 447 - 463, (2012).

[11] C.T. Haas and W.R. Fagerlund, Preliminary Research on Prefabrication, Preassembly, Modularization, an Offsite Fabrication in Construction: A Report to the Construction Industry Institute, University of Texas, (2002).

[12] J. Neelamkavil, Automation in the prefab and modular construction Industry, *Proc. International Symposium on Automation and Robotics in Construction* , Austin, (2009).

[13] L. Mei and J. Yang, Analysis of interrelationships between critical waste factors in office building retrofit projects using interpretive structural modelling, International Journal of Construction Management, **14**(1), 15-27, (2014).

[14] S. Azevedo, H. Carvalho and V. Cruz-Machado, Using interpretive structural modelling to identify and rank performance measures: An application in the automotive supply chain, Baltic Journal of Management, . **8**(2), 208-230, (2013).

MATEC Web of Conferences **47**, 04018 (2016)
DOI: 10.1051/matecconf/20164704018

Innovative Approach for IBS Vendor Selection Problem

Mohd Faizal Omar[1,a], Mohd Nasrun Mohd Nawi[2], Adi Irfan Che-Ani[2], Nor Intan Saniah Sulaiman[3] and Kai Chen Goh[4]

[1]*School of Quantitative Sciences, Universiti Utara Malaysia, 06010 Sintok, Kedah, Malaysia*
[2]*School of Technology Management and Logistics, Universiti Utara Malaysia, 06010 Sintok, Kedah, Malaysia*
[3]*Pusat Citra Universiti, Universiti Kebangsaan Malaysia, 43600 Bangi, Selangor, Malaysia*
[4]*Faculty of Business and Technology Management, Universiti Tun Hussein Onn Malaysia, 86400 Batu Pahat, Johor, Malaysia*

Abstract. Supply chain management in Industrialised Building System (IBS) construction management has significantly determined the successful of company and project performance. Due to the wide variety of criteria and vendor available, the vendor selection process for a specific project needs is becoming more difficult. The need of decision aid for vendor selection in other areas is widely discussed in previous research. However, study on vendor selection for IBS project is largely neglected. Decision Support System (DSS) is proposed for this purpose. Yet, most of the DSS models are impractical since they are complicated and difficult for a layman such as project managers to use. Research indicates that the rapid development of ICT has highly potential towards simple and effective DSS. Thus, this paper highlights the importance and research approach for vendor selection in IBS project management. The study is based on Design Science Research Methodology with combination of case studies. It is anticipates that this study will yield an effective value-for-money decision making platform to manage vendor selection process.

1 Introduction

Construction industry can be considered as one of the foundation in Malaysia economy. Due to the high demand of housing and infrastructure, Malaysia has progressively moved toward industrialisation. However, the traditional approach in construction is not parallel with this objective. Traditional approach is always led to construction problem such as delay, overrun cost, redesign, and low quality of project outcome. Thus, new approach and method of construction is significant to fulfil the need of housing and infrastructure development. Thus, the introduction of new approach in Malaysia construction industry such as IBS has been considered as solution to the construction problems [1-3]. IBS is a construction system that is built using pre-fabricated components. The manufacturing of the components is systematically done using machine, formworks, and other forms of mechanical equipment. The components are manufactured off-site and once completed will be delivered to construction sites for assembly and erection. Building system can be classified with four main categories [3]:

[a] Corresponding author : mfaizal.omar@gmail.com

a. Conventional building system
b. Cast-in-situ formwork system – table or tunnel formwork
c. Prefabricated system and
d. Composite system

IBS attempts to expedite the execution of the project by providing materials to be produced and constructed readily at the optimum time. The role of purchasing in supply management has been acknowledged as one of the foundation towards the successful of construction companies particularly in IBS construction project. Purchasing from the right vendor can enhances efficiency and competitiveness of IBS project. In prefabrication process, problems related to supplier such as material shortage or late delivery of material can negatively affect the IBS project process. Thus, there is a need to assess supplier's suitability for a specific IBS project. Client may need to consider supplier selection criteria such as quality, delivery, cost/price, financial, technology, etc. For IBS project, other related criteria might also be considered such as ISO certification, product development, flexibility, reliability, and management ability. This scenario increases the challenges to select the right and effective vendor. Effectiveness refers to vendor who can supply the right amount of materials or services at the right time, at right price and the right quality.

2 Problem Statement

One of the important processes during planning is vendor selection. Little attention is given in literature for IBS vendor selection problem. Based in the literature, there are multiple characteristics and criteria for industrialized building materials vendor to be considered in the project; such as delivery time, logistic issues, erection process, cost of purchasing etc. It imposed a complex scenario where it requires the right decision to be made before selecting an appropriate vendor. Poor decision in vendor selection would negatively affect not only company performance, yet affecting project performance as well. The performance of a few IBS vendor in Malaysian construction industry is an issue where poor decision were made in awarding the contract [5]. Thus, there is a need to aid decision making for IBS vendor selection. Till date, the decision making tools for this problem area is still far immature as most were ad-hoc decision. Thus, we aim to develop a Knowledge Based Multi Criteria Decision Support System for IBS Vendor Selection Problem. This study will attempt to fill the gap by developing a decision making framework and prototype to foster better decision making in IBS vendor selection.

3 Related Work

3.1 Decision in supplier selection

Vendor or Supplier selection has gained much attention from researchers in numerous fields worldwide. This is due to the importance of an effective supplier selection in supply chain management. For example, purchasing share would influence 40-80% of the total turnover of company [6]. In addition, in IBS construction management, supplier plays significantly importance towards project performance. An ineffective supply chain management would negatively affect project performance and led to construction problem such as delay, overrun time and, cost [7]. Furthermore, selection of the right vendor is crucial process due to the vagueness, imprecision of goal and constraint in decision process [8, 9] . Due to the diversity of purchasing context, vendor selection also deals with tremendous criteria. For instance, a study has been conducted to compare the importance of supplier selection based on 110 research paper within 1991-2001 [9]. Their study revealed that there were some criteria were used consistently in vendor selection such as price, delivery, quality, production facilities and capacity, technical capability, geographical location, management, reputation and position in industry and, financial position. Other study also mentioned that criteria such as reliability, flexibility, consistency and long term relationship were also used frequently in other field [21]. However, in recent years, the new criteria emerge such as e-transaction ability [10], integrity, professionalism, commitment, economic, political situation [8], and environmental and social criteria [11]. The increasing of criteria in vendor selection has increase the complexity for this problem. As a result, various analytical techniques has been proposed. Previous research categorize vendor selection method into two groups i.e. single model and integrated model (Figure 1).

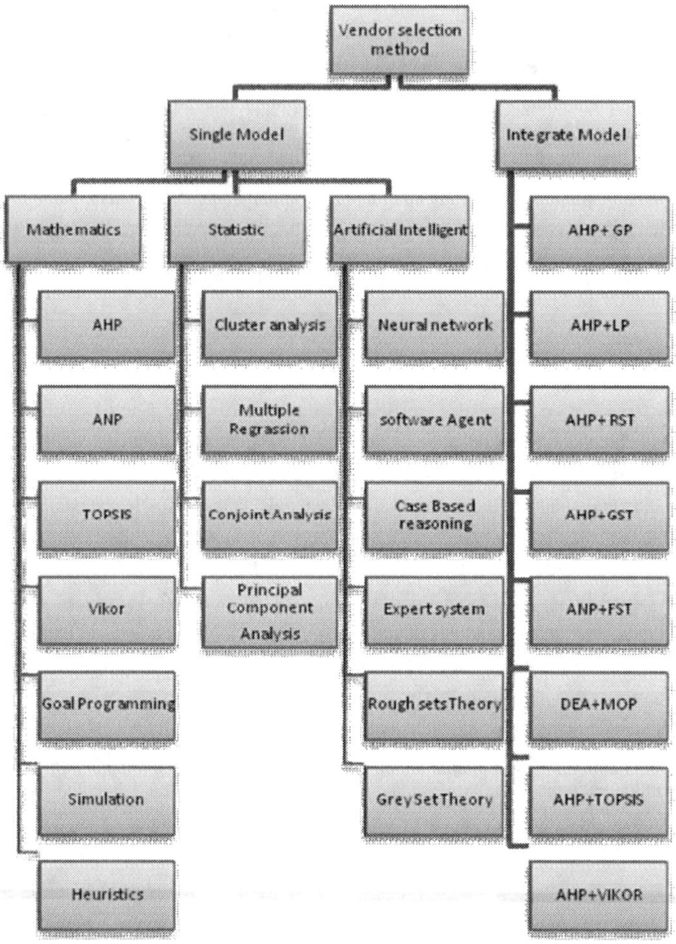

Figure 1. Method in supplier selection [12].

The above figure shows that Multi Attributes Decision Making (MADM) technique such as AHP, ANP, DEA, TOPSIS and, VIKOR dominates and capable of dealing with vendor selection problem. Thus, in order to enhance the vendor selection in IBS project management, we propose the integration of MADM method and DSS to be carried out.

3.2 Decision support system and applications in construction project management

DSS are computer programs that aid users in problem solving or decision-making environment. The system has detailed knowledge, data, models, algorithms, user interfaces, and control mechanisms to support a specific decision problem [13]. They are especially valuable in situations in which the amount of available information is prohibitive for the intuition of an unaided human decision maker. Furthermore, DSS can aid human cognitive deficiencies by integrating various sources of information, providing intelligent access to relevant knowledge, and aiding the process of structuring decisions. A lot of research on DSS in planning has been conducted to date in construction management [14-16]. However, most of the models are impractical since they are complicated and difficult for a layman such as project managers to use.

It is desirable to suppress those complexities in the models by an advanced DSS which can offer simple, effective and efficient tools towards better decision making. Some researchers only concentrate on model development and discards important computer science fundamentals, e.g. software engineering, information management and human-centred computing [14, 16, 17]. Some researchers have attempted to adopt computing essence in their model development [18]. However, they often imposed incorrect terms which led to erroneous concepts of software development modelling [19]. As a result, most of the models and methods are not being used or have limited impact for real-world decision making [20]. Nowadays, the trend of the cutting edge technology in ICT on cloud computing. The technology was named after the representation as a cloud, often described on a graphic representation of a network. Its concept on a large scale implies the use of the Internet in order to allow those interested to access virtual resources through which users can manage and control the services they have acquired. Thus, our proposed DSS will leverage the features in cloud technology to support mathematical approach for vendor selection problem In IBS project.

4 Methodology

Driven by the motivation and research objectives, Design Science Research Methodology (DSRM) will be utilised. Considering its capabilities to address issues pertaining to the design and development of an IT artefact, DSRM is well suited for the research involving system development. DSRM can also be used with case study approach in order to understand certain event or phenomena in depth [21]. In this research, we attempt to utilise a mix of case studies technique as strategy to demonstrates and develop DSS. Case study is chosen as a method to justify and validate the IT artefact. Hence, two case studies and software development process will be performed. Multiple of case study can enhance the development and validation of research rather than single case study. The descriptions of few main research activities are as follows.

4.1 Case study 1: Requirement gathering for DSS

The purpose of this activity is to deeply understand the requirements needed to develop DSS for a MCDM problem. Data will be collected through literature review. This phase is divided into three sub activities i.e. requirement elicitation, requirement analysis and System Design. The process in Case Study 1 mainly focuses on the aspect of Requirement Engineering. It concerned on the identification of the goals to be achieved by the envisioned system, the operationalisation of such goals into services and constraints, and the assignment of responsibilities for the resulting requirements to agents such as humans, devices, and software. There are six main categories of requirements that should be addressed which consist of (1) individual functionality, (2) business flow, (3) data, formats and information needs, (4) systems with other interfaces, (5) user interfaces, and (6) other constraints such as performance, reliability, and security

Requirements of DSS will be further analysed by using Case Study technique. To demonstrate a generic DSS architecture, it is proposed that a paper based decision modelling is conducted for vendor selection problem. A real case of IBS vendor selection problem in Malaysian construction industry will be selected to illustrate the modelling technique. As a result, mathematical decision model will be developed and validated by decision makers.

4.2 DSS development

Based on the previous software requirement in Case 1, DSS is proposed to incorporate model management, database and user interface.

4.3 Case study 2: Usability evaluation and validation of DSS

Case Study 2 focuses on the evaluation of DSS for IBS vendor selection by decision makers. In this phase, the current practice result will be compared to DSS prototype output. Then a usability testing will be performed to assess its usability and utility. A set of heuristics questionnaire will be established. This iterative process will be followed by unstructured interview to capture stakeholder and enable them to "think aloud". The DSS prototype will be revised based on experts recommendations. Figure 2 below illustrates the conceptual concept of proposed DSS as mentioned in case 1 and 2 above.

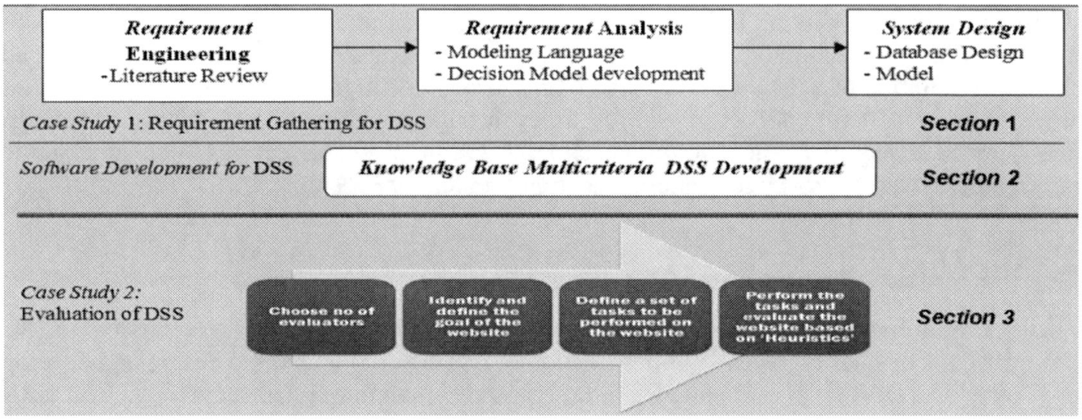

Figure 2. Conceptual framework for the proposed DSS.

5 Discussion and Conclusion

Vendor selection has been addressed as foundation towards an effective supply chain in numerous fields including construction. In recent years, the role of vendor selection has gained much attention. In Malaysian Construction Industry particularly for IBS, supplier/manufacture is one of the main project stakeholders who determined the successful of project. This is because IBS project is based on prefabrications method. Any disruption in supply chain such as material shortage or delay can automatically affect project process. In addition, issues regarding poor vendor performance are also raised. Hence, there is a need to select the right vendor to expedite the execution of the project by providing the materials for the project to be produced and constructed readily at the optimum time. Our study aims to develop knowledge base multi-criteria Decision Support System to profile and select the right IBS vendor to cater the need of Malaysian construction industry. Driven by the motivation and research aims, this study will base on case study and DSRM. Real case projects will be used to demonstrate the decision making framework. The data collection will include documental analysis and interviews with stakeholders. On the other hand, the implementation process involves the design of vendor profiling, implementation and evaluation of cloud based multi-criteria DSS. Unlike other DSS implementation in the literature, this DSS toolkit will be developed by using the latest cloud computing technology to solve IBS vendor selection in an innovative way. The artefact can be generalized to solve other type of multi criteria decision making problem which potentially foster a better quality, value-for-money and transparent decision making process.

Acknowledgment

The authors gratefully acknowledge the support by the Ministry of Education Malaysia for providing the funding under Research Acculturation Collaborative Effort (RACE). We also thank the contribution by other members in Construction Research Group and Analytical Minds CoE in UUM.

References

[1] B. López-Mesa, Á. Pitarch, A. Tomás and T. Gallego, Comparison of environmental impacts of building structures with in situ cast floors and with precast concrete floors, Build. Environ., **44**(4), 699-712, (2009).

[2] L. Jaillon, C.S. Poon and Y.H. Chiang, Quantifying the waste reduction potential of using prefabrication in building construction in Hong Kong, Waste Manag., **29**(1), 309–320, (2009).

[3] Y.F. Badir, M.R.A. Kadir and A.H. Hashim, Industrialized building systems construction in Malaysia, J. Arhit. Eng., **8**(1), 19-23, (2002).

[4] A. Cox, Power, value and SCM, Supply Chain Management, **4**(4), 167–175, (1999).

[5] S.A.A.H.S.S. Hassim and M.S. Jaafar, The contractor perception towers Industrialised building system risk in construction projects in Malaysia, Am. J. Appl. Sci., **6**, 937–942, (2009) .

[6] P.S. Tan, S.S.G. Lee and A.E.S. Goh, Multi-criteria decision techniques for context-aware B2B collaboration in supply chains, Decis. Support Syst., **52**(4), 779-789, (2012).

[7] M.N.M. Nawi, A. Lee, M.N.A. Azman and K.A.M. Kamar, Fragmentation issue in Malaysia industrialised building system (IBS) project, J. Eng. Sience Technol., **9**(1), 97-106, (2014).

[8] S. Thiruchelvam and J.E. Tookey, Evolving trends of supplier selection criteria and methods, Int. J. Automot. Mech. Eng., **4**(1), 437-454, (2011).

[9] S.H. Cheraghi, M. Dadashzadeh and M. Subramanian, For supplier selection : An update, J. Appl. Bus. Res., **20**(2), 91-108, (2004).

[10] A.K. Kar and A.K. Pani, Exploring the importance of different supplier selection criteria, Manag. Res. Rev., **37**(1), 89-105, (2014).

[11] B.A. Ozturk and F. Ozcelik, Sustainable supplier selection with a fuzzy multi-criteria decision making method based on triple bottom line, **5**(3), 129,147, (2014).

[12] K. Mukherjee, Supplier selection criteria and method: past, present and future, Int. Oper. Res., 39-44, (2014).

[13] H.K. Bhargava and C. Tettelbach, A web-based decision support system for waste disposal and recycling, Comput. Environ. Urban Syst., **21**(1), 47-65, (1997).

[14] I.M. Mahdi, A multi-criteria approach to contractor selection, Eng. Constr. Archit. Manag., **9**(1), 29–37, (2002).

[15] C. Kahraman, U. Cebeci and Z. Ulukan, Multi-criteria supplier selection using fuzzy AHP, Logist. Inf. Manag.,**16**(6), 382-394, (2003).

[16] A. Shapira and M. Goldenberg, AHP-based equipment selection model for construction projects, J. Constr. Eng. Manag., **131**(12), 1263-1273, (2005).

[17] R. Manoharan, Subcontractor selection method using analythic hierarchy process, Universiti Teknology Malaysia, (2005).

[18] A.K. McCowan and S. Mohamed, A classification of Decision Support System for the analysis and evaluation of concession project investment, J. Financ. Manag. Prop. Constr., **7**(2), 127-137, (2002).

[19] M.F. Omar, B. Trigunarsyah and J. Wong, The state of art of decision support system development: Progress of requirement engineering in construction, *International Building and Infrastructure Technology Conference*, (2011).

[20] Q. Tian, J. Ma and O. Liu, A hybrid knowledge and model system for R&D project selection, Expert Syst. Appl., **23**(3), 265-271, (2002).

[21] C.A. Weber, J.R. Current and W.C. Benton, Vendor selection criteria and methods, Eur. J. Oper. Res., **50**(1), 2-18, (1991

MATEC Web of Conferences **47**, 04019 (2016)
DOI: 10.1051/matecconf/20164704019

A Review on Industrialised Building System Issues in Malaysia

Farah Deba Ayeop Abdul Khalil[1,a], Farah Nora Aznieta Abd Aziz[1], Salihuddin Hassim[1] and Mohd Saleh Jaafar[1]

[1]*Department of Civil Engineering, Faculty Engineering, Universiti Putra Malaysia, 43400 Serdang, Selangor Malaysia*

Abstract. Industrialised Building System (IBS) has been identified as a potential method to improve overall construction performance in term of quality, cost effectiveness, safety and health, waste reduction and productivity. There are several issues in managing IBS construction projects which lead to delays, poor qualities and cost overrun. This paper aims to analyze the issues in managing the construction phases of IBS projects. Based on this review, it shows that less effective use of IBS in Malaysia construction industry is due to unwillingness of the players to change from conventional system because they are not convinced with the IBS system. Therefore, appropriate bodies should help to increase the contractor's understanding and prepareness in using IBS system in the construction industry.

1 Introduction

An Industrialised Building System (IBS) is a construction system that is built using pre-fabricated components. The manufacturing of the components is systematically done using machine, formworks and other forms of mechanical equipment. The components are manufactured off-site and once completed will be delivered to construction sites for assembly and erection. IBS can be defined as a construction system which components are manufactured in a factory, on or off-site, positioned and assemble into structures with minimal additional site work [1]. The construction industry has started to embrace IBS as a method of realizing better construction quality and productivity, reducing risks related to occupational safety and health, alleviating issues for skilled workers and dependency on manual foreign labour, and achieving the ultimate goal of reducing the overall cost of construction. Apart from this, it offers minimal wastage, fewer site materials, a cleaner and neater environment, controlled quality, and lower total construction costs [2]. The experiences in some developed countries such as Japan, Germany and the United Kingdom indicate that there is a great potential for IBS to progress, as shown by their market share [3]. Among the most successful IBS implementations in the world are Sekisui Home (Japan), Living Solution (United Kingdom), Open House (Sweden) and Wenswonen (Netherlands) [4].

IBS consists of precast component systems, fabricated steel structures, innovative mould systems, modular block systems and prefabricated timber structures as construction components which mainly are parts of the building that are difficult and time consuming if cast on-site hence, are designed

[a] Corresponding author : faralaiqa@gmail.com

detailed and cast as standardized components at the factory and transported to the site for assembly technologies [5]. According to Blismas [6] and CIDB [7], construction industry in developed countries has started to adapt IBS as a method of attaining better construction quality and productivity, reducing risks related to occupational safety and health, alleviating issues for skilled workers and dependency on manual foreign labor, and achieving the ultimate goal of reducing the overall cost of construction in order to minimal wastage, fewer site materials, a cleaner and neater environment, controlled quality and lower total construction costs [2]. On the same note, IBS adoption in Malaysia has long historical note since the 1960s to overcome shortage of houses such as Flat Pekeliling for Police Diraja Malaysia. However, the use of IBS was not sustained even though some of the international IBS systems were introduced in late 1960s and 1970s. It was found that those imported IBS systems were not compatible with Malaysia's climate and social practices resulting to closing down of precast concrete factories and continuing use of conventional system in the industry [8].

However, due to shortcoming of local skilled workers and overflow of unskilled workers from the neighboring countries, in 1999 country has recognised and enforcing use of IBS as one way of overcoming the problems. The IBS Strategic Plan was launched to promote IBS and followed by the IBS Roadmap 2003-2010 and IBS Roadmap 2011-2015, blueprints to industrialized construction by 2015 to chart and guide the awareness programmes, incentives, vendor scheme development, training, quality control and research and development programmes [9]. Since then, IBS is made compulsory in goverment building project with mimimum of 70% of the components should be made of IBS. However, implementation in private projects are non-traceble by the government and therefore the main problems are not solved althought the government already take initiative in using IBS system. Recently, enforcement of using IBS system are highlighted in the Economic Transformation Plan (ETP) particularlly to helps the country in managing a fast, better quality and environmental friendly construction methos in developing 'Greater Kuala Lumpur' in mega project such as Mass Rapid Transit (MRT), 1 Malaysia Financial District, Sungai Buloh and Sungai Besi mixed development project. The application of IBS in the construction industry plays an important role in supporting 12 National Key Economic Areas (NKEAs) in providing infrastructures and buildings. Therefore, the policies on IBS should be established in a sustainable manner to create high income society and economy and also creates more job opportunities. The potential usage of automation and robotic in IBS shall be explore by local industries so that the system becomes more effective and practical in the real application.

2 IBS Construction Scenario

Various attempts have been made to face the diverse challenges in the building sector by the government and related agency in the industry. The conventional construction method, which is commonly being practiced, is high cost and unable to respond to the huge demand in a short time with standard quality [10]. This traditional construction process has also been widely criticized for its fragmented approach to project delivery and its failure to form effective teams thus created a number of issues such as rework, time delay, rising cost, lack of communication and coordination, and wastages [10]. This traditional process which is segregated in nature are known as 'over the wall' syndrome, its breaking the project into a piece of smaller sub-project which is pass among different group in sequential manner throughout the entire project delivery and it contributed toward failure in forming effective teams [10].

In addition to that, the conventional construction method consists of a reinforced concrete frame and brick, beam, column, wall, and roof, which are cast in site using the timber framework while steel reinforcement is fabricated off-site. This method is labour intensive, involving formwork fabrication, steel bending and concreting. Thus, it requires many wet trades on-site, such as skill carpenters, plasterers, and brick workers. This process can be impeded and ruin by quality issues, unfavourable site conditions, a skilled labour shortage, bad weather conditions and material wastage [11]. The application of the conventional method to meet the huge demand of today with the certain quality standard, added up with the rising cost of labour and material will give a negative impact and hampered the development toward the construction industry.

According to Pan et al. [2], construction industry in developed countries has started to adapt IBS as a method of attaining better construction quality and productivity, reducing risks related to occupational safety and health, alleviating issues for skilled workers and dependency on manual foreign labour, and achieving the ultimate goal of reducing the overall cost of construction. It also had been highlighted as important use of technology and modern construction method in improving the efficiency of the implementation of construction projects [8]. According to CIDB [9], the use of IBS and innovation in the construction process also stimulated the growth of construction output and is seen as one of the most important sectors that contribute to the Gross Domestic Product (GDP) of a country. It is noted that the IBS can contribute significantly to fulfilling the demand and boosting it performance of construction industry.

Despite the change of policy, promotion and benefit that can be attain through the adoption of IBS construction industry, it is being hampered and still in its early stages. According to Hamid et al. [8] and Nawi et al. [10], although the IBS system improves the quality of projects, easier to control, reducing rectification work and lowering the total cost of construction, the trend of IBS usage in Malaysian construction projects are still below the national target. None the less, there are still limited numbers of contractors in Malaysia specialising in IBS that can undertake IBS jobs. Up until today, the IBS construction project still composed with the conventional construction method. Due to the combination of these two methods, IBS project can be assumed as multi-complex task and project [12] which will then contribute to other problems along the chain and the process of completing the project. This is agreed by several studies that opined that to reach the expertise in IBS, necessary knowledge and changes to organization process, practices, procedure and administration are required. Therefore it is clear that the IBS requires a shift in paradigm and a new set of management practise in undertaking the IBS project to its desirable outcome.

In addition toward the adoption and integration of IBS in the construction industry, it is revealed that the IBS supply chains are facing with the problem of fragmentation and adversarial relationships among players. It has been identified as the primary impediment toward the IBS construction project delivery. It should be noted that the construction process consists of five project management process which involved with various participants in a long duration that start with initiating, planning, executing, monitoring and closing phase [12]. The construction industry is unique compared to the other industries as it is a project oriented industry that produces unique projects [12] that are complex in nature thus, a proper and systematic management in needed. Management factors are domineering to ensure successful execution and determining the whole supply chain systems in IBS project [13].

The challenges and barriers in IBS implementation in Malaysia is basically due to the unwillingness of the players to convert from the conventional system of construction. The reasons for this could be due to lack of knowledge, unwilling to learn new things, comfortable and secure with current system and insufficient budget for mechanised system construction method [8]. Consequently, the awarded IBS projects are not well managed delayed and produced bad quality buildings [11].Another possible reason for this bad reputation of IBS system is that manufacture of IBS components is involved only after the tender stage of the value chain. This lack of integration among relevant players in the design stage has resulted in a need for redesign and additional costs to be incurred if IBS is adopted [8].

3 Summary

Based on above discussion, it can be summarised that applying IBS system in Malaysia construction industry is still in unacceptable stage although it has been 16 years since it is introduced in 1999. The players, particularly privates companies are not convinced that IBS can contribute to significant cost savings, especially with the small volume of building constructed. Also, there is no guidance in terms of finance and cost control in IBS projects [14]. Therefore despite a need of a new approach of procurement using strategic partnering in the construction delivery system, support and professional training from government agencies such as CIDB and CREAM should be continue and proactively share the knowledge with the key players. This in turn perhaps will produce more IBS players and more coordinated and comprehensible guidelines in terms of applications, elements and cost control of IBS projects.

References

[1] CIDB, IBS Roadmap 2003–2010, Construction Industry Development Board Malaysia, Kuala Lumpur, (2003).

[2] W. Pan, A.G.F Gibb and A.R.J. Dainty, Strategies for integrating the use of offsite production technologies in housebuilding, Journal of Construction Engineering and Management, **138**(11), 1331-1340, (2012).

[3] L.S. Peng and C.J. Chuan, a study of the readiness of precasters in just in time construction work study, International Journal of Productivity and Performance Management, **50**(4), 131-140, (2001).

[4] M. Oostra and C. Claeson-Jonsson, *Best Practices: Lesson Learnedon Building Concept in Open Building Manufacturing- Core Concept and Industrial Requirement*, TNO Bouw , Delft, (2007)

[5] CIDB, Malaysian Construction Outlook 2007, Construction Industry Development Board Malaysia, Kuala Lumpur, (2007).

[6] N. Blismas and R. Wakefield, Drivers, constraints and the future of offsite manufacture in Australia, Construction Innovation, **9**(1), 72-83, (2009).

[7] CIDB, IBS Roadmap 2003–2010, Construction Industry Development Board Malaysia, Kuala Lumpur, (2003).

[8] Z.A. Hamid, K.A.M. Kamar, M.Z.M. Zain, M.K. Ghani and A.H.A. Rahim, Industrialised building system (IBS) in Malaysia: The current state and R&D initiatives, Malaysian Construction Research Journal, **2**(1), 1–11, (2008).

[9] CIDB, The current state of industrialised building system (IBS) construction forward, *Proceedings of the 1st IBS Roundtable Workshop*, Kuala Lumpur, (2011).

[10] M.N.M. Nawi, A. Lee and K.M. Nor, Barriers to implementation of the industrialised building system (IBS) in Malaysia, The Built and Human Environment Review, **4**, 22-35, (2011).

[11] E.C.W. Lou and K.A.M Kamar, Industrialized building systems: Strategic outlook for manufactured construction in Malaysia, Journal of Architectural Engineering, **18**(2), 69-74, (2012).

[12] I.L. Jabar, F. Ismail, A.A. Mustafa, Issues in managing construction phase of IBS projects. Procedia - Social and Behavioral Sciences, **101**, 81–89, (2013).

[13] F. Ismail, N.M. Yusuwan, H.E.A. Baharuddin, Management factors for successful IBS projects implementation, Procedia - Social and Behavioral Sciences, 68, 99–107, (2012).

[14] A.B.A. Rahman and W. Omar, Issues and challenges in the implementation of industrialised building systems in Malaysia, *Proceedings of the 6th Asia-Pacific structural Engineering and Construction Conference*, Kuala Lumpur, (2006).

MATEC Web of Conferences 47, 04020 (2016)
DOI: 10.1051/matecconf/20164704020

Properties of Agro-Industrial Aerated Concrete as Potential Thermal Insulation for Building

Eeydzah Aminudin[1,a], Mohd Fadhil Md Din[2], Mohd Warid Hussin[3], Abd Halid Abdullah[1], Kenzo Iwao[4] and Yo Ichikawa[4]

[1]*Faculty of Civil and Environmental Engineering, Universiti Tun Hussein Onn Malaysia, 86400 Parit Raja, Johor, Malaysia*
[2]*Institute of Environmental Water Resources and Management, Universiti Teknologi Malaysia, 81310 Skudai, Johor, Malaysia*
[3]*Construction Research Centre, Faculty of Civil Engineering, Universiti Teknologi Malaysia, 81310 Skudai, Johor, Malaysia*
[4]*Nagoya Institutes of Technology, Gokiso, Showa, Nagoya, Aichi 466-8555, Japan*

Abstract. The present paper is aiming to provide an actual state of the development of non autoclaved Agro-Industrial aerated concrete by using multiple ashes as improvement of thermal behavior for building. The feasibility of Agro-Industrial wastes as lightweight concrete by utilizing the Palm Oil Fuel Ash (POFA) as binder replacement and bottom ash as fine aggregate was investigated in this paper. Portland cement, bottom ash, aluminum powder and lime (Ca(OH)2) were used in this study. The POFA was used to replace Portland cement and Hydrated Lime at 0%, 5%, 10% and 15% by weight and aluminum powder was added at 0.75% dry weight in order to form bubbles. The compressive strength, water absorption, porosity and the thermal conductivity test were carried out after the concrete were water cured for 7 days and later being exposed to the air and water until 28days. The results show that the 20% replacements give the optimum strength of 7.143MPa and 30% give the best thermal conductivity with 0.48W/mK. Hence, this study aim, was to develop an agro-industrial aerated concrete good in insulation but having an optimum strength. Hence, it has been found that the more the percentage of POFA is added the lower the thermal conductivity since the pore structure is increasing and by the optimization done, 30% replacement has been chosen as the best mix design for Agro-Industrial Aerated Concrete.

1 Introduction

In the world of urbanization, lightweight concrete has become a perfect product as construction materials since it offers advantages such as light in densities ranging from 1000 to 1600 kg/m^3, having a great fire protection, good in thermal and sound insulations compared to normal weight concrete [1]. Normally aerated concrete consists of a mixture of sand, lime, cement, gypsum, water, and an expanding agent [15]. [17] classified there are three divisions on the basic of strength and unit weight: low density or low strength concrete (0.7–2.0MPa, 300–800kg/m3) used for insulation; moderate-strength lightweight concrete (7–14MPa, 800–1360kg/m3) used for concrete block together with other

[a] Corresponding author : eeydzah@uthm.edu.my

applications for strength and structure lightweight concrete (17–63MPa, 1300–1900kg/m^3) [2]. This study focusing on the moderate-strength lightweight concrete as it is main application for non-load bearing wall that comes with insulations properties. It is important to develop such materials for mitigation due to climate change or known as Urban Heat Island (UHI) issues.

Lightweight concrete or also known as thermal insulation is one of key factor for assessing the material composition of building envelops [21]. This is because the behavior of materials will determine the properties which are important for urban thermal balance. Besides that the long term benefit of insulation is that it may reduce the cost of cooling as well as reducing the pollution of the environment [3]. Heat from solar radiation will be absorbed and the infrared will radiate and dissipate part of the accumulated heat through convective, conduction and radiation. Some of heat will reradiate back to the atmosphere causing the increasing of ambient temperature. Thus, technical characteristics of materials used need to comply with determination degree of energy consumption and comfort conditions of individual buildings as well as of open spaces [4].

Aerated concrete known as lightweight concrete has been used in current construction materials. Such invention, lots of other extensive research progress aiming in improving the thermal, hygric and other structure properties. Majority, researched innovation conducted in Table 1 proved that either the replacement of cement or fine aggregate proven that the it might lower the density as well as thermal conductivity hence create a better insulation. As been cited by [5] the reduction of density of mortar means probably due to high air content. This has been agreed in Fu and Chung, 1997 cited in [5] study showed that low thermal conductivity is related to high air void. However, study showed by [2], proved that the thermal conductivity (Tc) differ, depends upon the pore structure of the lightweight aggregates, density of concrete and the cement paste matrix. Thus, the pore structure is one of the main factors affecting Tc.

Still the limitation shows that non load bearing structures rarely proves the percentage of wastage can improve thermal behavior including the value of Tcand the improvement in heat flux. Hence for this segment the researcher would like to discuss the behavior of adding and combining the wastes from Agricultural and Industry can improve the behavior of thermal conductivity, as improving the thermal comfort. Develop an efficient building materials is important as to minimize the consumption of electricity [8].

2 Materials and Methods

2.1 Materials used for the experiment

Materials used in this study consist of Ordinary Portland Cement, Calcium hydroxide ($Ca(OH)_2$), Bottom Ash from Power Plant as replacement of sand, POFA, aluminum powder, super plasticizer and water. Bottom ash was first pretreated by crushing and particle size classification method to decrease its higher unburned carbon to desirable level or that gen. The bottom ash was from Tanjung Bin Power Plant, Malaysia and has been oven dried at the temperature of 110˚C ± 5 for 24 hour which later has been sieved before stored in airtight container. The Bottom Ash used in this study conforms with BS 882: 1983 and ASTM C 33-84, standard grading requirements for fine aggregate. The POFA used is a by-product obtained from burning the remaining of palm oil fibers and shells from Kahang Palm Oil Mill located in southern Malaysia. The collected ashes were dried in the oven at the temperature of 110˚C ± 5 for 24 hour to remove all the moisture content before sieved and ground to obtain finer particles. Aluminum powder was used as a pore-foaming agent which reacts with calcium hydroxide to form hydrogen gas (gas bubbles) in cement slurry. The reaction in making aerated concrete was as follow [22] as shown in Equation (1).

$$2Al(s)+3Ca(OH)_2(s)+6H2O \rightarrow 3CaO\cdot Al_2O_3\cdot 6H_2O+3H_2(gas) \quad (1)$$

Besides that powder form super plasticizer classified as type superplasticizers sulfonated naphthalene in powder form confirming to ASTM C-494-05 (ASTM 2005a) type A and ASTM type B, D respectively were employed in all agro-industry aerated concrete. Superplasticizer was being introduced as admixtures as it may add fluidity to the concrete while less water is being used. In addition, it also plays a significant role to reduce the flocculation of cement grains [23]. In the case of aerated concrete production, insufficient presence of superplasticizer may fail the dispersion of cement particles thus producing a mix with a very low workability. Hence, created medium cause's disruption to the aeration process which is crucial for decreasing the density of aerated concrete. Besides that, the use of BA in fine aggregate will also create a higher expansion as compared to sand with the increment percentage of superplasticizer. This might due to the porous structure of bottom ash which requires a lot water content hence leads to the reduction of compressive strength [9].

Table 1. Previous study on the additional of pozzolonic material for thermal conductivity.

Wastes Materials	Authors	Applications
Fly Ash from the Top Ash at Coal Power Plant	[6]	Effect of snow and Fly ash (type C) as additive of (FI10, FI20, and FI30) by weight. FI10: 0.284 W/m.K FI20: 0.255 W/m.K FI30: 0.272 W/m.K
	[5]	The more the replacement of fly ash in cement the lower thermal conductivity. Thermal conductivity: FA(%) 10: 1.016 W/m.K FA(%) 20: 0.879 W/m.K FA(%) 30: 0.792 W/m.K
	[2]	The Tc value increase as the percentage of BA is increase in light weight aggregate. Thermal conductivity: 28 Days normal cured BA(%) 10: 0.53 W/m.K BA(%) 20: 0.56 W/m.K BA(%) 30: 0.62 W/m.K Autoclaved BA(%) 10: 0.63 W/m.K BA(%) 20: 0.64 W/m.K BA(%) 30: 0.66 W/m.K
	[7]	As stated by Kurama, in his study proved that the more addition of BA as fine aggregate the lower the thermal conductivity hence the better the insulation. Thermal conductivity: Reference(%) : 0.3610 W/m.K BA(%) 25: 0.3060 W/m.K BA(%) 50: 0.2826 W/m.K BA(%) 75: 0.2569 W/m.K BA(%) 100: 0.2202W/m.K
Ashes from Palm Oil Mill (POFA)	[1]	The usage of POFA in foamed concrete provide a better insulation as stated in the results which shown: LFC-CM-0.56(0.65 W/mK) LFC-PF10-0.54(0.74 W/mK) LFC-PF20-0.56(0.67 W/mK)

2.2 Chemical composition of materials

Based on data presented in Table 2, shows the percentage of chemical compositions of materials used in this research. The sum of SiO_2, Al_2O_3 and Fe_2O_3 in POFA is 62.28% with CaO 5.70%. It has shown that the type of fly ash is Class C Fly Ash with the min percentage of SO3 of 1.06%. Based on the previous research that had done by [24] the quality of burning is depends on the burning process. The silica or silicon oxide (SiO_2) is perhaps the most essential substance which is found in POFA and contains most like more then 40% [18]. The amount of crystalline phase also can be found in POFA [10].

Table 2. Chemical composition of materials used in this research.

Oxide	Portland Cement (%)	POFA (%)	Calcium Hydroxide (%)
SiO_2	18.45	52.73	0.02
Al_2O_3	4.07	2.13	0.01
CaO	60.85	5.70	31.39
Fe_2O_3	3.19	7.42	0.00
MgO	2.85	3.94	0.55
K_2O	0.22	6.87	0.1
P_2O_5	0.06	3.18	0.00
TiO_2	0.12	0.17	0.17
MnO	0.12	0.33	0.18
SO_3	4.04	1.06	13.04

2.3 Preparation of mix proportions

In this research, cement/calcium hydroxide (lime) /sand ratio (C: CH: S ratio) were used based on Ing, E, 2010 in [25]. The novelty for this study was bottom ash has been used 100% replacing the sand. The 100% bottom ash as the fine aggregate is been set as the control specimen. The specimen has been mixed with three different stages which the first stage was the mixing preparation of cement, hydrated lime, POFA, and Bottom Ash. Next, the additives were added with aluminum powder and super plasticizer. The water dry mix ratio has been varies with different percentage of POFA (0, 10, 20, 30, 50%) in order to determine best mix ratio. Table 3 represents the mix proportion of Argo-Industrial Aerated Concrete. At the end of the stage the water is added before cast in the form of cubes (70 × 70 × 70 mm). All specimens were left in the mold for 24 h before being demolded and subjected to water curing for first 7 days continued with different curing regimes until the time is ready to test.

Table 3. Mix proportion of agro-industrial aerated concrete.

Materials	Mix design				
	C:CH:BA = 25:25:50				
	CS	0%	10%	20%	30%
Cement (kg/m³)	214	214	203.3	192.6	183.5
Hydrated Lime (kg/m³)	214	214	203.3	192.6	183.5
POFA (kg/m³)	-	-	21.4	42.8	61
Sand (kg/m³)	428	-	-	-	-
Bottom Ash (kg/m³)	-	428	428	428	428

All specimens were tested to determine density and compressive strength at the age of 7 and 28 days. Compressive strength test was conducted in accordance to BS 1881:116 (British Standard Institution 1983). The aerated concrete which having the lower dense with an adequate strength and lower in thermal conductivity suitable for non load bearing application as stated in ASTM C129-85 (ASTM 1990) was chosen as the optimum design mix for control. In this research focused on the non autoclaved aerated concrete due to the awareness on reducing the effect on global warming even though as been stated by Tada in [26] claim that non autoclaved will produced a lot of macro-size distribution which will not influence the strength. At the end of the experiment, the thermal conductivity test has been identified, using BS EN 12664:2001 standard.

3 Results and Discussion

3.1 Influence of water dry mix ratio in agro-industrial aerated concrete

Maintaining correct water content of the aerated concrete mix is important to control the early stages of the manufacturing process. If the water / dry mix ratio is too low the mix will be very viscous at casting and will make the expansion become impeded and the mix will not reach the correct final height. Vise verse, if the water/dry mix content is too high the cast for the aerated concrete will become low in viscosity and the stability will may be reduced. As reported on the action of compressive strength value and density with respect to water dry mix ratio in Figure 1 different percentage of POFA give different results on the strength. Based on Figure 1, it is reported that water dry mix ratio of 0.2% give the best strength for all mixes. This might due to the effect of POFA which has a pozzolanic properties besides having portlandite, that may enhance the strength of concrete at 28 Days [27]. Basically at optimum percentage of 20% replacement of POFA will give a higher strength [14]. As water depleting, the lower the creation of macropores entrapped within the hardened mix making the specimen not only heavier but possess high strength and a denser weight recorded. Still, insufficient water will create lots of internal stress (internal capillary) between the matrix and it might have the capability to be burst off and create a larger crack mode on the surface. The internal stress is much related to the capacity of cement paste produced.

Hence, adequate water is important due to the formation of cellular internal structure. The cellular structure is very important as it will depend on the level of expansion which offers explanation for the gradual drop of density and strength when water increased. In general, the level of water dry mix ratio can be selected depending on the targeted density and strength of aerated concrete.

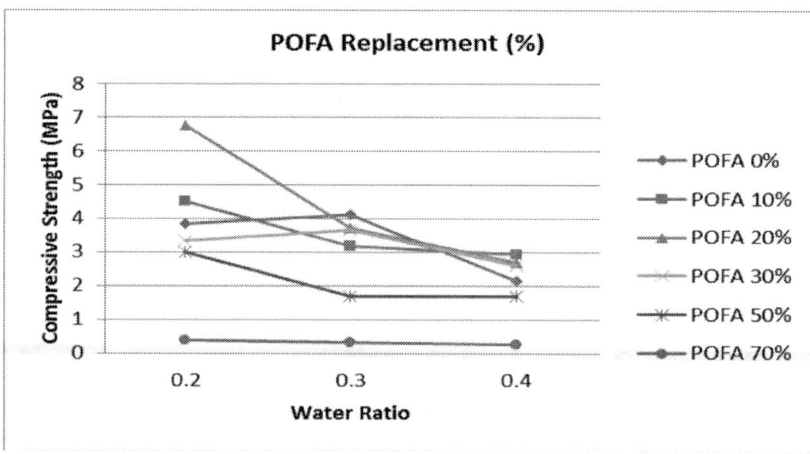

Figure 1. Variation of compressive strength with water dry mix ratio for mixes containing different amounts of POFA.

3.2 Relationship between mechanical properties of POFA in aerated concrete

Majority, researched innovation conducted in Table 1 proved that either the replacement of cement or fine aggregate proven that the higher the replacement when the lower the density, the lower the thermal conductivity hence a better insulation created. As been cited by [5] the reduction of density of mortar means probably due to high air content. This has been agreed in Fu and Chung, cited in [5] study showed that low thermal conductivity is related to high air void. However, study showed by [2] proved that the thermal conductivity differ, depends upon the pore structure of the lightweight aggregates, density of concrete and the cement paste matrix. Thus, the pore structure is one of the main factors affecting thermal conductivity.

Figure 2 indicates the mechanical properties including compressive strength, density, thermal conductivity and also the porosity value which narrow even precisely the indication for the best selection of mix based on the characteristic on reducing the thermal comfort. Overall, the higher percentage of POFA means that the more total of fines in matrix structure of aerated concrete. [28] findings shows that higher amount of fly ash leads towards lower density in aerated concrete. 20% replacement of POFA showed the highest strength among others at 28 days with 7 MPa. This might due to the pozzolanicity of POFA which plays the role in densifying internal structure through formation of C-S-H gel that fills the existing pores. This finding has been agreed by [29] which found that replacement of POFA up to 30% give the highest strength among the others. Pozzolanic reaction is a secondary reaction which forms the Calcium Hydroxide crystal from hydrating cement grains. Too much of POFA will not generate a higher strength development since lower content of calcium hydroxide produced from hydration process.

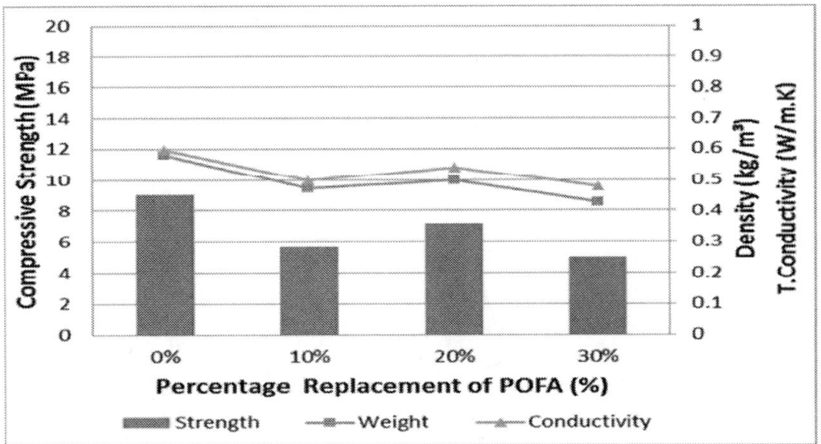

Figure 2. Overall properties for mixes containing different amounts of POFA subjected to different curing regimes at the age of 28 days.

But since, the knowledge of this study is more concerning on the behavior of thermal in the concrete hence, the thermal insulation has became the prior of the art of knowledge. The infrared thermograph was used as the technique to detect the behavior of the thermal load for particular brick at different time duration. By comparing three replacement percentages POFA for thermal insulation, it was found that 30 per cent gave better insulation properties yet with an adequate strength. The thermal conductivity from 30 per cent was 0.48W/m.K lesser even compared to the control aerated concrete with 0.5965W/m.K. Hence, the values of thermal conductivity decrease with the increasing percentage of POFA replacement. Again, it has been agreed in Fu and Chung, 1997 cited in [5] study showed that low thermal conductivity is related to high air void. Study showed by [6] stated that 20 per cent replacement gave the lowest value of thermal conductivity compared to control,10 and 30 per cent. The optimum proved that the samples have a highest amount of isolated voids compared to others. The basic idea behind using aerated concrete was to attain elongation of the heat flow path through the wall. The direct heat flow through vertical transverse joints was avoided by cancelling vertical joints fully filled with [3]. This caused to a high value of thermal conductivity.

3.3 Selected design mix

Based on the results obtained from the trial and error method, the mix having a lower density and exhibit an adequate strength for non load bearing wall with lower in thermal conductivity value has been chosen. It has been discovered that the replacement of 20% POFA exhibit the greatest strength at later age. As for 30% POFA it gave a bit lower strength but having an adequate strength with low thermal conductivity. As been stated in previous discussion, lower density, will create a more pores created between matrixes hence resulting a lower thermal conductivity value. Hence, in this study the aim investigates the lowest thermal conductivity in order to produce a high in time lagged which later will reduce the heat to transfer into a building. Table 4 shows the best mix design selected based on exhibit an adequate strength for non-load bearing specified in ASTM C129-85 (1990). As a control, 3 different mixes has been studied. First, the control having sand replacement in aerated concrete, control mortar based on C117 and commercial brick.

Table 4. Mix proportion of Agro-Industrial aerated concrete.

Cement: Lime: Bottom Ash	1:1:2
Ordinary Portland cement (%)	70
POFA Replacement (%)	30
Sand Fineness	Passing 4.75mm
Water Dry Mix Ratio	0.29
Alum Powder (% of dry mix)	1.5
Superplasticizer (%)	1

3.4 Microstructure

In aerated concrete, pore system normally divided into 3 regions. Based on [30] air pores with a radius of 50 to 500μm was introduced by hydrogen gas during manufacturing process and as for micro-capillaries of 50nm or less created between the gap of hydration developed in the wall between the air pores. It is also observed that the strength performance also influenced by this characteristic. It has been revealed that the porosity is not the unique parameter affecting the strength of concrete and many other factors such as pore size distribution, micro cracks, interface and other are also important factors for its properties. However, porosity which can semi-empirically describe the relationship between strength and microstructure for porous materials was still studied in this research. For the study of structure-property correspondence of aerated concrete, specimen having different porosity has been compared in Figure 3. From the results obtain, the regression line, R2= 0.8. This indicates that the correlation in between compressive strength and porosity is strong. Compressive strength is very important in order to catter a suitable load for a particular mixing.

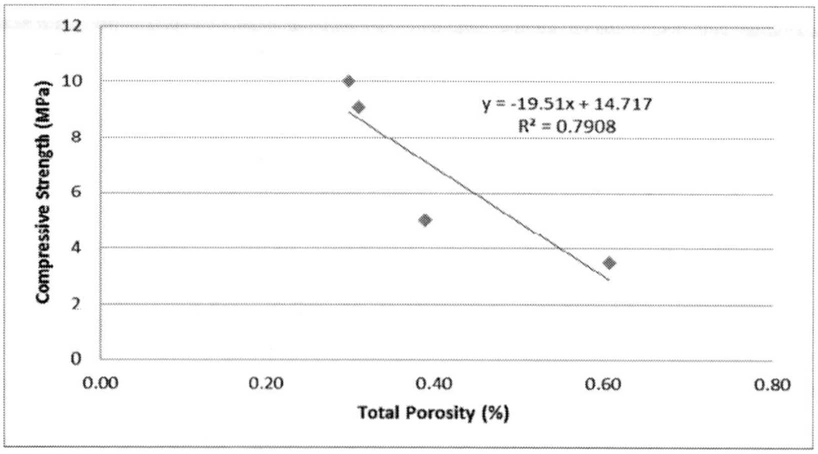

Figure 3. Relationship between total porosity and compressive strength of AC.

Besides that, variation of strength and density alter to different characteristic of air void or particle size distribution. It shows that the compressive strength increases with the decreasing of total porosity. According to [28] fine pore structure and narrow air void size caused increase in strength. A narrow air void will have well-structured cell in matrix form. The matrix link produces strengthen composition between one another. Figure 4 shows the pore size distribution diagarms and the mean diameters of bubbles for 30percent replacement. This can be understood as the fact that the increase of POFA will increase the total porosity and lower the bulk density. In other words, the larger volume of fine pore, thicker pore wall will have a uniform pore distribution from finest pore to macro capillary resulted from excessive process water. This can be proved by 30POFA having a pore diameter of 0.45 μm and density of 1225kg/m^3 with the porosity of 39%.

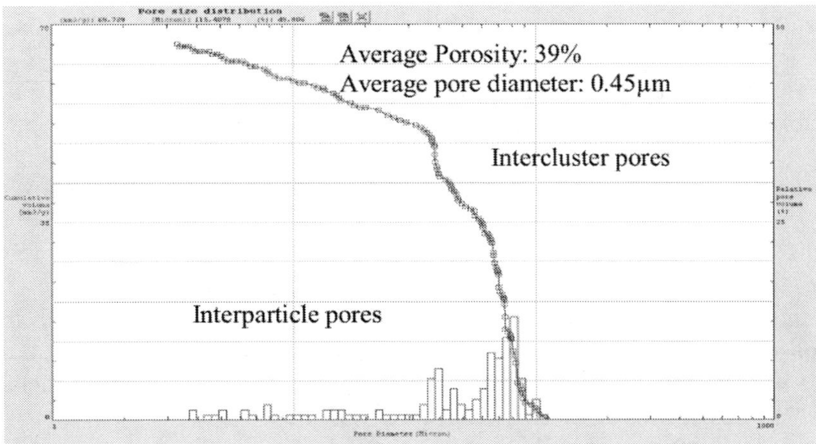

Figure 4. Pore size distribution diagrams and the mean diameters of bubbles for 30POFA.

4 Conclusion

On the basic of the results, the conclusion can be drawn: (i) POFA physical characteristic may exhibit the strength development for aerated concrete. The results show that the 20% replacements give the optimum strength of 7.143MPa and 30% give the best thermal conductivity with 0.48W/mK, (ii) The 100% replacement of bottom ash (0.52 W/m.K) as the sand replacement may enhance the thermal conductivity value compared to the control aerated concrete by sand (0.68 W/m.K), (iii) The optimum mix design which give an adequate result in strength but improve the thermal properties which can act as insulation and can be utilize for materials especially non-load bearing wall as one of indicator to overcome the Urban heat island issues. It shows that 30% POFA gave a better thermal comfort at night because it does not released back the heat at night, and (iv) The Agro-Industrial aerated concrete will produces initial needle-shaped hydrates which will change the structure into hexagonal and continues in block-shaped calcite crystals after the several years. Meaning to said, it will give a constant structure within years.

References

[1] S.K. Lim, C.S. Tan, O.Y. Lim and Y.L. Lee, Fresh and hardened properties of lightweight foamed concrete with palm oil fuel ash as filler, Constr.Build. Mater., **46**, 39–47, (2013).

[2] W. Wongkeo and A. Chaipanich, Compressive strength, microstructure and thermal analysis of autoclaved and air cured structural lightweight concrete made with coal bottom ash and silica fume, Mater. Sci. Eng. A, **527**(16-17), 3676–3684, (2010).

[3] K.S. Al-Jabri, A.W. Hago, A.S. Al-Nuaimi and A.H. Al-Saidy, Concrete blocks for thermal insulation in hot climate, Cem. Concr. Res., **35**(8), 1472–1479, (2005).

[4] M. Santamouris, A. Synnefa and T. Karlessi, Using advanced cool materials in the urban built environment to mitigate heat islands and improve thermal comfort conditions, Sol. Energy, **85**(12), 3085–3102, (2011).

[5] R. Demirbog, Influence of mineral admixtures on thermal conductivity and compressive strength of mortar, **35**, 189–192, (2003).
[6] A. Saygılı and G. Baykal, A new method for improving the thermal insulation properties of fly ash, **43**, 3236–3242, (2011).
[7] H. Kurama, İ.B. Topçu and C. Karakurt, Properties of the autoclaved aerated concrete produced from coal bottom ash, J. Mater. Process. Technol., **209**(2), 767–773, (2009).
[8] L.M. Al-Hadhrami and A. Ahmad, Assessment of thermal performance of different types of masonry bricks used in Saudi Arabia, Appl. Therm. Eng., **29**(5-6), 1123–1130, (2009).
[9] M. Singh and R. Siddique, Effect of coal bottom ash as partial replacement of sand on properties of concrete, Resour. Conserv. Recycl., **72**, 20–32, (2013).
[10] C. Chandara, K.A.M. Azizli, Z.A. Ahmad, S.F.S. Hashim and E. Sakai, Analysis of mineralogical component of palm oil fuel ash with or without unburned carbon, Adv. Mater. Res., **173**, 7–11, (2010).
[11] F.H. Wittman, *Autoclaved aerated concrete, moisture and Properties. Developments in civil Engineering*, Elsevier, (1983).
[12] M.W. Hussin, K. Muthusamy and F. Zakaria, Effect of mixing constituent toward engineering properties of POFA cement- based aerated concrete, J. of Materials in Civil Eng., **22**(4), (2010).
[13] V. Sata,C. Jaturapitakkul and K. Kiattikomol, Utilization of palm oil fuel ash in high – strength concrete, Journal of Materials in Civil Engineering, **16**, 623-628, (2004).
[14] K. Muthusamy, Properties of Palm Oil Palm Boiler Ash Fuel Ash Cement Based Aerated Concrete, PhD Thesis, Universiti Teknologi Malaysia, (2009),
[15] C. Karakurt, H. Kurama and I.B. Topçu, Utilization of natural zeolite in aerated concrete production, Journal of Cement and Concrete Composite, **32**, 1-8, (2010).
[16] I. Kula, A. Olgun, V. Sevinc and Y. Erdogan, An investigation on the use of tincal ore waste, fly ash, and coal bottom ash as Portland cement replacement materials, Cem. Concr. Res., **32**, 227–232, (2002).
[17] ACI Committee 213, Guide for structural Lightweight Aggregate Concrete, (2001).
[18] N.F. Zainudin, K.T. Lee, A.H. Kamaruddin, S. Bhatia and A.R. Mohamed, Study of adsorbent prepared from oil palm ash (OPA) for flue gas desulfurization, Journal of Separation and Purification Technology, **45**(1), 50–60, (2005).
[19] U.J. Alengaram, B.A. Muhit and M.Z. Jumaat, Utilization of oil palm kernel shell as lightweight aggregate in concrete – a review, Construction and Building Materials, **38**, 161–172, (2013).
[20] W.T.L. Chow and M. Roth, Temporal dynamics of the urban heat island of Singapore, Int. J. Clim., **26**, 2243-2260, (2006).
[21] V. Koc˘í, J. Made˘ra, R. C˘erny′ and P. Rovnaníková, *Application of a combined Computational-Experimental Approach for Service Life Estimate of Exterior Plasters of Historical Buildings*, WIT Press, Southampton, (2009).
[22] Short and Kinniburgh, *Lightweight Concrete*, CR Books, (1963).
[23] A.S.M.A. Awal and M.W. Hussin, Some aspects of durability performances of concrete incorporating palm oil fuel ash, *Proc. of the 5th Int. Conf. on Structural Failure, Durability and Retrofitting*, Singapore, (1997).
[24] N.K. Bansal, S.N. Garg and S. Kothari, Effect of exterior surface colour on the thermal performance of buildings, Building and Environment, **27**(1), 31–37, (1992).
[25] I.A. Shehu and A.S.M. Awal, Mechanical properties of concrete incorporating high volume palm oil fuel ash, Advances Materials Research, **599**, 537–540, (2012).
[26] K. Ramamurthy, E.K.K. Nambiar and G.I.S. Ranjani, A classification of studies on properties of foam concrete, Cement and Concrete Composite, **31**, 388–396, (2009).
[27] A.S.M.A Awal, A Study of Strength and Durability Performances of Concrete Containing Palm Oil Fuel Ash, PhD Thesis,Universiti Teknologi Malaysia, (1998).
[28] N. Arreshvhina, Application of Slag Cement Based Aerated Lightweight Concrete In Non-Load Bearing Wall Panels, Master Thesis, Universiti Teknologi Malaysia, (2002).

MATEC Web of Conferences **47**, 04021 (2016)
DOI: 10.1051/matecconf/20164704021

Determining the Technical Competencies of Construction Managers in the Malaysia's Construction Industry

Hairuddin Mohammad[1,a], Siti Khalijah Yaman[1], Fadzil Hassan[2] and Zulhabri Ismail[2]

[1]*Faculty of Civil and Environmental Engineering, Universiti Tun Hussein Onn Malaysia, 86400 Parit Raja, Johor, Malaysia*
[2]*Centre for Construction Project and Infrastructure Management, Universiti Teknologi Mara, 40450 Shah Alam, Selangor, Malaysia*

Abstract. It is a common conviction that project management, construction management, and site management are the tools for improving the performance of the construction project as a whole. Notwithstanding the similarities and differences between all of them, construction management which consists of off-site and on-site activities is deemed to have the major influence on the succession of a project. Consequently, a competent construction manager is required throughout the particular phase to shape the construction stability and geared up to complete the proposed development as according to several common objectives. Nonetheless, a competent construction manager will not just appear "out of the blue", they need to be teach, train, and develop in order to equip them with comprehensive competency sets. For that reason, education, training, and continuing professional development (CPD) are in place to provide the required knowledge, skills, and abilities. However, the debates on the inadequacy and problems on current construction management programs are continued to impede Malaysia's construction environment. It has been observed that lack of term of reference on the technical competency for construction manager is became the major setback. Hence, this paper tries to unfold the circumstances by offering a preliminary countermeasure to alleviate the situation through analysis of literatures. It was found that, the associated central phenomenon is believed to be explored holistically with the inclusion of multi layered thematic analysis as a tool to assist the exploration processes. Then, the respective outcomes from the analysis are deemed to be ready for subsequent phases, for instance the validation phase.

1 Introduction

Construction industry became renowned as one of the important industry throughout the globe. It was one of the contributors for a country's growth domestic product (GDP). Further, it appeared that construction industry act as a multiplier for other important industries such as manufacturing, agriculture, mining and even services sector to move forward especially in term of providing completed functional facilities (i.e. buildings and infrastructures). Their spills over effects are no doubt benefited many facets of construction related personnel across all level of positions. Regrettably, the progression of Malaysia's construction industry was hampered by a considerable number of problems. It was observed that the frequent theme of exceeding time and budget, not

[a] Corresponding author : hairuddin81@gmail.com

according to the specifications, excessive wastages, and subpar quality was particularly due to the contractors' fault which primarily occurred during the physical implementations of a construction project [1-4].

Currently, there are 67,833 of contractors (as of 2014) in the construction industry and 57,785 of construction managers (CM) (as of 2011) registered with the Construction Industry Development Board (CIDB) [5,6]. Since the CM is the key person in construction and generally representing the contractors, their contributions towards the successfulness of construction projects were undeniable, particularly during the construction phase. Given that construction phase exploits a lot of resources (manpower, materials, money, and machines), a competent CM is needed in order to manage those resources and geared up to achieve organizational objectives. Unfortunately, it was believed that only a few is qualified and fully trained CM since their (the competent CM) existence is in the dark. Even within nearly 58,000 construction managers [7], their level of competency and qualification is unknown [8]. Adding up salt to the wound, continuous critiques from past scholars on untiring degradation towards construction education and training does not helped to alleviate the previous problems [9-11]. Gaps particularly on knowledge and skills are still persisting, notwithstanding the level of offerings [12]. Furthermore, the absence of CM competency standard in the latest National Occupational Skills Standard (NOSS) indirectly impeded the appalling situation [13]. Hence, the particular mismatches together with problems of poor site management has resulted in many construction projects' failures and end up derailed from their true goals. Appropriately, the issue underpinning the lack of training and competent construction managers is the dearth of recognized and established term of reference for construction managers' competency provided within the industry.

Therefore, the author tries to unfold the predicament by focusing on CM technical competency. Since it is clear that faster proactive action should be in the training and education's pipeline, a spotlight into visible and ease of teachable technical competency is paramount [8,14-16]. Treating it as a central phenomenon, critical preliminary countermeasures are pursued to give ways for subsequent phases of the research exploration. Thus, culmination of views from past researchers were taken seriously by peeling off the CM technical competency's term, reviewing what others have done, and it will be followed by building up the relevant maiden stage through acceptable multiple layered thematic analysis.

2 Construction Managers Technical Competencies

2.1 Construction management

In reflection to Malaysia's preferred contractual approach, a general type of construction project life cycle will be taken into consideration. A general construction project will undergo several important phases throughout its life cycle (see Figure 1) [17-21]. Each phase has its own target and they may use different resources, together with different management approach. On top of that, generally each phase falls under different term of management, namely project management, construction management, and site management. Notwithstanding the similarities and differences between all of them, construction management which consists of off-site and on-site activities is deemed to have the major influence on the succession of a project.

As according to Koskela [20], construction management process is "where the detail designed is transformed into a construction/fabrication plan and into day-to-day coordination and control of processes on site or in a factory". Likewise, due to the consumption of a lot of resources (money, materials, machines, and manpower) during that phase alone [23], the determination of profit or losses are more prevalent; either on behalf of client or contractor.

Moreover, by referring to a general construction delivery and beside the existence of client and designer as a key player during the construction phase, contractor plays a significant role. Their contribution may start as early as the tendering or bidding process until the physical product complete as planned. Aware on the demanding nature of the construction phase, therefore it necessitates an all-around capable managers to lead and serve as moderators between each parties involved [24-26].

Thus, it requires a good management effort which normally leads by a construction manager in order to integrate all manageable items and transform it into an end solid product. For that reason, it was agreed that all construction managers should comprehensively have adequate competencies; a determinant to the successful of a construction project. In general, construction manager should have hard (technical) and soft (behavioural) competency to ensure that they are professionally fit for the construction world [27]. It is a 'must' because they are among the forefront in their respective field, where a lot of responsibilities is at stake in today's fast changing construction pace. Therefore, in order to summarise the culmination of common characteristics of construction management, subsequent table (refer Table 1) might be sufficient.

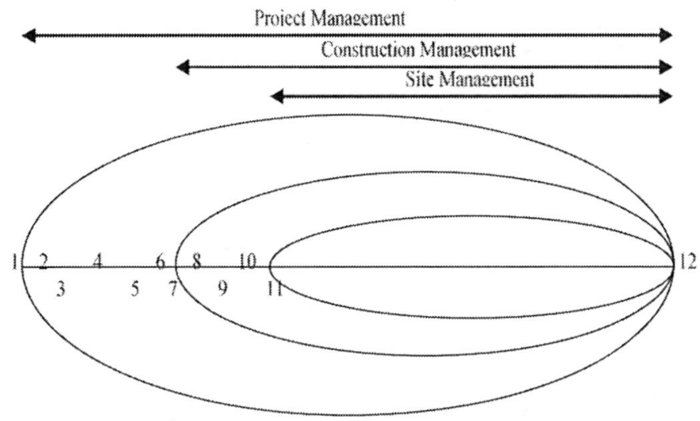

Legends: Process			
1.	Idea	7.	Advertise notice to bidders
2.	Conceptual design	8.	Bid period & receipt of proposals
3.	Approval of conceptual design	9.	Select contractor
4.	Preliminary & final design	10.	Notice to proceed
5.	Bid package complete	11.	On-site construction period
6.	Decision to release for bid	12.	Inspection & acceptance of project

Figure 1. Overall phases of a general construction project life cycle.

2.2 Competency

Competency or competence for some others is getting much attention in a past few decades, especially in the literature of management strategy during 1990s [28]. It was understood that the term is usually associated with training and educational fields. Hassan et al. [12] and Stoof et al. [27] were both agreed that the concept of competency is frequently used in the area of education including training and Continuing Professional Development (CPD). The particular concept is deemed to be the underlying formula to develop a new or improvisation of an existing educational method. In his work, McDaniel [28] came out with a model which shows the magnitude of competency identification as a critical consideration for CM curriculum design. Besides, Stoof et al. [27] added that competency is not limited only for educational purposes but the significance of the concept was also appreciated by business organizations through their human resource development (HRD) programs.

A preliminary study shows that skills, knowledge, and abilities (including attitudes) became the primary elements when discussing what is inside the "competency" [12, 29, 32, 33]. Additionally, based on the communal findings, it is obvious that the concept of competency underpinned the definition propounded by researchers. Besides, most of the researchers agreed that competency need to be discussed along with its intended usage, function, and amid its appropriate contexts, in order to have the most suitable meanings. Thus, in reaching the mutual concept of competency amid numerous deliberations, the attribute-based approach (work oriented) seems spot-on and was agreed by most of the researchers. Thus, the following diagram (see Figure 2) illustrates the concept of competency as according to Hassan et al. [12] and several others, and it will became the concern of this paper.

Table 1. Summary of several characteristics and approach to CM.

No.	Characteristics	Details	Approaches	Leads by
1	Period	Have defined start and end	Need to be managed properly	Construction Manager
2	Product/service	One of its kind, unique	According to client's requirements	
3	Critical components	Time, cost, quality, scope, resources	Need to be balance and in equilibrium	
4	Resources	Money, manpower, material, machines	Need to be managed properly	
5	Environment	Complex, uncertain, demanding	Inclusion of risk management, change management, etc.	
6	Phases	From bidding to acceptance of project	Need to be managed properly	

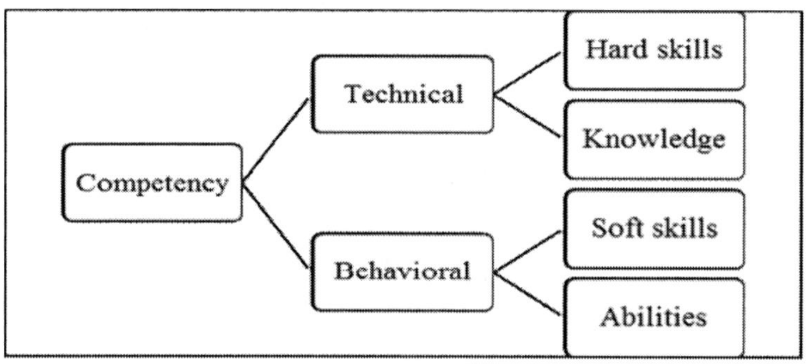

Figure 2. Concepts of competency as according to Hassan et al. [12].

Therefore, it is understandable that competency itself covers a large area. So as the identification of comprehensive technical competency and behavioural competency for construction manager, it requires a considerable work endeavour. For that reason, in accordance with the work of Stoof et al. [27] and Le Deist et al. [26] where they lend support from Spencer and Spencer in 1993, Hartle in 1995, and Parry in 1996, they were mutually agreed on the "visible" part of competency which comprises of skills and knowledge. It is accepted that skills and knowledge are noticeable and also teachable, where skills and knowledge are categorized as hard competency and can be develop through training and education. In a serious note, between both types of competency (i.e. technical and non-technical), technical competency is a vital consideration in order to match individual to their job scopes, but regrettably it was often being ignored by the common competency approaches [16]. To a certain extent, the superfluous focus made towards non-technical competency (i.e. behaviours and attitudes) by early developers of competency models were gradually diminished the important fixtures of technical competency. On the other hand, due to differentiate between skills and knowledge, Egbu's definitions were brought forward [34] :

a) On knowledge: "the ideas, wisdom and facts managers acquire through experience, theory and practice; the acquisition of which gives them an ability to understand".
b) On skills: "the activities or patterns of behaviour which managers undertake in order to accomplish a given desired outcome".

Based on the Egbu's definitions and the construction manager's realm, knowledge is everything to be learned and known [34]. It is an overall information gathering paradigm, where every associated knowledge elements need to be embedded (within the contexts of training and education). Further, for a certain knowledge that associated with a kind of activity, basically the learners knew about "know-how", but physical realisation-wise is not yet implemented. Moreover, "know-how" might be a combination of a number of knowledge in order to complete the entire "know-how". Meanwhile, skills required practices, actions and conducts, where basically uses human anatomy (i.e. physical structure such as hand and leg) and things. Similarly to knowledge, a mixture of skills may need in order to accomplish a given task. Here, it is clear that knowledge and skills compliments each other. Knowledge on most occasions became a primary element and as a basis for establishment of skills.

Therefore, with regard to technical competency of construction manager, both contexts (i.e. knowledge and skills) were utmost important. Thus, drawing from the works of several researchers, technical competency attributes which related to construction management were identified. Generally, the construction management functions such as planning, organising, staffing, coordinating, and controlling were considered as technical attributes. It was able to be derived from analysis of the key tasks or work outputs, where a pure construction activity that involves process, methods, materials, systems, and equipment comes into perspective (see summary on Figure 3).

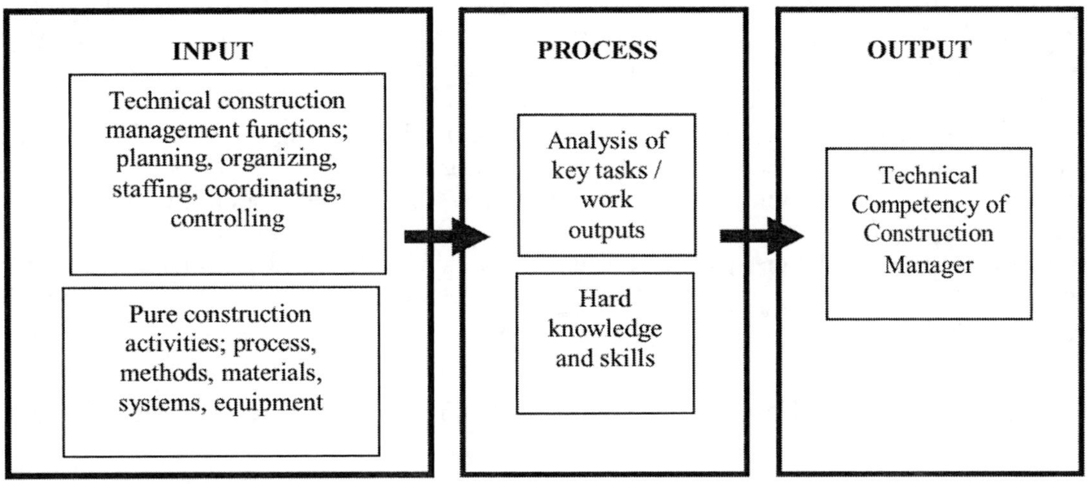

Figure 3. Identifying processes of technical competency for construction manager.

3 Past Research on Construction Manager's Competency

Research on CM competency is not new and much likely it will keep evolved from time to time. Drawing from the works of [32, 8-41] their main motive to reconsider the competency of construction manager is quite similar. Based on the problems, purposes, and actions taken, most of them stressed on the inadequacy of the construction management training/education in coping with fast-changing nature of the construction industry, which in turn resulting the damaging nature of construction project as a whole. However, the authors do not intend to deliberate on the issue of training/education as a whole, but only embarked on the quest of establishing the preliminary stage, which are identifying the technical competency for CM. Based on the previously cited researchers (as in the beginning of the topic), reviewing the similar quest is of paramount concern of this paper. Set aside their full context of epistemology's anchorage, their moves to choose positivism, constructivism, or pragmatism approaches is not aptly judged by the authors.

Collectively, all researchers were agreed that CM is representing the contractors. Only a tiny discrepancy was observed on behalf of [24, 38]. Riggs [38] postulated that the CM are representing the clients, meanwhile Love et al. [24] did not specify the exact organisation should the CM worked with, but stressed that CM may represents all facet of construction practitioners' organisations. This issue are debatable, and to some extent involves the semantic inquiry in term of the position title itself. Not favourable to the issue, the paper follows the majority of past researchers' standpoint; by insisting the CM representing the contractors.

As far as educational research is concerned, literature review is a must, where cases were built from ground up in order to fully understand the research topic (as a preliminary stage). Mutually, the authors observed that all researchers were compiling related competency for CM based on previous literatures. By referring Riggs [38] and Egbu [32] the move was apparent. Meanwhile, the rest was compiling from related institutional documents such as the American Council for Construction Education (ACCE), the American Institute of Constructors' (AIC), the California State University, and the Aalborg University [30, 36, 41, 43]. On the other hand, Arditi et al. [42] are compiling from a combination of past literatures and the Construction Management Association of America (CMAA), the American Institute of Architects (AIA), and the Associated General Contractors of America (AGC). Aside, there are two researches that use focus group interviews to gather CM's competency in the first place [42, 45]. Besides compiling activities, there were some attempt to cluster the competency from the ground up through inclusion of construction project stages [40, 45], abiding to predetermine cluster from institutional documents or past research [24, 30], and differentiation according to the nature of the competency itself [25].

In a nutshell, the authors believed that the previous research was providing a context rich competency for CM as according to their predetermined scopes. However, there was also evidence that the compilation was way beyond the boundary of construction management that being discussed previously by the authors. Additionally, proper iteration of competency either technical or non-technical was nowhere to be found or rather bland. Thus, pondering if the similar approach were taken by the authors and were applied towards Malaysia's construction industry that yet having established competency for CM, devastating effects would be around the corner. Confusion on the underlying generic competency for CM (especially the technical competency) might appear and it will further mess up the curricular design of related education/training. Thus, the authors assert appropriate compilation of technical competency for CM through preliminary stage of the research. Inclusion of stringent screening process is deemed to help the process.

4 Multi Layered Thematic Analysis and Clustering of Technical Competency

Guided by the written texts of Creswell [44, 45], preliminary identification of CM's technical competency is ideally follows the central phenomenon approach. It is also the approach taken by vast majority of past researchers. Literatures derived mostly by books, journals, and legitimate documents provide established listing of competency, where context-rich information is gathered. It may later on accompanied by qualitative (i.e. interview as validation and value added information), and/or quantitative (questionnaire survey as testing of theory and generalisation). Fully aware on the nature of compiling the CM's technical competency, the authors posits that credible qualitative-like approach is deemed appropriate Thus, multi layered thematic analysis is considered suitable as an approach to collect, sieve, and analysis towards the appropriate literatures by focusing on preliminary-selected themes before moving on to latter processes. Every layer regarding the analysis will become screening net, which at the end of the layer, comprehensive and meaningful data were extracted.

On the other hand, the authors also see the integration of clustering efforts might as well helps in term of organisation and significantly ease the latter processes of data. However, these were also became an issue, especially if their substance is questionable. Basically, there are two approaches that usually being considered by past researchers; clustering from the ground up (already being touched previously), and clustering through subsequent analysis of data (mostly by quantitative inquiry). The former is at ease since following the established predetermined clusters which maintaining meaningful clusters' title, meanwhile the latter are relying on computational mathematical formula to segregate the data into new found clusters. For instance, attempts to categorise competency into several clusters by using Principal Component Analysis (through Varimax Rotation Method with Kaiser Normalization). More often than not, the authors observed newly created clusters' title at the end of the particular analysis. Without doubting the approach, however the authors believed that the move was to a certain extent unleashed incomprehensiveness and requiring more translational works (e.g.

tweaking) especially during education/training design development (i.e. from competency sets to realisation of education/training module).

Therefore, as far as the paper is concerned, adoptions of multi layered thematic analysis together with appropriate clustering efforts were indeed beneficial towards preliminary identification of the said central phenomena. Multiple considerations' loop accompanying the process while gathering information from literatures is believed to produce exhaustive, holistic, and reliable outcomes (i.e. CM's technical competency) [7]. Thus, the following figure (see Figure 4) outlined the overall concept adopted by the authors. It can be seen that there are three major layers consists of construction management phase's boundary, identification of CM's technical competency, and preliminary clustering efforts according to valid past literature. All layers had been discussed appropriately in the preceding section of the paper. Nevertheless, there are added quality measures being imposed to each and every one of competency items before it was eventually finalised. Things such as reassessment and improvisation of cluster's categories might plausibly conduct in order to maintain cohesive and conciseness, as long as meaningful clustering approach were attained.

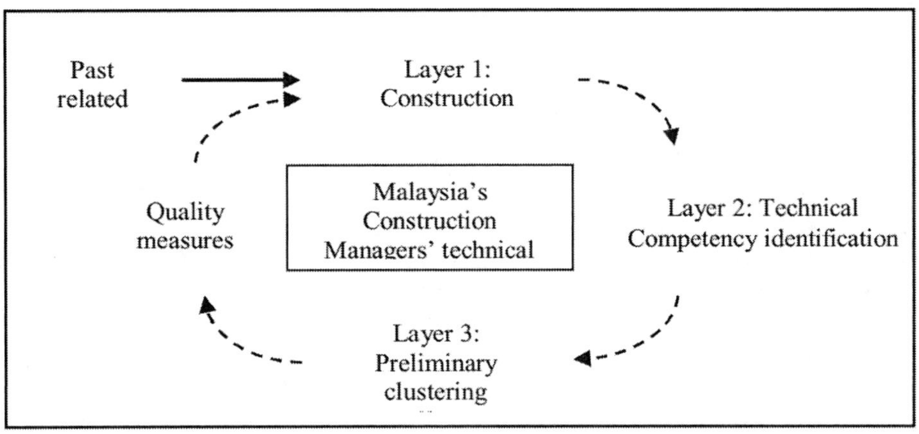

Figure 4. Multiple loop for assisting multi-layered thematic processes.

5 Conclusion

It was observed that in order to offer primary counter measure for problems depicted in the beginning of the paper, a thorough investigation on technical competency of construction manager would be essential. Due to the fact that the competency of construction manager is commonly being researched form time to time, abundance of technical competency was expected. Helped by delving into vast availability of literatures, the central phenomenon is believed to be explored holistically. Additionally, the inclusion of multi layered thematic analysis as a tool to assist the exploration may indeed fruitful. Purposely-built nets of construction management phase, technical competency identification, and proper clustering guidance is reckoned to efficiently sieve the information. Nevertheless, since the nature of required technical competency which may not come with "one size fits all" necessitates the need of local inputs. This is where subsequent undertaking follows, for instance, consideration of inputs from "academia-industry" namely the academicians and construction practitioners through other inductive or deductive approaches (e.g. interviews and surveys). It was apt since our construction environment is changing very fast and they are the ones who encountered those phenomena the first hand. To sum up, both competency sets (i.e. technical and non-technical) are indeed needed for a construction manager to be competent, despite different technique of construction methods were applied. However, these pragmatic moves to give a focus on the generic technical aspect will positively bridge the knowledge gaps and further assisting succeeding endeavours, at least in Malaysia.

References

[1] H.A. Rahman, M.A. Berawi, A.R. Berawi, M.O.O. Mohamed and I.A. Yahya, Delay mitigation in the malaysian construction industry, J. of Const. Eng. and Manag., 132, 125-133, (2006).

[2] M. Sambasivan and Y.W. Soon, Causes and effects of delays in malaysian construction industry, International Journal of Project Management, 25, 517-526, (2007).

[3] Unit Penyelarasan Pelaksanaan, Pengurusan Projek Sektor Awam, Jabatan Perdana Menteri, Kuala Lumpur, (2013).

[4] National Audit Department Malaysia, Auditor General's Report 2014: Activities of the Federal Ministries/Departments and Management of the Government Companies (Series One), Kuala Lumpur, (2014).

[5] Construction Industry Development Board Malaysia, Buletin Statistik Pembinaan Suku Tahunan 2014, Kuala Lumpur, (2014).

[6] Construction Industry Development Board Malaysia, Maklumat: Statistik Pendaftaran Personel, CIDB News, 23, (2011).

[7] Construction Industry Development Board Malaysia, Maklumat: Statistik Pendaftaran Personel. CIDB News, 1, (2011).

[8] S. Yaman, A. Abdullah, H. Mohammad and F. Hassan, Technical competency of construction manager in malaysian construction industry, Applied Mechanics and Materials, 773-774, 1053-1059, (2015).

[9] A.H.A. Bakar, Construction in the developing world: Issues on indigenous contractors, *Contextual Issues of the Built Environment in Malaysia, Pulau Pinang*, Penerbit Universiti Sains Malaysia, (2009).

[10] A.R. Ibrahim, M.H. Roy, Z. Ahmed and G. Imtiaz, An investigation of the status of the Malaysian construction industry, Benchmarking, 294-308, (2010).

[11] L.F. Tong, Identifying essential learning skills in students' Engineering education, *HERDSA Conference Proceedings*, (2003).

[12] F. Hassan, M. Maisham, S.M.I. Khan, S.A.S. Alwi and H. Ramli, Competencies, skills, and knowledge of quantity surveying graduates, *The Quantity Surveying International Convention*, (2010).

[13] NOSS Division, Dept. of Skills Development, National Occupational Skills Standard (NOSS) Registry 2014, Ministry of Human Resource, Malaysia, Kuala Lumpur, (2014).

[14] G.T. Haugan, *Project Management Fundamentals: Key Concept and Methodology, Management Concepts*, Vienna, (2006).

[15] H.S. Cooke and K. Tate, *Project Management*, McGraw-Hill, New York, (2005).

[16] R.S Mansfield, Building competency models: Approaches for HR professionals, Human Resource Management, 35, 7-18, (1996).

[17] C.F. Gray and E.W. Larson, *Project Management: The Managerial Process*, McGraw-Hill, New York, (2008).

[18] R. Burke, *Project Management: Planning and Control Techniques*, John Wiley and Sons, London, (2003).

[19] D.W. Halpin, *Construction Management*, John Wiley and Sons, Hoboken, (2006).

[20] L. Koskela, Application of the New Production Philosophy to Construction, Technical Report No. 72, Stanford University, (1992).

[21] G.J. Ritz, *Total Construction Project Management*, McGraw-Hill, New York, (1994).

[22] S.S.M. Isa, *A Study on Roles and Tasks of Construction Managers in Malaysian Construction Industry*, Universiti Teknologi Mara Malaysia, Shah Alam, (2007).

[23] R.U. Farooqui and M. Saqib, Desirable attributes and skills for graduating construction management students, *Associated Schools of Construction Proceedings*, (2010).

[24] P.E. Love and N.S. Haynes, Construction managers' expectations and observations of graduates, Journal of Managerial Psychology, 16, 579-593, (2001).

[25] S.U.R. Toor and G. Ofori, Developing Construction professionals of the 21st century: Renewed vision for leadership, Journal of Profes. Issues in Eng. Edu. and Practice, 279-286, (2008).

[26] F.O.D.L. Deist and J. Winterton, What is competence?, Human Resource Development International, 8, 27-46, (2005).
[27] A. Stoof, R.L. Martens, J J.G.V. Merriënboer and T. J. Bastiaens, The boundary approach of competence: a constructivist aid for understanding and using the concept of competence, Human Resource Development Review, 1, 345, (2002).
[28] D. McDaniel, Developing an integrated curriculum model for construction management education, Journal for Workforce Education and Development, 1, Article 2, (2005).
[29] P. Hager and A. Gonczi, What is competence?, Medical Teacher, 18, (1996).
[30] S. Hackett, Educating for competency and reflective practice: Fostering a cojoint approach in education and training, Journal of Workplace Learning, 13, 103-112, (2001).
[31] R. Dogbegah, D. Owusu-Manu and K. Omoteso, A principal component analysis of project management competencies for the ghanaian construction industry, Australasian Journal of Construction Economics and Building, 11, 26-40, (2011).
[32] C.O. Egbu, Skills, knowledge and competencies for managing construction refurbishment works, Construction Management and Economics, 17, 29-43, (1999).
[33] R.S. Mansfield, Building competency models: Approaches for HR professionals, Human Resource Management, 35, 7-18, (1996).
[34] S.O. Wandahl and L.W. Ussing, construction management engineer education with focus on competencies required by the construction industry, *Conference Proceeding W089 - Special Track 18th CIB World Building Congress*, (2010).
[35] P.T. Mullin and A.N. Williams, using skills gap analysis in construction management to stimulate a demand led model of curriculum, *Conference Proceeding W089 - Special Track 18th CIB World Building Congress*, (2010).
[36] Y.H. Ahn, H. Kwon, A.R. Pearce and H. Shin, Key competencies for US construction graduates: an exploratory factor analysis, *ASC Proceedings of the 46th Annual Conference*, (2010).
[37] N. Lee, R. Ponton, A. Jeffreys and R. Cohn, Analysis of industry trends for improving undergraduate curriculum in construction management education, *47th ASC Annual International Conference Proceedings*, (2011).
[38] L.S. Riggs, educating construction managers, J. Constr. Eng. Manage., 114, 279–285, (1998).
[39] D.E. Gunderson, J.W. Ra, H. Schroeder and H.R. Holland, needs assessment – a construction management Bachelor of Science degree program in Alaska, Journal of Construction Education, 7, 86 – 96, (2002).
[40] A.J. Hauck and B.J. Jackson, Design and implementation of an integrated construction management curriculum, *ASC Proceedings of the 41st Annual Conference*, (2005).
[41] C. Souder and D.M. Gier, What does the construction industry expect from recent construction management graduates?, *ASC Proceedings of the 42nd Annual Conference*, (2006).
[42] D. Arditi and D. Ongkasuwan, Technical notes: Duties and responsibilities of construction managers: Perceptions of parties involved in construction, Journal of Construction Engineering and Management, 135, (2009).
[43] B.L. Benhart and M. Shaurette, Establishing new graduate competencies: ensuring that construction management curriculums are delivering "job-ready" employees, *47th ASC Annual International Conference Proceedings*, (2011).
[44] J.W. Creswell, *Qualitative Inquiry and Research Design: Choosing Among Five Approaches*, Sage Publications, United Kingdom, (2007).
[45] J.W. Creswell, *Educational Research: Planning, Conducting, and Evaluating Quantitative and Qualitative Research*, Pearson, Ohio, (2008).

MATEC Web of Conferences 47, 05001 (2016)
DOI: 10.1051/matecconf/20164705001

Short Term Prediction of PM_{10} Concentrations Using Seasonal Time Series Analysis

Hazrul Abdul Hamid[1,a], Ahmad Shukri Yahaya[2], Nor Azam Ramli[2], Ahmad Zia Ul-Saufie[3] and Mohd Norazam Yasin[1]

[1]*Faculty of Civil and Environmental Engineering, Universiti Tun Hussein Onn Malaysia, 86400 Parit Raja, Johor, Malaysia*
[2]*School of Civil Engineering, Universiti Sains Malaysia, 14300 Nibong Tebal, Penang, Malaysia*
[3]*Faculty of Computer and Mathematical Sciences, Universiti Teknologi MARA, 40450 Shah Alam, Selangor, Malaysia*

Abstract. Air pollution modelling is one of an important tool that usually used to make short term and long term prediction. Since air pollution gives a big impact especially to human health, prediction of air pollutants concentration is needed to help the local authorities to give an early warning to people who are in risk of acute and chronic health effects from air pollution. Finding the best time series model would allow prediction to be made accurately. This research was carried out to find the best time series model to predict the PM_{10} concentrations in Nilai, Negeri Sembilan, Malaysia. By considering two seasons which is wet season (north east monsoon) and dry season (south west monsoon), seasonal autoregressive integrated moving average model were used to find the most suitable model to predict the PM_{10} concentrations in Nilai, Negeri Sembilan by using three error measures. Based on AIC statistics, results show that ARIMA $(1, 1, 1) \times (1, 0, 0)_{12}$ is the most suitable model to predict PM_{10} concentrations in Nilai, Negeri Sembilan.

1 Introduction

There are many sources of air pollution such as stationary sources, mobile sources, and open burning sources [1]. Air pollution emissions degrade air quality whether in urban or rural settings. An issue of great concern has been the detrimental effect of low air quality onto human health, chronically or acutely. Understanding the behaviour of air pollution statistically would allow predictions to be made accurately. In the early days of abundant resources and minimal development pressures, little attention was paid to growing environmental concern in Malaysia [1]. The Department of Environment is one of the bodies in Malaysia that is responsible in monitoring the status of air quality throughout the country to perceive any significant change which may cause harm to human health and environment.

In Malaysia, there are 52 monitoring locations throughout the country that belong to the Department of Environment [2]. The parameters monitored include Total Suspended Particulates, Particulate Matter (PM_{10}), Sulphur Dioxide (SO_2) and several airborne heavy metals. This 52 monitoring station was categorized in to 4 categories which is industrial, urban, sub-urban and background station. This research will focus on PM_{10} concentrations since previous researches have

[a] Corresponding author : hazrul@uthm.edu.my

shown that particles larger than 10 micrometer in aerodynamic diameter did not penetrate the body's defences in nose, mouth, and upper airways so it is unlikely to cause respiratory effects [3]. PM_{10} is also a useful indicator of several sources of outdoor air pollution, such as fossil-fuel combustion and also contaminants resulting from motor vehicle emissions [4]. Increase in ambient PM_{10} concentration can lead to significant impacts on the respiratory health especially for children, the elderly and susceptible individuals, which are normally associated with reduced lung function, asthma, pneumonia, bronchitis and emphysema [5]. At extremely high levels and long term exposure, it may even cause death. Malaysia Ambient Air Quality Guidelines state that the 24-hour mean for PM_{10} concentration should not exceed 150μg/m^3 and 12-month average should not exceed 50 μg/m^3.

Selecting appropriate time series models for the data is an important step to make prediction. These time series models may become the basis for estimating the parameters to meet the evolving information needs of environmental quality management. The developed models also can be used by related bodies to provide an early warning to the respective population. Since Malaysia has two seasons which is wet season (north east monsoon) and dry season (south west monsoon) [6], this research will focused on seasonal autoregressive integrated moving average models (Seasonal ARIMA). This research was carried out to find and proposed the most suitable time series model to predict the PM_{10} concentrations in Nilai, Negeri Sembilan, Malaysia using seasonal time series model.

2 Materials and Method

2.1 Study area

Nilai is a town located in Negeri Sembilan, Malaysia and this location classified as an industrial area by Department of Environment Malaysia. Geographically located at latitude 2^{0}45' N of the equator and longitude102^{0}15' E of the prime meridian, Nilai is a rapidly growing town due to its proximity and easy connection to Kuala Lumpur using the existing highway. The data used in this study is hourly PM_{10} concentration in Nilai, Negeri Sembilan taken from April 2008 to March 2009. Mean top bottom method were used to replace the missing values where the data were filled with the average of data available above and below the missing values.

2.2 Procedure of time series analysis

A time series is a sequence of values $\{y_1, y_2, y_3, ..., y_{t-1}, y_t, ...\}$ observed through time. Time series analysis involves the statistical method of the analysis of a sequence data. There are three phases in time series modelling which is identification, estimation and testing [7].

2.3 Seasonal ARIMA (*p*, *d*, *q*) x (*P*, *D*, *Q*)$_s$

Time series having a trend or a seasonal pattern are not stationary in mean. So, ARIMA models cannot really cope with seasonal behaviour. Since this study focus on wet season and dry season in Malaysia, seasonal ARIMA time series model has been used. Dry season period is from April to September (south west monsoon) and dry season is from September to March (north east monsoon) [6]. Seasonal ARIMA $(p, d, q) \times (P, D, Q)_s$ are defined by six parameters as follow:

$$\underbrace{\left(1-\varphi_1 B-\varphi_2 B^2-\ldots-\varphi_p B^p\right)}_{AR(p)}\underbrace{\left(1-\beta_1 B^s-\beta_2 B^{2s}-\ldots-\beta_P B^{Ps}\right)}_{SAR(P)}\underbrace{(1-B)^d}_{I(d)}\underbrace{\left(1-B^s\right)^D}_{I_s(D)} y_t = c+\underbrace{\left(1-\psi_1 B-\psi_2 B^2-\ldots-\psi_q B^q\right)}_{MA(q)}\underbrace{\left(1-\theta_1 B^s-\theta_2 B^{2s}-\ldots-\theta_Q B^{Qs}\right)}_{SMA(Q)}\varepsilon_t \quad (1)$$

where:

AR(p)	= autoregressive part of order p
MA(q)	= moving average part of order q
I(d)	= differencing of order d
SAR(P)	= seasonal autoregressive part of order P
SMA(Q)	= seasonal moving average part of order Q
$I_s(D)$	= seasonal differencing of order D
s	= the period of the seasonal pattern appearing

2.4 Stationarity of time series

Time series methods are typically based on stationarity. For stationary time series, the value of mean and variance are assumed constant with time [8]. There are many types of test that can be used to determine the stationarity of the time series. The most commonly used test for stationarity is augmented Dickey-Fuller test [9]. The Augmented Dickey-Fuller (ADF) test uses the following equation:

$$\text{ADF} = \alpha_0 + p_1 y_{t-1} + \sum_{j=2}^{p-1} \beta_j \nabla y_{t-j} + e_t \tag{2}$$

where:

α_0 = drift components

e_t = error term

Abdel-Aziz and Frey [10] stated that the hypothesis to determine the stationary of the data is as follows:

H_o : the time series data is non-stationary

H_1 : the time series data is stationary

The null hypothesis will be rejected if the ADF value is greater than the critical value.

2.5 Model identification and estimation

This study used Akaike Information Criterion (AIC) to identify the suitable model. AIC is the most commonly used criteria to select the best model in time series analysis [11]. The smallest AIC statistics indicates the best time series model. The Akaike Information Criterion (AIC) is define as follow:

$$\text{AIC} = -2\ln \hat{\sigma}_a^{\,2} + 2M \tag{3}$$

where,

$\hat{\sigma}_a^{\,2}$ = maximum likelihood estimate of $\sigma_a^{\,2}$

M = effective number of observations that is equivalent to the number of residuals that can be calculated from the series

2.6 Validation

The error measures were used to judge the developed model. Three performance indicators which are mean absolute percentage error (MAPE), normalized absolute error (NAE) and root mean square error (RMSE) as shown in Table 1 has been used to evaluate and validate the time series model [12, 13].

Table 1. Error measures.

Error Measure	Formula	Description
Mean Absolute Percentage Error (MAPE)	$MAPE = \frac{\sum_{i=1}^{N} \left\| \frac{(O_i - P_i)}{O_i} \right\|}{N} \times 100$	The smallest value of mean absolute percentage error indicates the best model
Normalized Absolute Error (NAE)	$NAE = \sum_{i=1}^{N} \|P_i - O_i\| / \sum_{i=1}^{N} O_i$	A small value for the normalized absolute error means that the model is appropriate.
Root Mean Square Error (RMSE)	$RMSE = \sqrt{\frac{1}{N} \sum_{i=1}^{N} (P_i - O_i)^2}$	A smaller RMSE value means that the model is more appropriate.

* N is the number of monitoring records, O_i is the observed monitoring records and P_i is the predicted monitoring records.

3 Results and Discussion

Descriptive statistics for hourly PM_{10} concentration in Nilai, Negeri Sembilan from April 2008 to March 2009 shown in Table 2. As an industrial area, the maximum concentration of PM_{10} is 96μg/m^3, less than Malaysia Ambient Air Quality Guideline (MAAQG) and these data is positively skewed which means that most of the readings are below the mean value.

Table 2. Descriptive statistics for PM_{10} concentration for Nilai.

Mean	55.56
Std deviation	14.51
Median	55
Mode	51
Maximum	96
Skewness	0.18
Kurtosis	0.05

Since this study considered the wet season and dry season in Malaysia, each set of data was start at 1st of April and end at 31st of March. Dry season period is from April to September which is during south west monsoon and wet season is from September to March which is during north east monsoon. Figure 1 show the time series plot for PM_{10} concentration. Since the observation from time series plot sometimes may not able to give clear information about the stationarity, the augmented Dickey-Fuller (ADF) test was performed to check the stationarity of this series.

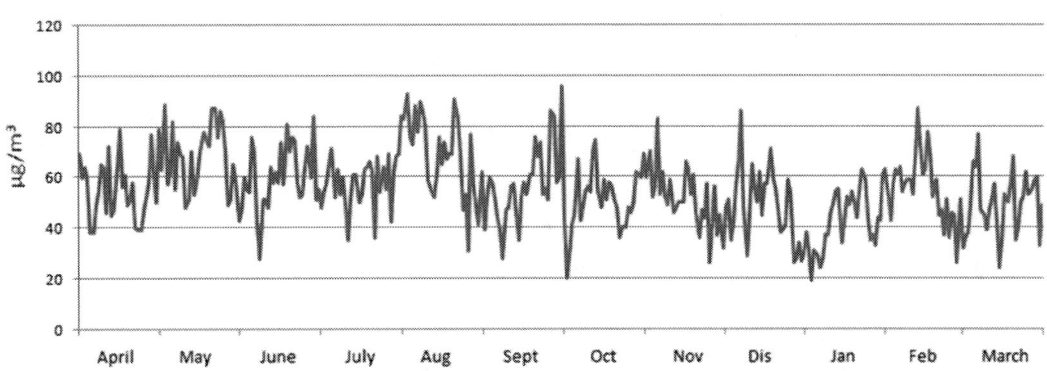

Figure 1. Time series plot of PM10 concentration for Nilai, Negeri Sembilan.

Result of Augmented Dickey Fuller test shows that the monitoring records of PM_{10} concentration from April 2008 to March 2009 are stationary with ADF statistic –4.9718 (p-value is less than 0.001). Seasonal Autoregressive Integrated Moving Average model (Seasonal ARIMA model) has been considered to find the most appropriate model for these set of data by comparing the AIC statistics values. Seasonal ARIMA (1, 0, 1) x $(0, 0, 1)_{12}$ with the smallest AIC statistics value 7.789 is the most appropriate model that give the most accurate forecasting for PM_{10} concentration in Nilai, Negeri Sembilan for composite years from April 2008 to March 2009. Time series equation based on the most appropriate time series model for Nilai is as follows.

$$y_t = 57.15 + 1.27y_{t-1} - 0.31y_{t-2} + 0.04y_{t-3} - 0.009y_{t-4} + 0.88e_{t-1} + e_t \tag{4}$$

The performance indicators for the time series model are summarized in Table 3. For the purpose of seeing more clearly, the 15 days observed and forecasted of PM_{10} concentration using the most appropriate time series model were plotted in Figure 2.

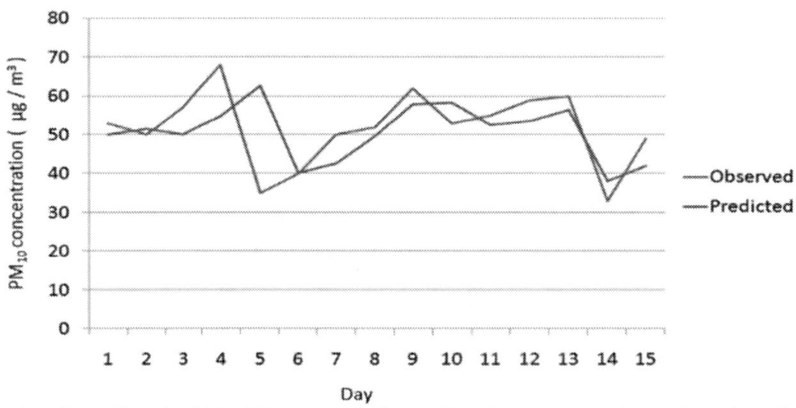

Figure 2. Observed and predicted of PM10 concentration using Seasonal ARIMA (1,0,1) x $(0,0,1)_{12}$.

Table 3. Model performance indicator.

Monitoring Site	Performance indicator (Error measure)		
	RMSE	NAE	MAPE
Nilai	11.9920	0.89053	10.2130

4 Summary

The hourly average actual monitoring record of PM_{10} concentration was used to find the best time series models. By using three error measure, seasonal ARIMA (1, 0, 1) x $(0, 0, 1)_{12}$ model is the most suitable model to forecasting the PM_{10} concentration in Nilai, Negeri Sembilan. The results of this study provide useful information on air quality status in Nilai Negeri Sembilan and can be used for prediction of future reading and also for air quality management.

References

[1] R. Afroz, M.N. Hassan and N. A. Ibrahim, Review of air pollution and health impacts in Malaysia, Environment Research, **92**(2), 71-77, (2003).

[2] Department of Environment Malaysia, Malaysia Environmental Quality Report 2013, Kuala Lumpur: Department of Environment, Ministry of Natural Resources and Environment, Malaysia, (2013).

[3] D.W. Dockery, Health effects of particulate air pollution, Ann Epidemiology, **19**, 257-263, (2009).

[4] N. Kunzli, R. Kaiser, S. Medina, M. Studnicka, O. Chanel, P. Filinger, M. Herry, J.F. Horak, T.V. Puybonnieux, P. Quenel, J. Schneider, R. Seethaler, J.C. Vergnaud and H. Sommer, Public health impact of outdoor and traffic related air pollution: a European assessment, The Lancet, **356**, 795-801, (2000)

[5] B.M.T. Shamsul, Paras pendedahan kepada PM_{10} dan hubungannya dengan simptom masalah pernafasan di kalangan pekerja Majlis Perbandaran Petaling Jaya. Master of Medical Sciences Thesis. Universiti Kebangsaan Malaysia, (2002)

[6] L.H.L. Oliver, S. Ahmad, K. Aiyub, Y.M. Jani and T.K. Hwa, Urban environmental health : Respiratory illness and urban factors in Kuala Lumpur city, Malaysia. Environment Asia, **4**(1), 39 – 46, (2011)

[7] T.M.J.A. Cooray, *Applied Time Series Analysis and Forecasting*, Alpha Science, Oxford, (2008).

[8] W.S.W. William, *Time Series Analysis-Univariate and Multivariate Methods,* Pearson Education, United States, (2006)/

[9] R.A. Carmona, *Statistical Analysis of Financial Data in S-Plus*, Springer, United States, (2004).

[10] A. Abdel-Aziz and H.C. Frey, Development of hourly probabilistic utility NOx emission inventories using time series techniques: Part I-univariate approach, Atmospheric Environment, **37**, 5379-5389, (2003).

[11] H. Bozdogan, Model selection and Akaike information criterion (AIC): The general theory and its analytical extensions, Psychometrika, **52**(3), 345-370, (1987).

[12] J. Tayman and D.A. Swanson, On the validity of MAPE as a measure of population forecast accuracy, Population Research and Policy Review, **18**, 299 - 322, (1999).

[13] H. Junninen, H. Niska, K. Tuppurainen, J. Ruuskanen and M. Kolehmainen, Method for impotation of missing values in air quality data sets, Atmospheric Environment, **38**(9), 2895-2907, (2004).

MATEC Web of Conferences **47**, 05002 (2016)
DOI: 10.1051/matecconf/20164705002

Experimental and Computational Investigations of Baffle Location Effect on the Performance of Oil and Water Separator Tanks

Rozi Abdullah[1,a], Haitham A. Hussein[1,2] and Md Azlin Md Said[1]

[1]*School of civil engineering, Universiti Sains Malaysia, 14300 Nibong Tebal, P. Penang, Malaysia*
[2]*Civil Engineering Department, Al-Nahrain University, Baghdad, Iraq*

Abstract. Gravity separator tanks are used to separate oil from water in treatment units. Achieving the best flow uniformity in a separator tank will improve the maximum removal efficiency of oil globules from water. In this study, the effect on hydraulic performance of different baffle structure positions inside a tank was investigated. Experimental data and 2D computation fluid dynamics were used for analysis. In the numerical model, two-phase flow (drift flux model) was used to validate one-phase flow. For laboratory measurements, the velocity fields were measured using an acoustic Doppler velocimeter. The measurements were compared with the result of the computational model. The results of the experimental and computational simulations indicate that the best location of a baffle structure is achieved when the standard deviation of the velocity profile and the volume of the circulation zone inside the tank are minimized.

1 Introduction

The water pollution control ordinance imposes an absolute prohibition on the discharge of oil or oily mixtures to public foul sewers. Fat, oil, and grease (FOG) blockages are the primary cause of 40% to 50% of all sanitary system overflows [1]. FOG may accumulate inside sanitary systems and eventually reduce the effective size of such systems [2]. These waste materials include source waste from petroleum and petrochemical refining and processing; tramp oil from mechanical repair stores, utility operations, restaurants, sanitary sewage, and bilge and ballast water; and contaminated surface runoff [3].

A gravity separation tank operates based on the specific gravity difference between water and immiscible oil globules. The effectiveness of a gravity separator depends on appropriate hydraulic design and period of wastewater detention for a given rise velocity. Construction baffle structure is one of the most effective ways to improve the separation of oil and water by enhancing the flow pattern and increasing the hydraulic retention time inside separator tanks.

Scientific studies conducted to establish design guidelines for oil and water separator tanks are limited. Septic tanks that provide at least 10 min to 30 min hydraulic retention time are suggested for adequate oil and grease removal [4]. Chu and Ng [2] investigated whether installing tube settlers in a grease intercepter design can improve performance. They used peanut oil with water and measured the

[a] Corresponding author : cerozi@usm.my

effluent oil and grease as well as the chemical oxygen demand. Their research showed that an 8% to 10% improvement in removal efficiency can be achieved for oil and grease by adding tube settlers.

The present research mainly aims to determine the best location of a single baffle inside rectangular gravity separation tanks. Such location must correspond to the optimum flotation of oil droplet performance in a tank. In this study, the effect of baffle position on separation efficiency is investigated through experimental and numerical simulations using two-phase flows, oil, and water. The appropriate location of the baffle in a separator tank is determined using computational fluid dynamics CFD. The volume of fluid (VOF) method is used for numerical method. In the laboratory, the velocity fields in the flume tank are measured by using an Acoustic Doppler Velocimeter (ADV), and the measurements are used to validate the numerical models. The ADV uses a technique known as pulse to pulse coherent Doppler sonar to measure the velocity vectors in 3D.

2 Laboratory Model

2.1 Laboratory setup details

A set of laboratory measurements is conducted for four baffle distances from the tank inlet over the length of basin ratios of a thin baffle in a rectangular oil and water separator tank (with water depth-to-tank length ratio of 0.31) to validate the performance of oil and water separator tanks via computation simulation [5, 6]. Figure1 illustrates the experimental setup and measurement system. This figure shows a rectangular oil and water separator tank with length L=130 cm, width W=50 cm, height H=42 cm, height of inlet opening H_{in}=10 cm, height of weir H_w=40 cm, and height of baffle H_b=31.8 cm. The distance from the tank bottom to the inlet opening H_1=15 cm. The laboratory experiments are conducted for four baffle positions at a flow rate equal to Q=2 L/s. In cases 1 to 4, a baffle is located at various baffle distances from the inlet-to-tank length ratios, that is, d/L=0.46, 0.61, 0.77, and 0.93. Experimental measurements of the cases are conducted, and the measured values of dimensionless x- and z-velocity profiles are recorded. The three velocity components are measured using an ADV. The ADV is relatively rugged, easy to operate, and can be readily mounted and maneuvered with the flow field. A 10 MHz Nortek acoustic Doppler velocity meter (Nortek AS, Norway) is used to measure the instantaneous velocities of liquid flow at different points in the oil and water separator tank. In this study, seven profiles are used to measure the velocity in the separator tanks for each case of baffle location. The velocity of each profile is then measured at eight points along the vertical line.

2.2 CFD model

2.2.1 *Time-averaged flow equations*

In hydraulic numerical modeling, the incompressible steady state flow with viscous effect is often considered to determine flow conditions. The mass continuity and momentum equations are generally used as governing equations, and the turbulence flow model is used with these equations to calculate the Reynolds stresses. The general mass continuity equation is [7, 8]:

$$V_f \frac{\partial \rho}{\partial t} + \frac{\partial}{\partial x}(\rho u A_x) + \frac{\partial}{\partial z}(\rho w A_z) = 0 \tag{1}$$

where V_f is the fractional volume open to flow in the calculation cell; ρ is the fluid density; and u and w are the velocity components in the length and height (x, z) directions, respectively. The momentum equation for the fluid velocity components in the two directions are the Navier–Stokes equations expressed as:

$$\frac{\partial u}{\partial t}+\frac{1}{V_f}\left\{uA_X\frac{\partial u}{\partial x}+wA_Z\frac{\partial u}{\partial z}\right\}=-\frac{1}{\rho}\frac{\partial P}{\partial x}+G_X+f_X \tag{2}$$

$$\frac{\partial w}{\partial t}+\frac{1}{V_f}\left\{uA_X\frac{\partial w}{\partial x}+wA_Z\frac{\partial w}{\partial z}\right\}=-\frac{1}{\rho}\frac{\partial P}{\partial z}+G_Z+f_Z \tag{3}$$

where G_x, G_z are the body accelerations, and f_x, f_z are the viscous accelerations that form a variable dynamic viscosity.

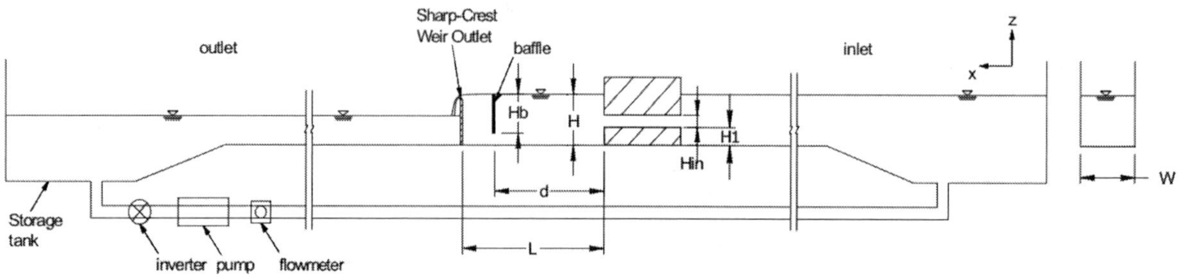

(a)

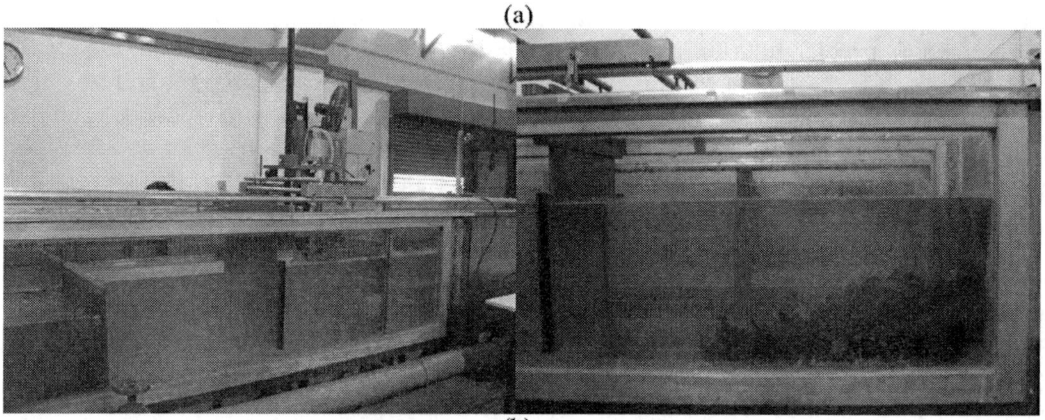

(b)

Figure 1. (a) Schematic diagram of the tank; (b) Photos of laboratory setup.

2.3 Drift flux model

In a fluid consisting of multiphase components, such as oil and water, wherein the components have different densities, the components also have different flow velocities. The two-phase (oil and water) numerical simulation is based on the drift flux model. The drift flux formulation of relative velocity assumes that flow is composed of two discrete phases: the continuous phase (water) and the dispersed phase (oil). The volume fractions of the two components that make up the mixture are denoted by f_1 and f_2, respectively, where :

$$f_1+f_2=1 \tag{4}$$

The momentum balance for the water phase is

$$\frac{\partial u_1}{\partial t}+u_1\cdot\nabla u_1=-\frac{1}{\rho_1}\nabla P+F+\frac{K}{f\rho_1}u_r \tag{5}$$

The momentum balance for the oil phase is

$$\frac{\partial u_2}{\partial t} + u_2 \cdot \nabla u_2 = -\frac{1}{\rho_2}\nabla P + F - \frac{K}{(1-f)\rho_2}u_r \tag{6}$$

where u_1 and u_2 are the microscope velocities of the water (continuous) and oil (dispersed) phases, respectively; and f is the volume fraction of the continuous phase. The microscopic velocity refers to the velocity of each phase with a small but finite volume of fluid. K is a drag coefficient that relates to the interaction of the two phases, F is the bodily force, and u_r is the relative velocity difference between the dispersed and continuous phases.

$$u_r = u_2 - u_1 \tag{7}$$

The aim of the drift flux model is to compute for the motion of the two phases relative to the volume average velocity $\bar{u}$. The volume weighted average velocity is

$$\bar{\mathrm{u}} = f\mathrm{u}_1 + (1-f)\,\mathrm{u}_2 \tag{8}$$

2.4 Verification test

The flow in an oil and water separator tank has two phases. In this study, one and two phase numerical models in the computation simulation of separation tanks are used. The oil phase is in the form of drops with an average diameter of 150 μm and a density of 910 kg/m^3. Water is discharged through the inlet slot at a flow rate of 2 L/s, whereas the inlet slot of 5% by volume oil [9] in water mixture has a constant discharge rate equal to 30 mg/L. 300×122 grids were chosen for the computation. Figure 2 shows a comparison of the velocity profile of the numerical model using the FLOW-3D program between the one-phase and two-phase flow models (drift flux model).

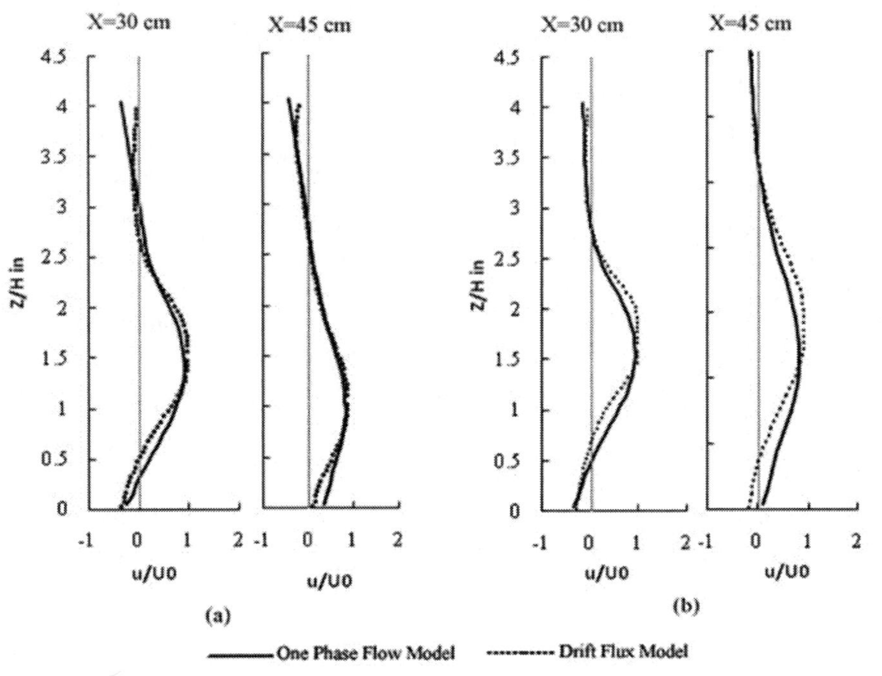

Figure 2. x-Velocity profile for one and two phase flow model for baffle height (Hb/H= 0.76) at d/L=0.77.

The velocity profiles of the two models are computed at 30 cm and 45 cm from the inlet for cases of baffle location at *d/L*=0.46 and 0.77. The velocity is computed in the *x* and *z* direction. The aforementioned profiles show that the velocities in the numerical simulation are in good agreement between one-phase flow (water) and two-phase flow (oil and water). Consequently, for these velocity profiles, the flow in the experimental model used pure water without oil droplet. Figure 3 shows the *x* computed velocity profiles compared with the experimental measurements for cases of baffle location *d/L*= 0.77 inside the oil and water separator tanks. The 2D CFD model is a good descriptor of the experimental measurements.

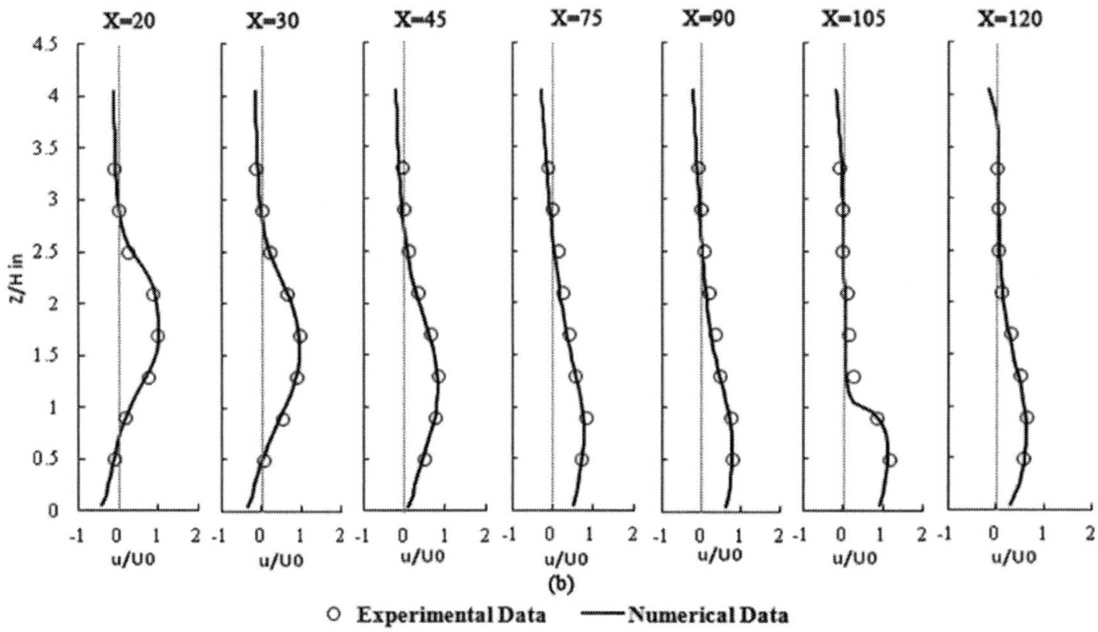

Figure 3. x-Velocity profile for for baffle height (Hb/H= 0.76) at d/L= 0.77.

3 Results and Discussions

3.1 Velocity profile

In two phase flow (oil and water) the rise of oil droplets (vertical velocity) in water generally depends on the density and viscosity of water. Moreover, the vertical velocity of oil droplets is highly dependent on droplet diameter, with small droplets rising significantly more slowly than large droplets. Consequently, the separator tank must be designed to reduce the inlet horizontal velocity as much as possible and to achieve more uniformity. This situation allows oil droplets to coalesce with other droplets and to rise rapidly toward the separator surface. The effectiveness of baffle location in improving the uniformity of the flow pattern has been assessed by comparing the standard deviation (SD) of *x*-velocity across the cross section [10]. For each baffle location case, the axial velocity is retrieved at 82 equally spaced points along five vertical positions in the separator tank, and then, the SD is calculated. Figure 4 shows the SD of x-velocity across the separator tank with four end baffle position (*d/L*). It can be seen that the baffle location at *d/L*=0.77 has the minimum magnitude (SD) along the separator tank from 0.1 to 0.5m, and thus, it exhibits the best performance (a better separator).

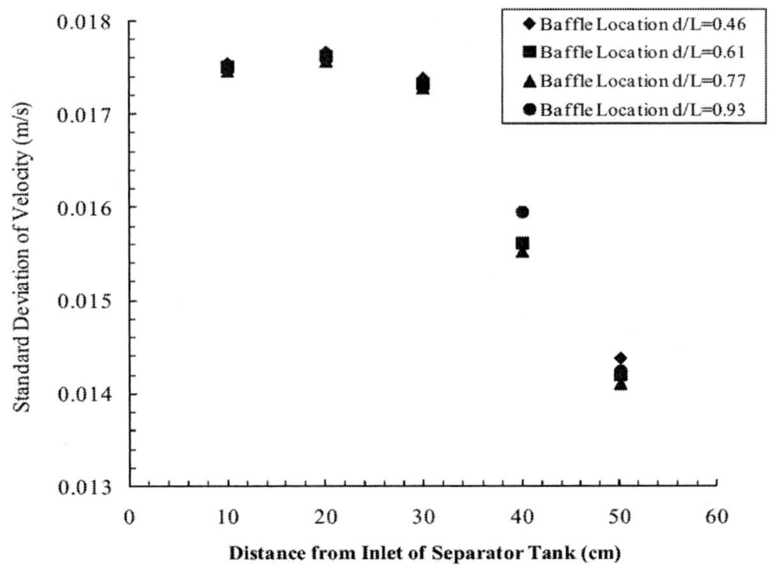

Figure 4. Standard deviation of velocity across the Separator tanks with different baffle locations d/L two phase flow.

3.2 Flow field

The high performance of separator tanks depend on the flow pattern. Therefore the baffle structure in the separator tanks is used as to reduce the energy dissipaters, circulation zone volume inside the tank, and in additional to collect the oil droplet from the water surface level. The best position of the baffle is acquired when the volume of circulation zone is minimized. Thus, the proper position for the baffle may achieve to obtain more uniform distribution of velocity in the separator tanks and minimize dead zone. Four different cases of baffle positions were modeled in this study. Figure 5 illustrate shows the baffles locations with the volume of vortices or circulation zone, which is intended by the total water volume in separator tanks and calculated by the numerical method. This figure illustrate The baffle position at d/L = 0.46, 0.61 and 0.93 has 66%, 65% and 60% circulation volume rate respectively, while the minimum circulation volume rate magnitude in case of d/L= 0.77 is 59% from total tank volume.

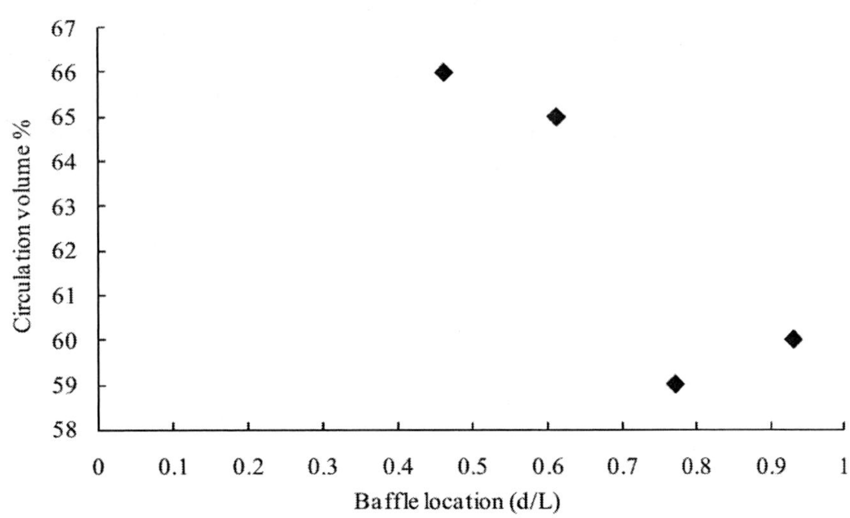

Figure 5. Circulation volume percentages in different baffle location (d/L).

4 Conclusion

This study was performed to evaluate the performance of oil and water gravity separator tanks. Laboratory experiments and CFD analysis were performed to assess the effect of baffle positions *d/L*=0.46, 0.61, 0.77, and 0.93 on the removal efficiency of oil and water separator tanks. A numerical model was applied to determine the effect of baffle position on velocity profile. Standard deviation served as an indicator of flow uniformity along separator tank sections. The minimum SD resulted in the best uniformity (a better separator). Moreover, the volume of vortices or the circulation zone was normalized to determine the smallest circulation zone in each baffle position case in the separator tank. The results showed that the baffle constructed on *d/L*=0.77 achieves the minimum SD and volume of circulation zone percentage.

References

The authors would like to acknowledge Universiti Sains Malaysia for funding this research through RUI grant (1001/PAWAM/814194).

References

[1] R. Southerland, Sewer fitness: Cutting the fat, American City and Country, **117**, 27-31, (2002).

[2] W. Chu and F.L. Ng, Upgrading the conventional grease trap using a tube settler, Environment International, **26**, 17-22, (2000).

[3] A.O. Abass, A.T. Jameel, S.A. Muyubi and M.I. Abdul Karim, Removal of oil and grease as emerging pollutants of concern (EPC) in wastewater stream, IIUM Engineering J., **12**(4), 161-169, (2011)

[4] I. Metcalf, *Wastewater Engineerin: Treatment and Reuse*: McGraw-Hill, (2003)

[5] API, Design and operation of oil-water separators, American Petroleum Institute, Washington, (1990).

[6] T. Morrow and F. Dodget, Fluid flow modelling of gravity separators, Multi-Phase Production, 364-380, (1991).

[7] C. Hirt and J. Sicilian, A porosity technique for the definition of obstacles in rectangular cell meshes, *Proc. of 4th Int. Conf. on Numerical Ship Hydrodynamics*, Washington, (1985).

[8] C.W. Hirt and B.D. Nichols, Volume of fluid (VOF) method for the dynamics of free boundaries, J. of Computational Physics, **39**, 201-225, (1981).

[9] T.N. Aziz, L.M. Holt, K.M. Keener, J.W. Groninger and J.J. Ducoste, Performance of grease abatement devices for removal of fat, oil, and grease, J. of Environmental Engineering, **137**, 84-92, (2011).

[10] D. Wilkinson, B. Waldie, M.I. Mohamad Nor and H.Y. Lee, Baffle plate configurations to enhance separation in horizontal primary separators, Chemical Engineering J., **77**, 221-226, (2000).

MATEC Web of Conferences **47**, 05003 (2016)
DOI: 10.1051/matecconf/20164705003

Physiochemicals and Heavy Metal Removal from Domestic Wastewater via Phycoremediation

Abdul Rafiq Ab Razak[1,a], Norshuhaila Mohamed Sunar[2] Nurul Azira Alias[1], Paran Gani[1] and Menega Subramaniam[1]

[1]*Faculty of Civil and Environmental Engineering, Universiti Tun Hussein Onn Malaysia, 86400 Parit Raja, Johor, Malaysia*
[2]*Faculty of Engineering Technology, Universiti Tun Hussein Onn Malaysia, 86400 Parit Raja, Johor, Malaysia*

Abstract. The common sources of water pollution in Malaysia are domestic sewage and industrial waste. Therefore, domestic wastewater quality effluent should be improved before discharged through the outlets. The alternative method of treatment uses microalgae for water remediation which is known as phycoremediation was applied. This technique is to remove or reduce nutrients and harmful pollutants in domestic wastewater. Thus, objective of the present study is to bioremediate the physiochemical and heavy metal from domestic wastewater using freshwater green microalgae *Botryococcus* sp. A photobioreactor is used to treat the wastewater by employing the microalgae *Botryococcus* sp. as a vital part of the treatment system. The results show that several nutrients have been reduced successfully such as phosphate and total phosphorus of 100% removal, inorganic carbon of 99% removal, total carbon of 42% removal, and nitrate of 10%. The most prominent heavy metal content that has been removed is Aluminium of 41%. At the same time, the growth of microalgae *Botryococcus* sp. in this wastewater has achieved the maximum value at Day 4 with 2.58 x 10^5 cell/ml only. These results show the potential of *Botryococcus* sp. cultivation as an alternative method to treat domestic wastewater and any other biotechnology works in the future.

1 Introduction

Wastewater treatment using microalgae is considered an environmentally sound approach to minimize the excessive nutrients and has been applied for almost 50 year ago in special reactors termed high-rate ponds system[1]. However, dewatering and harvesting of microalgae is assumed high in term of operation cost, especially where a limit on suspended solid is imposed by law [2].

In addition, the chosen microalgae species to be used in wastewater treatment normally determined based on their robustness to survive in wastewater against excessive nutrients such as nitrogen and phosphorus [3]. To breed 1kg algal biomass, it requires 1.8 kg of CO_2 [4]. This requirement helps to solve greenhouse problem[5]. Microalgae can also reduce the contents of pollutant and toxic metal in wastewater called phycoremediation[6][7]. Remediation is commonly refer to an array of regulatory requirements, and also can be based on assessments of human health and ecological risks where no legislative standards exist [8].

[a] Corresponding author : rafiqrazak@yahoo.com

Phycoremediation is a process of employing microalgae or macroalgae for improving wastewater quality which able to fix carbon dioxide by using photosynthesis process and to eliminate excess nutrients effectively at minimal cost of operation[9-11]. This sustainable technique had been applied and studied widely using different type of wastewater such as domestic wastewater, greywater, dairywater, piggery wastewater and textile wastewater [12]. Microalgae growth is able to remove phosphorus and nitrate while able to present itself as an alternative to avoid water contamination[13].

The quality determining factor of pollutants is coming from domestic wastewater composing three major elements which are carbon, nitrogen and phosphorus. This is due to the content of the wastewater has organic matter and human waste such as urine, garbage, food waste and other solid material[14]. Moreover, domestic wastewater also contain mixed of bacteria which can give serious impact on human health and environment if uncontrolled discharged happened [15]. Conventionally, physical, chemical and biological methods are used to remove contaminants from wastewater and in order to achieve different level of contaminants treatment in wastewater, individual wastewater treatment procedures are combined into variety systems [16].

Basically, higher BOD levels mean the water is at poor quality [17]. The term of microalgae refer to microscopic biological plant which unable to be seen by naked eyes without microscope, and categorized as eukaryotic and prokaryotic cyanobacteria [18]. Cultivation of microalgae on wastewater can absorb CO2 and release O_2 in the air through photosynthesis process. On global scale, microalgae be able to produce up to 75% of oxygen [19]. Other than that, others advantageous of phycoremedition is potentially valuable biomass produced for bio-based product such as bio-fertilizer, biofuel feedstock, food supplement and pharmaceutical industry [15]. Therefore, objective of the present study is to reduce physiochemical and heavy metal elements from domestic wastewater using microalgae.

2 Materials and Methods

2.1 Wastewater preparation

The domestic wastewater sample was collected at Universiti Tun Hussein Onn Malaysia domestic wastewater treatment plant. This plant caters the domestic waste such as sewage and rain water effluent. It has 2 stages of wastewater treatment which are primary and secondary treatment. The sample is collected before it reaches primary treatment. The wastewater sample is preserved at a temperature less than 4^0C prior to the experimental. The physic-chemical parameters (Ph, dissolved oxygen (DO), biochemical oxygen demand (BOD), chemical oxygen demand (COD), total organic carbon (TOC), total carbon (TC), inorganic carbon (IC), total nitrogen (TN), nitrate, phosphate and total phosphorus) and heavy metal elements (zinc (Zn), ferum (Fe), cadmium (Cd), copper (Cu), aluminium (Al), lead (Pb) and manganese (Mn) were tested immediately based on standard methods [20]. Before commence the experiment, wastewater was filtered using membrane filter 0.45um pore size to remove other microorganism and suspended solid.

2.2 Microalgae preparation

Microalgae *Botryococcus* sp. has been prepared at the microbiological experimental, *Botryococcus* sp. was cultured and the stock was maintained in the bold basal medium (BBM) for 14 days. The inoculums were counted using haemocytometer and initial cell concentration 10^3 cell/mL which is based on the standard method and examination of water and wastewater (2009) and was transferred to the treatment medium for phycoremediation process.

2.3 Experimental set-up

The pre-treated domestic wastewater was poured into a photobioreactor which can hold about 17L of water sample with aeration provided at 1L/min to homogenize the wastewater and microalgae during phycoremediation. After that, the sample was inculated with the microalgae seed at inoculum of 1.0 x 10^3 cell/ml and the cell count was performed using haemocytometer with the help of light compound microscope according to standard method [21]. The phycoremediation was done for 20 days under outdoor natural conditions.

3 Results and Discussion

3.1 Characteristic of wastewater

These parameters are important for next analysis which is analysis for physiochemical and heavy metal after phycoremediation process. The parameter chosen has shown at Table 1 below with the wastewater sample and compared the value with Standard A and Standard B of Malaysia Wastewater Effluent Discharge (2000) guidelines.

Table 1. The raw characteristic for wastewater sample (n=3).

Physiochemical	Domestic wastewater	Malaysia Wastewater Effluent Discharge Standard (2000)	
		A	B
Chemical oxygen demand (mg/L)	34	50	100
Biochemical oxygen demand (mg/L)	93	20	50
Total phosphorus (mg/L)	3.3	-	-
Total nitrogen (mg/L)	15.79	-	-
Total suspended solid (mg/L)	2267	50	100
Total dissolved solid (mg/L)	12600	-	-
Total organic carbon (mg/L)	5.53	-	-
Total carbon (mg/L)	24.39	-	-
Inorganic carbon (mg/L)	18.86	-	-
Nitrate (mg/L)	3.12	-	-
Phosphate (mg/L)	267.49	-	-
Dissolved oxygen (mg/L)	10.91	-	-
pH	2.5	6.0-9.0	5.5-9.0
Turbidity (mg/L)	38	-	-

Table 2. The heavy metal characteristic for wastewater sample (n=3).

Heavy metal	Zinc, mg/L	Ferum, mg/L	Cadmium, mg/L	Copper, mg/L	Lead, mg/L	Aluminium, mg/L	Manganese, mg/L
Domestic wastewater	0.662	0.907	0.0008	0.0132	0.447	0.0062	0.342
Standard A	1.0	1.0	0.01	0.2	10	0.1	0.2
Standard B	1.0	5.0	0.02	0.1	15	0.05	1.0

3.2 Removal of physiochemical properties

Phosphate, nitrate, TC, TOC, IC and TC classified as a nutrient in wastewater. The analysis has measured all the parameter proved the nutrient content can be treated effectively by *Botryococcus* sp. According to the Figure 1, the most effective nutrient removal are total phosphorus and phosphate with 100 % removal from Day 4 until Day 20 followed by inorganic carbon with 83.4% at Day 4 increase to 99.9% at Day 20. However, the value of total carbon removal is 73.1% at Day 4 decrease to 42.6% at Day 20. Phosphorus has been removed fully as microalgae uses phosphorus and nitrate for its growth[1]. The nitrate value also reduces from 95.2% to 10.1% at the last day. *Botryococcus* sp. consumes nitrogen quickly in ammonium form [22].The conclusion is both of these nutrients had been removed after treatment using microalgae (*Botryococcus* sp.).

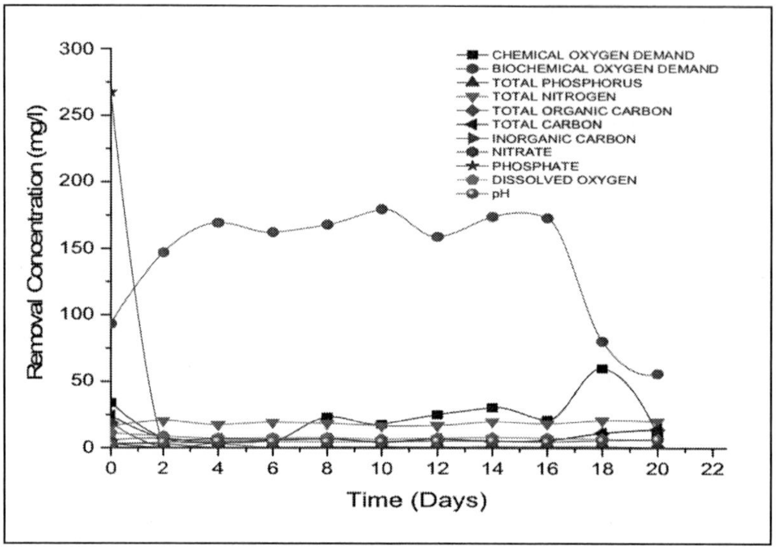

Figure 1. The nutrients removal for domestic wastewater during phycoremediation.

3.3 Removal of heavy metal properties

The highest removal of heavy metal at the end of research is ferum. However, after 20 days the higher value is still zinc and the lower value is cadmium. Some of the elements are not effectively removed from wastewater because the result does not achieve Standard of Malaysian Effluents Guidelines. The conclusion is aluminium value can be acceptable for the standard guideline which is the one and only heavy metal in domestic wastewater that can be treated by using microalgae (*Botryococcus* sp.).

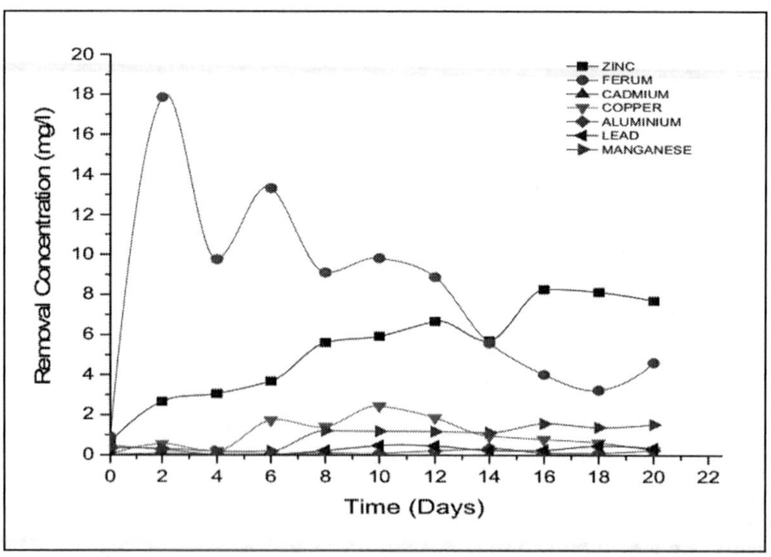

Figure 2. Graph of heavy metal in domestic wastewater during phycoremediation.

4 Conclusions

The growth of microalgae *Botryococcus* sp. in domestic wastewater has been achieved the maximum value at Day 4 with 2.58 x105 cell/ml. The nutrients had been removed at the domestic wastewaters which are phosphate, total phosphorus, inorganic carbon, total carbon and nitrate. The heavy metal shows high removal efficiency using phycoremediation. This study needs to be explored to unveil its potential. It is recommended that more studies be carried out to investigate the retention time required to efficiently treat the wastewater.

Acknowledgements

The authors would like to acknowledge Research and Innovation Centre (R&D) Research, Innovation, Commercialization and Consultancy Office (ORICC) Universiti Tun Hussein Onn Malaysia for funding this project under grant FRGS vot 1479.

References

[1] N. Abdel-Raouf, A.A. Al-Homaidan and I.B.M. Ibraheem, Microalgae and wastewater treatment, Saudi J. Biol. Sci., **19**(3), 257-275, (2012).

[2] J. Shi, B. Podola and M. Melkonian, Removal of nitrogen and phosphorus from wastewater using microalgae immobilized on twin layers: An experimental study, J. Appl. Phycol., **19**(5), 417-423, (2007).

[3] H.Azarpira, P. Behdarvand, K. Dhumal and G. Pondhe, Potential use of cyanobacteria species in phycoremediation of municipal wastewater, Int. J. of Biosciences, **4**(4), 105-111, (2014).

[4] U. Emeka, G.I. Ndukwe, K.B. Mustapha and R.I. Ayo, Constraints to large scale algae biomass production and utilization, J. of Algal Biomass Utilization, **3**(2), 14-32, (2012).

[5] F.M. Salih, Microalgae tolerance to high concentrations of carbon dioxide: A review, J. Environ. Prot., 2(5), 648–654, (2011).

[6] P.H. Rao, R.R. Kumar, B. Raghavan, V. Subramanian and V. Sivasubramanian, Application of phycoremediation technology in the treatment of wastewater from a leather-processing chemical manufacturing facility, Water SA, **37**(1), 7-14, (2011).

[7] P. Gani, N.M. Sunar, A.A.A. Latiff, I.T.K. Joo, U.K. Parjo, Q. Emparan and C.M. Er, Phycoremediation of dairy wastewater by using green microlgae: *Botryococcus* sp., Appl. Mech. Mater., **773-774**, 1318-1323, (2015).

[8] J. Chotwattanasak and U. Puetpaiboon, Full scale anaerobic digester for treating palm oil mill wastewater, J. Sustain. Energy Environ., **2**(3), 133–136, (2013).

[9] P.O. Souza, L.R. Ferreira, N.R.X. Pires, P.J.S. Filho, F.A. Duarte, C.M.P. Pereira and M.F. Mesko, Algae of economic importance that accumulate cadmium and lead: A review, Brazilian J. Pharmacogn., **22**(4), 825–837, (2012).

[10] S. Chinnasamy, A. Bhatnagar, R.W. Hunt and K.C. Das, Microalgae cultivation in a wastewater dominated by carpet mill effluents for biofuel applications, Bioresour. Technol., **101**(9), 3097-3105, (2010).

[11] P. Gani, N.M. Sunar, H. Matias-Peralta, A.A.A. Latiff, U. Kalthsom, A. Rafiq and A. Razak, Phycoremediation of wastewaters and potential hydrocarbon from microalgae: A review, Adv. Environ. Biol., **9**(20), 1-8, (2015).

[12] S. Elumalai, S. Ramganesh and T. Sangeetha, Phycoremediation for leather industrial effluent - treatment and recycling using green microalgae and its consortia, Int. J. of Current Phycoremediation, **2**(10), 1–9, (2014).

[13] P. Gani, N.M. Sunar, A.A.A Latiff, N.S. Kamaludin, U.K. Parjo, Q. Emparan and C.M. Er, Experimental study for phycoremediation of *Botryococcus* sp . on greywater, Appl. Mech. Mater., **773-774**, 1312-1317, (2015).

[14] M.A. Fulazzaky, Carbonaceous, nitrogenous and phosphorus matters removal from domestic wastewater by an activated sludge reactor of nitrification-denitrification type, J. Eng. Sci. Technol., **4**(1), 69-80, (2009).
[15] H. Kamyab, M. Soltani, M. Ponraj, M.F. Din and E.V. Putri, A review on microalgae as potential lipid container with wastewater treating functions, J. of Environmental Treatment Techniques, **1**(2), 76–80, (2013).
[16] A.L.A. Zinatizadeh, Biological treatment of palm oil mill effluent (Pome) using an up-flow anaerobic sludge fixed film (Uasff) bioreactor, PhD Thesis, Universiti Sains Malaysia, Penang, (2006).
[17] J.R. Votano, M. Parham, L.H. Hall, L.B. Kier and L.M. Hall, Microalgae: The green gold of the future?, Chem. Biodivers., **1**(11), 1829–1841, (2004).
[18] P. Metzger and C. Largeau, Botryococcus braunii: A rich source for hydrocarbons and related ether lipids, Appl. Microbiol. Biotechnol., **66**(5), 486–496, (2005).
[19] Z. Yaakob, K.F. Kamarudin and R. Rajkumar, The Current Methods for the biomass production of the microalgae from wastewaters: An overview, World Applied Sciences J., 31(10), 1744-1758, (2014).
[20] APHA, AWWA, and WEF, *Standard Methods for the Examination of Water and Wastewater*, American Public Health Association, American Water Works Association and Water Environment Federation, (2005).
[21] O. Bastidas, Cell counting with neubauer chamber (basic hemocytometer usage), Technical Note - Neubauer Chamber Cell Counting, Celeromics, (2014).
[22] R. Órpez, M.E. Martínez, G. Hodaifa, F. El-Yousfi, N. Jbari and S. Sánchez, Growth of the microalga botryococcus braunii in secondarily treated sewage, Desalination, 246(1), 625-630, (2009).

MATEC Web of Conferences **47**, 05004 (2016)
DOI: 10.1051/matecconf/20164705004

Nutrient Concentration Distribution in Sediment and Overlying Water at Bukit Merah Reservoir, Perak

S. H. A. Talib[1,a], M. S. Yusoff[2], Z. A. Hasan[3], W. R. Ismail[4] and M. S. Abustan[1]

[1]*Faculty of Civil and Environmental Engineering, Universiti Tun Hussein Onn Malaysia, 86400 Batu Pahat, Johor, Malaysia*
[2]*School of Civil Engineering, Universiti Sains Malaysia, 14300 Nibong Tebal, P.Pinang, Malaysia*
[3]*Engineering and Urban Drainage Research Centre, Universiti Sains Malaysia,14300 Nibong Tebal, P. Pinang, Malaysia*
[4]*HydroGeomorphology Research Group, Section of Geography, School of Humanities, Universiti Sains Malaysia,11800 Minden, Penang, Malaysia*

Abstract. Bukit Merah Reservoir (BMR) is the oldest reservoir in Malaysia constructed to supply irrigation water to the Kerian Irrigation Scheme. Depletion of storage capacity due to sedimentation process is the most concerning issue of the reservoir recently. Sediment analysis is also very important when monitoring the substances that appear in the reservoir water. Environmental toxins, like heavy metals and hydrophobic organic components, as well as nutrients such as nitrogen and phosphorus are readily bond to the particulate matter. Factors in the reservoir that affect the particulate matter will aid the distribution of contaminants in the sediments. The contaminants in the sediments could have higher concentrations compared to those found in the overlying water. This means that the sediment plays an important role in the cycling of nutrients and distribution of contaminants in the ecosystem. The objectives of this research are to determine the relationship of nutrient and heavy metal content for sediment and overlying water. Nutrient profile information is provided from sediment sample analysis. Based on the data from the horizontal distribution of surface sediment phosphorus in BMR, it was apparent that the highest concentration occurred in the reservoir inlet from Sungai Kurau catchment area. Phosphorus and nitrogen in BMR have a similar trend of decreasing concentration from upstream to downstream. The phosphorus and nitrogen concentrations of surface sediment were in fact significantly correlated with phosphorus and nitrogen concentrations of surface water because of the high value of determination of correlation (R^2). As a conclusion, the deposition of sediments was found to bring along external nutrients. Variability of phosphorus and nitrogen concentrations in the sediment directly affects the quality of water which is very important for irrigation and domestic uses.

1 Introduction

A reservoir is one of the key sources of water supply as it provides hydroelectric power and provides flood protection. However, reservoir storage is often affected by sedimentation due to soil erosion in the catchment area [1]. Reservoir sedimentation is a threat to the optimal use of water resources.

[a] Corresponding author : hidayaht@uthm.edu.my

Reservoir sedimentation is the process of sediment deposition that occurs after a dam construction [2]. Removal of vegetation and indiscriminate land development in the catchment area can increase the volume of stream flows. As a result of runoff from rainfall, soil particles on the surface of catchment area can be eroded and transported [3]. Erosion will decrease the soil water storage capacity and also reduce infiltration on the eroded sites. The eroded soil will then be transported along the water stream and deposited upon arriving at the reservoir. Repetition of transportation and deposition processes will result in the occurrence of reservoir sedimentation problems.

Sediment analysis is also very important when monitoring the substances that appear in the reservoir water. Environmental toxins, like heavy metals and hydrophobic organic components, as well as nutrients such as nitrogen and phosphorus are readily bond to the particulate matter. Factors in the reservoir that affect the particulate matter will aid the distribution of contaminants in the sediments. The contaminants in the sediments could have higher concentrations compared to those found in the overlying water. This means that the sediment plays an important role in the cycling of nutrients and distribution of contaminants in the ecosystem. Therefore, the objectives of this research are to determine the relationship of nutrient and heavy metal content for sediment and overlying water.

Phosphorus was chosen as it is a key nutrient. This is because phosphorus often functions as the limiting nutrient and thereby determines phytoplankton abundance [4]. Therefore, the availability of phosphorus is commonly considered as the most important factor for the overall environmental state of lakes. Iron indirectly has a strong influence on the nutrient yield in lake. This is due to the formation of compounds for example, iron hydroxids, which are important for inorganic binding of phosphorus in lake. Retention and availability of phosphorus are therefore closely linked with the capacity of iron to release or take up phosphorus.

Nitrogen is recognised to play an important role in some lake ecosystems [5]. This is true for most of the lake where nitrogen leaves the system through denitrification, while phosphorus remains [6]. However, the concentration of nitrogen is often strongly correlated with the phosphorus level, but the differentiation of impacts between these two substances may be difficult within the empirical context. However, all these sediment nutrient concentrations are high at the surface and fall to a more constant value about 20 cm below the sediment-water interface as explained by Fish and Andrew [7].

2 Methodology

2.1 Description of study area

BMR is the oldest artificial lake in Malaysia. It is located at the north-western corner of Perak state in Peninsular Malaysia, which is between longitude of 04° 48' and 05° 13' and latitude 100° 33' and 101° 40'. Figure 1 shows the location of BMR. BMR was constructed in 1906 on Sungai Kurau and located at the upstream of Kerian Irrigation area in order to supply water to the Kerian Irrigation scheme.

BMR has a distinctive unique shape compared with other manmade lake. Typically, shape of the reservoir tends to be narrow and elongated as it is formed in the river valley. The current shape of BMR is in crescent shape, where width longer than the length. The measurement of BMR is about 13.88km long running from north to south direction and a width of 4.5km from east to west. Its operation level is 8.5m above sea level with a maximum depth of 5.3m. The uniqueness of this shape tends to cause different types of flow and may reduce the retention time. Due to this factor BMR was chosen as the study area for this research.

Bukit Merah dam is an earth filled embankment constructed at the upstream of Sungai Kurau and Sungai Merah confluences in 1906. As reported by Anwar (2010), the embankment level is at reference level (RL) 8.08 m above the mean sea level and it is further improved to RL 10.67 m under the 2nd Malaysian Plan (1961-1965). In 1984, under Kerian-Sg.Manik Intergrated Agricultural Development Project (IADP KSM), the embankment level was again raised to the present level of

11.28 m. At present level, water can be stored to a maximum level of RL 9.14 m to enhance the double cropping planting intensity to Kerian Irrigation Scheme.

BMR functions as an irrigation supply to the Kerian Agricultural Scheme. Approximately 10,000 farmers rely on rice cultivation industry. In addition to the irrigation supply, it also provides fresh water to meet the domestic and industrial demands of Kerian District as well as Larut Matang District.

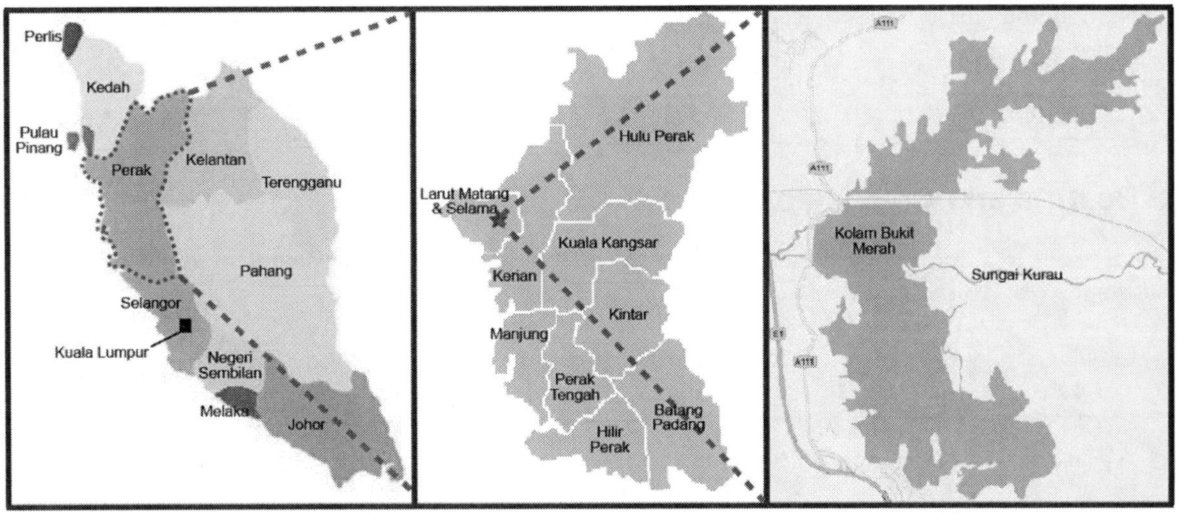

Figure 1. Reservoir Location of Bukit Merah.

2.2 Sampling equipment

Due to difficulty of existing corer, an attempt was made to preserve the best qualities of the existing corers while reducing the overall weight and size of the device. Newly invented gravity corer using transparent acrylic tubes with 5.0cm outer diameter was used to take out the sediment cores [8]. Each of these tubes is 0.5m long. During sampling activities, the tubes were covered with stainless steel casing which is fitted snug to the tubes. It was used to avoid any damage to the tubes and samples. This casing can go up to 6.5m length (length can be adjusted according to the water depth) with attached weights of up to 5kg. It was also equipped with a serrated edge to facilitate the penetration process.

The design consists of a corer covered with casing that uses the attached weight sink into the sediment when released from the surface. From the surface, a weight then was dropped and hammered to the steel casing. The corer was hammered into the sediment as deep as possible until it could not go any further. Once the corer reaches above the water, the tube was immediately taken out from its casing and then a cork was placed at the bottom of the tube.

2.3 Laboratory test

2.3.1 Sediment digestion

The sliced core sections were digested using USEPA Method 3050B (Acid Digestion for Sediments, Sludge and Soils) for Total Phosphorus (TP) and Total Nitrogen (TN) determination. The purpose of acid digestion process was to crush and dilute the sediment. At the end of the process, it turned the sediment into liquid phase. There are three chemical solutions needed in this process which are Nitrate Acid (HNO_3), Hydrogen Peroxide Solution (H_2O_2) and Hydrochloric Acid (HCl).

2.3.2 Total phosphorus and total nitrogen

Chemical analysis was carried out according to the method outlined in the Standard Method for the Examination of Water and Wastewater [9]. Total phosphorus was determined using Method 10127 Molybdovanadate Method with Acid Persulfate Digestion while, total nitrogen was measured according to Method 10072 Persulfate Digestion Method and were tested using DR 2010 Spectrophotometer by adding certain reagents to the samples. A spectrophotometer was used to measure the amount of light transmitted or absorbed by the solutions. The detection limits for total phosphorus and total nitrogen methods were 0.0 to 100.0 mg/L PO_4^{3-} and 10.0 to 150.0 mg/L N respectively.

3 Results and Discussion

Analyses of total nitrogen and total phosphorus concentration for surface sediment and water samples across the reservoir are shown in Figure 2 and Figure 3 respectively.

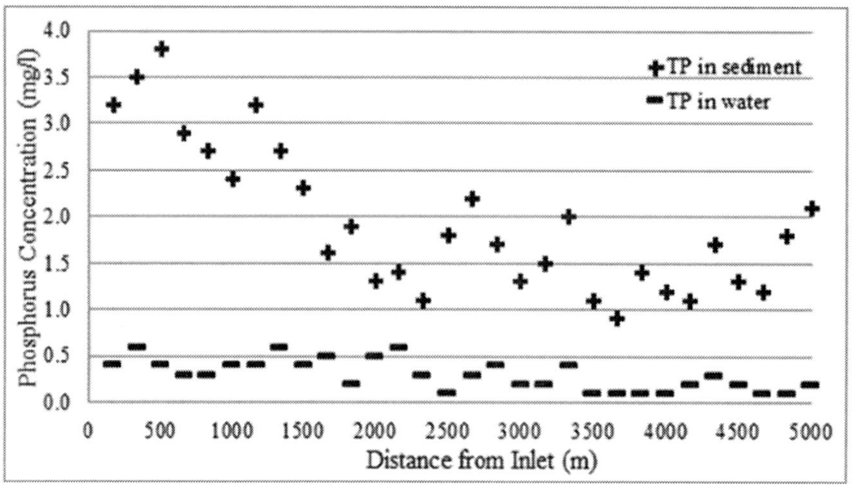

Figure 2. Total phosphorus concentration in surface sediment and water for 30 sampling points.

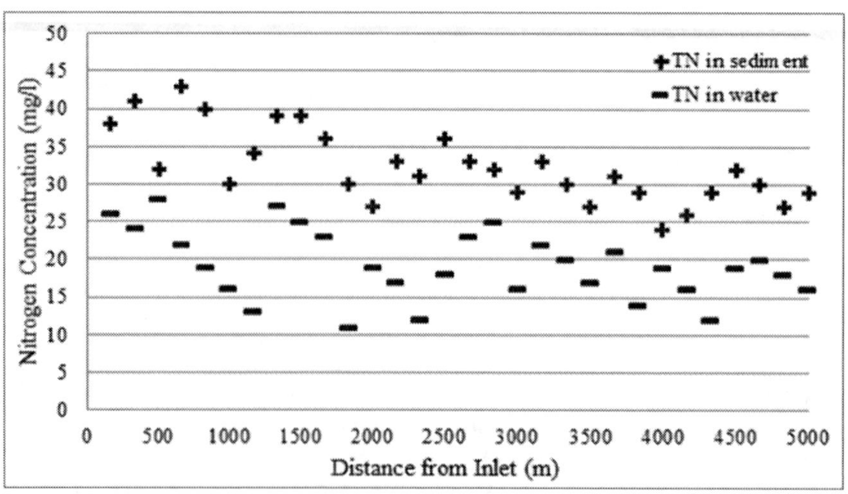

Figure 3. Total nitrogen concentration in surface sediment and water for 30 sampling points.

The nitrogen and phosphorus concentrations at surface sediment represented by the first discrete horizontal sediment sample slice (0 to 2cm) from each sediment core varied between 900 and 380 mg P kg^{-1} and 2400 and 4300 mg N kg^{-1} dry weight, respectively. The horizontal distribution of annual mean phosphorus and nitrogen concentrations in the surface water varied from 0.1 to 0.6 mg P l^{-1} and 11 to 28 mg N l^{-1}, respectively. The phosphorus relative variation across the lake tended to be higher for surface sediment than for surface water.

Phosphorus and nitrogen concentrations in both surface sediment and surface water are significantly correlated (Table 1). The sediment phosphorus were significantly correlated with surface water phosphorus ($r = 0.517$, $p < 0.003$, $n = 30$) as well as nitrogen ($r = 0.448$, $p < 0.013$, $n = 30$).

For the sediment, there was a clear incline of increasing nitrogen towards the outlet area of the reservoir, similar to the surface water phosphorus concentrations (Figure 2 and Figure 3). Due to that, there was high correlation coefficient between nitrogen and phosphorus concentration in surface sediment ($r = 0.93$, $p < 0.001$, $n = 32$). The ratios of N:P in surface water is 30. This suggests P is more limiting than N for primary production in BMR. The average N:P mass ratio in water of 64.2:1 during the survey. By contrast, the N:P mass ratio in the surface sediments was only 16.6:1.

Table 1. Correlation coefficients of phosphorus and nitrogen in sediment and water.

	N water	P water	N sediment	P sediment
P sediment	0.448	0.517	0.648	-
N sediment	0.517	0.503	-	
P water	0.403	-		
N water	-			

***All significant at p<0.05; n=30*

It has been shown above that the horizontal nutrient concentration gradients can be presented in surface sediments of lakes, induced by high external nutrient loadings and often associated with elevated sediment nutrient concentrations near river mouths [10]. Therefore, it is reasonable to assume that the horizontal distributions of sediment and water variables presented in Figure 1 and 2 gave a realistic representation of the actual horizontal distributions. Based on data from the horizontal distributions of surface sediment phosphorus, it was apparent that the highest concentration occurred in the reservoir inlet from Sungai Kurau catchment area. This area has the major inflows and sources of external phosphorus entered the reservoir. The phosphorus concentrations in the surface water were also high in this area.

Nitrogen concentrations tended to be higher in sediments of BMR. However, compared with other studies, including shallow eutrophic lakes in northern Europe [11-13], New Zealand [14] and Estonia and Russia [15], phosphorus and nitrogen in sediments are still low. This may be caused by the large surface area of BMR. Trolle et al. [16] found that in large surface areas, regardless of wind direction, energy is available for mixing and resuspension. Therefore, fine organic particles, which have high phosphorus and nitrogen contents, may be resuspended most of the time, and partly mineralised in the water rather than buried in the sediments.

The average N:P ratios in the surface water were higher than the N:P ratios found in the surface sediment (Table 1). The difference between the N:P ratios of surface water and sediment suggest that up to 70% of the nitrogen that was absorbed by the sediment vanished through denitrification. During denitrification, nitrate converts to ammonium or free nitrogen (N_2) that may diffuse into the water phase and the atmosphere. Thus, it is lost from the system. Furthermore, a BMR nutrient budget for year 2008 [17], reported that over 60 % and 30% of nitrate and nitrite respectively that entered the reservoir were retained and permanently buried in the sediment, while the rest are estimated to be removed via denitrification. In Denmark, Jensen et al. [18] found that nitrogen retention was indeed caused by denitrification. Akinbile et al. [19] also did a study on addition to the analysis of phosphorus and nitrogen contents of the sediment in BMR and found out that the water was slightly polluted (Class III according to Malaysia Water Quality Index) at the highest concentration of nitrogen and phosphorus. And it decreases to Class I when the concentration decrease.

Phosphorus and nitrogen in BMR have a similar trend of decreasing concentration from upstream to downstream. Figure 4 and Figure 5 show the percentage of total nitrogen and total phosphorus reductions in both surface sediment and water, respectively. The phosphorus and nitrogen concentrations of surface sediment were in fact significantly correlated with phosphorus and nitrogen concentrations of surface water because of the high value of determination of correlation (R^2). Previous study by Trolle et al. [16], showed that the nutrient concentrations of surface sediment were generally not significantly correlated with lake water quality (represented by surface water concentrations of phosphorus and nitrogen). However, this study shows otherwise.

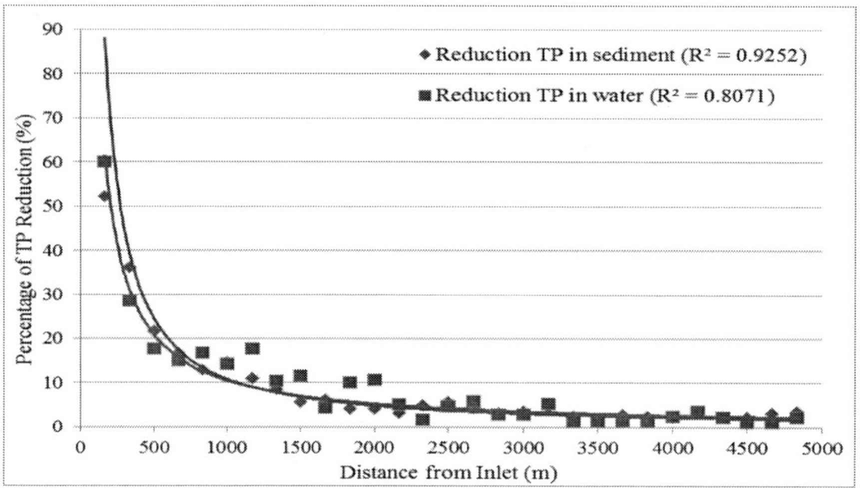

Figure 4. Relationship between phosphorus concentration in water and sediment with distance from inlet.

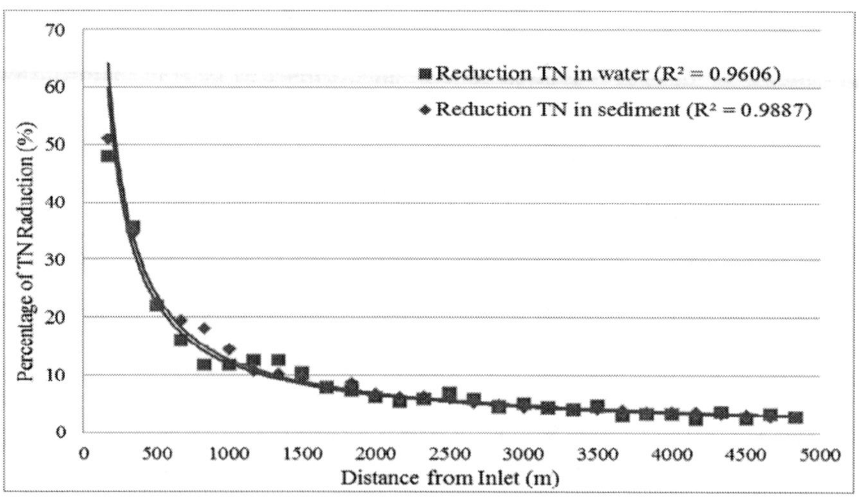

Figure 5. Relationship between nitrogen concentration in water and sediment with distance from inlet.

4 Conclusion

Interrelations between spatial variability in phosphorus and nitrogen concentrations of water and sediment were examined for 30 different sampling locations in Bukit Merah reservoir. Noticeable horizontal phosphorus and nitrogen concentrations in water have been shown to occur. It is apparent that the highest concentrations occurred in the reservoir inlet from Sungai Kurau catchment area. However, phosphorus and nitrogen concentration in surface sediment gave higher magnitude than the concentration in water. The phosphorus and nitrogen concentrations of surface sediment were in fact significantly correlated with phosphorus and nitrogen concentrations of surface water because of the high value of determination of correlation (R^2). As a conclusion, the deposition of sediments was found to bring along external nutrients. Variability of phosphorus and nitrogen concentrations in the sediment directly affects the quality of water which is very important for irrigation and domestic uses..

References

[1] S.H.A. Talib, M.S. Yusoff and Z.A. Hasan, Modeling of sedimentation pattern in Bukit Merah Reservoir, Perak, Malaysia, Procedia Engineering, **50**, 201-210, (2012).

[2] G.L. Mamede, Reservoir sedimentation in dryland catchments: Modelling and management, Doctoral Thesis, Universitätsbibliothek, (2008).

[3] W. Saenyi and H. Holzmann, An integrated approach of water erosion, sediment transport, and reservoir sedimentation, *ERB and Northern European FRIEND Project 5 Conf.*, Slovakia, (2002).

[4] P.J. Dillon and F.H. Rigler, The phosphorus-chlorophyll relationship in lakes, Limnol. Oceanogr, **19**(5), 767-773, (1974).

[5] S.C. Maberly, L. King, M.M. Dent, R.I. Jones and C.E. Gibson,. Nutrient limitation of phytoplankton and periphyton growth in upland lakes, Freshwater Biology, **47**(11), 2136-2152. (2002).

[6] W. Ripl, *Restoration of Eutrophic Lakes by Sediment Treatment, Restoration of Lakes, Streams, Floodplains and Bogs in Europe*, Springer, Netherlands, 77-84, (2010).

[7] G.R. Fish and I.A. Andrew, Nitrogen and phosphorus in the sediments of Lake Rotorua, New Zealand J. of Marine and Freshwater Research, **14**(2), 121-128, (1980).

[8] S.H.A. Talib, Sedimentation modelling and nutrient profile at Bukit Merah Reservoir, Perak, Malaysia. PhD Thesis, Universiti Sains Malaysia, Penang, (2014).

[9] APHA, Standard Methods for the Examination of Water and Wastewater, American Public Health Association, Washington, (2012).

[10] B. Qin, Z. Liu and K. Havens, *Eutrophication of Shallow Lakes with Special Reference to Lake Taihu, China*, Springer, Netherlands, (2007).

[11] L. Håkanson, On the relationship between lake trophic level and lake sediments, Water Research, **18**(3), 303-314, (1984).

[12] E. Jeppesen, M. Sãndergaard, N. Mazzeo, M. Meerhoff, C. Branco, V. Huszar and F. Scasso, Lake restoration and biomanipulation in temperate lakes: relevance for subtropical and tropical lakes, Restoration and Management of Tropical Eutrophic Lakes, 331-349, (2005).

[13] G.A. Weyhenmeyer, E. Jeppesen, R. Adrian, L. Arvola, T. Blenckner, T. Jankowski, E. Jennings, P. Noges, T. Noges and D. Straile, Nitrate depleted conditions on the increase in shallow northern European lakes, Limnology and Oceanography, **52**(4), 1346-1353, (2007).

[14] D. Hamilton, and S. Mitchell, Waves induced shear stresses, plant nutrients and chlorophyll in seven shallow lakes, Freshwater Biology, **38**, 159-168, (2003).

[15] T. Nõges, A. Järvet, A. Kisand, R. Laugaste, E. Loigu, B. Skakalski and P. Nõges, Reaction of large and shallow lakes Peipsi and Võrtsjärv to the changes of nutrient loading, *Shallow Lakes in a Changing World*, Springer, Netherlands, 253-264, (2007).

[16] D. Trolle, G. Zhu, D. Hamilton, L. Luo, C. Mcbride and L. Zhang, The influence of water quality and sediment geochemistry on the horizontal and vertical distribution of phosphorus and nitrogen in sediments of a large, shallow lake, Hydrobiologia, **627**, 31-44. (2009).

[17] W.R.Ismail and S.A.M. Najib, Sediment and nutrients balance of Bukit Merah Reservoir, Perak, Malaysia, Lakes and Reservoirs: Research and Management, **16**, 179-184, (2011).
[18] J.P. Jensen, E. Jeppesen, P. Kristensen, P.B. Christensen and M. Sãndergaard, Nitrogen loss and denitrification as studied in relation to reductions in nitrogen loading in a shallow, hypertrophic lake (Lake Sãbygãrd, Denmark). Internationale Revue der gesamten Hydrobiologie und Hydrographie, **77**, 29-42. (2007).
[19] C.O. Akinbile, M.S. Yusoff, S.H. Abu Talib, Z. Abu Hasan, W.R. Ismail and U. Sansudin, Qualitative analysis and classification of surface water in Bukit Merah Reservoir in Malaysia, Water Science and Technology Water Supply, **13**(4), 1138-1145, (2013).

MATEC Web of Conferences **47**, 05005 (2016)
DOI: 10.1051/matecconf/20164705005

Evaluating Chlorophyll-*a* Changes During Algal Bloom in Three Gorges Reservoir Using an Extended WASP Model

Li Jian[1,2], Qin Xiaosheng[2,a] and Chen Ming[2]

[1]*Yangtze River Scientific Research Institute, No.23 Huangpu Road, Wuhan, China*
[2]*School of Civil and Environmental Engineering, Nanyang Technological University, 50 Nanyang Ave., Singapore*

Abstract. Algal bloom in Three Gorges Reservoir (TGR) and one of its tributaries, Xiangxi River (XR), have become major concerns and the dynamic changes of such events were investigated using the hydrodynamic model SELFE and the extended Water Quality Analysis Simulation Program (WASP) model to address nutrient and biomass dynamics. The model has taken into consideration the surface wind, heat fluxes, oxygen exchanges, solar radiations and boundary conditions from main river channel and tributaries. As an extension to our previous work, this study aimed to report in more detailed the result of chlorophyll-*a* simulations, where the field observed data of algal blooms in TGR in 2007 was used for calibration and the horizontal and vertical distributions of phytoplankton biomass (based on chlorophyll-*a*) were presented. It was found that the chlorophyll-a concentration characterized as algal biomass was influenced by many complex factors. Further study results are yet to be reported.

1 Introduction

The Three Gorges Reservoir (TGR), with a storage capacity around 39.3 billion m^3, is located in the upstream of the Yangtze River, China. The Yangtze River and its tributaries in the reservoir area are featured by both narrow and deep boundaries. Meanwhile, obvious exchanges of water body and various solutes (including nutrients and suspended sediments) occur between the mainstream and the tributaries under frequent reservoir operations (for the purpose of flood control and hydroelectricity generation). Various pollutants flow into the reservoir especially the tributary of Xiangxi River (XR) from many chemical factories and sanitary sewage systems, which caused frequent the algal bloom events in the tributaries after water impoundment in 2003.

Over the past years, many researchers have conducted field observations and numerical simulations for studying problems of water quality deterioration and ecosystem evolution in TGR and its tributaries. Particularly, many hydrodynamic and water quality models have been applied for examining the TGR algal blooms, which include 1-Dimensional [1], vertical 2-Dimensional [2] and 3-Dimensional [3] models. From the previous works, it was found that the flowing field, solute transport and phytoplankton biomass evolution in TGR present significant spatial and temporal variations [4]; hence, a 3-dimensional model is considered better suited for the modeling purpose. Generally, the 3-dimensioal models can be divided into structured and unstructured meshes

[a] Corresponding author: xsqin@ntu.edu.sg

considering the topographical structure. For example, Dai et al. (2013) used a structured-mesh model, Delft3D, for modeling the TGR water quality [3]; however, the model cannot well follow the complex boundary and subaqueous relief in TGR. Therefore, in this study, an unstructured mesh water quality model was used based on the hydrodynamic model SELFE [5] and the Water Quality Analysis Simulation Program (WASP) model [6]. The detailed structure and general performance of this model was initially reported in Li [7]. The WASP model used here has been extended considering the updated research literatures [8] and the local characteristics about TGR water quality problem [4]. As an continual work to the previous study [7], the algal bloom occurred in 2007 was more extensively simulated and reported by the coupled models aiming at understanding the changing process of the horizontal and vertical distribution of phytoplankton biomass (chlorophyll-*a*).

2 Methodology

The coupled model was based on SELFE [5] and the extend WASP model [6] considering the most updated water quality research works [8]. The coupled hydrodynamic and extended WASP model covers the essential physiochemical and biological processes in the aquatic ecosystem in TGR in addition to the effect of sediments including suspended and bottom layer sediment on the water quality dynamics. The model also considers the interactions among many factors like hydrodynamics, water temperature, underwater light intensity, nutrients, dissolved oxygen (DO), biochemical oxygen demand (BOD) and phytoplankton biomass (denoted by chlorophyll-*a* concentration). The author has added more modules into the original WASP model including nonlinear adsorption-desorption interaction between suspended sediment and phosphate, solutes exchange between bottom sediment layer and overlying water considering bottom stress [7-9]. Meanwhile, the Nitrogen and Phosphorus nutrients as Total Nitrogen (TN) and Total Phosphorus (TP) used in the model have been divided into organic and inorganic nutrients including nitrate (NO_3), ammonia (NH_4), organic nitrogen (ON), phosphate (PO_4) and organic phosphorous (OP) [9, 10]. For technical details, readers are referred to literatures [6-10].

3 Study Case

The XR (with a length about 94 km, flowing through Xingshan and Zigui) is situated in the Hubei Province of China. Many other tributaries exist in TGR along the Yangtze River including Duanfangxi, Tongzhuanghe, and Qingshuihe, Yuanshuihe [7]. Since the discharge and pollutant load into XR are predominantly larger [11], other tributaries were considered as closed bay in the simulation. Serious water quality problems had occurred in these tributaries due to insufficient water mixing or flowing based on the in-situ observation [4]. Therefore, a high initial concentration of nutrients and chlorophyll-*a* is used in the calculation. The hydrodynamic condition at local tributaries is closely linked to the Yangtze River's major channel flow. For instance, there existed backflow mixing and vertical water exchange for the period of water impoundment and water discharging processes of TGR. Hence, the physiochemical and biological parameters close to the XR outlet would act together with the main river flow. Therefore, the water body exchange between Yangtze River and the tributaries influencing the algal bloom process was also simulated.

The computational domain including Yangtze River, Xiangxi River, Gaolan River (the second-class tributary of Yangtze River) and some other first-class tributaries in TGR is discretized with 15,282 horizontal triangle elements [7]. A total of 11 observation stations in XR and 6 observation stations in Yangtze River were established along the river channel to measure parameters like nitrogen, phosphate, suspended sediment, DO, BOD and chlorophyll-*a* in surface water; in addition, the biochemical and hydrological data during algal blooms occurring in 2007 were also collected. In numerical modeling, the input boundary of the Yangtze River, XR and Gaolan River were defined following the hydrologic stations data (i.e. discharge) during the events of actual algal blooms; the TGR dam boundary condition (i.e. water levels) was based on the actual reservoir operation data.

4 Results

The SELFE-WASP model was calibrated using the measured data in 2007 when the algal blooms, hydrodynamics, and water quality were observed together and the data were suitable for the calibration of the 3D coupled modeling. The model calibration efforts for flow velocity were based on implementation of SELFE modeling. The main parameters of WASP were estimated many times in order to generate a reasonable fit of the observed data for algal bloom process. Other model parameters were selected based on the works from the previous publications [1-3]. The estimated or selected parameter values were found within the intervals reported by Justić and Wang [8]. And the calibrated parameters used in modeling (not listed here) can reproduce the algal bloom process (chlorophyll-*a*) well. One example in Gaoyang Station is shown in Fig. 1, where the simulated chlorophyll-*a* concentration agreed well with the observed data at the 5-m depth under the water surface. Note the simulated and observed vertical profiles of chlorophyll-*a* concentration at Gaoyang Station was not shown here.

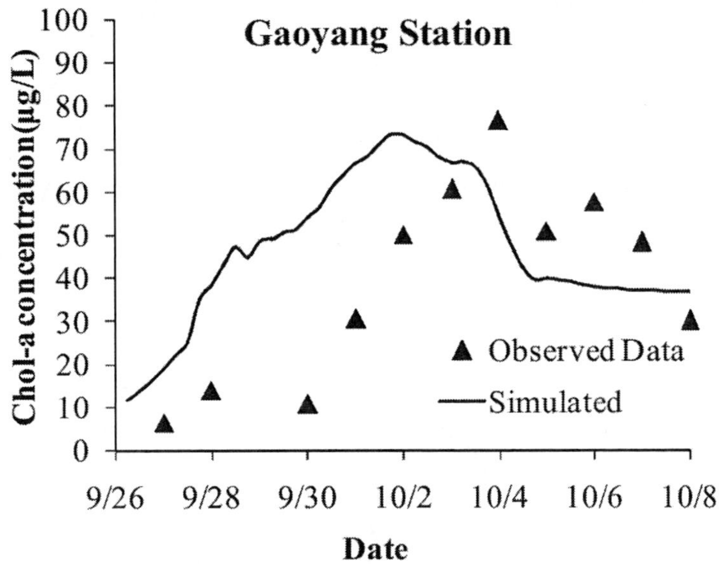

Figure 1. Observed vs. simulated concentrations in calibration of chlorophyll-*a*.

Figure 2 presents the results of horizontal distributions of flowing velocity and chlorophyll-*a* concentration. The flowing velocity in Yangtze River was obviously higher (the maximum velocity can reach up to 0.3 m/s) than that in the Xiangxi tributary (the velocity was around 0 m/s) as shown in Figure 2(a) which would lead to different algal reproduction phenomenon, the backflow happened in tributary Xiangxi River would also decrease the nutrients and phytoplankton transportations into the Yangtze River. Meanwhile, the vortex at the confluence could also be seen in Figure 2(a) which was caused by different flowing directions from Yangtze River and Xiangxi River. The chlorophyll-*a* concentration characterized as phytoplankton biomass was preliminarily simulated considering many factors influencing the growth, death and settling down of phytoplankton cells. The horizontal distribution of chlorophyll-*a* concentration at both water surface and bottom can provide the information about the algal bloom locations in the TGR area in Figure 2(b). Obviously, the algal bloom event mainly occurs in the upper reach of XR from Pingyikou to the confluence of XR and Gaolan River (Xiakou) [7]. Meanwhile, the bottom Chlorophyll-*a* concentration was also obviously high at the shallow water zone like Pingyikou and Tongzhuanghe [7]. The phytoplankton biomass in the Yangtze River was relatively lower than that in the tributaries. Generally, the coupled model could provide reasonable reproduction of the algal bloom event, and the event occurring locations could be preliminarily analyzed through the proposed modeling study.

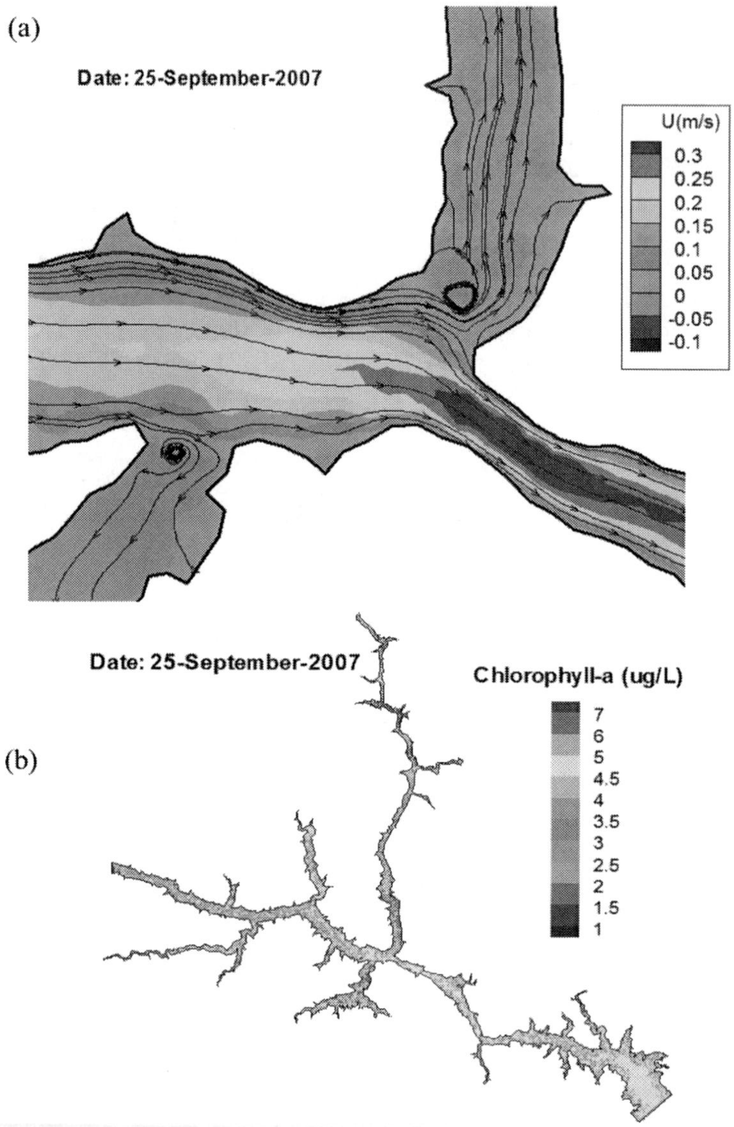

Figure 2. Horizontal distribution of variables (a) Flowing velocity at the confluence of Xiangxi River and Yangtze River and (b) Chlorophyll-a concentration near riverbed.

Figure 3 shows the results of vertical distribution of chlorophyll-*a* concentration. The phytoplankton cells at water surface can receive more lighting for photosynthesis than that in the deeper water column, which would lead to obvious differences of vertical chlorophyll-*a* concentration. Because of over-saturated phenomenon caused by solar irradiation that will inhibit the photosynthesis, the chlorophyll-*a* concentration beneath the water surface around water depth of 5.0~20.0 m reaches the maximum value. The chlorophyll-*a* concentration is not the highest at the water surface as shown in Figure 3. The vertical concentration profiles of chlorophyll-*a* show stratifications in both the XR tributary (Fig. 3a) and the Yangtze River mainstream (Figure 3b). The chlorophyll-*a* concentration in XR tributary was obviously higher than that in Yangtze River mainstream because the relatively high flowing speed could increase the self-cleaning capability and also more physical transport would decrease the phytoplankton biomass in Yangtze River mainstream as shown in Figure 3(a). The relatively stable flow in XR provides a suitable condition for the algal reproduction that has led to locally high chlorophyll-*a* concentrations; while the chlorophyll-*a* concentration in the Yangtze River is more related to the inflow boundary condition.

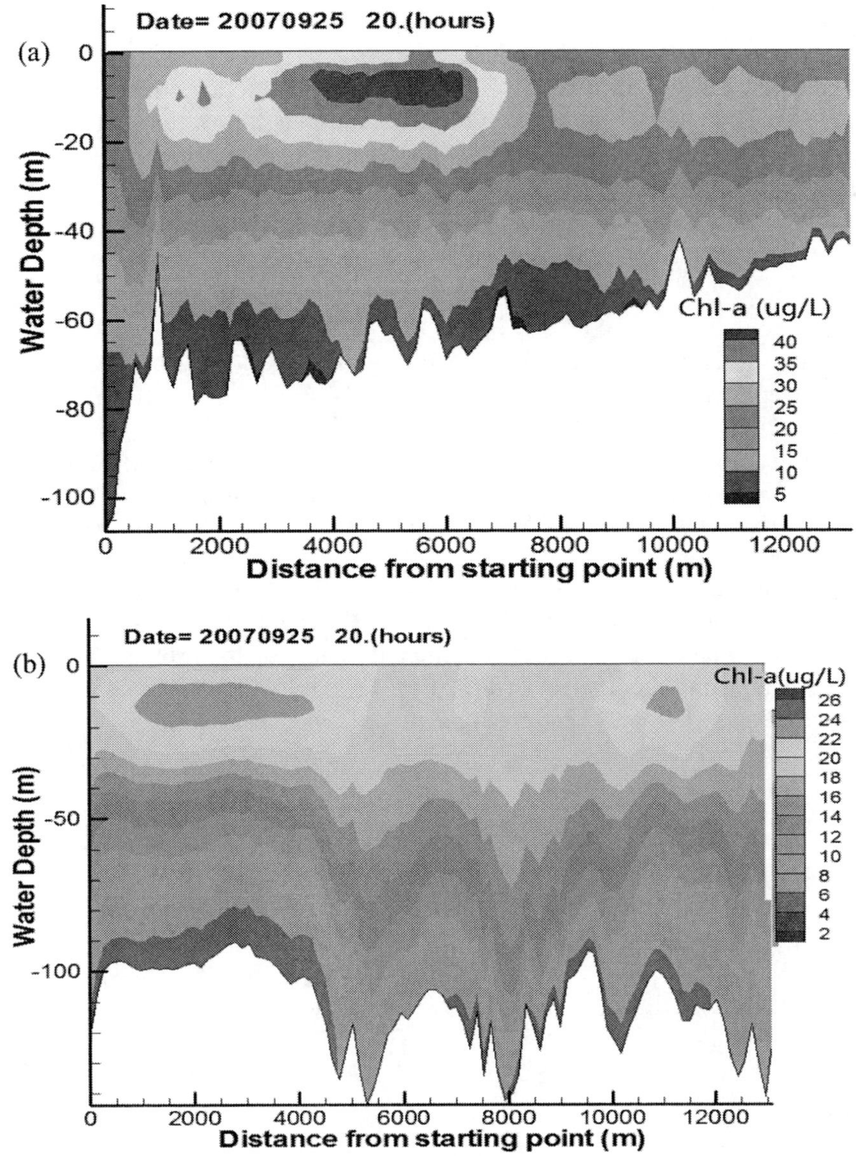

Figure 3. Vertical concentration profile of chlorophyll-*a* at: (a) XR and (b) Yangtze River.

5 Conclusion

This study presents a preliminary numerical simulation on the spatial-temporal variation of chlorophyll-*a* by using a numerical model coupling SELFE and an extended WASP module. The model was demonstrated effective for the case study of TGR and XR algal bloom event in 2007 [7]. The simulation domain covered Yangtze River and many tributaries which was different from the only simulation about XR algal bloom in the author's previous works [8, 9]. Reasonable tendencies of chlorophyll-*a* changes and its concentration were preliminarily examined from the numerical modeling. The algal bloom occurring locations could be analyzed successfully through the proposed model. Further studies on other water quality constituents are yet to be reported more thoroughly in the near future.

Acknowledgement

This material is based on research/work supported by the National Science Foundation of China (51309021) and in part by the Land and Liveability National Innovation Challenge under L2 NIC Award No. L2NICCFP1-2013-3 (WBS no.: M4061545.D63). We are also grateful to the Xiangxi River Ecological Observation Station of China, Three Gorges University for providing the field observation data.

References

[1] L. Wang, Z. Yu and H. Dai, Water Science and Engineering, **2**, 16-24, (2009).
[2] J. Lian, Y. Yao and C. Ma, Water, **6**, 3200-3223, (2014).
[3] H. Dai, J. Mao and D. Jiang, PLOS ONE, **8**, 1-14, (2013).
[4] T. Zheng, J. Mao and H. Dai, Science China Technological Sciences, **54**, 1588-1598, (2011).
[5] Y. Zhang and M. B. António, Ocean Modelling, **21**, 71-96, (2008).
[6] T.M. Wool, *Water Quality Analysis Simulation Program (WASP) Version 6 User's Manual*, US Environmental Protection Agency, Atlanta, (2001).
[7] J. Li, *E-proceedings of the 36th IAHR World Congress,* The Hague, Netherlands, (2015).
[8] D. Justić and L. Wang, Continental Shelf Research, **72**, 163-179, (2014).
[9] J. Li, D. Li and X. Wang, Journal of Environmental Sciences, **24**, 1569-1578, (2012).
[10] J. Li, Z. Jin and W. Yang, Ecological Informatics, **22**, 23-35, (2014).
[11] K. Zhu, Doctoral dissertation of Chinese Academy of Sciences, (2013).

MATEC Web of Conferences **47**, 05006 (2016)
DOI: 10.1051/matecconf/20164705006

Removal of Iron and Manganese Using Cascade Aerator and Limestone Roughing Filter

Azrin Mohd Sanusi[1], Mohd Nordin Adlan[1], M. A. Z. Mohd Remy Rozainy[1] and Rhahimi Jamil[1,a]

[1]*Faculty of Civil Engineering, Universiti Sains Malaysia 14300 Nibong Tebal, Pulau Pinang, Malaysia*

Abstract. Combination between oxidation and filtration can be used for removing iron and manganese from groundwater especially when the concentrations of these metals were high. This study focused on the effectiveness of the cascade aerator and the size of the limestone filter media to remove iron and manganese from groundwater. Water samples used for this study were collected from orphanage home, Rumah Nur Kasih, Taiping. Universiti Sains Malaysia (USM) has provided a tube well of 15 m depth and 150 mm diameter for the orphanage home. However, the water cannot be used for domestic consumption due to high amount of iron and manganese at 6.48 and 1.9 mg/L which exceeded the drinking water standard of 0.3 and 0.1 mg/L respectively. Using laboratory physical model, the study has shown that the removals of iron and manganese have reduce the concentration until 0.17 and 0.2 mg/L respectively. Thus, the results from this study which utilize cascade aerator and limestone roughing filter could be implemented on site for the community to use the ground water for domestic purposes.

1 Introduction

Groundwater is one of the sources of water that can be used for daily needs such as bathing, cooking and washing where it represents about 97% of the freshwater resources on the Earth that are avilable for use by humans [1]. Iron and mangenese present in drinking water are not health treatening but it's a problem when the baceria exists in soil aquifers and some surface water [2]. Conventionally, iron is removed from groundwater by the processes of aeration and rapid filtration. Different mechanisms may contribute to the iron removal in filters; flock filtration, adsorptive iron removal and biological iron removal. Water containing iron can be divided into two main groups: Waters which separate iron just after aeration and waters where iron remains in the solution after aeration for endlessly long period [3]. Roughing filters are primarily used to separate fine solids from the water that are only partly or not retained at all by stilling basin or sedimentation tanks. The large filter surface area available for sedimentation and relatively small filtration rates also supports absorption as well as chemical and biological processes [4]. Universiti Sains Malaysia (USM) has provided a tube well of 15 meter depth and 150 mm diameter for the orphanage home called Rumah Nur Kasih, Taiping. The objective of this study is to determine the effects of the flow rates and limestone sizes on the removal of iron and manganese and to determine the amount of oxygen dissolved using the cascade aerator based on different flow rates. The scope of work for this study includes details on the parameters of

---

[a] Corresponding author : nu_eiman82@yahoo.com

iron, manganese and dissolved oxygen that determined through experiment on samples of water from the tube well at Rumah Nur Kasih, Taiping.

2 Materials and Methodology

2.1 Study area

Orphanage house, Rumah Nur Kasih, Taiping is inspired by the committee to create a home that is perfect for placing orphans and the poor. A committee was established to achieve the objective, and an old mosque has been applied for and received approval from religion and treasury offices to house all the selected orphans and the poor. After obtaining a consensus, then on 1st June 2010 the committee agreed to name this organization with the name of Nur Kasih Welfare Organisation, Taiping and named orphanages and poor in the name of Rumah Nur Kasih, Taiping. Location to take water samples is at Rumah Anak-Anak Yatim dan Miskin Nur Kasih, Taiping, Perak.

2.2 Description of instruments

Samples for analysis in a pilot study should be collected carefully to make sure the most representative sample possible is obtained. Water samples were taken once a week using five clean bottles. Equipment was used to take the water samples is bottles and hosepipe. The measurement of total iron and manganese was made using Method Hach DR 2800 with standard 3500FeB (APHA, 2014), where for iron the DR 2800 entered the 265 program and for manganese selected 296 program. The DR2800 were installed into the cascde aerator shown at Figure 1 to measure the dissolved oxygen and also the total iron and manganese.

Figure 1. Cascade aerator model.

2.3 Measurement technique

Pilot study has been done to make the most representative sample that possible for the study obtained. Water samples were taken once a week using five clean bottles. After water sample arriving at the laboratory, experiment at cascade aerator shown at Figure 1, must continue to avoid the water exposed to air for long time. The measurement of total iron was made using Method Hach DR 2800 with standard 3500FeB (APHA, 2014). The DR 2800 entered the 265 programs. The two clean cells, one cell was filled with distilled water (blank) and another one with a water sample. The cell with sample was added with one Ferro Ver Iron Reagent Powder Pillow. When the colour of the regent turns to orange its show the iron was presented. DR2800 also used to determine manganese, where after stored the programs, 296 Manganese test was selected. A violet colour will develop if manganese is present. Roughing filtration can be considered as a major pre-treatment process for wastewater, since they could efficiently separate fine solid particles over prolonged periods without addition of chemicals. The roughing filter size in this research is 29 x 6.5 x 6 inches. There three sizes of filtration media had been chosen are 12.5 mm - 10 mm, 5 mm - 3.35 mm, and 2.36 mm - 0.6 mm, suggested by Aziz and Smith [5, 6] the limestone was sieved a size between 2.36 and 4.75 mm. Figure 2 shows the scematic diagram of filtration media that has been built as a model.

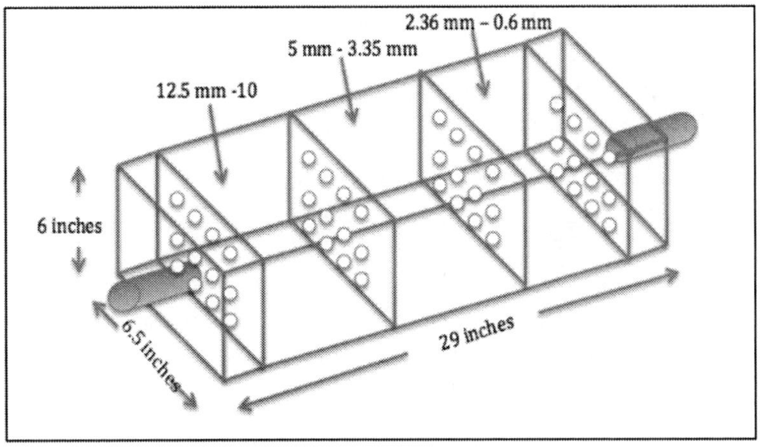

Figure 2. Scematic diagram for limestone roughing filter.

3 Results and Discussion

The results obtained for iron and manganese test were using DR2800. The test results are based on samples taken from Rumah Nur Kasih, before and after filtration. For iron and manganese, flow rates used were 25, 10 and 40 L/min. According to Rathinakumar et al. [2], the flowrate that had chosen between 0.003 m3/s until 0.021 m3/s, means the flow rate increment ie proportional to the aeration improvement . The graph is divided into two according to the mix trial. First graph the data analysed between percentage removals after filter against limestone size and the second graph is percentage removals after filter against flow rate. Based on results mix trial, the experiment continued using flow rate 25 L/min and limestone size 2.36 mm – 0.6 mm and the results is satisfactory and achieve the standard referred to the DOE Water Quality Index Classification. Figures 3 and 4 respectively for iron and manganese represents the aeration efficiency of a stepped channel with flow rate 25 L/min while constant step rise.

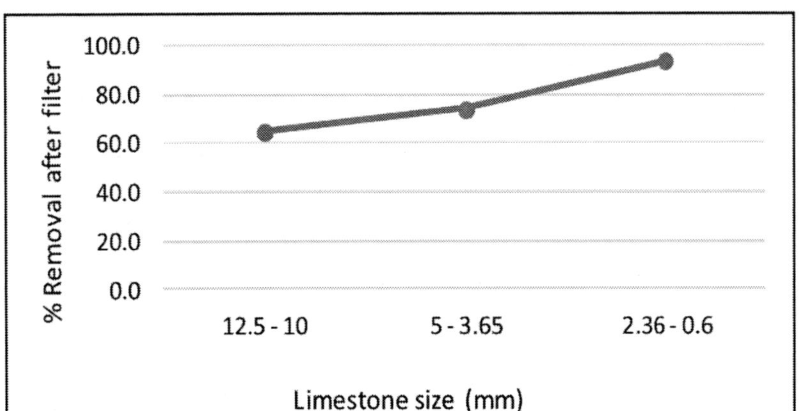

Figure 3. Percentage removal of iron against limestone size with flow rate 25 L/min.

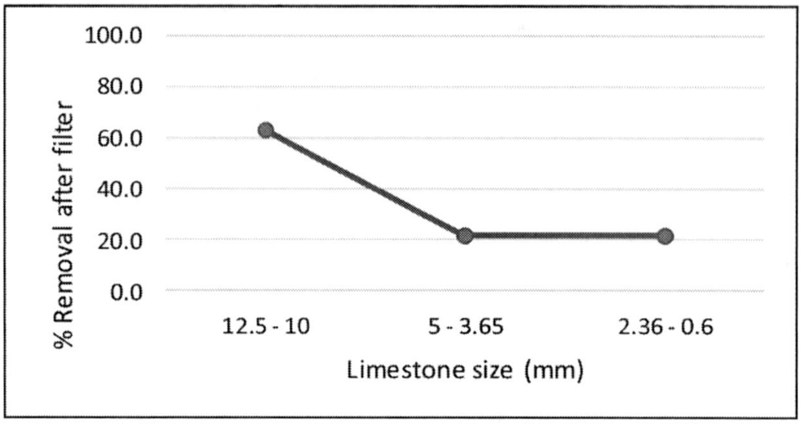

Figure 4. Percentage removal of manganese against limestone size with flow rate 25 L/min.

Figures 5 and 6 respectively for iron and manganese show the filtration efficiency with limestone size 2.36 mm – 0.6 mm with different flow rates. The flow rates are 10, 25 and 40 L/min. From graph, the highest iron removal is at flow rate 25 L/min.

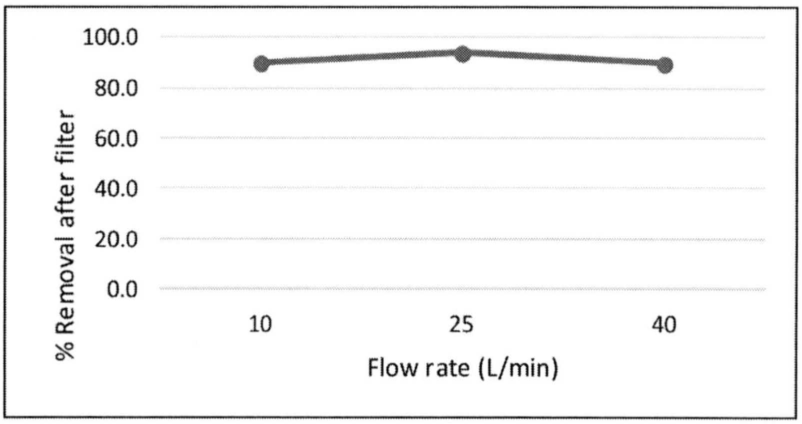

Figure 5. Percentage removal of iron against flow rate with limestone size 2.36 mm – 0.6 mm.

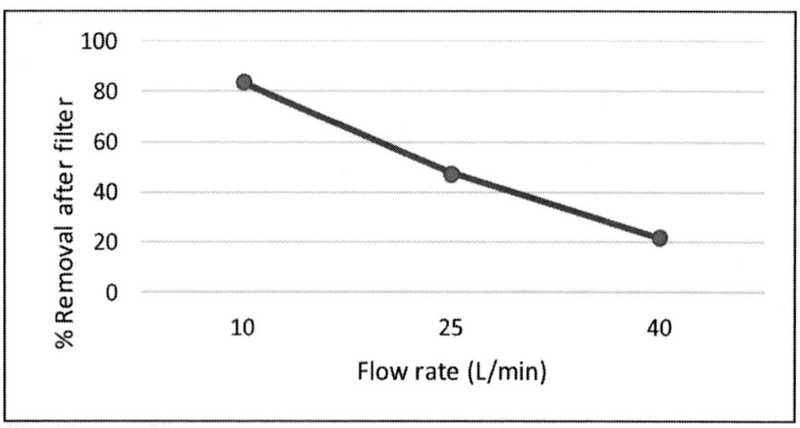

Figure 6. Percentage removal of manganese against flow rate with limestone size 2.36 mm – 0.6 mm.

Table 1 shows the increasing DO obtained from the tank 1 to tank 4. DO value obtained at the bottom of the tank is at a level II according to standard shown of DOE Water Quality Index Qualitification. In theory, the dissolved oxygen is an indicator of the degree of contamination of water because the lower the dissolved oxygen, the more polluted the water.

Table 1. The results of DO obtained.

Dissolved Oxygen, DO (mg/L)					
Sample	**1**	**2**	**3**	**4**	**Flowrate (L/min)**
1	4.0	5.4	5.9	6.2	12
2	3.5	5.1	6.0	6.4	22
3	6.5	6.9	7.3	7.9	25
4	3.5	5.3	6.4	6.8	32

4 Conclusions

This study was utilized the cascade aerator and limestone roughing filter could be implemented on site for the community to use the ground water for domestic purposes. Based on the results its shows that from iron parameter, the highest removal is 0.17 mg/L while the highest removal for manganese is 0.2 mg/L and the highest DO value is 6.8 mg/L at 32 L/min after pumping the water sample by using different flow rates.

Acknowledgement

The authors would like to acknowledge Ministry of Education Malaysia for providing LRGS grant No. 203/PKT/6726001 – Riverbank Filtration for Drinking Water Source Abstraction to fund this research.

References

[1] E. Lopez-Gun and W.T. Jarvis, Groundwater governance and the low of the hidden sea, Water Policy, **11**(6), 742-762, (2009).

[2] V. Rathinakumar, G. Dhinakaran and C.R. Suribabu, Assessment of aeration capacity of stepped cascade system for selected geometry, Int. J. of ChemTech Research, **6**(1), 254-262, (2014).

[3] R. Munter, H. Ojaste and J. Sutt, Complexed iron removal from groundwater, J. of Environmental Engineering, **131**, 1014-1020, (2005).

[4] O. Nkwonta and G. Ochieng, Roughing filter for water pre-treatment technology in developing countries: A review, Int. J. of Physical Sciences, **4**(9), 455-463, (2009).

[5] H.A. Aziz and P.G. Smith, The Influence of pH and coarse media on manganese precipitation from water, Water Research, **26**(6), 853-855, (1992).

[6] H.A. Aziz and P.G. Smith, Removal of manganese from water using crushed dolomite filtration technique, Water Research, **30**(2), 489-492,1996.

MATEC Web of Conferences **47**, 05007 (2016)
DOI: 10.1051/matecconf/20164705007

The Growth of Aspergillus Niger on a Wood Based Material with 4 Types of Wall Finishing

Menega Subramaniam[1,a], Norshuhaila Mohamed Sunar[2], Adnan Latif[1], Umi Kalthsom Parjo[2], Chin Ming Er[1] and Abdul Rafiq Ab Razak[1]

[1]*Faculty of Civil and Environmental Engineering, Universiti Tun Hussein Onn Malaysia, 86400 Batu Pahat, Johor, Malaysia*
[2]*Faculty of Engineering Technology, Universiti Tun Hussein Onn Malaysia, 86400 Batu Pahat, Johor, Malaysia*

Abstract. Buildings are a vital component in a human's daily life. It provides shelter from the environment, weather and animals. Mold growth within the building might be caused by the moisture problems which directly act on it such as water leaks or indirect factor such as high humidity levels. This growth causes esthetic problems and deterioration of its wall coatings. Spores from the fungi also cause health problems to humans. The fungus species studied in this research is *Aspergillus niger*. The material is made of wood and its finishing is thick wallpaper, thin wallpaper, acrylic paint and glycerol based paint. ASTMD5590-00 standard was used to evaluate fungal growth and to determine if non antifungal agent was effective in inhibiting the amount of fungal growth on four types of wall finishing used on wooden walls. This research was conducted without using any antifungal agent. Highest percentage of growth of the fungi was found on acrylic paint, followed by glycerol based paint and thin wallpaper. Thick wall paper shows the least growth of fungi. The maximum growth is visible on day 12 which is more than 60% by all the wall finishing.

1 Introduction

Mold causes significant problems when it comes to aesthetic value within a building. It may lead to surface problems such as degradation and spalling. Fungi are organisms that colonize on building coatings and thrive on them [1, 2]. Using wood as the building material is also the main cause of mold and fungi colonization in buildings. Controls on the pollutant sources are considered because the design, efficiency and economics of the IAQ control can be influenced [3]. Most people work in an office environment [4]. This is an area where a hazard such as chemicals, lighting and indoor air quality rises. Health problems such as asthma, irritations and other issues are spread due to airborne fungal spore which effect humans drastically [5-7]. These signs are uncommon in buildings that are well maintained but water leaks and inappropriate humidity levels causes high fungal spore loads [8, 9]. A fungus is able to grow almost on all natural and synthetic materials especially if they are wet [10, 11]. Wood is vulnerable to fungal attack [12]. Kiln dried wood surfaces are more susceptible to fungi infection compared to other wood types [13]. Acylated wooden furniture's, wood polyethylene composites, plywood and modified wood products are susceptible to infestation by *Aspergillus*, *Trichoderma* and *Penicillium*. Paper and glue used on indoor surfaces are very good growth substrates

[a] Corresponding author : menegasubramaniam@gmail.com

for most of the indoor fungi as they provide a wet environment for the fungi. *Aspergillus* and *Penicillium* grow superficially on painted surfaces, but *Aureobasidiumpullulans* was found to deteriorate the paints. Acrylic painted surfaces are attacked by *Alternaria*, *Cladosporium*, and *Aspergillus*. There are 3 main factors that influence the growth of fungi in buildings. They are moisture, nutrients and temperature [14]. Material types and humidity are also factors that control the fungi growth in structures. Material such as gypsum board, wallpaper, mortar and paint are common in the construction industry. *Aspergillus* species are commonly found on ceramic-type materials (concrete, mortar) and paints and glues. Cellulose based materials are much sensitive to contamination by fungi compared to inorganic materials such as gypsum, mortar and concrete. Cellulose can be metabolized by the microorganisms [15]. A large variety of construction materials are sensitive to the development of mold. Nevertheless, it seems clear that the materials/substrate does influence the type of mould that develops on the surface. A few studies have demonstrated that the fungal genera detected on construction plaster are predominantly *Penicillium* or *Aspergillus* [16]. Studies have shown materials with strong concentrations of organic carbon like cellulose or carbonates (wall paper, wood-based building materials) are more favourable to the development of mould than those with lower carbon content (plaster, glass wool) [16]. Thus the objective of this study is to determine the effectiveness of different wall finishing (thick wallpaper, thin wallpaper, glycerol based paint and acrylic paint) on wood based material to reduce the growth of *Aspergillus niger* without applying any type of antifungal agent.

2 Materials and Methods

2.1 Preparation of bio-resistance test

The wall finishing which is the acrylic paint (AP), glycerol based-paint (GBP), thin wall paper (TnW) and thick wallpaper (TkW) are used to cover the supports (woods). The weight of the thick wall paper is 240 g/m^2 and the thin wall paper is 80 g/m^2. According to Vacher et al. [16] the supports (woods) were cut into samples size of 50mm x 50mm. Then, 70% of ethanol was used to wash each surface of the samples. The samples were left to dry for a few hours. After drying was completed, the acrylic paint and the glycerol based paint were each applied on different surface of wood samples to cover the wood. The thin and thick wall paper was glued on the surface of the supports. Approximately 24 to 48 hours were taken to dry the samples. Next, the fungal spores were inoculated. The samples are proceeded to test in a germicide UV-lamp for 40 minutes before the samples were kept in individual dishes. 50 µl of the fungus spore suspension was applied on top of the wall finishing. Finally, the petri dish was sealed using parafilm. The samples were placed into the incubator for 27 days for observation. Relative humidity (RH) was set up to 55% ± 2% and the temperature at 28°C ± 1°C as stated in ASTM D5590-00 Guidelines [17]. Regular observations were done for 3, 6, 9, 12, 15, 18, 21, 24, and 27 days for each test at the same time.

2.2 Measurement of result

American Society for Testing and Materials (ASTM) D5590-00 Standard Scale was referred for analyzing the data [17]. The analyzing of the data was carried out for the free-antifungal substrate and consists of different wall finishing cover surface. As the result, the efficiency of the various wall finishing in treating the indoor fungal is observed by the rate of the fungal growth (at 27th days). For observation, the decrease rate of mold growth on the supports, graph of fungal growth versus four types of wall finishing are plotted. These graphs can show the reduction in rate of indoor fungi on the support. The measurement also includes the percentage of fungi growth and the diameter of the fungi.

Table 1. Scale for evaluation fungal growth (ASTM D5590-00 Standard Scale).

Scale	Percentage of Fungi Growth	Degree of Fungi Growth
0	0	None trace of growth
1	(<10%)	Trace Growth
2	(10 – 30%)	Light growth
3	(30 – 60%)	Moderate growth
4	(60 – 100%)	Heavy growth

Three graphs are tabulated according to percentage of fungi growth, rating scale and also the diameter of fungi growth.

3 Results and Discussion

The growth of fungi on different wall finishing which is not treated with any type of antifungal agent was shown (Figure 1). Acrylic paint (AP) and glycerol based paint (GBP) have higher percentage of fungi growth compared to thick wallpapers (TkW) and thin wallpapers (TnW). The highest percentage of fungi growth was by AP which reached its maximum by day 12 at 100%. This shows that AP and GBP is unsuitable as wall finishing on wood unless it was treated or applied with a layer of antifungal [13]. The least growth of fungi was found on control sample which was the TkW. It took almost 27 days to reach 90% compared to other types of wall finishing. The thickness of the wall finishing influences the growth of the fungi. Method is as shown above.

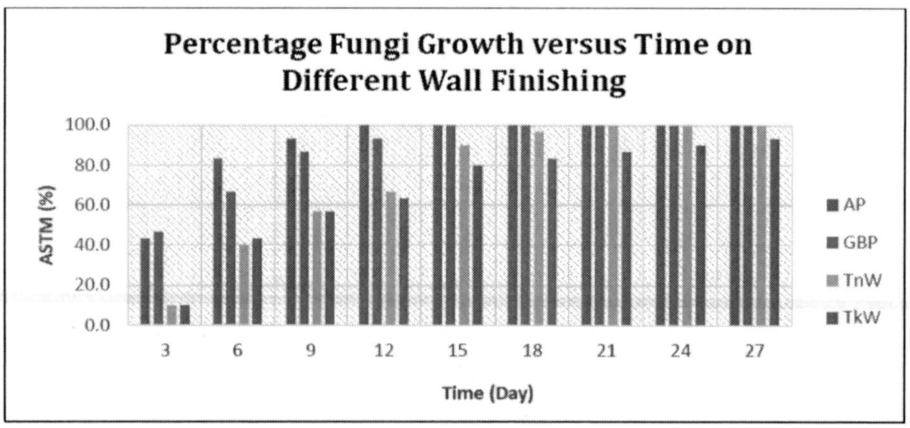

Figure 1. Percentage of fungi growth on different wall finishing without antifungal agent.

The Rating Scale for control sample that is sample without antifungal agent was shown (Figure 2). On day 3 as the first reading, acrylic paint shows the highest rating on fungi growth. It reached rating as 2.6 followed by 2.4 by glycerol based paint. Both wall papers had the same rating on day 3 which is 1.0. Indoor fungal growth of this sample increased drastically until Day 15 then remained stable where all of them reached 4.0 rating scale.

The diameter of fungi growth on different types of wall finishing is as shown (Figure 3). The highest diameter of fungi recorded on day 3 was the sample with acrylic paint. It was 2.5 cm of fungi formed on acrylic paint as wall finishing on wood. The minimum diameter at day 3 was stated on thick wallpaper. It is proved that in this study, wallpapers also as equal as the paints on fungi growth beyond day 18. Thick wallpaper is more effective to resist fungal growth compared to thin wallpaper. This might be due to the difference in thickness of wallpaper that can prevent the spore from getting the nutrient from the agar.

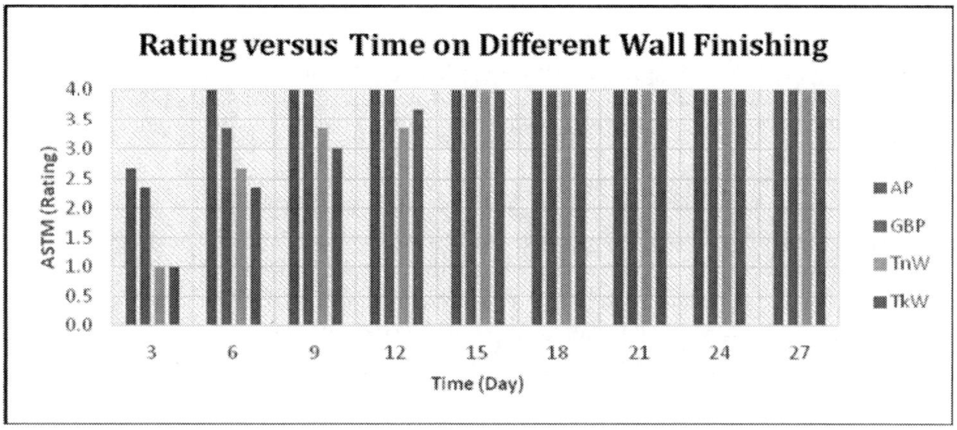

Figure 2. Rating scale versus time on different wall finishing without antifungal agent.

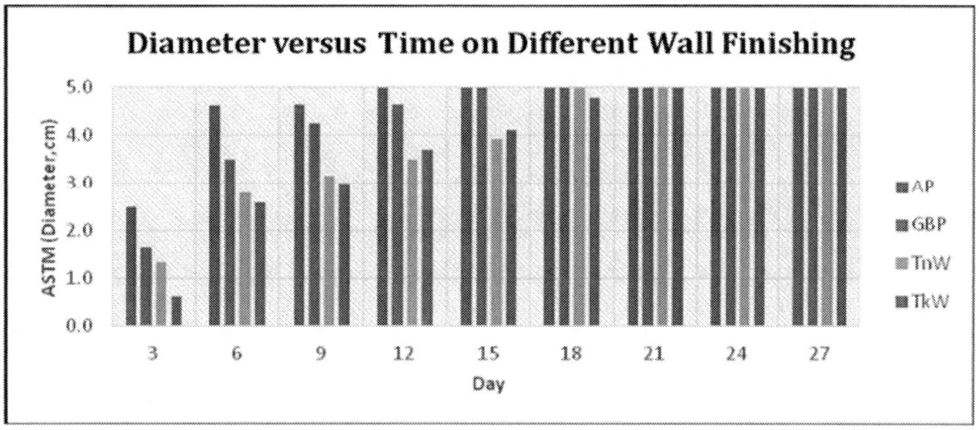

Figure 3. Diameter of fungi growth versus time.

4 Conclusions

The fastest growth of *Aspergillus niger* on wood based material with 4 types of wall finishing was found on acrylic paint, followed by glycerol based paint and thin wallpaper. The slowest fungi growth for the sample was found on thick wallpaper. It is suggested that antifungal agent should be added to avoid indoor fungi growth and to act as a protective layer for the wall finishing.

Acknowledgements

The authors would like to acknowledge Research and Innovation Centre (R&D) Research, Innovation, Commercialization & Consultancy Office (ORICC) Universiti Tun Hussein Onn Malaysia for funding this project under grant FRGS vot 1479.

References

[1] N. Bellotti, L. Salvatore, C. Deyá, M.T. Del Panno, B. del Amo and R. Romagnoli, The application of bioactive compounds from the food industry to control mold growth in indoor waterborne coatings, Colloids Surf. B. Biointerfaces, **104**, 140-144, (2013).

[2] C.M. Er, N.M. Sunar, A.M. Leman and N. Othman, Direct growth inhibition assay of total airborne fungi with application of biocide-treated malt extract agar, MethodsX, **2**, 340-344, (2015).

[3] S. Lee, G. Kwon, J. Joo, J.T. Kim and S. Kim, A finish material management system for indoor air quality of apartment buildings (FinIAQ), Energy and Buildings, **46**, 68-79, 2012.

[4] C.M. Er, N.M. Sunar, A.M. Leman and O. Norzila, Evaluation of potassium sorbate as a biocide to reduce viability of total airborne fungi in a higher educational building of computer studies, J. of Advanced Research in Materials Science, **2**(1), 15-19, (2014).

[5] C.M. Er, N.M. Sunar A.M. Leman, N. Othman, Q. Emparan, U.K. Parjo, P. Gani, N.A. Jamal and N.A. Ideris, The evaluation of indoor microbial air quality in two new commissioning higher educational buildings in Johor , Malaysia, Applied Mechanics and Materials, **774**, 1068-1072, (2015).

[6] J.A. Crawford, P.F. Rosenbaum, S.E. Anagnost, A. Hunt and J.L. Abraham, Indicators of airborne fungal concentrations in urban homes: Understanding the conditions that affect indoor fungal exposures, Sci. Total Environ., **517**, 113-124, (2015).

[7] C.M. Er, N.M. Sunar, A.M. Leman, N. Othman, U. Kalthsom, N.A. Jamal and N.A. Ideris, The biocidal effect of potassium sorbate for indoor airborne fungi remediation, Desalin. Water Treat., **57**(1), 1–6, (2015).

[8] U.K. Parjo, N.M. Sunar, A.M. Leman, N.A. Ideris, P. Gani, Q. Emparan and C.M. Er, Indoor fungal treatment by using potassium sorbate as bio-resistance coating for different plasterboard wall finishings, Applied Mechanics and Materials, **774**, 1116-1120, (2015).

[9] K. Parjo, N.M. Sunar, A.M. Leman and C.M. Er, Effect of fungal growth on the surface of painted plasterboards, Advances in Environmental Biology, **9**(20), 15-19, (2015).

[10] C.M. Er, N.M. Sunar, A.M. Leman, N. Othman, P. Gani, N.A. Jamal, Q.A. Emparan and U.K. Parjo, In vitro inhibitory assay of an isolated indoor airborne fungus from an institutional building of computer education by using potassium sorbate, Applied Mechanics and Materials, **774**, 1091-1095, (2015).

[11] X. Zhang, B. Sahlberg, G. Wieslander, C. Janson, T. Gislason and D. Norback, Dampness and moulds in workplace buildings: Associations with incidence and remission of sick building syndrome (SBS) and biomarkers of inflammation in a 10year follow-up study, Sci. Total Environ., **430**, 75-81, (2012).

[12] G. Ayanbimpe, W. Danjuma and M. Okolo, Relationship between fungal contamination of indoor air and health problems of some residents in Jos, InTech, 1-19, (2003).

[13] C.A. Clausen and V.W. Yang, Multi-component biocide protects wood from fungi and insects in UC2 applications, American Wood Protection Association, **91**, 31-35, (1998).

[14] U.K. Parjo, N.M. Sunar, A.M. Leman, P. Gani, Q. Emparan and C.M. Er, Coating bio-resistance test of different wall finishing for isolated indoor fungal treatment by using potassium sorbate biocide on wood, Applied Mechanics and Materials, **774**(2), 1384-1388, (2015).

[15] T. Verdier, M. Coutand, A. Bertron and C. Roques, A review of indoor microbial growth across building materials and sampling and analysis methods, Build. Environ., **80**, 136-149, (2014).

[16] S. Vacher, C. Hernandez, C. Bärtschi and N. Poussereau, Impact of paint and wall-paper on mould growth on plasterboards and aluminum, Build. Environ., **45**(4), 916–921, (2010).

[17] ASTM ASTM D5590-00, *Standard Test Method for Determining the Resistance of Paint Films and Related Coatings to Algal Defacement*, ASTM International, (2013).

MATEC Web of Conferences **47**, 05008 (2016)
DOI: 10.1051/matecconf/20164705008

Pollutants Characterization of Car Wash Wastewater

Nor Haslina Hashim[1,a] and Nadzirah Zayadi[2]

[1]*Faculty of Engineering Technology, Universiti Tun Hussein Onn Malaysia, 86400 Batu Pahat, Johor, Malaysia*
[2]*Faculty of Civil and Environmental Engineering, Universiti Tun Hussein Onn Malaysia, 86400 Batu Pahat, Johor, Malaysia*

Abstract. The huge quantity of water consumed per car during washing cars yields the untreated effluents discharged to the stormwater system. Wastewater samples from snow car wash and two full hand service car wash station were analyzed for pH and the presence of PO_4^{3-},TP, O&G, alkalinity, TSS, NO_3^-, NO_2^-, COD and surfactant in accordance Standard Method of Water and Wastewater 2012. Two full hand wash service stations and one station of snow foam service were investigated in this study. Amongst the stations, snow foam car wash station indicates the highest concentration of PO_4^{3-}, TP, O&G, TSS, COD and surfactant with the average value of 10.18 ± 0.87 mg/L, 30.93 ± 0.31 mg/L , 85.00 ± 0.64 mg/L 325.0 ± 0.6 mg/L, 485.0 ± 0.3 mg/L and 54.00 ± 2.50 mg/L as MBAS, respectively. Whereas, in parameters characterization in different stages throughout the car wash process, O&G was found to be the highest in pre soak stage, PO_4^{3-}, TP, TSS and COD in washing stage and NO_3^- and NO_2^- in rinse stage. All parameters were compared to Environmental Quality (Industrial Effluent) Regulations, 2009. There is a strong need to study on the characterization of car wash water in order to suggest the suitable treatment need for this type of wastewater.

1 Introduction

A car wash is defined as a non-domestic installation for external cleaning of cars [1]. Car wash generates high amount of wastewater approximately 150 to 350 litres per day being released in environment. Car washing in the street consists of oil and grease leaking during engine washing, high suspended solids from brake linings, with sand and dust being washed away and trace amounts of surfactant which are hardly degraded in environment. Commonly, parts of cars such as engine block, intake manifold, and brake drums have exceeded permissible limit discharges of heavy metal including iron, copper, zinc, and lead [2]. All the accumulated sediments will flow to the gutter system and go into storm water system without undergo treatment. Hence, endangered the ecosystem when washes away to the drainage system, streams and to the receiving waters, leads to the increased environment pollution [3]. Malaysia have their own environmental quality regulations, but the regulation is seldom enforced to the car wash industry. Generally, public perceived that the wastewater from car washing is not severely contaminated compared with other industrial wastewaters [4]. Hence, little attention is given to the car wash industries [5]. Presents, Malaysia has increased in car volumes on the road together with the existing number of cars. It has boosted the car wash industry, leading to the increase in car wash service stations, particularly in rapid development and

[a] Corresponding author : haslina@uthm.edu.my

high population of the focal point area. The aim of this study was to provide basic data required for the characteristics of car wash wastewater in Malaysia.

2 Sampling Sites

Three car wash stations with different type of washing were selected in two districts in Johor which are Parit Raja (a small town in Batu Pahat district) and Johor Bahru based on the geographical location and most common type of car wash industries in Malaysia. The type of washing, approximate number of washed cars and the volume of water used per washed car are tabulated in Table 1 below. All the information was obtained from the interviewing session with the car wash station owner. In full hand service, it required full labour intensive of cleaning, with desired amount of surfactant being applied with buckets full of water. While, in a snow foam car wash the cleaning processes required high amount foams of surfactant with high pressure trigger guns and hoses connected to a water pump (Figure 1). These three stations finally discharged the untreated wastewater to the drains downstream of a water bodies system (Figure 2).

Table 1. Operation time, number of washed car and volume of water used for car wash stations.

	Car Wash Station		
	1	**2**	**3**
Location	Parit Raja	Johor Bahru	Johor Bahru
Type	Full hand service	Full hand service	Snow Foam
Operation time per day (hours)	12	24	24
Approximate number of washed cars per day (unit)	50	100-150	100-150
Average volume of water used per car (L)	40	67	120

Figure 1. Snow car wash activity at Station 3.

Figure 2. Outlet drainage of carwash wastewater.

3 Sampling Method

Samples were collected for three car wash stations for 8 weeks. The samples were collected based on a composite samples collected when a car was being washed (approximately 1L per 5 minutes). Station 1 has better drainage system and different outlet for each washing processes, leads to the determination of pre-soak, wash and rinse stages. Pre-soak defined as the first process of car washes, where an amount of waters being applied during the exterior cleaning of cars before washing. The wash processes including of cleaning with surfactants was applied to wash away the dirts and grimes. Whereas, the rinse processes required washing proceses with clean water to remove the surfactants and dirts. The sample for each processes was differentiate and collected at the outlet point of drainage during the different stages were demonstrated. For Station 2 and Station 3, composite samples were obtained for three vehicles at same outlet, therefore the samples produced represent the whole pre-soak, wash and rinse stages. The sampling and sample preservation were based on Standard Methods for the Examination of Water and Wastewater (APHA, 2012) and stored at 4 °C until analysis.

4 Analysis

All parameters were measured in accordance Standard Method of Water and Wastewater 2012. pH was measured using HACH pH meter. Dionex ICS-2000 of Ion Chromatography system was used to analyse anions such as phosphate, PO_4^{3-}(4500-PE), total phosphorus, TP (4500-PE), nitrate, NO_3^- (4500 B) and nitrite,NO_2^- (4500 E) in filtered samples. Trace of total suspended solids (TSS) was analysed by filtration standard of APHA 2540 C. The organic compounds of chemical oxygen demand (COD) was followed APHA 5220 D procedure and digested using COD reactor of HACH DRB 200 for 2 hours before undergoes titration method on samples. Tests of oil and grease (O&G) concentration were measured according to modified APHA 5520B using extraction followed byevaporation method. The determination of alkalinity on raw samples were analysed using titration method of APHA 2320 B. The concentration of anionic surfactants experimented was modified from APHA 5540 C, Methylene Blue Absorbing Substances (MBAS). Cationic dye of Methylene blue (MB) formed complex with AR grade: Sodium Luryl Sulphate (SLS), EMORY of MW: 288.38 g/mol. The results gives the colour intensity of chloroform layer, measured by absorbance, at λmax = 652 nm. A series of SLS stock solution ranges within 0.5 mg/L - 5 mg/L were prepared, as a calibration curve graph. All the measurement of the absorbance were obtained from the spectrophotometer (Model: UV-VIS Scanning Spectrophotometer: Spectro UV-2650 by Labomed). All samples determination were done in triplicate homogenious sample.

5 Results and Discussion

Table 2 present the results of the characterization for the eight samples collected at each stations. Overall, station 3 reported the highest concentrations of PO_4^{3-}, TP, O&G, TSS and COD (on average, 10.180 ± 0.87 mg/L PO_4^{3-}, 30.933 ± 0.31 mg/L TP, 85 ± 0.64 mg/L O&G, 325 ± 0.6 mg/L TSS, 485 ± 0.33 mg/L COD and 54 ± 2.5 mg/L surfactant as MBAS compared to other stations. However, no determination of surfactant (as MBAS) was conducted in Station 2 due to unavailability of samples during the sampling time.

Surfactant was found in greater concentration (as MBAS) at Station 3 more than Station 1 due to the uses of greater nozzles hoses of foams that were applied during washing cars which yielded more foams containing surfactant. Historically, most car wash cleaning products based on detergents, containing mild alkalis [6]. However, as the global car wash industries grew, there had also been improvement on the ingredients compounds of the surfactants used to make cars more clean and shines forced more chemicals being used as the ingredients, resulting greater amount of pollutants being washed away during cleaning.

Table 2. Pollutant characteristics for accumulation processes for Station 1, Station 2 and Station 3 (n=8).

Parameter	Values	Unit	Station 1	Station 2	Station 3
pH	min	-	7.40	7.38	7.96
	max		7.88	7.90	8.33
	mean		7.73 ± 0.85	7.43 ± 0.37	8.20 ± 1.66
Phosphate (PO_4^{3-})	min	mg/L	0	0	8.35
	max		2.31	3.48	10.23
	mean		2.02 ± 1.4	3.40 ± 2.05	10.18± 0.87
Total phosphorus (TP)	min	mg/L	0	0	25.6
	max		7.09	10.66	33.70
	mean		7.05 ± 2.2	9.40 ± 1.55	30.93 ± 0.31
Oil and grease (O&G)	min	mg/L	55	78	79
	max		62	86	89
	mean		68.0 ± 0.4	80.0 ± 0.3	85.0 ± 0.6
Alkalinity (Alk)	min	mg/L	0	0	108
	max		23.12	34.77	300
	mean		21.53 ± 0.64	31.94 ± 1.22	293.00 ± 0.12
Nitrate (NO_3^-)	min	mg/L	0	0	2.16
	max		0.40	0.42	2.57
	mean		0.29 ± 0.58	0.36 ± 0.76	2.86 ± 0.25
Nitrite (NO_2^-)	min	mg/L	0	0	0
	max		0.15	0.19	0.01
	mean		0.13 ± 0.31	0.15 ± 0.47	0.01 ± 0.72
Total suspended solid (TSS)	min	mg/L	56	95	268
	max		89	104	333
	mean		85.00± 1.20	100.00 ± 0.62	325.00 ± 0.6.00
Chemical oxygen demand (COD)	min	mg/L	89	180	398
	max		210	266	490
	mean		190.0 ± 1.00	232.0 ± 0.6	485.0 ± 0.3
Surfactant (as MBAS)	min	mg/L	7.75	-	51.20
	max		33.16		60.50
	mean		17.90 ± 8.09		54.00 ± 2.50

For other parameters in Station 3, slightly alkaline pH of 8.20 ± 1.66, low concentration of NO_3^- (2.860 ± 0.25 mg/L) and NO_2^- (0.009 ± 0.72 mg/L) were within the typical ranges found in the industrial effluents of standard B. For comparison purposes, the mean concentration of contaminant levels reported for COD shows concentration within range of 75 mg/L - 738 mg/L [4]. The results revealed that the concentration of car washes for different car washes have greater COD, and most of the time exceeded the Standard B in environmental regulations (200 mg/L of COD) with Station 1 (190 ± 1.00 mg/L), Station 2 (232 ± 0.59 mg/L), and Station 3 (485 ± 0.33 mg/L). Generally, COD comes from the use of non biodegradable surfactant with dirts during car washing [4]. The permissible limit of Malaysia Effluent Discharge Standards of Standard B for O&G, is 10 mg/L. In this study O&G at all stations have exceeded the permissible limit in three stations, where Station 1 recorded (68 ± 0.42 mg/L), Station 2 (80 ± 0.31mg/L), and Station 3 (85 ± 0.64 mg/L), above the ranges of standard regulation. The results indicated the different types of car wash stations are not the determinant in the variation of oil and grease that was observed. The age of equipment, type and concentration of cleaning solutions, size of separation tanks, and time since last emptied could all have contributed to the differences observed in the O&G concentrations [7]. Whereas, the concentration of NO_3^- and NO_2^- were not considered as major pollutants to environments, since the ranges detected by all the car wash stations below the detection limit of standard B, 50 mg/L. The elevated concentrations of TSS for Station 2 (100 ± 0.62 mg/L TSS) and Station 3 (325 ± 0.6 mg/L TSS) were above the limit of 100 mg/L TSS regulated for standard B. The results of the higher TSS could be from the tire dust, rust stain, dirt, films from brakes, and grimes being washed away during cleaning processes.

Among the three car wash stations, Station 1 was the one that can be collected the samples and differentiate between stage processes of pre soak, wash and rinses. From Table 3, PO_4^{3-} (3.00 ± 1.22 mg/L) and TP (8.88 ± 2.0 mg/L) were higher during the washing process. The pollutants could be come from the dirt washed off vehicles, nature and concentration of contaminants with the chemicals and procedures used in the wash process during car wash activities [4, 5]. In this study, the TSS concentration for wash processes (90.00 ± 0.76 mg/L) was slightly near to the range of standard B, 100 mg/L. High amount of TSS were being washed away during washing process in removing the dirts and grimes on body parts and tire of cars. For COD, the values were complying with the regulation standard B for pre soak, wash and rinse processes. Based on the results of the characterization, the pollutant mean for O&G during pre soak (79.0 ± 1.10 mg/L), wash (50.0 ± 0.38 mg/L) and rinse (18.0 ± 1.26 mg/L) does not meet with standard B regulation. However, the higher concentration was generated by the pre soak process since it was washed away together with the crude oil and lubricating oil comes from motor oil of car wash stations or might the leaking of engine parts resulting high amount of O&G were detected [10].

Table 3. Pollutant characteristics for different stages processes for Station 1.

Parameter	Values	Unit	Pre-soak	Wash	Rinse
pH	min		7.48	7.80	7.96
	max		7.80	7.98	7.99
	mean		7.59 ± 0.55	7.81 ± 0.99	7.60 ± 0.53
Phosphate (PO_4^{3-})	min	mg/L	0.00	0.35	0.13
	max		0.89	3.55	2.60
	mean		0.41 ± 1.23	3.00 ± 1.22	2.00 ± 1.80
Total phosphorus (TP)	min	mg/L	0.00	0.98	0.93
	max		2.35	9.33	6.93
	mean		2.09 ± 1.32	8.88 ± 2.0	5.57 ± 1.08
Oil and grease (O&G)	min	mg/L	64.00	31.00	20.00
	max		96.00	59.00	29.00
	mean		79.00 ± 1.10	50.00 ± 0.38	18.00 ± 1.26
Alkalinity (Alk)	min	mg/L	0.00	18.80	9.21
	max		22.00	25.56	19.68
	mean		20.00 ± 1.53	23.00 ± 0.20	18.50 ± 1.31
Nitrate (NO_3^-)	min	mg/L	0.00	0.41	0
	max		0.40	1.59	1.61
	mean		0.30 ± 1.09	1.23 ± 1.22	1.45 ± 1.10
Nitrite (NO_2^-)	min	mg/L	0	5.8988	6.76
	max		0.009	6.146	7.230
	mean		0.002 ± 0.260	4.905 ± 0.640	5.981 ± 0.930
Total suspended solid (TSS)	min	mg/L	75.00	89.00	44.00
	max		97.00	100.00	52.00
	mean		80.99 ± 1.10	90.00 ± 0.76	36.00 ± 1.25
Chemical oxygen demand (COD)	min	mg/L	68.0	71.0	32.0
	max		126.0	192.0	75.0
	mean		100.0 ± 1.5	166.0 ± 1.8	48.50 ± 0.3

6 Conclusion

The car wash stations were analyzed regarding the characterization of car wash wastewaters with parameter of pH, PO_4^{3-}, TP, O&G, Alk, TSS, NO_3^-, NO_2^-, COD, and surfactant as MBAS. The characteristics of car wash wastewater discharges was dependent on the type of car washes. Snow foam car wash station, Station 3 presents the high concentration of pollutants due to the greater foam of surfactants. It yields high concentration of PO_4^{3-}, TP, O&G, TSS, and COD where the values were above the permissible limit of effluent standard B and not safe to be discharged to water body. The characterization between different phases of car washing processes including of pre soak, wash and rinse processes of Station 1 have been characterized. The high concentrations were obtained for PO_4^{3-}, TP, TSS and COD (wash), NO_3^- and NO_2^- in (pre soak), and O&G in (pre soak, wash and rinse) processes. All the parameters were following the permissible limit of standard B, except for the concentration of O&G.

Acknowledgement

This work is funded by University Tun Hussein Onn Malaysia of Postgraduate Incentive Research Grant (GIPS), Vot No. 1404.

References

[1] K. Boussu, C. Kindts, C. Vandecasteele and B.V.D. Bruggen, Applicability of nanofiltration in the carwash industry, Separation and Purification Technology, **54**, 139-146, (2007).
[2] B.S. Shete and N.P. Shinkar, Use of membrane to treat car wash wastewater, Int. J. of Research in Science and Advanced Technologies, **3**(1), 13-19, (2014).
[3] J. Perkowski, S. Bzdon, A. Bulska and W.K. Jozwiak, Decomposition of detergents present in car-wash sewage by titania photo-assisted oxidation, Polish J. of Environmental Studies, **15**(3), 457-465, (2006).
[4] W.J. Lau, A.F. Ismail and S. Firdaus, Car wash industry in Malaysia: Treatment of car wash effluent using ultrafiltration and nanofiltration membranes, Separation and Purification Technology, **104**, 26-31, (2013).
[5] F.K. Abagale, D.A. Sarpong, J.O. Ojediran, R. Osei-Agyemang, A.G. Shaibu and P.T. Birteeb, Heavy metal concentration in wastewater from car washing bays used for agriculture in the Tamale Metropolis, Ghana, Int. J. of Current Research, **5**(6), 1571-1576, (2013).
[6] H.C. Genuino, N.N. Opembe, E.C. Njagi, S. McClain and S.L. Suib, A Review of hydrofluoric acid and its use in the car wash industry, J. of Industrial and Engineering Chemistry, **18**, 1529-1539, (2012).
[7] C. Brown, Water effluent and solid waste characteristics in the professional car wash industry, Int. Car Wash Association, Toronto, (2002).
[8] A. Al-Odwani, A. Ahmed and S.B. Hamad, Carwash water reclamation in Kuwait, Desalination, **206**(1-3), 17-28, (2007).
[9] J. Oknich, *Handbook for the Perceived Environmental Impact of Car Washing*, Ramsey-Washington Metro Watershed District, Washington, (2002).
[10] R.S. Rao, C.P. Jyothi, R.S. Prakasham, P.N. Sarma and L.V. Rao, Xylitol production from corn fiber and sugarcane bagasse hydrolysates by Candida tropicalis, Bioresource Technology, **97**(15), 1974-1978, (2006).

MATEC Web of Conferences **47**, 05009 (2016)
DOI: 10.1051/matecconf/20164705009

Optimization of Sorption Parameters for Color Removal of Textile Dye by Cross-linked Chitosan Beads Using Box-Behnken Design

Ali H. Jawad[1,a], Nur Shazwani Abdul Mubarak[1] and W. I. Nawawi[1,2]

[1]*Faculty of Applied Sciences, Universiti Teknologi MARA, 40450 Shah Alam, Selangor, Malaysia*
[2]*Coal and Biomass Energy Research Group, Faculty of Applied Sciences, Universiti Teknologi MARA, 02600 Arau, Perlis, Malaysia*

Abstract. Cross-linked chitosan-epichlorohydrin beads (CEB) was prepared and studied for the removal of Reactive Red 4 (RR4) from textile industry. The effect of parameters such as adsorbent dosage (A: 0.5-1.5 g/L), pH (B: 4-10) and time of contact (C: 30-80 min) on the color removal of RR4 was investigated using response surface methodology (RSM) based on Box–Behnken surface statistical design at an initial RR4 concentration, Co = 100 mg/L as a fixed input parameter. The results of RSM indicated that the proposed model predicted the responses adequately within the limits of input parameters being used. Regression analysis showed good fit of the experimental data to the second-order polynomial model with coefficient of determination (R^2) value of 0.9958 and Fisher F-value of 185.36. Optimization of adsorbent dose (1.0 g/L), pH (4.0), and contact time (80 min) gave a maximum of 70.53% RR4 removal by CEB.

1 Introduction

Colored effluents from textile industries contain several types of synthetic dyestuffs [1]. These dyes are considered hazardous, toxic, and non-biodegradable and tend to suppress photosynthetic activity in aquatic systems by preventing the penetration of sunlight and oxygen. Therefore it posed a serious environmental problem and direct destruction to the aquatic system when discharged into water stream [2, 3]. Among the various treatments of effluents that contain dyes, the adsorption method is the best alternative and has been widely used to remove pollutants from effluents [4] because of its high performance, flexibility and simplicity of design, convenience of operation, and selectivity [5]. Adsorption also does not result in formation of harmful by-products [6]. Chitosan is an abundant cationic biopolymer produced through a deacetylation process of the naturally occurring biopolymer chitin [7]. Chitosan is an excellent material compared to other polyssacharides due to the presence of active adsorption sites such as amino ($-NH_2$) and hydroxyl (–OH) groups. Unfortunately, drawbacks of natural chitosan such as low mechanical properties, high swelling and shrinkage index, compressible at high operating pressure, and soluble in many organic acids limits the adsorption process [8]. Therefore, chemical modification through a cross-linking process is an important step to enhance the mechanical resistance, stabilize chitosan at low drastic pH and increase the adsorption

[a] Corresponding author : ali288@perlis.uitm.edu.my

abilities of chitosan [9]. The application of chitosan and its cross-linked derivatives in the form of flakes and/or powder is significantly limited in scaling up the adsorption system for industrial applications due to its small size particle that leads to serious clogging effect [10]. Therefore formation of chitosan beads together with crosslinking reaction can enhance the diffusion mechanism and improve the access to internal adsorption sites.

The application of experimental design such as response surface methodology (RSM) was employed which result in improved decolorization and the most importantly can reduce the process variability, cost and time consuming [11]. In addition RSM can be applied in different types of disciplined and provides a maximum number of information with a minimal number of experiments. Moreover, the classical way of changing one variable at a time and keeping the others constant are incapable of revealing the optimal combination of the parameters involved [12]. Thus, the present study aims to prepare cross-linked chitosan-epichlorohydrine beads (CEB) as a promising synthetic biopolymer for removal of reactive dyes from polluted water by using three level factorial of Box–Behnken design (BBD). The effect of operational parameters such as adsorbent dosage, pH and contact time of CEB for the removal of RR4 dye were evaluated and optimized. Reactive red 4 (RR4) was chosen as a model pollutant because it facilitates electrostatic interaction between polymer chain and the negatively charged anionic dye.

2 Material and Methods

Crosslinked chitosan-epichlorohydrine beads (CEB) of particles size (<250 μm) was prepared according to the same procedure described previously [13]. Reactive Red 4 (RR4) or otherwise commonly known as Cibacron Brilliant Red (Colour Index Number: 18105, chemical formula: $C_{32}H_{23}ClN_8Na_4O_{14}S_4$, MW: 995.23 gmol^{-1}, and with 50% dye content was chosen as adsorbate. The stock solution of RR4 was prepared (1000 mg/L), and the desired concentration can be obtained by diluting stock solution with ultrapure water (18.2 MΩ cm^{-1}). Initial pH of the RR4 solution was adjusted with 0.1 mol/L HCl or 0.1 mol/L NaOH using a Metrohm, 827 pH lab. Colour removal was determined by using a direct reading spectrophotometer (DR 2800 HACH, USA) at λ_{max} 517 nm. The percentage of RR4 colour removal in the aqueous solutions by CEB was calculated by the following Equation (1):

$$\text{Color removal (CR\%)} = (C_o\text{-}C_t/C_o) \times 100 \quad (1)$$

where C_o and C_t both in (mg/ L) are, respectively, the initial and remaining concentration of RR4 in solution at time t (min).

In the present study, the 3-level, 3-factor Box–Behnken experimental design is applied to investigate and validate adsorption process parameters affecting the removal of RR4 dye by CEB. Adsorbent dose (A: 0.5-1.5 g/L), pH (B: 4-10) and time of contact (C: 30-80 min), while dye concentration of 100 mg/L was kept as a constant input parameter. A total of 17 experiments have been conducted in this work to evaluate the effects of the three main independent parameters on the removal of RR4 dye. The actual experimental design matrix is given in Table 1. The results were analyzed using the coefficient of determination (R^2), Pareto analysis of variance (ANOVA) and statistical and response plots.

3 Results and Discussion

The statistical significance of the BBD model was assessed by analysis of variance (ANOVA). The results of second-order response surface model in the form of analysis of variance (ANOVA) are shown in Table 2. Values of model terms Prob> F <0.0500 designate that factors are significant under selected conditions. Significant model terms for response (RR4 removal %) are A, B, C, AB, AC, A^2, B^2, and C^2, whereas another remaining term BC had less significance for the response and can be neglected to improve the model and thus the equation developed is expressed by the following Equation (2):

$$CR\,\% = +51.69 +10.13A -13.80B +10.78C -5.15AB +2.93AC -12.85A^2 -6.20B^2 +1.90C^2 \qquad (2)$$

The goodness of fit of the model was also tested by the multiple correlation coefficients (R^2). In this case, the value of the multiple correlation coefficients was 0.9958, which implied that this regression is statistically significant. This can be observed in Figure 1 by comparing the actual values against the predicted responses by the model for the percentage of RR4 removal. The value of predicted multiple correlation coefficient (pred. R^2 = 0.8778) is in reasonable agreement with the value of the adjusted multiple correlation coefficient (adj. R^2 = 0.9527). The individual effect of the studied variables including adsorbent dosage (A), pH (B), contact time (C) and color removal efficiency (%) was investigated by perturbation plots as shown in Figure 2. The steep curvature in A curve indicates that the color removal was sensitive to this variable. The relatively sharp line of C shows the high sensitivity to change in operation time levels. The pH curve shows a slow curvature indicating that this factor has slight effect on the response. From Figure 2, it can be seen that the RR4 dye color removal efficiency increases as the adsorbent dosage increases and decreases with increasing pH.

Table 1. The 3-factors BBD matrix and the value of the response function (color removal, CR (%)).

Run	Adsorbent dose (g/L)	pH	Time (min)	CR%	
				Actual	Predicted
1	0.5	4	55	31.28	31.16
2	1.5	4	55	60.19	61.72
3	0.5	10	55	15.40	13.87
4	1.5	10	55	23.69	23.81
5	0.5	7	30	20.95	22.76
6	1.5	7	30	37.00	37.16
7	0.5	7	80	38.63	38.47
8	1.5	7	80	66.39	64.58
9	1.0	4	30	53.28	51.59
10	1.0	10	30	21.92	21.64
11	1.0	4	80	70.53	70.81
12	1.0	10	80	43.85	45.54
13	1.0	7	55	51.08	51.69
14	1.0	7	55	51.99	51.69
15	1.0	7	55	52.26	51.69
16	1.0	7	55	51.72	51.69
17	1.0	7	55	51.40	51.69

Table 2. Analysis of variance (ANOVA) for response surface quadratic models.

Source	Sum of Squares	df	Mean Square	F Value	p-value Prob > F
Model	4318.32	9	479.81	185.36	< 0.0001
A-Adsorbent dose	820.33	1	820.33	316.91	< 0.0001
B-pH	1524.07	1	1524	588.79	< 0.0001
C-Time	929.88	1	929.88	359.24	< 0.0001
Adsorbent dose x pH	106.3	1	106.3	41.06	0.0004
Adsorbent dose x Time	34.28	1	34.28	13.24	0.0083
pH x Time	5.48	1	5.48	2.12	0.1892
Adsorbent dose^2	695.39	1	695.39	286.65	<0.0001
pH^2	161.79	1	161.79	62.5	<0.0001
Time^2	15.56	1	15.56	5.9	0.0456
Residual	18.12	7	2.59		
Core Total	4336.44	16			

ANOVA results for the response parameter (Table 2) indicated that the maximum RR4 removal (%) can be obtained under following interaction terms: AB and AC. Figures 3 and 4 show the three dimensional (3-D) and contour plots of the response at the optimal interaction terms of AB and AC respectively, and simultaneously keeping other two variables constant at central level. The combined effect of adsorbent dose (A) and pH (B) on RR4 removal efficiency is shown in 3-D plot of Figure 3. In fact, adsorbent dose reflects the capability of the CEB adsorbent to uptake the RR4 molecules from a given solution, while initial solution pH is responsible for surface charge of adsorbent, hence their interaction effect is influential in the model.

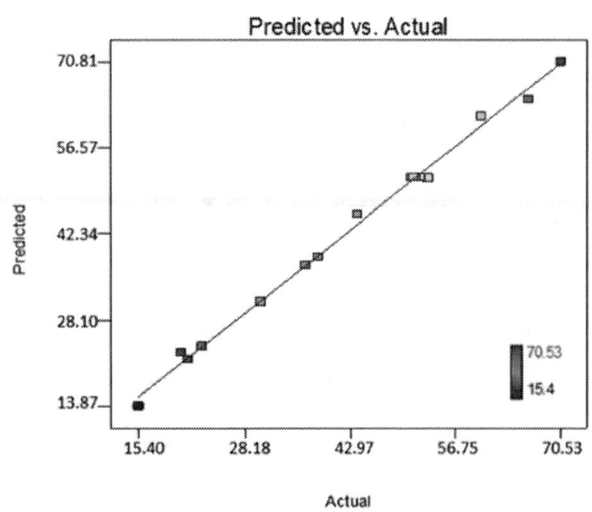

Figure 1. Comparison of actual results of RR4 removal % with the predicted values by BBD model.

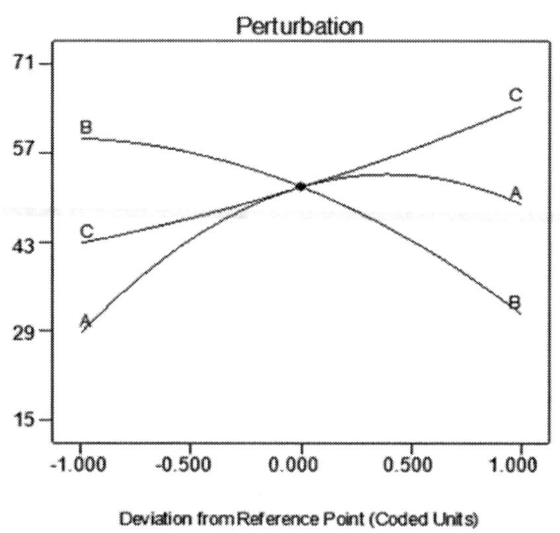

Figure 2. Perturbation plots for the removal of RR4 using CEB.

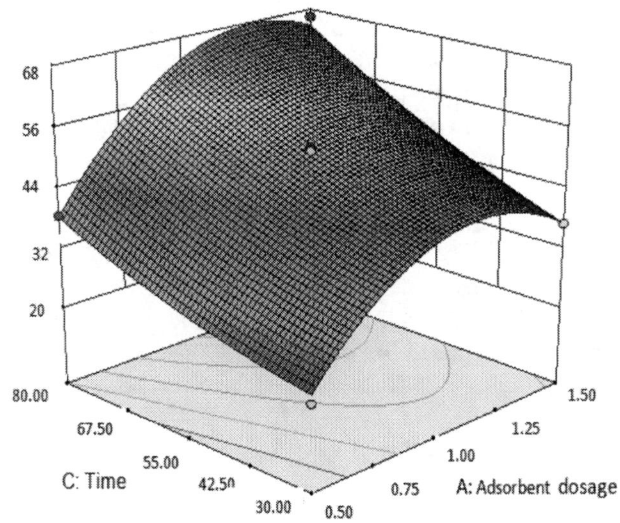

Figure 3. The 3-D response surface of the RR4 removal efficiency as the function of adsorbent dose and pH $[RR4]_o$ = 100 mg/L and agitation time = 55.00 min).

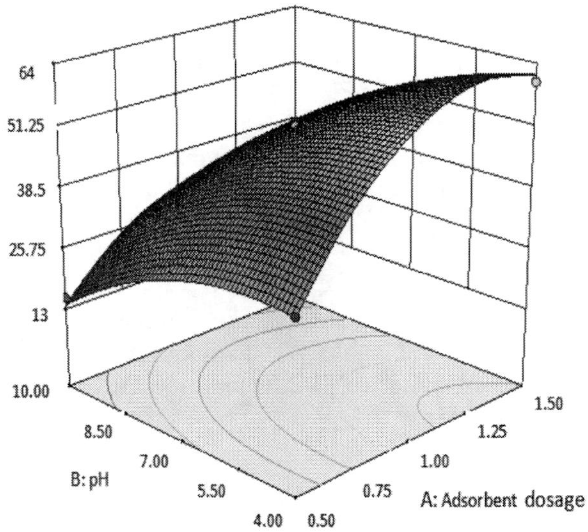

Figure 4. The 3-D response surface of the RR4 removal efficiency as the function of adsorbent dose and agitation time $[RR4]_o$ = 100 mg/L and pH = 7.00).

As for pH, the explanation of this phenomenon is that the hydrogen atom (H^+) available in the solution could protonate the amine groups ($-NH_2$) of chitosan. Consequently, the surface of the CEB would be positively charged, especially in acidic environment at pH lower than 6.5, as expressed by Equation (3).

$$R\text{–}NH_2 + H^+ \leftrightarrow R\text{–}NH_3^+ \quad (pH < 6.5) \tag{3}$$

In addition, the intensity of the positive charge on the chitosan polymeric surface can be increased by decreasing the pH value and making the sorption process pH sensitive [14]. The combined effect of (A) and (C) on RR4 removal efficiency is shown in 3-D of Figure 4. Time of contact offers better contact between the RR4 and CEB. As can be seen, the RR4 removal efficiency increases up to 70.53 % by increasing time of contact and CEB dose simultaneously. However, it was also observed that there is no remarkable change in RR4 removal after reaching 1g/L CEB adsorbent dose even with longer contact time.

4 Conclusion

In this work, the feasibility of color removal of RR4 onto CEB as a biosorbent was investigated. RSM and BBD were appreciable in determining the optimal conditions for color removal such as adsorbent dose, pH and contact time. The results obtained indicated that maximum RR4 color removal of 70.53% by CEB can be achieved at adsorbent dose (1.0 g/L), pH (4.0), and contact time (80.0 min).

References

[1] B. Balci, O. Keskinkan, M. Avci, Expert Systems with Applications, **38**, 949–956, (2011).

[2] R. Huang and Q. Liu, Arabian J. Chem, (2013), online available 17.05.2013: http://dx.doi.org/10.1016/j.arabjc.2013.05.017

[3] L.Wang, J. Li, Ind. Crop Prod., **42**, 153–158, (2013).
[4] M. Vakili, M. Rafatullah, B. Salamatinia, A.Z. Abdullah, M.H. Ibrahim, K.B. Tan, Carbohydr Polym., **113**, 115-130, (2014).
[5] M. Rafatullah, O. Sulaiman, R. Hashim and A. Ahmad, J. Hazar. Mater., **177**, 70–80, (2010).
[6] C. Crini and P.M. Badot, Prog. Polym. Sci., 33399-447, (2008).
[7] A.C. Rodrigues, M. Boroski and N.S. Shimada, J. Photochem. Photobiol., A **194**, 1-10, (2008).
[8] W.S.W. Ngah, C.S. Endud and R. Mayanar, React. Funct. Polym., **50**,181-190, (2002).
[9] J. Wang and C. Chen, Bioresour. Technol., **160**, 129-141, (2014).
[10] K. Papadopoulou, I.M Kalagona, A. Philippoussis and F. Rigas, Int. Biodet and Biodeg., **77**,31-38, (2013).
[11] P. Sharma, H. Kaur and M. Sharma, Environ. Monit. Assess., **183**, 151–195, (2011).
[12] A.A. Oladipo, M. Gazi, E. Yilmaz, Chem. Eng. Res. Design, **104**, 264–279, (2015).
[13] W.S.W. Ngah, S.A. Ghani and A. Kamari, Bioresour. Technol., **96**, 443-450, (2005).
[14] S. Chatterjee, B.P. Chatterjee, A.R. Das and A.K. Guha, J. Colloid Interface Sci., **288**, 30–35, (2005).

MATEC Web of Conferences **47**, 05010 (2016)
DOI: 10.1051/matecconf/20164705010

Pleurotus Spent Mushroom Compost as Green Supplementary Nutrient in Tissue Culture

Chia Chay Tay[1,2,a], Mohamad Shariman Shaari[1], Wan Nurul Hidayah Wan Anuar[3] and Saiyidah Nafisah Hashim[3]

[1]*Faculty of Applied Sciences, University Teknologi MARA, 02600 Arau, Perlis, Malaysia*
[2]*Faculty of Civil Engineering, Universiti Teknologi MARA, 40450 Shah Alam, Malaysia*
[3]*Faculty of Applied Sciences, Universiti Teknologi MARA, 35400 Tapah, Perak, Malaysia*

Abstract. This study investigates the effect of *Pleurotus* spent mushroom compost as green supplementary nutrient in tissue culture. Different concentration of *Pleurotus* spent mushroom compost in powder (SMC) and activated carbon (SMCAC) were added into Murashige and Skoog media (MS) with 0.1 mg/L 6-benzylaminopurine (BAP). The shoot regeneration of *Clinacantus nutans* plant such as number of shoots, number of leaves and length of leaves were recorded for four weeks. Collected data was analyzed using one way analysis of variance (ANOVA) and Tukey test through the SPSS Statistics 21 software. The 0.1 g/L SMC recorded the highest leaves numbers. Meanwhile, 0.2 g/L SMC showed highest performance in length of shoot. The SMCAC showed a negative response in number of leaves and shoot as well as length of shoot. In summary, the SMC showed significant performance in number of leave and shoot length but less effective on number of shoot. In contradictory, the SMCAC reported poor performance in shoot regeneration but contribute in absorption of nutrient from environment and storage of the nutrient asfunction of biochar. Therefore, the SMC has a high potential as a green supplementary nutrient for tissue culture. The application of this material has contributes into the green technology via convert waste to product, which is in-line with zero waste concept.

1 Introduction

Tissue culture technique improves shoot regeneration of plants' growth with high proliferation rates when compared to on the soil. But, some of plants that grow in the culture media still giving low shoot regeneration rate because of low absorption of nutrients [1]. In order to increase shoots proliferation, the supplementary nutrient is added in culture media in order to reduce the use of exogenous hormones, improve performance and minimize the cost [2, 3]. Generally, the supplementary nutrient is based on synthetic chemicals [4, 5]. On the other hand, agricultural waste which has high organic compounds has been neglectedin tissue culture application [6].

Pleurotus ostreatus spent mushroom compost (SMC) is an agriculture waste that obtained from mushroom cultivation farm. For traditional on soil stem cutting study, SMC supplied nutrients to the crops and thus potential to replace inorganic fertilizer [7]. However, the activated carbon showed no detectable relationship between the activated carbon and the productivity of plant, but promised a win-win solution to energy, carbon storage and ecosystem function [8]. To authors' knowledge, there is

[a] Corresponding author : taychiay@gmail.com

lack of study in SMC and SMCAC as supplementary nutrient in tissue culture and comparison of performance.

Clinacantus nutans plant has been chosen in this study due to its medicinal properties and high demand in market [9]. The traditional on soil stem cutting technique hardly supply consistent quality and quantity ofplant. Therefore, this study focuses on the effects of SMC and SMCAC in culture medium as supplementary nutrient in tissue culture media to the *C. nutans*. Moreover, this study compares their shoot regeneration performance on the *C. nutans* through statistical analysis.

2 Methodology

2.1 Samples collection

The sample of *Pleurotus* spent mushroom compost was collected from C & C Mushroom Cultivation Farm Sdn. Bhd, Johor, Malaysia. Meanwhile, the stems of *C. nutans* as tissue culture plant were taken from plant nursery in Arau and surface sterilized by using 30% Clorox and 70% ethanol prior to subculture and plant regeneration study.

2.2 Shoot regeneration study

Nodal segments of *C. nutans* explants were cultured on full strength MS media supplemented with SMC at 0.05, 0.1, 0.2 g/L and SMCAC at 0.3, 1.25, 2.5 g/L, respectively. The media contained standardized 0.1 mg/L BAP hormone in order to induce plant regeneration. The culture jars have been incubated in growth room at adjusted temperature of 26 ± 1 °C and under initial photoperiod of 16 hours light and 8 hours dark period. The number of shoots, number of leaves and length of shoots were observed and recorded every week for four weeks. The experiment was repeated with two nodal segments of *C. nutans* per jam jar.

2.3 Statistical analysis

Data was analyzed using SPSS Statistics 21 software for Windows 7. The values were expressed in mean ± standard error and analyzed by one-way analysis of variance (ANOVA) and tukey test.

3 Results and Discussion

Effects of *Pleurotus* spent mushroom compost (SMC) and *Pleurotus* spent mushroom compost activated carbon (SMCAC) on *C. nutans* shoot regeneration were examined through number of leaves, number of shoot and length of shoot. In addition, comparison of them were evaluated through statistical analysis.

3.1 Effect of SMC

Figure 1(a) and (b) illustrate the effects of SMC and SMCAC on *C. nutans* shoot proliferation through number of leaves and shoots as well as shoot length. In shoot proliferation, the 0.1 g/L SMC recorded the highest leaves numbers of 7 compared to 5 in other treatments. While, similar number of shoot of 2 - 3 was recorded for all treatments. Briefly, SMC enhanced number of shoot but not showed significant different performance in shoot proliferation for number of leaves and shoots amongst treatments. This revealed that the addition of SMC increased the pH values, salt contents, macro and micronutrient concentrations, thus improved shoot regeneration [10]. Wang et al. [11] supported this observation as reported that *Agaricus bisporus* SMC increased number of leaves of green cauliflower and cabbage in soil.

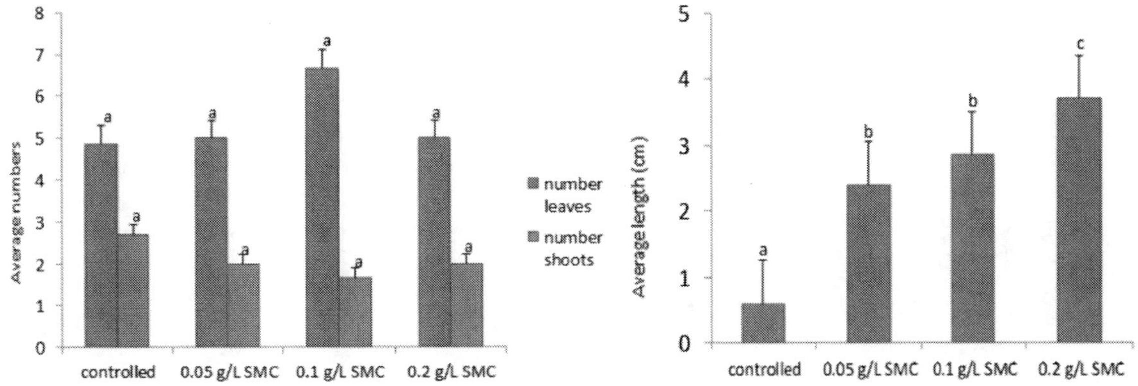

Figure 1. Effects of different concentrations of SMC on the growth performance of *C. nutans* (a) number of leaves and shoots (b) length of shoot.

For shoot length in *C. nutans*, SMC at 0.2 g/L exhibited the highest length of shoots (Figure 1 (b)). The SMC showed significant effect on length of shoot among control, 0.05-0.1 g/L SMC and 0.2 mg/L SMC. Basically, the SMC serves as organic soil amendments to improve plant growth in agriculture [12]. Thus, the SMC improved the shoots length of *C. nutans* and directly increased the height of the plant. Increase in shoot length contributes to quantity production through higher yield of leaves obtained in order to meet market demand in a shorter time with consistent quality of culture. Similar positive observation also reported by Jonathan et al. [13] on the effects of SMC as soil conditioner for the growth of soybean plant.

3.2 Effect of SMCAC

The shoots regeneration patterns of *C. nutans* nodes segments in different concentration of SMCAC are shown in Figure 2 (a) and (b). Generally, the explants were survived in 0.3-2.5 g/L of SMCAC but showed slower in shoot regeneration than control treatment. Control and treatments of SMCAC showed significantly different in number of leaves and shoots. Meanwhile, there is no significant difference amongst treatments and control statistically for shoot length. This is because activated carbon absorbed high concentration of nutrients and growth hormone in both liquid and solid media [14]. Generally, activated carbon, a stable and recalcitrant carbon, could function as biochar which is reported as slow sequestration in carbon source [15]. Thus, such circumstances of fast absorption and slow in released of nutrients caused slower cell division which led to slower shoot regeneration. This observation was supported by Zimmerman et al. [16] who reported that the soil organic matter stimulated the co-mineralization of the labile components of biochar over the short term, while over the long term, biochar-soil interaction enhanced carbon storage through the process of organic matter sorption to biochar and physical protection.

3.3 Comparison study

The different growth performance of *C. nutans* using SMC and SMCAC at different concentrations compared to control were demonstrated in Figure 3 (a) and (b). A 0.1 mg/L SMC showed the highest in number of leaves and significantly different statistically with control and other SMC treatments as well as SMCAC treatments. Generally, a decrease in number of shoots when adding SMC and SMCAC. The controlled treatment showed the highest shoots number among other treatments (Figure 3 (a)).

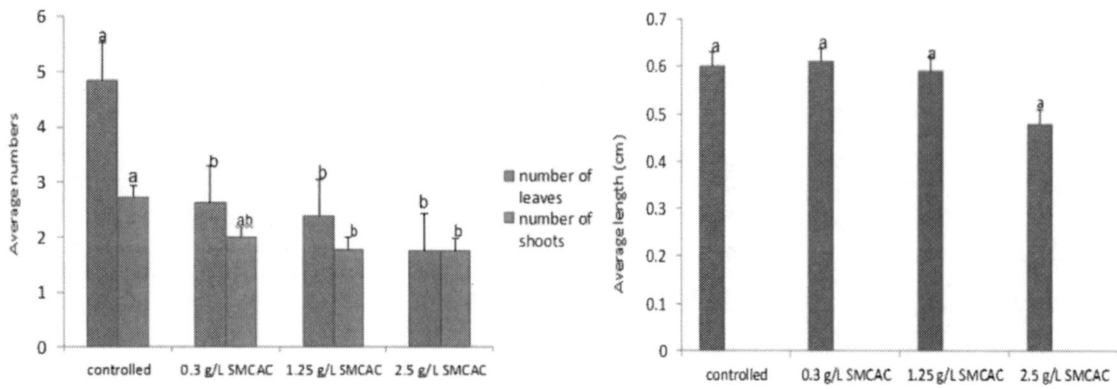

(a) (b)

Figure 2. Effects of different concentrations of SMCAC on the growth performance of *C. nutans* (a) number of leaves and shoots (b) length of shoot.

Overall, the control showed significantly different with SMC and SMCAC in number of shoot. The explants in 0.2 g/L of SMC showed the highest length of shoots with increment of two times when compared to the control and SMCAC treatments (Figure 3 (b)). It can be summarized that SMC is an excellent candidate as supplementary nutrient in tissue culture technique for *C. nutans* as it exhibited highest and significant performance in number of leaves and length of shoots compared to other treatments. The application of SMC in tissue culture technique not only significantly improves plant shoot regeneration and plant growth but also convert agricultural waste to product which is in-line with green technology development and zero waste concept.

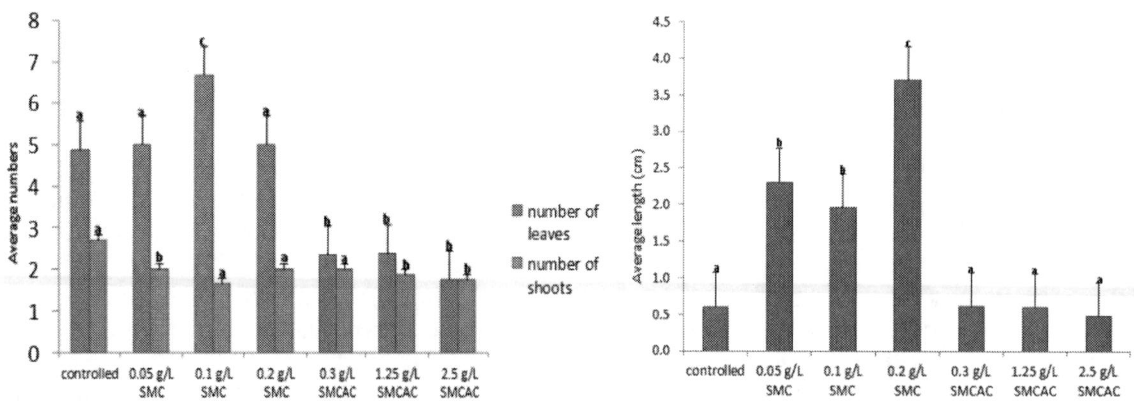

(a) (b)

Figure 3. Effects of different concentrations of SMC and SMCAC on the growth performance of *C. nutans* (a) number of leaves and shoots (b) length of shoot.

4 Conclusion

The potential of SMC and SMCAC as green supplementary nutrient in tissue culture were evaluated. This study exhibit that SMC was found to be increased number of leaves and length of shoot of *C. nutans*. The highest number of leaves and length of shoots were recorded at 0.1 /L SMC and 0.2 g/L SMC, respectively. But, SMC showed a reverse relationship for number of shoots with increase of SMC concentration. On the other hand, SMCAC revealed a negative performance in shoot regeneration in terms of number of leaves, number of shoots and length of shoots. In conclusion, SMC is a potential green supplementary nutrient in tissue culture with significant performance in number of leaves and length of shoots. Meanwhile, SMCAC was found to be a good absorbent as potential activated carbon functions as biochar for nutrient adsorption from environment. The application of SMC and SMCAC not only develop green biotechnology but also apply zero waste concept by convert waste to product.

References

[1] A. Ebert, F. Taylor and J. Blake, Changes of 6-benzylaminopurine and 2, 4-Ddichlorophenoxyacetic acid concentrations in plant tissue culture media in the presence of activated charcoal, Plant Cell Tiss. Org., **33**,157-162, (1993).

[2] M. La Habi, B. Prasetya, S. Prijono and Z. Kusuma, The effect of sago pith waste granule compost and inorganic fertilizer on soil physical characteristics and corn (Zea mays L.) production in inceptisol, IOSR J. of Environmental Science, Toxicology and Food Technology, **8**, 32-40, (2014).

[3] N.S. Karam, K.I. Ereifej, R.A. Shibli, H. AbuKudais, A. Alkofahi and Y. Malkawi, Metals concentration, growth, and yield of potatoproduced from in vitro plantlets or microtubers and grown in municipal solid-waste-amended substrates, J. Plant Nutr., **21**, 725-739, (1998).

[4] J. Singh, F. Sabir, R.S. Sangwan, L.K. Narnoliya, S. Saxena and N.S. Sangwan, Enhanced secondary metabolite production and pathway gene expression by leaf explants-induced direct root morphotypes are regulated by combination of growth regulators and culture conditionsin centella asiatica (L.) urban, Plant Growth Regul., **75**, 55-66, (2015).

[5] F. Trevisan, E.P. Chu, M. Gaspar and M.A.M. Carvalho, In vitro culture and fructan production by Vernonia herbacea (Asteraceae), Acta Physiol. Plant, **36**, 2299-307, (2014).

[6] R. Us-Camas, G. Rivera-Solis, F, Duarte-Ake and C. De-la-Peria, In vitro culture: an epigenetic challenge for plants, Plant Cell Tiss. Cult., **118**, 187-201, (2014).

[7] D.P.C. Stewart, K.C. Cameron, I.S. Cornforth and J.R. Sedcole, Effects of spent mushroom substrate on soil physical conditions and plant growth in an intensive horticulture system, Aust. J. Soil Res., **36**, 899-912, (1998).

[8] L.A. Biederman and W.S. Harpole, Biochar and its effects on plant productivity and nutrient cycling: a meta-analysis, GCB Bioenergy, **5**, 202-214,(2013).

[9] Y.K. Yong, J.J. Tan, S.S. The, S.H. Mah, G.C.L. Ee, H.S. Chiong and Z. Ahmad, Clinacanthus nutans extracts are antioxidant with antipoliferative effect on cultured human cancer cell lines, Evid. Based Complementary Altern.Med., **2013**, 1-8, (2013).

[10] E. Medina, C. Paredes, M.D. Perez-Murcia, M.A. Bustamante and R. Moral, Spent mushroom substrate as component of growing media for germination and growth of hoticultureal plants, Bioresour. Technol., **100**, 4227-4232, (2009).

[11] Y.J. Wang, R.X. Chen, D.Q. Lu, P.J. Zhang and B.Q. Gu, Recycle experiment on flower and vegetable with spent mushroom substrate, Edible Fungi, **1**, 45-46, (2006).

[12] R.I. Lopez-Castro, S. Delmastro and N.R. Curvetto, Spent oyster mushroom substrate in a mix with organic soil for plant pot cultivation, Micol. Aplicada Int., **20**, 17-26, (2008).

[13] S.G. Jonathan, O.J. Oyetunji, O.J. Olawuyi and P.O. Uwukhor, Application of Pleurotus ostreatus SMC as soil conditioner for the growth of soybean (Glycine max), Academia Arena, **5**, 55-61, (2013).

[14] M.J. Pan and J. Van-Staden, The use of charcoal in vitro culture – a review, Plant Growth Regul., **26** ,155-163, (1998).

[15] K.A. Spokas, K.B. Cantrell, J.M. Novak, D.W. Archer, J.A. Ippolito, H.P. Collins and A.A. Nichols, Biochar: A synthesis of its agronomic impact beyond carbon sequestration, J. Environ. Qual., **41**, 973-989, (2012).

[16] A.R. Zimmerman, B. Gao and M. Ahn, Positive and negative carbon mineralization priming effects among a variety of biochar-amended soils, Soil Biol. Biochem., **43**, 1169-1179, (2011).

MATEC Web of Conferences **47**, 05011 (2016)
DOI: 10.1051/matecconf/20164705011

The Strength, Leachabilty and Microstructure Characteristics of Chromium Contaminated Stabilised Sediment

Mohammed Kabir Aliyu [1,a], Ahmad Tarmizi Abd Karim[1] and Chan Chee Ming[1]

[1]*Faculty of Civil and Environmental Engineering, Universiti Tun Hussein Onn Malaysia, 86400 Parit Raja, Johor, Malaysia*

Abstract. The effect of replacing cement with rice husk ash (RHA) on compressive strength and leachability of Chromium stabilised sediments was investigated. Artificially contaminated sediment was prepared by spiking sediment with solution of Potassium dichromate (K2Cr2O7) to achieve an average of 1000 ppm target concentration. Cement was added at 10% and rice husk ash at 5%, 10%, 15% and 20% to the total dry weight of the mixture, which were then cured at room temperature (27 ± 3°C) and humidity of 75 ± 5 % for 7, 14 and 28 days. The addition of RHA with cement to the spiked sediment increased the compressive strength (UCS) of solidified sample with increase in curing period. The reaction products and crystalline mineral phases were identified using X-ray diffraction (XRD) which shows the formation of various crystalline phases of the treated sediment after 28 days which explain the mechanisms responsible for immobilization of chromium under study. The UCS values of solidified samples at 7, 14 and 28 days exceeded the minimum landfill disposal limits of 0.34N/mm2 (340 kPa). It was observed that leaching of Cr (VI) was higher in acetic acid medium (TCLP) as compared to deionized water (DIW) medium. The leached chromium concentrations in the two different leachates (acetic acid, and deionized water) were all below the regulatory limits of 5 mg/l of U.S EPA. Results have indicated that the partial replacement of cement with RHA in the binder system has increased the strength and reduced leachability of chromium in the treated compared to untreated sediment samples.

1 Introduction

Solidification/Stabilisation (S/S) is a treatment technique by which contaminated soils, sediments or waste are mixed with a binder and/or specific additives with the aim of decreasing the mobility of the toxic contaminants by increasing the pH and fully or partially binding the contaminants in the solid matrix (stabilisation), but also of improving the physical properties (strength, compressibility, permeability and durability) of the final treatment products (solidification) [1].

The wide spread use of chromium has resulted in the contamination of soils and water. Chromium contamination is of great concern due to its toxic, mutagenic, and carcinogenic nature. The chromium is generated from steel and other alloy's production, pigments, chrome plating and leather tanning industries. Among the various forms of chromium, Cr (VI) is the form which is commonly found at contaminated sites and most important one because of its toxicity, solubility, and mobility

[a] Corresponding author : hf110153@siswa.uthm.edu.my

characteristics [2, 3]. It can cause skin rashes, nose irritation, stomach upsets, respiratory problems, weakened immune systems, kidney and liver damage, lung cancer and sometimes even death.

X-ray diffraction (XRD) has been found to be a useful application to characterize the crystalline phases of the solidified wastes [4, 5]. Information from such studies can be used successfully to obtain the immobilization mechanisms and reactions that occur in an S/S process [4, 6]. At present, Toxicity characteristic leaching procedure (TCLP) is used by the US EPA to evaluate whether a particular S/S process is effective in treating a given waste. The TCLP limit for chromium is 5 mg/l. Factors such as waste form, binder system, curing time and cement-to-waste ratio have been identified to affect the leachability of a waste after an S/S treatment [4, 7-9].

2. Experimental Program

2.1 Materials

TiO_2 The main source of data for this study are sediments taken from Sembrong river at a site located between geographical coordinates of latitude 1° 52'.18.44" N and longitude 103° 06' 15.71" E, at Parit Sempadan, Parit Raja near Universiti Tun Hussein Onn Malaysia (UTHM). Holcim Top standard cement was used for this study. This type of cement is one of the cement that its quality complies with the Malaysian standard MS 522:2007. This type of cement is usually used as the main binder for ordinary construction and is readily available and also is environmentally friendly due to the reduced CO_2 emission when it is produced. The rice husk used was burnt at a controlled temperature of 700°C in the furnace for a period of 6 hours, with a heating rate of 5°C /min and then it was left overnight to cool. The burnt ash was then ground to produce ash size less than 75μm.

2.2 Sediment contamination procedure

The natural sediment contains very low (below detection limits) concentration of chromium. Therefore to obtain synthetic contaminated sediment, the dried sediment samples was spiked with potassium dichromate ($K_2Cr_2O_7$) to obtain approximately 1000 ppm target concentration to simulate worst case contamination scenario at a site prior to treatment. The contaminated sediment was prepared by mixing pre - determined amount of the potassium dichromate ($K_2Cr_2O_7$) with deionized water prior to addition into the sediment. The sediment sample was then mixed thoroughly and allowed to mellow foe a period of 7 days to ensure homogeneous distribution of the contaminant in the sediment.

2.3 Solidification/stabilisation sample preparation

The chromium contaminated sediment samples were mixed with the required amount of cement and/or RHA. Although based on current practice, cement has been added to soil from 5% to 20% [10]. It was decided to experiment with 10% cement to total mixture while rice husk ash was added at 5% to 20% to the total dry weight of the mixture. The sediment, cement and RHA were then thoroughly mixed in a clean Baker stand mixer. The water to cement ratio used was 0.4 for all the mixing process. The mixture was then transferred to a custom-made plastic cylindrical mould of 25mm diameter and 50mm height and compacted in 3 layers of 10g each, each layer was compressed and compacted with a custom made miniature compaction tool and compressed for 25 blows at each layer. The sample specimen was then gently pushed out of the mould and wrapped with a cling film and kept at relative humidity of not less than 95% in tightly closed plastic curing box at 27 ± 2°C for different periods of up to one month, the specimens were left to cure for 7, 14 and 28 days and UCS test was carried out on the specimens at the above mentioned curing days. A detail of the mix ratios of the specimens for the unconfined compressive strength (UCS) test is given in Table 1.

Table 1. Mix ratios for unconfined compression test (cement + RHA + spiked sediments).

Mix Ratios Symbol	Sediment (%)	Cement (%)	RHA (%)	w/c Ratio
100:0:0	100	0	0	0.4
90:10:0	90	10	0	0.4
85:10:5	85	10	5	0.4
80:10:10	80	10	10	0.4
75:10:15	75	10	15	0.4
70:10:20	70	10	20	0.4

2.4 Compressive strength

Unconfined compressive strength test (UCS) was conducted to evaluate the strength of the mixture of rice husk ash with cement and the chromium spiked sediment samples at higher stress. The test was conducted in accordance with BS 1377 (1990), part 7[11]. The compressive strength of the solidified samples was determined after 7, 14 and 28 days of moisture curing. Specimen tested for UCS were collected and dried in the oven at 105°C for 24 hours before being crushed to pass through a 1 mm sieve prior to TCLP and deionised water leaching tests.

2.5 Leachability and XRD analysis

The leaching test was carried out in acidic as well as in distilled water medium at 7, 14 and 28 days of curing using the toxicity characteristics leaching procedure (TCLP) in accordance with EPA method 1311 (USEPA, 1992). Samples were ground and passed through a No. 40 sieve (0.425mm) prior to the TCLP tests. Precisely, 5g of the crushed stabilised sediments was placed in 500-ml plastic bottles and mixed with a designated 100ml of TCLP extraction fluid, For this research TCLP extraction fluid No.2 was used because the pH of all the samples were above 5.0 even after the addition of 3.5 ml 1M HCl and heated for 10 minutes at 50 ± 5°C. The leaching solution was designed to simulate the worst case leaching conditions on disintegrated landfill waste due to prolonged aging effects. The extraction fluids used are acetic acid (pH 2.88 ± 0.05), and distilled water (DIW) (pH= 7.0 ± 0.1) was used to examine the impact of a non- aggressive solution, and it is used here as a control. It is useful to evaluate the potential of a material when exposed to normal precipitation, all at a solid - liquid ratio of 1:20 by weight of the crushed sediment.

XRD was used to investigate the crystalline mineral phases responsible for Chromium immobilisation in the stabilised sediment samples. Samples were ground to fine powders and investigated for the reflection angle (2θ) in the range 10° to 90° with a step size of 0.02° and a count time of 15.4 s per step. Step-scanned X-ray diffraction data was collected with XRD Bruker D8 Advance machine. The XRD analysis was conducted at 40 kV and 40 mA using a diffracted beam graphite-monochromator with Cu radiation. XRD patterns were qualitatively analysed and interpreted using Eva software and reference data base from the International Centre for Diffraction Data (ICDD) data base.

3 Results and Discussion

The effect of adding RHA and cement to the chromium contaminated sediment on the compressive strength and leachability of chromium was investigated. The results are reported and discussed in the following sections.

3.1 Unconfined compressive strength

The strength of the solidified sediment samples was measured after 7, 14 and 28 days of curing. The results of the UCS are shown in Table 2. The strength was dependent on the amount of RHA added in the binder system. The highest strength of 4780 kPa was obtained with mix ratio Cr28 80:10:10 (80% Chromium spiked sediment, 10% cement and 10% RHA at 28 days of curing). The least compressive strength was indicated by the control sample Cr7.100:0:0 (100% Chromium spiked sediment, 0% Cement and 0% RHA at 7 days of curing) which was 234.6 kPa. However, except for the control sample which the strength was 234.6 kPa, the strength in each case was more than the minimum strength required 340 kPa as per the landfill disposal limit. This is because the calcium hydroxide crystal produced during cement hydration was consumed during the pozzolanic reaction of RHA, which occurred mostly during these curing durations, and as a result, the secondary calcium silicate hydrate (CSH) gel was formed. It is believed that this secondary CSH gel contributes to the additional strength for the mixture other than the primary CSH gel obtained from the hydration of cement [12].

Table 2. UCS of solidified/stabilised chromium spiked samples throughout 28 days of curing.

Mix Ratios	**Unconfined Compressive Strength (kPa)**		
	7 days	**14 days**	**28 days**
100:0:0	234.6	377.2	523
90:10:0	1614	2179	2379
85:10:5	2341	2844	3407
80:10:10	3003	3071	4780
75:10:15	3484	3639	4505
70:10:20	3075	4142	3297

3.2 Leachability and XRD analysis

The leachability tests of solidified samples containing Cr (VI) were carried out at various mix ratios and curing time (7, 14 and 28 days). As can be seen in Table 3, it will be observed that leaching of Cr (VI) was higher in acetic acid (TLCP) medium as compared to deionized water (DIW) medium. The leachability of chromium was reduced from 2.197 mg/l to 0 (below detection limits) for the TCLP, and 0.259 mg/l to 0 (below detection limits) for the DIW leachates. The leached chromium concentrations in the two different leachates (Acetic acid, and deionized water) were all below the regulatory limits of 5 mg/l. The chromium concentration in the leachates of cement/RHA was less compared to the control sample which may be due to binding of Cr (VI) by cement/RHA [13]. The percentage of chromium leached out in the TCLP was between 1.94 to 0% for all the mix ratios and at curing days of 7, 14 and 28 days, while for the deionized water the percentage of chromium leached out was between 0.2 to 0%.

Table 3. The concentration of chromium in the TCLP and DIW leachates at 7, 14 and 28 days.

Mix ratios	**Conc. at 7 days for TCLP (mg/l)**	**Conc. at 7 days for DIW(mg/l)**	**Conc. at 14 days for TCLP (mg/l)**	**Conc. at 14 days for DIW(mg/l)**	**Conc. at 28 days for TCLP (mg/l)**	**Conc. at 28 days for DIW(mg/l)**
100:0:0	1.224	0.043	1.121	0.019	1.482	0.043
90:10:0	0.011	0	2.048	0.259	1.806	0.106
85:10:5	2.197	0.200	1.942	0	1.821	0.079
80:10:10	1.888	0	1.344	0	1.474	0.040
75:10:15	0.0180	0	1.174	0	0	0
70:10:20	0.0184	0	1.391	0	1.118	0.009

The results of X-ray diffraction show the formation of various crystalline mineral phases after 28 days of curing, it was performed on the treated chromium spiked sediment after 28 days of curing for the sample Cr28d.85:10:5 (85% Chromium spiked sediment, 10% cement and 5% RHA at 28 days) and Cr28d.75:10:15. (75% Chromium spiked sediment, 10% cement and 15% RHA at 28 days). It can be observed in Figure 1 which shows the XRD pattern of Cr28.85:10:5 and Cr28.75:10:15 after 28 days of curing. The most prominent peaks in the Chromium spiked sample (Cr28.85.10.5) were those of Kaolin at 2θ of 12.41° and 24.90°, Quartz at 2θ of 20.86°, 26.65°, 50.14°, 59.96°, Pyrite at 2θ of 33.02°, 40.74°, 47.39° and 56.27°, Montmorillonite at 2θ of 64.36° and 73.57°, Muscovite at 2θ of 34.98°, 42.38°, 45.83° and 61.79° and for the chromium phase were observed at 2θ 19.93°, 26.91°, 29.48° and 37.76°. Subsequent to treatment with the Cement/RHA (Cr28.75:10:15) the peaks of Chromium were not detected, but only for Chromium oxide, Portlandite $Ca(OH)_2$, one of the cement hydration products, was identified at 2θ of 17.69° and 36.50°. The peaks of ettringite (E) are also observed at 15.77° and 34.53°. The chromium ettringite phase is a low-solubility product which restricts Cr (VI) leaching in the (Cr28.75:10:15) and may be due to addition, substitution or precipitation of new components as observed by earlier workers [14]. Kaolinite peaks were observed at 2θ of 12.35°, 24.86° and 38.34°, Pyrite at 2θ of 33.04°, 47.36° and 56.13°, montmorillonite peak at 2θ 61.82°. 2θ of 33.04°, 47.36° and 56.13°, montmorillonite peak at 2θ 61.82°.

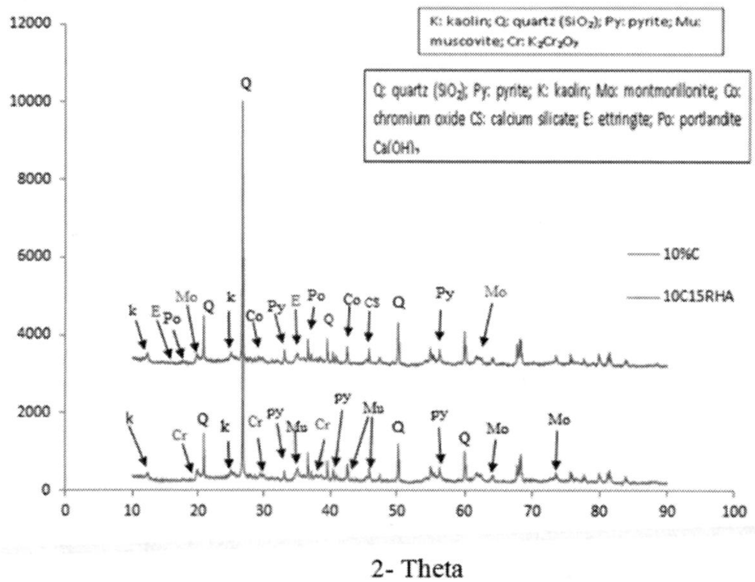

Figure 1. XRD patterns of chromium spiked sediment for cr28d.85:10:5 and cr28d.75:10:15 after 28 days.

4 Conclusions

The following conclusions can be drawn based on the experimental study:

i. The compressive strength of chromium spiked sediment was in the range of 234.6 kPa to 4780 kPa, there was enormous increase in the compressive strength with increase in the rice husk ash content of 5 - 20%, as the number of days of curing increases. The maximum strength was obtained between 15-20% RHA additions. While minimum strength was obtained for the control samples (0%cement / 0% RHA). All the UCS values obtained for all the solidified sediment samples exceeded the Landfill disposal limit of 0.34 Nmm2 (340 kPa) for a waste, except for the control sample at 7 days which has the value of 234.6 kPa slightly below the regulatory limit.

ii. Leachability tests (TCLP and Deionised water) conducted on the crushed Stabilized/Solidified sediment samples indicated that chromium in the leachates of acetic acid and deionized water were either undetectable or appreciably below the leachability limits. The retention capacity of chromium spiked sediment sample was in the range of 97% to 100% for the TCLP and 100% for Deionised water leaching test.

iii. The XRD analysis of the stabilised chromium spiked sediment showed reaction products and various peaks of Cr28d.75:10:15 and Cr28d.85:10:5 samples. The retention of chromium by the S/S in the cementitious matrix of cement-RHA mixture is due to addition, substitution (Cr-ettringite, Cr-CSH) or precipitation of new components. Portlandite $Ca(OH)_2$ was identified for the treated samples after 28 days indicating pozzolanic reactions were not over.

References

[1] A. Antemir, C.D. Hills, P.J. Carey, M.C. Magnie and A. Polettini, Investigation of 4-year-old stabilised/solidified and accelerated carbonated contaminated soil, J. of Hazardous Materials, **181**(1), 543-555, (2010).

[2] S.A. Katz and H. Salem, *The Biological and Environmental Chemistry of Chromium*, VCH Publishers, New York, (1994).

[3] S. Wang and C. Vipulanandan, Solidification/stabilization of Cr (VI) with cement: Leachability and XRD analyses, Cement and Concrete Research, **30**(3), 385-389, (2000).

[4] C. Vipulanandan and S. Wang, Remediation of hexavalent chromium contaminated soil, *Proc. of In-Situ Remediation of the Geoenvironment*, ASCE, New York, 363-373, (1997).

[5] C. Vipulanandan and S. Krishnan, Leachability and biodegradation of high concentrations of phenol and o-chlorophenol, Hazardous Waste and Hazardous Materials, **10**(1), 27-47, (1993).

[6] A. Roy, H.C. Eaton, F.K. Cartledge and M.E. Tittlebaum, Solidification/stabilization of a heavy metal sludge by a Portland cement/fly ash binding mixture, Hazardous Waste and Hazardous Materials, **8**(1), 33-41, (1991).

[7] P.L. Bishop, Contaminant leaching from solidified-stabilized wastes. *Emerging Technologies in Hazardous Waste Management II*, American Chemical Society, Washington, (1991).

[8] S.Y. Wang and C. Vipulanandan, Leachability of lead from solidified cement-fly ash binders, Cement and Concrete Research, **26**(6), 895-905, (1996).

[9] C.K. Lin, J.N. Chen and C.C Lin, An NMR and XRD study of solidification/stabilization of chromium with Portland cement and β-C 2 S, J. of Hazardous Materials, **48**(1), 137-147 (1996).

[10] R. Malviya, and R. Chaudhary, Factors affecting hazardous waste solidification/stabilization: A review, J. of Hazardous Materials, **137**(1), 267-276 (2006).

[11] BS 1377-7, Methods of Tests for Soils for Civil Engineering Purposes, British Standards Institution, London, (1990).

[12] R.J. Detwiler, J.I. Bhatty and S. Battacharja, *Supplementary Cementing Materials for Use in Blended Cements*, Portland Cement Association, Illinois, (1996).

[13] N. Jain, Solidification and Leachability of Cr (VI) in rice husk ash-blended cement, ISRN Civil Engineering, **2011**, 1-6, (2011).

[14] V.L. Bonavetti, V.F. Rahhal and E.F. Irassar, Studies on the carboaluminate formation in limestone filler-blended cements, Cement and Concrete Research, **31**(6), 853-859, (2001).

MATEC Web of Conferences **47**, 05012 (2016)
DOI: 10.1051/matecconf/20164705012

Steam-Enhanced Extraction Experiments, Simulations and Field Studies for Dense Non-Aqueous Phase Liquid Removal: A Review

Nor Asni Azizan[1,a], Samira Albati Kamaruddin[2] and Shreeshivadasan Chelliapan[2]

[1]*Faculty of Civil Engineering, Universiti Teknologi Malaysia, 81310 Johor Bahru, Malaysia*
[2]*Department of Engineering, Razak School of Engineering and Advanced Technology, Universiti Teknologi Malaysia Kuala Lumpur, 54100 Kuala Lumpur, Malaysia*

Abstract. Experiments, simulations and field studies for dense non-aqueous phase liquid (DNAPL) removal have demonstrated successful recovery through steam-enhanced extraction. Steam-enhanced extraction is an innovative technology for soil and groundwater remediation to remove as much contamination as possible. Most of researchers study the main DNAPL recovery mechanisms such as physical displacement by vaporization, evaporation and condensation, reduction in interfacial tension and DNAPL viscosity influenced by temperature. Other removal mechanism such as steam distillation and steam stripping also has been studied. The removal of DNAPL using steam-enhanced extraction shall be investigated to identify, acquire, analyze, visualize, and evaluate the effectiveness of the remediation. Several parameters can be controlled to justify the successful of the remediation. A comprehensive understanding of the subsurface environment, multiphase fluid flow and the physical processes is required to prevent remediation failure. Thus, it will avoid continuous contamination of the subsurface environment. The researcher can quantify the reduction in contamination remediation and acquire high quality data sets to validate future numerical model. Aim of this paper is to review and to summarize the existing laboratory experiment, simulations and field studies from other researchers regarding steam-enhanced extraction for dense non-aqueous phase liquid removal.

1 Introduction

Steam-enhanced extraction (SEE) was applied as environmental application from its original developed purpose to enhanced oil recovery. Since 1980s, extensive experiments, simulations and field studies has led to the point which proven the SEE technology can successfully remove the volatile contaminant from unsaturated subsurface [1]. Recently, these studies have been extending the SEE method to recoveries of saturated subsurface with more complex heterogeneity conditions and fractured geologic media. The SEE method involves continuous injection of steam through injection well into subsurface. The steam flows depend mostly to heterogeneity condition of subsurface. Three temperature zones was form which is steam-temperature, variable temperature and ambient temperature. These temperature influence the DNAPL viscosity thus effect the overall process of SEE

[a] Corresponding author : asnie86@yahoo.com

[2-4]. Several processes occur during SEE such as condensation, evaporation and volatilization. This process will continuously occur until it reaches the extraction well [5].

2 Steam-Enhanced Extraction

2.1 Selection of steam-enhanced extraction

Factors influence the application of steam-enhanced extraction as DNAPL remediation depend on the DNAPL properties such as volatility and viscosity. Volatility of DNAPL can increase the DNAPL saturation from residual saturation. Higher temperature can enhance volatilization thus increase the DNAPL removal rates [6, 7]. Viscosity of DNAPL determines the amount of DNAPL saturation displaced by steam front. The steam was capable in mobilizing DNAPL saturation until the ratio of viscosity of DNAPL and water equal to three. If larger, fingering may occur thus reducing the displacement efficiency [8]. Other factors that can influence the application of steam-enhanced extraction as DNAPL remediation is the porous media properties. The porous media types affect the DNAPL removal rate from porous media [9-11]. The DNAPL removal rate is higher in coarser soil than fine soil. The DNAPL removal rate is lower for soil with wider range of grain size due to lower permeability in fine grain. The DNAPL removal rate is the same for variable range of sand size but with longer remediation time as the grain size decrease [12].

The porous media heterogeneity influenced the time of steam move and directly affects cost. Steam movement is drive more towards the permeable layer instead of low permeable layer. The heterogeneity of subsurface in small scale cause fingering at higher flow rates [13]. The layered heterogeneity can cause channeling to occur due to natural beds or manmade such as buried utilities [14, 15]. Other factors to be considered are DNAPL location and amount of DNAPL. Small amount of DNAPL with location near to ground surface more economical to treat by excavation and incineration. The efficient ways to utilize steam-enhanced extraction is to treat contaminated soil to depth greater than 100 feet [5].

The most effective ways to increase the efficiency of steam-enhanced extraction is to consider the location of injection and extraction well placement and operation. The most effective ways is to surround the contaminated area with 4 to 6 injections well and place the extraction well at center. Thus, the propagation of steam will be balance during steam-enhanced extraction. The distance between each injection well is reported to be between 1.5m to 18 m and made from steel [5].

The efficiency of steam-enhanced extraction also influence by soil permeability. Soil permeability must be sufficient to allow steam heat. The injection rate is depending on pressure. The pressure must be not more than soil fracture pressure of 1.65 pounds per square inch per meter of depth to avoid overburden pressure [16]. Low soil permeability with low injection rate will increase time of heat lost to overburden and underburden thus is ineffective. Injection pressure influences the steam injection rate as well as temperature and steam quality. Increase temperature does not increase the heating rate. It is because the increase temperature will cause increase of heat loss. The steam generation will be slow causes longer recovery time [17]. Steam quality is the total water proportion in vapor phase. 100 % steam quality is when no liquid water present. Steam quality influence the ratio of oil and steam at extraction well [18, 19].

2.2 Recent research using steam-enhanced extraction

Recent laboratory, numerical simulation studies and field studies have contribute major affect in remediation of DNAPL on contaminated sites. From laboratory and numerical simulation studies, researcher can study and investigate how steam and air act and react to remediate DNAPL. Laboratory and numerical simulation can provide accurate data sets to study DNAPL remediation issues due to possible conduct of controlled experiment and model under know condition of porous media heterogeneity. The summary of recent research using steam-enhanced extraction is shown in Table 1.

Table 1. Recent research of DNAPL remediation through steam-enhanced extraction technology or thermal remediation from 2011 to 2015.

Research Type	Size	DNAPL	Remediation Method	Research Issue	Media/ Condition	Ref.
Field demonstration	$1210m^2$ with 14m to 50m depth	PCE, TCE, 1,1,1,1-TCA	Steam-enhanced extraction and soil vapor extraction	Savannah River serious contamination problem	Heterogeneous Saturated / Unsaturated	[20]
Experiment 2-D Simulation T2VOC	40cm height, 50cm width, 5cm thickness	TCE	Steam-enhanced extraction injection and air injection	TCE removal in variable porous media	Heterogeneous / Homogeneous Saturated / Unsaturated	[21]
Simulation 3-D TMVOC	109m height, 204m width, 218m length	TCE	Thermal*	TCE removal in fractured geologic	Fractured Geologic	[22]
Simulation 3-D TMVOC	20m depth, 8m radius	1,2-DCA	Thermal*	1,2-DCA removal in fractured geologic	Fractured Geologic	[23]
Simulation 3-D TMVOC	27.7m depth, 1.5m radius	TCE	Thermal*	TCE removal in fractured geologic at jet engine testing facilities	Fracture Geologic	[24]
Experiment 2-D	40cm height, 50cm width, 5cm thickness	TCE	Steam-enhanced extraction	TCE removal in variable heterogeneous media	Heterogeneous Unsaturated	[25]
Field demonstration Experiment Mathematical	15.2m depth, volume of $696m^3$	TCE, DCE and VC	Thermal*	TCE, DCE and VC in variable fractured geologic	Fracture Geologic	[26]
Field demonstration	6.2m depth, volume $5097m^3$	PCE	Thermal*	PCE removal from tight clays	Homogeneous Saturated	[27]
Experiment 1-D	0.3m depth, 5cm diameter	1,2-DCA	Thermal*	1,2-DCA removal characterization from fractured clay	Homogeneous Saturated	[28]
Experiment 2-D	38.1cm height, 13.97cm width, 73.66cm length	TCE	Thermal*	Observation of trapped gas during TCE removal	Heterogeneous Saturated	[29]

*Thermal means in situ thermal remediation either being stated as other than steam-enhanced extraction or no specific method stated.

From 1952 to 1978, multiple volatile organic compound (VOC) such as TCE, PCE and 1,1,1-TCA (1,1,1-Trichloroethane) has been identified causing the Savannah River in serious contamination problem. Total amount of 400,000 pounds VOC was successfully removed to date and 75% removed within first 18 month [20]. Peng, et al. [21] investigate the TCE remediation using co-injection of steam and air in unsaturated heterogeneous porous media and further research to saturated homogenous porous media. The results show that higher removal rate for coarse sand compared to fine sand. Chen, et al. [22] has study TCE removal in 3-D fractured geologic media using thermal remediation. The study was aim to investigate the heat transfer behavior and evaluate the sensitivity of parameter through numerical simulation using TMVOC TOUGH2 program. The results show that the TCE removal rate influence by matrix permeability and encouraged by increased fracture density.

Chen, et al. [23] has research the removal of 1,2-DCA (1,2-Dichloroethane) from fractured geologic media of sandstone using thermal remediation. The results shows that more than 99% 1,2-DCA are successfully removed. McKenzie [24] has investigate the remediation of TCE in fractured geologic media using thermal conductive heating at jet engine testing facilities of Naval Air Warfare Center (NAWC) in West Trenton, New Jersey. The results from numerical simulation shows 67% TCE was successfully removed compared to 63.5% TCE removal reported on site. The heterogeneity effect of porous media on TCE removal using steam-enhanced extraction (SEE) was studied by Wang, et al. [25]. The results show that the TCE removal percentage for three experiments is 89.5%, 88.2% and 85.2%.

Rodriguez [26] research the thermal heating application to removed chlorinated solvents such TCE, cis-1,2-dichloroethylene (DCE) and vinyl chloride (VC) from fractured geologic. The results show that the distance between point of compliance and upgradient domain face influence pore water concentration of fracture rock. The PCE remediation using in situ thermal desorption has been investigated by Heron, et al. [27] on site at San Francisco, California. After 110 days of heating, 2540kg of PCE was successfully recovered. Liu, et al. [28] investigate the chlorinated volatile organic compound removal characteristic from fractured clay using boiling. The results reverses the multiphase flow code predictions which state that pressure will increase when heating low permeable material, leading to overturn the boiling. The trapped gas behavior under passive venting conditions was investigate by Martin and Kueper [29] during application of electrical resistance heating in removal of TCE. The results show the influence of gas phase venting during remediation process of electrical resistance heating.

3 Discussions

The physical displacement is the main important removal mechanism for DNAPL remediation involves volatilization, evaporation, steam distillation and steam stripping [8]. The physical displacement is influenced by steam boiling point. Contaminants with low boiling point than water show greater recoveries [30-34]. The steam temperature, variable temperature and ambient temperature plays an important factor in the performance of SEE [14, 18, 35]. These temperature also influence the oil residual saturation [2-4] vapor pressure and interfacial effects [7]. Heterogeneity of subsurface shows greater impact on DNAPL recoveries rate. The DNAPL removal rate is higher in coarse sand compared to fine sand [21, 25]. In fractured geologic media, the DNAPL removal rate is influence by matrix permeability and fracture density [22]. In addition, the distance between point of compliance and upgradient domain face directly affect the pore water concentration of fracture geologic media which consequently affect the DNAPL removal [26]. The DNAPL removal can be accelerated from low permeability clay and occur when half or less pore water are recovered [28]. The SEE application shows successful application in DNAPL remediation [23, 24, 27].

4 Conclusions

Steam-enhanced extraction has shown greatest impact to DNAPL removal from saturated/unsaturated subsurface with complex heterogeneity and fractured geologic media. More subsurface characterization effort is required and comparison to other technology with the same condition shall be made in terms of performance, cost and remediation time. More experiments, simulations and field studies shall be performed to develop advanced steam-enhanced extraction technology to ensure continuance of environmental quality protection.

Acknowledgements

This work was supported by of the Ministry of Education Malaysia and sponsor by Universiti Teknologi Malaysia (UTM). The sponsorship is through TIER 2 Encouragement Grant UTM with cost center number of Q.K130000.2640.10J93.

References

[1] E.L. Davis, Steam injection for soil and aquifer remediation, Environmental Protection Agency, United States, (1998).

[2] T.A. Edmondson, Effect of temperature on water flooding, J. of Canadian Petroleum Technology, **4**, 236-242, (1965).

[3] S.W. Poston, S. Ysrael, A. Hossain and E.F. Montgomery III, The effect of temperature on irreducible water saturation and relative permeability of unconsolidated sands, Society of Petroleum Engineers, **10**,171-180, (1970).

[4] E.L. Davis and B.K. Lien, Laboratory study on the use of hot water to recover light oily wastes from sands, Environmental Protection Agency, United States, (1993).

[5] Engineering Forum Issue Paper, In situ treatment technologies for contaminated soil, Environmental Protection Agency, United States, (2006).

[6] K. Udell and L. Stewart, Combined steam injection and vacuum extraction for aquifer cleanup, *Proc. of the Int. Conf. on Subsurface Contamination by Immiscible Fluids*, Calgary, (1990).

[7] S. Lingineni and V.K. Dhir, An experimental and theoretical study of remediation of multicomponent organic contaminants in the unsaturated soil by venting, ASME-HTD, **193**, 99-99, (1992).

[8] L.D. Stewart and K.S. Udell, Mechanisms of residual oil displacement by steam injection, SPE Reservoir Engineering, **3**, 233-1,242, (1988).

[9] P.R de Percin, Demonstration of in situ steam and hot-air stripping technology, J. of the Air and Waste Management Association, **41**(6), 873-877, (1991).

[10] J.S. Lighty, G.D. Silcox, D.W. Pershing, V.A. Cundy and D.G. Linz, Fundamentals for the thermal remediation of contaminated soils, Particle and bed desorption models, Environmental Science and Technology, **24**, 750-757, (1990).

[11] L. Tognotti, M. Flytzani-Stephanopoulos, A.F. Sarofim, H. Kopsinis and M. Stoukides, Study of adsorption-desorption of contaminants on single soil particles using the electrodynamic thermogravimetric analyzer, Environmental Science and Technology, **25**, 104-109, (1991).

[12] A. Hadim, F. Shah and G. Korfiatis, Laboratory studies of steam stripping of LNAPL-contaminated soils, Soil and Sediment Contamination, **2**, 37-58, (1993).

[13] M.D. Basel and K.S. Udell, Two-dimensional study of steam injection into porous media, Multiphase Transport in Porous Media, ASME-HTD, **127**, 39-46, (1989).

[14] K. Cook, In situ steam enhanced recovery process, Environmental Protection Agency, United States, (1995).

[15] M.D. Basel and K.S. Udell, Effect of heterogeneities on the shape of condensation fronts in porous media, Heat Transfer in Geophysical Media, ASME HTD, **172**, 63-70, (1991).

[16] K.S. Udell, M. Itamura, L. Alvarez-Cohen and M. Hernandez, NAS Lemoore JP-5 cleanup demonstration, Berkeley Environmental Restoration Center, University of California, Berkeley, (1994).

[17] F.S. Johnson, C.J. Walker and A.F. Bayazeed, *Oil Vaporization During Steamflooding*, Society of Petroleum Engineers, Dallas, (1968).

[18] N.A. Myhill and G.L. Stegemeier, Steam-drive correlation and prediction, J. of Petroleum Technology, **30**(2), 173-182, (1978).
[19] A.K. Singhal, Physical model study of inverted seven-spot steamfloods in a pool containing conventional heavy oil, J. of Canadian Petroleum Technology, **19**, 12-124, (1980).
[20] B. Kramer, J. Kupar, J. Ross, J. Cardoso-Neto, D.G. Jackson and B.B. Looney, Concluding a steam injection remediation project at a DNAPL Source Zone at the Savannah River Site, *Waste Management Symposia*, Arizona, (2015).
[21] S. Peng, N. Wang and J. Chen, Steam and air co-injection in removing residual TCE in unsaturated layered sandy porous media, J. of Contaminant Hydrology, **153**, 24-36, (2013).
[22] F. Chen, R.W. Falta and L.C. Murdoch, Numerical analysis of thermal remediation in 3D field-scale fractured geologic media, Groundwater, **53(**4), 572-587, (2014).
[23] F. Chen, R.W. Falta and L.C. Murdoch, Numerical analysis of contaminant removal from fractured rock during boiling, J. of Contaminant Hydrology, **134–135**, 12-21, (2012).
[24] A. McKenzie, Simulating remediation of trichloroethylene in fractured bedrock by thermal conductive heating using the numerical model TMVOC, Master Thesis, Queen's University, Canada, (2013).
[25] N. Wang, S. Peng and J. Chen, 2D-box study on effects of media heterogeneity on remediation of TCE by steam enhanced extraction, Advances in Environmental Protection, **3**, 45, (2013).
[26] D. Rodriguez, Assessment of thermal heating for the removal of chlorinated solvents from fractured bedrock, PhD Thesis, Queen's University, Canada, (2012).
[27] G. Heron, J. Lachance and R. Baker, Removal of PCE DNAPL from tight clays using in situ thermal desorption, Groundwater Monitoring and Remediation, **33**, 31-43, (2013).
[28] X. Liu, L.C. Murdoch, R.W. Falta and T. Tan, Experimental characterization of CVOC removal from fractured clay during boiling, Int. J. of Heat and Mass Transfer, **70**, 764-778, (2014).
[29] E.J. Martin and B.H. Kueper, Observation of trapped gas during electrical resistance heating of trichloroethylene under passive venting conditions, J. of Contaminant Hydrology, **126**, 291-300, (2011).
[30] J.R. Hunt, N. Sitar and K.S. Udell, Nonaqueous phase liquid transport and cleanup: 1. analysis of mechanisms, Water Resources Research, **24**, 1247-1258, (1988).
[31] P.W. Atkins, *Physical Chemistry*, Oxford University Press, (1998).
[32] G. Heron, S. Carroll and S.G. Nielsen, Full-scale removal of DNAPL constituents using steam-enhanced extraction and electrical resistance heating, Groundwater Monitoring and Remediation, **25**, 92-107, 2005.
[33] G. Heron, K. Parker, J. Galligan and T.C. Holmes, Thermal treatment of eight CVOC source zones to near nondetect concentrations, Groundwater Monitoring and Remediation, **29**, 56-65, (2009).
[34] K.S. Udell, Application of in situ thermal remediation technologies for DNAPL removal, International Association of Hydrological Sciences, (1998).
[35] P.E. Baker, An experimental study of heat flow in steam flooding, Society of Petroleum Engineers J., **9**, 89-99, (1969).

MATEC Web of Conferences **47**, 05013 (2016)
DOI: 10.1051/matecconf/20164705013

Isolation and Characterization of Microcrystalline Cellulose (MCC) from Rice Husk (RH)

Zuliahani Ahmad[1,a], Nurul Nadhirah Roziaizan[1], Rozyanty Rahman[2], Ahmad Faiza Mohamad[3] and Wan Izhan Nawawi Wan Ismail[1]

[1]*Faculty of Applied Sciences, Universiti Teknologi Mara, 02600 Arau, Perlis, Malaysia*
[2]*School of Material Engineering, Universiti Malaysia Perlis, 02600 Arau, Perlis, Malaysia*
[3]*Faculty of Industrial Technology, Universiti Teknologi Mara, 40450 Shah Alam, Selangor, Malaysia*

Abstract. Microcrystalline cellulose (MCC) was extracted from local agricultural residues, rice husk (RH). RH undergone alkaline, bleaching and acid hydrolysis treatments. Nitric acid was used in the hydrolysis process as a potential acid to replace strong acids such as hydrochloric and sulphuric acids. Hydrolysis of RH was carried out by using different molarity of nitric and hydrochloric acids (control). The properties of MCC obtained such as % yield, crystallinity index (CrI) and functional group present were studied. % yield shown a comparable result regardless of different acid used. Whilst Fourier transform infrared (FTIR) spectroscopy showed the progressive removal of non-cellulosic constituents. X-ray diffraction (XRD) analysis revealed that the crystallinity increased with successive treatments regardless different molarity and acids used. The results showed potential of nitric acid to be used in MCC isolation process.

1 Introduction

A lot of rice husk (RH) is generated during the milling process in rice milling industries. 78% of the paddy is received as bran, broken rice and rice. Other 22% of the paddy received as husk. Generally, RH made up of 33% cellulose, 26% hemicelluloses and 7% lignin [1]. Most of the RH widely used in building material, fertilizer, insulation material, oil spill absorbent, heat insulators as well as a fillers in composite preparation [2,3]. Nevertheless, large quantities of RH are usually burn in an open air, which creates environmental issues. Therefore, it is necessary to reduce environmental problem by fully utilize RH as cellulose derivatives source.

Cellulose is one of the biodegradable and renewable structural plant polymer which can be processed into whisker-like micro fibrils, microcrystalline cellulose (MCC), nanocrystalline cellulose (NCC) and many more [4,5]. It can be obtained from agricultural by-products such as cotton fiber, kenaf, hemp, corn stalk, bagasse and RH. MCC exhibits excellent properties in composite fabrication such as renewability, biodegradability, high surface area for bonding with resins [4-6].

The isolation technique is called hydrolysis method where the amorphous region of cellulosic material is disintegrated resulting with different degrees of crystallinity index. Many researchers reported on using strong minerals acids such as sulphuric acid (H2SO4) and hydrochloric acid (HCl) hydrolyzed the amorphous regions [6-7]. In this study, HNO3 has been chosen as an alternative to the

[a] Corresponding author : zuliahani@perlis.uitm.edu.my

strong acid because it is completely miscible with water, only small amounts needed to catalyze the hydrolysis reaction and less corrosive. The significance of this research is to encourage the utilization of MCC isolated from RH in composite preparation via acid hydrolysis using HNO3 in comparison with HCl.

2 Materials

The rice husk was collected from BERNAS, a local rice mills in Simpang Empat, Perlis. NaOCl and NaOH were used as bleaching agents while HNO3 and HCl were used for hydrolysis. All the chemicals was purchased from Fluka and was used without further purification.

2.1 Pre-treatment of RH

The alkaline treatment was performed to solubilise the hemicelluloses. 10 g of RH were refluxed with 120 mL of 1M NaOH at 80°C for 1h and 36 min. Following alkali treatment, the alkaline treated RH were refluxed with 140 mL of 5% sodium hypochlorite (NaOCl) at 80 °C for 18 min [8] in the bleaching process. The lignin become soluble in the alkaline medium and oxidized during bleaching.

2.2 Acid hydrolysis process

10 g of pre-treatment RH were treated with 61 mL of 1M and 2M HNO3 at room temperature for 30 min under continuous stirring. It was washed repeatedly with distilled water until pH 7 is achieved. The resultant MCC were grounded into a fine powder by using rotary ball mill after the washing and drying process. The mass of dried RH were recorded and the same steps were used for 1M and 2M HCl.

2.3 Percentage yield

Percentage yield was calculated using the following equation:

$$\text{Percent yield (\%)} = \frac{\text{weight initial}}{\text{weight after hydrolysis}} \times 100\% \quad (1)$$

2.4 Fourier transform infrared (FTIR) spectroscopy

FTIR analysis was carried out using Perkin Elmer FTIR spectrometer 1650 with scanning ranges of 4000.0 cm^{-1} to 400.0 cm^{-1} at room temperature.

2.5 X-ray diffraction (XRD)

X-ray diffraction was performed to determine crystallinity index at different treatments. Diffraction patterns were obtained using a PANalytical X'PertPRO Multi-Purpose Diffractometer with Cu Kα radiation. The crystallinity index (CrI) were calculated via Equation (2) [9]:

$$\text{CrI} = \left(\frac{I_{002} - I_{am}}{I_{002}} \right) \quad (2)$$

where I_{002}=intensity of the 002 peak (at about 2θ=26)I_{am} = intensity corresponds to the peak at about 2θ = 8.

3 Results and Discussion

3.1 Percentage yield

The yield of MCC sample was determined and the data was summarized in Table 1. Different molarity and acids were used to extract MCC from RH which were 1M HNO_3, 1M HCl, 2M HNO_3 and 2M HCl. 1M and 2M HCl gave 60.58% and 69.24% while 1M and 2M HNO_3 gave 80.64% and 83.26% of MCC yield respectively. Higher molarity of acids gave higher MCC yield. This is due to to at higher acid concentration, more solubilised amorphous region of cellulose contribute to increase % of MCC yield. The percentage yield of MCC varied from sample to sample. Also, the difference was depending on the type of paddy, climatic, geographical conditions, sample preparation and method of analysis which could be the reason for this variation [7]. As a conclusion, 1M HNO_3 and 2M HNO_3 gave higher MCC yield as compared to HCl regardless of molarity.

Table 1. Yield of MCC sample at different molarity and acids used.

Samples	% Yield
1M HNO_3	80.64
1M HCl	60.58
2M HNO_3	83.26
2M HCl	69.24

3.2 FTIR

Figure 1 shows the FTIR spectra recorded for untreated RH and RH treated with different acids. All samples display two main absorbance regions at 2900-3500 cm^{-1} and 450-1750 cm^{-1} in line with Rosa et. al. [10] corresponding to OH bending, CH_2 groups, C-O-C stretching and C-H. The peak at 3350-3440 cm^{-1}, which was observed in all spectras, represents of the C-H and O-H groups. The absorption at region from 1637 until 1651 cm^{-1} in all samples are corresponding to the absorption of water. According to Nurain Johar et. al. [7], this band is related to bending modes of water molecules due to a strong interaction between cellulose and water. This indicates that the cellulose components were present after the chemical treatment carried out on the rice husk fibres.

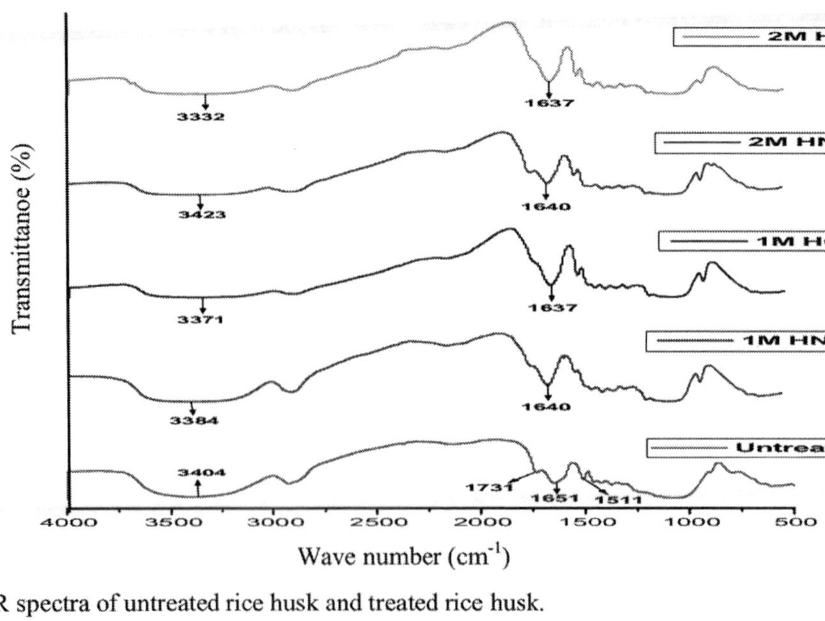

Figure 1. FTIR spectra of untreated rice husk and treated rice husk.

In addition, Rosa et. al. [10] found the absence of peaks located in the range of 1509-1609 cm^{-1}, which correspond to the C=C aromatic skeletal vibrations indicate the complete removal of lignin. The absorption band which corresponds to the acetyl or ester groups of hemicelluloses appears in the region 1700-1740 cm^{-1} as can be seen in untreated RH. Whilst the absence of this peak was found in treated RH indicates the removal of hemicelluloses was successfully done [10-11]. The peak located at 1049-1079cm^{-1} was observed in five spectras referring to the C-H stretching vibration of C-O [12]. Based on FTIR spectras, lignin and hemicelluloses have been removed via bleaching treatment, alkali treatment and acid hydrolysis. In addition, no significance difference were observed in all samples at the cellulose region, which indicates the cellulose molecular structure remains unchanged from the treatment.

3.3 XRD

Figure 2 shows the diffraction patterns obtained for untreated and treated rice husk with different molarities and acid used. The peak at 2θ= 16°, 22.5° and 36° is correspond to cellulose I region for all samples. This is supported by Rosa et al. [10] and Satyamurthy et al. [13] which there are typical of cellulose I with three well-defined crystalline peaks around 2θ = 16°, 22.5° and 36° and indicated by the absence of the doublet located at 22.5°.

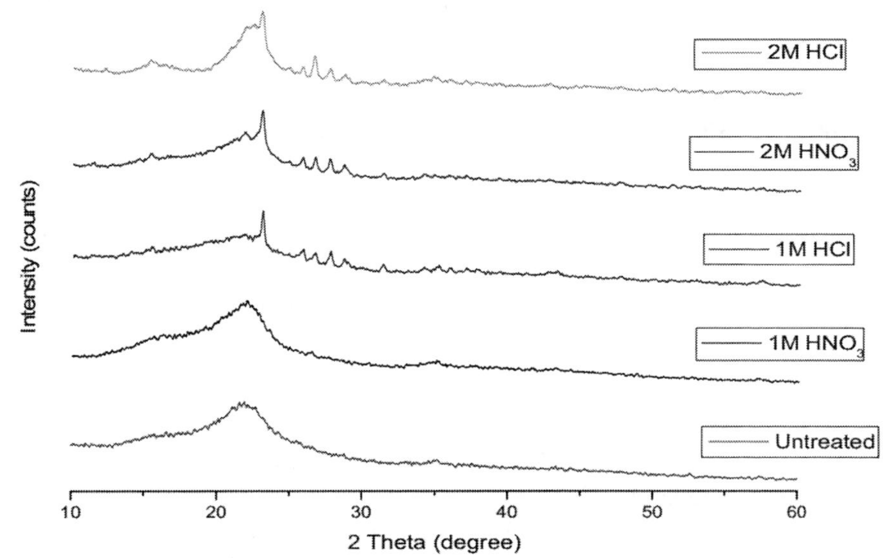

Figure 2. X-ray diffraction patterns of untreated and treated rice husk with different molarities.

The crystallinity index was determined for various samples and the results are summarized in Table 2. The crystallinity values of treated RH with 2M HNO_3 is 63.9% which is higher compared to others. It is due to the removal of amorphous regions of cellulose by acid hydrolysis, which prompts the hydrolytic cleavage of glycosidic bonds, finally releasing individual crystallites [6]. In comparison, 2M HNO_3 and 2M HCl shows comparable result and 1M HNO_3 gives higher percentage crystallinity compared to 1M HCl.

During the hydrolysis process, hydronium ions can penetrate the more accessible amorphous regions of cellulose and allow the hydrolytic cleavage of glycosidic bonds, which eventually releases individual crystallites [7]. In addition, during the preparation of cellulose the growth and realignment of monocrystals may occur in parallel and thus can improve the cellulose crystallinity [7]. An increase in crystallinity is related to increases in the rigidity of the cellulose structure, which can lead to higher tensile strength to fibers. This increase would be expected to enhance the mechanical properties of composites [10].

Table 2. Crystallinity index of treated and untreated RH.

Sample	Crystallinity (%)
Untreated RH	39.3
MCC-1M HNO_3	42.0
MCC-1M HCl	37.0
MCC-2M HNO_3	63.9
MCC-2M HCl	52.2

4 Conclusion

MCC has been successfully isolated from RH using chemical treatments involving alkali, bleaching and acid hydrolysis treatments. The results obtained from FTIR analysis confirmed that chemical structure of cellulosic fragments is not influenced by the acid hydrolysis. The crystallinity analysis shows that MCC give high crystallinity providing evidence that acid hydrolysis does not alter crystal structure from cellulose I of RH. The proposed HNO_3 gives comparable % yield and % crystallinity as compared to HCl making it a potential acid to be used in acid hydrolysis process.

References

[1] A.M. Adel, H.Z. Abd El-Wahab, A.A. Ibrahim and M.T. Al-Shemy, Characterization of microcrystalline cellulose prepared from lignocellulosic materials, Part II: Physicochemical properties. Carbohydrate Polymers, **83**, 676-687, (2011).

[2] O. Olamide and F.A. Oyawale, Characterization of rice husk via atomic absorption spectrophotometer for optimal silica production, Int. J. of Science and Technology, **2**(4), 210-213, (2012).

[3] C.P. Azubuike and A.O. Okhamafe, Physicochemical, spectroscopic and thermal properties of microcrystalline cellulose derived from corn cobs, Int. J. of Recycling of Organic Waste in Agriculture,**1**(9), 1-7, (2012).

[4] E. Abraham, B. Deepa, L.A. Pothen, J. Cintil, S. Thomas, M.J. John, R. Anandjiwala and S.S. Narine, Enviromental friendly method for the extraction of coir fibre and isolation of nanofibre, Carbohydrate Polymers, **92**, 1477-1483, (2013).

[5] S.J. Eichhorn, Cellulose nanowhiskers: Promising materials for advanced applications, Soft Matter, **7**, 303-315, (2011).

[6] M.K. Mohamad Haafiz, S.J. Eichhorn, A. Hassan and M. Jawaid, Isolation and characterization of microcrystalline cellulose from oil palm biomass residue, Carbohydrate Polymers, **93**, 628-634, (2013).

[7] N. Johar, A. Dufresne and I. Ahmad, Extraction, preparation and characterization of cellulose fibers and nanocrystals from rice husk, Industrial Crops and Products, **37**, 93-99, (2012).

[8] A.P. Mario, D.W. Marseno and Haryadi, Synthesis and characterization of sodium carboxymethylcellulose from cavendish banana pseudo stem (Musam cavendishii Lambert), Carbohydrate Polymers, **62**, 164-169, (2005).

[9] A.M. Adel, H.Z. Abd El-Wahab, A.A. Ibrahim and M.T. Al-Shemy, Characterization of microcrystalline cellulose prepared from lignocellulosic materials, Part I: Acid Catalyzed Hydrolysis, Biosource Technology, **101**, 4446-4455, (2010).

[10] S.M.L. Rosa, N. Rehman, M.I.G. De Miranda, S.M.B. Nachtigall and C.I.D. Bica, Chlorine-free extraction of cellulose from rice husk and whisker isolation, Carbohydrate Polymers, **87**, 1131-1138, (2012).

[11] M. Nuruddin, A. Chowdhury, S.A. Haque, M.,Rahman, S. F. Farhad, M. Sarwar Jahan and A. Quaiyyum, Extraction and characterization of cellulose nanofibrils from agricultural wastes in an integrated biorefinery initiative, Cellulose Chemical and Technology, **45**, 347-354, (2011).

[12] A. Alemdar and M. Sain, Biocomposites from wheat straw nanofibres: Morphology, thermal and mechanical properties, Composites Science and Technology, **68**, 557-565, (2008).

[13] P. Satyamurthy, P. Jain, R.H. Balasubramaya and N. Vigneshwaran, Preparation and characterization of cellulose nanowhiskers from cotton fibres by controlled microbial hydrolysis, Carbohydrate Polymers, **83**, 122-129, (2011).

MATEC Web of Conferences **47**, 05014 (2016)
DOI: 10.1051/matecconf/20164705014

The Effect of Reactives Diluents to the Physical Properties of Acrylated Palm Oil Based Polyurethane Coatings

Munirah Onn[1,a], Ahmad Faiza Mohd[2] and Mohd Firdaus Yhaya[3]

[1]*Department of Polymer Technology, Faculty of Applied Sciences, Universiti Teknologi Mara, 02600 Arau, Perlis, Malaysia*
[2]*Department of Polymer Technology, Faculty of Applied Sciences, Universiti Teknologi Mara, 40450 Shah Alam, Selangor, Malaysia.*
[3]*School of Industrial Technology, Universiti Sains Malaysia, 11800 Minden, Penang, Malaysia*

Abstract. The development of polyurethane with hydroxyl access in a molecule leads to a new alternative of low toxicity green product. Palm oil is one of the major commodities in Malaysia. The potential of palm oil to be used as coatings raw material such as alkyd is limited due to low unsaturated side on fatty acid chains. To overcome this limitation, palm oil was modified through transesterification process to produce polyol. Acrylated isocyanate (urethane oligomer) was then grafted onto polyol to produce polyurethane with vinylic ends. The polyurethane was formulated with different cross-linkers (reactive diluents) and cured under UV radiation. The effect of three different diluents; monoacrylate, diacrylate and triacrylate on the properties of cured polymer were studied in this research. Fourier Transform Infrared (FTIR), Hydroxyl Value Titration, Gel Content, and Volatile Organic Compound (VOC) were used for characterization. Physical testing performed were Pencil Hardness and Pull-Off Adhesion test. Novel palm oil-based polyurethane coatings have been found to have good properties with mono acrylate functionality.

1 Introduction

Polyurethanes are versatile class of polymers with variety of applications, because their properties can easily be modified by varying the components from which they are constructed [1]. Polyurethanes have been used in adhesive and coating industry for a long time. Since the conventional polyurethanes are petroleum based, the utilization of renewable resources to produce environmental friendly coatings is gaining popularity [2]. Oils and fats of vegetables and animals can be utilized as alternative raw materials. They offer huge number of possibilities in chemistry which rarely can be met by petrochemicals [3]. Vegetable oils contain unsaturation on fatty acids bond which have potential in synthesis of biopolymers [4].

This research, palm oil was used as the renewable raw material. Unfortunately, its limitation was due to low unsaturation on fatty acid chains [5]. To overcome this limitation, palm oil was modified through transesterification process with glycerine to produce polyol (Scheme 1). Glycerine used in this process contributed to primary and secondary hydroxyl groups on the polyol. This is an effort to have alternative hydroxyl functionality similar to castor oil. Isocyanate is one of the basic components

[a] Corresponding author : munirah591@perlis.uitm.edu.my

in producing polyurethane. Isophorone diisocyanate (IPDI) was monoacrylated with 2-hydroxylethyl acrylate (HEA) to form urethane oligomer, which was later grafted onto polyol to produce polyurethane with vinylic ends. Different types of reactive diluents; monoacrylate, diacrylate and triacrylate were used as cross-linkers.

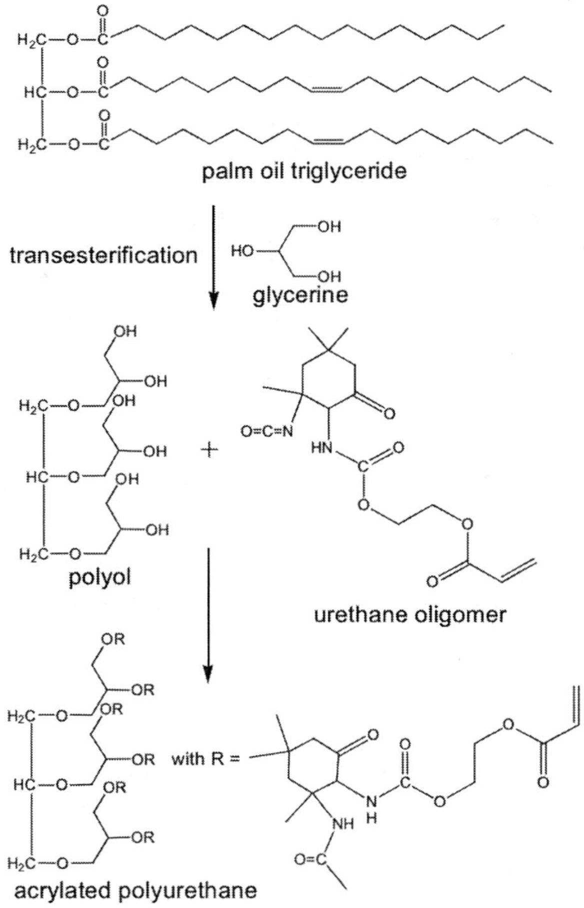

Scheme 1. Synthesis of polyurethane.

2 Experimental

2.1 Materials

Triglyceride palm oil (824 g/mol) from Sri Murni brand was used. Fatty acid ester was found to be around 2.0. Glycerol (92.09 g/mol) was supplied by May & Baker Ltd. Dagenham, England were used. Isopohorone diisocyanate (IPDI) and 2-hydroxylethyl acrylate (HEA) were supplied by Sigma-Aldrich. *p*-toluenesulfonic acid (PTSA), dibutyltin dilaurate (DBTL) and hydroquinone (as catalysts) were used as received. Acetone as solvent in polymerization process was supplied by Sigma-Aldrich. Three (3) types of reactive diluents were used; butyl-methacrylate, tripropyleneglycol diacrylate (TPGDA), Trimethylolpropane triacrylate (TMPTA). Irgacure 500 supplied by CIBA was used as photoinitiator and *N-methylpyrrolidone* (NMP) from Sigma-Aldrich was used as received.

2.2 Instrumentation

FTIR spectroscopic measurements using KBr pellets were carried out using Perkin-Elmer IR spectrometer and the samples were cured under Fusion UV radiation curing machine (80 kW) lamp.

2.3 Synthesis

The process began with transesterification reaction of palm oil to produce polyols. The transesterification process was done with the access of one mole triglyceride, 4 mole of glycerine and 6 mole of methanol. 5% PTSA was used as catalyst. Palm oil, glycerin and methanol were charged into 500 mL multi-neck reaction flask equipped with a stirrer, thermometer, water cooled condenser, and nitrogen inlet. Catalyst was added after the reaction temperature reached 50°C and stir rate was set to 550 – 650 rpm. Hydroxyl value test was done during the reaction to obtain the optimum hydroxyl value of polyol from palm oil.

The formation of urethane oligomer was done by charging one mole IPDI and one mole HEA into multi-neck reaction flask. 1% DBTL was used as catalyst and 0.5% hydroquinone was added to inhibit further reaction. FTIR was used to monitor the progress of urethane oligomer formation.Polyol from palm oil was then physically blended with urethane oligomer to form polyurethane resin. The reactions occur at room temperature and the resin was characterized by FTIR. 3 types of reactive diluents were used as cross-linkers; monoacrylate, diacrylate and triacrylate. Polyurethane was mixed well with reactive diluents, 3% of photoinitiator (Irgacure 500) and 1% of film formation agents (NMP). The ratio between polyurethane resin and reactive diluents were varied from 50% to 100%. The coatings were applied on the glass substrates as thin films (40μm) by using hand coater and cured under Fusion UV radiation light conveyer at constant speed 3.2 m/min, 80 kW bulb with 15 cm light intensity with one pass.

2.4 Physical testing

Gel content test was done on the cured samples. The cured films were carefully peeled off from the glass substrates and then weighed. A known weight of the cured films was extracted in acetone at 32°C for 24 hours. The residues were dried at 50°C to a constant weight. The gel content of the cured film was then determined by the following equation.

$$\text{Gel Content } (\%) = W_t / W_o \text{ X } 100 \quad (1)$$

where W_t is the weight after extraction and W_o is the weight before extraction.

Volatile organic compound test was conducted to determine the amount of volatile compounds in cured samples. These compounds have low boiling point and high vapor pressure. The cured samples were weighed and put under 120°C for 2 hours. After 2 hours samples were taken out and weighed. The percent of volatile organic compounds was calculated by following equation :

$$\%VOC = (W_f\text{-}W_i) / W_i \text{ X } 100) \quad (2)$$

where W_i is initial weight and W_f is final weight.

The hardness test was done by pencil hardness and referring to ASTM D3363 as standard method. This test uses special pencils with different degrees of hardness to scratch the coating as indication of hardness. The amount of pressure and the angle of the pencil to the substrate affect the results. By referring to ASTM D3363 standard method, pull-off adhesion test was done using a portable Defelsco posi test MP 2571 adhesion tester. A load was increasingly applied to the surface until the dolly was pulled off. The force required to pull the dolly off yielded the tensile strength in MegaPascals (MPa).

3 Experimental

3.1 Synthesis and characterization test

Three different types of acrylates (reactive diluents/cross-linkers) were used in this research: butyl acrylate, tripropyleneglycol diacrylate (TPGDA), and trimethylolpropane triacrylate (TMPTA) with acrylate functionality of one, two, and three, respectively. In theory, more acrylates should translate to better physical properties due to the higher cross-linking density. Hydroxyl value of polyol obtained from palm oil was calculated to be 1331.12 mg NaOH/g using hydroxyl value titration method. Higher hydroxyl values indicate better properties by the polyurethane resin, because more urethane linkages can be formed. FTIR spectra showed the formation of polyol from palm oil. The formation of hydroxyl in polyol was detected at 3307 cm^{-1} (Figure 1).

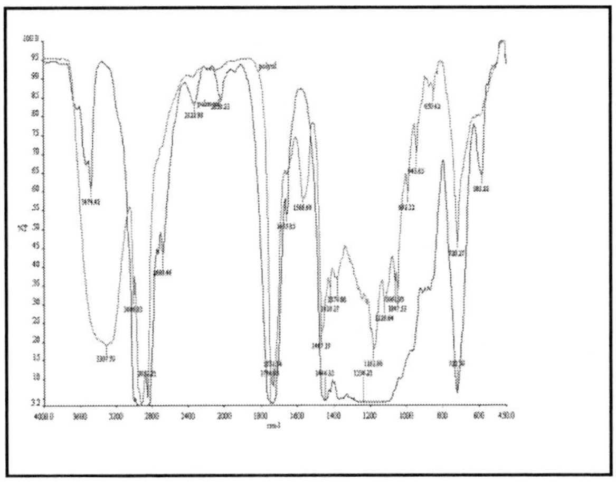

Figure 1. FTIR spectra of polyol.

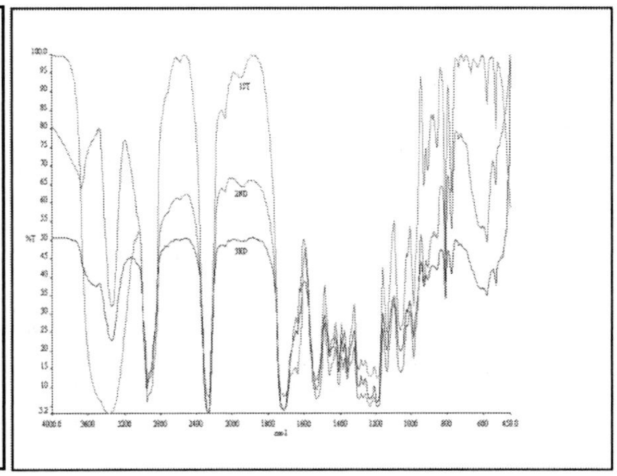

Figure 2. FTIR-spectra of urethane oligomer.

The formation of urethane oligomer was determined by FTIR spectroscopy. The presence of isocyanate and hydroxyl was detected at 2260 cm^{-1} and a broad peak at 3200 cm^{-1}, respectively. The spectrum shows the peak at three time interval (1st, 2nd and 3rd). By measuring the peak area of NCO peak, it was possible to give an estimate of the degree of urethane formation. Intensity of absorption bands at 1709 cm^{-1} (C=O) and 1120 (C-O) shows the present of urethane linkage. The hydrogen-bonded NH peaks can also be observed at 3200 cm^{-1}. However this peak overlapped with hydroxyl broad peak as shown in Figure 2. The process was stopped when half of the initial NCO contents were eliminated. The remaining NCO groups were used in the formation process of acrylated polyurethane. Acrylated polyurethane was synthesized by mixing the polyol from palm oil with the urethane oligomer. The full formation of acrylated polyurethane took about 9 hours as indicated by the NCO peak at 2260 cm^{-1} (Figure 3). The peak intensity for isocyanate decreased during the reaction and totally depleted at the end of the reaction as shown in Figure 4. The formation of acrylated polyurethane was further judged through the reduction of peak intensity of OH at 3300-3400 cm^{-1}.

3.2 Gel content test

Gel content values obtained from the test can be used to predict the degree of cross-linking after UV curing process because the gel content is proportional to degree of cross-linking. The measurements indicate the amount of insoluble material in a cross-linked sample. The higher the percentage of gel content, the higher the degree of cross-linking. Gel content of control sample was found to be 83%. Degree of cross-linking in cured coatings can be varied by adding the different types of reactive diluents; acrylates and varying the ratio between resin and acrylates. The percentages obtained were relatively high, indicating that the coatings were highly cross-linked material. When acrylates were added into the formulation, the acrylates cross-linked the long polyurethane chains together which altered the chemical and physical characteristics.

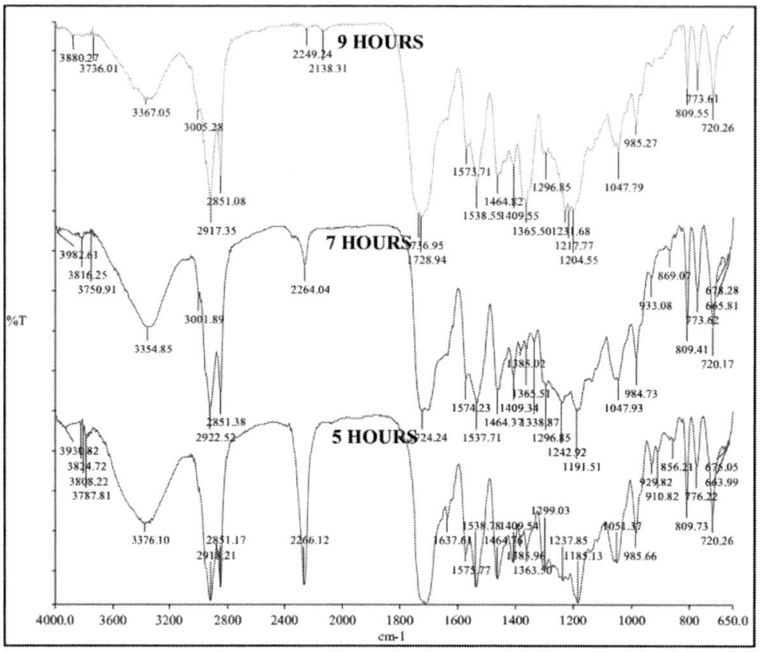

Figure 3. FTIR spectra of acrylated polyurethane.

Table 1. Effect of types of acrylates on coating properties.

		Gel Content (%)	VOC (%)	Pencil Hardness	Pull-Off Adhesion (MPa)
CONTROL		**83**	**10**	**2B**	**0.797**
MONO	SAMPLE 1	72	11	2B	1.457
	SAMPLE 2	76	7	2B	1.247
	SAMPLE 3	80	6	2B	1.263
	SAMPLE 4	82	7	2B	1.290
	SAMPLE 5	84	9	2B	1.257
TPGDA	SAMPLE 1	87	14	2B	1.000
	SAMPLE 2	90	12	2B	0.853
	SAMPLE 3	82	13	2B	0.673
	SAMPLE 4	81	11	B	0.617
	SAMPLE 5	86	7	2B	1.093
TMPTA	SAMPLE 1	90	5	2B	0.573
	SAMPLE 2	85	4	2B	0.593
	SAMPLE 3	85	7	2B	0.653
	SAMPLE 4	86	9	B	0.550
	SAMPLE 5	87	15	B	0.947

3.3 Volatile organic compound (VOC) test

Volatile organic compound usually refers to solvent content in the system. The lower the amount of VOC in the coating, the safer it is to be used. The present of VOC will increase the degree of flammability of the coating and also increase the health risk associated with application of the coating. Therefore, there is a maximum level of VOC emission fixed for certain applications. According to the result of VOC test, usage of TPGDA gives relatively high percentage of VOC as compared to another two acrylates; butyl acrylate and TMPTA. When it has less functionality, the acrylate was diluted with more solvent to achieve the volume needed. These lead to faster curing due to solvent evaporation and lead to lower solid content.

3.4 Pencil hardness

Hardness is the resistance of a coating to indentation or scratching. 'H' stands for hardness and 'B' stands for blackness. The hardness of these coatings was found to be between B and 2B. It shows that the coating has a soft feeling surface compared to conventional polyurethane which range from H to 2H. It also been influence by fewer airborne impurities which have time to contact the wet film and lead to softness.

3.5 Pull-off adhesion test

Pull-off adhesion test indicates the coating resistance towards the applied perpendicular tensile force when the coatings are forced to be separated from the substrate. According to Figure 6, it shows butyl acrylate gave relatively high coating resistance towards applied force as compared to TPGDA and TMPTA. TPGDA and TMPTA caused higher cross-linking density, which in turn contributed to more rigid coatings. Rigid coatings have higher cohesion between themselves rather than adhesion to the substrates.

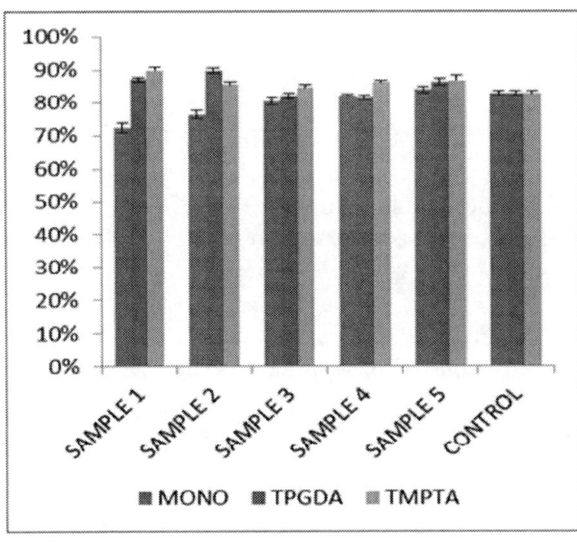

Figure 4. Gel content of cured film with different types of acrylate loading.

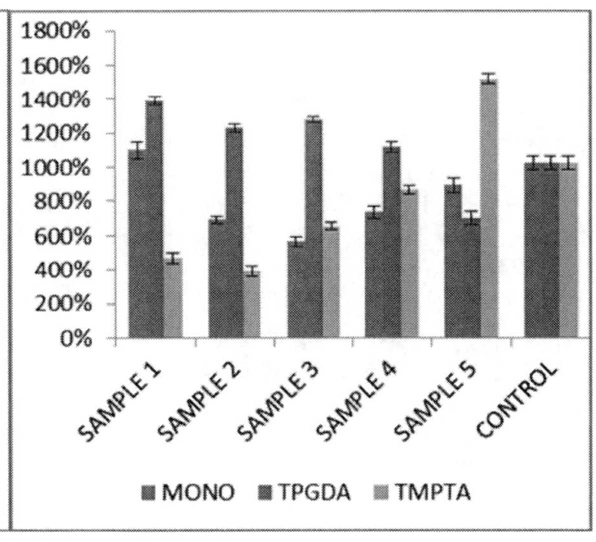

Figure 5. VOC of cured films with different types of acrylate loading.

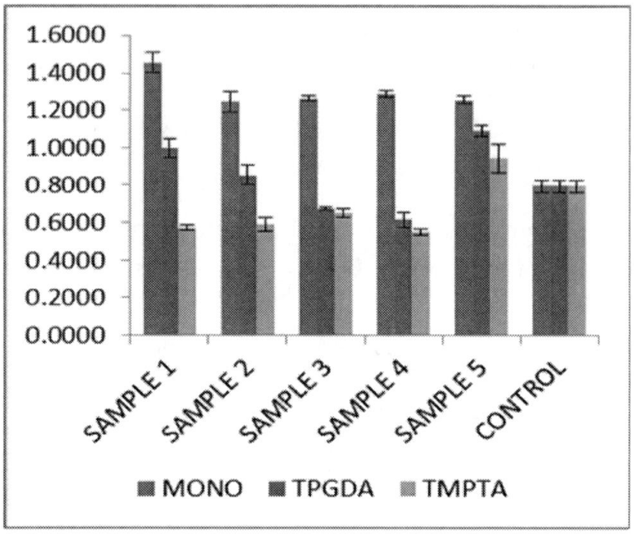

Graph 6. Pull-off adhesion of cured film with different types of acrylate loading.

4 Conclusions

Synthesis of renewable, environmental friendly polyurethane coatings were successfully done through transesterification of palm oil and subsequent modification with acrylated urethane oligomer. Among the reactive diluents, Butyl acrylate is the best reactive diluents to be crosslink with the acrylated palm oil oligomer.

References

[1] Y.C. Lai, E.T. Quinn, and L.Valint. Control of hard segment size in polyurethane formation. Journal of Polymer Science. **33**, 1767–1772, (1995).

[2] R.I. Muhammad, D.H. Beg Mohammad and S.J. Saidatul. The effects of increasing unsaturation in palm oil properties. World Academy of Science, Engineering and Technology. International Journal of Chemical, Nuclear, Materials and Metallurgical Engineering, **9**, 3-9, (2015).

[3] G. Lligadas, I.C. Ronda, M. Galia, V. Cadiz, Novel silicon-containing polyurethanes from vegetable oils as renewable resources. Synthesis and properties, Biomacromolecules **7**, 2420-2426 ,(2006).

[4] C.T. Yuan,K. Pimphan, S. Galen and H.H. Fu, Physical properties of water-blown rigid polyurethane foams from vegetable oil-based polyols, Journal of Applied Polymer Science, **105**, 453-459, (2007).

[5] G. Yanchang and J. Xuesong, Photo-cured materials from vegetable oils, The Royal Society of Chemistry, **1**, 1-27, (2015).

MATEC Web of Conferences **47**, 05015 (2016)
DOI: 10.1051/matecconf/20164705015

Immobilized/P25/DSAT and Immobilized/Kronos/DSAT on Photocatalytic Degradation of Reactive Red 4 Under Fluorescent Light

M. S. Azami[1, 2,a], S. K. Ain[1,2,], R. Zaharudin[1,2], F. Bakar[1,2] and W. I. Nawawi[2]

[1]*Faculty of Applied Sciences, Universiti Teknologi MARA, 40450 Shah Alam, Selangor, Malaysia*
[2]*Faculty of Applied Sciences, Universiti Teknologi MARA, 02600 Arau, Perlis, Malaysia*

Abstract. In this work, photocatalytic degradation of Reactive Red 4 (RR4) using immobilized P25 and kronos were performed under fluorescent light sources. The photocatalysis activity for both catalysts was investigated under fluorescent lamp source which consist UV and Visible light. The effect of various parameters such as initial concentration, initial pH and strenght of immobilized plate were studied. The result showed that 90% of RR4 dye was degrade in 1 hr using immobilized/kronos/DSAT at 100 mg L^{-1} of RR4 dye while 81% degradation was achieved by immobilized/P25/DSAT at the same condition. The lowest pH showed the higher photocatalytic activity. Hence, the effect of dye concentration and pH on the photocatalysis study can be related with the behavior of environmental pollution. The low strength showed by immobilized/P25/DSAT where it remain 37 % as compared with strength of immobilized/kronos/DSAT (52 wt.%). For the future work, the polymer binder like Polyvinyl alcohol (PVA), Polyethylene glycol (PEG), and others polymers can be apply in immobilized study to overcome the strength problem.

1 Introduction

The development of rapid industrialization in this world will contribute toward the major pollutant. Furthermore, organic dyes are one of the main industrial wastewater pollution. Despite the industrial are already schemes the various treatment to combat the pollutant unfortunately, the decrease in levels for specific pollutants has not been achieved. This environment problem of wastewater is significant focus by researcher to study toward effective removal and degradation processes [1-3]. The common biological processes of degradation and discoloration on modern dyes are ineffective because of high degree of aromatic groups in dye molecules [4]. The traditional physical methods such as using filtration, reverse osmosis, active carbon and coagulation are costly. Recently, Heterogenous photocatalysis have gained much attention toward the degradation of water pollution through complete mineralization. heterogenous photocatalysis is the process basically based on generates the hydroxyl radical which is highly oxidizing agents. This hydroxyl radical will oxidize the organic pollutants into harmless product. The combination between semiconductor photocatalyst and light sources is the key process of photodegradation. Among of the semiconductor, titanium dioxide (TiO_2) is the most popular and widely used by researchers due to the chemically stable, low cost, non-toxic,

[a] Corresponding author : saifulddinazami@gmail.com

reusable and its high photocatalytic activity [5, 6]. This material consist a wide band gaps between 3.0 eV to 3.2 eV which require the UV light irradiation. In this work, photocatalytic degradation of RR4 was studied using immobilized P25 and kronos VLP 7000 under fluorescent light illumination. Herein, RR4 was selected as model pollutant for degradation study. Subsequent experiments were conducted to investigate the effect of parameter of P25 and kronos on the photodegradation of RR4. The effect of initial dye concentration, initial pH on degradation of RR4 was investigated. Otherwise, strength of P25 and kronos on glass plate was investigated.

2 Experimental

2.1 Materials

P 25 TiO_2 (particle size: 20 nm, BET surface: 60 m^2 g^{-1}). P 25 was supplied by Evonic-Aeroxide. Kronos VLP 7000 TiO_2 (BET surface: 250 m^2 g^{-1}) was provided by Kronos International Inc, Hydrochloric acid (HCl) and sodium hydroxide (NaOH) from R&M chemical were used in this study. Reactive red (RR4) dye or commonly known as Cibacron Brilliant Red (Colour Index Number: 18105, chemical formula: $C_{32}H_{23}ClN_8Na_4O_{14}S_4$, molecular weight: 995.23 g mol^{-1}, λ max: 517 nm). Ultra-pure water was used to prepare all solutions in this work.

2.2 Methods

2.2.1 Preparation of immobilized P25/DSAT and kronos/DSAT and photodegradation study

Typically, TiO2-P25 solution was prepared by mixing 13 g of N doped TiO2 with 100 ml of distilled water in 250 ml reagent bottle. The solution was undergoing shaking process for 30 minutes using orbital shaker model PSU-20i, Grant-bio to make it homogenize. P25 solution was then coated (immobilized) by using brush technique and DSAT was stacked onto glass plate with a dimension of 13 x 4.8 cm (L x H) prior coated with N-TiO2. The glass cell with immobilized/P25/DSAT was then dried in the oven at 100 °C for 20 minutes. A dried immobilized/PS25/DSAT sample was cleaned by using distilled water under irradiated of 55-W fluorescent lamp model Qusun E27, 6400K in aerated condition for 1 hour prior to photocatalytic degradation.

The experimental was repeated by using kronos VLP 7000. Photocatalytic degradation study for immobilized was carried out by 25 ml of 100 mg L-1 of anionic RR4 dye was poured into a glass cell with a dimension of 150 cm × 10 mm × 80 mm (L×B×H). Immobilized TiO2 was then transfer into glass cell containing RR4 dye and irradiated with a 55 Watt fluorescent lamp at specific time interval until it turns colourless. An aquarium pump model NS 7200 was used as an aeration source to supply oxygen.

2.2.2 Effect of initial dye concentration, initial pH, on the immobilized kronos

Four sets of RR4 dye with different concentrations (100, 200, 300 and 400 mg L^{-1}) were used as the model pollutant in order to study the effect of concentration of the RR4 dye on the photocatalytic efficiency of the photocatalyst. For pH study, a series of 100 mg L-1 RR4 dye solutions at different initial pH values were used. The pH of each solution was adjusted to values of about 1, 3, 8 and 10 using 1 M NaOH and 1 M HCl solutions. The respective immobilized of the catalyst within the dye solutions was then irradiated with 55 Watt fluorescent lamp for 1 hour. Aliquots were taken at every 15 minutes time interval. The concentration of the remaining RR4 dye was measured spectrophotometrically using HACH DR1900 spectrophotometer.

2.2.3 Effect of strength immobilized P25 and kronos

The strength study was carried out by measuring the remaining weight in immobilized/P25/DSAT and immobilized/kronos/DSAT sample after sonication process with ultrasonic bath Cress Ultrasonic, Model 4HT-1014-6.

3 Results and Discussion

3.1 Photocatalytic activity of immobilized/P25/DSAT and immobilized/kronos/DSAT

Figure 1 shows plot of ln (C/C_0) vs. time of immobilized/P25/DSAT and immobilized/kronos/DSAT. This investigation was carried out by using immobilized P25 and kronos under fluorescent lamp. Based on result, it was observed that 90% of degradation of the RR4 was achieved by kronos/DSAT in 1 hr. Meanwhile, about 81% degradation of RR4 was contributed by P25/DSAT in 1 hr. Kronos VLP 7000 contain C-doped TiO_2 and it also known as visible light photocatalyst. Previous study claimed that the band gap of kronos VLP 7000 is around 2.32 eV meanwhile band gap for TiO_2 P25 is around 3.0-3.2 eV [7, 8].

Thus, it can be claimed that the immobilized kronos was degrade the RR4 dye under visible light region and UV light region resulting in more photogenerated electrons and holes thereby will increase the photocatalytic activity. The surface area also play an important role toward the decolorized the RR4 dye since kronos consist high surface area 250 m^2 g^{-1} resulting more absorption of energy enhancing the photocatalytic activity .

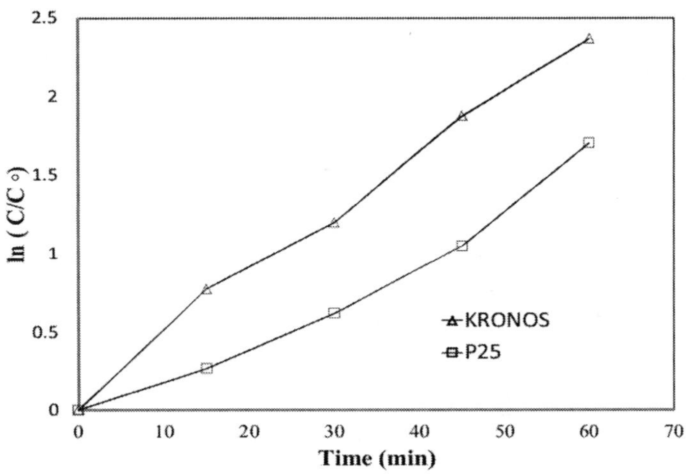

Figure 1. ln (C/C_0) vs. time of immobilized/P25/DSAT and immobilized/kronos/DSAT.

3.2 The effect of initial dye concentration and initial pH on the photocatalytic efficiency of immobilized/kronos/DSAT.

The effect of initial dye concentration on the photocatalytic efficiency of immobilized sample is shown in Figure 2(a). As expected, the photocatalytic activity of RR4 dye decreased with increasing RR4 dye concentration where the highest photocatalytic activity occurred for 100 mg L^{-1} concentration of RR4 dye solution. This phenomenon caused by the color of solution becomes more intense while increasing the concentration of solution. The light of fluorescent lamp was retarded to penetrate into the surface of TiO_2 resulting in reducing the formation of OH• and O_2• radicals thereby decreasing the photogenerated of electrons and holes. Figure 2(b) shows the rate of the photodegradation of RR4 under immobilized/kronos/DSAT at different pH conditions as a function of the irradiation time. As a result showed that the photocatalytic degradation increase with decreasing the pH value. At pH 1-3, a significantly high electrostatic attraction exists between negatively charged anionic dyes which is RR4 and positively charged surface of the TiO_2 kronos . This phenomenon is due to the amphoteric behavior of the TiO_2 semiconductor. The potential at zero charge of TiO_2 surface is at pH_{pzc}= 6.80 [9, 10]. The surface charged of TiO_2 is positively charged in acidic PH (PH<6.8) and negatively charged in alkaline pH (pH>6.80). As a consequence of TiO_2 surface above pH ~ 6.8 is negatively charged hence repulsive force dominates between negative TiO_2 surface and RR4 anions. So the basic pH (pH 8-10) will cause poor degradation of RR4 molecules.

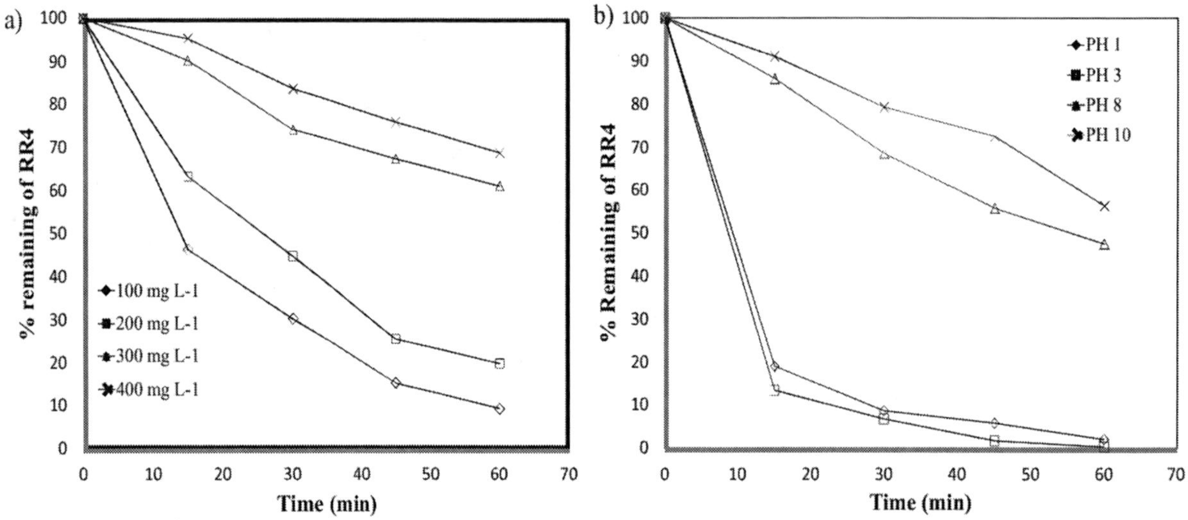

Figure 2. Photocatalytic degradation at different initial concentration: (a) and initial pH, (b) of RR4 using immobilized/kronos/DSAT sample irradiated with 55 Watt fluorescent lamp.

3.3 Effect of strength of immobilized/P25/DSAT and immobilized/kronos/DSAT

TiO_2 layer in immobilized/P25/DSAT is only remain 52 wt.% meanwhile, the result showed by immobilized/kronos/DSAT is only remain almost 37 wt.% after sonication process at strength test as can be seen in Figure 3. It is well expected because no binder was applied on entire immobilization system. But, the immobilized TiO_2 by using DSAT as a thin layer binder significantly improved the photocatalytic degradation rate of RR4.

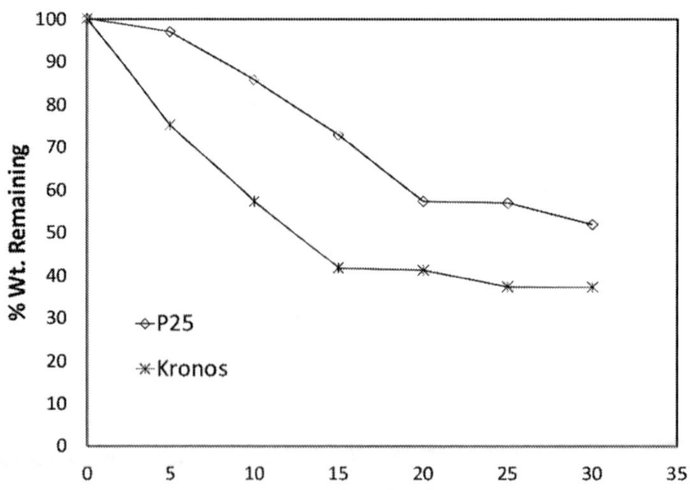

Figure 3. Effect of strength on immobilized/P25/DSAT and immobilized/kronos/DSAT under sonication process.

4 Conclusions

In this work,the comparative study was done for photocatalytic degradation of RR4 using TiO_2kronos VLP 7000 and TiO_2 P25. The immobilized technique was successfully applied this work. The initial pH has significant effect on photocatalytic degradation. The result showed that the photodegradation anionic RR4 is more active toward the acidic pH due to the amphoteric behavior of TiO_2. The ability of kronos VLP 7000 to enhanced photocatalytic degradation under visible light that related with high surface area was showed higher photocatalytic activity than P25. It was observed that 90% and 81 % of degradation was achieved by kronos VLP 7000 and P25 in 1 hr respectively under fluorescent lamp. Less strength showed by immobilized/kronos/DSAT when test on sonication process compared to the layer of coated P25. This research studies will contributed toward the eco-environmental that can be great potential to becoming preliminary treatment of wastewater especially for dyes industries. This strength study of immobilized can be proved that its can be purify contaminated water continuosly since the P25 and kronos can intact even not strong enough intact on the glass plate. Hence, the polymer binder like Polyvinyl alcohol (PVA), Polyethylene glycol (PEG), and others polymers can be apply for future work of immobilized study to overcome the strength problem.

References

[1] X.P. Wang, and T.T. Lim, Solvothermal synthesis of C-N codoped TiO_2 and photocatalytic evaluation for bisphenol a degradation using a visible-light irradiated led photoreactor, Appl. Catal., **100**, 355-364,(2010).

[2] S. Bangkedphol, H.E. Keenan, C.M. Davidson and A.W. Sakultantimetha, Enhancement of tributyltin degradation under natural light by N-doped TiO2 photocatalyst, J. Hazard. Mater., **184,** 533-537,(2010).

[3] D. Chen, Z. Jiang, J. Geng, Q. Wang and D. Yang, Carbon and nitrogen co-doped TiO_2 with enhanced visible light photocatalytic activity, Ind. Eng. Chem. Res., **46**, 2471-2746, (2007).

[4] M. Cristina Fernandez, M. Soledad Larrecshi, M. Pillar Callao, An analytical overview of processes for removing organic dyes from wastewater effluents, Trac-Trend Anal. Chem., **29**, 1201-1211,(2010).

[5] M.A. Nawi and S.S. Sheilatina, Photocatalytic decolourisation of reactive red 4 dyes by an immobilised TiO_2/chitosan layer by layer system, J. Colloid Interface Sci., **372**, 80-87, (2012).

[6] T.K. Tseng, Y.S. Lin, Y.J. Chen and H. Chu, A Review of photocatalysts prepared by sol-gel method for VOCs removal, Int. J. Mol. Sci., **11**, 2336-2361, (2010).

[7] T. Fotiou, T. Triantis, T. Kaloudis and A. Hiskia, Photocatalytic degradation of cylindrospermopsin under UV-A, solar and visible light using TiO_2, Mineralization and Intermediate Products, Chemosphere, **119**, 89-94, (2015).

[8] A. Manassero, M. Lucila Satuf and O. Mario Alfano, Evaluation of UV and visible light activity of TiO_2 catalysts for water remediation, Chem. Eng. J., **225**, 378–386, (2013).

[9] C. Hu, J.C. Yu, Z. Hao and P.K. Wong, Effects of acidity and inorganic ions on the photocatalytic degradation of different azo dyes, Appl. Catal., **46**, 35-47, (2003).

[10] K. Bubacz, J. Choina, D. Dolat and A. W. Morawski, Methylene blue and phenol photocatalytic degradation on nanoparticles of anatase TiO_2, J. of Environ. Stud., **4**, 685-691, (2010).

MATEC Web of Conferences **47**, 05016 (2016)
DOI: 10.1051/matecconf/20164705016

An Overview of Composting Based on Variable Feedstock Material

Aeslina Abdul Kadir[1,a], Siti Noratifah Jamaludin[1] and Nur Wahidah Azhari[1]

[1]*Faculty of Civil and Environmental Engineering, University Tun Hussein Onn Malaysia, 86400 Parit Raja, Johor, Malaysia*

Abstract. Composting is a biological treatment method that provides a potential sustainable way to convert food waste into organic compost. In composting, the feedstock material is an important item to ensure the success of the composting process. This paper reviewed the process of composting based on implementation different types of feedstock, namely: 1) animal waste such as cow dung, poultry litter, swine manure and chicken manure; and 2) agricultural waste such as sawdust, rice straw, bran, bagasse, banana waste and pine chip. The result for poultry litter, cow manure, swine manure, sawdust and rice straw has C/N ratio lower than 20 at final composting process which is considered as satisfactory level for compost maturity. As a conclusion, the selection of the feedstock material is based on the characteristics of the material itself and the selection of materials is important for the quality of compost.

1 Introduction

In Malaysia, solid waste generation has increased more than 91% over the past 10 years, due to the rapid development of urban areas [1]. Currently, Malaysia produced approximately 30,000 metric tons of waste every day or 1.3 kg/person/day [2]. The waste consists of domestic and industrial refuse. Waste management due to rapid urbanisation has become a serious concern for government departments, pollution control agencies, regulatory bodies and also public.

Proper waste management is very important to prevent further environmental destruction and to promote sustainable development. Sustainable development aims to meet human needs while preserving the environment so that these needs can be met not only in the present, but also for future generations. The 3.6 % annual increase in solid waste generation required appropriate facilities and technologies which are unfortunately not available to match the requirement for sustainable waste management [3]. Sanitary landfills are only 3% of the total existing disposal sites in the country while the remaining are non-sanitary landfills [4]. This makes the management of waste in Malaysia is more delicate matter to tackle and constrain to establish a sustainable waste management system. The characteristics of municipal solid waste (MSW) components also play important roles to determine the suitability of the disposal systems. Solid waste composition in most Asian countries is highly biodegradable with high moisture content such as food waste, paper, agriculture waste and textiles [5]. In Malaysia, the average components of MSW are quite similar with the largest categories consist of food waste (43.5%), plastic (25.2%), followed by paper (22.7%), glass (2.6%), ferrous (2.1%), textiles

[a] Corresponding author : aeslina@uthm.edu.my

(0.9%) and lastly 1.8% for others [6]. Therefore, a continuous effort is required to identify the most suitable alternative for long terms solutions to reduce the burden of existing MSW disposal systems such as open dumping and landfilling specially on food waste.

The issues of food waste that represent the highest percentage in MSW component are now starting to become a regional issue even for developed countries, thus alternative method is in need. For countries such as Korea and Japan, they practice the separation of food waste from other solid waste because of the biodegradability of the food waste, lack of landfill space as well as transport and disposal of food waste problem that occurs in many combustion processes [7].

Food waste is one of the important components of MSW generated and the disposal in landfills has critical environmental effects [8]. Furthermore, a constantly increasing population makes the land available to use as landfills is more difficult to find and local governments are facing with the need to extend the life of their landfills and avoid the cost of developing a new landfill.

In Malaysia, the solution to solve the problem of solid waste management based on the concept of environmentally acceptable to the public is in need. The commingled in Malaysia without any separation need to be resolved immediately because food waste accounts for almost 45% of municipal solid waste. Total generation of food waste is a major cause of many problems associated with the landfill, such as odor, toxic leachate, and greenhouse gas emissions [9].

Food waste recycling in Malaysia is rarely implemented. There are several options for implementing the method of recycling food waste such as composting, anaerobic digestion or biogass technology and recycling of yellow grease. Few numbers of countries have implemented the recycling method including Germany and Britain [10] that perform the anaerobic decomposition by nurturing methane gas to be used as fuel. Countries like Japan, China, Indonesia and Philippines are known for using backyard composting and successfully reduce their food waste produced.

2 Composting

Composting is a natural recycling system that breaks organic matter down into valuable nutrition to the soils. The benefit of composting is that it can reform soil structure, texture and aeration and improve the capacity of the soil to hold water. Furthermore, utilizing composting as a fertilizer could develops good root formation in soil. Composting also has the ability to improve soil conditions and can contribute to the reduction of greenhouse gas emission and groundwater pollution as well as reducing the total amount of waste [11].

There are many types of composting and each method has different features. Generally, there are two major types of composting namely anaerobic and aerobic. The first method is processed in the absence of oxygen and the second one is processed in the presence of oxygen. Anaerobic composting is less labour intensive as it does not require frequent turning of the pile. It is a low-temperature process because the process of decomposition by anaerobic bacteria requires less energy released [12]. However, unlike aerobic composting, anaerobic compost maintains both seeds of weeds and pathogenic microorganisms because the temperature produced is not high enough to kill them. Hence, it is more unsanitary and there might be higher incidence of serious diseases occurred. Moreover, anaerobic composting decomposes at a very slow rate and produces a strong odour.

Conversely, aerobic composting is a high-temperature process because the process of oxidizing compounds requires more energy. However, the temperature will decrease gradually after intense activity of microorganisms. Therefore, occasional monitoring of moisture and temperature are needed to maintain air circulation during this process. Therefore, it will need more labours to mix the pile thoroughly. However, the process of decomposing is faster than anaerobic methods and there is less unpleasant odour [13]. People usually show their resistance towards home composting due to the particular smell or the presence of flies [14]. Hence, aerobic composting may achieve higher public acceptance by local people. Composting method continues to attract more users not only due to being an environmentally sustainable product but also for its great qualities as soil amendment. Compost utilization will assist to improve soil quality and reduces soil loss, and also increases soil water retention and reduce the need for extra inputs [15].

3 Overview of Composting Based on Variable Feedstock Material

Composting is the best process to reduce the amount of waste in landfill, since up to 30% of the materials send to the landfill is organic and could be composted. Compost application returns nutrients to the land, hold moisture in gardens and on lawns, contributes to watershed health by controlling run-off and naturally fertilize and provide structure to the soil. There are many types of animals and agricultural waste that can be used on compost making. It is essential to mix carbon-rich materials such as dried leaves, woody materials and straw with nitrogen-rich materials such as grass, fresh plant cuttings and manure because different types of organic matter contain different proportions of carbon (C) and nitrogen (N). C/N ratio for compost should be maintain at 25-50 because if the C/N ratio is in lower ratios, ammonia is given off and biological activity is also impeded at lower ratios. On the other hands, at higher ratios nitrogen may be a limiting nutrient. This review will consist of two main feedstock materials which are animal waste and agricultural waste.

3.1 Animal waste

Compost made from animal waste and other organic refuse can serve as a valuable nutrient resource to the agricultural fields, and decrease environmental load if they are treated properly. Frank et al. [16] used cow dung and poultry litter along with banana peels in their study. Poultry litter and cow dung are wastes that may fit in the category of nitrogen supplements (low C/N ratio <12). The characteristics of cow dung shows that pH value with 6.6, electrical conductivity (EC) with 11.1 dS m-1, C/N ratio with 13.4, phosphorus (P) value with 2.41 g/kg and potassium (K) value with 38.61 g/kg. On the other hand, the characteristics of poultry litter is different with pH value of 7.7, EC value with 10.01 dS m-1, C/N ratio with 7.6, P value with 14.8 g/kg and K value with 17.62 g/kg. The C/N ratio of the cow dung and poultry litter itself is in low ratio which is good to help the compost in the maturity process. The result of the compost with cow dung and banana peels started with C/N ratio of 15.5 and after 12 weeks the value drop to 10.5. While the C/N ratio value for compost with poultry litter and banana peels is start with 14.6 and becoming 10.2 after 12 weeks. The final compost C/N ratio of the compost was within the range of 10-15, which is ideal for final stabilized composts [17].

Cow manure also have been used by Nonglak et al. [18] along with urea fertilizer, tobacco plant waste and compost inoculum in their study. The cow manure were air dried at 80°C for 24 hours and pressed trough a 0.2 mm sieve. The result shows that the highest temperature of the compost is 62°C, pH value with 11.78 to 7.78 and initial value of C/N ratio with 17.76 and then decreased to 12.27 at day 50. C/N ratio in this study also cannot be used as an absolute indicator of compost maturation same as Frank et al. [16] study since the initial C/N ratio was below 20. However, the compost in this study is apparently reached maturity faster within 30 to 40 days as also indicated by the decreased of temperature. Cow manure can be concluded as a good feedstock material as it help to speed up the maturity process of compost.

Swine manure can be an excellent source of nutrients for crop production and to be known as an economical fertilizer. Jishao et al. [19] used fresh swine manure as the raw materials along with wheat straw at a ratio of 10.5: 1 in aerated static composting box with dimension 0.65 m x 0.50 m x 0.40 m. The moisture content of the swine manure is 70.9% with C/N values of 12.8. The result shows that the initial value of the compost with pH value of 6.5, EC value with 3.5 and C/N ratio with 12.9. The maximum temperature in the compost was 65.2°C at day 8. The final value of the compost is pH with 8.9, EC value with 2.2 ms.cm-1 and C/N ratio with 9.4. In this study, C/N could not be as an indicator because of the low initial C/N (10-13), so the ratio of the ending value of C/N to the initial C/N (T) was used to estimate the maturity. The compost reached maturity when T value was ranged from 0.53 to 0.72 [20]. The value of T in this study is 0.73 which is almost reached the requirement of maturity. So, the study period should be longer to let the compost achieve the requirement of maturity. Therefore, swine manure is still a good feedstock material as it help the maturity process but it need a bit more time to achieve it.

Chicken manure is also full of nutrients that will benefit the gardening plot. It has been used by Naser et al. [21] as one of the feedstock for their co-composting study along with sawdust. The EC of chicken manure is 7.49 ds m-1 and the pH value is 7.99. The characteristics of the chicken manure shows that the C value is 33.04%, N value is 3.73%, P value is 2.02% and K value is 2.69%. The result shows that the C/N ratio started with 24 and ended with 34. The C/N values were high which is higher than the optimum value of 21 for matured compost. This is due to the high amount of recalcitrant carbon present in the feedstock mixture. Thus, chicken manure is not a good help for compost to achieve the maturity by depending on the C/N value

3.2 Agricultural waste

In general, most of agricultural waste such as crop residues, sawdust, rice straw and others contain large amounts of carbon but low amounts of nitrogen and will cause slower decomposes process. So the organic material of the compost needs to contain the correct proportions of carbon and nitrogen. This is because the microorganisms that decompose the organic matter need both carbon and nitrogen to function well.

Many of previous researchers used sawdust as one of the material in their study. Sawdust is the most widely used bulking agent in the composting systems. Ying et al. [22] used sawdust along with food waste in their study at 1:1 ratio and the characteristics shows that the moisture content of the sawdust is 7.24%, the TOC value is 52.9%, the TKN value is 0.59% and the C/N ratio is at 89.8 respectively. The result shows that the pH reached 8.0 after 21 days then was stable during the rest of the composting process. As for C/N ratio the value decreasing from 25.5 to 18.1 after 56 days. The reduction of total carbon content but an increase in total nitrogen content resulted in a decrease of C/N ratio indicating that the composts were becoming more stable and mature. C/N ratio less than or equal to 20 is to be considered as satisfactory level for compost maturity [23], which is prove that sawdust is a good feedstock material if the mixture used is suitable.

Jiachao et al. [24] used rice straw which is difficult to be decomposed along with bran, soil and vegetables at ratio of 11:2:8:3. The rice straw was air-dried and cut into 10 to 20 mm lengths. The characteristics for rice straw are determined. The moisture content is 11.73%, TOC value is 428.0 g kg-1, TN value is 8.8 g kg-1 and C/N ratio is 48.8 accordingly. As for bran, the characteristics are 14.06% for moisture content, TOC value with 474.1 g/kg, TN value with 41.2 g/kg and C/N ratio with 11.5 respectively. The result shows that the temperature of the pile compost increased significantly to 56°C at the first 4 days and decreased gradually to the ambient level after a few days. On the other hand, C/N ratio value decreased from 31.15 to 16.62 due to the decomposition of the organic matter by the microorganisms during the composting process and reached the maturity level of the compost. Thus, rice straw is a good feedstock material as the compost final result achieve the maturity level at satisfactory time.

In another research, bagasse, pine chip and banana waste was used by Brito et al. [25] as the materials for their composting at a ratio of 5:2.5:2.5. The characteristics for baggase indicates that the pH value is 4.9, EC value with 4.72 dS m-1, organic matter (OM) with 81.5%, TOC value with 40.4% and N value with 3.4%. As for banana waste the pH value is 5.5, EC value is 1.34 dS m-1, OM value is 42.1%, TOC with 1.63% and N value is 1.63%. Meanwhile, the characteristics for pine chip is pH with 5.3, EC value with 0.20 dS m-1, OM value with 86.6%, TOC value with 53.9% and TN 0.28%. The result shows that the pH value is around 4.7 to 6.7, the TN value increased from 2.36% to 4.07% and the C/N ratio also decreasing from the beginning of the process until day 126 with the value of 9.6 which is showed an acceptable degree of maturation [26]. Therefore, the mixture of bagasse, pine chip and banana waste as feedstock material in composting is also proven to be good.

4 Conclusions

Based on the literature review, we can see that the selection of the feedstock materials is based on the characteristics of the material itself. The selection of the organic material can be any type of plants or animals with suitable proportions of C and N. The microorganisms decompose the organic matter need both carbon and nitrogen to function well. Nitrogen-rich material such as fruit, vegetables peels and young leaves that expedite decomposes process that contains low levels of carbon but high levels of nitrogen. Carbon-rich material such as crop residues and small twigs that decomposes slowly contains large amounts of carbon but low amounts of nitrogen. Too little nitrogen in material means the composting process will be slow, too much of it will result in the material becoming acid and smelly. The ideal ratio of carbon and nitrogen for starting a compost pile is 25-50: 1. Therefore, the selection and combination of feedstock materials is important for the quality of compost. Most of the previous study successfully reached the standard of maturity with the compost feedstock material their chosen.

References

[1] P. Agamuthu and S.H. Fauziah, MSW disposal in Malaysia: Landfill management, *Proc. of the 2nd Expert Meeting on Solid Waste Management in Asia and the Pacific Islands*, Kitakyushu, (2006).

[2] P. Agamuthu, K.M. Khidzir and F.S. Hamid, Drivers of sustainable waste management in Asia, Waste Management and Research, **27**, 625–633, (2009).

[3] P. Agamuthu and S.H. Fauziah, Sustainable management of wet market waste, *Proc. of the Int. Conf. on Sustainable Solid Waste Management*, 239-243, (2007).

[4] S.H. Fauziah, Municipal Solid Waste Management: A Comprehensive Study in Selangor, PhD Thesis, University of Malaya, Kuala Lumpur, (2009).

[5] C. Visvanathan, O. Tubtimthai and P. Kuruparan, Influence of landfill top cover design on methane oxidation: Pilot scale lysimeter experiments under tropical conditions, *Proc. of the 3rd Asia Pasific Landfill Symposium*, Kitakyushu, (2004).

[6] K.H. Chua, E.J.M. Sahid and Y.P. Leong, Sustainable Municipal Solid Waste Management and GHG Abatement in Malaysia, *ST-4: Green and Energy Management*, 1-8, (2011).

[7] M.H. Kim and J.W. Kim, Comparison through a LCA evaluation analysis of food waste disposal options from the perspective of global warming and resource recovery, Science of the Total Environment, **408** (19), 3998-4006, (2010).

[8] Environmental Protection Agency, Basic Information about Food Waste, (2009) <retrieved from http://www.epa.gov/osw/conserve/materials/organics/food/fd-basic.htm>

[9] S.H. Lee, K.I. Choi, M. Osako and J.I. Dong, Evaluation of environmental burdens caused by changes of food waste management systems in Seoul, Korea, Science of the Total Environment, **1**(3), 42-53 (2007).

[10] A.D. Knipe, The management of household food waste, *Environmental Research and Consultancy*, West Morden, United Kingdom, (2005).

[11] C. Muller, Decentralised composting in developing countries: Financial and technical evaluation in the case of Asmara City, Diploma Thesis of Christian Müller, Swiss Federal Institute of Technology, Zurich, (2005).

[12] D. Gaschk, T. Ayaka, V. Hanh, Dominique Wisniewski, Waste Management in Devikulam. Sustainability in Developing Communities, (2009)

[13] C. Henry, Chapter VII: Composting, (2005). <retrieved from http://www.unep.or.jp/ietc/publications/spc/solid_waste_management/Vol_I/14Chapter.pdf >

[14] M. Ali, Sustainable composting: Case studies and guidelines for developing countries, Water, Engineering and Development Centre, Loughborough University, Leicestershire, (2004).

[15] United States Composting Council, Greenhouse gases and the role of composting: A primer for compost producers, (2008). <retrieved from http://www.compostingcouncil.org>

[16] K. Frank,N. Eiji, E. Tsuneyoshi, A. Zahoor, Y. Rumana, M.T. Moses and Y. Sadahiro, An evaluation of aerobic and anaerobic composting of banana peels treated with different inoculums for soil nutrient replenishment, Bioresource Technology, **126**, 375-382, (2012).

[17] A. Boldrin, I. Korner, U. Krogman and T.H. Cristensen, Composting:mass balances and product quality, Solid Waste Management, **978**, 1-3, (2010).
[18] S. Nonglak, D. Srisulak, S. Panalak, J. Somsak, B. Sukanya, W. Janewit and J. Chaiwat, Composting of tobacco plant waste by manual turning and forced aeration system, Maejo Int. J. Sci. Technol., **3**(02),248-260, (2009).
[19] J. Jishao, H. Yimei, L. Xueling and H. Hua, The effects of apple pomace, bentonite and calcium superphosphate on swine manure aerobic composting, Waste Management, **34**, 1595-1602, (2014).
[20] M. Itavaara, M. Vikman and O. Venelampi, Windrow composting of biodegradable packaging materials, Compos. Sci. Utiliz., **5**, 84-92, (1997).
[21] K. Naser, C. Ian, A.S. Miguel, S. Syd, M. Sebastian and B. Nanthi, Maturity indices in co-composting of chicken manure and sawdust with biochar, Bioresource Technology, **168**, 245-251, (2014).
[22] Z. Ying, S. Ammaiyappan and W.C.W. Jonathan, Evalution of humic substances during co-composting of food waste, sawdust and chinese medicinal herbal residues, Bioresource Technology, **168**, 229-234, (2014).
[23] M.F. Hirai, V. Chanyasak and H. Kubota, A standard method for measurement of compost maturity, Biocycle, **24**, 54-56, (1983).
[24] Z. Jiachao, Z. Guangming, C. Yaoning, Y. Man, Y. Zhen, L. Hui, Y. Yong and H. Hongli, Effects of physico-chemical parameters on the bacterial and fungal communities during agricultural waste composting, Bioresource Technology, **102**, 2950-2956, (2011).
[25] E. Brito, M.A. Bustamante, C. Paredes, J. Moreno-Caselles, M.D. Perez-Murcia, A. Perez-Espinosa and R. Moral, Composting of brewery wastes with agricultural and forest residues. <retrieved from http://www.ramiran.net/ramiran2010/docs/Ramiran2010_0114_final.pdf>
[26] M.P. Bernal, C. Paredes, M.A. Sanchez- Monedero and J. Cegarra, Maturity and stability parameters of compost prepared with a wide range of organic waste, Bioresource Technol., **63**, 91-99, (1998).

MATEC Web of Conferences **47**, 05017 (2016)
DOI: 10.1051/matecconf/20164705017

A Comparison Study of New TiO_2/PEG Immobilized Techniques Under Normal and Visible Light Irradiations

R. Zaharudin[1,2,a], S. K. Ain[1,2], F. Bakar[1,2], M. S. Azami[1,2] and W. I. Nawawi[2]

[1]*Faculty of Applied Sciences, Universiti Teknologi MARA, 40450 Shah Alam, Selangor, Malaysia*
[2]*Faculty of Applied Sciences, Universiti Teknologi MARA, 02600 Arau, Perlis, Malaysia*

Abstract. Novel photocatalysts of TiO_2 and TiO_2/PEG were immobilized using new technique which is double sided adhesive tape (DSAT) as a thin layer binder onto glass plate. The photocatalytic activity study of immobilized/TiO_2 and immobilized/TiO_2/PEG were carried out by irradiating of 36 mg L^{-1} MB dye in the presence of normal and visible light with rate of decolourization was estimated from aliquot concentration spectrophotometrically. The high photocatalytic activity from immobilized/TiO_2/PEG was observed with the rate constants of ca 0.06 and 0.02 min^{-1} under normal and visible light irradiation respectively. An active photocatalytic activity was observed for immobilized/TiO_2/PEG sample while no photocatalytic activity detected under immobilized/TiO_2. The high photoactivity was also detected under normal light for immobilized/TiO_2/PEG sample. This might be due to the presence of electron conjugate in PEG that acted as electron donor for TiO_2 thus increase the photocatalysis process. Moreover, this electron donor is reported to be able to become active under low energy visible light hence the immobilized/TiO_2/PEG can perfomed its photoactivity under visible light irradiation.

1 Introduction

TiO_2 is commonly known for its use in water purification since it is cheap, chemically inert and has high photo activity [1-3]. TiO_2 with a band gap of of 3.2 eV, is capable to degrade a variety of organic compounds via photo catalysis [4-5]. Nevertheless, it can only absorb UV (which constitutes 4% of the solar light) due to its wide energy band gap. Several methods have been assessed to shift its optical response to visible light region such as TiO_2 doping with metal and non-metal compound [6-8]. Even though its optical response could be shifted to Visible Light (VL) region, the preparations for samples were complicated. Currently, unsaturated polymers were applied as photosensitizer to produce the optical response of VL. For instance, Su et al. [9] had developed TiO_2/polymer by calcination process and the optical response was shifted to VL region. In addition, Zhu et al. [10] suggested that conjugated polymer such as poly (fluorene-co-bithiophene) (PFB) combined with TiO_2 to be an effective photosensitizer since 75 % of phenol was removed under visible light irradiation. For degradation of methyl orange [11], TiO_2/PVA (polyvinyl alcohol) showed a significant visible-active photodegradation. Other conjugated polymers [12] such as polypyrrole, polythiophene and polyaniline have also been reported and exhibited similar photodegradation pattern.

[a] Corresponding author : nurraihanzaharudin@gmail.com

In this research, a novel photocatalyst (TiO_2/PEG DSAT) was successfully prepared using brush coating method. More importantly, DSAT method was firstly reported to produce TiO_2/polymer film. The prepared composite film could effectively degrade Methylene Blue (MB) under normal and VL irradiation. Due to its cheapness and conveniences, TiO_2/PEG may contribute to application to the disposal of industrial effluents.

2 Experimental

2.1 Materials and preparation of photocatalyst

The sample solution was prepared by adding 6.5 g of titanium dioxide (TiO_2) Degussa P25 powder in 50 mL of deionised (DI) water added with polyethylene glycol (Merck, MW: 6000). The sample solution is thoroughly mixed through mechanical stirring. The compositions of the solution for obtaining immobilized/TiO_2/PEG films and molecular structure for PEG were presented in Table 1 and Figure 1 respectively. The composition presented in Table 1 has been selected based on the previous works that has established the experimental conditions to prepare pure (unmodified TiO_2) and modified TiO_2/PEG films. The sample films were prepared by using a brush-coating method applied onto glass substrates, which was cleaned and taped with Double Sided Adhesive Tape (DSAT) beforehand. The coated glass was then dried in the oven at 100 °C for 15 minutes.

Table 1. The composition of the solution for obtaining immobilized/TiO2/PEG DSAT films.

Samples	TiO_2 (g)	DI water (mL)	PEG (8g in 100 mL of DI water)
Immobilized/TiO_2	6.5	50	-
Immibilized/TiO_2/PEG	6.5	50	1 mL

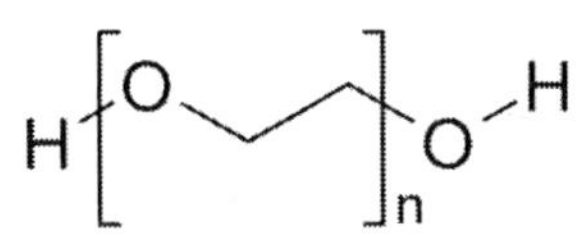

Figure 1. Molecular structure for poly ethylene glycol (PEG).

2.2 Photocatalytic degradation study

The activity of the catalyst was tested by degradation of methylene blue (MB; Fluka Analytical). The experimental setup used was to flow a coated glass with 20 mL of 36 mg L-1 concentration of MB dye inside a glass cell reactor (dimension: 50 mm Length x 10 mm Width x 80 mm Height). An NS 7200 aquarium pump was used as an aeration source. Area of photo catalytic activity equals to Length X Width of DSAT applied which is 6.6 cm X 4.6 cm.The light intensity of the lamp used were 461 W m^{-2} for Visible and 6.7 wt/m^2 for UV respectively. Direct photocatalysis and adsorption reactions were also performed for the photodegradation efficiency of MB. Turning on the 55-W compact fluorescent light initiated the photocatalytic reaction and outlet absorbance was measured. A 4 mL aliquot was extracted from the glass cell reactor at pre-specified intervals. The absorbance of the aliquot was measured by an UV spectrophotometer (Varian UV-Vis detector model HACH DR 1900) set to 661 nm which corresponds to the MB absorbance. Absorbance was converted to MB concentration using a Pseudo First Order plot. With the Pseudo First Order plot, absorbance was converted to concentration and rate of degradation (k) was calculated. The experimental design was modified from previous works [13-15].

3 Results and Discussion

3.1 Photocatalytic study

Photocatalytic degration of MB dye by using immobilized/TiO_2/PEG and immobilized/TiO_2 under normal and visible light irradiations are shown in Figure 2(a) and (b) respectively. A complete MB decolorization for both immobilized photocatalysts with high photocatalytic activity was observed under immobilized/TiO_2/PEG which can be seen in Figure 2(a). It is evident that under normal light irradiation PEG gives a high photoactivity. Since PEG acts as a pore forming agent, [16-18] thus, this characteristic displays increased porosity, which resulted in high numbers of reaction sites. Therefore, a lot of photocatalytic reaction may take place onto immobilized/TiO_2/PEG surface. The degradation rate obtained at 0.06 min-1 was found to be higher than ca 0.024 of other immobilized TiO_2/polymer [19]. The lower degradation rate is subjected to less reaction sites available due to utilization of various cross-linking methods. A complete MB decolorization takes approximately 90 minutes for immobilized/TiO_2/PEG while immobilized/TiO_2 takes almost 2 hours. In Figure 2(b), decolorization of MB was observed in immobilized/TiO_2 under visible light irradiation. However, it is just observed as a result for adsorption of immobilized/TiO_2. Moreover, it is generally accepted that pristine TiO_2-has wide bandgap energy thus make visible light disable to activate electron-hole pair. Hence, no photocatalytic activity will happen under visible light as well as observed on immobilized/TiO_2 sample in Figure 2(b).

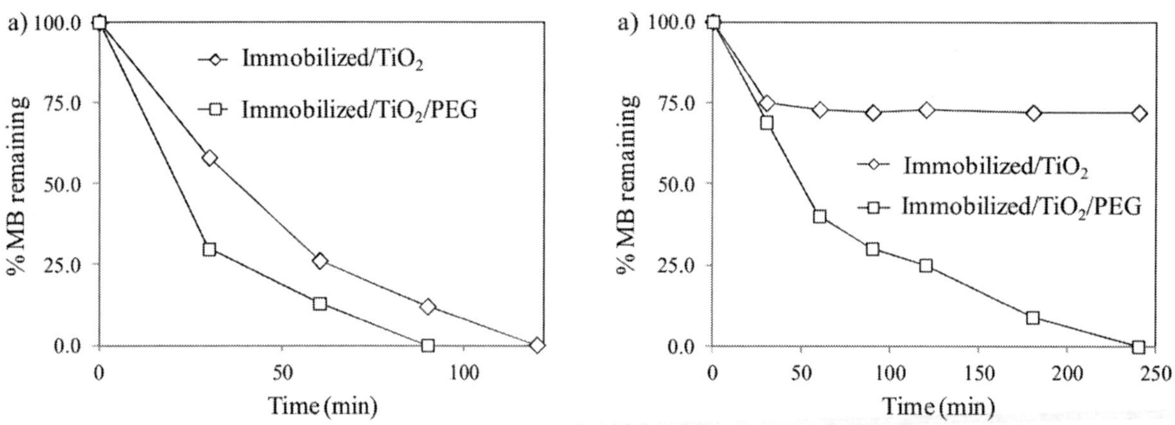

Figure 2. Decolorization of 36 mg L^{-1} MB dye using immobilized/TiO_2 and immobilized/TiO_2/PEG under a) normal and b) visible irradiations.

In contrary, an active photocatalytic activity was observed on immobilized/TiO_2/PEG sample under visible light irradiation. As can be seen in Figure 2(b), immobilized/TiO_2/PEG sample can perform a complete decolorization of 36 mg L^{-1} MB which takes around 240 minutes. An active visible light of immobilized/TiO_2/PEG might be due to the conjugated unsaturated polymer from PEG which is similar in photosensitized mechanism to the dye on TiO_2 under visible light. The sensitizer molecules adsorbed on TiO_2 are excited by visible light and electrons are subsequently injected to conduction band (CB) of TiO_2. While the CB acts as a mediator for transferring electrons from the sensitizer to substrate electron acceptors on TiO_2 surface, the valence band (VB) remains unaffected in a typical photosensitization. The excited electrons in CB of TiO_2 create a series of chain reactions and this explained visible light active for immobilized/TiO_2 sample. Subsequently, this behavior is expected as PEG is proven to enhance the transmittance value from 380-540 nm. According to Calderon et al. [20], the optical response has already shifted into VL region. For this reason, immobilized/TiO_2/PEG produced is visible-active by using PEG [21]. Nevertheless, through this research paper, the enhancement has been made better since less power and less chemical are consumed, as compared to the previous work.

3.2 Strength study

Effect of strength for immobilized photocatalysts were determined by % remaining of photocatalysts after exposed under ultrasonic vibration as testing media. Figure 3 shows the % remaining of immobilized/TiO_2 and immobilized/TiO_2/PEG under sonication process. Weak adherence strength was observed on immobilized/TiO_2 sample since 56 % of photocatalyst remained on to the glass plate when applied under 20 seconds of ultrasonic vibration as opposed to TiO_2/PEG DSAT. Interestingly, coating adherence was proven to be better with PEG polymer supports with 92 % of catalyst on immobilized/TiO_2/PEG remained after 20 seconds of sonication. The strong intact on immobilized/TiO_2/PEG was due to the presence of PEG polymer to bind the TiO_2 particles. Hence, the developed immobilized/TiO_2/PEG prepared via brush coating method was proven to be strong, intact and durable. The evidence suggests that strong surface of immobilized/TiO_2/PEG is devoted to application of DSAT. It can be noticed that the photocatalyst film with PEG6000 m.wt showed a uniform, crack-free, porous and good adhesion with glass substrates [22]. In comparison to the previous work, the application of DSAT gives superior characteristic which greatly improves the durability of immobilized TiO_2 system. By far, this finding has confirmed that immobilized/TiO_2/PEG has already suppressed other immobilized TiO_2 system showing prominent strength in binding TiO_2/PEG surface.

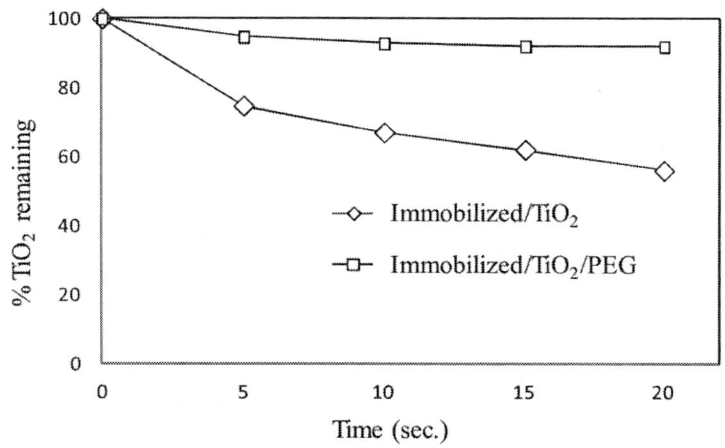

Figure 3. Percentage of immobilized/TiO_2/PEG and immobilized/TiO_2 weight remained after sonication.

4 Conclusions

The immobilized/TiO_2/PEG samples were successfully prepared by means of brush coating method. Immobilized/TiO_2/PEG exhibited higher visible light photocatalytic activity for the MB degradation in contrast to immobilized/TiO_2/DSAT. The photocatalytic activity under visible light was obviously depended on PEG dosage. The TiO_2/PEG DSAT exhibited higher activity for MB degradation which took 240 minutes to complete decolorization of MB dye under visible light irradiation. Through brush coating method, the photocatalyst surface was strong and intact in comparison to TiO_2/DSAT surface. In brief, the visible light photocatalytic activity in immobilized/TiO_2/PEG was attributed to the conjugated unsaturated polymer from PEG which is similar in photosensitized mechanism to the dye on TiO_2 under visible light.

References

[1] D. Mukherjee, Development of A Novel TiO_2-Polymeric Photocatalyst for Water Purification Both under UV and Solar Illuminations, PhD Thesis, The University of Western Ontario, Canada, (2011).

[2] N. Negishi, T. Iyoda, K. Hashimoto and A. Fujishima, Preparation of transparent TiO_2 thin film photocatalyst and its photocatalytic activity, Chem. Lett., **24**(9), 841-842, (1995).

[3] M.R. Hoffamnn, S.T. Martin, W. Choi and D.W. Bahnemann, Environmental applications of semiconductor photocatalysis, Chem. Rev., **95**, 69-96, (1995).

[4] K. Jatinder, dual effect of photocatalysis and adsorption in degradation of azorubine dye using nanosized TiO_2 and activated carbon immobilized with different techniques. Int. J. of ChemTech Research, **3**, 1537-1543, (2010).

[5] W. Zhao, Z. Bai, A. Ren, B. Guo and C. Wu, Sunlight photocatalytic activity of CdS modified TiO_2 loaded on activated carbon fibers, Appl. Surf. Sci., **256**(11), 3493–3498, (2010).

[6] V. Iliev, D. Tomova, R. Todorovska, D. Oliver, L. Petrov, D. Todorovsky and M. Uzunova-Bujnova, Photocatalytic properties of TiO_2 modified with gold nanoparticles in the degradation of oxalic acid in aqueous solution, Appl. Catal A: Gen., **313**(2), 115–121, (2006).

[7] J. Li, D. Shieh, D. Li, C. Ho, S. Yang and J. Lin, Photodegradation of CH3I on mesoporous TiO_2-B nanofibers with Au nanoparticles, Appl. Surf. Sci., **254**(15), 4655–4664, (2008).

[8] M. Bowker and E. Fourre, Direct interactions between metal nanoparticles and support: STM studies of Pd on TiO_2(1 1 0), Appl. Surf. Sci., **254**(14), 4225–4229 (2008).

[9] B.T. Su, X.H. Liu, X.X. Peng, T. Xiao and Z.X. Su, *Functional Nanostructures: Processing, Characterization and Applications*, Spinger, New York, (2003).

[10] J.F. Zhu, F. Chen, J.L. Zhang, H.J. Chen and M. Anpo, Visible light responsive titania-based nanostructures for photocatalytic, photovoltaic and photoelectrochemical applications, J. Photochem. Photobiol. A: Chem., **180**, 196–204, (2006).

[11] Y.Z. Wang and M.Q. Zhong, Photocatalytic activities of TiO_2 modified by poly (fluorene-co-bithiophene) under visible light, J. Applied Catalysis B: Environmental, **90**, 249–254, (2009).

[12] P.Lei, F. Wang, X. Gao, Y. Ding, S. Zhang, J. Zhao, S. Liu and M. Yang, Immobilization of TiO_2 nanoparticles in polymeric substrates by chemical bonding for multi-cycle photodegradation of organic pollutants, J. of Hazardous Materials, **227–228**, 185-194, (2012).

[13] C.Y Kuo, C.H. Wu and H.Y. Lin, Photocatalytic degradation of bisphenol A in a visible light/TiO_2 system, Desalination, **256**(1-3), 37-42, (2010).

[14] R. Ali and B.S Ooi, Photodegradation of New Methylene Blue in aqueous solution using zinc oxide and titanium dioxide as catalyst, J. of Technology, **45**, 31-42, (2006).

[15] W.I.N. Wan Ismail, S.K. Ain, R. Zaharudin, A.H. Jawad, M.A.M. Ishak, K. Ismail and S. Sahid, New TiO_2/DSAT immobilization system for photodegradation of anionic and cationic dyes, Int. J. of Photoenergy, **2015**, 1-6, (2015).

[16] S. Suzana, C. Lidija, D. Ljubas, V. Svetličić, I.F. Houra and N. Tomašić, Synthesis, characterization and photocatalytic properties of sol–gel TiO_2 films, Ceramics Int., **37**(4), 1153-1160, (2011).

[17] S.J Bu, Z.G Jin, X.X Liu, L. Yang and Z. Cheng, Fabrication of TiO_2 porous thin films using peg templates and chemistry of the process, Materials Chemistry and Physics, **88**(2-3), 273-279, (2004).

[18] A.A. Ramírez-Santos, P. Acevedo-Peña and E.M. Córdoba, Enhanced photocatalytic activity of TiO_2 films by modification with polyethylene glycol, Química Nova, **35**(10), 1931-1935, (2012).

[19] D. Mukherjee, S. Barghi and A.K. Ray. Preparation and characterization of the TiO_2 immobilized polymeric photocatalyst for degradation of aspirin under UV and solar light, Processes, **2**(1), 12-23, (2014).

[20] J.M. Calderon-Moreno, S. Preda, L. Predoana, M. Zaharescu, M. Anastasescu, M. Nicolescu, M. Stoica, H. Stroescu, M. Gartner, O. Buiu, M. Mihaila and B. Serban, Effect of polyethylene glycol on porous transparent TiO_2 films prepared by sol–gel method, Ceramics Int., **40**(1), 2209-2220, (2014).

[21] L.C. Chen, F.R. Tsai and C.M. Huang, Photocatalytic decolorization of methyl orange in aqueous medium of TiO_2 and Ag–TiO_2 immobilized on γ-Al_2O_3, J. Photochem. Photobiol. A: Chem., **170**(1), 7–14 (2005).
[22] N. Tasić , Z. Branković, Z. Marinković-Stanojević and G. Branković, Effect of binder molecular weight on morphology of TiO_2 films prepared by tape casting and their photovoltaic performance, Science of Sintering, **44**(3), 365-372, (2012).

MATEC Web of Conferences **47**, 05018 (2016)
DOI: 10.1051/matecconf/20164705018

Carbon Nitrogen Co-Doped P25: Parameter Study on Photodegradation of Reactive Red 4

M. S. Azami[1,a], W. I. Nawawi [2], M. A. M. Ishak [2], K. Ismail [1], Z. Ahmad [2] and Ali H. Jawad [2]

[1]*Faculty of Applied Sciences, Universiti Teknologi MARA, 40450 Shah Alam, Selangor, Malaysia*
[2]*Faculty of Applied Sciences, Universiti Teknologi MARA, 02600 Arau, Perlis, Malaysia*

Abstract. Photocatalytic degradation rate of reactive red 4 (RR4) using carbon coated nitrogen doped TiO_2 (C N co-doped TiO_2) in photocatalysis process is main goal on this research. The main operating the parameters such as effect of initial dye concentration, catalyst loading, aeration flow rate and initial pH on degradation of RR4 under 45 W fluorescent lamp was investigated. photocatalytic activity of RR4 dye decreased with increasing RR4 dye concentration. The optimum loading is around 0.04 g and optimum aeration rate is about 25 mL min^{-1} of C N co-doped TiO_2. Effect of pH was conducted based on the optimum loading and conclude that the photocatalytic degradation of RR4 became faster at pH 2 - 7. For the future work, the modification of doping with others element like non-metal or metal with C N co-doped TiO_2 can be enhanced toward the higher efficieny of photodegradation under visible light. Moreover, the immobilized technique can be used in future to overcome the difficulty of filtration on suspension.

1 Introduction

Heterogenous photocatalysis is the process basically based on the oxidation process of photogenerated holes with hydroxide ion or water to form hydroxyl radical. This hydroxyl radical which is highly oxidizing agents will oxidized the organic wastewater pollutants into harmless product. The combination between semiconductor photocatalyst and light sources is the key process of photodegradation. Among of the semiconductor, titanium dioxide (TiO_2) is the most popular and widely used by researchers due to its chemical stability, low cost, atoxic, capability for the reusable and its high photocatalytic activity [1-4]. This material consist a wide band gaps between 3.0 eV to 3.2 eV which require the UV light irradiation which means that only (4-6%) of sunlight has the required energy to activate the photocatalyst [5]. Thus, the doping techniques have been explored to enhance the photocatalysis toward visible light. The non-metal element (N, C, S, F and P) have been discovered as great potential for enhance the visible light in TiO_2 materials [6-8]. Among all the non-metal doped TiO_2 materials, nitrogen doped carbon coated TiO_2 have been discerned to exhibit superior photocatalytic activity under visible light irradiation [9]. Several advantages of carbon coated can be listed such as improved the adsorption of pollutants, improvement of anatase crystalline structure, better rate of sedimentation of the photocatalyst particles and suppression of phase transformation from anatase to rutile at high temperature [10]. Meanwhile, nitrogen doped TiO_2 can

[a] Corresponding author : saifulddinazami@gmail.com

be supreme potential toward the visible-light absorption due to metastable center formation, small ionization energy and its comparable atomic size in oxygen [11]. In this work, The C, N co-doped P25 sample was already preparad by Nawawi [10]. Subsequent experiments were conducted to investigate the effect of parameter of C N co-doped TiO_2 on the photodegradation of RR4. The effect of initial dye concentration, catalyst loading, aeration flow rate and initial pH on degradation of RR4 was investigated.

2 Experimental

2.1 Materials

C N co-doped TiO_2 preparad by Nawawi [10], Sodium Hydroxide (NaOH), Hydrochloric acid (HCl), Potassium Chloride (KCl), Reactive red (RR4) dye or commonly known as Cibacron Brilliant Red (Colour Index Number: 18105, chemical formula: $C_{32}H_{23}ClN_8Na_4O_{14}S_4$, molecular weight: 995.23 g mol^{-1}, λ max: 517 nm). Ultra-pure water was used to prepare all solutions in this work, 45 Watt fluorescent lamp, aeration pump.

2.2 Methods

2.2.1 Effect of initial dye concentration on the photocatalytic efficiency of the modified photocatalyst

Four sets of RR4 dye with different concentrations (5, 10, 30 and 60 mg L^{-1}) were used as the model pollutant in order to study the effect of concentration of the RR4 dye on the photocatalytic efficiency of the modified photocatalyst. 0.024 g of a selected modified photocatalyst sample was dispersed in 20 ml of RR4 dye solution at different concentrations under ultrasonic sonication for 30 seconds. The respective aqueous suspension of the catalyst within the dye solutions was then irradiated with 45 Watt compact visible light fluorescent lamp for 1 hour. Aliquots were taken at every 15 minutes time interval and the solution was filtered through a 0.45 μm cellulose acetate membrane filter to remove the catalyst particles. The concentration of the remaining RR4 dye was measured spectrophotometrically using HACH DR2000 spectrophotometer.

2.2.2 The Effect of the photocatalyst loading

A series of 20 mL of 30 mg L^{-1} RR4 dye solution with respective loading of 0.012, 0.024, 0.04, and 0.08 g of the optimum photocatalyst were prepared separately and dispersed under ultrasonic vibration for 30 seconds. Each aqueous suspension was then irradiated with 45 Watt visible light fluorescent lamp for one hour. The remaining RR4 in the sample solution was measured using HACH DR2000 spectrophotometer after every 15 minutes interval of irradiation. The percent degradation of RR4 dye solution was plotted against the dosage of the photocatalyst.

2.2.3 The effect of the aeration flow rate

Various aeration flow rates were applied to the system since air can provide oxygen for the degradation of RR4 dye. For the effect of aeration study, the flow rate of 0, 15, 25, 50 and 100 mL min-1 was respectively supplied into the systems containing the mixture of 0.024 g of the optimum photocatalyst in 20 ml of 30 mg L^{-1} RR4 dye solution. The RR4 solution was irradiated with a 45 Watt fluorescent lamp at each respective aeration rate.

2.2.4 The effect of initial pH values of the RR4 dye solutions

For this experiment, a series of 20 mL, 30 mg L^{-1} RR4 dye solutions at different initial pH values were used. The pH of each solution was adjusted to values of about 2, 4, 7 and 10 respectively before being photocatalytically decolorized using 0.024 g of optimum photocatalyst under a 45 W fluorescent lamp. The pH values of every solution were adjusted using 1 M NaOH and 1 M HCl solutions

2.2.5 Determination of point of the zero charge (pHzpc) for the unmodified and modified photocatalyst

pH of the zero charge (pHzpc) for photocatalysts were determined according to the procedures outlined by Machado and Santana [12]. A series of 50 mL ultra-pure water with different pH ranging from pH 2 to 10 was prepared. The pH of the solution was adjusted using 1 M NaOH and 1 M HCl solutions. 0.01 g of the unmodified and modified catalyst was added into each solution followed by addition of 10 ml of 0.1 M KCl. The solutions were stirred and the final pH values of the solutions were recorded after 24 hours of stirring. The discrepancies of the pH were obtained by subtracting the initial pH with final pH. A plot of pH discrepancies versus the initial pH was constructed and the point of zero charge was obtained from the values which cuts the X-axis.

3 Results and Discussion

3.1 The effect of initial dye concentration on the photocatalytic efficiency of C N co-doped TiO_2˙P25

The effect of initial dye concentration on the photocatalytic efficiency of PU1-350 sample is shown in Figure 1. As expected, the photocatalytic activity of RR4 dye decreased with increasing RR4 dye concentration where the highest photocatalytic activity occurred for 5 mg L^{-1} concentration of RR4 dye solution. Pseudo first order rate constants for the photocatalytic degradation of different concentration of RR4 dye by PU1-350 photocatalyst were 0.290, 0.235, 0.160 and 0.085 min^{-1} for 5, 10, 15, 30 and 60 mg L^{-1} respectively. Accordingly, this is due to the color of the solution that becomes more intense with increasing concentration. Photons from the light irradiation were absorbed by the dye molecules rather than by TiO_2 particles which caused the light-screening effect that retarded the penetration of light into the surface of the photocatalyst and this eventually decreased the efficiency of the photocatalyst and thereby reduced the formation of OH• and O_2• radicals.

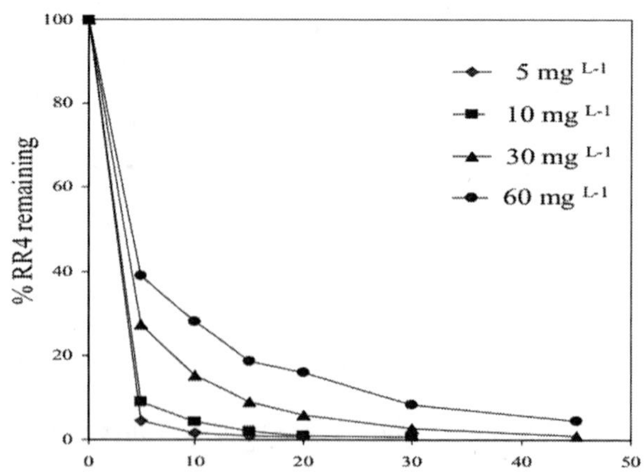

Figure 1. Photocatalytic degradation at different initial concentration of RR4 using C N co-doped TiO_2˙P25 sample (PU1-350).

3.2 Effect of catalyst loading on the photocatalytic efficiency of C N co-doped TiO_2-P25

Figure 2 shows the rate of the photocatalytic degradation of 30 mgL^{-1} RR4 dye by various amount of C N co-doped TiO_2-P25 (PU1-350) loading under a 45 W compact fluorescent lamp. As observed with TC1-560 and PP0.6-450, the degradation rate of RR4 dye by PU1-350 increased with increasing catalyst loading until 0.04 g beyond which the rate began to be slightly reduced (Figure 2). The increasing rate of RR4 removal for the loading of 0.012 to 0.04g PU1-350 catalyst was due to increasing amount of PU1-350 particles in the system. By increasing the PU1-350 particles, the number of photon and the dye molecules absorbed on the surface of the active site would also increase which eventually caused the photocatalytic activity of PU1-350 to increase. However, by further increasing the amount of PU1-350 beyond 0.04g, the rate decreased instead of becoming constant. There are two possibilities that influence this phenomenon might be due to the scattering of light and reduction in light penetration through the solution as a result of the excess catalyst particles and the aggregation of TiO_2 particles at high concentration which caused a decrease in the number of surface active sites, thus bringing little stimulation to the catalytic reaction [13].

3.3 Effect of aeration flow rate on the photocatalytic efficiency of C N co-doped TiO_2 P25

Aeration was supplied continuously to the solution throughout the experiment by using an aquarium air pump as the source of aeration. The presence of air bubbles from the aeration in the system promoted a significance agitation of the aqueous suspension for homogeneous distribution of the catalyst in the solution and to provide oxygen for the oxidative photocatalytic degradation. Figure 3 shows the effect of flow rates towards photocatalytic degradation of RR4 dye using C N co-doped TiO_2-P25 (PU1-350) as photocatalyst. It was found that 25 mL min^{-1} was the optimum aeration rate for a (PU1-350) where the the degradation rate of RR4 was almost two times faster than the system running without aeration (stirring process) with pseudo first order rate constant (k) for 25 mL min^{-1} aeration and stirring process ca. 0.058 and 0.027 min^{-1} respectively. All photocatalytic and adsorption experiments in this work were therefore carried out under this optimum aeration rate.

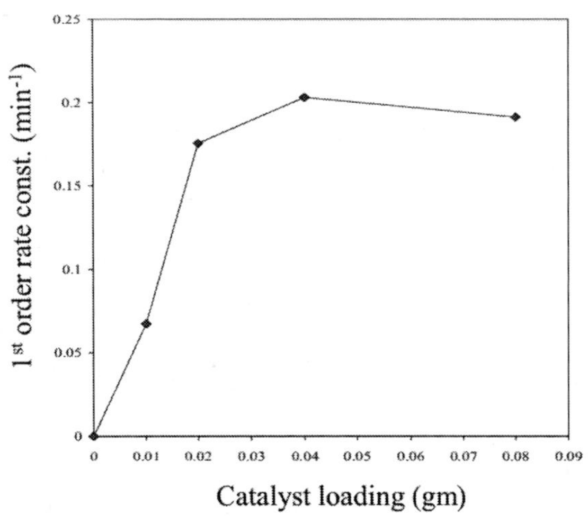

Figure 2. Photocatalytic degradation of RR4 dye by different amount of PU1-350 loading irradiated with a 45 Watt fluorescent lamp.

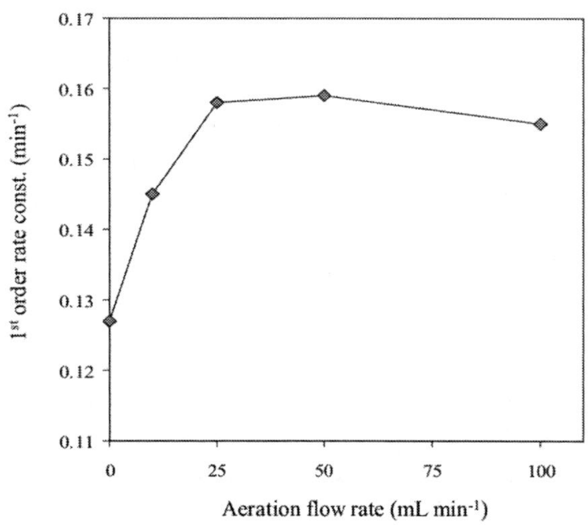

Figure 3. Effect of aeration flow rate on the photocatalytic degradation of RR4 under a 45 W fluorescent lamp.

3.4 Effect of initial pH of the RR4 dye solution on the photocatalytic efficiency of C N co-doped TiO_2-P25

Figure 4a shows the rate of the photodegradation of RR4 under C N co-doped TiO_2-P25 (PU1-350) at different pH conditions as a function of the irradiation time. The results indicated that the highest rate of photocatalytic degradation of RR4 was within the range of pH 2 - 7 while at pH 10, the rate was significantly slower where less than 30% of RR4 removed after 10 minutes of irradiation. Slower photocatalytic degradation of RR4 at pH 10 was due to the repulsion between the dye molecules and PU1-350 particles which led to decreased adsorption efficiency. The photocatalytic degradation of RR4 became faster at pH 2 - 7 due to the PZC value for PP0.6-450 to be c.a. 4.5 (Figure 4b) thus making it's particles to be positively charged with less agglomeration thus increased adsorption process.

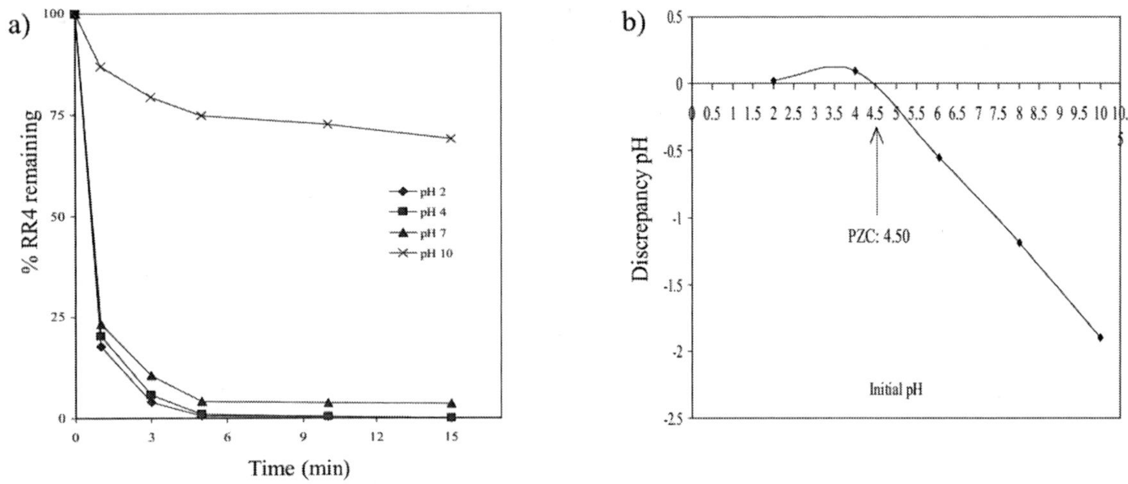

Figure 4. a) Photocatalytic degradation at different pH of RR4 dye under PP0.6-450 sample irradiated with 45 Watt fluorescent lamp, and b) Point of zero charge for C N co- doped TiO_2-P25, PU1-350.

4 Conclusion

The irradiation of C N co-doped TiO_2 was conduct using four parameters namely initial concentration,effect of loading, effect of aeration flow rate and effect of pH. It can be observed the photocatalytic activity of RR4 dye decreased with increasing RR4 dye concentration where the highest photocatalytic activity occurred for 5 mg L^{-1} meanwhile 0.04g of catalyst loading give optimum photocatalytic degradation. It was found that 25 mL min^{-1} was the optimum aeration rate for a (PU1-350). Effect of pH was conducted based on the optimum loading and conclude that the photocatalytic degradation of RR4 became faster at pH 2 - 7 due to the PZC value for PP0.6-450 to be c.a. 4.5. This research studies will generate towards the green technology that can be destructive the wastewater especially from dyes industries. This parameters study can be proved that C N co-doped TiO_2 can be purify contaminated water in various situation. For the future work, the modification of doping with others element like non-metal or metal with C N co-doped TiO_2 can be increase the efficiency of photodegradation under visible light. Moreover, the immobilized technique can be used in future to overcome the difficulty of filtration on suspension.

References

[1] M.T. Yagub, T K. Sen, S. Afroze and H.M. Ang, Dye and its removal from aqueous solution by adsorption: A review, Adv. Colloid. Interface. Sci., **209**, 172–184, (2014).

[2] V. Vaiano, O. Sacco, D. Sannino and P. Ciambelli, Nanostructured N-doped TiO_2 coated on glass spheres for the, Appl. Catal. B., **170-171**, 153–161, (2015).
[3] M.A. Nawi, and S.S. Sheilatina, Photocatalytic decolourisation of reactive red 4 dyes by an immobilised TiO2/chitosan layer by layer system, J. Colloid Interface Sci., **372**, 80-87, (2012).
[4] T.K. Tseng, Y.S. Lin, Y.J. Chen and H. Chu, A review of photocatalysts prepared by sol-gel method for VOCs removal, Int. J. Mol. Sci., **11**, 2336-2361, (2010).
[5] J. Yuan, M. Chen, J. Shi and W. Shangguan, Preparation and photocatalyst hydrogen evolution of N-doped TiO2 from urea and titanium tetrachloride, Int. J. Hydrogen Energy, **31**, 1326-1331, (2006).
[6] P. Li, G. Zhao, K. Zhao, J. Gao and T. Wu, An efficient and energy saving approach to photocatalytic degradation of opaque high-chroma methylene blue wastewater by electrocatalytic pre-oxidation, Dye Pigments, **92**, 923-928, (2012).
[7] J.P. Michael, W.D. Charles and P.P. Ivan, N-doped TiO_2 visible light photocatalyst films via a sol– gel route using TMEDA as the nitrogen source, J. Photochem. Photobiol. A., **281**, 27-34, (2014).
[8] X. Hou, C.W. Wang, W.D. Zhu, X.Q. Wang, Y. Li, J. Wang, J.B. Chen, T. Gan, H.Y. Hu and F. Zhou, Preparation of nitrogen-doped anatase TiO_2 nanoworm/nanotube hierarchical structures and its photocatalytic effect, Solid State Sci., **29**, 27-33, (2014).
[9] D. Chen, Z. Jiang, J. Geng, Q. Wang and D. Yang, Carbon and nitrogen co-doped TiO_2 with enhanced visible-light photocatalytic activity, Ind. Eng. Chem. Res., **46**, 2741-2746, (2007).
[10] W.I. Nawawi and M.A. Nawi, Carbon coated nitrogen doped P25 for the photocatalytic removal of organic pollutants under solar and low energy visible light irradiations, J. Mol. Catal, A: Chem., **383-384**, 83-93, (2014).
[11] S.H. Shin, H.H. Chun and W.K. Jo, Enhanced photocatalytic efficiency of N–F-co-embedded titania under visible light exposure for removal of indoor-level pollutants, Materials, **8**, 31-41, (2015).
[12] N. R. C. Fernandes MacHado and V. S. Santana, Influence of thermal treatment on the structure and photocatalytic activity of TiO_2 P25, Today, **107-108**, 595–601, (2005).
[13] M.I. Badawy, F.El. Gohary, M.Y. Ghaly and M.E.M. Ali, Enhancement of olive mill wastewater biodegradation by homogeneous and heterogeneous photocatalytic oxidation, J. Hazard. Matter., **169,** 673–679, (2009).

MATEC Web of Conferences **47**, 05019 (2016)
DOI: 10.1051/matecconf/20164705019

Photocatalytic Study of New Immobilized TiO_2 Technique Towards Degradation of Reactive Red 4 Dye

S. K. Ain[1,2,a], M. S. Azami[1,2], R. Zaharudin[1,2], F. Bakar[1,2] and W. I. Nawawi[2]

[1]*Faculty of Applied Sciences, Universiti Teknologi MARA, 40450 Shah Alam, Selangor, Malaysia*
[2]*Faculty of Applied Sciences, Universiti Teknologi MARA, 02600 Arau, Perlis, Malaysia*

Abstract. The study on TiO_2 for wastewater remediation has gained interest among researchers. However, the application of this photocatalyst is limited due to non-recyclability of conventional TiO_2. Thus, immobilization technique has been developed to solve this issue. Hence, a comparison study between two types of immobilized photocatalysts namely titanium dioxide (TiO_2) and TiO_2 mixed with polyvinyl alcohol (PVA) has been conducted in this work to observe the significant effect of PVA polymer in photocatalysis reaction of reactive red 4 (RR4) dye. Double sided adhesive tape (DSAT) was used as thin layer binder in this immobilization system. The result shows that the photocatalytic performance of TiO_2-PVA/DSAT was higher than that of TiO_2/DSAT under both normal UV and visible light irradiations due to the conjugated unsaturated polymer from PVA serve as electron donor for TiO_2 thus increase the photocatalysis process. Besides, TiO_2-PVA/DSAT was also found to possess much better adhesion strength to the support material compared to TiO_2/DSAT. Based on the findings, this TiO_2 immobilization system is expected to be beneficial in the industrial wastewater treatment. Thus, further study to improve the photocatalytic activity of this immobilized TiO_2 will be in our future work.

1 Introduction

Advanced oxidation processes (AOPs) have seen to be growing interest recently. The practical application of AOPs is commercially known as photocatalysis. Titanium dioxide (TiO_2) is the best known phototcatalyst and has been employed in almost studies. Immobilization of TiO_2 has been introduced to replace the existing suspension system. This is due to the main drawback from the suspension system that is formation of sludge which requires further purification steps thus resulting in difficulties to recycle the photocatalyst. Since past decades, researchers have taken various approaches to modify the TiO_2 for a better performance of photocatalysis reaction. Thus, many different methods that have been developed to prepare the immobilized TiO_2.

Adding polymer as matrices has been widely used by researchers in immobilized TiO_2 preparation to produce photocatalyst with better strength, adsorption capability and surface morphology. Some examples of the widely used polymers are polyvinyl chloride (PVC) [1], polyethylene glycol (PEG) [2] and polyvinyl alcohol (PVA) [3]. Under certain state, polymers with carboxyl or hydroxyl groups particularly, will bond chemically with the hydroxyl groups on the surface of TiO_2.

[a] Corresponding author : ain.syazwani@hotmail.com

PVA which possess good film-forming ability is a hydrophilic polymer and potentially useful in immobilization of TiO_2 [4-6]. Its low market price is another factor that many researchers employed this polymer in their work. Yang et al. [7] prepare C-PVA/TiO_2 composites by calcinating the mixture of TiO_2 sol and C-PVA solution for photocatalytic degradation of rhodamine B. Meanwhile, H-PVA/ TiO_2 composite was synthesized by Song et al. [8] under sol–gel and thermal treatment process with tetrabutyl titanate as titanium source.

In this paper, a comparison study between two types of immobilized photocatalyst which were TiO_2/DSAT and TiO_2-PVA/DSAT has been carried out successfully. The photocatalytic activities of both photocatalysts were investigated by studying the degradation of reactive red 4 (RR4) dye under two different light sources namely normal and visible light. Besides, the adhesion strength of the photocatalysts was also compared through sonication procedure.

2 Experimental

2.1 Materials

TiO_2 powder used (P25, 20% rutile and 80% anatase) was supplied by Evonic-Aeroxide. PVA was provided by R&M Chemicals. The model pollutant used was Reactive red 4 (RR4) dye which was purchased from Sigma-Aldrich. Deionized water was used to prepare all experimental solutions throughout this study.

2.2 Fabrication of immobilized TiO_2/DSAT and TiO_2-PVA/DSAT plates

The TiO_2 coating formulation was prepared by mixing TiO_2 with water while the TiO_2-PVA coating formulation was prepared by adding a specific amount of PVA solution into the TiO_2 formulation. Clean glass plates (130 mm × 50 mm) were prepared as support material to immobilize both photocatalyst formulations by brush-coating method. DSAT was stuck onto one side of the plates to cover an area of 65 mm × 47 mm prior to coating. The plates were each coated with fixed amount of both immobilize photocatalysts.

2.3 Photocatalytic study

A custom made glass cell with dimension of 50 mm × 10 mm × 80 mm (L × B × H) was used as a container to fill in 20 mL of the 200 mg L−1 RR4 sample together with the coated plate. An aerator made from aquarium pump model NS 7200 was connected to the glass cell via PVC tubing and a glass Pasteur pipette to supply oxygen and bubbles constantly throughout the experiment. The plate was placed inside the cell. A 55 W Qusun E27, 6400K fluorescent lamp was used to irradiate the normal light source on to the plate. Hoya ultraviolet (UV) filter was used to the normal light to get visible light source. Samples of the irradiated solutions were taken at every 15 min interval to determine their concentration by using HACH DR1900 ultraviolet-visible (UV-vis) spectrophotometer.

2.4 Immobilized TiO_2/PVA adhesion strength test

To test the adhesion strength of the immobilized TiO_2/PVA, the plate was sonicated for 30 seconds with Cress Ultrasonic, Model 4HT-1014-6. The sonicated plate was then taken out after every 5 seconds interval and its weight was recorded. The differences of weight before and after every 5 seconds intervals for up to 30 seconds sonication of the catalyst were measured in order to evaluate the percentage of the catalyst particles adhered onto the surface of the glass plates.

3 Results and Discussion

3.1 Effect of light sources

Figure 1 shows the comparison between photocatalytic activity of TiO_2-PVA/DSAT and TiO_2/DSAT when irradiate under normal and visible light. Photocatalytic efficiency of TiO_2-PVA/DSAT was found to be greater than TiO_2/DSAT under normal light irradiation with first order rate constant was ca. 0.0485 and 0.0309 min^{-1} respectively. Meanwhile, TiO_2-PVA/DSAT shows a more active visible light photocatalysis towards the degradation of RR4. Both types of immobilized photocatalysts are able to degrade RR4 under visible light rather great with 26% and 16% of RR4 colour remaining for TiO_2/DSAT and TiO_2-PVA/DSAT.

This happened due to the RR4 self-sensitization feature and the adsorption properties of the TiO_2 itself since other researchers reported certain kind of dyes may become a sensitizer during the photodegradation and the process could happened under visible light even the bandgap energy of the involved photocatalyst is higher than 3.0 eV [9-10]. Hence, photodegradation study using other pollutants is compulsory.

Methylene blue (MB) dye is used as a model pollutant to affirm the photoactivity of TiO_2/DSAT and TiO_2-PVA/DSAT sample under visible light. Figure 2 shows the photodegradation of 36 mg L^{-1} MB using TiO_2/DSAT and TiO_2-PVA/DSAT under visible light. As expected, no photocatalytic degradation of MB observed under TiO_2/DSAT since the bandgap energy of pristine TiO_2 sample is insufficient for energy from visible light to activate an excitation of electron hole pair.

In contrary, photodegradation activity was observed under under TiO_2-PVA/DSAT with complete decolorization were recorded at 300 minutes, it was clearly proven that TiO_2-PVA/DSAT sample is active under visible light irradiation. The conjugated unsaturated structure of PVA might be the main reason for the activity of the immobilized TiO_2-PVA/DSAT under visible light.

The mechanism is similar with the photosensitized mechanism of the dye sample towards TiO_2 under visible light. Visible light will cause the excitation of the sensitizer molecules adsorbed onto the surface of TiO_2 while consequently excite electrons to the conduction band (CB) of TiO_2. Electrons will be transferred from the sensitizer to the electron acceptor substrate present on TiO_2 surface by the CB which acts as a media while the valence band (VB) will remains unaffected in a standard photosensitization process. The visible light activity of the immobilized polymer/TiO_2 sample happens due to series of chain reactions that will be formed by the excited electrons in CB of TiO_2.

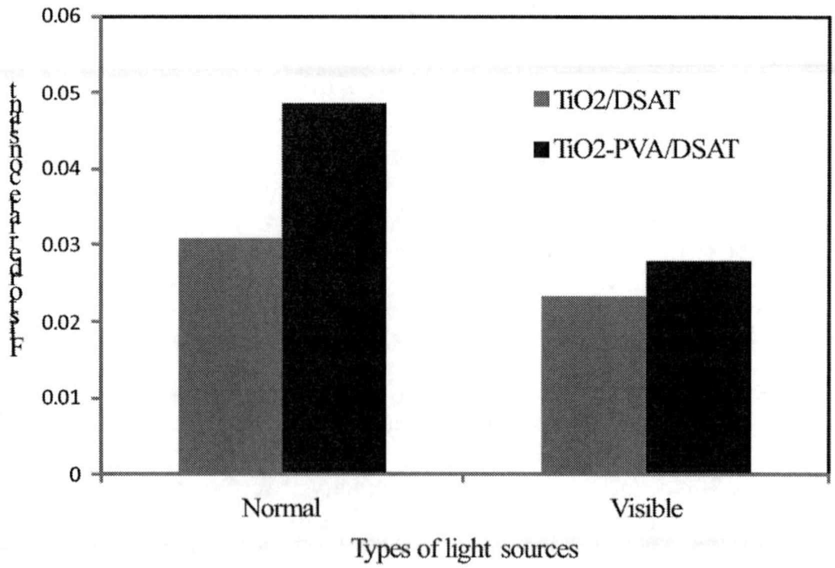

Figure 1. First order rate constant for immobilized TiO_2-PVA/DSAT and TiO_2/DSAT under normal and visible light.

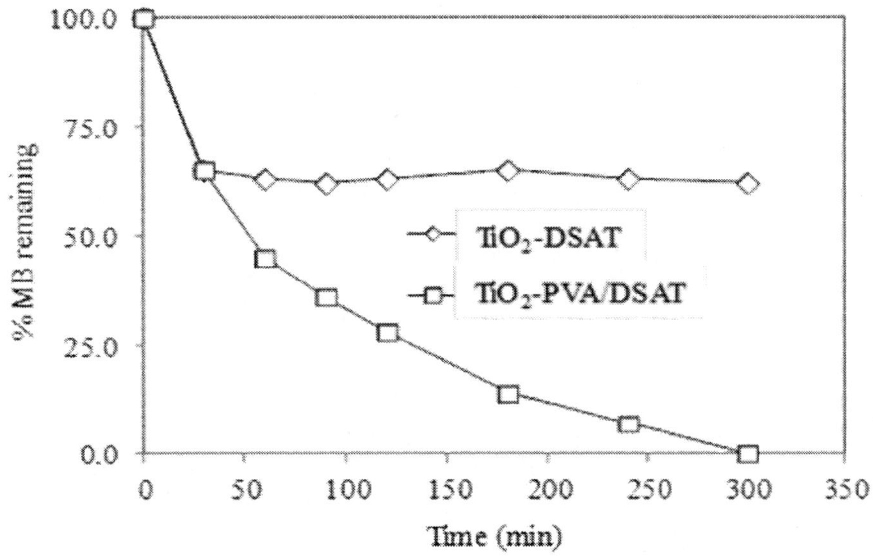

Figure 2. Decolorization of 36 mg L^{-1} MB dye using immobilized TiO_2-DSAT and TiO_2-PVA/DSAT under visible irradiation.

3.2 Adhesion strength test

This test identifies the adhesion strength of the immobilized TiO_2-PVA/DSAT and TiO_2/DSAT onto the support material. Figure 3 shows percentage of each photocatalyst weight remained after 30 seconds of sonication. Immobilized TiO_2-PVA/DSAT shows much better adhesion strength towards the glass plate compared to the TiO_2/DSAT. There is still as much as 87% of catalyst remaining for immobilized TiO_2-PVA/DSAT while only 47% catalyst left on TiO_2/DSAT after sonication. This was due to the significant effect of PVA that binds well chemically with TiO_2 matrices which results in stronger adhesion strength toward the support material. Since there is no binder present in immobilized TiO_2/DSAT, a great amount of its photocatalyst was removed during sonication.

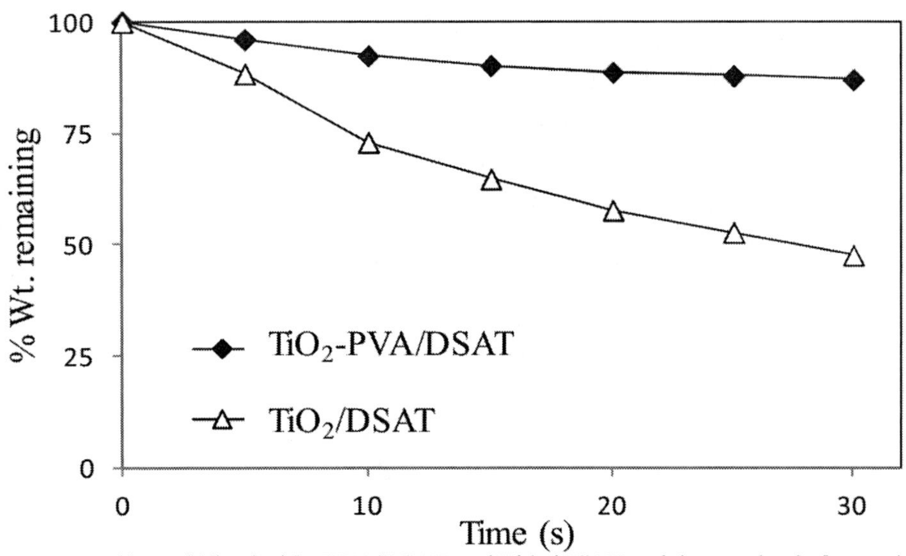

Figure 3. Percentage of immobilized TiO_2-PVA/DSAT and TiO_2/DSAT weight remained after sonication.

4 Conclusion

Photocatalytic efficiency of immobilized TiO_2-PVA/DSAT and TiO_2/DSAT has been successfully carried out under normal and visible light with RR4 as the model pollutant. It was found that photocatalytic activity of immobilized TiO_2-PVA/DSAT was not affected by different light sources. When test under ultrasonication, the adhesion of immobilized TiO2-PVA/DSAT was a lot better than TiO_2/DSAT towards the support material. Thus, this polymer bound immobilized TiO_2 photocatalyst was expected to be employed for industrial application for its durability. The study towards improvement of TiO_2-PVA/DSAT immobilized photocatalyst will be further investigated.

References

[1] S. Cho and W. Choi, Solid-phase photocatalytic degradation of PVC–TiO_2 polymer composites, J. Photochem. Photobiol. A, **143**, 221–228, (2001).

[2] Y. Zhang, C. Han, G. Zhang, D.D. Dionysiou and M.N. Nadagouda, PEG-assisted synthesis of crystal TiO2 nanowires with high specific surface area for enhanced photocatalytic degradation of atrazine, Chem. Eng. J., **268**, 170–179, (2015).

[3] M. Ren, F. Frimmel and G. Abbt-Braun, Multi-cycle photocatalytic degradation of bezafibrate by a cast polyvinyl alcohol/titanium dioxide (PVA/TiO2) hybrid film, J. Mol. Catal. A: Chem., **400**, 42–48, (2015).

[4] C.C. Yang, S.J. Chiu, K.T. Lee, W.C. Chien, C.T. Lin and C.A. Huang, Study of poly(vinyl alcohol)/titanium oxide composite polymer membranes and their application on alkaline direct alcohol fuel cell, J. Power Sources, **184**, 44–51, (2008).

[5] B. Bolto, T. Tran, M. Hoang and Z.L. Xie, Crosslinked poly(vinyl alcohol) membranes, Prog. Polym. Sci., **34**, 969–981, (2009).

[6] C.C. Yang, Synthesis and characterization of the cross-linked PVA/TiO2 composite polymer membrane for alkaline DMFC, J. Membr. Sci., **288**, 51–60, (2007).

[7] H. Yang, J. Zhang, Y. Song, S. Xu, L. Jiang and Y. Dan, Visible light photocatalytic activity of C-PVA/TiO2 composites for degrading rhodamine B, Appl. Surf. Sci., **324**, 645–651, (2015).

[8] Y. Song, J. Zhang, H. Yang, S. Xu, L. Jiang and Y. Dan, Preparation and visible light-induced photocatalytic activity of H-PVA/TiO_2 composite loaded on glass via sol–gel method, Appl. Surf. Sci., **292**, 978– 985, (2014).

[9] L. Xiangzhong, Z. Wei and Z. Jincai, Visible light-sensitized semiconductor photocatalytic, J. Sci. China, **45,** 421-425, (2002).

[10] R. Malka, P. Sagi and P. Yaron, Using dyes for evaluating photocatalytic properties: A critical review, Molecules, **20**, 88-110, (2015).

MATEC Web of Conferences **47**, 05020 (2016)
DOI: 10.1051/matecconf/20164705020

A Study on RR4 Dye as a Sensitizer in Enhancing Photoactivity of Immobilized Photocatalysts

F. Bakar[1,2,a], M. S. Azami[1,2], S. K. Ain[1,2], R. Zaharudin[1,2] and W. I. Nawawi[2]

[1]*Faculty of Applied Sciences, Universiti Teknologi MARA, 40450 Shah Alam, Selangor, Malaysia*
[2]*Faculty of Applied Sciences, Universiti Teknologi MARA, 02600 Arau, Perlis, Malaysia*

Abstract. In this work, anionic RR4 dye was used to sensitize TiO_2/PVA and TiO_2/PEG immobilized system in enhancing photocatalytic degradation of cationic methylene blue (MB) dye. 0.3g of TiO_2 and polymer binder was coated onto a clean glass plate by using brush technique to develop optimum immobilize TiO_2 system. A comparison study between immobilized TiO_2/PVA (Im/TiO_2/PVA) and Immobilized TiO_2/PEG (Im/TiO_2/PEG) system with and without RR4 sensitizer were carried out under 45 W fluorescent lamp and visible light irradiation. The photocatalytic degradation of MB was significantly enhanced for both RR4 dye sensitized Im/TiO_2/PVA and Im/TiO_2/PEG with 1st order rate constant was ca. 0.035 and 0.030 min-1 respectively under 45-W fluorescent lamp. Same observation as well under visible light irradiation whereby enhanced of those RR4 sensitized immobilized photocatalysts were recorded as compared with immobilized photocatalysts without RR4 as sensitizer. The photocatalytic enhancement under Im/TiO_2/PVA/RR4 and Im/TiO_2/PEG/RR4 are due to the ability of RR4 dye to become electron (e-) donor for conduction band (CB) of TiO_2, thus making TiO_2 CB riches with electron, eventually this e- is used to remove MB by producing hydroxyl radical.

1 Introduction

The growth of industry in worldwide has discharge variety of contaminants and tremendously increase the generation of byproducts which lead to environment severe. One of the biggest effluents generated from the industries are dyes. These dyes are toxic and give serious hazard to human and aquatic life [1]. Therefore, researcher from all over the world has been working on various approaches in order to solve these issues. Advance oxidation process (AOP) [2], ozonation [3], Fenton technology [4], and photocatalysis [5] has been extensively used by previous researcher to mitigate variety of pollutants in various environmental condition. Recently, photocatalysis has grabbed many attentions from researcher to overcome these issues. Among all of semiconductor photocatalyst, titanium dioxide (TiO_2) is considered very close to the ideal semiconductor for photocatalysis due to its high stability, low cost, readily available and most important it is safe toward both humans and environment [6].

However there are two main disadvantages of TiO_2, firstly it has relatively wide band gap which allow it to absorb only 3-4% energy of the solar spectrum [7] and secondly the suspension mode application of TiO_2 only can be used in a limited time of application, leeched out easily and none

[a] Corresponding author : faezah_bakar@ymail.com

recyclable [7]. Therefore, many efforts have been done in order to increase the sensitivity of TiO_2 towards visible light region and improve its ability as photocatalyst such as by using dye sensitizer, doping with metallic and non-metallic species, heterojunction system and utilization thin film [7]. Mital et al. [8] reported that dye sensitization is expected to increase the efficiency of the excitation process and also expand the wavelength range of the excitation for the photocatalysis. This is done through excitation of the sensitizer followed by charge transfer to the semiconductor. Charge transfer can be formed in the semiconductor particles by exciting the dye attached to the photocatalyst surface. The excited state of the dye will eject e- to the semiconductor particles. Thus, this sensitization will increase the range of wavelength response of the photocatalyst in order to operate under normal sunlight.

Recently, immobilization of TiO_2 powder has been used by researcher in order to overcome issues arise from suspension mode of TiO_2 as it can be used in a long-term applications [7]. Despite, immobilize TiO_2 system also has its drawbacks such as less surface area for photocatalytic reactions as well as cracking and peeling off of the catalyst layer thus reduce the degree of degradation [7]. Therefore, water soluble polymer such as PEG and PVA is used as a binder to the TiO_2 system to improve its adhesiveness. From the best of knowledge, no study observed on using RR4 as dye sensitizer for MB removal and to compare with PVA and PEG as binder, moreover DSAT method is a new approach used to overcome some issues arise from previous method.

Therefore the aim of this study is to determine the effectiveness of RR4 as sensitizer for photodegradation of MB dye under different polymer binders in immobilized TiO_2. Cationic MB dye was used to evaluate the photocatalytic activity of immobilized photocatalysts under 45 W fluorescent lamp and visible light irradiations.

2 Experimental

2.1 Preparation of TiO_2 immobilized system

TiO_2 solution was prepared by adding 6.5g of TiO_2 powders into 50mL distilled water. The solution will then added with 1mL of PVA polymer in a reagent bottle followed by mechanical stirring until homogenized. The white solution of TiO_2 was then coated onto a clean glass plate with dimension of 50mm x 10mm x 80mm (Length x Wide x Height), which has been taped before with double sided adhesive tape (DSAT). The formulation was then dried in the oven for 10 minutes forming immobilized/TiO2/PVA (Im/TiO_2/PVA), coating process was repeated until 0.3 g loading of Im/TiO_2/PVA was achieved. Im/TiO_2/PVA samples was undergoing washing process by exposing immobilized sample with 45 W of fluorescent lamp under aerated condition for 1 hour prior sensitized with RR4 dye. The preparation procedure of immobilized sample will be repeated using other polymer which is PEG forming Im/TiO_2/PEG sample. This experimental method was modified from Wan et al. [8].

2.2 Preparation of RR4 sensitizer TiO_2 immobilized and photocatalytic study

Both of Im/TiO_2/PVA and Im/TiO_2/PEG plates were immersed into glass cell contained distilled water for washing purpose in 40 minutes prior sensitized with RR4 dye for about 2 hours forming Im/TiO_2/PVA/RR4 and Im/TiO_2/PEG/RR4. Then, those immobilized sample were tested for its photocatalytic activity by applying the plate in a glass cell containing methylene blue (MB). The process was carried out by immersing immobilized sample plate into a glass cell filled with 12 mg L^{-1} of MB dye and irradiated with 45-W fluorescent lamp named as normal light and 45-W fluorescent lamp covered by UV cut-off filter for visible light irradiation. An aquarium pump model NS 750 is used as an aeration source to ensure enough oxygen supply. The dimension of glass cell used in this study is 50mm x 10mm x 80mm (Length x Wide x Height). The decolourization degree of MB dye was determined at 15 minutes time interval until complete decolourization is achieved. The color reduction value of MB was determined by measuring the absorbance using HACH DR 1900 spectrophotometer with 661 nm wavelength.

3 Results and Discussions

3.1 Place Photocatalytic study of Im/TiO_2/PVA and Im/TiO_2/PEG under 45-W fluorescent lamp

The photocatalytic degradation of 12 mg L^{-1} MB using Im/TiO_2/PVA and Im/TiO_2/PEG with and without RR4 sesitized was illustrated in Figure 1a. Previously, limited study has been conducted on this subject matter. Therefore, this results demonstrate that the increment of photocatalytic activity for RR4 sensitized immobilized samples were observed where the complete decolorization of MB dye can be shorten into 45 and 60 minutes for Im/TiO_2/PEG and Im/TiO_2/PVA respectively, compared to the irradiation time for sample without RR4 sensitizer which both samples are takes more than 90 minutes to complete the decolorization process. This phenomenon supposed to be from the additional of RR4 dye as sensitizer where a significant photocatalytic activity enhancement might be due to the two major factors from RR4 dye. Firstly, RR4 was observed can increased the adsorption ability of the immobilized photocatalyst. As can be seen in Figure 1a, the big different percentage decolorization of MB was observed at first 15 minutes for both Im/TiO_2/PVA/RR4 and Im/TiO_2/PEG/RR4 where the trend for MB color removal was in line with the adsorption study of MB using those immobilized photocatalysts as can be seen in Figure 1b.

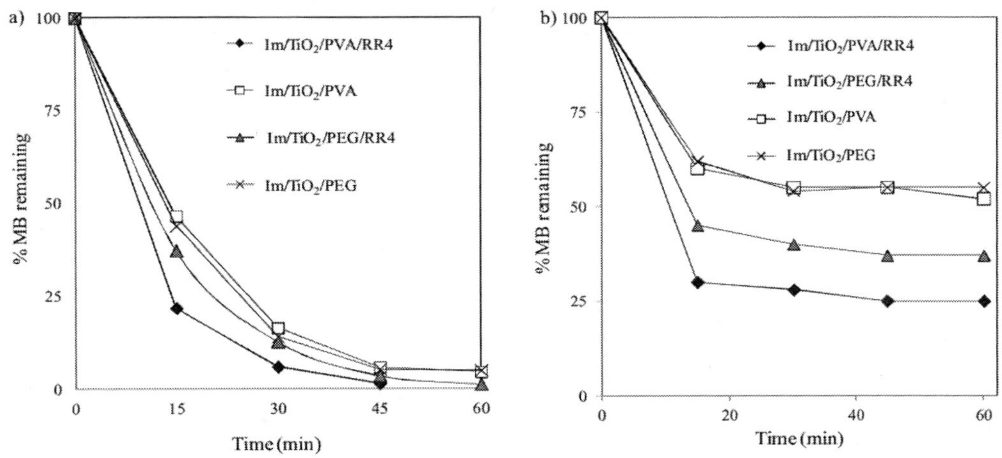

Figure 1. a) Photocatalytic degradation of MB under 45 W fluorescent lamp and b) adsorption study using various immobilized photocatalysts.

Secondly, RR4 dye was observed to become electron (e-) donor thus make the wide range of lights (UV and visible lights) to excite an e-hole pair and become e- donor for conduction band (CB) of TiO_2, thus making TiO_2 CB riches with electron, eventually this e- is used to remove MB by producing hydroxyl radical. Figure 2a shows the photodegradation study of MB under visible light irradiation. Those Im/TiO_2/PEG/RR4 and Im/TiO_2/PVA/RR4 shows a good photo response under visible light where complete decolorization of 12 mg L^{-1} MB was 60 and 75 minutes respectively while no photocatalytic activity was observed under Im/TiO_2/PEG and Im/TiO_2/PVA. According to Mital et al. [9] in their study showed that dye sensitization is expected to increase the efficiency of the excitation process of e-hole pair and also expand the wavelength range towards visible light region. This is done through excitation of the sensitizer followed by charge transfer to the semiconductor. Charge transfer can be formed in the semiconductor particles by exciting the dye attached to the photocatalyst surface. The excited state of the dye will eject e- to the semiconductor particles. Thus, this sensitization will increase the range of wavelength response of the photocatalyst in order to operate under normal sunlight. Therefore as a surface modification for TiO_2, dye sensitizer can be effective in broadening the range of absorption of this semiconductor. Figure 2b shows the schematic diagram the propose mechanism of photocatalysis process in presence of RR4 dye as sensitizer.

a)

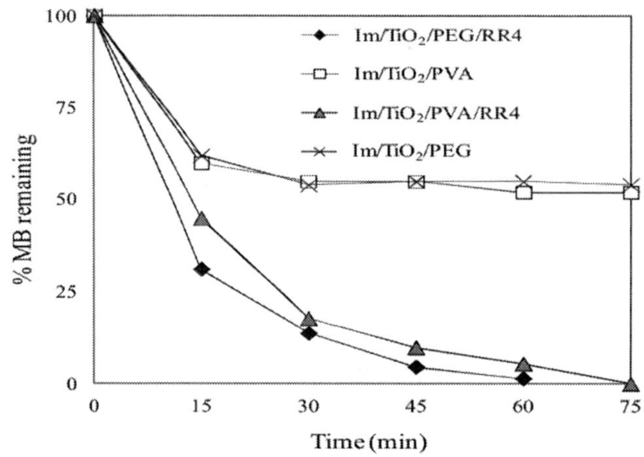

b)

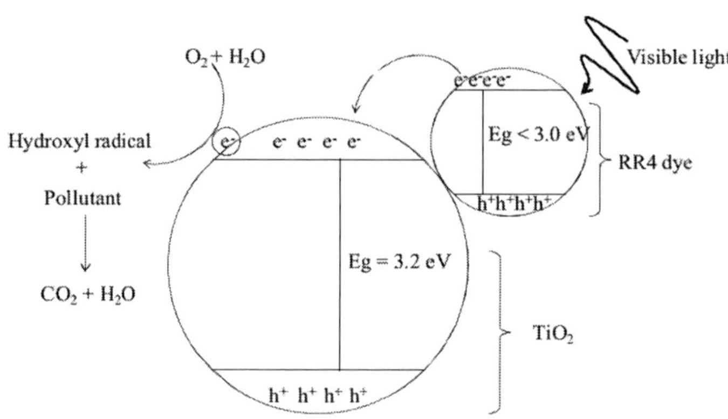

Figure 2. a) Photocatalytic degradation of MB dye under visible light irradiation and b) propose mechanism of photocatalysis in presence of RR4 dye sensitizer.

4 Conclusions

A study on RR4 dye as sensitizer for TiO_2 using immobilization technique was successfully applied namely Im/TiO_2/PVA/RR4 and Im/TiO_2/PEG/RR4. The photocatalytic activity of samples under RR4 sensitizer is significantly enhanced due to the ability of RR4 to become absorbance and electron donor for visible light irradiation. No photocatalytic activity under visible light irradiation was observed for those immobilized photocatalysts without presence of RR4 sensitizer. Hence, the influent of sensitize RR4 in immobilized TiO_2 makes this photocatalyst become more effective.

References

[1] M.A. Nawi, S. Sabar, and Sheilatina, Photocatalytic decolourisation of Reactive Red 4 dye by an immobilised TiO2/chitosan layer by layer system, J. of Colloid and Interface Science, **372**, 80–87 (2012).

[2] M. Pera-Titus, V. Garcia-Molina, M.A. Banos, J. Gimenez and S. Esplugas. Degradation of chlorophenols by means of advance oxidation process: A general review, Appl. Catal. B., **47**, 219–256, (2004).
[3] K. Ikehata and M.G. El-Din, Aqueous pesticides degradation by ozonation and ozone-based advances oxidation processes: A review (part II), Ozone-Sci. Eng., **27**, 173–202, (2005).
[4] K. Ikehata and M.G. El-Din, Aqueous pesticides degradation by hydrogen peroxide/ultraviolet irradiation and fenton type advance oxidation processes: A review, J. Environ. Eng. Sci., **5**, 81–135, (2006).
[5] J.M. Britto and M.D. Rangel, Advanced oxidation process of phenolic compounds in industrial wastewater, Quim. Nova, **31**, 114–122, (2008).
[6] M.N. Chong, B. Jin, C.W.K. Chow and C. Saint, Recent developments in photocatalytic water treatment technology: A review, Water Res., **44**, 2997–3027, (2010).
[7] M.A. Nawi, A.H. Jawad, S. Sabar and W.S. Wan Ngah, Immobilized bilayer TiO_2/chitosan system for the removal of phenol under irradiation by a 45 watt compact fluorescent lamp, Desalination, **280**, 288–296, (2011).
[8] W.I. Nawawi, S.K. Ain, R. Zaharudin, A.H. Jawad, M.A.M Ishak, K. Ismail and S. Sahid, New TiO_2/DSAT immobilization system for photodegradation of anionic and cationic dyes, International Journal of Photoenergy, **2015**, 1-7, (2015).
[9] G. S. Mital and T. Manoj, A review of TiO_2 nanoparticles, Chinese Science Bulletin, **56**, 1639-1657, (2011)

MATEC Web of Conferences **47**, 05021 (2016)
DOI: 10.1051/matecconf/20164705021

Challenges in Selecting a Sustainable Landfill Site in Malaysia

Tey Jia Sin[1,a], Goh Kai Chen[1] and Goh Hui Hwang[2]

[1]*Faculty of Technology Management and Business, Universiti Tun Hussein Onn Malaysia, 86400 Parit Raja, Johor, Malaysia*
[2]*Faculty of Electrical and Electronic Engineering, Universiti Tun Hussein Onn Malaysia, 86400 Parit Raja, Johor, Malaysia*

Abstract. Landfill is essential in the management of municipal solid wastes in Malaysia. The role of a landfill site has become prominent with the increasing quantity of municipal solid waste. However, the selection of a landfill site has always been one of the crucial yet complicated parts in constructing a landfill. Due to the negative impacts that caused by a landfill site, an appropriate landfill site is required in order to minimize the negative impacts to the lower. Therefore, this paper probes into the challenges in landfill site selection in Peninsular Malaysia. This research is conducted through the reviewing of past similar research and case study with eight different government departments that have play a role in evaluating the suitability of a site to be used as landfill. The results obtained show the challenges in selecting a landfill site. Through the understanding of the landfill site selection, it enables future research to improve the current landfill site selection system and ensure its sustainability.

1 Introduction

As the population and the generation of municipal solid waste increases simultaneously, Malaysia starts to realize that the need of a proper waste management system is required to cope with the daily quantity of municipal solid wastes produced. In 2007, the government established a new act which is known as Solid Waste and Public Cleansing Management Act, 2007. This act is used to provide proper guidelines and regulations, and a holistic approach to waste management [1]. The purpose of establishment of the act is for the transferring of executive authority from the local authority to the Federal Government. Through this act, the Department of National Solid Waste Management and the Solid Waste Management and Public Cleansing Corporation are operating throughout peninsular Malaysia and it imposes higher responsibilities on the stakeholders [2].

The generation of municipal solid wastes is more than 30,000 tonnes per day [3]. However, Malaysia shows a statistic that about 60% of the wastes generated are disposed into landfill in which at present, more than half of the existing landfills has reached its maximum capacity but the number of landfills are limited [4, 5]. Hence, a landfill site that is selected according to the concept of sustainability is necessary to mitigate the negative impacts towards human health and the environment. As a result, this paper focuses on revealing the challenges in selection of a landfill site in peninsular Malaysia.

[a] Corresponding author : jiasin.star@gmail.com

2 Waste Management

As the waste management has been given more attention, many techniques are then utilized in the industry in managing wastes. Among the waste management techniques are, recycling, incineration and composting are the common approaches. Wastes generated must be reduced or reused, unless the waste generated cannot be recycled, then applying incineration or landfill is the last resolution [6, 7].

The sanitary landfill method for the ultimate disposal of solid waste material continues to be widely accepted and used in Malaysia at the present stage [8, 9]. This is due to its economic advantages. Furthermore, landfill as compared to other disposal options, it requires less costs [8, 9].

Besides, government faces the problem in selecting an appropriate landfill site. An appropriate site has to be environmental friendly, acceptable by the public and the least cost that require in managing wastes. There are tipping fees to be charged from the imported waste [10]. Thus, a simpler way in selecting a landfill site has to be identified so as to increase the accuracy of the end result.

2.1 Development of municipal solid waste management

Solid waste is defined as any scrap material or other unwanted surplus substance or rejected products arising from the application of any process or any substance required to be disposed of as being broken worn out, contaminate or otherwise spoiled [11].

As year pass by, Malaysia government has been putting effort in the improving of managing municipal solid waste that created daily by establishing Department of National Solid Waste Management and Public Management Corporation. The highly increasing of solid wastes as compare to the past, thus a holistic system and law in managing municipal solid waste need to be enforced in order to ensure the environment being protected.

2.2 Landfill site selection

Both Siting a landfill site plays an important role in contributing a sustainable landfill. The decision of optimizing landfill siting have gained considerable importance so the least damage impacts could be ensured towards to the various environmental aspects as well as to the individual who are living surround the landfill site, thereby enhancing overall sustainability associated with the life cycle of a landfill [12]. By virtue of these, the Malaysia government recognizes that the appropriate waste management is essential tool in achieving sustainable development [13].

In terms of economics aspects, the price of the land depreciates as it getting nearer to the landfill site [14]. The effects generated through the landfill are such as noise pollution, surrounding area are aesthetical unattractive, and also, air pollution. Other than that, construction and operation of landfill requires high capital cost [15].

In Malaysia, the landfill site selection process for the land which is owned by the federal government is done by the National Solid Waste Department. However the rest is owned by the state government or private companies. Once the site being selected, it will be executed by Solid Waste and Public Cleansing Management Department (PPSPPA). National Solid Waste Department holds the authority in the planning for the locating process whereby it has to be matched with the requirement set by the Department of Environment and Town Country Planning.

Currently, Malaysia is using geographic information system in the evaluation of a landfill site. It is a tool to speed up the site selection process by showing a clearer presentation of suitable and unsuitable sites based on the set of criteria given. However, geographic information system is not comprehensive enough; it is mainly due to the fact that evaluation criteria produced ultimately weak evidence to support the selection [16]. Thus, it leads to the need to develop a proper tool in the evaluation process for a site and ensure the criteria uses are fulfil sustainable concept.

2.3 Landfill site selection criteria

Under Department of Environment in Malaysia, prior for a landfill site to be selected, it has to fulfill a series of criteria listed out to reduce the impacts towards the environment to the minimum. In relation to this, Department of Environment has produce a guideline namely, Guidelines for Development of Solid Wastes Sanitary Landfill. In this guideline, it stated six major criteria that should be taken into consideration when selecting a landfill site [17]. Among the six criteria are available area, surrounding environment and conditions, topography and geological conditions, transport infrastructure and access, post closure and land use plan, and availability of supplies.

Table 1. Impact of environmental impact assessment criteria towards landfill site selection.

Criteria	Influence towards Landfill Site Selection
Available area	An appropriate landfill site is the one having sufficient area to cater municipal solid wastes up to 20 to 25 years, besides, the area have to include with landfill facilities and buffer zone [17]. A landfill with short lifespan will cause the search of a new landfill site to be done in short period of time. Due to the land in Malaysia is getting scarce, a high cost is needed in acquisition a land [18], it becomes tougher for the searching of an appropriate land to be used as landfill site.
Surrounding environment/ Conditions	The environment of the surrounding have to take into consideration. It is necessary to ensure over the place has basic utilities supply, and the proximity to rivers and water intake point [17]. The bacis utilities is requires to make sure the landfill can be operating well while avoiding water area is to prevent leachate contaminate the water. The negative effects left from the landfill site such as noise, pollution will caused the land price surround the landfill site depreciate [14]. Thus, it is highly important to select a landfill where it can lower the possible chance to bring negative effect the surrounding area.
Topography and geological conditions	The geology of an area will directly control the types of soil created from the parent material, load bearing capacity of the landfill's foundation soil, and the mitigation of leachate [10]. A good landfill site is the one that can reduce the need in of doing serface earthworks. The area which is prone to flood should be avoid in use as landfill site as well [17].
Transport infrastructure and access	A good landfill site has to locate away from the residential area, but not isolated. The easy accessible of the waste vehicles will enable emergency helps in case of accidents happan on site [17]. On the other hand, the distance of the municipal solid wastes generation and landfill site will affect the transport cost directly [10].
Post closure land use plan	It is essential for the authority to plan for the post closure land use for the landfill site before a land is gazetted for landfill use. This is a close landfill site will still produce gases which is not good for human body and it is ideally not suitable for residential purpose [17]. To select a good landfill site has to consider the impact not only during the landfill site operating but also post closure land use plan.
Availability of supplies	Observing and identifying the materials that can be provided from the surrounding of the land is necessary as it can helps to reduce huge cost of transporting and purchasing of the materials from afar [17]. One of the example is cover soil that is use in landfill site.

From Table 1, it can be seen that the environmental impact assessment criteria in landfill site selection is still focusing more on the environmental criteria whereas on the economic and social part is giving lesser emphasis. This makes the whole evaluation of a selection of a landfill site become not holistic. Hence, the site that selected might be able to reduce the negative impacts towards the environment but it may incur higher cost for the people in charge or has the authority.

3 Method

This paper employed literature review and qualitative case study as the research methods to gather a more in- depth information from eight of the government sectors which involves in the evaluation of landfill site selection. This research was done through the qualitative case studies in which it has eight respondents from different government departments. Among the departments are Department of National Solid Waste Management, local authorities, State Economic Planning Unit, Public Work Department, Department of Environment, Town and Country Planning Department, Department of Irrigation and Drainage, and Solid Waste Management and Public Cleansing Corporation.

The current waste management methods adopted by Malaysia were identified. Besides, the problems faced during the landfill site selection are being discussed. This research is expected to improve current condition in landfill site selection by identifying the challenges. The concept of sustainability is desire to be implemented in the consideration of a suitable land to turn it into a landfill site. Thus, the perspective of each respondent which represents each government sector is determined in this research as well.

4 Results and Discussion

4.1 Challenges in landfill site selection

According to El Haggar [15], selecting a site for landfill purpose is one of the very significant processes to reduce the negative impacts to the surrounding. In Malaysia, the landfill site selection process for the land which is owned by the federal government is done by the National Solid Waste Department. However, the rest is owned by the state government or private companies. Once the site being selected, it will be executed by Solid Waste and Public Cleansing Management Department.

4.1.1 Limited area

According to the respondents, there are many criteria has to consider for a site to be chosen as the landfill site. A thorough examination needs to carry out and the potential landfill site is limited as not all the lands are suitable to be used as landfill site. Thus, although there are many lands in Malaysia, however, not every land is suitable for the development of a landfill site. Furthermore, some authorities in Malaysia hardly select a land to be used as landfill because the land is getting scarce and high cost is required in acquisition a land [18].

4.1.2 Non standardisation in selection criteria

Moreover, the identification of landfill sites for municipal solid waste disposal becomes a critical management issue in which it requires a number of considerations before the implementation to reduce the impacts it may cause to the individual [19]. This makes the decision makers face difficulties in weighing the factors in the selection of a landfill site as to reduce the negative impacts that may bring to the surrounding. Factors that are in the consideration are such as environmental, social and economic aspects.

However, according to the respondent, currently Malaysia has yet to come out general sustainable criteria as guideline for the selection of a landfill site to be sustainable. Most of the criteria that have been taking as consideration is environmental aspect because of the leachate as a by- product from landfill. It may be a challenge for a landfill site that is selected fulfill the concept of sustainability.

4.1.3 Politics (public against landfill)

According to respondent, politics in social aspect plays an important role in determining the construction of a landfill at a selected site. There are times where the suitable site has been selected, but the objection by the public will cause the project unable to be carried out.

In addition, in Malaysia, it is getting tougher to locate a suitable landfill site as each individual is possessing the attitude of not in my backyard and not willing to accept the implementation of a new landfill site, especially the landfill is near to the residential area with the reason of pollution, unpleasant smell and littering may happen in the future [20].

4.1.4 Technology selection

The adoption of an incinerator in Malaysia is still not being widely used based on the respondents. Some of the incinerators constructed by the Ministry of Urban Wellbeing, Housing and Local Government are yet to operate as well as they fail in terms and conditions. Besides the capital expenditures and operate expenses are very high as compare to landfill. Thus, landfill is still required in managing the solid wastes generated and a sustainable site should be selected.

On the other hand, currently Malaysia is only selects a landfill site through geographic information system. Geographic information system was employed by consultants under the Department of National Solid Waste in the evaluation of a site base on the criteria given. However, geographic information system is not comprehensive enough [16]. It leads to the needs of a more effective model to analyse the criteria and ensure a site that is sustainable to be chosen.

In addition, some of the respondents also highlighted that Malaysia is still lacking of the selection technology as the characteristic of municipal solid wastes are very different with other countries. In Malaysia, the wastes collected is damp and it unable to be decompose using the same technique as other countries. The more municipal solid wastes dump into a landfill, the faster the land become full and increase the need to select a new land for new landfill.

4.1.5 The application of sustainable concept into landfill site selection

Sustainable landfill is crucial in a waste management system as it can reduce the potential risk to the environment [21]. According to the respondents, not all aspects in sustainable will be considered in the evaluation of a site to become a landfill. Some of the government sector is only emphasizing on environmental aspects such as on the buffer zone however, the cost of transportation for sending municipal solid wastes to landfill from the origin has been put it aside. It caused a great problem as the site selected may be bringing an advantage for the environment and social aspects. On the other hand, it may incur high cost on the landfill management, which is not sustainable concept in landfill site selection. Hence, the results have shown a need in generating a better way of landfill site selection as the practitioner has yet to fully apply the concept of sustainability.

5 Conclusion

This paper has pointed out the best suitable Malaysia waste management treatment technique, landfill in the near future, and its challenges in the selection of the site. Waste management is an important sector because the municipal solid wastes will not stop generating due to the waste management system is not ready in a country. The challenges mentioned in previous lead to the need of having a decision model in evaluating the most suitable land base on a set of sustainable landfill site selection criteria. As the area for the land to be used as landfill is limited, the improvement in landfill site selection through a better system which considers the sustainable criteria will definitely brings a great benefit not only to the government sector but also to each individual in Malaysia. The upgrading of the landfill site selection system can speed up and increase accuracy in choosing a sustainable site which emphasing at all environmental, social and economic aspects.

Acknowledgement

The authors express gratefully acknowledgement to the industry stakeholders for their valuable contributions towards success of this research. This research is fully funded by UTHM.

References

[1] N.S. Romali, N. Mokhtar, W.F.W. Ishak and M.A.A. Samah, Solid waste management: Development of AHP model for application of landfill sites selection in Kuantan, Pahang, Malaysia, *Proc. of the Int. Symposium on the Analytic Hierarchy Process*, Kuala Lumpur, (2013).

[2] N. Yahaya and I. Larsen, Federalising solid waste management in peninsular Malaysia, *Proc. of Int. Solid Waste Association (ISWA) World Congress*, Singapore, (2008).

[3] P. Agamuthu and S. Fauziah, Challenges and issues in moving towards sustainable landfilling in a transitory country-Malaysia, Waste Management and Research, **29**,13-19, (2011).

[4] F.M. Ghazali, S. Syafalni and S.M. Noor, Public perception on the current solid waste management system in Malaysia: A comparative study of Matang Landfill and Bukit Tagar Sanitary Landfill (BTSL), World Applied Sciences J., **32**, 872-883, (2014).

[5] A. Johari, S.I. Ahmed, H. Hashim, H. Alkali and M. Ramli, Economic and environmental benefits of landfill gas from municipal solid waste in Malaysia, Renewable and Sustainable Energy Reviews, **16**, 2907-2912, (2012).

[6] M.O. Saeed, M.N. Hassan and M.A. Mujeebu, Assessment of municipal solid waste generation and recyclable materials potential in Kuala Lumpur, Malaysia, Waste Management, **29**, 2209-2213, (2009).

[7] O. Ortiz, J. C. Pasqualino and F. Castells, Environmental performance of construction waste: Comparing three scenarios from a case study in Catalonia, Spain, Waste management, **30**, 646-654, (2010).

[8] Z.Z. Noor, R.O. Yusuf, A.H. Abba, M.A. Abu Hassan and M.F. Mohd Din, An overview for energy recovery from municipal solid wastes in Malaysia scenario, Renewable and Sustainable Energy Reviews, **20**, 378-384, (2013).

[9] N.P. Cheremisinoff, *Handbook of Solid Waste Management and Waste Minimization Technologies*, Butterworth-Heinemann, Elsevier Science, (2003).

[10] B. Şener, M.L. Süzen and V. Doyuran, Landfill site selection by using geographic information systems, Environmental Geology, **49**, 376-388, (2006).

[11] Solid Waste and Public Cleansing Management Act 2007, Putrajaya, (2007).

[12] V.R. Sumathi, U. Natesan and C. Sarkar, GIS-based approach for optimized siting of municipal solid waste landfill, Waste Management, **28**, 2146-2160, (2008).

[13] E. Papargyropoulou, D.C. Preece, D.R. Padfield and A.A.B. Abdullah, Sustainable construction waste management in Malaysia: A contractor's perspective, *Proc. of the Int. Conf. on Management and Innovation for a Sustainable Built Environment,* Amsterdam, (2011).

[14] S.K. Jung, An integrative approach for municipal solid waste (MSW) landfill selection: A case study of O'ahu, Hawai'i, PhD Thesis, University of Hawai'i at Manoa, Ann Arbor, (2011).

[15] S. El Haggar, *Sustainable Industrial Design and Waste Management: Cradle-to-Cradle for Sustainable Development*, Academic Press, Elsevier, (2010).

[16] S.N.S. Ismail and L.A. Manaf, The challenge of future landfill: A case study of Malaysia, J. of Toxicology and Environmental Health Sciences, **5**, 86-96, (2013).

[17] Department of Environment, *Environmental Impact Assesement-Guidelines for Development of Solid Waste Sanitary Landfill*, Department of Environment, Malaysia, (2012).

[18] Z. Sakawi, Municipal solid waste management in Malaysia: Solution for sustainable waste management, J. of Applied Sciences in Environmental Sanitation, **6**, 29-38, (2011).

[19] E. Suresh and N. Usha, GIS modelling of land suitability for solid waste disposal, Int. J. of Geoinformatics, **4**(3), (2008).

[20] U.N. Ngoc and H. Schnitzer, Sustainable solutions for solid waste management in Southeast Asian countries, Waste Management, **29**, 1982-1995, (2009).
[21] S. Fauziah and P. Agamuthu, Trends in sustainable landfilling in Malaysia, a developing country, Waste Management and Research, **30**(7), 656-663, (2012).

MATEC Web of Conferences **47**, 05022 (2016)
DOI: 10.1051/matecconf/20164705022

A Monitoring of Air Pollutants (CO, SO_2 and NO) in Ambient Air Near an Industrial Area

Radin Maya Saphira Radin Mohamed[1,a], Ahmad Faizol Hakim Rahim[1] and Amir Hashim Mohd Kassim[1]

[1]*Department of Water and Environmental Engineering, Faculty of Civil & Environmental Engineering, University Tun Hussein Onn Malaysia.*

Abstract. A monitoring assessment was carried out to measure the concentration of air pollutants in ambient air in the university campus, which is located adjacent to the industrial area. The air pollutants were monitored for CO (Carbon monoxide),SO_2 (Sulfur dioxide) and NO (Nitrous oxide) at the three sampling points, with distance reference based from the industrial area. Air pollutant gases were sampled from the I-Brid Toxic Gases Analyzer with the sampling hour referred to the Recommended Malaysian Air Quality Guidelines (RMAQG) during October 2013 to Jun 2014. Meteorological data was collected from the E-Sampler device for 24 hours. It was found that the CO concentrations were fall within the RMAQG at all stations monitored. The SO_2 concentration was high at Station 3 (Material lab), with 0.66 ppm which was exceeded the RMAQG of 0.13 ppm. All three stations recorded high concentration of NO, which the peak concentration occurred at the afternoon sampling. The nearest Station 3 (Material lab) has recorded the highest level of NO, SO_2 and CO compared to the other stations. The monitoring data has contributed some highlights to the authority and awareness about possible long risk effect of the air pollutants at the case study .

1 Introduction

Air pollution is one of the most serious problems in the world. It occurs when the air contains gases, dust, fumes or odor in harmful amounts, which could be harmful to the health or comfort of humans and animals. For the past 50 years, air pollution was a widely recognized problem that comes with long and short term impact on human health and the environment [1]. With high rate of urbanization, industrialization and increase in motorized transport from the rapid population growth has resulted in increased concentration of various air pollutants sulphur dioxide (SO_2), nitrogen oxide (NOx), carbon monoxide (CO), lead (Pb), ozone (O_3) [2].SO_2 and CO concentrations in industrial areas and housing in Senai, Johor Malaysia were reported high concentration which were 2 ppm and 37 ppm, respectively, and the concentration had exceeded the limit SO_2 (0.13 ppm) and CO (30 ppm) [3].

The Universiti Tun Hussein Onn Malaysia (UTHM) is located in Batu Pahat, Johor, Malaysia (1°51'32"N 103°5'3"E) and has a student population of approximately 14,638 (Jun 18, 2015). Due to its proximity to a nearby industrial park (activities include data centers, electrical appliances manufacturing, wood processing plants, corrugated carton maker and packaging facilities), events of air pollution and its persistence, especially that of air pollutant gases is of particular concern to the

[a] Corresponding author : maya@uthm.edu.my

health and well-being of the student and faculty population. The ambient air across the UTHM campus is polluted by emission from vehicles (motocycles and cars) as the data showing the rise of number 11, 037 (2014) compared to 10, 232 in the year 2012 as recorded by the Safety Department UTHM. Pollution from motor vehicle is the single largest source of air pollution emissions, which composition depends on fuel used, and type and operating condition of the engine. The major constituents of motor vehicle pollutants are carbon monoxide (CO), hydrocarbon (HC), 8.5% nitrogen oxides (NOx), 0.8% particulate matter and 0.6% sulfur oxide (SOx). Cannistraro and Ponterio [4] stated that the pollution from motor vehicles is of great importance in developing countries with problems such as increasing vehicle fleets, infrastructural limitations, absence of emission control technologies of vehicle engines and poor maintenance or vehicle regulation. The wood processing involves the use of various chemicals (adhesives, thinners, paints, preservatives) [5]. These activities emit many gaseous pollutants including CO that may cause irritation of respiratory tracts and lungs and elevate the risk of respiratory tract infections.

Previous study on analysis of NO stated that, the combination of nitrous oxide causes paralysis of the nervous system, and concentrations greater than 100 $\mu g/m^3$ cause death for animals and humans. NO_2 is a deep lung irritant causing bronchial neutrophilic infiltration, pro inflammatory cytokine production, and responses to inhaled allergens in patients with asthma, both alone, and with SO_2 [6].To address the problem of air pollution, it is important to measure the level of air pollutants in the risk ambient air surrounding. The information of the status of air pollutants are essential in preventing the long term effects. Therefore, this monitoring data of the concentration of CO, NO_2 and SO_2 in ambient air was performed. As UTHM campus is located nearby an industrial area, it has raised the concern of air quality towards students and university staff. The concentration of air pollutants acts as an indicator in assessing the level of ambient air in UTHM campus.

2 Materials and Method

The air pollutants; carbon monoxide (CO), nitrogen monoxide (NO) and sulphur dioxide (SO_2) were measured during weekend and weekdays in October 2013 to Jun 2014. The sampling points were set up according to the distance from industrial area; Station 1-Stadium (1,000 m), Station 2-Kolej Kediaman Tun Dr. Ismail (500 m) and Station 3- Material Laboratory (300 m) as shown in Figure 1.

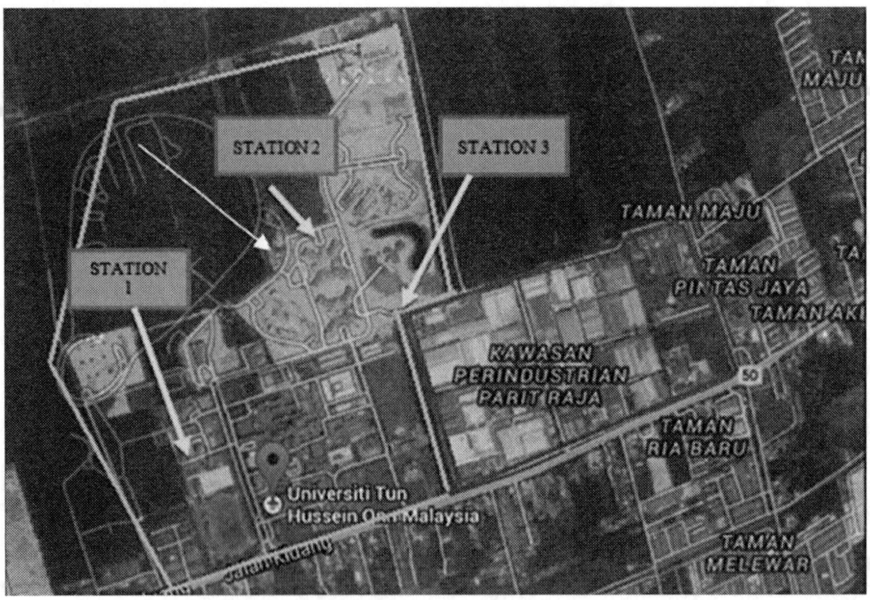

Figure 1. Location of air pollutants sampling station in UTHM and their proximity with the industrial area (adopted from Google Map, retrieve on 25th Dec 2014).

Air pollutants wasmonitored by using toxic gas analysis (Toxic Gas Monitor I-Bret Mx6) and E-Sampler Particulate Matter (Met One Instruments, Inc) for meteorological data. Sampling time was performed according to the standard Malaysia Ambient Air Quality as in Table 1. The Statistical Package for Social Sciences (SPSS) Software was used to obtain the significant value or not significant valueto determine whether the influence of meteorology was one of the factors that contribute to the incrementof concentration ofair pollutant. A result was considered 'significant' if the probability of the null hypothesis was equal to or less than 0.05 or ($P \leq 0.05$).

3 Air Pollutants Quality Standard

The air quality standards have been set by the United States Environmental Protection Agency (US EPA) and the Department of Environment Malaysia which is to determine the concentration of polluted air that is considered harmful to public health and environment. In this study, the Recommended Malaysian Air Quality Guidelines (RMAQG) which set up by DOE was used to configure the concentration of contaminants in polluted air. Table 1 shows the guidelines for the concentration of air pollutants in Malaysia.

Table 1. Malaysia Ambient Air Quality Guidelines (DOE, 2014).

Air pollutant	**Average time**	**Malaysia guidelines**	
		(ppm)	($\mu g/m^3$)
Carbon Monoxide (CO)	1 Hour	30	35 mg/m^3
	8 Hour	9	10 mg/m^3
Nitrogen Monoxide (NO)	1 Hour	0.17	320
Sulphur Dioxide (SO_2)	10 Minute	0.19	500
	1 Hour	0.13	350
	24 Hour	0.04	105

4 Results and Discussion

Concentration of air pollutants gases carbon monoxide (CO), sulphur dioxide (SO_2) and nitrogen monoxide (NO) were analysed and compared with the Recommended Malaysian Air Quality Guideline (RMAQG). Table 2 shows a summary data concentration of CO, SO_2 and NO between sampling stations during weekdays and weekend. The monitoring result shows that the concentration of CO at three sampling area was not exceeding the standard 30 ppm. The highest concentration of CO was at Station 3 was below than 10 ppm. Meanwhile the concentration of CO on weekend at all station was in range 0-2 ppm.It shows that CO concentration is still within safe levels in weekend. It is seconded by Cannistraro and Ponterio [4] stated that primary pollutants concentrations, benzene, carbon monoxide and particulate, decreased during weekends, this is due to the low vehicular traffic circulating in this period.

The highest concentrationSO_2 was 0.66 ppm at the Station 3 occurred during weekdays from 8.00-10.00 am, which exceeded the RMAQG of 0.13pm. Humidity and wind speed from the meteorological data on the sampling days had identified the factors that influenced the high concentration of SO_2 ($p<0.05$). Nevertheless, there was no SO_2 detected during weekend. Meanwhile, most of the NO concentration at three stations exceeded the standard which was 0.17 ppm occurred during weekdays and weekends especially at 8.00-10.00 am and 12.00-2.00 pm. The highest concentration was clearly recorded as being at afternoon peak between 12.00-2.00pm at Station 2

(KKTDI) which was 3.2 ppm. Besides that, the range for concentration during weekdays was 0.1-3.2 ppm and during weekends was 0.1-1.96 ppm.

Table 2. The concentration of CO, SO_2 and NO at three sampling stations during weekdays and weekend.

Station	Time	Toxic Gases (ppm)								
		CO			SO_2			NO		
		Weekday	Weekend	Standard (1hr)	Weekday	Weekend	Standard (1hr)	Weekday	Weekend	Standard (1hr)
Stadium (Stn 1)	8.00-10.00 am	0	0	30	0	0	0.13	0.66	0.66	0.17
	12.00-2.00 pm	0	0		0	0		0.46	0.23	
	4.00-6.00 pm	0	0		0.1	0		0	0.06	
	8.00-10.00 pm	0	0		0	0		0	0	
KKTDI (Stn 2)	8.00-10.00 am	0	0		0	0		0.26	0.26	
	12.00-2.00 pm	0	0		0	0		0.66	1.2	
	4.00-6.00 pm	0	0		0	0		0.43	0.33	
	8.00-10.00 pm	0	0		0	0		0	0	
Material laboratory (Stn 3)	8.00-10.00 am	0	2		0.66	0		0.16	1.96	
	12.00-2.00 pm	7	0		0	0		3.2	1.06	
	4.00-6.00 pm	0.66	0		0.1	0		0.16	0.1	
	8.00-10.00 pm	0	0		0.03	0		0.1	0	

5 Influence of Sampling Stations Distance to the Level of Air Pollutants

The levels of NO, CO and SO_2 were compared according to the distance from the industrial area as in Figures 2 to 4. It is shown that the distance has given a different level of air pollutants. Figure 2 shows the concentration of CO on weekday was recorded at three stations. For station 1 (Stadium) and Station 2 (KKTDI), the average reading recorded was 0 ppm. Comparatively at Station 3 (Material Laboratory) can be seen that there was obvious differences at 12:00 to 2:00 pm which the average reading was sharply increased from 0 ppm to 7 ppm and a sudden drop to 0.66 ppm at 4:00 to 600 pm.

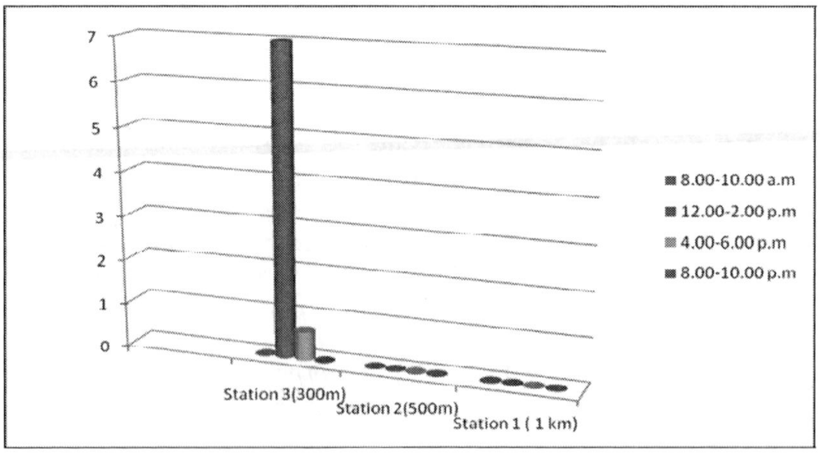

Figure 2. CO concentration in sampling stationsbased on the distance from industrial area.

Figure 3 refers to the average concentration of SO_2 readings at three sampling point which is have a different distance with the industrial plant. At 8:00 to 10:00 am, the highest reading was recorded at station 3 (Material Laboratory) which is 0.66 ppm. This indicates that the SO_2 concentration at that time has exceeded the standards and unsafe. While, the concentrations readings at the other stations was not change from 0 ppm. In the afternoon, starting from 4.00-6.00 pm only 0.1 ppm readings were recorded at station 1 (Stadium) and Station 2 (KKTDI), while in station 3 does not record any reading. At night, the air quality was safe where the reading of SO_2 concentrations was 0.03 ppm at station 3 only and it were below the standard of 0.13 ppm.

Figure 4 shows the NO concentration between the three stations based on distance from industrial area. The concentration at Station 3 (Material Laboratory) recorded the highest concentration at 12.00 to 2.00 pm, which the concentration rose dramatically from 0.16 ppm to 3.2 ppm. From the graph, it clearly seen the station 3 was the nearest to the industrial area.The concentration also was exceeded the standard RMAQG which was 0.17 ppm. This result shows that the industrial activities were contribute to the increased concentration of NO could jeopardize human health if exposed to concentrations of NO continuously.

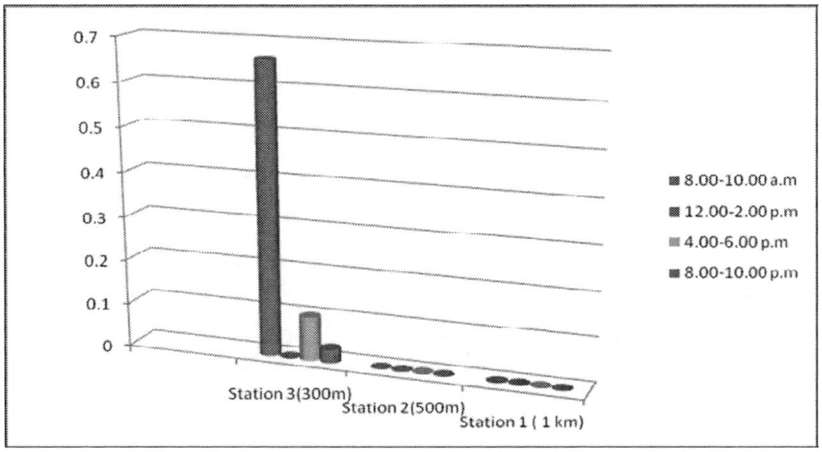

Figure 3. SO_2 concentration in sampling stations based on the distance from industrial area.

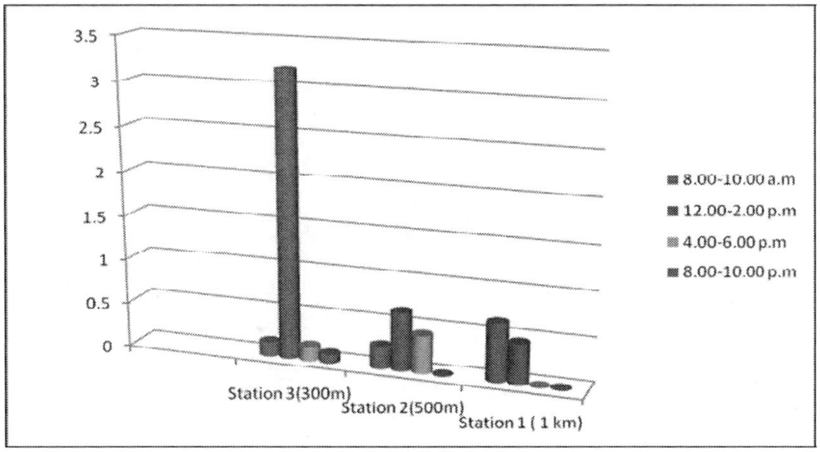

Figure 4. NO concentration in sampling stations based on the distance from industrial area.

6 Conclusion

The air pollutants concentrations were found high in the Station 3 (Material Lab) and Station 2 (KKTDI), with the concentration of SO_2 and NO were 0.66 ppm and 3.2 ppm, respectively and did not comply with the RMAQG of 0.13 ppm and 0.17 ppm. The results obtained from the distance between stations with industrial plant also play a role where near the station increased potential for exposure to toxic gas. The highest concentration was identified have occurred at Station 3 (Materials Laboratory) where that sampling station was closest station to the industrial area.

This monitoring work has suggested the links between certain meteorological factors and air pollutants concentration. This was derived through observations of air quality and postulation of wind speed and humidity. A more detailed analysis, correlating air quality with measured factors (such as human activities, an expanded set of meteorological factors) is necessary. Hence, the extensive monitoring of ambient air quality is currently conducted to provide a clear dispersion of air pollutants in ambient air of UTHM campus.
when preparing them.

Acknowledgement

The authors are grateful for the financial support provided by the Contract Grant UTHM Vot C006 and U396, Ministry of Education, Malaysia to cater this monitoring works.

References

[1] O. Ozden, T. Dogeroglu and S.Kara, Assessment of ambient air quality in Eskişehir, Turkey, Environment Int., **34**, 678–687, (2008).
[2] S.K. Goyal, S.V. Ghatge, P. Nema and S.M. Tamhane, Understanding urban vehicular pollution problem vis-a-vis ambient air quality - case study of a Megacity (Delhi, India), Environmental Monitoring and Assessment, **119**, 557–569, (2006).
[3] F. Mohmad, Pencemaran udara di kawasan perindustrian dan perumahan di Senai Johor, Degree Thesis, Universiti Teknologi Malaysia, (2003).
[4] G. Cannistraro and L. Ponterio, Analysis of air quality in the outdoor environment of the City of Messina by an application of the pollution index method, Int. J. of Civil and Environmental Engineering, **3**(6), 214-220, (2009).
[5] R.M.S.R. Mohamed, F.H.B.C. Aziz and A.H.M. Kassim, An assessment of selected heavy metal concentrations (Pb, Cu, Cr, Cd, Ni, Zn) in university campus located in industrial area. ARPN J. of Engineering and Applied Science, **9**(12), 2724-2730, (2014).
[6] P. Lazar, R. Zbořil, M. Pumera and M. Otyepka, Chemical nature of boron and nitrogen dopant atoms in graphene strongly influences its electronic properties, Phys. Chem. Chem. Phys., **16**, 14231-14235, (2014).

MATEC Web of Conferences **47**, 05023 (2016)
DOI: 10.1051/matecconf/20164705023

Tackling Carbon Emission with Nature: Effectiveness of Indigenous Microalgae Mixed Culture

M. Nazry Chik[1,a], Liyana Yahya[1], Afifi Zainal[1], Kamrul F. Kamarudin[2], Ding G. Tao[2] and M. Sobri Takriff[2]

[1]*Emission and Waste Management Technology Group, TNB Research Sdn Bhd, 43000 Kajang, Malaysia*
[2]*Department of Chemical and Process Engineering, Faculty of Engineering and Built Environment, Universiti Kebangsaan Malaysia, 43600 Bangi, Malaysia*

Abstract. Marine microalgae species was isolated and identified from its native condition of sea water discharge canals at the TNB Janamanjung Sdn Bhd's coal-fired power station. The species was expected to be a robust one based on the prevalent harsh upstream conditions and processes the survived ones endured. The isolation involves streaking-plating method as well as serial dilution and liquid media culture propagation. Morphological and molecular identification were both carried out before satisfactory identification of *Chlorella sp.* be made. This species was later mixed with marine *Isochrysis sp.* with varying volumetric proportions and cultured in five separate air-lift bubbling column photobioreactors, as part of the experiment to determine which optimum volumetric culture ratio is best to fix CO_2 from the power plant. Results indicate that a more dominant ratio of *Chlorella sp.* 75% and *Isochrysis sp.* 25% by volume provide gives a better growth profile and indicates a better CO_2 fixation rates.

1 Introduction

Any utility power company in a developing country is on constant pressure to deliver electricity to consumers reliably, safely and affordably, amidst the ever-increasing power demand and the current carbon-constraint world. Nowadays, curbing CO_2 emissions has been a new agenda, internationally and nationally - as in our Prime Minister's pledge to reduce 40% carbon intensity, from 2005 level [1]. Utilizing renewable energy and embarking on energy efficiency are some of the major approaches in curbing CO_2 emissions, and they have been promoted and implemented in Malaysia with some level of success. Of late, tackling the issue at the point source through Carbon Capture, Utilization and Storage Technologies (CCUS) is getting its foot in Malaysia. Being part of it, many literatures and works have identified phototropic microalgae as an alternative or complementary method among available CCS technologies [1]. The use of its natural, more efficient photosynthesis process, and the potential of valuable downstream products it can generate, make algal bio-fixation process eligible and attractive to be a more sustainable method in curbing CO_2 emissions [3]. Using this biological method, the CO_2 is fixed bio-chemically into C_x-H_y-O_z compound in product of photosysnthesis reaction originating from CO_2, water (H_2O) and photon energy from the sun as the reactants. By mass ratio, approximately 2 tonnes of CO_2 can be fixed by 1 ton of microalgae [5]. The baseline parameter

[a] Corresponding author : mnazry.chik@tnbr.com.my

of utmost concern is the CO_2 fixation rate by microalgae species, often expressed as grams of CO_2 fixed per culture volume per culture duration (e.g. $g_{CO2}/(L.day)$).

However, there are few issues and challenges regarding algae-based fixation and one of them is low productivity. Productivity of microalgae needs to be enhanced in order to portray a respected amount of CO_2 removal capability. One of the means to improve productivity is by using microalgae consortia, as it was reported to have better growth rate as it complement the metabolic activities of one another to grow and reproduce [6]. The growth characteristics, biochemical compositions, nutritional compositions and other growth factors of mixed culture can also be substantially different from those of monocultures [7]. Besides that, a mixed culture composed of various similarly productive strains with various optimal growing conditions would be less susceptible to environmental shifts than a monoculture [8]. Thus quests on finding a robust and high fixation rate microalgae species has been one of research focuses in this biological endeavour. This work expresses some results out of the effort in identifying robust indigenous microalgae consortia from the 3 x 700 MW coal-fired Sultan Azlan Shah Power Station in Manjung, Sitiawan, Perak Darul Ridzuan (TNBJ).

2 Materials and Methods

2.1 Species sampling and isolation of microalgae species

Samples were collected from the sea water outfall at Stesen Janakuasa Sultan Azlan Shah, Manjung Perak, west coast of Malaysia Peninsular, 4° 9' 31.158"S latitude and 100° 38' 30.3396" E longitude. This coal-fired power plant is among the major power generation plant in Malaysia, having an installed capacity of 3 units of 700 MW. This power station has been equipped with the state-of-the-art emission mitigation equipment to meet the stringent World Bank's emission standard for NO_x, SO_2 and particulate matters [9]. The outfall is a discharge point of spent sea water used in the power generation process in the upstream. Among processes undergone by the sea water at this point include high velocity spray, disinfection with hypochlorite, water-spray as part of flue gas desulphurisation and aerated (air bubbled). Thus, any microalgae species that survive at this point shall experience varying degrees of hydrodynamic stresses, disinfection and temperature difference based on the process described above, and thus any ones which found dominant in the outfall shall be the ones who are robust enough to dwell further in bioreactors. The samples were cultured using both F/2 and TNBR medium. The samples were maintained in 50 mL centrifuge tube, at room temperature at 28°C and kept with continuous light exposure at 12,000 lux by 50 W fluorescent daylight type lamps. Observation of microalgae growth in the liquid medium were continued with plating method by using respective medium by using 1.25% agar plate. In the parallel of liquid and 1.25% agar cultivation, the growth of microalgae were measured by biomass accumulation, cell density and amounts of colonies on the plate respectively.

2.2 Carbon fixation experiment

Optical densities were measured daily using HACH's 2700 Spectrophotometer at 560 nm wavelength. Algal biomass were extracted daily using vacuum filtration pump and its dry weight measured and recorded. The carbon fixation calculation was adopted by taking into considerations of the ratio between the moles of CO_2 and the moles of typical molecular formula of biomass [11]. By taking into considerations of a balanced photosynthesis formula as in Equation (1) below:

$$4CO_2 + \text{nutrient} + H_2O + \text{light} \rightarrow 4CO_{0.48}H_{1.83}N_{0.11}P_{0.01} + 3(1/2)O_2 \quad (1)$$

The ratio of molecular weight of CO_2 and biomass is 1.882, which upon various elemental analysis performed on microalgae biomass, the ratio ranges from 1.81 to 2.37 [11].

2.3 Morphological identification

Growing microalgae species were observed and morphological screening was done by comparing to the online catalogue as per University of Texas (UTEX), Culture Collection of Microalgae and The Culture Collection of Algae at Goettingen University (international acronym SAG). Morphological identification was done by determining the characteristic of isolated microalgae, which include size, availability of membrane wall and flagella and the shape of chloroplast.

2.4 Molecular identification

Single/pure strain of microalgae obtained in isolation process was used in molecular identification. Genomic DNA was extracted from the wet biomass by using liquid nitrogen and was further purified using Promega Wizard Genomic DNA Purification Kit. 18s rRNA genes Polymeric Chain Reaction (PCR) amplified using a primer designed by FirstBase Sdn Bhd. The sequenced products were compared with the National Center for Biotechnology Information (NCBI) data bank using the BLAST Program. (NCBI BLAST, USA). Top ten similar results were achieved and compared for the similarity in the percentage value.

2.5 Mixed culture growth profile experiment

Healthy culture of microalgae species isolated and identified earlier were later propagated through series of volume scaling-up before being tested in five 250-liter air-lift bubbling column photobioreactors located at TNBJ. Each of the reactors were cultured with varying volumetric ratio composition between the new species and *Isochrysis sp.*. *Isochrysis sp.* was isolated from the open sea off-coast of the same power plant and its selection in this work to form a consortium of species was based on previous works [12]. Table 1 below lists species volumetric ratio distribution across each of photobioreactor columns in use.

Table 1. Species volumetric ratio distribution.

Column Label	Volume New Species	Volume *Isochrysis sp.*
2	25%	75%
3	50%	50%
4	0%	100%
5	100%	0%
6	75%	25%

Each of the columns is bubbled using actual flue gas from a unit of power plant's flue gas duct. Fine bubbles were made possible by BLOWTAC/Disc-170 diffusers. The flue gas was extracted from a sampling point using PermaPure Baldwin's (USA) stainless steel flue gas probe system, through KNF's (Germany) vacuum pump. The flowrate was maintained at 1 liter per minute using Dwyer's rotameters, while pH was monitored using ThermoScientific's (USA) pH sensors. In order to avoid excessive acidic culture condition, the admission of flue gas was controlled using actuator valves of which their operation (open or close) were dictated by miliampere signals from pH transmitters. The flue gas composition varied throughout the experiment duration, in relation to varying coal used and regimes of plant's operation. Average composition of carbon dioxide (CO_2) content in the flue gas was between 4 – 6 vol. % with maximum of 8 vol. %, alongside other major gas components e.g. carbon monoxide (CO), nitrogen oxides (NO_x), sulfur dioxide (SO_2), etc. Daily growth monitoring was done by reading the optical density of 25 mL sample using HACH's D4000 Spectrophotometer at 560 nm wavelength and the readings were plotted.

3 Results and Discussions

3.1 Morphological and molecular identification

From microscope observation, the size of the purified microalgae ranges around 4.24 ~10 μm with the area around 56.48μm^2. Notable characteristic that can be observed was round or ellipsoidal shape with transparent outer membrane, with thin layer of membrane without any extension of flagella. By these observations, the suspected species was *Chlorella sp.* By comparison with *Chlorella sp.* from UTEX. *Chlorella* is blue-green algae species which belongs to phylum *Chlorophyta* division. Report from Fott and Novakova (1969) showed the versatility of *Chlorella* species to adapt to marine environment, despite its abundance in freshwater environment. Other results from NCBI also showed that several microalgae which were isolated from South China Sea were confirmed to be *Chlorella* species. Next confirmation step requires molecular level analysis.

In Table 2 depicted below, top ten hits blast result were observed and the list suggests that seven out of ten confirm up to 99% that an isolated, purified and sequenced microalgae is *Chlorella* sp. Thus, from molecular result plus with similar characteristic by morphological identification, it was confirmed that isolated microalgae were *Chlorella* sp.

Table 2. Top 10 Hits blast results against NCBI nucleotide collection (nr/nt) database excluding bacteria.

Description	Identification	Accession
Chlorella sp. KAS001 gene for 18s rRNA, partial sequence	100%	AB176663.1
Chlorella sp. KAS002 gene for 18s rRNA, partial sequence	100%	AB176666.1
Chlorella sp. SAG211-18 8s rRNA gene	99%	X73992.1
Chlorella sp. MDL5-18 18S ribosomal RNA gene, partial sequence	99%	AY197632.1
Chlorella sp. KAS005 gene for 18s rRNA, partial sequence	99%	AB176664.1
Chlorella sp. KMMCC C-137 gene for 18S rRNA gene, partial sequence	99%	GQ122349.1
Chlorella sp. MBIC10088 gene for 18S rRNA, partial sequence	99%	AB058308.1
Dictyosphaerium sp. NIES 453 18S ribosomal RNA gene, partial sequence; internal transcribed spacer 1,5.8S ribosomal RNA gene, and internal transcribed spacer 2	99%	GQ487248.1
Dictyosphaerium sp. CCAP 222/4 ribosomal RNA gene, partial sequence; internal transcribed spacer 1,5.8S ribosomal RNA gene, and internal transcribed spacer 2	99%	GQ487246.1
Kalenjinia gelatinosa strain CCAP 222/8 18S ribosomal RNA gene, partial sequence; internal transcribed spacer 1,5.8S ribosomal RNA gene, and internal transcribed spacer 2	99%	GQ477061.1

3.2 Growth profile of mixed culture

In discussing this, reference is made to the growth profile by optical density measurements of mixed culture between *Isochrysis sp.* and *Chlorella sp.* as depicted in Figure 1(a) for the first run and Figure 1(b) for the second run. General growth profiles for both runs are observed to be in accordance to the typical growth profile of microalgae, with distinct lag, exponential and decay phases especially for the first run. From Figure 1(a), the best growth curve – having the highest and steadiest optical density - occurs in column no. 6 where the mixture of *Isochrysis* 25% and *Chlorella* 75% by volume reside. The trend, in descending order then followed by *Isochrysis* 50% and *Chlorella* 50%, *Isochrysis* 0% and *Chlorella* 100%, *Isochrysis* 75% and *Chlorella* 25% and finally *Isochrysis* 100% and *Chlorella* 0%. Similar trend was also observed in the second run where *Isochrysis* 25% and *Chlorella* 75% tops the chart. The species in the second run also seemed to be more acclimatized to the experiment conditions where no lag phases were observed in early culture stage, steeper exponential phase and longer period of stationary phase. However, the second and third place here was interchanged with that from the second run, which warrants for a more finer volumetric ratio around that region to be performed in the future. It can be concluded that better growth and fixation rates occur in cultures where there is higher proportion of consortia. The presence of *Isochrysis sp.*, together with the consortia, acting as symbiont and complements the metabolic activities of one another to grow and reproduce. The volumetric ratio of *Isochrysis* 25% and consortia 75% gives the best growth and fixation rates, now determined to be 0.965 g.CO2/L.day, a 34% improvement from our previous effort.

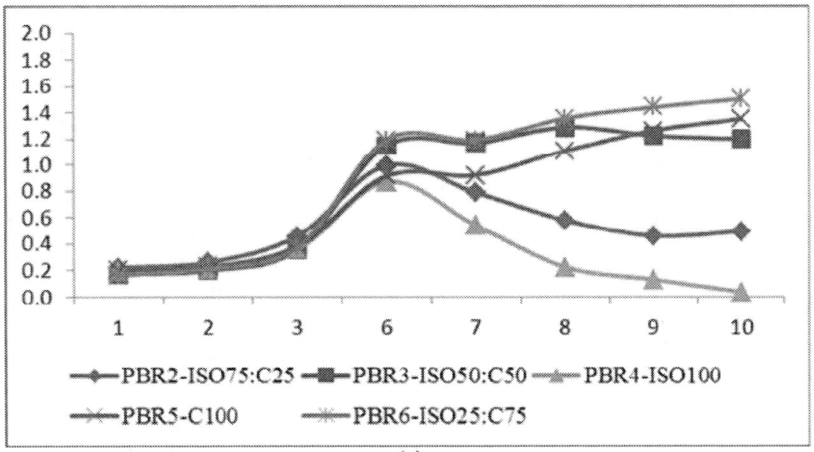

(a)

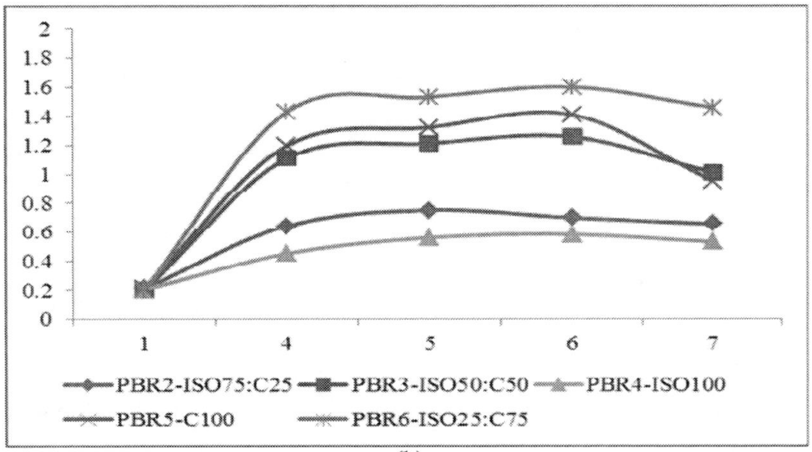

(b)

Figure 1. Optical density for mixed culture (a) first run, (b) second run.

4 Conclusions

The observations obtained from the above suggest two important findings – *Chlorella sp.* isolated from the power plant's outfall is a robust species and meet hypothesis made earlier based on the environmental condition and upstream processes they have endured. Its robustness is evidenced by the better growth profile observed in the columns where its volume is more dominant. Second interesting finding lies on the observed dependency of *Chlorella sp.* on *Isochrysis sp.* as illustrated by better growth profile on mixed culture against mono-culture. This suggests that *Isochrysis sp.* can act as a symbiont and complements *Chlorella sp.* A better data to support this can be made available through population dynamics study to be performed later.

Thus, this work has demonstrated the basic findings on the importance of searching an indigenous species in the locality where the CO_2 that is going to be fixed originates. *Chlorella's* acclimatization to the surrounding level of CO_2 and other flue gas components and its readiness to co-exist with other species has made it a good mixed-culture species for biological fixation at this power plant.

The authors would like to thank TNB Research Technical Board Committee for approving a seeding fund no. TNBR/SF0157/2014 and Inbiosis UKM in supporting the research financially and educationally between 2014 and 2015. Special thanks also extended to the Management team of TNB Janamanjung Sdn.Bhd. for their co-operation and support.

References

[1] Demotix online news, Malaysian Prime Minister at COP15, (2011). <retrieved from http://www.demotix.com>

[2] M. Packer, Algal capture of carbon dioxide: biomass generation as a tool for greenhouse gas mitigation with reference to New Zealand energy strategy and policy, Energy Policy, **37**, 3428–3437, (2009).

[3] S.H. Hoa, C.Y. Chena, D.J. Leed and J.S. Chang, Perspectives on microalgal CO_2-emission mitigation systems – A review, Biotechnology Advances, **29**, 189-198, (2011).

[4] B. Wang, Y. Li, N. Wu and C. Q. Lan, CO_2 bio-mitigation using microalgae, Applied Microbiol Biotechnology, **79**, 707–718, (2008).

[5] P. Spolaore, C. Joannis-Cassan, E. Duran, and A. Isambert, Commercial applications of microalgae, J. of Bioscience and Bioengineering, **101**(2), 87–96, (2006).

[6] Y. Chisti, Biodiesel from microalgae, Biotechnology Advances, **25**, 294–306, (2007).

[7] T. Papone, S. Kookkhunthod and R. Leesing, Microbial oil production by monoculture and mixed cultures of microalgae and oleaginous yeasts using sugarcane juice as substrate, World Academy of Science, Engineering and Technology, **6**(2), 195-199, (2012).

[8] W.W. Huang, B.Z. Dong, Z.P. Cai and S.S. Duan, Growth effects on mixed culture of Dunaliella salina and Phaeodactylum tricornutum under different inoculation densities and nitrogen concentrations, African J. of Biotechnology, **10**(61), 13164-13174, (2011).

[9] K.R. Johnson and W. Admassu, Mixed algae cultures for low cost environmental compensation in cultures grown for lipid production and wastewater remediation, J. of Chemical Technology and Biotechnology, **88**(6), 992-998, (2013).

[10] A. Mahmud and E.W. Baker, Environmental considerations for the 3 x 700MW coal-fired Manjung project, Jurutera, (2004).

[11] I. Suryata, H.G. Svavarsson and S. Einarsson, Geothermal CO_2 bio-mitigation techniques by utilizing microalgae at the blue lagoon, Iceland, *Proc. of the 34th Workshop on Geothermal Reservoir Engineering*, Stanford, (2010).

[12] S. Van Den Hende, H. Vervaeren and N. Boon, Flue gas compounds and microalgae: (bio-) chemical interactions leading to biotechnological opportunities, Biotechnology Advances, **30**(6), 1405–1424, (2012).

[13] M.N. Chik, L. Yahya, M.H. Boosroh and L. Shamsuddin, Scanning of microalgae species for biological CO_2 fixation at a Malaysian coastal coal-fired power station, *Proc. of the Int. Conf. on Sustainable Development*, Putrajaya, (2011).

[14] L. Yahya, M.N. Chik and M.H. Boosroh, Carbon fixation ability between marine Isochrysis sp. and NannochloropsisOculata. *Proc. of the 12th Malaysian Society of Applied Biology,* Kuala Terengganu, (2012).

[15] L. Yahya, M.N. Chik and M.H. Boosroh, Ability of Isochrysis sp. as the biological carbon fixer under actual coal-fired power plant's flue gas. *Proc. of the 6th Int. Conf. on Clean Coal Technology,* Greece, (2013).

MATEC Web of Conferences **47**, 05024 (2016)
DOI: 10.1051/matecconf/20164705024

Low Thermal Pretreatment as Method for Increasing the Bioavailability of Organic Matters in Domestic Mixed Sludge

Roslinda Seswoya[1,a] and Ahmad Tarmizi Abdul Karim[1]

[1]*Faculty of Civil and Environmental Engineering, Universiti Tun Hussein Onn Malaysia, 86400 Parit Raja, Johor, Malaysia*

Abstract. In practice, primary and secondary sludge are fed into anaerobic digestion. However, the microbial cell exists in secondary sludge are an unfavorable substrate for biodegradation. Thermal pretreatment is proved to increase the bioavailability of organic and improve the biodegradation subsequently. During low thermal pretreatment, both intracellular (within the microbial cell) and extracellular (within the polymeric network) materials were extracted. This process increases the bioavailability meaning that organic compounds are accessible to the microorganisms for their degradation. This research aims to investigate the effect of thermal pretreatment on domestic mixed sludge disintegration. Domestic mixed sludge was thermally treated at 70^{0}C for various holding times. The pre-thermally treated domestic mixed sludge was measured for protein and carbohydrates following the Lowry Method, and Phenol-Sulphuric Acid Method respectively. DR 6000 UV-Vis spectrophotometer, DRB200 Reactor (digester) and COD vial (TNT plus 822) were used for COD determination, based on Reactor Digestion Method approved by USEPA. The results showed that the organic matter in domestic mixed sludge is efficiently solubilised during thermal treatment organic matter. The higher soluble yield for each monitored parameter determined in this study indicated that low thermal pretreatment improve bioavailability.

1 Introduction

Sewage was treated using anaerobic digestion widely [1]. About 55% to 70 % methane rich biogas will be produced during the anaerobic digestion of sewage sludge, due to the presence of easily biodegradable organic in sewage sludge [2]. Typically, methane production potential from sewage sludge is approximately 300–400 m^3/ VS_{added} [3].

The pretreatment or sludge disintegration can be used to accelerate the solubilization of particulate organic matter in sewage. During pre-treatment, both intracellular (within the microbial cell) and extracellular (within the polymeric network) materials were extracted, increased the bioavailability subsequently. In comparison with other methods of pre-treatment, thermal pre-treatment has been considered to be a more promising technique due to less energy consumption and no chemical addition. More recently, the low and mild thermal pretreatment, has been gaining more attention because of similar efficiency compared with the conventional thermal pretreatment, which operated at high temperature [4 - 9]. The low and mild thermal pretreatment is operated at temperature less than

[a] Corresponding author : roslinda@uthm.edu.my

120^0C[6]. The low thermal pretreatment (70^0C) showed positive effects on bioavailability of organic matter from municipal sewage sludge [10, 11] . This paper seeks to evaluate the effect of the low thermal treatment of domestic sewage sludge disintegration. The influence of treatment time and treatment temperature is determined through variation of both parameters. In addition, the types of organic matters that are solubilized during the low thermal pretreatment are also discussed.

2 Material and Methods

2.1 Sludge sampling

For the experiments, sludge samples were taken from the full scale sewage treatment plant (STP) in Malaysia. The conventional wastewater treatment used in this plant consists of preliminary, primary treatment and secondary treatment. Activated sludge is a process applied in the secondary treatment. The sample is a mixture of primary and secondary sludge. The sludge was stored at 4^0C in the laboratory until use.

2.2 Low temperature (700C) pre-treatment

The low temperature pretreatment was carried out at 70^0C in order to enhance thermal solubilization of particulate material. The effect of thermal pretreatments depends both on treatment temperature and time. Therefore, in this work the effect of pretreatment time were evaluated by identifying the samples characteristics at different pretreatment times (20 min, 40 min and 60 min) in order to study the combined effect. Two (2) numbers of 0.5 L duran bottles, each containing 0.2 L of domestic mixed sludge were submerged in a thermostatic bath at 70^0C during 20 min, 40 min, and 60 min. The duran bottles were covered with household aluminium foil, to avoid evaporation. Firstly, the thermostatic water bath was heated at the desired temperature, and the duran bottles were immersed in the water bath only when the temperature has reached the desired temperature. The thermal treatment duration was defined as the elapsed time between the moment at which the sample was immersed in the thermostatic water bath and the moment at which the sample was removed from the thermostatic water bath. After treatment, the sample was placed on the table and cooled passively to room temperature and stored at 4^0C until use.

2.3 Analytical method

The following components were analyzed before and after thermal treatment on both the sludge and the supernatant to identify total and soluble fractions: COD, carbohydrate, and protein. The supernatant samples were obtained after centrifugation at 10,000 rpm for 10 min and subsequent filtration through 0.45 μm cellulose acetate filter [12] .The analysis performed on the supernatant will be called 'soluble', and the analysis on the sludge will be called 'total'. DR 6000 UV-Vis spectrophotometer, DRB200 Reactor (digester) and COD vial (TNT plus 822) were used for total and soluble COD determination, based on Reactor Digestion Method which was approved by USEPA. The protein and carbohydrate determination were conducted according to the Lowry Method [13] and Phenol-Sulphuric Acid Method [14] respectively. Bovine Serum Albumin (BSA) and glucose stock solutions were used as the protein and carbohydrate standards respectively. According to [10] the degree of solubilisation (%) of the organic components is calculated by the following equation:

$$\text{degree of solubilisation}\,(\%) = \frac{\text{sC treated} - \text{sC untreated}}{\text{tC untreated}} \tag{1}$$

where: sC treated = soluble concentration treated, sC untreated = soluble concentration untreated, and tC untreated = total concentration untreated.

3 Results and Discussion

3.1 Effects of thermal treatment on the sludge solubilisation

Thermal treatment on the domestic mixed sludge was performed at 70^0C for three different treatment durations. An overview of the results is shown in Table 1. From the table, it is clear that a thermal treatment (70^0C) effectively releases the studied components (COD, protein, and carbohydrate) to the water phase, as the degree of solubilisation (%) is higher at any treatment time. This means that the bioavailability of low thermally treated domestic mixed sludge is improved consequently. This finding is similar with the result obtained by [15], indicating that the sludge disintegration improves COD bioavailability. However, in this study the ultrasonic (low frequency of 31 kHz) was used for sludge disintegration and the municipal sewage sludge was studied instead of domestic sewage sludge.

Table 1. Concentration and degradation of the organic components of the untreated (raw) and pre-thermally treated domestic mixed sludge.

Parameter	Blank (Raw)	70^0C		
		20 min	40 min	60 min
COD total, TCOD (g/L)	16.83 ± 0.25	8.05 ± 0.07	20.85 ± 0.92	21.45 ± 0.21
COD soluble, SCOD(g/L)	2.55 ± 0.02	3.19 ± 0.01	3.43 ± 0.01	3.54 ± 0.02
Solubilisation %	0.00	3.80	5.23	5.88
Protein total (mgBSA-eq /L)	3568.50 ± 27.78	6634.00 ± 33.81	8249.67 ± 19.69	7934.17 ± 15.67
Protein soluble (mg BSA -eq/L)	47.9 ± 0.89	110.97 ± 5.35	163.37 ± 2.30	176.83 ± 1.04
Solubilisation %	0.00	1.77	3.24	3.61
Carbohydrate total (mg Glu- eq /L)	1279.83 ± 3.06	2461.83 ± 8.39	4787.33 ± 23.14	5314.17 ± 35.65
Carbohydrate solubl e(mg Glu-eq/L)	52.77 ± 0.31	139.73 ± 1.08	169.73 ± 0.75	202.63 ± 2.48
Solubilisation %	0.00	6.79	9.14	11.79

Note :

1) 16.83 ± 0.25 = average ± stdev,
2) Triplicate sample were prepared for all measurement except duplicate sample for COD
3) Coefficient of variation (CV) for each data is less than 5%

3.2 Organic matter solubilisation

Table 2 shows the ratio of SCOD/TCOD of untreated domestic mixed sludge and pre-thermally treated domestic mixed sludge at 70^0C. The value for ratio of SCOD/TCOD is higher at each treatment time. However, that treatment time of 20 min increased the ratio of SCOD/TCOD to almost 3 times higher than the SCOD/TCOD value of untreated domestic mixed sludge. It seems that 20 min treatment time is able to release the particulate COD into the water phase.

Treatment time at 40 min shows the biggest solubilisation, and no significant solubilisation increment observed at treatment time of 60 min. There is a significant difference of COD solubilisation at 70^0C of 60 min for both pre-thermally treated domestic sewage sludge and municipal sewage sludge. As for pre-thermally treated municipal sewage sludge, the COD solubilisation at 60 min is almost 3 times higher than the untreated sample [10]. On the contrary, the degree of solubilisation (in term of COD at 70^0C of 60 min) for pre-thermally treated domestic sludge is four times higher than pre-thermally treated municipal sewage sludge (1.36 %). This showed that that the low thermal pretreatment have a great potential for improving the bioavailability of domestic mixed sludge.

Table 2. The changes of SCOD/TCOD ratio at different treatment time (700C).

Parameters	Blank (raw)	70^0C		
		20 min	40 min	60 min
COD total, TCOD (g/L)	16.83 ± 0.25	8.05 ± 0.07	20.85 ± 0.92	21.45 ± 0.21
COD soluble,SCOD (g/L)	2.55 ± 0.02	3.19 ± 0.01	3.43 ± 0.01	3.54 ± 0.02
SCOD/TCOD (%)	15.1	39.6	16.5	16.5

3.3 Carbohydrate and protein solubilisation

Carbohydrates and proteins were solubilised due to thermal treatment at 70^0C. It was observed that the carbohydrate soluble and protein soluble concentration increased gradually with treatment time, indicating that more and more particulate organic matters in domestic mixed sludge become the soluble ones with the increase of treatment time. The soluble protein and carbohydrate is almost similar at treatment time 40 min. While, at treatment time of 20 min and 60 min the carbohydrate is soluble more than protein. There are no significant changes of solubilisation (%) for protein at 40 min and 60 min treatment time. On the other hand, the higher solubilisation (%) observed for carbohydrate at higher treatment time. Referring to Table 1, the degree of solubilisation of carbohydrate is always higher compared to protein at any treatment time. This showed that carbohydrate in domestic mixed sludge was more soluble than protein when it was treated at 70^0C. This result contradicted with the findings from thermal pretreatment at 70^0C for municipal sewage sludge as observed by [10], where carbohydrate was less soluble. The carbohydrate was also less soluble from the municipal sewage sludge with low organic content (VS/TS = 40%) after undergoing pretreatment at 70^0C [6]. The percentage solubilisation (%) observed from this study (domestic sewage sludge) for both carbohydrate and protein are higher than what was observed when municipal sewage sludge was used as shown by [10]. At the applied temperatures, a reaction of soluble carbohydrates with themselves or soluble proteins (forming e.g. Amadori compounds or melanoidins) was not expected in this study since the applied temperature is low. This hypothesis was confirmed visually because the sludge supernatant remained transparent and did not turn brown to a large extent after treatment.

4 Conclusion

The present paper studied the application of low temperature thermal treatment (70^0C) for domestic mixed sewage sludge disintegration. It was seen that organic compounds are efficiently solubilised during thermal treatment as indicated that higher organic release ina longer treatment time.This proved that sludge disintegration is able to improve the bioavailability of organic matter. This would promote the use of thermal pretreatment to enhance the anaerobic digestion of domestic mixed sewage sludge especially for efficient methane recovery. Future research will focus on the methane production from the pre-thermally treated domestic mixed sludge at various low temperature (< 100^0C).

Acknowledgement

The authors would like to thank Universiti Tun Hussein Onn Malaysia (UTHM) for their financial support to make these research possible. The authors also would like to thank Unit Inovasi, IWK for the technical support.

References

[1] A.F. Morgan-Sagastume, S. Pratt, A. Karlsson, D. Cirne, P. Lant and A. Werker, Production of volatile fatty acids by fermentation of waste activated sludge pre-treated in full-scale thermal hydrolysis plants, Bioresour. Technol., **102**,3089–97, (2011).

[2] L. Appels, J. Baeyens, J. Degrève, and R. Dewil, Principles and potential of the anaerobic digestion of waste-activated sludge, Prog. Energy Combustion Science, **34**(6), 755–781, (2008).

[3] Å. Davidsson, C. Lövstedt, J. Ia Cour Jansen, C. Gruvberger and H. Aspegren, Co-digestion of grease trap sludge and sewage sludge, Waste Manag., **28**(6), 986–992, (2008).

[4] L. Appels, J. Lauwers, J. Degrève, L. Helsen, B. Lievens, K. Willems, J. Van Impe and R. Dewil, Anaerobic digestion in global bio-energy production: Potential and research challenges, Renew Sustainable Energy, **15**(9), 4295–4301, (2011).

[5] W.J. Park and J.H. Ahn, Effects of microwave pretreatment on mesophilic anaerobic digestion for mixture of primary and secondary sludges compared with thermal pretreatment, Environ. Eng. Res., **16**(2),103–109, (2011).

[6] Y. Yan, H. Chen, W. Xu, Q. He and Q. Zhou, Enhancement of biochemical methane potential from excess sludge with low organic content by mild thermal pretreatment, Biochemical Engineering J., **70**, 127–134, (2013).

[7] D. Bolzonella, P. Pavan, P. Battistoni and F. Cecchi, Mesophilic anaerobic digestion of waste activated sludge: influence of the solid retention time in the wastewater treatment process, Process Biochemical, **40,** 1453–1460, (2005).

[8] M. Kuglarz, D. Karakashev and I. Angelidaki, Microwave and thermal pretreatment as methods for increasing the biogas potential of secondary sludge from municipal wastewater treatment plants, Bioresource Technologies, **134**, 290–297, (2013).

[9] I.A. Nges and J. Liu, Effects of anaerobic pre-treatment on the degradation of dewatered-sewage sludge, Renewable Energy, **34**(7), 1795–1800, (2009).

[10] L. Appels, J. Degrève, B. Van der Bruggen, J. Van Impe and R. Dewil, Influence of low temperature thermal pre-treatment on sludge solubilisation, heavy metal release and anaerobic digestion, Bioresource Technolgy, **101**(15), 5743–5748, (2010).

[11] I. Ferrer, S. Ponsá, F. Vázquez and X. Font, Increasing biogas production by thermal (70°C) sludge pre-treatment prior to thermophilic anaerobic digestion, Biochemical Engineering Journal, **42**(2), 186–192, (2008).

[12] H. Yi, Y. Han and Y. Zhuo, Effect of combined pretreatment of waste activated sludge for anaerobic digestion process, Procedia Environmental Science, **18**, 716–721, (2013).

[13] O.H. Lowry, N.J. Rosebrough, A.L. Farr, R.J. Randall and A. Lewis, Protein measurement with the folin phenol reagent, J. of Biological Chemistry, **193**, 265–275, (1951).

[14] M. Dubois, K.A. Gilles, J.K. Hamilton, P.A. Rebers and F. Smith, Colorimetric method for determination of sugars and related substances, American Chemical Society, **28**(3), 350–356, (1956).

[15] U. Neis, K. Nickel, and A. Tiehm, Enhancement of anaerobic sludge digestion by ultrasonic disintegration, Water Science Technologies, **42**(9), 73–80, (2000).

MATEC Web of Conferences **47**, 05025 (2016)
DOI: 10.1051/matecconf/20164705025

An Overview of Organic Waste in Composting

Aeslina Abdul Kadir[1,a], Nur Wahidah Azhari[1] and Siti Noratifah Jamaludin[1]

[1]*Faculty of Civil and Environmental Engineering, Universiti Tun Hussein Onn Malaysia, 86400 Parit Raja, Johor, Malaysia*

Abstract. This paper reviewed studies on the composting process of organic waste. Organic wastes are wastes that easily biodegradable. These wastes are produced from many sources such as agricultural waste, market waste, kitchen waste, urban solid food wastes and municipal solid waste. Without proper management, these waste could create several environment problem. Therefore, composting is the best low cost alternative solution to overcome this problem. Composting method can degrade all types of organic wastes like fruits, vegetables, plants, yard wastes and others. The composition from organic waste that could be used as nutrients for crops, soil additive and for environmental management. However, many factors can contribute to the quality of the compost products as different types of organic wastes have different concentrations of nutrients, nitrogen, phosphorous and potassium (N, P, K) which are the common macro nutrients present in fertilizers. The presences of heavy metals show how composts can be applied to soils without contributing any ill effect. In term of the factor affecting the composting process, temperature, pH, moisture contents and carbon nitrogen ratio (C:N) are the main parameters that contribute to the efficiency of the composting process.

1 Introduction

In Malaysia, the percentage of organic waste in the environment was between 48 to 68% [1]. Due to the large amount of organic waste in the environment, it has become one of the main global issues. Among the various treatments in managing the organic waste such as the use of landfill and incineration, decaying organic wastes by using biological processes is considered as more suitable solution method [2, 3]. Composting is one of low cost biological decomposition process. The composting process is circuited by microbial activity. The physical-chemical parameter affected by this process include temperature, aeration, moisture content, C:N ratio and pH [4].

Composting is an alternative solid waste management system (SWM), it can be used for the recycling of organic matters into useful products. In addition, it can also be used to control the increase of waste [5]. This process is considered to be the most efficient, environmentally safe and most agronomically sound, where compost can be used as a soil conditioner, organic fertilizer as well as containing high nutrients for the soil [6]. The microbial community in compost, which are bacteria, fungi and worms can also stabilize the degradable organic matters.

Besides that, the characteristics of the microbial population depend on the substrate and physical conditions, which are the moisture, temperature and aeration of where composting take place. The performance of the composting process also will depend on the characteristics of the waste because

[a] Corresponding author : aeslina@uthm.edu.my

composting is only suitable for waste that is biodegradable [7]. There are many advantages that can arise from composting, including the reduction of volume, weight and water content of the waste as well as producing inactive pathogenic organisms [8, 9]. Therefore, the compost can contribute to the improvement of soil contents and nutrients required for harvesting plants, and will significantly reduce the use of synthetic fertilizers [10]. The application of compost can improve soil properties that are badly in need of renewal as it can increase the organic carbon contents in the soil. Meanwhile, compost also act as a soil intervention in improving soil structure, water infiltration rate, water holding capacity and tilth [12].

2 Overview of Organic Waste in Composting

The composting of organic waste in Malaysia is highly encouraged. As a huge amount of wastes are filling our overfilled landfills. Therefore, this paper will review composting as an alternative method for waste management. The physical and chemical properties of organic wastes during the composting processes had been discussed in details by many researchers. This paper will review studies on how the organic wastes were disposed using the composting method. The studies reviewed were divided into three classifications of organic wastes, which are agriculture waste, municipal solid waste and kitchen waste.

2.1 Agriculture wastes

Agricultural Wastes (AW) are wastes produced from agricultural activities in agricultural premises. Agricultural wastes are often managed poorly because of the limited access to disposal facilities, hence most of agricultural wastes are burned or incinerated. Taleb et al., [13] studied the compost of plant residues and sheep manure with soil (1:1). The compost pile was put in a box with the dimension of 1.7 m x 5 m x 5 m. The initial temperature of the compost was 25℃, the value increased after 30 days to 30℃ but declined to 24℃ in day 50 at the end of the composting process. The increasing of the temperature had affected the moisture content. The moisture content slowly decreased at the end of 50 days, the pH value also decreased from 8.38 to 6.02. While for N, P and K contents, 0.65%, 0.00295% and 0.3640% were recorded at the end of the composting process. Unfortunately, the availability of P and K content was less compared to the compost fundamental by Washington State University (WSU).

Furthermore, Ch'ng et al. [14] studied the compost of pineapple leaves (PL) wastes. In this study, Polystyrene boxes were used. The ambient temperature was recorded between 25 to 32.5℃. The compost seed, was a combination of PL (3.5 kg), chicken manure (2.8 L), chicken feeds (350 g) and molasses (175 g). Chicken manure and molasses were used as microbial seedings and to supply carbohydrates for the microbes. During the composting process, the physical-chemical tested were temperature, moisture content, pH, C:N ratio, and nutrients contents for N, P and K. The temperature on the first day rose to 50.9℃ in day 13 and decreased to the ambient temperature of 33.5℃. The moisture content of the compost, was maintained at 50 to 53.3%. For the C:N ratio, the initial compost process was 30 and decreased to 19.8 at the end. The content of nutrient of N, P and K were recorded at 2.3%, 0.46% and 2.67% and recorded low heavy metal content.

Composting of rice straws (RS), wheat straws (WS), potato plants (PP) and mustard stovers (MS) with fish pond bottom was conducted by Karak et al. [15]. This composting process was conducted for 56 days by using a heap as compost box, with the dimension of 2.5 m x 1.5 m x 1.5 m. For all of the compost preparations, the temperature of compost at the first day ranged from 24 to 26.8℃ and increased to 81℃ before becoming constant. The initial values of the pH ranged from 6.76 to 7.68 while the total N concentrations were in the range between 14.56 g/kg to 21.57 g/kg; the heavy metal concentration was below the standard set by the Indian Ministry of Agriculture and Cooperation. At the end of the composting, the C:N ratio varied from 11 to 18.

In addition, Tripetckul et al. [16] conducted the composting of wastes that were produced from the coconut plants which are coir pitch and coconut juice. The ratio was 3 (coir pitch):1 (rice bran) :0.08

(Mollases) with cow manure. Three composts were prepared for this study, the initial C:N ratios were 30, 25 and 20 and the moisture content was retained between 50 to 60%. Coconut juice was used to adjust the moisture content. During the composting process, the textures, color, odor and microbial evaluation were observed by visualizations. On day 1, white fungi appeared. At the 2nd and 3rd week, the composting material was the smallest in size and brownish in color and at the end of the composting process, the compost became blackish brown in color and turned into a humus-like texture. In early days of composting, the temperature reached 56 to 58°C and the temperature maintained for 9 to 11 days before dropping to the ambient temperature. The pile was turned consistently to increase the oxygen supplies that can help to reactivate the composting process and the availability to microorganisms present during the composting process. The increase of pH in the beginning, was due to the metabolic degradation of organic acids and ammonification process during organic matter degradations and this slowly declined because of the production of organic acids during the biodegradation of lignocellulose in the compost pile. At the end of the composting process, the C:N ratio decreased, this showed that the degree of humification of organic matter had increased through the composting process, while the total N content increased due to the weight loss of the compost pile.

Other researcher, Janakiram and Sridevi [17] added manure in the composting process and studied the compost of succulent plant, shrubs and trees, mixed with cow dung. The physical and chemical parameters were tested. The moisture contents of the compost were between 12.30 to 18.82% and pH values were 7.23 to 7.59. On the other hand, the N contents were 1.22 to 1.50%, P 0.32 to 0.40% and K 1.70 to 1.8% and the C:N ratios were at 10:1, 11:1 and 12:1. This study suggested that to increase the presence of K, fibrous materials like straws and wood chips which can absorb the water and prevent loss of K can be added into the compost.

The study by Saithep et al. [18] used tobacco plants (TP) as their composting wastes. The compost seed was made out of TP (100), cow manure (10), urea fertilizer (2) and compost inoculums (0.1). The moisture content was maintained at 60%. The Bacillus, Streptomyces and Aspergillus bacterias were present in the compost inoculum. The composting process was done by pilling the compost and the aeration of the compost was supplied by manual turning and forced aeration. During the composting process, the temperature ranged between 29 to 62℃ and decreased slowly to the ambient temperature of 30℃. The pH values of the composts during the manual turning and forced aeration were 9.45 to 11.78 and 9.10 to 10.97 and decreased to 7.78 and 7.80 respectively after 10 days. The moisture contents for the compost was initially at 60% and decreased to 50%, the decrease of the moisture contents in the compost occurred because of the water evaporation due to the heat generated by microbial reactions. In this study, at the end of composting, the C:N ratio after 50 days of composting was recorded at 13 and the C:N ratio of a matured compost in general is below 20.

Furthermore, Pagnakorn [11] used banana peels that increasingly becoming a common agricultural waste in Thailand. Banana peels were used as a fermented organic liquid. During this compost run, the pH, moisture, organic carbons and nutrients (N, P, K) were tested and compared with the standard level set by the Ministry of Agriculture and Cooperative. The pH value obtained was higher than 7, which was 9.91, this is unsuitable and not recommended because it may cause the soil to loss its efficient nutrients such as iron and manganese. However, this is suitable for the soils in Thailand. The moisture content recorded was 15.55%, while the N, P and K levels were higher than than standard level; 1.69%, 2.92% and 0.84% respectively.

2.2 Municipal solid waste

Manohara and Belagali [19] study the composting quality during three of the monsoons which are pre-monsoon, monsoon and post-monsoon. During the composting, water and bio-inoculum were added to maintain the moisture contents of the pile. The maximum temperature was recorded at 68℃, high temperature is important to remove the weed seeds and eliminating the presence of pathogens in the solid wastes. The pH values were 6.88 to 7.67 as for the moisture content 44 to 64%, organic carbon and nutrients (N, P, K) were 1.75%, 3.4%, and 0.57% respectively. As for the heavy metal, copper was 700 ppm, zinc was 630 ppm, lead was 34 ppm, chromium was 69 ppm and nickel was 23 ppm.

On the other hand, composted wast0es was also Rawat et al. [20] in Ahmadebad, Bangalore and Delhi in India. The compost values were compared to the Indian Standard. Through the composting, the moisture contents were between 23.83 to 31.61%. The moisture content for the compost was insufficient, but turning the compost periodically had fasten the process of composting. The pH values were 7.82 to 8.19 and the C:N ratios were 18.99 to 25.36. The N and P contents were 0.85 to 1.13% and 2.52 to 2.9% respectively. While the heavy metal contents, copper, chromium, nickel, lead and cadmium were recorded to be less than the standard provided which were 300mg/kg, 50mg/kg, 50 mg/kg, 100 mg/kg and 5 mg/ kg respectively. It is suggested that to increase the quality of the compost, adding cowdung, bagasse and garden wastes was recommended.

In addition, studied compost as the mean for waste management Syamala and Belagali [5] at MySore, Karnataka, India. During the composting, the pH values recorded were between 7.13 to 8.76 and the moisture content decreased from 85.19 to 29.87% due to high evaporation rates. The pH values were between 7.13 to 8.76, which were higher than normal range due to production of carbon dioxide (CO_2) from organic acids and the loss of N. The total organic carbon (TOC) decreased throughout the composting process and this was related to the decomposition of waste by microbial populations. The heavy metal concentration was compared to Ohai EPA. There are presence of copper, cadmium, lead and mercury varied between 146.8 to 325 mg/kg, 0.1 to 1.4 mg/kg, 1 to 8 mg/kg and 0 to 3 mg/kg respectively.

2.3 Kitchen wastes

Kitchen waste mostly generates food scraps for disposal. Throwing this scraps can create odor problems and adds to the volume of waste going to landfills. There are three methods of composting kitchen wastes were reviewed. As for the composting of kitchen waste, Pathak et al. [21] used bio-composter as a composted and the study was conducted for 4 years, from 2008 to 2011 for 135 days per year.

The highest temperature recorded was 64℃ and decreased to 32.7℃ at the ambient temperature, the moisture content was observed at 55.8% and decreased to 21.7% at day 36. The loss of moisture content was due to the high temperature. For the pH, an acid pH was recorded during the early days of composting due to the production of organic acids, the pH rose up to 8.6 but later dropped to 6.3. This was caused by the ammonification and mineralization of the organic matters through the activities of microorganisms. For composition of the nutrients N, P, K were between 1.16 to 1.20%, 0.03 to 0.053% and 0.30 to 0.38% respectively, as for heavy metals, the content were, copper (45.25 to 48.39 mg/kg), zinc (51.1 to 54.4 mg/kg) and iron (1134.8 to1274.2 mg/kg). On the other hand, Gautam et al. [22] mixed vegetables and fruit wastes for their studies. A heap 4' high and 8' long was use as a composted area.

The ambient temperature was recorded, which was between 35 to 45℃. The initial moisture content was maintained at 50 to 60%. Periodical turning was conducted to aerate the compost heap every 3 to 5 days. The maximum temperature recorded at was between 48 to 50℃ on the 3 days of the composting process. The moisture content and pH were between 25 to 41% and 7.75 to 7.84 respectively. For N, P and K contents, they ranged from 0.03 to 0.07%, 0.002 to 0.005% and 0.32 to 0.36% respectively. However, the N concentration was inadequate when compared to the standard concentration. Hence, this study suggested adding phosphoric acid to avoid the unnecessary volatilization of ammonia.

In another study, Arslan [8] used in-vessel composter as a method to compost kitchen wastes. The composting process was conducted for 22 days. The compost was mixed with 2 kg of sludges as inoculum and 3.5 kg of sawdust. The temperature was recorded at 55℃ on day 2 and this temperature maintained at day 7. The initial pH value was 5.5 and moisture contents were between 48 to 53%. The C:N ratio decreased from 35.92 to 19.69 while the total Kjeldahl Nitrogen (TKN) had increased at the end of composting from 1.43 to 2.45%. As for the heavy metal contents- chromium, cadmium, zinc, copper, iron and nickel contents were tested. The cadmium concentration was below the detection limit and the others were 22.4 mg/kg, 190.7 mg/kg, 35 mg/kg, 2641.75 mg/kg and 15.33 mg/kg respectively.

3 Conclusion

Based on extensive literature review, composting research of different types of organic wastes shown different performance on the effectiveness of the composting process. Composting, as a treatment of organic waste, had been proven to significantly reduce the volume of wastes in the country. In addition, composting can also provide nutrients that are suitable for agriculture and can be used as fertilizer to replace chemical fertilizer. Furthermore, compost can also be used as soil amendments as well as being eco-friendly, hygienic economical and toxic free. In conclusion, during the composting of agricultural wastes the addition of animal manure can enhance the degradation process, whilst in the composting of municipal solid waste and kitchen waste it is important to measure the heavy metal content because of its toxicity and different method of composting influenced the nutrient status of compost. Nevertheless, the compost provided must comply with the standard limit to ensure the quality of the compost.

References

[1] A. Abd Hamid, A. Ahmad, M.H. Ibrahim, N. Nik Abdul Rahman, Food waste management in Malaysia – current situation and future management option, J. of Industrial Research and Technology, **2**(1),36-39,(2012).

[2] Y. Dhokhikah, Y. Trihadiningrum, Solid waste management in asian developing countries: Challenges and opportunities, J. of Applied Environmental and Biological Sciences, **2**(7), 329-335, (2012).

[3] M. Rawat, A.L. Ramanathan and T. Kuriakose, Characteristics of municipal solid waste compost (mswc) from selected indian cities – a case study for its sustainable utilisation, J. of Environmental Protection, **4**, 163-171, (2013).

[4] H. Fathi, A. Zangane, H. Fathi, H. Moradi, Municipal solid waste characterization and it is assessment for potential compost production: A Case Study in Zanjan City, Iran, American J. of Agriculture and Forestry, **2**(2), 39-44, (2014).

[5] D.C. Shymala, S.L. Belagali, Studies on variations in physico-chemical and biological characteristics at different maturity stages of municipal solid waste compost, Int. J. of Environmental Sciences, **2**(4), 1984-1997, (2012).

[6] L. Rama and M. Dr. Vasanthy, Market waste management using compost technology, Int. J. of Plant, Animal and Environmental Sciences, **4**(4), 57-61, (2014).

[7] R. Ahmad, G. Jilani, M. Arshad, Z.A. Zahir and A. Khalid, Bio-conversion of organic wastes for their recycling in agriculture: An overview of perspectives and prospects, Annals of Microbiology, **57**(4), 471-479, (2007).

[8] E. Işıl Arslan, Ayhan Ünlü and Murat Topal, determination of the effect of aeration rate on composting of vegetables-fruit wastes, CLEAN – Soil, Air, Water, **39**(11), 1014-1021, (2011).

[9] J.C. Hangreaves, M.S. Adl and P.R. Warman, A review of the use of composted municipal solid waste in agriculture, Agriculture, Ecosystems and Environment, **123**, 1-14, (2008).

[10] A. Hernandez, H. Castillo, D. Ojeda, A. Arras, J. Lopez and E. Sanchez, Effect of vermicompost and compost on lettuce production, Chilean J. of Agricultural Research, **70**(4), 583-589, (2010).

[11] U. Pangnakorn, Valuable added the agriculture waste for farmers using in organic farming groups in Phitsanulok, Thailand, *Tropentag 2006 Prosperity and Poverty in a Globalized World Challenges for Agricultural Research*, Born, (2006).

[12] M. Risse, Food waste composing: Institutional and industrial applications, (2012) <retrieved from http://extension.uga.edu/publications/files/pdf/B%201189_3.PDF.>

[13] R.T. Abu-Zahra, A. R.Ta'any and A. R. Arabiyyat, Changes in compost physical and chemical properties during aerobic decomposition, Int. J. of Current Microbiology and Applied Sciences, **3**(10), 479-486, (2014).
[14] H.Y. Ch'ng, O.H. Ahmed, S. Kassim and N.K. Ab Majid, Co-composting of pineapple leaves and chicken manure slurry, Int. J. of Recycling of Organic Waste in Agriculture, **2**(23) ,1-8, (2013).
[15] T. Karak, P. Bhattacharya, R.K. Paul, T. Das and S.K. Scha, Evaluation of compost from agriculture waste with fish pond sediment as bulking agent to improve compost quality, Enviromental Chemistry, **41**(7), 711-723, (2013).
[16] S. Tripetchkul, K. Pundee, S. Koonrisuk and S. Akeprathumchai, Co-composting of coir pitch and cow manure: initial c/n ratio vs physico-chemical changes, Int. J. of Recycling of Organic Waste in Agriculture, **1**(15), 1-8, (2012).
[17] T. Janakiam and K. Sridev, Conversion of waste into wealth: A study in solid waste management, E-J. of Chemistry, **7**(4), 1340-1345, (2010).
[18] N. Saithep, S. Dheeranupatana, P. Sumrit, S. Jeerat, S. Boonchalearmkit, J. Wongsanoon, and C. Jatisatienr, Composting of tobacco plant waste by manual turning a forced aeration system., Journal of Science and Technology., **3**(2), 248-260, (2009)
[19] B. Manohara, S.L. Belagali, Characterization of essential nutrients and heavy metals during municipal solid waste composting, Int. J. of Innovative Research in Science, Engineering and Technology, **3**(2) , 9664-9672, (2014).
[20] M. Rawat, A.L. Ramanathan and T. Kuriakose, Characteristics of municipal solid waste compost (mswc) from selected Indian cities - a case study for its sustainable utilisation, J. of Environmental Protection, **4**,163-171, (2013).
[21] A.K. Pathak, M.M. Singh, V. Kumara, S. Arya and A.K. Trivedi, Assessment of physico-chemical properties and microbial community during composting of municipal solid waste (Viz. Kitchen Waste) at Jhansi City, U.P (India), Recent Research in Science and Technology, **4**(4), 10-14, (2012).
[22] S.P. Gautam, P.S. Bundela, A.K. Pandey, M.K. Awasthi and S. Sarsaiya, Composting of municipal solid waste of Jabalpur City, Global J. of Environmental Research, **4**(1), 43-46, (2010).

MATEC Web of Conferences **47**, 05026 (2016)
DOI: 10.1051/matecconf/20164705026

Emerging Pollutant of Concern: Occurrence of Pharmaceutical Compounds in Asia with Particular Preference to Southeast Asia Countries

Nor Haslina Hashim[1,a], Muhammad Hafiz Nasir[2] and Muhammad Syamim Ramlee[2]

[1]*Faculty of Engineering Technology, Universiti Tun Hussein Onn Malaysia, 86400 Parit Raja, Malaysia*
[2]*Faculty of Civil and Environmental Engineering, Universiti Tun Hussein Onn Malaysia, 86400 Parit Raja, Johor, Malaysia*

Abstract. Southeast Asia countries were developing countries, with the rapid development for a better living condition leads to longer life expectancy, which increased total population. It may result in increased the demand of pharmaceutical in domestic use and or in hospital. Although most wastewater treatment plants (WWTPs) met the local authority's regulatory requirement, there are still many pharmaceutical compound incompletely removed and discharge to the water stream and enter the environment. Recently many studies and researches have published on the occurrence and source as well as the fate of pharmaceuticals all over the world including Asia. As part of Asia region, Southeast Asia countries (Malaysia, Philippines, Singapore, Thailand, Vietnam, Brunei, Cambodia, and Lao PDR) seem has lack of research knowledge and information regarding this emerging pollutant issues compared to developed Asia country as China, Korea and Hong Kong. In this paper, it will review recent studies occurrences (surface and wastewater) on most common pharmaceuticals with several recommendations to overcome and thus summarise the actual situation in Southeast Asia.

1 Introduction

Recently, the World Health Organization (WHO), authorities and researcher across Europe, North America, and elsewhere have emphasized the concern regarding potential aquatic ecosystems and human health impacts connected to contaminants of this emerging pollutant [1]. Pharmaceutical are part group of emerging, probably hazardous pollutants that have attract highly concern among researcher and scientist in recent years. These compounds among the emerging environment issue lately. Pharmaceutical compounds has detected in surface water, wastewater treatment plants (WWTPs) influent and effluent, and in the hospital effluent [2].

The Asian Development Bank (ADB), in a widely acclaimed paper published in 2001, accused Southeast Asia of having the world's dirtiest environment (land, sea and air). Urban corridors such as Kuala Lumpur's Klang Valley and Singapore Island seem to have the pollution situation more or less under control. Currently, there are few published reports of pharmaceutical compounds in this region. Alkylated musk recently has been measure of wastewater treatment in Guangdong, China [3]. In 2003, China was the world's largest producer of pharmaceutical products, with an annual production of

[a] Corresponding author : haslina@uthm.edu.my

28,000 tonnes of penicillin (60% of the world total). At present, little is known about the adverse effects of human and veterinary pharmaceuticals on aquatic organisms, although standard acute toxicity data have been reported for some pharmaceuticals [4].

2 Occurrences

The occurrence of pharmaceutical in surface water and wastewater is already published and well documented in European and North American countries [5-8]. A published research concerning the occurrence of pharmaceutical active compounds in aquatic environments in Southeast Asia countries reviewed by Richardson [9] and indicated has no environment data for the fate and effects have been reported in Southeast Asia before 2005. Other than that, developing countries generally has poor treatment to treat wastewater or less advanced than the developed countries because of the lower economic budget. Thus, effluent was released to surface stream water without specific treatment.

2.1 Surface water

The occurrence of pharmaceuticals in surface water has been summarised and shown in Table 1 (selected Southeast Asia countries) which is obtained from recent studies. In Asia, the antibiotic concentrations such as Oxytetracyclin, Trimethoprim and Sulfamethoxazole are high in both wastewater influent effluent and surface water [10]. In additional, researcher analyzed for 10 NSAIDs, 2 blood lipid regulators, and 2 antiepileptic drugs in the Pearl River system in China (Liuxi, Zhujiang, and Shijin Rivers) and four sewage effluents during the dry and wet seasons, results were reported and the environmental risks they posed were assessed [11].

Table 1. Detection of selected pharmaceutical compound in surface water of several Southeast Asia countries.

Country	Location	Compound	Conc. (ng/L)	Ref.
Taiwan	Taiwanese River	Ketoprofen	110–620	[12]
		Ibuprofen	<12–30	
		Diclofenac	24–62	
Vietnam	Hanoi	Ketoprofen	<40–330	[13]
		Ibuprofen	100–1100	
		Diclofenac	<140–310	
		Naproxen	80–380	
Singapore	Marina Bay	Naproxen	>12-24	[14]
		Ketoprofen	>3-23	
		Diclofenac	>4-26	
		Ibuprofen	>37-60	
	Rochor Canal	Naproxen	20	
		Ketoprofen	25	
		Diclofenac	Below Detection	
		Ibuprofen	139	
Malaysia	Langat and Muar Rivers	Caffeine	138–760	[15]
		Diclofenac	Non Detect–54	

Other in China, river water and tap water screened for the presence of β-lactam antibiotics. River water samples were collected from upstream to downstream, which a drainage river that connects with a drainage ditch of the pharmaceutical factories located in this area, and receives the effluent from two WWTPs. The β-lactam antibiotics were found within the range of 0–9.56 μg/L in river water at the downstream of antibiotics manufacturers, and none were detected in the tap water [16]. Further in Southeast Asia country, researcher identified a method to determine the risk of pharmaceutical compounds applied it to 10 of the human-use pharmaceuticals detected in Taiwan's Sindian River [17]. Additionally, they showed that in river water samples, NSAIDs, steroids, propranolol, and gemfibrozil were found at ng/L to mg/L levels with night-time concentrations of these compounds, except estrogens, doubled or even greater, compared to day-time levels. Table 1 provide an overview of the various pharmaceuticals in surface, with the concentrations recoded in Asia countries. Meanwhile, scientist developed and applied a method for quantification of 10 pharmaceuticals in surface water samples from the major rivers in South Korea. Results showed that overall, surface water contamination near livestock WWTPs was higher than near domestic WWTPs [18]. Experimenter collected samples from upstream/downstream and effluent dominated creeks along the Han River, Seoul (South Korea) and examined them for selected emerging pollutants. Most target compounds were detected in both the Han River samples (63%) and in the effluent dominated creek samples (79%). Both river and creek samples were frequently detected atenolol, musk ketone, naproxen, carbamazepine, caffeine, and benzophenone, although the mean concentrations in effluent dominated creek samples (102 – 3745 ng/L) were significantly higher than those in river samples (56–1013 ng/L). However, steroid hormones 17β-estradiol, 17α ethynylestradiol, progesterone, and testosterone, were not detected (< 1 ng/L) in both samples [19].

2.2 Wastewater

Many research reported that those WWTPs, which meet the regulatory requirement for waste water treatment are only moderately effective in removing the pharmaceuticals, expect the plants install with a tertiary treatment [20]. Occurrence and fate of selected pharmaceutical compound were investigated in three WWTPs in Beijing, China. All of the pharmaceutical compounds were detected in the influents of all three WWTPs with bisphenol-A being the most abundant compound detected [21]. In another study, four WWTPs in Beijing, China were investigated on occurrence and removal of 13 pharmaceuticals and 2 consumer products, including antibiotic, anti-lipidemic, anti-inflammatory, anti-hypertensive, insect repellent and antipsychotics [22].

Meanwhile in Malaysia, researcher developed a method for simultaneous determination and quantification of 23 pharmaceuticals and synthetic hormones from different therapeutic classes in water samples and applied the method to river water and WWTP effluents with results showing that among target compounds, chlorothiazide was found at the highest level, with concentrations reaching up to 865 ng/L in WWTP effluent, and 182 ng/L in river water [23]. In Korea, scientist studied the occurrence and fate of pharmaceuticals in WWTPs and rivers found that pharmaceuticals were present. The researched found biological processes in the WWTP showed the greatest efficacy for reducing the concentration of the pharmaceuticals. Overall, although conversional wastewater treatments only partially remove concentration pharmaceutical several compound, while some of the pharmaceuticals are resistant to degradation and persistent in the environment [24]. The result of WWTPs is still unsatisfactory.

Table 2. Detection of selected pharmaceutical compound in wastewater of several Asia countries.

Country	Compound	Ref.
Malaysia	Chlorothiazide	[23]
	Chloramphenicol	[28]
Pakistan	Oxytetracyclin	[29]
	Trimethoprim	[29]
China	bisphenol-A	[21]

Via search engine from Sciences Direct website, based on "Occurrence of pharmaceutical active compounds in Southeast Asia" and "Occurrence of pharmaceutical active compounds in Europe" in between 2010 until 2015 (recently- 1 July), a comparison are made to visualise the few amount of publication of journal and research about this emerging pollutant in Asia region. Only few studies are available regarding the occurrence of Pharmaceutical Active Compounds (PAC's) in all of environment matrices including industry wastewater, domestic wastewater, hospital wastewater, surface water, and sediments of Southeast Asia countries. Thus detailed and focused investigations specifically in Southeast Asia are required to determine the sources, pathway and fate of these PACs and their biological resistance in the entire environment.

3 Conclusion

The important action should be taken instantly by upgrading available system for treating wastewater in Southeast Asia countries because there are still high concentrations found in effluent WWTPs. According to research, the removal of pharmaceuticals is still incomplete except WWTPs includes tertiary treatment [25]. A study found that the removal efficiency is different with divergent filtration. Biological Activated Carbon (BAC) filtration is more effective to remove the pharmaceuticals compare to sand filtration [26]. Other than improving the treatment system process in WWTPs, it is important for government to set regulation globally. Each country has a government body that is responsible for the control of environmental pollution. The formulation of policy, enforcement and implementation of regulation vary from country to country depending on various factors. Some countries have stringent regulation and enforce it strongly; some have slow approach, where as some are still in primary stage of environmental protection. In some countries, the implementation part is very weak that it cannot enforce the existing regulation effectively due to inexperienced manpower and financial constraints. Some countries are still concentrating on rapid economic development through industrialization at the cost of environment damage. Meanwhile, government in each Southeast Asia countries need to create a module to educate their own citizen awareness this emerging pollutant issues.

It is well known that there has not been sufficient evaluation of pharmaceutical residue in environment from the Southeast Asian region. Evaluation and reporting inventories of production and usage of pharmaceutical compound products in all countries of Southeast Asia is essential needs. The volume of pharmaceutical industry and human population in these countries has increase significant of pharmaceutical contamination and its associated risk. The approach and framework of Southeast Asian Mussel Watch is appropriate plan to organise local scientists and researchers into common aim and focussed of contaminant measurement [27]. In response to this, suggested that government should establish more resources cooperate nationally with country between Asia region and globally with Europe and North America in order to know more about the effect and hence setting up with different pharmaceutical compounds. Therefore the well knows it toxicity; impact or monitoring data can achieve. Lastly each Southeast Asia countries should assemble regulation to control the use and disposal of this pharmaceutical to eliminate probable chance discharging the residue to the environment.

References

[1] World Health Organization, Pharmaceuticals in Drinking-Water, Water Sanitation Health, (2012).

[2] M.J. Gomez, M. Petrovic, A.R. Fernandez-Alba and D. Barcelo, Determination of pharmaceutical of various therapeutic classes by solid-phase extraction and liquid chromatography-tandem mass spectrometry analysis in hospital effluent wastewaters, J. Chromatogr. A111, **4**(2), 224-233, (2006).

[3] X. Zeng, G. Sheng, Y. Xiong and J. Fu, Determination of polycyclic musks in sewage from Guangdong China using gas chromatography–electron impact-mass spectrometry, Chemosphere, **60**(6), 817–823, (2005).

[4] D. Pascoe, W. Karntanut and C.T. Muller, Do pharmaceuticals affect freshwater invertebrates? A study with the cnidarian Hydra vulgaris, Chemosphere, **51**, 521–528, (2003).

[5] D.W. Koplin, E.T. Furlong, M.T. Meyer, E.M. Thurman, S.D. Zaugg, L.B. Barber and H.T. Buxton, Pharmaceuticals, hormones, and other organic wastewater contaminant in U.S. stream, Environmental Science and Technology, **36**, 1202-1211, (2002).

[6] K.V. Thomas and M.J Hilton, The occurrence of selected human pharmaceutical compounds in UK Estuaries, Marine Pollution Bulletin, **49**, 436-444, (2004).

[7] A. Nikolaou, S. Meric and D. Fatta, Occurrence patterns pharmaceuticals in water and wastewater environments, Analytical Bioanalytical Chemistry, **387**, 1225-1234, (2007).

[8] S. Zhang, Q. Zhang, S. Darisaw, O. Ehie, and G. Wang, Simultaneous quantification of polycyclic aromatic hydrocarbons (PAHs), polychlorinated biphenyls (PCBs), and pharmaceuticals and personal care products (PPCPs) in Mississippi river water in New Orleans, Louisiana, USA Chemosphere, **66**, 1057-1069, (2007).

[9] B.J. Richardson, P.K.S. Lam and M. Martin, Emerging chemical of concern: pharmaceutical and personal care products (PPCPs) in Asia, with particular reference to Southern China, Mar. Pollut. Bull., **50**, 913-920, (2005).

[10] P.T. Hoa, S. Managaki, N. Nakada, H. Takada, A. Shimizu, D.H. Anh, P.H. Viet and S. Suzuki, Antibiotic contaminant and occurrence of antibiotic-resistant bacteria in aquatic environment in northern Vietnam, Sci. Total Environ., **409**, 2894-2901, (2011).

[11] J.L. Zhao, G.G. Ying, Y.S. Liu, F. Chen, J.F. Yang, L. Wang, X.B. Yang, J.L. Stauber and M.S.J. Warne, Occurrence and a screening-level risk assessment of human pharmaceuticals in the Pearl River system, South China. Environ. Toxicol. Chem., **29**(6), 1377–1384, (2010).

[12] C.C. Hsin, L.W. Pi and H.S. Wang, Using liquid chromatography ion trap mass spectrometry to determine pharmaceutical residues in Taiwanese rivers and wastewaters, Chemosphere, **72**, 863-869, (2008).

[13] N.H. Tran, T. Urase and T.T. Ta, A preliminary study on the occurrence of pharmaceutically active compounds in hospital wastewater and surface water in Hanoi, Vietnam, CLEAN – Soil, Air, Water, **42**(3), 267–275, (2014).

[14] J. Wu, J. Yue, R. Hu, Z. Yang and L. Zhang, Use of caffeine and human pharmaceutical compounds to identify sewage contamination, Int. J. of Medical, Health, Biomedical, Bioengineering and Pharmaceutical Engineering, **2**(8), 289-293, (2008).

[15] F.A.Q. Fouad, M.P. Abdullaha, R. O. Mohamed, J. Latipa and Z. Zakaria, Multi-residue analytical methodology-based liquid chromatography-time-of-flight-mass spectrometry for the analysis of pharmaceutical residues in surface water and effluents from sewage treatment plants and hospitals, J. Chromatogr. A., **1345**, 139–153, (2014)

[16] J. Yin, Z. Meng, M. Du, C. Liu, M. Song and H. Wang, Pseudo-template molecularly imprinted polymer for selective screening of trace Betalactam antibiotics in river and tap water, J. Chromatogr., A. **1217**(33), 5420–5426, (2010).

[17] A.Y.C. Lin, C.F. Lin, Y.T. Tsai, H.H.H. Lin, J. Chen, X.H. Wang and T.H. Yu, Fate of selected pharmaceuticals and personal care products after secondary wastewater treatment processes in Taiwan, Water Sci. Technol., **62** (10), 2450–2458, (2010).

[18] S.H. Koo, C.H. Jo, S.K. Shin and S.W. Myung, Simultaneous determination and occurrences of pharmaceuticals solid-phase extraction and liquid chromatography-tandem mass spectrometry (LC-MS/MS) in environmental aqueous samples, Bull. Korean Chem. Soc., **31**(5), 1192–1198, (2010).

[19] Y. Yoon, J. Ryu, J. Oh, B.G. Choi and S.A. Snyder, Occurrence of endocrine disrupting compounds, pharmaceuticals, and personal care products in the Han River (Seoul, South Korea), Sci., Total Environ., **408**(3), 636–643, (2010).

[20] P. Gao, D. Mao, Y. Luo, L. Wang, B. Xu and L. Xu, Occurrence of sulphonamide and tetracycline-resistant bacteria and resistance genes in aquaculture environment, Water Res., **46**, 2355-66, (2012).

[21] L. Zhou, C. Gao and W. Xu, Magnetic dendritic materials for highly efficient adsorption of dyes and drugs, ACS Appl. Mater. Interfaces, **2**(5), 1483–1491, (2010).

[22] Q. Sui, J. Huang, S. Deng, G. Yu and Q. Fan, Occurrence and removal of pharmaceuticals, caffeine and DEET in wastewater treatment plants of Beijing, China, Water Res., **44**(2), 417–426, (2010).

[23] N.A. Al-Odaini, M.P. Zakaria, M.I. Yaziz and S. Surif, Multi-residue analytical method for human pharmaceuticals and synthetic hormones in river water and sewage effluents by solidphase

extraction and liquid chromatographytandem mass spectrometry, J. Chromatogr., A. **1217**(44), 6791–6806, (2010).

[24] W.J. Sim, J.W. Lee and J.E. Oh, Occurrence and fate of pharmaceuticals in wastewater treatment plants and rivers in Korea, Environ. Pollut. (Oxford), **158**(5), 1938–1947, (2010).

[25] A. Jelic, N. Gros, A. Ginebreda, R. Chespedes-Sancez, F. Ventura, M. Petrovic and D. Barcelo, Occurrence, partition, and removal of pharmaceuticals in sewage water and sludge during wastewater treatment, Water Res., **45**, 1165-1176, (2011).

[26] J. Reungoat, B.I. Escher, M. Macova and J. Keller, Biofiltaration of wastewater treatment plant effluent: Effective removal of pharmaceutical and personal care products and reduction of toxicity, Water Res., **45**, 2751-2762, (2011).

[27] L. Monirith, D. Ueno, S. Takahashi, H. Nakata, A. Sudaryanto, A. Subramanian, S. Karuppiah, A. Ismail, M. Muchtar, J. Zheng, Asia-Pacific mussel watch: monitoring contamination of persistent organochlorine compounds in coastal water of Asian countries, Mar. Pollut. Bull., **46**, 281-300, (2003).

[28] S.A. Marni, N.T. Malintan, I. Faridah, A. M. Mustafa, Chloramphenicol in Malaysia waste water and its residues in animal husbandaries products, Health and the Environment Journal, **1**(1), 41-45, (2010).

[29] M. Ahmad, A. U. Khan, A. Wahid, Z. A. Butt, M. Farhan, F. Ahmad, Role of hospital effluents in the contribution of antibiotics and antibiotic resistant bacteria to the aquatic environment, Pak. J. Nutr., **11**(12), 1177-1182, (2012).

AUTHOR INDEX

AUTHOR INDEX

AUTHOR INDEX

EDP Sciences – Web of Conferences
17, Avenue du Hoggar
Parc d'Activité de Courtabouf
BP 112
F-91944 Les Ulis Cedex A
France

ISBN 978-1-5108-2212-2

CURRAN ASSOCIATES INC.
proceedings.com

International Conference on Advances in Mechanical Engineering-2022 (ICAME-2022)

IOP Conference Series: Materials Science and Engineering
Volume 1259

Nagpur, India
26 – 28 May 2022

IOP
Institute of Physics